Plant Equipment Reference Guide

The McGraw-Hill Engineering Reference Guide Series

This series makes available to professionals and students a wide variety of engineering information and data available in McGraw-Hill's library of highly acclaimed books and publications. The books in the series are drawn directly from this vast resource of titles. Each one is either a condensation of a single title or a collection of sections culled from several titles. The Project Editors responsible for the books in the series are highly respected professionals in the engineering areas covered. Each Editor selected only the most relevant and current information available in the McGraw-Hill library, adding further details and commentary where necessary.

Hicks · CIVIL ENGINEERING CALCULATIONS REFERENCE GUIDE

Hicks • MACHINE DESIGN CALCULATIONS REFERENCE GUIDE

Hicks • PLUMBING DESIGN AND INSTALLATION REFERENCE GUIDE

Hicks • POWER GENERATION CALCULATIONS REFERENCE GUIDE

Hicks • POWER PLANT EVALUATION AND DESIGN REFERENCE GUIDE

Higgins • PRACTICAL CONSTRUCTION EQUIPMENT MAINTENANCE REFERENCE GUIDE

Johnson & Jasik • ANTENNA APPLICATIONS REFERENCE GUIDE

Markus and Weston • CLASSIC CIRCUITS REFERENCE GUIDE

Merritt • CIVIL ENGINEERING REFERENCE GUIDE

Perry • BUILDING SYSTEMS REFERENCE GUIDE

Rosaler & Rice • INDUSTRIAL MAINTENANCE REFERENCE GUIDE

Rosaler & Rice • PLANT EQUIPMENT REFERENCE GUIDE

Woodson • HUMAN FACTORS REFERENCE GUIDE FOR ELECTRONICS AND COMPUTER PROFESSIONALS

Woodson • HUMAN FACTORS REFERENCE GUIDE FOR PROCESS PLANTS

Plant Equipment Reference Guide

Robert C. Rosaler, P.E. Editor in Chief

James O. Rice Associates
New York

James O. Rice Associate Editor

James O. Rice Associates
New York

Tyler G. Hicks, P.E. Project Editor

McGraw-Hill Book Company

New York St. Louis San Francisco Auckland Bogotá
Hamburg London Madrid Mexico Milan
Montreal New Delhi Panama Paris São Paulo
Singapore Sydney Tokyo Toronto

Library of Congress Cataloging-in-Publication Data

Plant equipment reference guide.

(McGraw-Hill engineering reference guide series)
"The material in this volume has been published previously in Standard handbook of plant engineering, by Robert C. Rosaler and James O. Rice"—Verso of t.p.
Includes index.
1. Industrial equipment—Handbooks, manuals, etc.
I. Rosaler, Robert C. II. Rice, James O'Neill, 1910– . III. Hicks, Tyler Gregory, 1921– .
IV. Standard handbook of plant engineering. V. Title.
VI. Series.
TS191.P57 1987 621.8 87-3165

ISBN 0-07-052163-8

1234567890 DOCDOC 89210987

ISBN 0-07-052163-8

Printed and bound by R. R. Donnelley and Sons

Contents

Preface

This reference guide is a comprehensive coverage of the selection and maintenance of plant operating equipment of many types. A condensation of the Rosaler and Rice—*Standard Handbook of Plant Engineering,* this guide will be valuable to a variety of management, operating, and maintenance personnel in plants of all types.

The guide opens with a discussion of mechanical power transmission. Many different types of power transmission equipment are considered—bearings, drives, couplings, chains, etc. Each is discussed in terms of its selection for a specific set of operating conditions, along with expected maintenance requirements. The user of the guide will find this section especially useful since so much practical information is provided.

Next, the guide considers materials handling. The user is shown how to plan materials-handling systems. Once the plan is completed, specific equipment can be chosen to meet the needs of the plant. Typical equipment discussed includes containers, fixed-path units, mobile units, along with warehousing and storage.

Hydraulic and pneumatic systems are discussed next. In these chapters the guide user is shown how to chose and maintain such systems for best economy and greatest reliability.

Piping and valving for plants of many types are discussed in a series of seven chapters. Metallic, plastic, and asbestos-cement piping are analyzed for their suitability and maintenance requirements. Other topics discussed are pipe insulation, industrial hose, and valves.

Instrumentation and automatic controls are the next topics considered in the guide. A variety of instruments and controls are described and evaluated for various plant applications.

Noise and vibration control—important in all plants today—are discussed in the next two chapters. A variety of materials and techniques to control both noise and vibration are evaluated and assessed for use in different plants.

Pollution control and waste disposal—both of which are receiving intense attention currently from various regulatory bodies—are considered in the next three chapters. Topics include air pollution control, liquid-waste disposal, and solid-waste disposal.

The guide concludes with six chapters on plant safety and sanitation—important topics for every plant manager and operator. Since safety is so important in terms of human lives, it is discussed first. Next, fire and electrical hazard protection and prevention are considered. Security equipment is analyzed in light of typical requirements in today's plants. Toxic substances and radiation hazards are evaluated in terms of safety and security. Lastly, sanitation control and housekeeping are discussed in detail.

This reference guide is a valuable tool and information source for plant engineers, plant managers, plant operators, maintenance supervisors, cost estimators, schedulers, mechanical engineers, electrical engineers, civil engineers, consulting engineers, and a variety of other maintenance personnel in plants of all types throughout the world. The information it presents is practical, up-to-date, and highly useful. With this reference guide on hand, there are few daily problems that cannot be solved easily and quickly.

Tyler G. Hicks, P.E.

Contributors

Charles H. Artus
Product Application Engineer
Gates Rubber Co.

Richard Best
Sr. Fire Analysis Specialist, Research and Fire Information Services
National Fire Protection Association

Michael M. Calistrat
Manager, Mechanical Development
Metal Products Division
Koppers Company, Inc.

William Carter, Standards Engineer
The Torrington Company

Karl J. Danz, P.E.
Principal Engineer
Roy F. Weston, Inc.

A. J. Grosso
Vice President, Engineering
American District Telegraph Corporation

Michael R. Harrison
Manager, Engineering and Technical Services
Insulation Systems Dept.
Johns-Manville Sales Corp.

Leo J. Horvath
Director, Technical Affairs
Association of Asbestos Cement Pipe Producers

Paul Jensen
Manager, Industrial Services
Bolt Beranek and Newman Inc.

Charles Albert Johnson, Ph.D.
Technical Director
National Solid Wastes Management Association

Chris S. Louskos
O-Ring Division
Parker-Hannifin Corp.

Stanley A. Mruk
Technical Director
Plastics Pipe Institute

John B. Painter
Gasket Division
Parker-Hannifin Corp.

Ralph E. Peterson
Parker Packing
Parker-Hannifin Corp.

Richard G. Ramsdell
Seal Group
Parker-Hannifin Corp.

Ranjit S. Randhawa
The Foxoboro Company

Thomas C. Roberts
Product Engineering Manager
S.S. White Industrial Products Division
Pennwalt Corporation

Thomas J. Rost
Project Engineer
Chain Division
FMC Corporation

Richard B. Ruch, Jr.
Project Manager
Roy F. Weston, Inc.

Robert William Ryan
Assistant Director
Department of Environmental Safety
University of Maryland

Ken S. Satija, Ph.D., P.E.
United Engineers & Constructors Inc.

John B. Scannell
Parker Packing
Parker-Hannifin Corp.

Benjamin A. Schranze, P.E.
Roy F. Weston, Inc.

Jaswant Singh, Ph.D., CIH
Vice President and Technical Director
Clayton Environmental Consultants, Inc.

Robert L. Smith, Jr.
General Electric Co.

Arthur Spiegelman, P.E.
M & M Protection Consultants
Marsh & McLennan, Inc.

William M. Throop, P.E.
Manager, Market Development
Envirex Inc.

Eric E. Ungar
Principal Engineer
Bolt Beranek and Newman Inc.

George W. Walsh
General Electric Co.

Don Williams
Technical Service Director
Huntington Laboratories, Inc.

D. G. Wilson
Vice President and General Manager
Facility Plans Engineering and Construction
United States Steel Corp.

American Gear Manufacturers Association
Arlington, Virginia

***Machine Design* A Penton/IPC Publication**

K. W. Tunnell Company, Inc.
King of Prussia, Pennsylvania

The Valve Manufacturers Association (Communications Committee)
McLean, Virginia

section 1

Mechanical Power Transmission

chapter 1-1

Gearing and Enclosed Gear Drives

prepared by

American Gear Manufacturers Association*
Arlington, Virginia

GEARING

Functions of Gear Drives

The major functions of gears and gear drives are: reduction of speed, multiplication of torque, and positioning of shafts.

Speed Reducer

Economically, it is normally better to use a small, high-speed prime mover and gear-reducer combination than a larger, low-speed power source.

*Individual contributors:
James E. Gutzwiller, Boston Gear Division of INCOM International, Inc.
Harold O. Kron, Philadelphia Gear Corporation
Raymond J. Birck, The Falk Corporation

Speed Increaser

In some instances, it is impractical to operate a prime mover at a speed high enough to suit requirements of the driven equipment. For such applications, gears may be used as a speed increaser.

Shaft Orientation

Gears may provide desired orientation and relative rotation of shafts. Some common arrangements available are: in-line, parallel shaft, and right angle. Miter gears (1:1 ratio bevel gears), for example, serve the specific purpose of providing a 90° shaft orientation. Other angles can be supplied by specially designed gears of several types.

Gear Types

The most common types of gears are illustrated in Figs. 1-1 to 1-11. Other available types are generally modifications of the basic gears shown.

Spur Gears. These are cylindrical in form and operate on parallel axes. The teeth are straight and parallel to the axes (Fig. 1-1).

Spur Rack. A spur gear has straight teeth that are at right angles to the direction of motion (Fig. 1-1).

Helical Gears. A helical gear is cylindrical in form and has helical teeth. Parallel helical gears operate on parallel axes and, where both are external, the helices are of opposite hand (Figs. 1-2, 1-3, 1-12, and 1-13).

Double Helical Gears. Each of these has both right-hand and left-hand helical teeth, and operates on parallel axes. These also are known as *herringbone gears* (Fig. 1-4).

Crossed Helical Gears. These gears operate on crossed axes and may have teeth of the same or opposite hand. The term *crossed helical gears* has superseded the old term *spiral gears* (Fig. 1-5).

Worm Gear. This gear is the mate to a worm. A worm gear that mates with a cylindrical worm is said to be single-enveloping. Worm gearing operates on nonintersecting axes (Figs. 1-6, 1-7, 1-14, and 1-15).

Cylindrical Worm Gearing. This is a form of helical gearing that mates with a worm gear (Fig. 1-6).

Double-Enveloping Worm Gearing. This comprises hourglass worms mated with a worm gear (Fig. 1-7).

Bevel Gears. These are conical in form and operate on intersecting axes that usually are at right angles (Figs. 1-8 to 1-10).

Straight Bevel Gears. These gears have straight-tooth elements which, if extended, would pass through the point of intersection of their axes (Fig. 1-9).

Spiral Bevel Gears. These have teeth that are curved and oblique (Figs. 1-10 and 1-16).

Hypoid Gears. Similar in general form to bevel gears, hypoid gears operate on nonintersecting axes (Fig. 1-11).

Gear Geometry

Gear-Tooth Action

Gears of all types have the common characteristics of theoretically smooth transmission of motion through engagement of successive teeth. Certain design parameters must be met to ensure that one pair of teeth starts engagement before the preceding pair leaves off.

Standard Tooth Forms

The involute form is almost universally used for spur and helical gears and has some application for worm and bevel gears. The involute provides for accuracy of motion transmission, even when there is some change in center distance between gears. It also offers a number of manufacturing advantages. In worm and bevel gearing, conjugate forms are used to suit the manufacturing processes employed.

Modified and Special Tooth Forms

Common modifications to involute gear teeth are: long and short addenda, stub teeth, and tip and root relief. Special tooth forms are sometimes used to provide higher capacity. Normally, accurate mounting of the gears must be maintained to realize this advantage.

Gear Materials and Heat Treatment

Common Materials

Gears may be made up of a wide range of materials that may be ferrous, nonferrous, and nonmetallic. For industrial applications, steel and iron are most commonly used for spur, helical, and bevel gears. Worm-gear pairs generally are made up of bronze or iron for the gear and steel for the worm. Use of plastics is generally limited to light applications, particularly those in which minimal lubrication is available.

Hardened vs. Unhardened Material

Since the gear set rating is governed to a great extent by hardness of the teeth, heat treating often is used to provide higher capacity in the same space. Pinions, which endure more load cycles, generally are made slightly harder than their mating gears. Hardening adds to the cost, particularly where an additional finishing operation is required to correct distortion. For some applications, untreated gears may prove more economical. Obviously, this will be true when the equipment arrangement will not permit use of smaller hardened gears.

Material Selection

Information regarding material selection and properties is contained in AGMA 390, "Gear Handbook" (part II, sec. 2) and Ref. 1.* The selection of specific material and treatment combinations should be based on an analysis of the overall requirements and conditions. Some of the fundamental factors to be considered when making material and treatment selections for gearing are as follows:

- Factor of safety, loading, duty cycle, mounting, gearing enclosure, lubrication, and ambient atmospheric conditions.
- For replacement gearing, the life obtained from the previous gearing should be evaluated. If satisfactory, replace with similar material. If longer life is required, selection of a heat-treatment specification yielding a higher hardness and, if necessary, a better material, may provide the desired improvement.
- Annealed carbon steels, bar stock, forgings, or castings, are usually satisfactory for pinions and gears for uniform or moderate shock loads when the size of the gearing is not an important factor.
- Annealed carbon-steel pinions with cast-iron gears are sometimes used for the same reason mentioned above for annealed carbon steels.

*References are listed at the end of the chapter.

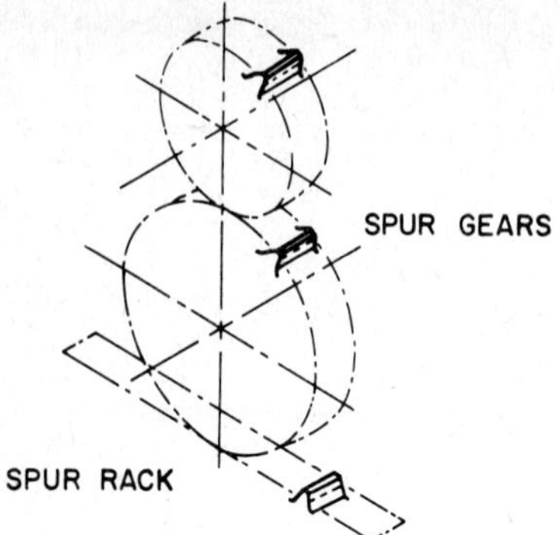

Figure 1-1 Spur gears and spur rack. *(Extracted from Ref. 2.)*

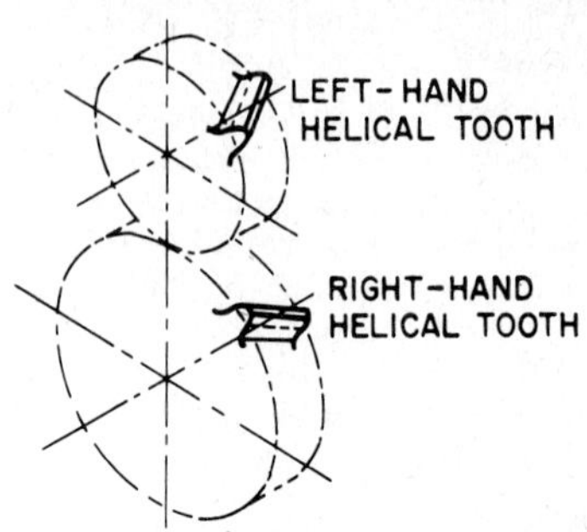

Figure 1-2 Parallel helical gears. *(Extracted from Ref. 2.)*

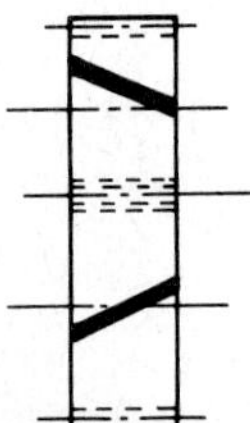

Figure 1-3 Single helical gears. *(Extracted from Ref. 2.)*

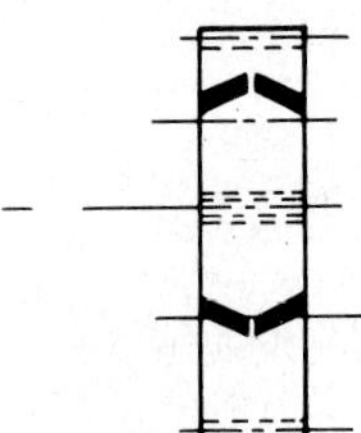

Figure 1-4 Double helical (herringbone) gears. *(Extracted from Ref. 2.)*

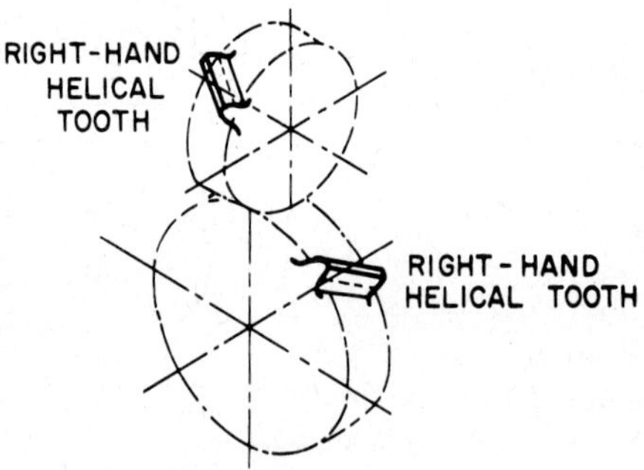

Figure 1-5 Crossed helical gears. *(Extracted from Ref. 2.)*

- Alloy-steel pinions are used when there are increased loads or greater life is desired. They may be used with cast iron or annealed (forged or cast) steel gears, usually when the ratio is about 6:1 or higher.
- Alloy-steel pinions and gears, heat-treated, should be used with the higher hardness ranges when space limitation is a factor, i.e., where smaller center distances and face widths may be necessary.

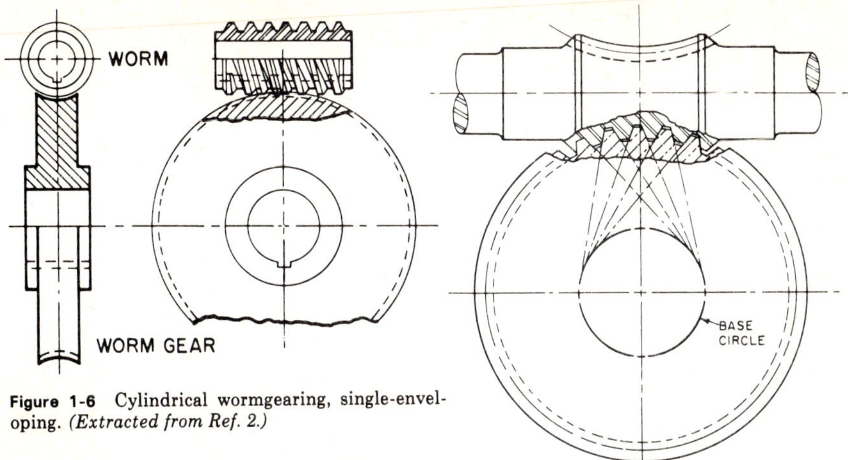

Figure 1-6 Cylindrical wormgearing, single-enveloping. *(Extracted from Ref. 2.)*

Figure 1-7 Double-enveloping Cone® worm gearing. *(Extracted from Ref. 2.)*

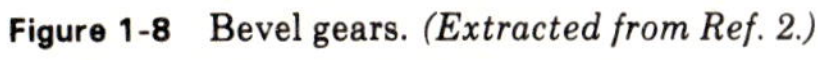

Figure 1-8 Bevel gears. *(Extracted from Ref. 2.)*

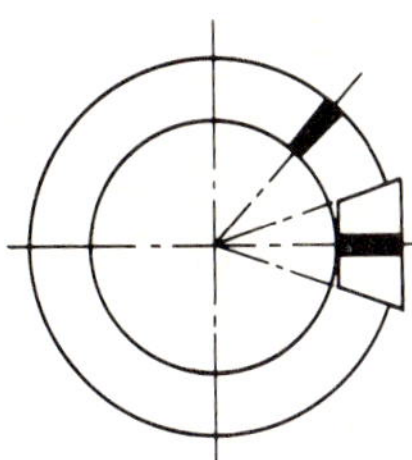

Figure 1-9 Straight bevel gears. *(Extracted from Ref. 2.)*

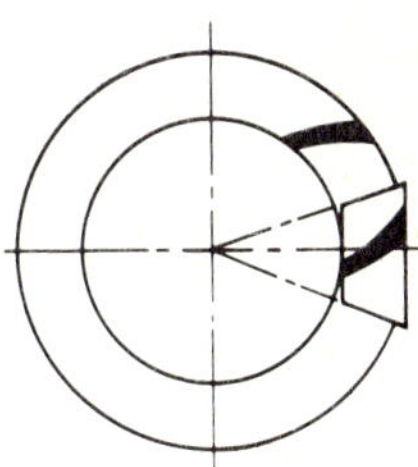

Figure 1-10 Spiral bevel gears. *(Extracted from Ref. 2.)*

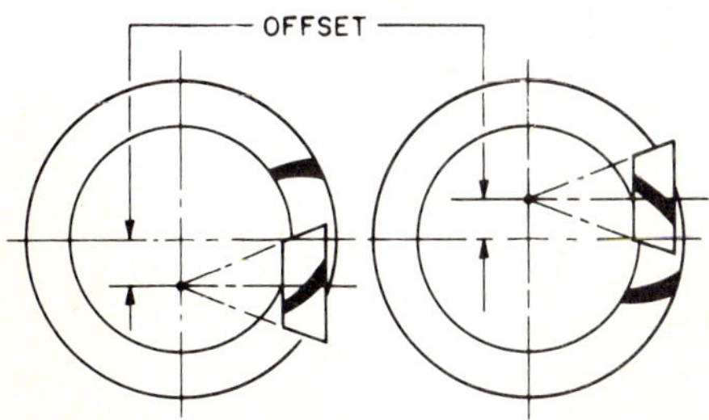

Figure 1-11 Hypoid gears. *(Extracted from Ref. 2.)*

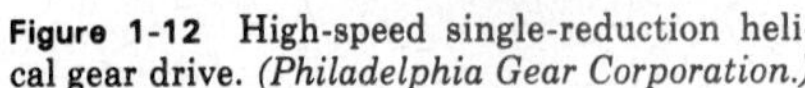
Figure 1-12 High-speed single-reduction helical gear drive. *(Philadelphia Gear Corporation.)*

Figure 1-13 Double-reduction helical gear drive. *(Philadelphia Gear Corporation.)*

- Steel pinions and gears which are to be machined and cut after heat treatment should have the pinion hardness specified as follows:
  Ratios up to 2:1; pinion and gear to be same minimum hardness
  Ratios from 2:1 to 8:1; minimum hardness of pinion to be 40 Bhn (Brinell hardness number) higher than minimum gear hardness
  Ratios of 8:1 and higher; pinion to be more than 40 Bhn harder at minimum than gear
- Steel pinions and gears hardened to 400 Bhn or higher after cutting are generally specified with the same hardness, unless extremely high hardness is desired for the pinion.

Figure 1-14 Double-reduction worm-gear drive. *(Boston Gear Division of INCOM International Inc.)*

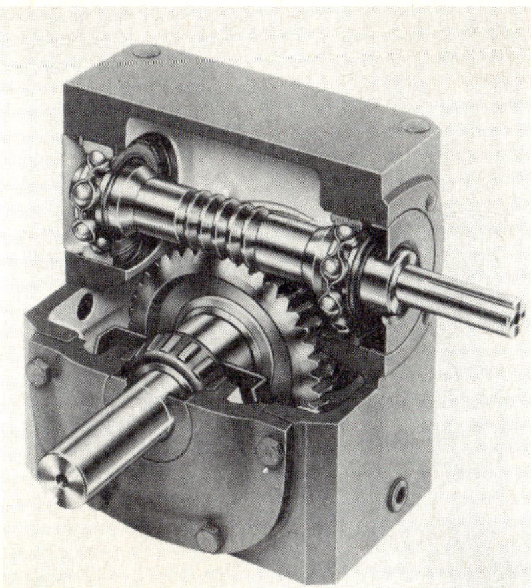

Figure 1-15 Single-reduction worm-gear drive. *(Boston Gear Division of INCOM International Inc.)*

- A range of 40 Bhn should be specified for minimum hardnesses up to 285 Bhn. A range of 50 Bhn should be specified for minimum hardnesses over 302 Bhn.[1] For example, a pinion might be specified with a range from 285 to 321 Bhn and the mating gear from 223 to 262 Bhn.
- When core hardness is a requirement, consideration should be given to the size and shape of the cross sections involved.
- Where impact loads exist, the use of alloys and the lowering of hardness are recommended for carburized gears and pinions.
- When accelerated wear is encountered in service, heat-treated gearing providing greater hardness will, in most cases, help to alleviate this condition. Consult with an AGMA company member for appropriate recommendations.

Figure 1-16 Single-reduction spiral bevel gear drive. *(The Falk Corporation.)*

- On flame- and induction-hardened gears, where hardness in the root area is a requirement, it should be specified.

Manufacturing Processes

Generating

Most gear teeth are produced by a generating process that takes into account the shape of the tool and relative motions between the tool and workpiece. Examples are: hobbing, shaping, rolling, grinding, and shaving.

Forming

The forming process produces gear teeth by a direct replication method. Examples are: casting, molding, and broaching.

Finishing

Finishing operations remove a relatively small amount of material from gear teeth. These may be used to improve accuracy and finish. Shaving is used to improve profile and finish on relatively soft gears. Grinding and honing are employed for harder gears.

Size Limitations

General limitations on gear diameters are listed below. A specific company may have the capability of producing larger gears of a particular type. Spur gears of very large diameter for low-speed application have been cut in arc sections and also fabricated from rack sections, bent to the proper radius of curvature.

	Approximate Maximum Diameter	
Gear Type	*in*	*m*
Spur	400	10
Helical	400	10
Straight bevel	100	2.5
Spiral bevel	90	2.3
Worm gear	150	4

Accuracy Requirements

Requirements for gear accuracy may be governed by several factors. Operating speed and gear size are probably the most important. From an application standpoint, specifications on sound level and smoothness of motion transmission are often controlling.

AGMA 390, "Gear Handbook," contains a classification system covering tolerances for common types of gears. In this system tolerances are tabulated for a series of *quality numbers* that range from Q3 to Q16. The higher the quality number the more precise are the tolerances. The handbook covers the gears themselves and does not apply to gears in enclosed drives.

By far the majority of applications can be satisfied by gears in the range of Q5 to Q8. Rarely does a job require a gear more precise than class Q14. AGMA 390 tabulates a range of suggested quality numbers for various applications. This is only a guide. Judgment should be used in making a final selection to meet the specific conditions involved.

Precise Motion Transmission

Many applications use gears as a means for precise transmission of motion. Usually, high AGMA quality numbers (Q12 to Q14) are required for these gears. Examples of such applications are: navigational tracking, telescopes, and index devices for machine tools.

High-Speed Considerations

Gear inaccuracies become more critical as operating speed increases. The most obvious problem may be a high noise level. In addition, the dynamic loads on the teeth caused by these errors may be a substantial part of the total transmitted load.

Pitch line velocity, which takes into account size as well as rotational speed, is the usual measure of gear speed. The unit normally used to measure this velocity at tooth mesh is feet per minute.

AGMA 390 suggests the following quality numbers for power drives in the machine tool industry.

Gear Pitch Line Velocity, ft/min	*Quality Number*
0–800	Q6–Q8
800–2000	Q8–Q10
2000–4000	Q10–Q12
Over 4000	Q12 and up

Accuracy vs. Cost

Cost of manufacturing and inspection escalates rapidly with increasing quality number. Therefore, the user should use care to avoid specifying higher classes than required for the application.

Mounting of Gears

Close attention must be given to the mounting of gears and the maintenance of alignment under operating conditions. Obviously, precision of mounting must match that of the gears.

Enclosed Gear Drives

Advantages

Enclosed gear drives marketed by gear manufacturers offer several advantages over open power-transmission devices.

- Safety—protection from moving parts
- Retention of lubricant
- Protection from environment
- Economics of quantity manufacture

Types and Features

Enclosed gear drives generally are classified by the principal type of gearing used. They may have a single set of gears or additional gears of either the same or different types to form multiple reductions. Figures 1-12 to 1-17 provide illustrations of common types. Table 1-1 covers important features of each.

Mounting

Gear drives may be designed for base, flange, or shaft mounting. The last type makes use of a hollow output shaft for direct mounting on the driven shaft (refer to Fig. 1-18). A reaction arm or similar device is required to secure the unit against rotation.

Gear Motors

A gear motor is an integral drive unit incorporating an electric motor and gear reducer, with the frame of one supporting the other. Some designs use motors with special shaft ends and/or mountings, while others adapt standard motors (refer to Fig. 1-19).

Figure 1-17 Double-reduction spiral bevel–helical gear drive. *(The Falk Corporation.)*

Normal Speed vs. High Speed

AGMA standards for enclosed gear drives, used in general industrial service, limit input speed to 3600 r/min. An additional limitation is imposed: 5000 ft/min pitch line velocity for helical and bevel units and 6000 ft/min sliding velocity for cylindrical worm gears. Above these limits, special consideration should be given to such items as gear quality, lubrication, cooling, bearings, etc. See Refs. 3 and 4.

TABLE 1-1 Enclosed Gear-Drive Units—Features
(Refer to Figs. 1-12 to 1-15.)

Type of unit	Ratio range*	Horsepower range*	Efficiency %†	Fig. no.
Helical & double helical				
Single reduction	to 7:1	to 20,000	96–98	1-12
Double reduction	5:1–40:1	10,000	95–97	1-13
Triple reduction	20:1–300:1	3,000	93–95	——
Quadruple reduction	150:1–1000:1	1,000	91–93	——
Bevel, helical	5:1–40:1	3,000	95–97	——
Bevel helical, helical	20:1–300:1	2,000	93–95	——
Bevel				
Single bevel	to 9:1	2,000	96–98	
Worm gear				
Single reduction	5:1–70:1	300	50–96	1-15
Double reduction	25:1–4900:1	100	20–92	1-14
Helical, worm	20:1–300:1	150	50–94	——

*The information on ratio and horsepower ranges is approximate and is for the product usually offered.

†Efficiency is based on transmission of full rated power.

Figure 1-18 Shaft-mounted double-reduction helical gear drive. *(The Falk Corporation.)*

INSTALLATION AND MAINTENANCE

The variety of types and sizes of gears and enclosed gear drives makes it impractical to cover installation and maintenance in specific detail. The user should refer to the manufacturer's literature, nameplate data, and warning tags. Such information should take precedence over the generalized comments that follow.

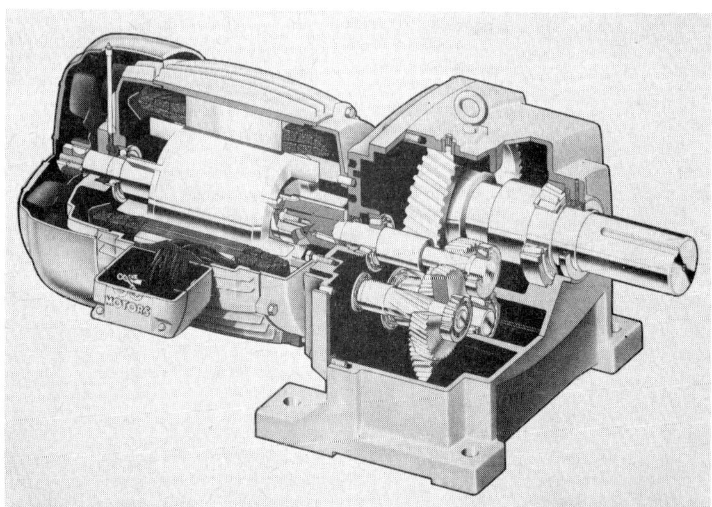

Figure 1-19 Triple-reduction helical gear motor. *(U.S. Electrical Motors.)*

Mounting and Installation of Gears

Accuracy of mounting must be commensurate with the quality of the gears themselves to obtain optimum results. Some types of gears require close endwise positioning of either or both members of a pair to obtain proper operation. Examples are bevel gears (both), cylindrical worm gears (gear only), double-enveloping worm gears (both). This positioning must be provided by bearings of suitable capacity to accommodate the thrust loads involved.

Provision must be made for adequate lubrication of open or semi-enclosed gears and guarding for safety. See Chap. 17-1, "Lubricants," for information on gear lubrication.* Reference 5 also provides information on lubricant types and methods of lubrication.

Installation and Start-Up of Enclosed Gear Drives

The handling, installation, and servicing of a new enclosed gear drive deserves close attention to avoid damage and to assure proper operation. A checklist of important items is provided in Table 1-2.

Lubrication of Enclosed Gear Drives

Improper lubrication is one of the major causes of failure of gear drives. The gear manufacturer's instructions must be followed to assure proper operation.

Reference 6 provides detailed information on recommended lubricants and maintenance procedures. Gear type, size, and speed, along with ambient temperature range, are major influencing factors on lubricant selection.

The gear unit should be drained and cleaned with a flushing oil after 4 weeks of initial operation. For refilling, either the filtered original lubricant or new lubricant may be used. For normal operation, oil changes should be made after every 2500 h of service.

Periodic checks must be made on oil levels, oil cups, and grease fittings. When pressure lubrication is used, proper functioning of pump, filter, and cooler should be frequently audited.

TABLE 1-2 Installation and Start Up Checklist

Step	Instruction
Storage	If necessary to store or maintain the gear unit in an inactive condition for more than a month, contact the manufacturer to determine need for protective action.
Handling	Observe manufacturer's instructions for unpacking and handling of gear drive.
Foundation	Provide adequate foundation commensurate with size and type of unit. Surface is to be level unless drive has been specifically designed for other positioning.
Accessibility	Provide adequate space to permit future maintenance.
Auxiliary parts	Assemble components, such as coupling hubs, sprockets, etc., to shafts with shrink or slip fits in accordance with instructions. Do not force fit.
Alignment	Align shafts and auxiliary drives accurately. Most couplings are designed to accommodate only minor misalignment.
Guards	Install suitable guards for safety in accordance with OSHA standards.
Lubrication	Observe manufacturer's instructions, using the specified lubricants. *Note: Most gear drives are shipped without lubricant.*
Rotation	Check for proper rotation and freedom from obstructions before start of full operation. For pressure lube systems, make certain pump is delivering oil.
Operation	Inspect for oil leaks and unusual noise or vibration immediately after start-up. Check oil temperature after several hours. A temperature of 180 to 200°F is not unusual for most gear drives. After first week, recheck alignment and tightness of all fasteners, fittings, and pipe plugs.

*See Rosaler and Rice: *Standard Handbook of Plant Engineering*, 1983.

TABLE 1-3 Trouble Chart

Trouble	What to inspect
Heating	Is unit and fan assembly covered with dirt? Is unit overloaded? Has recommended oil level been exceeded or is level too low? Are couplings in alignment? Have bearings been properly adjusted? Are oil seals or stuffing boxes the cause? Is oil clean or is sludge content high? Has oil filter been cleaned? Is oil pump functioning?
Shaft failure	Check alignment; most shafts fail owing to misalignment. Some troubles are caused by use of rigid couplings. Is overhung load beyond capacity of unit? Is unit subject to high energy loads or extreme repetitive shocks not previously considered?
Bearing Failure	Rust formation caused by high humidity or the entrance of water. Unsuitable lubricant. Abnormal loading causing excessive deflection results in flaking, cracks, and fractures. Improper adjustment causes abnormal loading if bearings are pinched or abnormal gear wear if bearings are too free. This is dependent upon type of bearing and the possible lack of lubrication.
Oil leakage	Check oil seals and replace if worn. Check stuffing boxes and adjust or replace packing. Check tightness of drain, level, and other plugs or fittings.
Wear	Backlash may be insufficient. Misalignment due to worn bearing. Incorrect lubrication. Insufficient lubrication. Lubricant carrying foreign matter, viz., abrasive dirt or particles of worn metal teeth.[7] Excessive temperature. Excessive speeds. Excessive loads.
Noise or vibration	Bad alignment. Loose or worn bearings. Insufficient lubrication. Excessive lubrication.

Troubleshooting

Alertness to changes in operating characteristics, such as increased temperature rise over ambient noise and vibration and oil leakage, can prevent costly shutdowns. Table 1-3, "Trouble Chart," provides a checklist to diagnose various problems in operation.

APPLICATION OF GEARING AND ENCLOSED GEAR DRIVES

Gear Ratings

AGMA has developed rating formulas for most types of gearing and enclosed gear drives. The ratings determined from these formulas are intended for applications where loads of a uniform nature are applied for no more than 10 h/day. It is these ratings that normally are tabulated in manufacturers' catalogs.

Service Factors

In selecting gearing or an enclosed gear drive for an application, the horsepower to be transmitted is multiplied by a service factor to determine an *equivalent horsepower*. Service factors have been developed from the experience of manufacturers and users to allow for the nature and duration of the transmitted load. Table 1-4 provides a tabulation of service factors extracted from Ref. 3 for enclosed speed reducers or increasers using spur,

TABLE 1-4 AGMA Standard Practice for Enclosed Speed Reducers or Increasers Using Spur, Helical, Herringbone, and Spiral Bevel Gears

Prime mover	Duration of service, h/day	Driven machine load classifications: Uniform	Moderate shock	Heavy shock
Electric motor,	Occasional, ½	0.50	0.80	1.25
steam turbine,	Intermittent, 3	0.80	1.00	1.50
or	Over 3 up to and incl. 10	1.00	1.25	1.75
hydraulic motor	Over 10	1.25	1.50	2.00
Multicylinder	Occasional, ½	0.80	1.00	1.50
internal	Intermittent, 3	1.00	1.25	1.75
combustion	Over 3 up to and incl. 10	1.25	1.50	2.00
engine	Over 10	1.50	1.75	2.25
Single-cylinder	Occasional, ½	1.00	1.25	1.75
internal	Intermittent, 3	1.25	1.50	2.00
combustion	Over 3 incl. 10	1.50	1.75	2.25
engine	Over 10	1.75	2.00	2.50

helical, herringbone, and spiral bevel gears. The factors for other types of gears vary slightly from those shown.

Application Classification

Most AGMA standards for enclosed gear drives provide tables for various applications as a guide for selecting service factors. This information is usually also contained in manufacturers' catalogs.

Product Selection

After the equivalent horsepower has been determined, selection of gearing or enclosed gear drives can be made by comparing this figure with the basic rating. It is necessary that the product selected have a rated load capacity equal to or in excess of the equivalent horsepower. An enclosed gear drive usually must also be checked for thermal rating. This is the horsepower that can be transmitted continuously for 3 h or more without causing a temperature of more than 100°F above ambient temperature. Should this limitation prevail, several alternatives are available, such as auxiliary cooling systems, oil pans to reduce churning, or selection of a larger unit.

Systems Considerations

An essential phase in the design of a system of rotating machinery is the analysis of the dynamic (vibration) response of a system to excitation forces.

The dynamic response of a system results in additional loads imposed on the system and relative motion between adjacent elements in the system. The vibratory loads are superimposed upon the mean running load in the system and, depending upon the dynamic behavior of the system, could lead to failure of the system components.

In a gear unit, these failures could occur as tooth breakage or pitting of the gear element, shaft breakage, or bearing failure.

Any vibration analysis must consider the complete system, including prime mover, gear unit, driven equipment, couplings, and foundations. The dynamic loads imposed upon a gear unit are the result of the dynamic behavior of the total system and not of the gear unit alone.

For further information, see Ref. 8.

Sound and Vibration

The greatest concern regarding sound and vibration of gear drives is the contribution to the industrial noise level. A second concern is that these may be sypmtomatic of abnormal wear and impending failure.[9–12] Refer to Sec. 6 of this Handbook, "Noise and Vibration Control," for additional information.

REFERENCES: AGMA STANDARDS APPLICABLE TO ENCLOSED GEAR DRIVES

1. "Gear Materials Manual," AGMA 240.01, 1972.
2. "Terms, Definitions, Symbols, and Abbreviations," ANSI/AGMA 112.05, 1976.
3. "Practice for Enclosed Speed Reducers or Increasers Using Spur, Helical, Herringbone and Spiral Bevel Gears," AGMA 420.04, 1975.
4. "Practice for High Speed Helical and Herringbone Gear Units," AGMA 421.06, 1969.
5. "Lubrication of Industrial Open Gearing," AGMA Specification 251.02, 1974.
6. "Lubrication of Industrial Enclosed Gear Drives," AGMA Specification 250.03, 1972.
7. "Gear-Tooth Wear and Failure," ANSI/AGMA 110.03, 1962.
8. "Systems Considerations for Critical Service Gear Drives," AGMA Information Sheet 427.01, 1976.
9. "Measurement of Sound on High Speed Helical and Herringbone Gear Units," AGMA Specification 295.04, 1977.
10. "Sound for Enclosed Helical, Herringbone and Spiral Bevel Gear Drives," AGMA 297.01, 1973.
11. "Sound for Gearmotors and In-Line Reducers and Increasers," AGMA 298.01, 1975.
12. "Fundamentals of Sound as Related to Gears," AGMA 299.01, sec. 1, 1978.
13. "Manual for Assembling Bevel and Hypoid Gears," AGMA 331.01 (R1976), 1969.
14. "Manual for Machine Tool Gearing," AGMA 360.02, 1971.
15. "Gear Classification, Materials and Measuring Methods for Unassembled Gears," "AGMA Gear Handbook," AGMA 390.03, Vol. I, 1973.
16. "Practice for Single- and Double-Reduction Cylindrical-Worm and Helical-Worm Speed Reducers," AGMA 440.04, 1971.
17. "Practice for Single, Double and Triple-Reduction, Double-Enveloping Worm and Helical-Worm Speed Reducers," AGMA 441.04, 1978.
18. "Practice for Worm Hollow Output Shaft Speed Reducers," AGMA 442.01 (R1974), 1965.
19. "Practice for Gearmotors Using Spur, Helical, Herringbone and Spiral Bevel Gears," AGMA 460.05, 1971.
20. "Practice for Worm Gearmotors," AGMA 461.01 (R1974), 1966.
21. "Practice for Spur, Helical and Herringbone Gear Shaft-Mounted Speed Reducers," AGMA Information Sheet 480.06, 1977.
22. "Spiral Bevel, Helical and Herringbone Gear Units for Water Cooling Tower Fans," AGMA Information Sheet 490.02, 1972.

chapter 1-2

Bearings

PART 1

Rolling Element Bearings

by
William H. Carter
Standards Engineer,
The Torrington Company
South Bend, Indiana

GLOSSARY*

Aligning bearing A bearing which, by virtue of its shape, is capable of considerable misalignment.

*Extracted with permission from Anti-Friction Bearing Manufacturers Association Standards.

Antifriction bearing A term commonly given to ball and roller bearings.

Average life The summation of all bearing lives in a series of life tests, divided by the number of life tests.

Axial clearance See end play.

Basic load rating The calculated, constant, radial, or centric thrust load which a group of apparently identical bearings can theoretically endure for 1 million (10^6) revolutions.

Basic static-load rating The static load which corresponds to a total permanent deformation of ball and race, or roller and race, at the most heavily stressed contact of 0.0001 in (0.00254 mm) of the ball or roller diameter.

Bearing series A graduated dimensional listing of a specific type of bearing.

Bore The area of the bearing making contact with the bearing seat on the shaft.

Boundary dimensions Dimensions for bore, outside diameter, width, and corners.

Cage A device which partly surrounds the rolling elements and travels with them. The main purpose of the cage is to space the rolling elements in ball bearings and space and guide the elements in roller bearings.

Combined load A combination of all radial and axial forces on a bearing.

Cone The inner ring of a tapered roller bearing.

Conrad ball bearing A nonfilling-slot type of ball bearing.

Cup The outer ring of a tapered roller bearing.

Diametral clearance See radial internal clearance.

Double-row bearing A bearing with two rows of rolling elements.

End play The measured maximum possible movement parallel to the bearing axis of the inner ring in relation to the outer ring.

Equivalent radial load The calculated, constant, stationary, radial load, which, if applied to a radial bearing, would give the same life as attained under actual conditions of load and rotation.

Equivalent thrust load The calculated, constant, centric axial load which, if applied to a thrust bearing, would give the same life as attained under actual conditions of load and rotation.

Fixed bearing A bearing which positions a part of the shaft against axial movement.

Floating bearing A bearing so designed or mounted as to permit axial displacement between shaft and housing.

Full-complement bearing A cageless bearing with a maximum number of rolling elements.

Housing bearing seat The part of the housing bore which contacts the outside diameter of the bearing.

Housing fit The amount of interference or clearance between the bearing outside diameter and the housing bore seat.

Inch bearing A bearing designed in inch dimensions.

Internal clearance See radial internal clearance.

Lateral travel See end play.

Life The life of an individual ball or roller bearing is the number of revolutions (or hours at some given constant speed) which the bearing runs before the first evidence of fatigue develops in the material of either the ring (or washer) or of any of the rolling elements.

Load rating Load ratings for specific speeds are based on a rating life of 500 h (AFBMA).

Loading groove A slot in the raceway shoulder that permits assembly of a maximum number of rolling elements.

Lubrication groove A continuous recess in a bearing for conveying lubricant.

Lubrication hole A hole in the rings to pass lubricant to rolling elements.

Metric bearing A bearing designed to metric dimensions.

Multirow bearing A bearing with more than two rows of rolling elements.

Needle roller A load-carrying rolling element generally understood to be long in relation to its diameter.

Outside diameter The area of the bearing making contact with the housing bearing seat.

Pitch diameter, rolling elements The diameter of the pitch circle generated by the center of a rolling element as it traverses the bearing's axis of rotation.

Pocket (cage) The portion of cage which is shaped to receive the rolling element.

Preload An internal loading characteristic in a bearing which is independent of any external radial and/or axial load carried by the bearing.

Races The inner ring or outer ring of a bearing.

Raceway The path of the rolling element on either ring of a bearing.

Radial bearing An antifriction bearing primarily designed to support a load perpendicular to the shaft axis.

Radial internal clearance For a single-row radial contact bearing, this is the average outer-ring raceway diameter, minus the average inner-ring raceway diameter, minus twice the rolling-element diameter.

Radial load Radial load is that load which may result from a single force or the *resultant* of several forces acting in a direction at right angles to the bearing axis.

Radial play See radial internal clearance

Rating life For a group of apparently identical bearings the rating life (L_{10}) is the life in millions of revolutions that 90 percent of the group will complete or exceed.

Retainer See cage.

Sealed bearing A ball or roller bearing protected against loss of lubricant and from outside contamination.

Self-aligning bearing A bearing with built-in compensation for shaft or housing deflection or misalignment.

Self-contained bearing Unit bearing assembly (nonseparable).

Separable bearing A bearing assembly that may be separated completely or partially into its component parts.

Separator See cage.

Shaft bearing seat The portion of the shaft upon which the bearing is mounted.

Shaft fit The amount of interference or clearance between the bearing inside diameter and the shaft bearing seat diameter.

Shield A circular part affixed to one bearing ring to cover the interspace, but not to run in contact with the other ring.

Single-row bearing A bearing having only one row of rolling elements.

Special bearing A bearing not meeting the requirements of standard or established line bearings.

Spherical roller bearing (radial) A bearing which, by virtue of the raceway or outer-ring construction, is capable of considerable misalignment.

Spherical roller thrust bearing An antifriction thrust bearing using spherical rollers as rolling elements.

Standard bearing An antifriction bearing conforming to AFBMA, "General Boundary Plans of Metric and Inch Dimensions."

Static equivalent load The calculated, static radial or static centric thrust load, which if applied to a bearing would cause the same total permanent deformation at the most heavily stressed rolling element and raceway contact as that which occurs under an actual condition of loading.

Static load Static load is a load acting on a nonrotating bearing.

Tapered roller A roller with one end smaller than the other (the frustum of a cone).

Thrust bearing A bearing designed primarily to support a load parallel to shaft axis.

Thrust load The load which results from a single force or the resultant of several forces acting in a direction parallel to the bearing axis.

Washer (thrust bearing) An annular ring upon which thrust bearing raceways are ground.

INTRODUCTION

Antifriction (rolling-element) bearings continue to find increased use wherever the reduction of friction is required at the interface of dynamic and static components in machinery. A principal advantage of these bearings is the ability to operate at friction levels considerably lower than those of plain or oil-film bearings, while maintaining coefficients of friction at start-up that are close to those in normal operation.

The complement of rolling elements in antifriction bearings comprises balls or rollers (or even a combination of both in some special designs). In concept, the balls or rollers are arranged within a bearing primarily to support either pure radial or pure thrust loading, but they sometimes have a capability for accomplishing both.

Standard (off-the-shelf) bearings usually consist of four essential components. The inner ring and outer ring with their raceways, and a complement of rolling elements (balls or rollers) contained and separated by the cage, retainer, or separator. In operation the hardened and ground surfaces of the raceways form the track for the rolling elements (load-supporting members) to follow while transmitting load from the dynamic members of an assembly to the stationary members.

APPLICATION INFORMATION

Bearing Life

The life of a bearing is expressed as the number of revolutions or the number of hours, at a given speed, for which the bearing will operate before any evidence of fatigue develops on the rolling elements or the raceways. Life may vary from one bearing to another, but stabilizes into a predictable pattern when considering a large group of the same size and type of bearings. The *rating life* of a group of such bearings is the number of hours or revolutions (at a given constant speed and load) that 90 percent of the tested bearings will exceed before the first evidence of fatigue develops. This is called L_{10} *life* or *minimum life.*

Average Life

The results of testing a large group of ball or roller bearings may be graphically illustrated. The distribution curve shown is obtained by plotting relative life vs. percent of bearings tested. See Fig. 2-1.

From the curve in Fig. 2-1 it is apparent that the average life is approximately five times the minimum life. About 50 percent of the bearings will exceed the average life. Since it is not possible to predict the exact life of a single bearing, a safety factor must

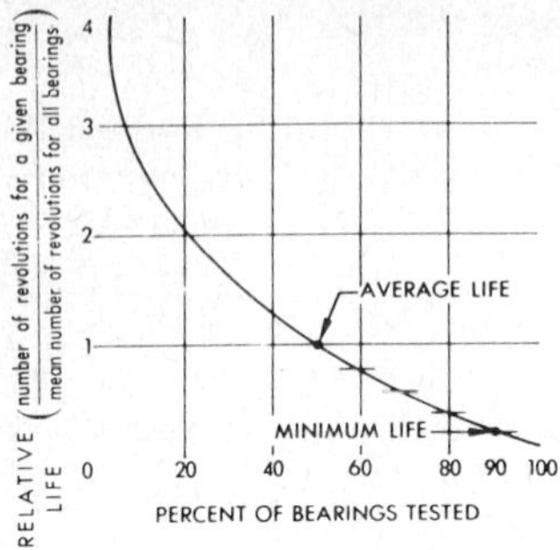

Figure 2-1 Average-life curve.

be allowed to minimize the chances of early failure. The cost of replacing a bearing plus the expense of machine downtime may greatly exceed the relatively low cost of the bearing. Therefore, most designers prefer to use minimum life as a basis for design. In some applications where safety or maintenance economy is not critical and low initial bearing cost is desirable, the average-life value may be used.

Life-and-Load Relationship

Empirical calculations and experimental data point to a predictable relationship between bearing load and life. This relationship may be expressed by formulas. In these empirical formulas the bearing life is found to vary inversely as the applied load to an exponential power. The assigned value of the exponent depends upon the basic type of rolling element.

For all types of roller bearings the formula is

$$\text{Life} = \left(\frac{\text{basic dynamic capacity}}{\text{load}}\right)^{3.33} \qquad 10^6 \text{ revolutions}$$

For all types of ball bearings the formula is

$$\text{Life} = \left(\frac{\text{basic dynamic capacity}}{\text{load}}\right)^{3} \qquad 10^6 \text{ revolutions}$$

Ring Rotation Factors (RF)

The basic dynamic capacity of a radial bearing is based on the inner ring rotating with respect to the applied load. Recent findings indicate that the bearing capacity does not have to be downgraded when the outer ring rotates. If the inner ring or outer ring rotates with respect to load, the rotation factor is 1.0.

Effect of Load

It is also evident from the exponential character of the basic life–load relationships, that for any given speed, a change in load may have a substantial effect on the life. For a roller bearing, if the load is doubled, the life is reduced to one-tenth its former calculated duration. Similarly, if the load is halved, the life is increased tenfold. For a ball bearing, if the load is doubled, the life is reduced to one-eighth its former value. Likewise, if the load is reduced one-half, the life is increased eightfold.

Effect of Speed

The preceding expressions are independent of the speed of the bearing and are valid for speeds ranging from 10 to 10,000 r/min. For speeds below 10 r/min, consult the engineering department of a bearing manufacturer.

If bearing life is measured in hours, an increase in speed results in a decrease in hours of life. The number of revolutions per unit time determines the hours of life available before the fatigue limit of the bearing is reached. If the speed is doubled, the hours of life are halved. Conversely, if the speed is reduced by 50 percent the hours of life will be doubled.

Selection of Bearing Type

Selection of bearing type is made after the general design concept of the machine has been established and the magnitude of the loads and speeds estimated. Special conditions can directly affect bearing operation and must be considered. These include ambient or localized temperatures, shock or vibration, dirt or abrasive contamination, difficulty in obtaining accurate alignment, space limitations, need for shaft rigidity, etc.

Selection of the proper type of bearing is not an exact science. The fields of application for many types of bearings overlap, and the value of experience in bearing applications cannot be overemphasized. Each type of bearing, however, has inherent features, which determine its relative suitability for a specific application. Careful analysis of these features and familiarization with the fundamental characteristics of each type of bearing will help in selecting the proper bearing.

As an aid to experienced designers and inexperienced bearing users alike, the similarities and differences of ball and roller bearings are outlined. Table 2-1 shows the relative operating characteristics of each cataloged bearing type.

TABLE 2-1 Relative Operating Characteristics

Bearing type	Radial capacity	Thrust capacity	Limiting speed	Resistance to elastic deformation	
				Radial	Axial
Ball radial					
Conrad	Moderate	Moderate in both directions	High	Moderate	Low
Maximum-capacity	Moderate +	Moderate in one direction	High	Moderate +	Low+
Angular-contact	Moderate	Moderate + in one direction	High −	Moderate	Moderate
Roller radial					
Cylindrical roller	High	None	Moderate +	High	None
	High	Light in one direction	Moderate +	High	Not recommended
	High	Light in both directions	Moderate +	High	Not recommended
Journal	High +	None	Low	High	None
Spherical roller	High	Moderate in both directions	Moderate	High −	Moderate
Tapered roller					
Single-row	High −	Moderate + in one direction	Moderate	High −	Moderate
Single-row, steep angle	Moderate +	High in one direction	Moderate −	Moderate	High
Double-row	High	Moderate + in both directions	Moderate	High	Moderate
Double-row, steep angle	Moderate +	High in both directions	Moderate −	Moderate	High
Four-row	High +	High in both directions	Moderate −	High +	High
Thrust					
Angular-contact ball thrust	Low +	High − one direction	Moderate	Low	High −
Ball thrust	None	High one direction	Moderate −	None	High
Roller thrust	None	High + one direction	Low	None	High +
Self-aligning roller thrust	None	High + one direction	Low	None	High +
Tapered roller thrust	Locational only	High + one direction	Low	None	High +

Ball Bearings vs. Roller Bearings

Using balls as the rolling elements in bearings offers certain performance advantages. Most of the advantages of ball bearings are derived from the small area of contact between ball and raceway.

Ball bearings may be operated at higher speeds, with less internal friction and with less heat generation. They have a greater inherent ability to accommodate slight misalignment. Under certain conditions of combined loading, ball bearings occupy less space than required for roller bearings of the same bore size.

Rollers are not limited to a single geometric shape. There are several types, such as tapered rollers, spherical rollers, and cylindrical rollers. For a given load and diameter of rolling element, rollers transmit load through a larger contact area than do balls. This allows roller bearings to support greater loads and accommodate far more shock than ball bearings of equivalent size. For a given applied load, contact area stresses for roller bearings are lower than for ball bearings, and, therefore, they produce lower elastic deformation. Since the larger contact areas create more friction, permissible operating speeds for roller bearings are lower than those for ball bearings.

Selection of Bearing Size

Once the designer selects a suitable type of bearing for a set of specific conditions, the size of the bearing needed to provide adequate service life is determined. Many cases exist in which more than one bearing type will satisfy the operating conditions. In these instances the designer should determine the most suitable size of each type considered and make the final selection on the basis of mounting simplicity, space considerations, and overall economy.

The basic parameters affecting the choice of bearing size are radial load, thrust load, speed, required life, ring rotation, and shock or vibration conditions. Other factors such as misalignment, abnormal temperature, contamination, and poor lubrication will seriously reduce service life, but their exact effect cannot be determined. These factors should be eliminated by proper mounting design rather than by attempting to estimate their effect on bearing life.

Limiting Speeds

The ability of a bearing to operate at high speeds is dependent upon the rate at which generated heat is dissipated. Maximum speed is governed by bearing type, size, bearing load, ambient temperature, and type of lubricant.

The geometric design of a bearing and the method of positioning the rolling elements basically determine the coefficient of friction. Since frictional losses are proportional to the peripheral speed, it follows that the smaller the bearing the greater the speed at which it may operate.

Elastic deformation of the raceways and rolling elements is increased by heavy loads, and this creates additional heat, thereby limiting the allowable speed.

Ambient temperature may affect the rate of heat dissipation. Applications having a high ambient temperature require careful selection of the method of lubrication.

The type of lubricant is a basic criterion for establishing limiting speeds. Lubricants with high viscosity offer more frictional resistance, therefore oil is preferable to grease for higher speeds. Even with the proper frequency and amount of lubrication, limiting speeds for grease are approximately 50 percent of the given values. The use of circulating oil or oil mist will allow higher speed limits than does oil-bath lubrication.

Fits of Shaft and Housing

To ensure the full utilization of bearing capacity under operating conditions, it is important to have the proper fit between inner ring and shaft, and outer ring and housing. The tolerances to which the bearing is made are standardized, so desired fits may be obtained by controlling the dimensions and tolerances for the shaft and housing.

Normally, the problem in fit determination is to make the rotating ring of the bearing and its associated shaft or housing rotate as a single unit by using an interference fit. The fit of the nonrotating ring should be loose, with minimum clearance, for ease of assembly and axial movement in the housing.

The amount of interference fit employed should not create in the bearing rings excessive stress that might result in early fatigue failure. Under conditions of light load, the interference fit can be small. As the loads increase or shock loading is introduced, the interference must be increased so that no clearance exists and none can be induced by the load. This is the only effective means of preventing "creep." As a rule, axial clamping cannot be relied on to prevent creep since the clamping force must be excessively high. Thus, the heavier the load, the tighter the fit.

The degree of fit is labeled in the Anti-Friction Bearing Manufacturers Association (AFBMA) tolerance system which has been adopted where applicable. This tolerance system is in accordance with that adopted by the American National Standards Institute (ANSI). This system applies to all ball thrust bearings and radial bearings (except tapered roller bearings). For tapered roller bearings, an adaptation of the recommended AFBMA fitting practice has been made for the convenience of designers.

Special Materials for Bearings

Industrial demands for special bearings to meet abnormal service requirements spur the continual search for new and improved bearing materials. High temperatures, corrosive atmospheres, massive size, marginal lubrication, complex design, and space and weight limitations are typical abnormal requirements.

Conventional bearing steels are often inadequate when these problems are present. Sustained high operating temperatures reduce hardness, wear resistance, yield strength, and, therefore, bearing life. Conventional bearing steels also lack resistance to the oxidation which takes place at elevated temperatures.

Several materials such as 440-C stainless; 18-4-1 high-speed steel; M-1, M-2, M-10, and MV-1 high-speed steels; and high-cobalt alloys are used for high-temperature applications. For extremely high temperatures, materials such as metallic carbides and ceramics are used.

The combination of bearing size, complexity of design, and space and weight limitations can be a governing factor in the selection of bearing material. For example, a large-diameter, thin-section bearing with integral gear teeth and bolt holes would require a material which could be selectively hardened.

Corrosion is also a factor to be considered. The corrosive agent, load, temperature, and lubrication all affect the final choice of material. Some high-temperature materials with capacities equivalent to conventional bearing steel are used in both the normal- and high-operating-temperature ranges because of their good corrosion resistance. Other corrosion-resistant materials such as Monel, 18-8 stainless steel, and beryllium-copper alloys have sharply limited capacity, but excellent resistance to specific kinds of corrosion.

Monel and beryllium copper are not as hardenable as bearing steels and are nonmagnetic and resist saltwater corrosion. These qualities make them excellent materials for marine applications.

A variety of materials is also used for cages. Synthetic resin–impregnated fabrics, aluminum, and nylon products are becoming popular in the normal temperature ranges. Ductile iron, certain stainless steels, and iron-silicon bronze are used for higher temperatures.

Mounting Design

Mounting design varies widely, depending upon the type of bearing used and the requirements of the application. Selection of bearing type and mounting design are closely related since many cases exist where selection of bearing type is influenced by mounting design considerations.

Most applications require the use of more than one bearing on a shaft. Two identical bearings or a combination of different types and sizes may be used on a common shaft. The advantages of each combination should be evaluated by the designer in selecting bearings.

A *fixed*-bearing mounting locates the shaft and carries any thrust loads which exist in the application. A *float*-bearing mounting accommodates relative axial movement between the shaft and housing. Various combinations of these two basic mountings are used:

1. Fixed-float mounting
2. Fixed-fixed (or opposed) mounting
3. Float-float mounting

The fixed and float arrangement is necessary when: (1) a long shaft is used, (2) thermal expansion or contraction of the shaft with respect to the housing occurs, or (3) separate housings are required for two or more bearings on a common shaft.

The fixed and float combination offers many desirable features for heavy industrial equipment. The upper half of Fig. 2-2 shows a typical arrangement. The axial location

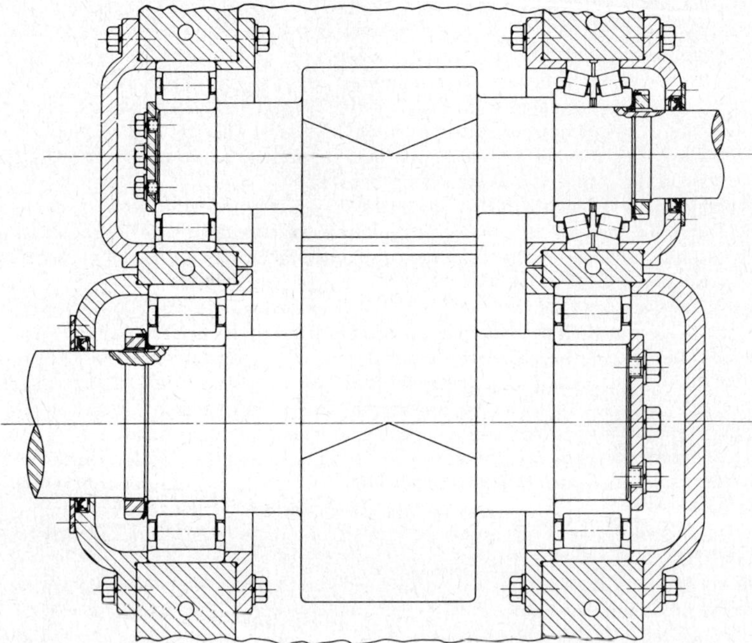

Figure 2-2 Typical fixed- and float-bearing arrangements.

is accomplished through the fixed bearing since it is clamped rigidly to the shaft and the housing. A two-row tapered roller bearing is shown in the fixed position. However, other radial bearings capable of taking thrust loads in both directions and applied radial loads may be used. For the float position the cylindrical roller bearing shown supports relatively heavy loads and permits free axial displacement. A type TDO or TNA tapered roller bearing may be used if the sliding pressures between the outer ring and the housing bore are not excessive.

The lower half of Fig. 2-2 demonstrates the float-float mounting. A pair of cylindrical roller bearings permits the entire shaft to float. A herringbone gear is used and the floating shaft allows the gears to mesh properly. The arrangement shown permits manufacturing economies since gear axial alignment is assured without very accurate machining or shimming of the bottom housing cover and end plate. A pair of type TDO or TNA tapered roller bearings, not fixed axially in their housings, may be used if the sliding pressures are not excessive.

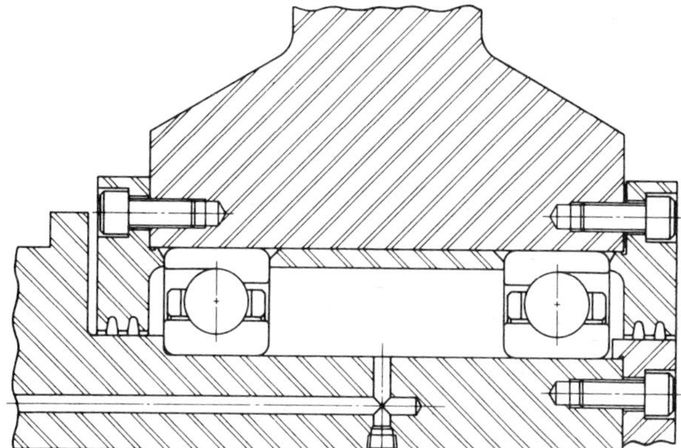

Figure 2-3 Typical fixed-fixed bearing arrangement.

Figure 2-3 illustrates a typical mounting arrangement where the bearings will be subjected to a radial load with some locational thrust load. A gap is left between the clamping ring and the face of the inner ring to provide for tolerance accumulation. Use of a spacer between the faces of the outer rings allows through-boring of the housing. Although maximum-capacity (types BH or BIH) ball bearings are shown, Conrad (types BC or BIC) ball radial bearings could also be used when there are moderate loads. Some applications, such as flywheels or spur gear hubs, require a tight fit of the outer ring in the housing with a loose fit of the inner ring on the shaft.

Seals and Closures

Seals are used to protect the bearing from contamination as well as to retain the lubricant. There are three basic types of seals:

1. **Lip Contact, Commercial Seals** These are usually standard components made by several manufacturers.
2. **Annulus and Labyrinth Seals** These are noncontact seals with slight clearance between stationary and rotating members depending on the lubricant to effect a frictionless closure.
3. **Slinger Seals** External types depend on centrifugal force to fling foreign matter away from the shaft. Internal slinger-type seals are used to distribute lubricant within the housing and to shield the bearing.

Basic types of sealing arrangements are shown in Fig. 2-4.

Type A. Commercial seal. The contact lip may be synthetic rubber or leather, spring-backed for more positive sealing. Consult manufacturers' literature for shaft finish and limiting speeds. May be used for either grease or oil.

Type B. Annular grooves. Shown here with drain slot at bottom. These may be in either the shaft or the housing, or in both. The effectiveness is increased by keeping the running clearance small and by using multiple grooves. Used for oil or grease lubrication.

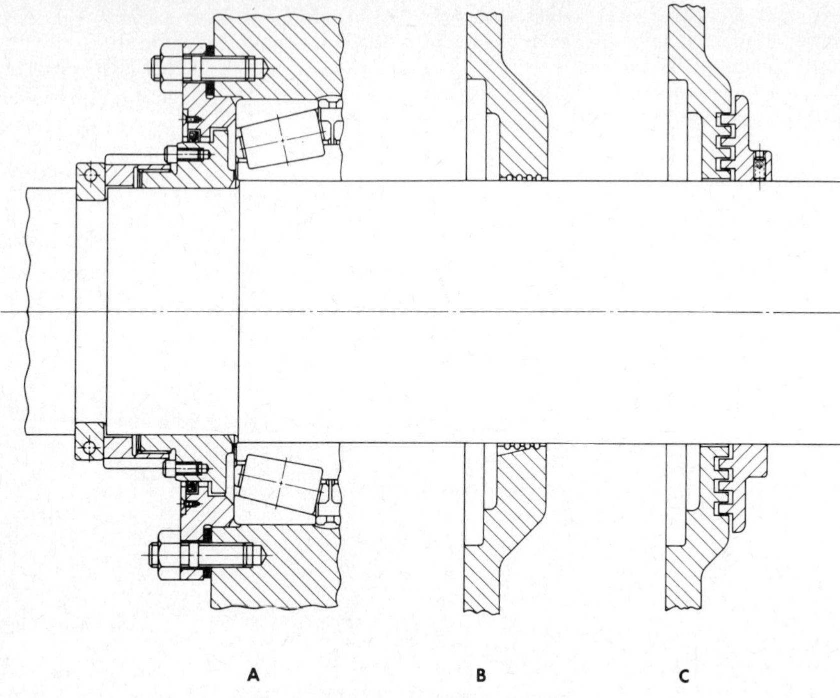

Figure 2-4 Basic sealing arrangements.

Type C. Axial labyrinth seal. Does not require a split housing. Clearance must be allowed for axial movement. Effective for abrasive environment. Use for oil or grease.

Type D. Radial labyrinth seal. Used with a split housing or end cap. Bore of grooved sealing ring is slightly larger than shaft, allowing it to float axially. Angular surface of housing groove reduces pumping action. Suitable for oil or grease.

Type E. Felt seal. Provides medium effectiveness at low speeds but it loses its effectiveness at high speeds and high temperatures. In most cases, it functions as a contact seal, but after "wearing in" often functions as a simple close-clearance annulus seal. Not suitable for an abrasive environment. Use for grease lubrication.

Type F. The piston ring seal is a modification of the labyrinth seal. The piston rings are stationary and are mounted under radial compression in the housing. The split rings have rabbetted joints. Clearances between the rings and grooves are slight. This seal accommodates axial displacement. It is easy to install or remove, and can be used with a spacer or in shaft grooves as shown. Accepting deflection and misalignment, it offers a relatively positive closure augmented by a grease annulus. A particularly effective seal for an abrasive environment, it is suitable for grease and oil.

Type G. This is a combination annular groove-axial labyrinth seal. Annular grooves retain lubricant in the housing. The external flinger serves as a shield and flings contaminants away from the seal. It is suitable for grease or oil.

Type H. This is a triple combination of lip contact seals. Two commercial seals are mounted and opposed with a spacer in between to allow relubrication of the contact surfaces. A face contact seal is also used to prevent the entrance of contaminants. It is used primarily for grease.

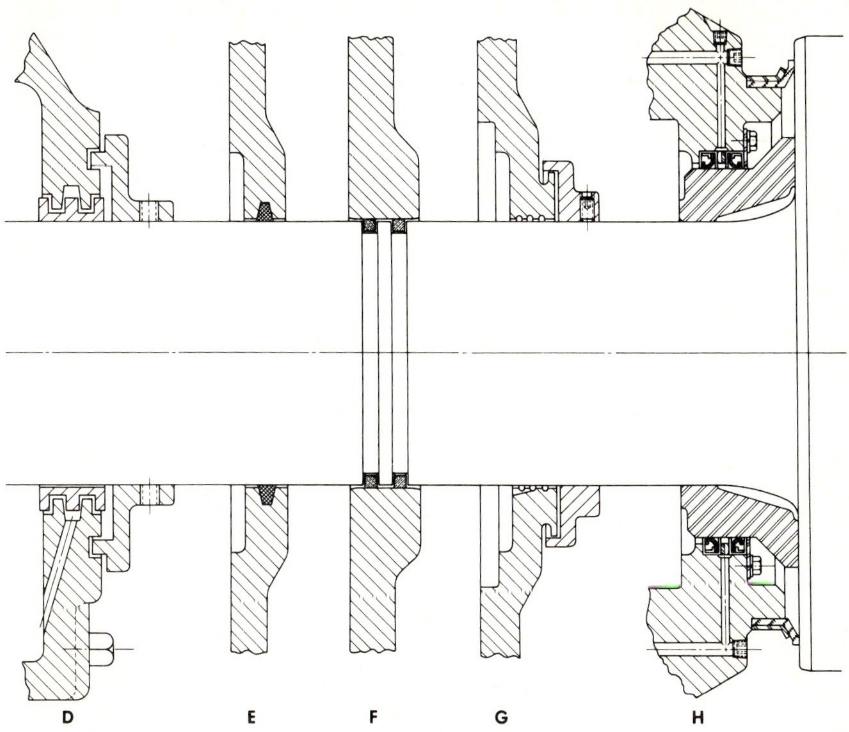

Figure 2-4 Basic sealing arrangements. (*Continued*)

PHYSICAL DESCRIPTION

Ball Bearings

Ball Radial Bearings

There are three major types of ball radial bearings with metric bore range of 100 to 320 mm and inch bore range of 4 to 40 in normally listed in bearing catalogs.

Conrad. See Fig. 2-5.

Type BC—metric sizes

Type BIC—inch sizes

Maximum Capacity. See Fig. 2-6.

Type BH—metric sizes

Type BIH—inch sizes

Angular Contact. See Fig. 2-7.

Type BA—metric sizes

Type BIA—inch sizes

For the convenience of the designer, metric and inch sizes of the three ball-bearing types are tabulated in order of increasing bore. Selection of the proper type of ball bearing is determined by consideration of the direction and magnitude of the bearing load. Ball bearings are particularly suited to high-speed operation because they have a lower coefficient of friction than roller bearings.

Figure 2-5 shows the most widely used ball-bearing type. Although it is primarily a

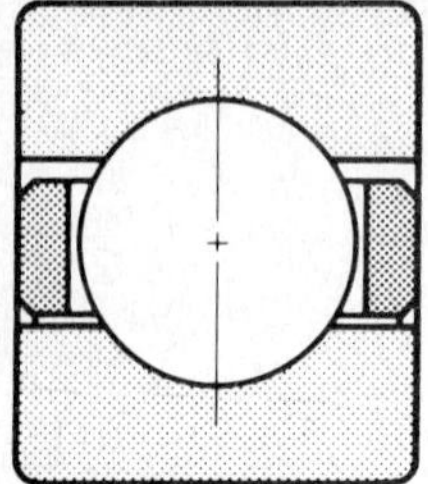

Figure 2-5 Conrad-type ball bearing.

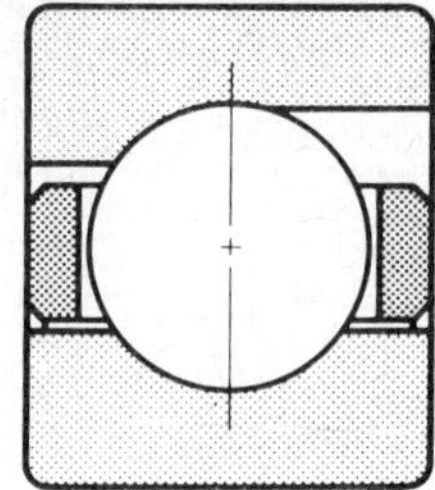

Figure 2-6 Maximum-capacity-type ball bearing. BH, BIH

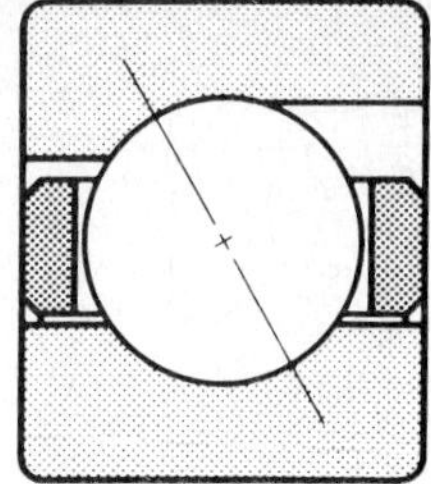

Figure 2-7 Angular-contact-type ball bearing.

radial bearing, it is capable of handling moderate thrust loads from either direction and operating at relatively high speeds.

The bearing rings have symmetrical, deep-grooved raceways without filling slots or a counterbore. The raceways are precision-ground to conform closely to ball curvature, consistent with minimum friction, maximum capacity, and practical manufacturing techniques. Balls are selected for uniformity to ensure optimum internal load distribution. The bearing utilizes the maximum number of balls which can be inserted between the raceways by eccentrically displacing the inner and outer rings. Balls are spaced by a two-piece, machined bronze cage. For higher speeds, and when other operating conditions warrant, the cage may be made of other materials. The nonseparable bearing construction facilitates handling.

The maximum-capacity ball bearing shown in Fig. 2-6 has the greatest radial capacity obtainable in a single-row ball bearing with cage, but it can take thrust in only one direction.

This type of bearing has an inner ring with a deep groove, such as the Conrad bearing, but the outer ring is counterbored; this substantially reduces the raceway shoulder on one side. The counterbore allows a maximum complement of balls in a one-piece, machined bronze cage to be assembled in the bearing. The outer ring is thermally expanded and slipped over the cage, ball, and inner-ring assembly. After cooling, the bearing is nonseparable.

Types BH and BIH are suited for higher radial loads than Conrad bearings of equivalent size as well as for combined radial and moderate unidirectional thrust loads. Since thrust capacity is unidirectional, the preferred mounting is in opposed pairs. A maximum-capacity bearing may also be opposed by a Conrad or other type of axially locating bearing. The center distance between bearings should be held to a minimum. Otherwise, axial shaft expansion may impose thrust on the shallow shoulder of the maximum-capacity bearing or cause preloading.

The design of angular-contact ball bearings, as shown in Fig. 2-7, is the same as that of the maximum-capacity type except the contact angle is 30°. This design feature allows the bearing to resist heavier thrust loads and minimizes axial deflection under load. The limiting speed of an angular-contact ball bearing is less than that of a Conrad or maximum-capacity type.

Angular-contact ball bearings are suited for applications where the thrust load is of the same order as or greater than the radial load. The preferred mounting arrangement is two bearings per shaft, with their contact angles opposed. As in the case of maximum-capacity bearings, the center distance between opposed bearings should be held to a minimum.

Angular-contact bearings can be furnished in matched pairs for duplex mounting (Fig. 2-8). When the bearings supplied for opposed mounting are placed together, there will be a small gap between the inner rings (type DB) or outer rings (type DF) before clamping. After clamping together, an internal preload is introduced in the set which

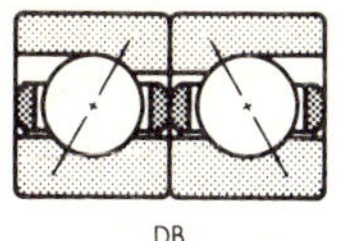

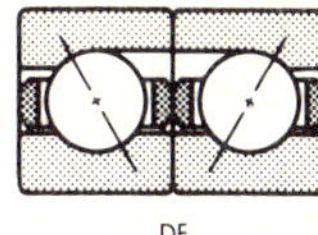

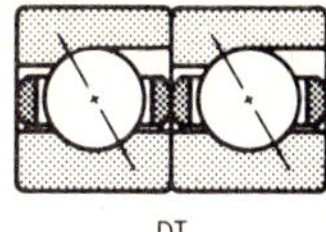

Figure 2-8 Duplex mounting of matched-pair angular-contact ball bearings.

increases the axial and radial rigidity of the duplexed pair. When supplied for tandem mounting (type DT), very heavy unidirectional thrust loads may be almost equally distributed among the bearings of the set. Details of application for DB and DF pairs should be submitted to the engineering department of the bearing manufacturer for preload recommendations.

Ball Thrust Bearings

Ball thrust bearings are used for lighter loads and higher speeds than roller thrust bearings.

The type TVB ball thrust bearing (Fig. 2-9) is separable, and consists of two hardened and ground steel washers with grooved raceways, and a cage which separates and retains precision-ground and -lapped balls. The standard cage material is bronze, but this may be varied according to the requirements of the application.

The type TVB bearing provides axial rigidity in one direction and its use to support radial loads is not recommended. It is very easily mounted. Usually the rotating washer is shaft-mounted. The stationary washer should be housed with sufficient outside diameter clearance to allow the bearing to assume its proper operating position. In most sizes both washers have the same bore and outside diameter. The housing must be designed to clear the outside diameter of the rotating washer, and it is necessary to step the shaft to clear the bore of the stationary washer.

Type TVL (Fig. 2-10) is a separable angular-contact ball bearing designed primarily for unidirectional-thrust loads. The angular contact design, however, will accommodate

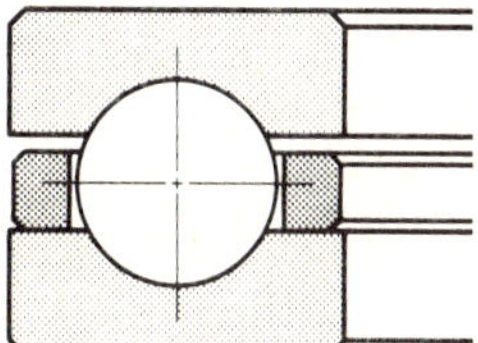

Figure 2-9 Ball thrust bearing.

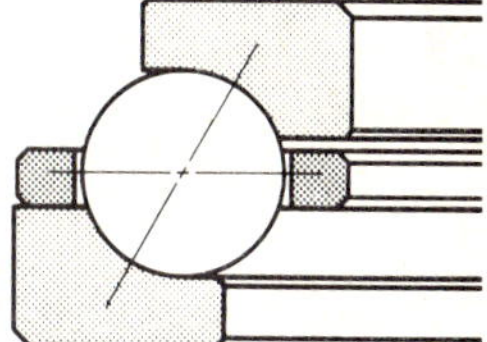

Figure 2-10 Angular-contact ball thrust bearing.

combined radial and thrust loads since the loads are transmitted angularly through the balls.

Tbe bearing has two hardened and ground steel rings with ball grooves and a one-piece bronze cage which spaces the ball complement. Although not strictly an annular ball bearing, the larger ring is called the outer ring, and the smaller the inner ring.

Usually the inner ring is the rotating member and is shaft-mounted. The outer ring is normally stationary and should be mounted with outside diameter clearance to allow the bearing to assume its proper operating position. If combined loads exist, the outer ring must be radially located in the housing.

The type TVL bearing should always be operated under thrust load. Normally, this presents no problem as the bearing is usually applied on vertical shafts in oil-field rotary tables and machine-tool indexing tables. If a constant-thrust load is not present, it should be imposed by springs or other built-in devices.

Low friction, cool running, and quiet operation are advantages of the type TVL bearing, which may be operated at relatively high speeds. The bearing is also less sensitive to misalignment than other types of rigid thrust bearings.

Roller Bearings

Tapered Roller Bearings

Tapered roller bearings are generally considered to offer the best support for combinations of heavy radial and thrust loads at moderate speeds.

A single-row tapered roller bearing consists of an inner ring (called a *cone*), an outer ring (called a *cup*), a bronze or steel cage, and a complement of controlled-contour rollers. In multiple-row tapered roller bearings one or more cones, cups, and cage assemblies may be used.

Tapered rollers and raceways are designed on the geometric principle of a cone (Fig. 2-11). Extensions of the lines of contact between the rollers and the raceways all meet at

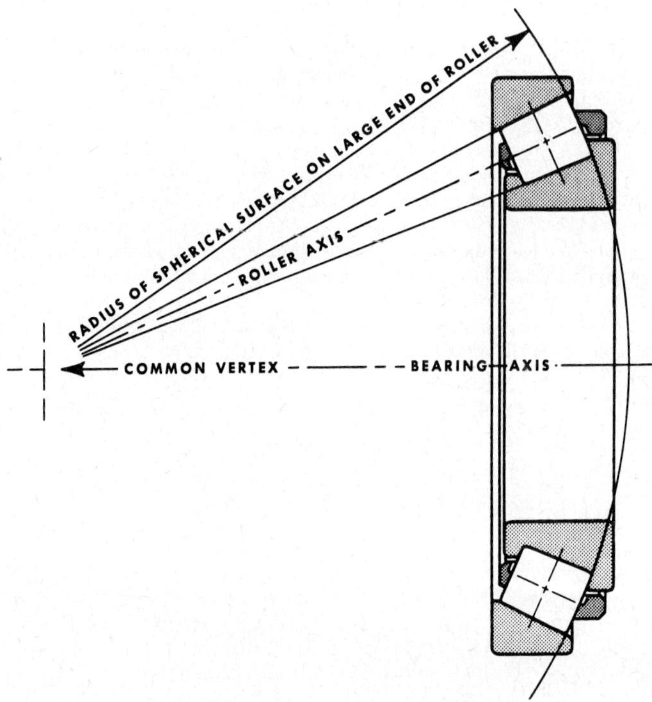

Figure 2-11 Geometric principle of tapered roller bearing.

a common point on the axis of the bearing. The design assures true geometric rolling. The large ends of the tapered rollers are spherically ground to match the spherically ground face on the guiding cone rib. Under load, the nominal pressure exerted between these two ground surfaces accurately positions the rollers within the load zone.

Single-Row Tapered Roller Bearings. Three types (TS, TSF, and TSS) of single-row tapered roller bearings are offered (Fig. 2-12). Each has a cup and a cone with a cage and roller assembly. Type TS serves as the basic design for the others.

Since a single-row tapered roller bearing supports thrust loads from only one direction, the preferred mounting is in opposed pairs. The proper internal clearance for the two bearings may be obtained by axial adjustment at the time of assembly.

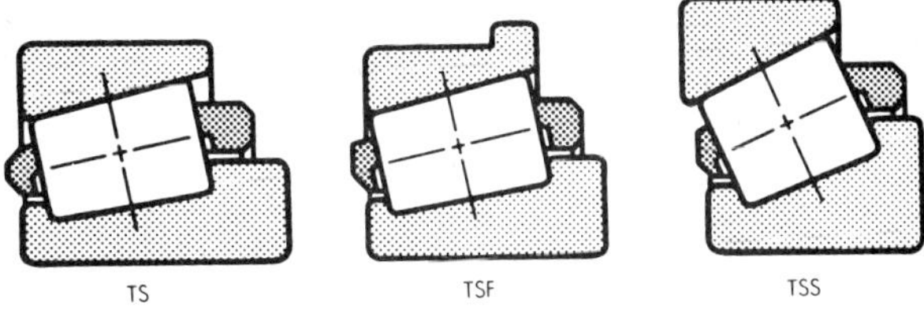

Figure 2-12 Single-row tapered roller bearings.

Type TSF bearings are identical with type TS except that the cup of the TSF type incorporates an external flange which, in some mountings, facilitates location and permits economies in design. When through-boring of the housing is advantageous, the use of the type TSF bearing is suggested.

Type TSS bearings are similar to type TS, but have a steeper angle of contact. These bearings are recommended for applications where the thrust load is predominant.

Two-Row Tapered Roller Bearings. Three basic types (TDI, TDO, and TNA) of two-row tapered roller bearings are available (Figs. 2-13 and 2-14). Types TDIS, TDIE,

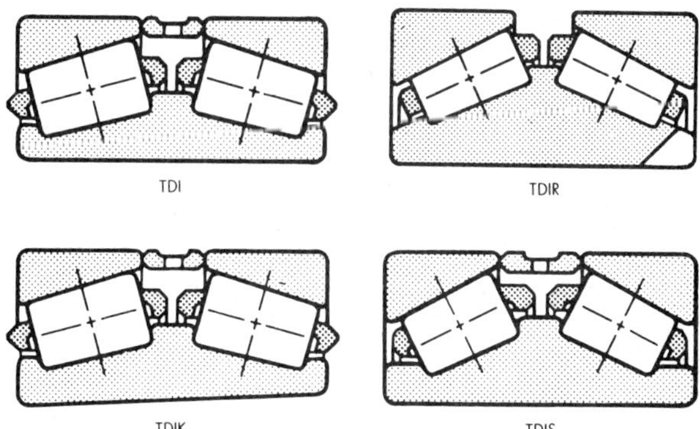

Figure 2-13 Two-row tapered roller bearings with converging angles of contact.

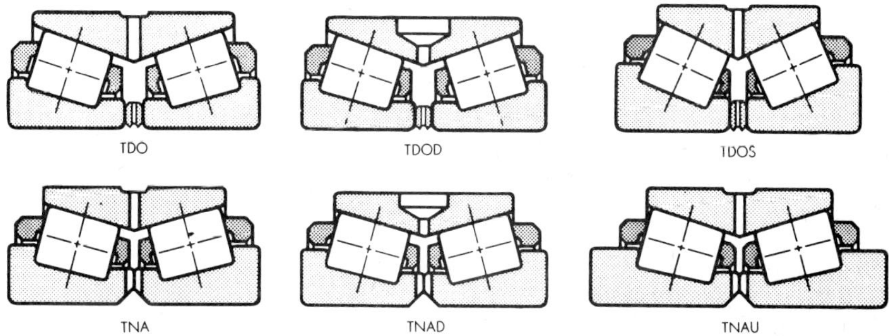

Figure 2-14 Two-row tapered roller bearings with diverging angles of contact.

and TDOS are steep-angle versions of the basic types. Type TDIE is supplied with a face keyway and without a cup spacer. Type TDIK is identical with type TDI except it has a tapered bore. Type TNAU is an extended-cone rib version of type TNA. Types TDOD and TNAD are identical, respectively, to types TDO and TNA, except they are supplied with a dowel hole in the outer ring. *All types (except TDIE) are normally furnished as preadjusted assemblies.*

Two-row tapered roller bearings have twice the radial capacity of single-row bearings of the same series and are used in positions where radial loading is too severe for single-row bearings. They have the further advantage that a two-row bearing can take thrust loads in both directions, thus allowing all applied thrust and shaft location to be taken at one position. This often simplifies design and reduces the danger of bearing clearance changes due to axial shaft expansion.

The steep-angle versions (TDIS and TDIE) offer greater thrust capacity with reduced radial capacity. These are widely used as backup thrust bearings in rolling mills and other applications where heavy thrust loads are encountered.

The design of types TDI, TDIS, and TDIE is such that the contact angles converge as they approach the axis of rotation (Fig. 2-15). Consequently, the use of these bearings will not appreciably increase the rigidity of the shaft mounting, and they should not be used singly on a shaft since they will not resist overturning moments.

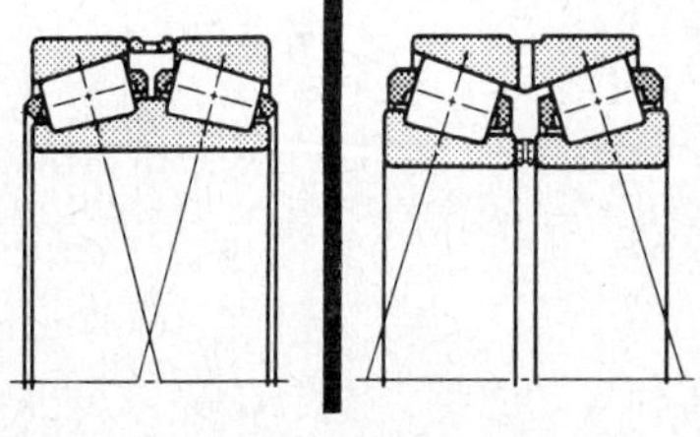

Figure 2-15 Comparison of contact angles for two-row tapered roller bearings.

In types TDO, TDOS, and TDOD, the contact angle lines diverge as they approach the axis of rotation, thus increasing the rigidity of the shaft mounting (Fig. 2-15). Therefore, these bearings are suited for resisting overturning moments. Due to the increased bearing ridigity, housing bore alignment is somewhat more critical than with types TDI, TDIS, and TDIE.

Four-Row Tapered Roller Bearings. The four-row tapered roller bearing (type TQO) (Fig. 2-16) is an extremely high capacity bearing designed primarily for heavy-duty roll-neck applications in metal rolling mills. The bearing is composed of two double-cone assemblies, one double cup, two single cups, and factory-adjusted cup and cone spacers. Spacers for each bearing are face-ground, after accurate measurement of the distance between adjacent cups and cones, to obtain the required initial internal clearance. Because spacers are ground to specific dimensions for each bearing, they are not to be interchanged and they are individually marked for proper assembly. It is important to submit sufficient information for proper internal clearance to be established.

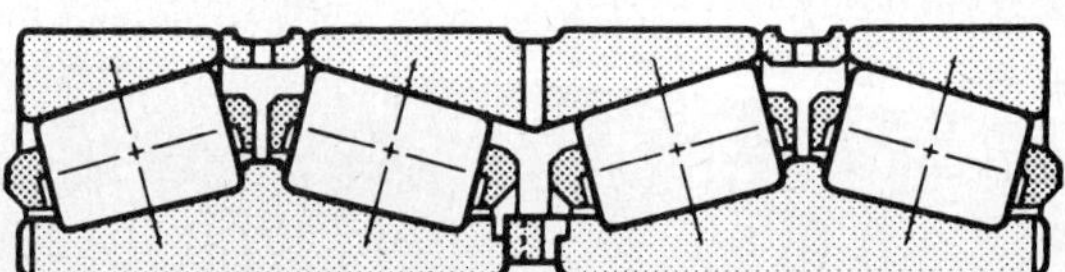

Figure 2-16 Four-row tapered roller bearing.

Lubrication grooves and oil holes are provided in the cup spacers and double cup. For rolling-mill applications, lubrication slots in the cone faces and cone spacer permit the lubricant to pass through the bearing to the roll neck.

Since wear- and shock-resistance under heavy rolling loads are requirements of roll-neck bearings, highest quality carburizing-grade bearing alloy steels are normally used in all four-row tapered roller bearings.

The type TQOK bearing (Fig. 2-17), with tapered bore for roll-neck mounting, is a development for modern high-speed rolling-mill requirements. The TQOK bearing features high capacity, compactness, and a mounting system that guarantees a positive interference fit on roll necks. The tightly-fitted twin-cone mounting of type TQOK provides extra mill rigidity and rolling accuracy by eliminating extraneous clearance. Roll-

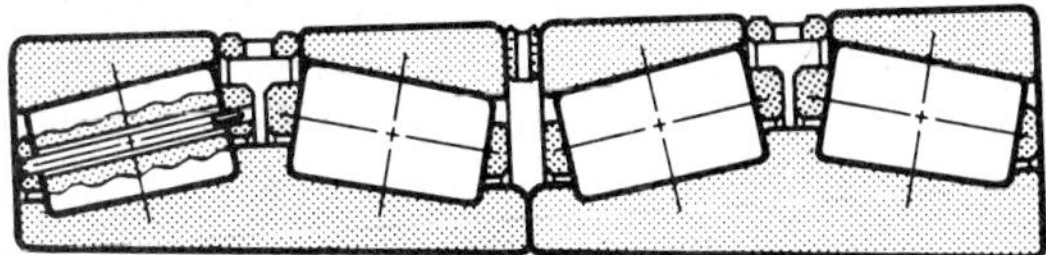

Figure 2-17 Four-row tapered roller bearing with tapered bore.

ing-mill speeds in excess of 5000 ft/min are permissible. Gauge uniformity is assured during the acceleration cycles of feeding the strip through the mill. The mill can be stopped and started without downtime for screw adjustment.

The advantages of compactness of the TQOK design are numerous, but the chief advantage is the short lever arm required on the roll neck with reduction in neck bending stress. Conversely, more capacity can be obtained for any given space limit than with other bearing designs. These and other design features greatly improve rolling-mill production and make the bearing ideal for backup and work-roll mountings in four-high and two-high mills.

Cylindrical Roller Bearings

Six standard types of cylindrical roller bearings are shown in Fig. 2-18. All six types have the same roller complements and, consequently, the same capacity for a given size. All

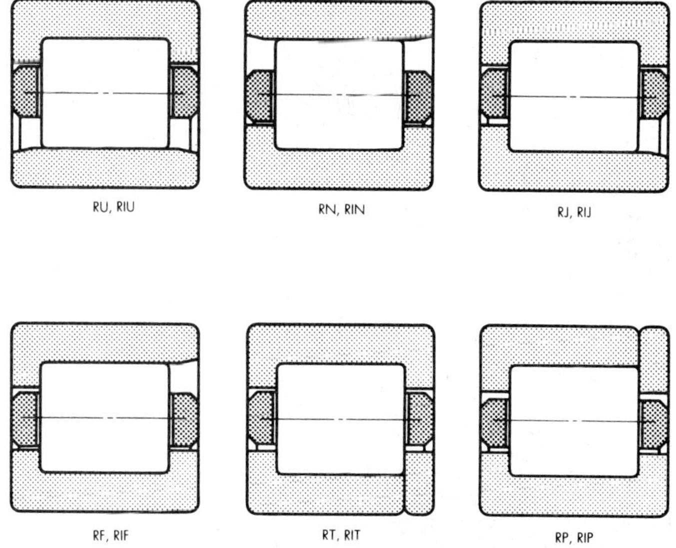

Figure 2-18 Standard types of cylindrical roller bearings.

types can be mounted with interference fits on either the inner or the outer ring, or both. In the latter case, a bearing with increased internal clearance must be specified to provide proper running clearance.

For convenience, bearings are listed according to bore, with both metric and inch bearings in the same figure. Inch bearings are identified by the letter "I" in the type code of the bearing number; thus where RN denotes a particular type of metric bearing, RIN denotes an inch bearing of the same type.

Types RU and RIU have double-ribbed outer and straight inner rings. Types RN and RIN have double-ribbed inner and straight outer rings. The use of either type at one position on a shaft is ideal for accommodating nominal expansion or contraction. The relative axial displacement of one ring to the other occurs with minimum friction while the bearing is rotating. These bearings may be used in two positions for shaft support if other means of axial location are provided.

Types RJ and RIJ have doubled-ribbed outer and single-ribbed inner rings. Types RF and RIF have double-ribbed inner and single-ribbed outer rings. Both types can support heavy radial loads, as well as light unidirectional thrust loads up to 10 percent of the radial load. The thrust load is transmitted between the diagonally opposed rib faces in a sliding action rather than a rolling action. Thus, when limiting thrust conditions are approached, lubrication can become critical. When thrust loads are very light, these bearings may be used in an opposed mounting to locate the shaft. In such cases, shaft end play should be adjusted at time of assembly.

Types RT and RIT have a doubled-ribbed outer ring and a single-ribbed inner ring with a loose rib which allows the bearing to provide axial location in both directions. Types RP and RIP have a double-ribbed inner ring and a single-ribbed outer ring with a loose rib.

Types RT and RP (as well as RIT and RIP) provide heavy radial capacity and light thrust capacity in both directions. Factors governing the thrust capacity are the same as for types RF and RJ bearings.

A type RT or RP bearing may be used in conjunction with a type RN or RU bearing for applications where axial shaft expansion is anticipated. In such cases the fixed bearing is usually placed nearest the drive end of the shaft to minimize alignment variations in the drive. Shaft end play (or float) is determined by the axial clearance in the bearing.

Spherical Roller Bearings

The self-aligning spherical roller bearing (Fig. 2-19) is a combination radial and thrust bearing designed for taking misalignment under load. When loads are heavy, alignment of housings difficult, and shaft deflections excessive, the use of spherical roller bearings assures best service life results.

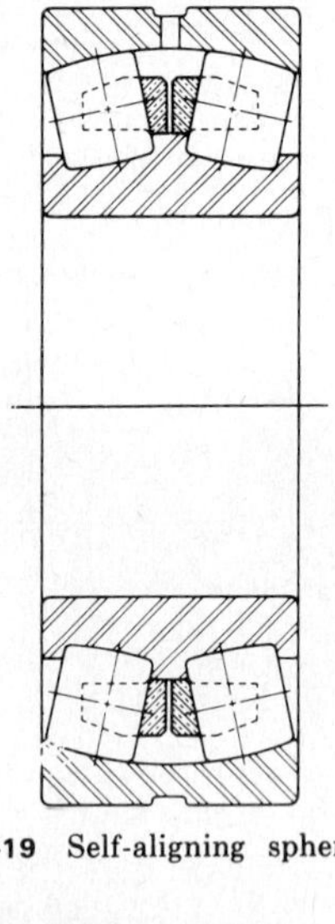

With the spherical rollers operating on the spherically shaped outer race, the assembly of the inner ring, retainers, and rollers may take up to $\pm 1\frac{1}{2}°$ of misalignment and continue to function with full capacity. High radial load capacity is secured by a large area of roller-to-race contact. Double-direction thrust capacity results from angular location of rollers relative to bearing axis.

Shaft deflections and housing distortions caused by shock loads are compensated for by the internal self-alignment of the free-rolling bearing elements. The binding stresses that limit service life of non-self-aligning bearings cannot develop in spherical roller bearings. The retainers ride on the center flange of the inner ring rather than on the rollers. Unit design and construction make the spherical roller bearing simple to handle at assembly point or during removal for maintenance.

Figure 2-19 Self-aligning spherical roller bearing.

Cylindrical Roller Thrust Bearings

Cylindrical roller thrust bearings withstand heavy loads at relatively moderate speeds. Standard bearings can be operated at bearing outside-diameter peripheral speeds of 3000 ft/min (1000 m/min). Special design features can be incorporated into the bearing and mounting to attain higher operating speeds.

Because loads are usually high, extreme pressure lubricants should be used with roller thrust bearings. Preferably, the lubricant should be introduced at the bearing bore and distributed by centrifugal force.

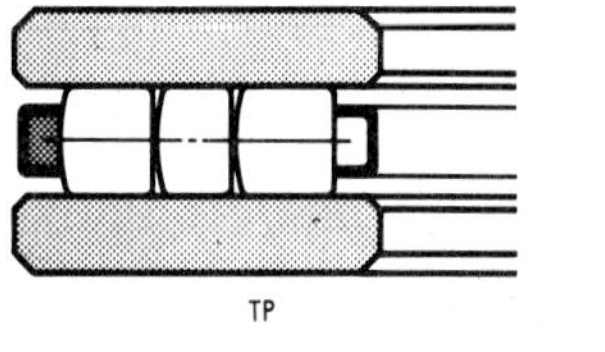

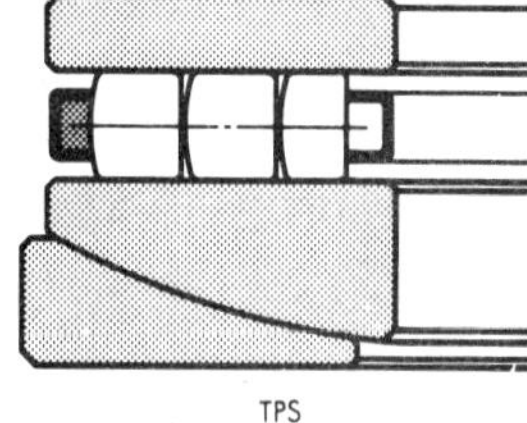

Figure 2-20 Cylindrical roller thrust bearings.

The type TP cylindrical roller thrust bearing (Fig. 2-20) has two hardened and ground steel washers, with a cage retaining one or more rollers in each pocket. When two or more rollers are used in a pocket they are of different lengths and are placed in staggered position in adjacent cage pockets to create overlapping roller paths. This prevents wearing grooves in the raceways and prolongs bearing life.

Because of the simplicity of their design, type TP bearings are economical. Since minor radial displacement of the raceways does not affect the operation of the bearing, its application is relatively simple and often results in manufacturing economies for the user. Shaft and housing seats, however, must be square to the axis of rotation to prevent initial misalignment problems.

Type TPS bearing (Fig. 2-20) is the same as type TP bearing, except one washer is spherically ground to seat against an aligning washer, thus making the bearing adaptable to initial misalignment. Its use is not recommended for operating conditions where alignment is continuously changing (dynamic misalignment).

The type TTHD tapered roller thrust bearing (Fig. 2-21) has an identical pair of hardened and ground steel washers with conical raceways, and a complement of tapered rollers equally spaced by a cage.

In the design of type TTHD, the raceways of both washers and the tapered rollers have a common vertex at the bearing center. This assures true rolling motion. The large end of each tapered roller is spherically ground to match the concave faces of both washer ribs. The pressure exerted under load by the roller ends at the rib surfaces accurately guides the rollers. The center of the large end of each roller is counterbored to improve lubrication at the guiding rib surfaces.

TTHD bearings are well suited for applications such as crane hooks, where extremely high thrust loads and heavy shock must be resisted and some measure of radial location obtained. For very low speed, extremely heavily loaded applications, these bearings are supplied with a full complement of rollers for maximum capacity. The TSR spherical roller thrust bearing design (Fig. 2-22) achieves a high thrust capacity with low friction and continuous roller alignment. The bearings can accommodate pure thrust loads as

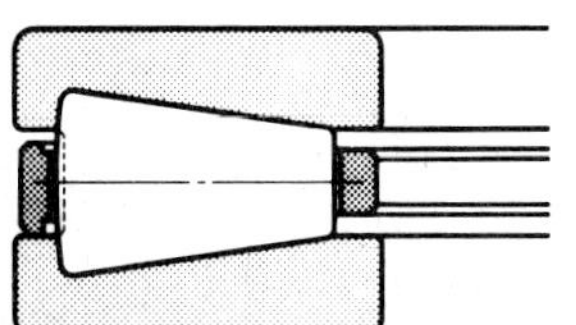

Figure 2-21 Tapered roller thrust bearing.

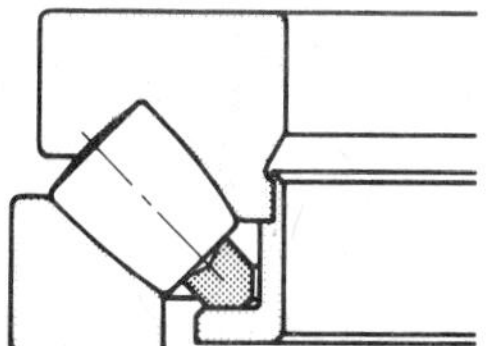

Figure 2-22 Spherical roller thrust bearing.

well as combined radial and thrust loads. Higher speeds are obtainable with this design than with any of the other roller thrust bearings. Typical applications are air regenerators, centrifugal pumps, and deep-well pumps.

Because the spherical roller thrust bearing must always carry some thrust load, it cannot be used as a float bearing. Most applications require the use of two or more bearings on a shaft: a fixed bearing and a float bearing.

As a rule, coil springs must be applied against the stationary ring of the bearing. Among the more common conditions requiring the use of springs are:

1. Applied thrust load is too low to overcome induced thrust load.
2. Load is temporarily reversed, and excessive lateral movement would be detrimental.
3. Thrust load varies from zero to a maximum and back to zero, and roller-to-raceway contact must be maintained.

Needle Roller Bearings

Size for size, needle bearings (Fig. 2-23) have more rollers and more lines of contact, and thus, generally, have higher capacities than roller bearings. This is particularly true

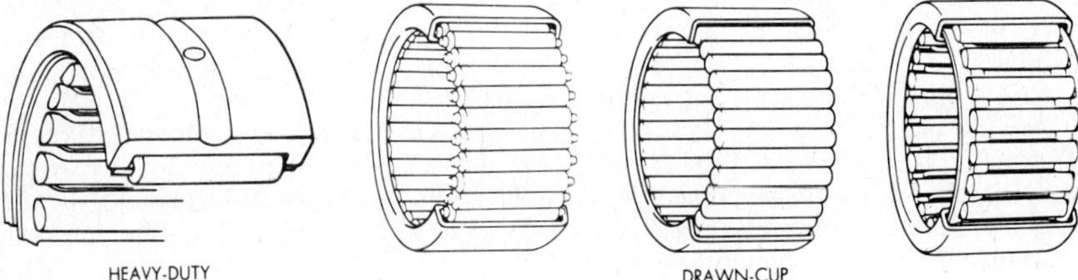

Figure 2-23 Needle roller bearings.

under static, slow-rotating, or oscillating conditions. Both types have far higher capacity in less radial space than other antifriction bearings.

The heavy-duty bearing has an outer race which is deeply hardened, while the race of the drawn-cup bearing has a thin, hardened case over a relatively ductile core. Thus, the heavy-duty bearing can withstand heavier shock or more continuous loads than the drawn-cup bearing, which should never be dynamically loaded, even momentarily, beyond the load limit given in the manufacturer's tabular data.

The smaller, more compact cross section is provided by the drawn-cup bearing with its smaller roller diameter and thin-steel outer race, as opposed to the larger roller diameter and thick outer race of the heavy-duty bearing.

For moderate speeds, both types are satisfactory. However, the cage in the roller bearing allows higher speeds for a given shaft size.

MOUNTING AND MAINTENANCE

General Installation Rules

Depending on the size of bearing and the application, there are different methods for mounting bearings. In all methods, however, certain basic rules must be observed.

Maintain Cleanliness

Choose a clean environment. Work in an atmosphere free from dust or moisture. If this is not obtainable, and sometimes in the field it isn't, the installer should make every effort to ensure cleanliness by use of protective screens, clean cloths, etc.

Plan the Work

Know in advance what you are going to do and have all necessary tools at hand. This reduces the amount of time for the job and lessens the chance for dirt to get into the bearing.

Preparation and Inspection

All component parts of the machine should be on hand and thoroughly cleaned before proceeding. Housings should be cleaned, even to blowing out the oil holes. Do not use an air hose on bearings. If blind holes are used, insert a magnetic rod to remove metal chips that might have become lodged during fabrication.

Shaft shoulders and spacer rings contacting the bearing should be square with the shaft axis. The shaft fillet must be small enough to clear the radius of the bearing.

On original installations, all component parts should be checked against the detailed specification prints for dimensional accuracy. Shaft and housing should be carefully checked for size and roundness.

Mounting Straight-Bore Bearings

Heat-Expansion Method

Most applications require a tight interference fit on the shaft. Mounting is simplified by heating the bearing to expand it sufficiently to slide easily onto the shaft. Two methods of heating are in common use:

1. Immersion in a tank of heated oil (Fig. 2-24)
2. Induction heating

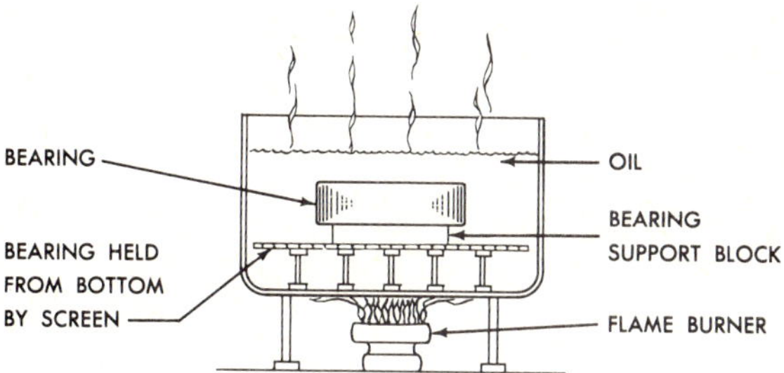

Figure 2-24 Typical arrangement for heat-expansion method.

The first is accomplished by heating the bearing in a tank of oil having a high flash point. The oil temperature should not exceed 250°F (120°C). A temperature of 200°F (93°C) is sufficient for most applications. The bearing should be heated at this temperature, generally for 20 or 30 min, until it is expanded sufficiently to slide onto the shaft very easily.

The induction-heating method is particularly suited for mounting small bearings in production-line assembly. Induction heating is rapid, and care must be taken to prevent bearing temperature from exceeding 200°F (93°C). Trial runs with the unit are usually necessary to obtain the proper timing. Thermal crayons which melt at predetermined temperatures can be used to check the bearing temperature.

While the bearing is still hot, it should be positioned squarely against the shoulder. Lock washers and lock nuts, or clamping plates, are then installed to hold the bearing against the shoulder of the shaft. As the bearing cools, the lock nut or clamping plate should be tightened.

The oil bath is shown in Fig. 2-24. The bearing should not be in direct contact with the heat source. The usual arrangement is to have a screen several inches off the bottom of the tank. Small support blocks separate the bearing from the screen. It is important to keep the bearing away from any localized high-heat source that may raise its temperature excessively, resulting in race hardness reduction.

Flame-type burners are commonly used, but may have to be replaced with some other heat source to comply with local safety regulations. An automatic device for temperature control is desirable. If safety regulations prevent the use of an open heated-oil bath, a mixture of 15 percent soluble oil in water may be used. This mixture may be heated to a maximum temperature of about 200°F (93°C) without being flammable. The bath should be checked from time to time to ensure its proper composition as the water evaporates. The bath leaves a thin film of oil on the bearing sufficient for temporary rust prevention, but normal lubrication should be supplied to the bearing as soon as possible after installation. Be sure all of the oil-in-water solution has been drained away from the bearing.

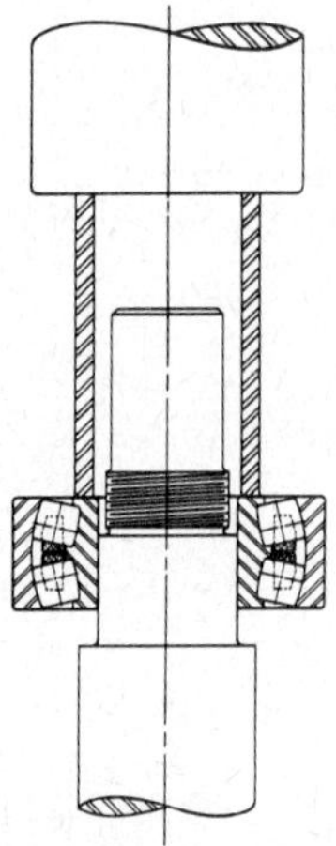

Figure 2-25 Arrangement for typical Arbor press bearing mounting.

Arbor Press Method

The alternative method of mounting, generally used only on smaller sizes (Fig. 2-25) is to press the bearing onto the shaft or into the housing. This can be done by using an arbor press and a mounting tube. The tube can be of soft steel with inside diameter slightly larger than the shaft. The outside diameter of the tube should not exceed the maximum shoulder height given in the tables of dimensions. The tube should be faced square at both ends, thoroughly clean inside and out, and long enough to clear the end of the shaft after the bearing is mounted.

If the outer ring is being pressed into the housing, the outside diameter of the mounting tube should be slightly smaller than the housing bore, and the inside diameter should not be less than the recommended housing-shoulder diameter in the tables of dimensions.

Coat the shaft with light machine oil to reduce the force needed for the press fit. Carefully place the bearing on the shaft, making sure it is square with the shaft axis. Apply steady pressure from the arbor ram to drive the bearing firmly against the shoulder.

Never attempt to make a press fit on a shaft by applying pressure to the outer ring, or to make a press fit in a housing by applying pressure to the inner ring.

Straight-Bore-Bearing Removal

Bearing pullers of various types and designs are available from several manufacturers. These pullers are useful in removing bearings up to about 10-in bore. The preferred method utilizes a split ring which is placed behind the inner ring. The pulling device is assembled so as to cause the split ring to push the bearing off the shaft. If machine elements interfere with this method, a prong-type puller may be used with a roller bearing which has side flanges on the inner ring. Insert the prongs of the puller behind the flange and pull the bearing off the shaft.

Hydraulic Method

Remove lock plate or lock nut. Provide a means for pulling the bearing off the shaft. A bearing puller applied to the outer ring may be used without fear of brinelling the races since only a nominal force is required for removal by this system. Or, the shaft may be placed in a vertical position, simply allowing gravity and slight manual force to remove the bearing. Pull the bearing all the way off the shaft quickly to prevent freezing after the loss of oil pressure as the bearing clears the oil groove.

Mounting Bearings in Housings

To facilitate installing the bearing in the housing and to minimize fretting corrosion during operation, coat the housing with a light-grade machine oil. Make sure the outer ring is square with the housing bore before inserting it. If the outer ring becomes misaligned and sticks, do not attempt to force it farther into the housing. Use a soft brass or steel bar and tap the outer ring until it becomes free and is realigned. An assembly plate for holding the outer ring in alignment with the inner ring is often useful for applications in which conditions tend to result in misalignment.

In cases of outer-ring rotation, where the outer ring is a tight fit in the housing, the housing member can be expanded by heating.

Cast housings should be annealed. Split housings present no alignment problems during installation, but the housings should be checked carefully for size and roundness.

Internal Clearance

Internal clearance is the amount of radial play within a bearing. This clearance accommodates the effects of tight shaft and housing fits, radial thermal expansion, speed, and other operating conditions.

The internal clearance in radial bearings other than tapered roller bearings is defined as *diametral clearance*. The diametral clearance is the amount of radial play built into the bearings (see Figs. 2-26 and 2-27).

Straight-bore or tapered-bore bearings installed with interference fits will result in a reduction of internal or diametral clearance. Several factors influence the reduction of internal clearance. When the bearing inner ring is pressed onto a solid steel shaft, reduction of internal clearance is approximately 80 percent of the shaft interference fit. For an outer ring pressed into a steel or cast-iron housing, the reduction of internal clearance is about 60 percent of the housing-interference fit. If the shaft is hollow, the housing walls are thin, or materials other than steel are used, clearance reduction may vary considerably from the above percentages.

Tapered-bore spherical roller bearings require a slightly greater interference fit with the shaft than straight-bore bearings of corresponding size. The resultant fit cannot conveniently be determined by direct shaft measurement. It can be checked by the distance the bearing is pressed onto a carefully gauged tapered shaft, or more easily by the effects of radial expansion of the inner ring as it is pushed onto the taper.

Lubrication

Since the lubricant affects bearing life and operation, selecting the proper lubricant is an important design function. The purpose of lubrication in bearing applications is to:

1. Minimize friction at points of contact within the bearings
2. Protect the highly finished bearing surfaces from corrosion
3. Dissipate heat generated within the bearings
4. Remove foreign matter or prevent its entry into the bearings

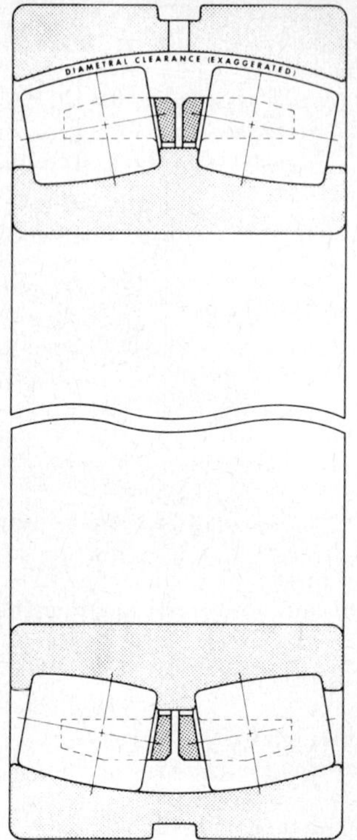

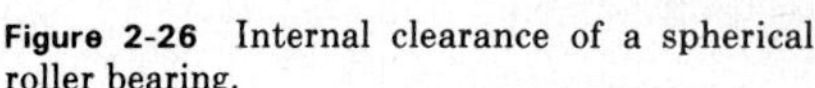

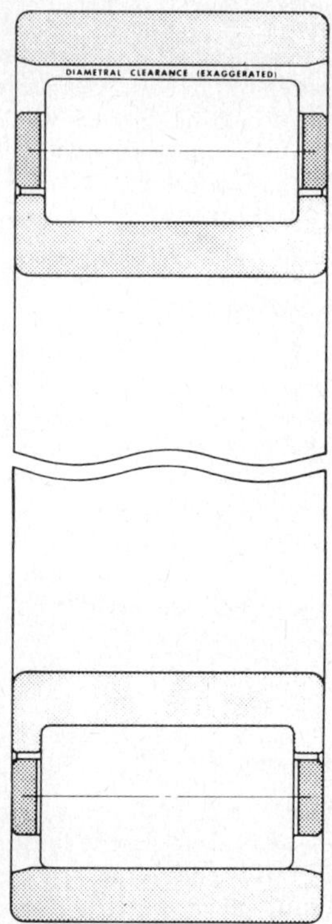

Figure 2-26 Internal clearance of a spherical roller bearing.

Figure 2-27 Internal clearance of a cylindrical roller bearing.

Two basic types of lubricants used with antifriction bearings are oils and greases. Each has its advantages and limitations.

Since oil is a liquid, it lubricates all surfaces and is able to dissipate heat from these surfaces more readily. Because oil retains its physical characteristics over a wider range of temperatures, it may be used for high-speed and high-temperature applications. The quantity of oil supplied to the bearing may be accurately controlled. Oil lubricants can be circulated, cleaned, and cooled for more effective lubrication.

Grease, which is easier to retain in the bearing housing, aids as a sealant against foreign matter and corrosive fumes.

Oil Lubrication

1. Oil is a better lubricant for high speeds or high temperatures. It can be cooled to help reduce bearing temperature.
2. Oil is easier to handle, and with oil it is easier to control the amount of lubricant reaching the bearing. It is harder to retain in the bearing. Lubricant losses may be higher than with grease.
3. As a liquid, oil can be introduced into the bearing in many ways, such as drip feed, wick feed, pressurized circulating systems, oil bath, or air-oil mist. Each is suited to certain types of applications.
4. Oil is easier to keep clean for recirculating systems.

Grease Lubrication

1. Restricted to lower-speed applications within operating-temperature limits of the grease.
2. Easily confined in the housing. This is important in the food, textile, and chemical industries.
3. Bearing enclosure and seal design simplified.
4. Improves the efficiency of mechanical seals to give better protection to the bearing.

For all new applications, a competent lubrication engineer should study the operating conditions and recommend the specific lubricant required.

BIBLIOGRAPHY

For additional or more detailed information on a specific bearing type or application, reference should be made to the bearing manufacturers' catalogs and or one of the following:

AFBMA Standards, The Anti-Friction Bearing Manufacturers Association, Inc., 2341 Jefferson Davis Highway, Arlington, VA 22202.

SKF Industries, Inc., Front Street and Erie Avenue, P. O. Box 6731, Philadelphia, PA 19132.

The Timken Company, 1835 Dueber Avenue S.W., Canton, OH 44706.

The Torrington Company, 3702 West Sample Street, P. O. Box 1684, South Bend, IN 46634.

PART 2

Plain Bearings

adapted with permission from

"Mechanical Drives" issue
Machine Design Magazine
A Penton/IPC Publication

INTRODUCTION

A *plain bearing* is any bearing that works by sliding action (with or without lubricant). This group encompasses essentially all types other than rolling-element bearings. Many of the basic principles dealing with plain-bearing loading and lubrication are explained in the prior section, and the reader is advised to review these basics before continuing with the more specific and detailed treatment here.

Plain bearings are often referred to as *sleeve bearings* or *thrust bearings,* terms that designate whether the bearing is loaded axially or radially.

Lubrication is critical to the operation of plain bearings, so their application and function is also often referred to according to the type of lubrication principle used. Thus, terms such as hydrodynamic, fluid-film, hydrostatic, boundary-lubricated, and self-lubricated are designations for particular plain bearings.

BEARING TYPES

Journal or Sleeve Bearings

These bearings (Fig. 2-28) are cylindrical or ring-shaped bearings designed to carry radial loads. The terms *sleeve* and *journal* are used more or less synonymously since sleeve refers to the general configuration while journal pertains to any portion of a shaft supported by a bearing. In another sense, however, the term journal may be reserved for two-piece bearings used to support the journals of an engine crankshaft. Sleeve bearings may be lubricated by either hydrostatic, hydrodynamic, or boundary means. In practice, most of them are designed to operate hydrodynamically, and hydrodynamic lubrication is normally implied when the terms sleeve or journal bearing are used.

The simplest and most widely used types of sleeve bearings are cast-bronze and porous-bronze (powdered-metal) cylindrical bearings. Cast-bronze bearings are oil- or grease-lubricated. Porous bearings are impregnated with oil and often have an oil reservoir in the housing.

Cast bearing bores are held within a few thousandths of an inch and are often finish-machined after installation to provide extreme accuracy of shaft position. Porous bearings are sized more accurately and are mounted with a special sizing tool so that no finish machining is needed. Machining poses the danger of smearing or closing the all-important pores.

Oscillating or slow-moving sleeve bearings cannot develop hydrodynamic lubrication, so their design criteria are different from those of other sleeve bearings. Oscillating bearings are generally designed on the basis of allowable wear or upon the load that produces "pounding out."

Engine bearings are designed to operate under hydrodynamic conditions. Their surfaces are smooth, and clearances are optimized to maximize load capacity while allowing enough oil flow to aid cooling.

Plastic bearings are being used increasingly in place of metal. Originally, plastic was used only in small, lightly loaded bearings where cost saving was the primary objective. More recently plastics with high load-carrying capacities have been developed for direct replacement of metals. Also, plastics are used because of functional advantages, including resistance to abrasion, and are being made in large sizes.

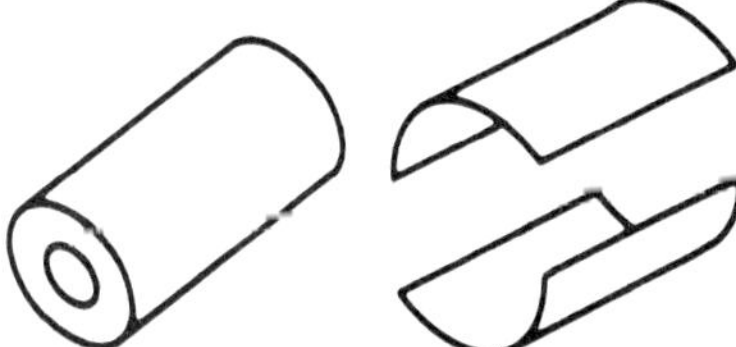

Figure 2-28 Sleeve or journal bearings.

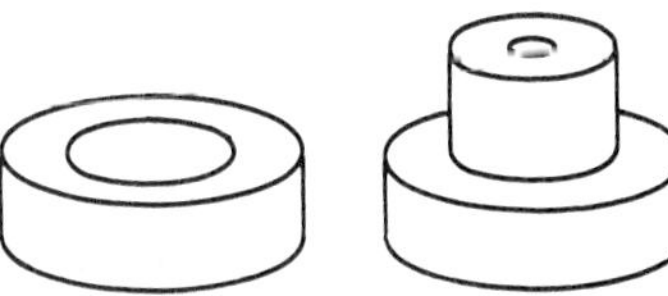

Figure 2-29 Thrust bearings.

Thrust Bearings

This type of bearing (Fig. 2-29) differs from a sleeve bearing in that loads are supported axially rather than radially. Thin, disklike thrust bearings are called thrust washers.

Hydrodynamic action in a thrust bearing can be improved by putting grooves in the surface, rather than having it flat. Small surface irregularities—either etched-in deliberately or resulting as a normal consequence of manufacturing—aid hydrodynamic action. Even assembly misalignment can help produce hydrodynamic behavior. Geometry of the grooves can be optimized for particular loads, speeds, and oil viscosities. For full hydrodynamic action, a thrust bearing must have tapered lands on tilting pads.

Spherical Bearings

The outside diameter of this type of bearing is spherical (or at least curved) to permit "wobble" of the bearing axis. This ball-and-socket action compensates for shaft and mount misalignment; thus these bearings (Fig. 2-30) are called *self-aligning*. Once the spherical tilt accommodates misalignment, subsequent bearing relative motion is normally between the shaft and the cylindrical bore in the ball.

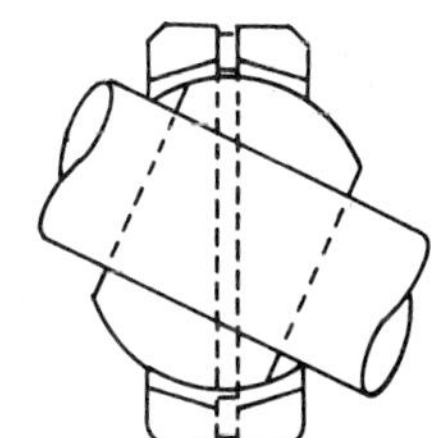

Figure 2-30 Spherical bearings.

When the ball is enclosed at the end of a link, the bearing is called a *rod-end bearing*. These bearings are normally used in oscillating linkages, and most of the relative motion is between the ball and the socket.

Hydrostatic Bearings

Hydrostatic bearings (Fig. 2-31) can be either of the thrust or sleeve (journal) type. Bearing characteristics are highly predictable, including such characteristics as stiffness, friction, and oil flow. Advantages include nearly zero starting friction, the ability to carry heavy loads at low speeds, and negligible wear. But hydrostatic bearings are expensive and require bulky equipment to provide pressurized oil. However, machines with hydrostatic bearings maintain their accuracy, so savings in depreciation can be significant. This type of lubrication is used most frequently in machine tools, but it is also employed in gyroscopes and other precision instruments where low friction is important.

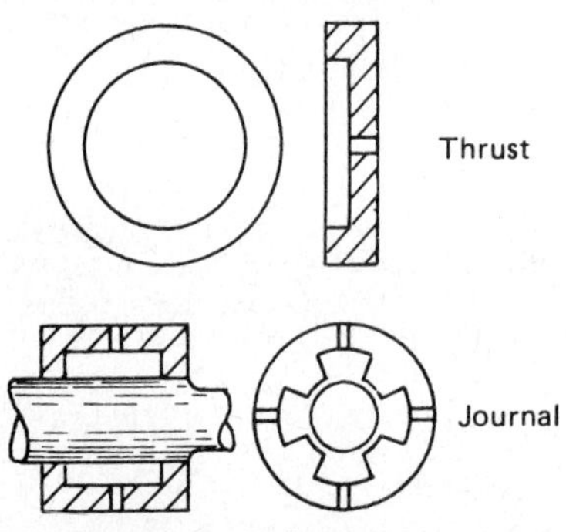

Figure 2-31 Hydrostatic bearings.

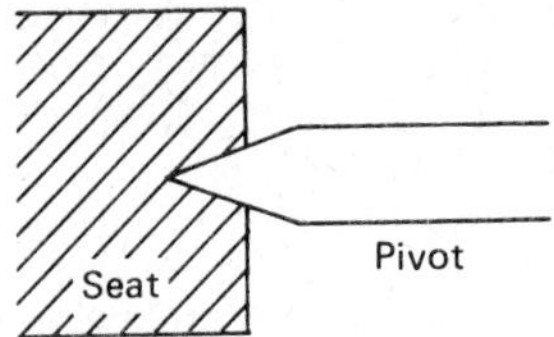

Figure 2-32 Pivot bearing.

Jewel and Pivot Bearings

These bearings (Fig. 2-32) are suitable for tiny, lightly loaded shafts in such applications as watches, meters, and instruments. The shaft does not pass through the bearing but rests in a cone or cup on each end. The pivots (pointed shaft ends) are usually hardened steel, and seats are usually sapphire, glass, beryllium copper, carbide, or carbon. The bearing normally is lubricated with a light nongumming oil.

These bearings operate in the boundary regime, but because of the small radius of contact, they have lower friction than rolling-element bearings. Jewel bearings come in a variety of standard designs.

Gas Bearings

Gas bearings are lubricated with a gas (usually air) instead of with an oil. They can either be externally pressurized (hydrostatic) or self-acting (hydrodynamic). A special type of self-acting gas bearing uses metal-foil interleaved to form a raceway and is thus called a *foil bearing*.

The operating principles and relative merits of hydrodynamic and hydrostatic gas bearings are similar to those of their oil-lubricated counterparts. The hydrostatic type can support loads at rest or low speed but requires an expensive and complex pumping system. The hydrodynamic type is simpler but must have significant relative motion to function and must be built to very close tolerances.

Gas bearings are used mainly for applications demanding extremely high rotational speeds or uniformly low frictional drag. Ambient atmospheres can be used as the working lubricant, so gas bearings also are used where possible contamination from oil lubricants must be avoided. The design of gas bearings is usually complex and requires sophisticated engineering. They usually are an integral part of the equipment in which they are used. However, some gas bearings are available as off-the-shelf components.

Most gas bearings demand small operating clearances, so they require precision manufacturing methods. They also are susceptible to overload, vibration, and dust contamination. Another problem is that they are subject to instabilities caused by the dynamics of a gas film or by unbalanced rotating masses.

HOW BEARINGS FAIL

The failure mode anticipated in a bearing influences decisions on the selection of a bearing type, bearing material, lubricant, and bearing surface finish. Typical failure modes include:

- Loss of shaft positioning from wear; seizure by overheating and loss of clearance
- Seizure by failure of lubricant and galling of contacting surfaces
- Destruction of journal by hard particles imbedded in the bearing surface
- Edge loading caused by dimensional instability of bearings and housings, or by shaft deflection
- Fatigue of bearing surface; and high friction

Friction

The frictional properties of any plain bearing depend on the lubrication system. Either hydrodynamic or hydrostatic lubrication can provide low friction. The lowest friction levels are attained in gas bearings or with magnetically supported bearings.

Friction in hydrodynamic and hydrostatic bearings is a function of lubricant viscosity and shear rate. Shear rate increases with increasing rotational speed and decreasing film thickness. The friction coefficient is generally below 0.001.

The friction in boundary and self-lubricated bearings varies widely and is difficult to predict for a given bearing-lubricant system. The range of coefficients of friction is 0.01 to 0.10 for boundary lubrication and 0.01 to 0.3 for self-lubrication.

Caution must be used when applying friction-coefficient handbook data. Conditions under which the values were measured should be known, and these conditions should be duplicated in the application. Coefficient of friction tends to increase with increasing surface roughness with dryness and cleanliness of surfaces, and with decreasing temperature.

Wear

Wear of plain bearings is strongly influenced by the state of lubrication and, conversely, wear characteristics influence the various lubrication states.

Hydrostatic bearings do not wear if operating properly because the bearing surfaces are separated by a film of oil, even when the bearing is stationary. Erosion of flow restrictors in hydrostatic bearings with high flow rates can and eventually does cause failure. Plugging of flow restrictors by wear debris can cause catastrophic failure. In gas-lubricated, externally pressurized bearings, occasional rubs and resulting wear can occur under impact and vibration loads.

Hydrodynamic bearings wear very slowly. Wear occurs during start-up and slowdown when the speed is too low to produce sufficient fluid pressure to support the bearing surfaces on a lubricant film. If hard debris (harder than the journal surface) imbeds in a babbitt or plastic bearing and protrudes above the bearing surface, the journal can wear seriously during start-up.

In a severe and catastrophic type of journal wear—known as *wire wooling*—the journal surface is machined by hard scabs of wear debris that pack into the babbit surface. In this failure the journals are deeply grooved and can no longer generate a hydrodynamic film.

Boundary-lubricated and self-lubricating bearings wear much faster than fluid-film bearings. Self-lubricating plastic bearings wear at higher rates than boundary-lubricated, metal-alloy bearings.

Babbitts are subject to fatigue damage in which pieces of the babbitt metal spall out of the surface. This might be considered as a catastrophic type of failure after a prolonged (but undetectable) period of wear buildup. Babbitt fatigue is most likely in bearings subjected to reciprocating or vibrating loads.

Babbitts and leaded bronzes are also susceptible to cavitation erosion. This is a form of wear in which shock waves, from gas bubbles collapsing in the lubricant, produce surface pits. Eventually, the removal of surface material impairs performance. Cavitation erosion has a characteristic lacy appearance and often is found around lubrication feed holes and slots in bearings subjected to vibratory or impact loading—such as in internal combustion engines.

Heat

Heat is generated either by shearing of the oil film or by rubbing contact. In hydrostatic and hydrodynamic bearings, heat generation at running speeds is the result of oil shear, and the amount of temperature rise can be estimated if oil viscosity and shear rates are known. Temperature can be regulated by controlling the oil flow through the bearing or by using external cooling.

High-speed and close-clearance fluid-film bearings are difficult to keep cool. The flow rate through a journal varies with the cube of oil-film thickness and linearly with supply pressure.

Boundary-lubricated and self-lubricating bearings are more sensitive to sliding velocity than fluid-film types because the coefficient of friction is as much as 10 times greater in the first two. Frictional heating is a function of bearing pressure, sliding velocity, and coefficient of friction. Therefore, if the coefficient of friction remains constant for a range of loads and speeds, a rough indication of heat load is provided by the *pressure-velocity* (PV) factor.

Plastic bearing materials are sensitive to PV because of their low thermal conductivities and high thermal-expansion rates.

Lubrication

Although some materials have an inherent lubricity or can be lubricated by virtue of a film of slippery solid, most bearings operate with a fluid film—generally oil but sometimes a gas.

By far the largest number of bearings are oil-lubricated. The oil film can be maintained through pumping by a pressurization system—in which case the lubrication is termed *hydrostatic*. Or it can be maintained by a squeezing or wedging of lubricant produced by the rolling action of the bearing itself—termed *hydrodynamic* lubrication. If loads are too high or speeds too slow, the hydrodynamic action begins to break down (allowing, in some cases, metal-to-metal contact), a condition referred to as *boundary* lubrication.

Hydrostatic Lubrication

The main virtue of hydrostatic lubrication is that it can accommodate heavy loads at low speeds because it does not depend upon relative motion to maintain the lubrication film. Instead, lubricant is supplied from a special pump and feed lines to the bearing or bearings. The oil is fed through flow restrictors, which generally are in the stationary part of the bearing. The flow restrictors automatically adjust the oil flow for the applied load. Another advantage of this lubrication system is low deflection in certain load ranges, making it preferred for many high-precision machine tools. The disadvantage of hydrostatic lubrication is its high cost and complexity.

Hydrodynamic Lubrication

This form of lubrication occurs more or less naturally in properly finished, sized, and lubricated holes and shafts. Essentially, rotation of the journal causes it to "roll up" one side of the bearing, thus creating a wedge-shaped channel into which lubricant is forced. The forcing of the lubricant into this wedge creates sufficient pressure to keep the journal riding on the oil film. This form of the lubricant is generally preferred because it is simple and dependable. Also, the lubrication action improves as speed increases, which in most applications goes hand-in-hand with an increase in loads experienced as speed increases. Its main drawbacks are an inability to carry heavy loads at low speed and appreciable wear under frequent stops or starts, or motion reversals.

The oil for hydrodynamic lubrication can be fed from an oil reservoir, or the bearing can be made of a porous metal that is impregnated with oil that "bleeds" to the bearing surface as the shaft rotates. (Most porous-metal bearings, however, operate under boundary or mixed-film conditions.)

Boundary Lubrication

This form of lubrication is essentially a breakdown of hydrodynamic action. At high loads or low speeds, the pressure of the hydrodynamic film cannot prevent metal-to-metal contact. So the opposing surfaces partially ride on an oil film and partially rub together as surface high points contact.

Lubrication is provided by lubricant decomposition products or surface-active additives which form a thin, soft, solid film on the metal surfaces and prevent metal-junction adhesion.

Boundary lubrication is not the most desirable operating mode, yet at times it is completely unavoidable. It is found mainly with slow-moving loads where the cost and expense of a hydrostatic system is not warranted. Hinge bearings in aircraft landing gears, for example, do not move fast enough to develop hydrodynamic films, yet hydrostatic systems would be too heavy, costly, and cumbersome.

Boundary-lubricated bearings using grease require special consideration to ensure adequate distribution of the lubricant. Grease travels only where is is carried on a moving surface, so the moving shaft carries a gob of grease at the center of the bearing, in a band only somewhat wider than the original gob. For this reason, sleeve bearings are grooved to distribute the grease over the entire load-bearing area, ensuring an adequate grease supply for boundary-film maintenance. In addition, the grooves collect wear debris, preventing it from damaging the bearing surface.

Many different groove designs are used (see Fig. 2-33). Some were developed for special applications; however, all are designed for easy manufacture. Basic principles to follow when selecting a groove design are:

- Grooves must carry the grease over the entire bearing width.
- Grooves must be placed so that the shaft contacts the grease within them. (A simple axial slot opposite the point of maximum load does not feed to the load zone because of internal clearances.)
- Groove edges should be rounded so that they do not act as scrapers that clean grease from the shaft.

Groove cross section is usually V-shaped, but it can be rectangular or semicircular. A flat-bottom V groove provides ample reservoir capacity and reduces edge

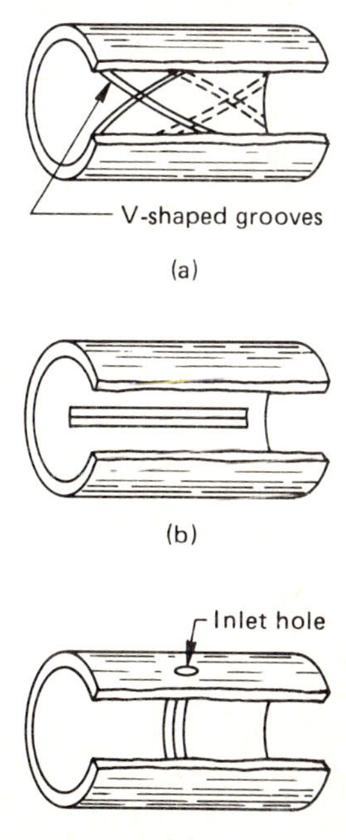

Figure 2-33 Adding grooves to distribute grease.

scraping. Groove width and depth are not critical as long as they do not weaken the bearing structurally or carve up too much of the bearing area. As a rule, groove depth should not exceed one-third the bearing wall thickness.

The 00 design is appropriate for continuous rotation under heavy loads. Conventional design calls for the grooves to be machined within about 0.125 in of the ends to retain the lubricant within the bearing area. However, under heavy loads where bearing temperature is high and debris becomes a problem, the grooves should exit through the ends, so that periodic flushing with fresh grease removes debris and lubricant decomposition products. Experiments with these *nonrecirculating* grooves indicate lower wear than with recirculating grooves.

The *straight axial* groove is a simple arrangement for moderate, unidirectional loads and continuous rotation. The axial slot should be located about 20° from the load line, counter to the direction of shaft rotation. This design spreads grease over the bearing width and still provides maximum bearing area. However, its lubricating effectiveness is not as good as that of the 00 design.

A circumferential slot in the center of the bearing is effective for oscillating bearings, especially in applications where load direction varies.

COMPARING PLAIN BEARINGS AND ROLLER-ELEMENT BEARINGS

Friction

The torque required to put a bearing into motion from rest is usually higher than the torque required to keep it running once it is in motion. *Starting friction* thus has an important influence on the power required in a drive system.

Externally pressurized bearings have very low starting torque. Roller bearings have a low starting torque. Unpressurized sleeve (fluid-film) bearings have substantially higher starting torque. The coefficient of friction at start-up for self-lubricated bearings is highly variable. It may range from 0.04 to 0.16.

The fluid-film bearing has a high starting torque because it must pass through boundary lubrication stages as it comes up to speed. Once running under a hydrodynamic film, the fluid-film bearing exhibits friction characteristics comparable to a rolling-element bearing.

At running speed, the friction characteristics of various bearing types cannot be summarized in a few words. The externally pressurized bearing runs with low friction. Friction in a self-lubricating sleeve bearing is quite variable, depending upon the application.

Running friction for a rolling-element bearing is lower than starting friction. If torque characteristics are critical to a design, recommended practice is to measure starting and running frictional characteristics experimentally.

If a bearing must be started repeatedly under heavy load, rolling-element bearings are a better choice than sleeve bearings. If the increased complexity is acceptable, externally pressurized (hydrostatic) bearings are the best choice of all. If starting load is light and load increases gradually with speed, the conventional hydrodynamic sleeve bearing usually is preferred.

Noise

Whenever a machine is subject to a noise-reduction program, bearings are normally involved in one way or another. Even if bearings are not generating noise, they often have much to do with noise transfer.

Generally, rolling-element bearings are noisier than fluid-film bearings. Minor inaccuracies in rollers or raceways can generate sounds that are amplified by the machine structure. Improving bearing quality can reduce this effect.

Fluid-film bearings under steady radial load generally do not produce any noise whatsoever. However, if this type of bearing is reversed frequently, it can generate considerable noise if it doesn't have enough lubricant to fill the bearing. Fluid-film bearings running unstable in a whirling mode can also produce noise.

Size

A bearing requiring a separate pressurized lubrication system requires more total space than a self-lubricating type. The relative space required at the actual load-support point is not a clear-cut matter. A pressurized bearing can conceivably be more compact than self-contained bearings at the load-support point.

For self-contained bearings, the sleeve bearing requires less radial space than a rolling-element bearing; however, the sleeve bearing requires slightly more axial space. Needle bearings require about as much space as journal bearings.

Cost

Hardware Costs

In high-volume lots, sleeve bearings and bushings are considerably less costly than rolling-element bearings. In mid-range volumes, prices for the two types are comparable. For special designs in small quantities, sliding bearings are usually more costly than rolling-element bearings. Dry-film and boundary-lubricated bearings usually employ proprietary materials that are expensive. Powdered-metal bearings, however, are inexpensive.

Design Costs

Rolling-element and dry-lubricated bearings normally require the least engineering cost to the end user. Manufacturers of rolling-element bearings, in particular, can provide considerable cost-saving assistance by virtue of well-documented design manuals.

Self-acting sleeve bearings, in contrast, may require considerable end-user design effort except for light-duty applications or where there is considerable application experience.

The behavior of externally pressurized bearings usually can be predicted easily by simple calculations, but considerable design effort may be required to verify the design completely.

Shop Costs

Rolling-element bearings normally require precise housings and shafts and thus require fairly costly machining for products in which they are used. Sleeve-bearings, in contrast, generally operate well with less finely prepared machine finishes. Many plain bearings operate satisfactorily with lathe-turned journals.

Maintenance Costs

If the bearing lubrication is self-contained, maintenance costs are normally determined by sealing requirements. If there is full-pressure lubrication, costs may be determined by the amount of filtration needed to keep out contaminants. Generally, rolling-element bearings have the lowest maintenance costs because of the lower lubrication requirements. The very minimum maintenance cost is associated with self-lubricating bearings—provided that they deliver sufficient service life.

Replacement Costs

These costs depend more on the specific design than on the type of bearing used. In general, however, sliding bearings normally are replaced more easily than rolling-element types. Both types can be damaged during installation if not handled properly. Sliding bearings can often be replaced quickly by machining from bar stock or by altering available stock sizes.

Cost of Failure

Rolling-element bearings usually give ample warning that they are approaching failure (by virtue of increasingly noisy operation). They usually fail from fatigue. Sliding bearings, on the other hand, usually perform well until just moments before violent failure.

If a rolling-element bearing fails at high speed, the failure is usually total and catastrophic. With a journal bearing, the effect is normally less drastic. Often, only a bit of polishing can put it back into service. However, sliding bearings are capable of total catastrophic failure.

chapter 1-3

Shaft Drives and Couplings

PART 1

Belt Drives

by
Wally Erickson
Chief Engineer, Product Application Department
The Gates Rubber Company
Denver, Colorado

BELT DRIVES

Belt drives are the most widely used method of power transmission. Improvements in the design and manufacture of belts have broadened their application and utility.

Belt drives employ V, flat, synchronous, or ribbed (poly V) belts. Although the different types of belt drives are considered interchangeable, this is really not the case. Each type is restricted to a well-defined application area.

For example, flat belts are satisfactory for drives operating at high speed and low horsepower. On the other hand, flat-belt drives become overly large when the primary consideration is high-power transmission, and they must give away to V-belt drives in such applications.

V belts provide the best overall power transmission and capability per dollar cost and unit of space. V belts also are the only type of belt that can be used on variable-speed drives.

Synchronous belts are used in applications requiring precise speed ratios and synchronization. These belts also offer lower static tensions, but operating tensions are about the same as on V-belt drives. The ability to operate over small sheave diameters often makes them attractive alternatives to V belts even when precise speed control is not important.

V Belts

The problems of high tension and instability with flat belts led to the development of V belts. These belts have deep, V-shaped cross sections that wedge into sheaves to provide the required traction. Because of this wedging action, V belts are highly stable and can operate at tensions considerably lower than those needed by flat belts. Thus, V-belt drives are more compact, allowing shafts and bearings to be smaller.

The load in a V belt is carried by a fiber tensile section located near the top of the belt. This section can contain one or several layers of cord, depending on the method of manufacture (Fig. 3-1).

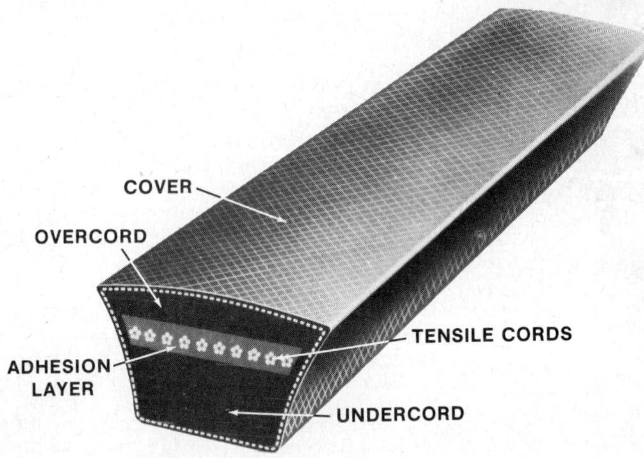

Figure 3-1 Typical V-belt construction.

V belts operate through a wide range of belt speeds. Standard sheaves are limited to 6500 ft/min (33 m/s) upper limit. Speed or peak capacity varies with the type of belt and section.

There are four basic types of V belts: industrial, light-duty, agricultural, and automotive.

Industrial V Belts

Classical. This line of belts includes five standard cross sections—A, B, C, D, E, (13C, 16C, 22C, 32C, 39C)*—ranging in width from ½ in (13 mm) for the A section to 1½ in (39

*Metric sizes are shown in parentheses.

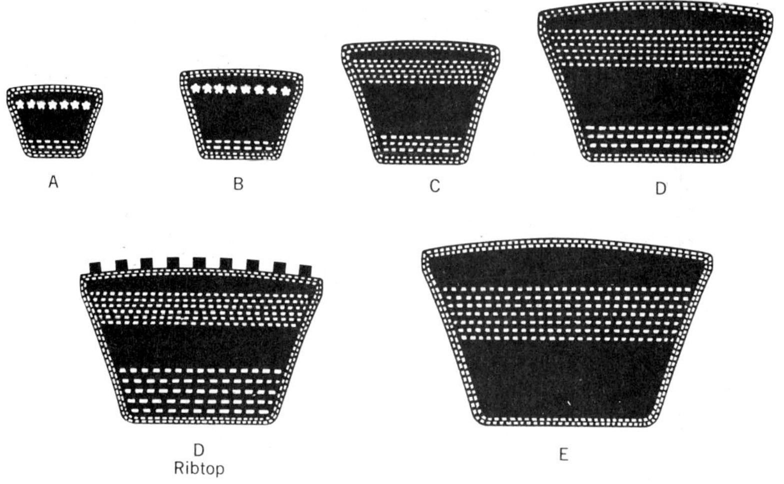

Figure 3-2 Classical V-belt cross sections.

mm) for the E section (Fig. 3-2). Classical belts can be teamed up in multiples of two or more belts, and multiple-belt drives can deliver up to several hundred horsepower continuously and can absorb reasonable shock loads. Temperature limits range from −30 to +140°F (−34 to +60°C).

Classical belts are available in the joined configuration, where several belt strands are connected by a tie band across the top. The tie band improves lateral stability and solves the problems of belts turning over and jumping off sheaves. The band rides above the sheave and does not interfere with the wedging action of any of the individual belt strands (Fig. 3-3).

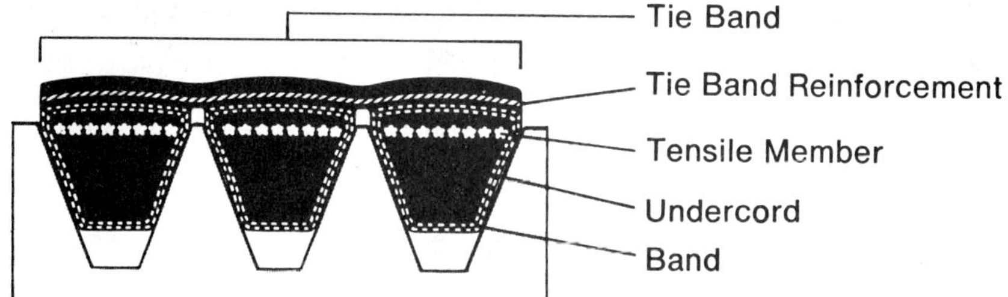

Figure 3-3 Joined V-belt tie-band configuration.

A widely used variation of the classical belt is the molded-notch or molded-cog belt. Being more flexible, the cog belt can operate on smaller-than-normal sheave diameters on drives with limited space for sheaves.

Narrow. These belts serve the same application area as multiple, classical V belts but with a lighter, more compact drive. Three cross sections—3V, 5V, 8V (9N, 15N, 25N)—replace the five classical cross sections (Fig. 3-4).

Narrow V belts usually provide substantial space and weight savings over classical belts. For instance, narrow belts can transmit up to three times the horsepower of conventional belts in the same drive space—or the same horsepower in one-half to two-thirds the space. In addition, the use of narrow belts results in a more aesthetically pleasing machine.

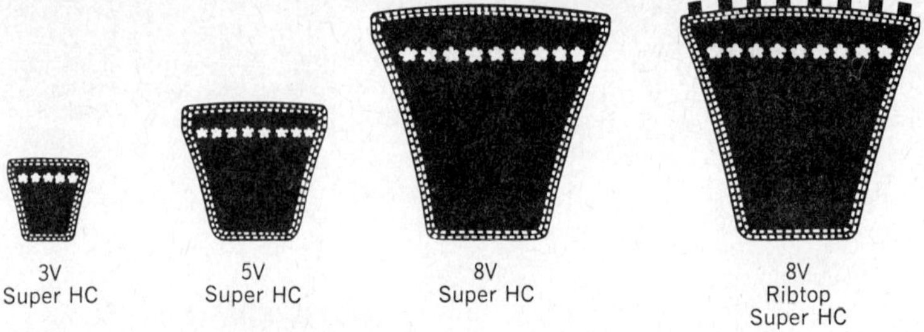

Figure 3-4 Narrow V-belt cross sections.

Variable-Speed Belts. Within certain limits, V belts are well suited to drives that must run at varying input and output speeds. Speed ratio on these drives is controlled by moving one sheave sidewall relative to the other so the belt rides at different pitch diameters.

Unlike gear transmissions where speed variation is limited to finite steps, variable-speed V-belt drives offer an infinite selection of speed ratios within a certain range. In simple applications, speed is varied manually by adjusting the sheave sidewalls while the drive is stopped. On other drives, speed ratio is varied while the drive is operating, either by manual control or by an automatic control sensitive to the speed and torque requirements of the machine.

But adjustable-speed operation generates high frictional heat and sidewall stresses. Until recently, belt fibers and molding compounds could not handle the strains of high-torque applications, and variable-speed belts were largely limited to industrial applications. In the past few years, manufacturers have found belt materials that withstand the heat better, and they have devised ways to optimize the belt configuration. As a result, variable-speed belts are moving into more demanding applications.

These belts are used in industrial, agricultural, automotive, and recreational vehicle applications where speed variation is a prerequisite.

Light-Duty V Belts

These belts are similar to classical belts, except they have slightly thinner cross sections. They are best suited for smaller sheaves, found on fractional-horsepower drives. Most drives designed for light-duty belts are single, rather than multiple drives. Usually these drives are operated intermittently. Service requirements may vary widely from 2 to 3 h/week for power lawn mowers to 40 or more hours per week for office machines.

Standard cross sections for light-duty V belts are 2L, 3L, 4L, 5L (6R, 9R, 12R, 16R) and range in width from ¼ to ⅝ in (6 to 16 mm).

Agricultural V Belts

While agricultural applications require heavy-duty drives, the requirements are significantly different. For example, industrial drives run at fairly constant loads, whereas agricultural drives are subjected to intermittent shock loads. In addition, most agricultural drives require the belt to bend in reverse over small-diameter sheaves; therefore, agricultural belts have more flexible jackets.

Agricultural V belts are available in joined, hexagonal, or double-V styles. Double-V belts are used on drives with more than one driven shaft where shaft location and direction of rotation make it necessary to drive from both sides of the shaft.

Standard cross sections for fixed-speed, classical agricultural V belts are HA, HB, HC, HD, HE (13F, 16F, 22F, 32F, 39F). Cross sections for the double-V belts are HAA, HBB, HCC and HDD (13FD, 16FD, 22FD, 32FD).

Automotive V Belts

Automotive accessory drives require special, narrow V belts that can fit in the limited space available in engine compartments. These belts must not only be able to transmit high horsepower, they must also do so over relatively small-diameter pulleys.

Flat Belts

The flat belt is still a widely used method of power transmission. These belts are flexible over small diameters, and some types adapt to almost any drive configuration. They provide good shock resistance, need no lubrication, are quiet, and offer good design flexibility.

One disadvantage of flat belts is they require higher belt tensions than V belts; this results in higher shaft and bearing loads. The need for higher tensions may cause more stretch, causing the belt to slip and require retensioning.

Synchronous Belts

V belts and flat belts "creep" somewhat in use, so they are not suitable for drives requiring synchronized input and output. Synchronous belts were designed to overcome this limitation. These belts eliminate slip by transmitting power through the positive engagement of belt teeth against pulley teeth. Thus, precise speed ratios and synchronization are possible in such applications as machine-tool indexing heads and automotive camshafts (Fig. 3-5).

Figure 3-5 Synchronous V belt.

Synchronous belt drives have an advantage over gears and chains in that they transmit reasonably high loads with low noise and without lubrication. The shock-absorbing characteristics of the rubber teeth against the metal pulley can be helpful. However, synchronous belts wear rapidly and fail permanently if pulleys are not aligned to close tolerances.

There is one situation (when synchronization is not important) in which synchronous belts have an advantage over V belts. These belts have an extremely high modulus, low-stretch tensile cord to maintain uniform spacing between teeth. This low growth under load results in minimal need for center-distance adjustment. Therefore, synchronous belts are often used on drives with limited space for center distance movement.

There are two basic configurations to synchronous belt teeth: the standard trapezoidal shape and a recently developed deeper, rounded tooth which increases belt torque capacity.

V-Ribbed Belts

Ribs on the bottom side of V-ribbed belts (Fig. 3-6) mate with corresponding grooves in the pulley. This tracking guides the belt and makes it more stable than a flat belt. The ribs fill the pulley grooves completely; therefore, V-ribbed belts do not have the wedging action of V belts, so they must operate at higher tensions.

Figure 3-6 V-ribbed belt.

V-ribbed belts have high lateral stability, much the same as joined V belts. However, the tensile section rides above the grooves and the ribs can be deflected.

While V-ribbed belts are not as flexible as flat belts, they still perform fairly well on small-diameter pulleys, provided that the shafts and bearings can handle high belt tensions. V-ribbed belts are typically used in light-duty applications requiring high speed ratios, e.g., in clothes dryers.

The belts come in five standard sections, ranging in rib width from approximately 1/16 in. to 3/8 in. (1.6 to 9.4 mm). The sections are H, J, K, L, M (PH, PJ, PK, PL, PM).

BELT SELECTION

To aid in selection of the proper belt for the proper application, manufacturers put technical and design information on their belts. In addition, the Rubber Manufacturers' Association and Mechanical Power Transmission Association have worked together to publish engineering standards for most types of belts (see references at end of the chapter). These standards contain information from which a drive can be designed.

In basic terms there are four questions that need to be answered in drive design:

1. What horsepower is required of the drive?
2. What is the speed of the *driver* shaft?
3. What is the speed of the *driven* shaft?
4. What is the approximate center distance?

Also, while selecting or evaluating the drive, consider the following points:

1. If you need to keep the width of the sheave face at a minimum, select the largest-diameter drive from the group.
2. Larger-diameter sheaves will keep drive tensions (and shaft pull) at a minimum.
3. Larger-diameter sheaves will generally give a more economical drive, but not so large that multiple-belt capability is sacrificed.
4. If space is limited, consider using the smallest-diameter drive from the group. However, sheaves on electric drives must be at least as large as the National Electrical Manufacturers' Association minimum standards.
5. When the belt is midway between two belt cross-sectional sizes, the larger cross section will usually be a more economical drive.

Selecting an optimum belt involves too many factors and belt types to be discussed here, but it can be readily accomplished using the reference material and manufacturers' literature.

MAINTENANCE AND SAFETY

Once belts have been selected, they require a minimum of maintenance, but certain procedures can help reduce equipment downtime and increase safety.

Installing Belts

When installing belts on any multiple-belt drive, always replace all the belts because older belts naturally become stretched or worn from use. If old and new belts are mixed, the new belts will be tighter, will carry more than their share of the load, and will probably fail before their time.

Be sure to use a set of belts from a single manufacturer. If brands are mixed, the belts may have different characteristics and they could work against each other. This would result in unusual strain and would reduce the life of the belts.

There are two ways to assure a well-matched drive system depending on the method of matching used by the individual belt manufacturer:

- If the belt manufacturer uses the older conventional matching system, each belt is measured under tension on V sheaves and marked with a match number designating the small increment of length within the overall belt length tolerance. Then the match numbers are grouped within the limits given in Table 3-1.
- Because of improved production methods, some manufacturers, like Gates, can build belts to overall length tolerances that are within the RMA standard matching limits. This means any belts of a specified length can be grouped together to form a matched set.

After the belts have been properly selected, it is time to put them on the drive.

The most important rule is never to pry or roll the belts onto the sheave groove. Prying the belt onto the sheave shortens the belt's life, even though damage may not be visible. Safety also dictates that belts not be rolled onto sheaves; this is one of the most dangerous acts that can be done around V-belt drives. If the sheaves turn, clothing or fingers can be caught and painfully damaged. Rolling is not the easiest way to install belts. The best way to install a belt is to use the drive adjustment to move the sheaves closer together and then drop the belt into the sheave groove.

A sturdy pry bar will help move the motor. Keep the take-up rails free of dirt, rust, and grit; lubricate them from time to time. This will make belt change easier and safer.

The final V-belt installation procedure involves properly tensioning the drive for trouble-free service. Here are three helpful tensioning tips:

1. The best tension for a V belt is the lowest tension at which the belt will not slip under a full load. Too much tension shortens belt life; too little causes slippage, which also leads to premature belt failure.
2. Take up the drive until the belts are snug in their grooves. Run the drive for 15 min to seat the belts; then impose the peak load. If the belts slip, tighten them until they no longer slip at peak loads.

TABLE 3-1 Belt Matching Limits

Belt length		
Inches	Millimeters	Matching limit
up to 100	up to 2540	Use only one number
100–200	2540–5080	Within two consecutive match numbers only
200–300	5080–7620	Within three consecutive match numbers only
300–400	7620–10,160	Within four consecutive match numbers only
400–500	10,160–12,700	Within five consecutive match numbers only
over 500	over 12,700	Within six consecutive match numbers only

3. Check the V-belt tension frequently during the first day of operation. Check periodically thereafter, and make necessary adjustments.

Most V-belt manufacturers publish tensioning procedures for stock drives which give more accurate values.

V-Belt Safety

Safety is a critical factor in the efficient operation of belt drives. The maintenance crews can take several positive steps to help keep drives running smoothly and safely:

1. Keep belted drives properly guarded. Virtually all regulatory agencies (particularly the Occupational Safety and Health Administration), insurance companies, and safety authorities require that drives be completely guarded. The guard should allow proper ventilation of the drive, but should have no gap where workers can reach inside and become caught in the drive.
   The guard is a positive safety feature, but also enhances maintenance by protecting the drive from environmental conditions, debris, and tampering.
2. Always turn the equipment off for maintenance, lock the controls, and place a "down for maintenance" sign on the machine whenever work is performed on the drive.
3. Never roll a belt onto a sheave. As mentioned earlier, use the drive take-up and simply drop the belts into the sheave grooves.

TABLE 3-2 Why V Belts Fail

Trouble area and observation	Cause	Remedy
Worn side patterns	Constant slip	Retension drive until belt stops slipping
	Misalignment	Realign sheaves
	Worn sheaves	Replace with new sheaves
	Incorrect belt	Replace with new belts
Bottom of belt cracking	Belt slipping, causing heat buildup and gradual hardening of undercord	Install new belt, tension to prevent slip
	Idler installed on wrong side of belt	Refer to a V-belt installation manual
	Improper storage	Refer to belt storage information available from manufacturer
Bottom and sides burned	Belt slipping under starting or stalling load	Replace belt and tighten drive until slipping stops
	Worn sheaves	Replace sheaves
Belt turnover	Foreign material in grooves	Remove material, shield drive
	Misaligned sheaves	Realign the drive
	Worn sheave grooves	Replace sheave
	Tensile member broken through improper installation	Replace with new belt(s)
	Incorrectly aligned idler pulley	Carefully align idler, checking alignment with drive loaded and unloaded
Belt pulled apart	Extreme shock load	Remove cause of shock load
	Belt came off drive	Check drive alignment, foreign material in drive; ensure proper tension and drive alignment

V-Belt Inspection

If proper V-belt inspection procedures are followed, 90 percent of all maintenance problems are eliminated. This is because a properly installed V belt is a remarkably trouble-free piece of equipment. But, to assure continued trouble-free service, a quick inspection of the belt drive should be made a part of the routine maintenance schedule.

Listen to the sounds that could mean trouble: *slapping* of belts against the drive guard could be caused by an improperly placed guard, loose belts, or excessive vibration; *squealing* can be caused by poorly tensioned belts or foreign material (grease, dirt, or paint) in the sheave grooves. (Never use a belt dressing that can cause belts to slip or collect dirt and grit.)

Watch the drive while it runs, removing the guard *temporarily* if necessary. With multiple-belt drives, there will be some variation in tension, but there should be a tight side and a slack side. If one or more belts are loose or tight, check one of the following problems:

1. **Improper Tension** Drive may be poorly positioned. Measure tension and adjust movable sheaves as necessary.
2. **Damaged Belt** Remove the belt and carefully inspect to see if the belt is broken internally.
3. **Improper Matching** Replace belts.

When the drive is turned off, inspect the belts for wear. If sidewall wear is excessive, check drive alignment.

V-Belt Failure

The five most common symptoms, causes, and remedies of belt failure are listed in Table 3-2.

BIBLIOGRAPHY

Publications of the Rubber Manufacturers Association, Washington, D.C.:

Number	Title
IP-3	*Power Transmission Belt Technical Information (PTBTI).* A complete set of IP-3-1 thru 3-13, listed below (also available separately).
IP-3-1	*V-Belt Heat Resistance* (1972).
IP-3-2	*V-Belt Oil Resistance* (1972).
IP-3-3	*Static Conductive V-Belts* (1972).
IP-3-4	*Storage of V-Belts (1972).*
IP-3-5	*A Drive Design Procedure for Double V-Belts* (1972).
IP-3-6	*Effect of Idlers on V-Belt Performance* (1972).
IP-3-7	*V-Flat Drives* (1972).
IP-3-8	*Steel Cable V-Belts* (1972).
IP-3-9	*Joined V-Belts* (1972).
IP-3-10	*V-Belt Drives with Twist* (1975).
IP-3-11	*Inspection Guide for Automotive V-Belt Drives* (1973).
IP-3-12	*Belt Tension Calculation Method for Automotive V-Belt Drives* (1975).
IP-3-13	*Mechanical Efficiency of Power Transmission Belt Drives* (1978).

Number	Title
IP-20	Specification: Joint MPTA/RMA/RAC *Classical Multiple V-Belts* (1977). A, B, C, D, and E Cross Sections. (An American National Standard.)
IP-21	Specification: Joint RMA/MPTA *Double V-Belts* (1965). AA, BB, CC, DD Cross Sections.
IP-22	Specification: Joint MPTA/RMA/RAC *Narrow Multiple V-Belts* (1978). 3V, 5V, and 8V Cross Sections.
IP-23	Specification: Joint RMA/MPTA *Single V-Belts* (1968). 2L, 3L, 4L, and 5L Cross Sections.
IP-24	Specification: Joint MPTA/RMA/RAC *Synchronous Belts* (1978). MXL, XL, L, H, XH, and XXH Belt Sections.
IP-25	Specification: Joint MPTA/RMA/RAC *Variable Speed Belts* (1979). Twelve Cross Sections.
IP-26	Specification: Joint MPTA/RMA/RAC *V-Ribbed Belts* (1977). H, J, K, L, and M Cross Sections

PART 2

Flexible Couplings

by

Michael M. Calistrat
Manager, Mechanical Development
Metal Products Division
Koppers Company, Inc.
Baltimore, Maryland

INTRODUCTION

If one could perfectly align two shafts, they could be connected with a muff or two flanged hubs bolted together. Once this is done, it must be ensured that neither of the machines will move on the foundation and that the foundation will not settle. Actually, some misalignment between a driving and driven shaft is a fact of life and this is why *flexible coupling must* be used. The American Gear Manufacturers Association (AGMA) defines flexible couplings as machine elements which transmit torque without slip and accommodate misalignment between the driving and driven shafts.

Flexible couplings are perhaps the most mistreated parts of any machinery, both at selection time and at installation time. Through proper coupling selection and good alignment procedure, high maintenance costs and lost production time can be avoided.

Different types of couplings can accommodate various misalignments; selecting the one that accommodates the largest misalignment is not always the wisest choice. Sometimes a larger misalignment is made possible by a reduction in the power transmitted or a reduction in the useful life of the couplings.

Coupling catalogs usually list the maximum misalignment for each coupling. At installation, one should strive for the best alignment possible, so that the coupling has enough reserve misalignment capacity to accommodate future worsening of the operating conditions.

The misalignment can change for many reasons such as foundation settling, bearing wear, and distortions caused by vibration and temperature changes as well as by the movement of the connected machines under external forces induced through pipes, belts, and chains.

SHAFT MISALIGNMENT

In order to perform a good alignment, one should understand the various relative positions two connected shafts can have.

Parallel Misalignment

This type of misalignment (Fig. 3-7) is the easiest to understand, measure, and correct. Assuming that the motor shaft is higher above the base than the pump shaft, how do we measure this offset? If the two shafts have the same diameter we can measure the offset directly using a straightedge and feeler gauges, as shown. If the two shafts do not have the same diameter, which is often the case, we must first measure dimension *E* and then calculate the offset:

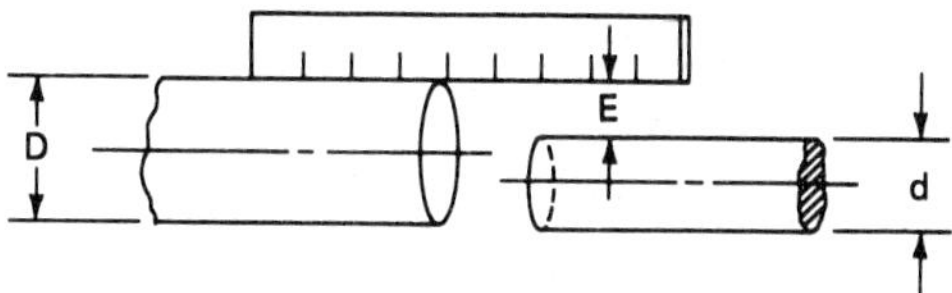

Figure 3-7 Parallel misalignment.

$$\text{Offset} = E - \frac{D - d}{2}$$

If the result is positive, then the large shaft is higher; if the result is negative then the small shaft is higher. The offset misalignment can be corrected easily by adding shims under the machine with the lower shaft.

Angular Misalignment

This is more difficult to measure and correct. Seldom can one see pure angular misalignment.

Combination Misalignment

This type (see Fig. 3-8) is what is most likely to be found between two shafts; it cannot be corrected in one step. When combined misalignment is discovered, the first step should be an attempt to correct the angular misalignment. This can be observed by using a straightedge and feeler gauges. It can be corrected by shimming either the front or the rear of the motor (or pump) base. Once the angular misalignment is corrected, the parallel misalignment can be treated as shown above.

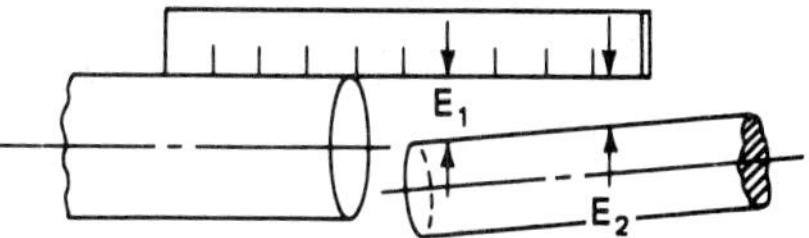

Figure 3-8 Combination misalignment.

In reality, aligning the two shafts is much more complicated, because in most cases the couplings cannot be installed on the shafts after the machines are in place. Sometimes the distance between the shafts is too large to be spanned with a straightedge. For proper alignment, the following steps are to be followed:

1. With a dial indicator point resting on the shaft, verify if the shafts are running true or if they are out of round. A machine with a bent shaft cannot be aligned!
2. The couplings should be installed on the shafts; they should then be checked for running true, both on the diameter and on the face. An eccentrically bored coupling will always operate in the misaligned condition and will have a short life.
3. For closely spaced shafts and moderate speeds, the method of aligning two machines using a straightedge and feeler gauges is satisfactory (Fig. 3-9).
4. The best alignment procedure, known as the *reverse-indicator* method, requires two readings: one with a dial indicator attached to one shaft and the point on the other,

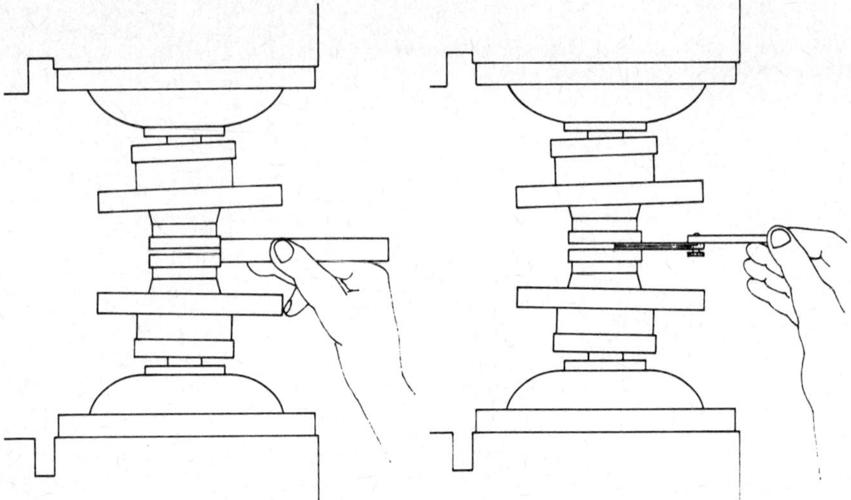

Figure 3-9 Coupling alignment.

and the second reading with the dial indicator reversed. For this method only small, light dial indicators and a sturdy support arm (rather than a magnetic base) should be used.

The importance of good alignment cannot be overemphasized. Fortunately many good articles on the subject are available, and some companies even specialize in selling courses in alignment as well as alignment tools.[1–8]

COUPLING INSTALLATION

Couplings are installed in two steps: first, each half-coupling is installed on its shaft; second, after the machines are aligned, the two coupling halves are bolted together either directly or through a spacer.[9]

The torque must be transmitted from the driving shaft to the first coupling hub, then through the coupling, and finally from the second hub to the driven shaft. In fractional horsepower applications the hubs are attached to the shafts through set screws. When the power is ½ hp (370 W) or more, then one or even two keys are used. Again, for small shafts a set screw over the key prevents the coupling from moving axially, but larger couplings are pressed on the shafts. Tapered shafts are often used, and they have a threaded extension. A nut and locking washer are used to retain the coupling on the shaft. The keys must fit perfectly in the keyway at the sides, but there *must* be a slight radial clearance.

It is important to have interference between the coupling hubs and the shafts because flexible couplings tend to resist misalignment, particularly under torque. The *restoring* forces generated by the couplings tend to rock the hubs on the shafts. Without interference, the movement between the hubs and shafts will wear the hub bore (fretting) and the coupling must be replaced. A safe rule is to use 0.0005 in per inch of diameter for interference.

Tapered shafts are often used because of the ease of installing and removing the hubs. First one should check to make sure there is good contact between the hub bore and the shaft. This is done by painting the bore with Prussian blue and then slightly rotating the hub on the shaft, without the key.

A minimum of 50 percent contact should be obtained. In no case should one attempt to obtain a better contact through lapping between the hub and the shaft. Lapping would generate a ridge on the shaft which in turn could cause the shaft to fail. A plug and ring should be used if lapping is necessary.

To calculate the *draw* of the hub on the shaft, use the formula

$$a = D \times i \times \frac{12}{t}$$

where

D = diameter, in
i = interference, in/in
t = taper, in/ft

For instance, for a 2-in shaft, 0.0005-in/in interference, and 1 ½-in/ft taper, the draw is

$$a = 2 \times 0.0005 \times \frac{12}{1.5} = 0.008 \quad \text{in}$$

For metric units:

$$a = D \times \frac{i}{t}$$

where

D = diameter, mm
i = interference, mm/mm
t = taper, mm/mm

For instance, for a 50-mm shaft, 0.00005-mm/mm interference, and 0.125 mm/mm taper, the draw is

$$a = 50 \times \frac{0.00005}{0.125} = 0.02 \quad \text{mm}$$

TYPES OF COUPLINGS

Depending on the method used to accommodate the misalignment, flexible couplings can be divided into: (1) sliding-element couplings, (2) flexing-element couplings, and (3) combination sliding and flexing couplings.

Until recently all sliding-element couplings required lubrication. With the advent of synthetic materials, some of these couplings can operate dry. For this reason, some people prefer to divide couplings into lubricated and dry categories; but for a better understanding of the mechanism of torque transmission and coupling dynamics, the division into the aforementioned three categories is preferable.

Whether one should use lubricated or dry couplings depends a lot on the type of maintenance available. A properly serviced lubricated coupling can last as long as the connected equipment. A dry coupling does not require periodic servicing, but the flexing elements, whether they are metal or elastomer, have to be replaced sooner or later. An interesting article on this subject can be found in Ref. 10.

Sliding-Element Couplings

These types of couplings accommodate misalignment by sliding between two or more of their components. This sliding, and the forces generated by the transmitted torque, generate wear. In order to provide adequate life, these couplings are either lubricated or use elements made of low-friction plastics. Sliding-element couplings have two half-couplings, because each sliding pair of elements can accommodate only angular misalignment; it takes two such pairs, properly spaced, to accommodate parallel misalignment. One can better understand this fact if each pair of sliding elements is assumed to be a hinge joint (Fig. 3-10). It can be seen that the maximum offset (parallel misalignment) between the two shafts is a function of the angle of misalignment α the coupling can accommodate *and* the distance L between the two joints. In analyzing Fig. 3-10, one can understand the advantage of using a coupling with a large shaft separation: it is easier to align, and changes in the positions of the connected machines have a small influence on the coupling alignment.

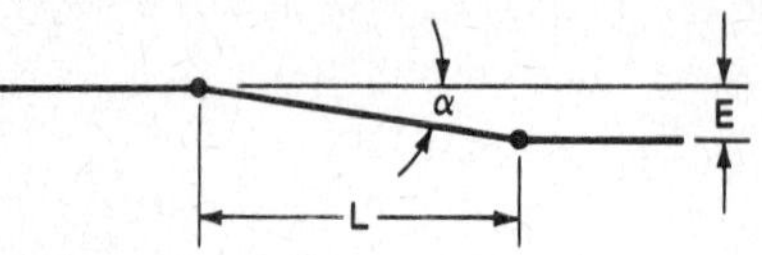

Figure 3-10 The effect of shaft separation.

Gear-Type Couplings

The gear-type couplings are probably the most versatile coupling design. They can be manufactured for almost any application, from a few horsepower to thousands of horsepower, from less than 1 r/min to more than 20,000 r/min. For a given application, a gear coupling is generally smaller and lighter than any other type. Gear couplings can be used on machines with closed coupled shafts or for long separations between the connected shafts. On the other hand, they require periodic relubrication (about every 6 months), they are rigid torsionally, and they are more expensive than other types of couplings.

A gear coupling for closed coupled shafts has two halves bolted together (Fig. 3-11). Each half-coupling has only three components: a hub, a sleeve, and a seal. The hub has a set of external teeth and is very similar to a pinion. The sleeve has a matching set of internal teeth cut in such a way that when the sleeve is slid over the hub there is play (backlash) between the meshing teeth. The seal is installed in a groove machined in the end plate of the sleeve and serves the double purpose of retaining the lubricant and preventing dirt or water from entering the coupling. The sleeves are also provided with one or two grease fittings or plugs. For long shaft separations, a spacer is introduced between the two sleeves. The flanges are connected with eight or more bolts, and a paper gasket, or an O ring, is installed between the flanges for sealing the joint.

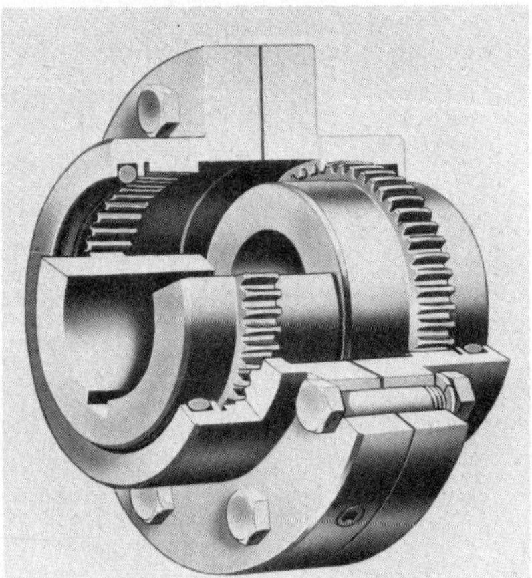

Figure 3-11 Gear-type coupling.

Figure 3-12 Chain coupling. *(Reproduced by permission of FMC Corporation.)*

Chain Couplings

The chain couplings excel through their simplicity; two sprockets and a length of double-row chain is all that is needed. They are generally used at low speeds, except when a special metal or plastic cover is used to contain the lubricant which otherwise will be thrown off by the centrifugal forces (Fig. 3-12). Chain couplings are used in closed coupled applications.

Steel-Grid Couplings

This type of coupling is in many respects similar to the gear coupling. It has two hubs with external teeth (Fig. 3-13) but of a special profile. Instead of the sleeves with internal teeth, it has a steel grid that loops through the spaces between the teeth. Because the steel grid flexes to some extent under torque, the steel-grid coupling is torsionally less rigid than the gear-type coupling.

Coupling Lubrication

Couplings that incorporate sliding elements require lubrication to minimize wear and thus increase their useful life. With few exceptions this type of coupling is grease-lubricated. Using the proper lubricants and procedures rewards the user with long, trouble-free service from the coupling. Not every grease is suitable for coupling lubrication. Manufacturers' catalogs list only a few greases. If these recommendations, or the greases listed in them, are not available, the following guidelines should be used:

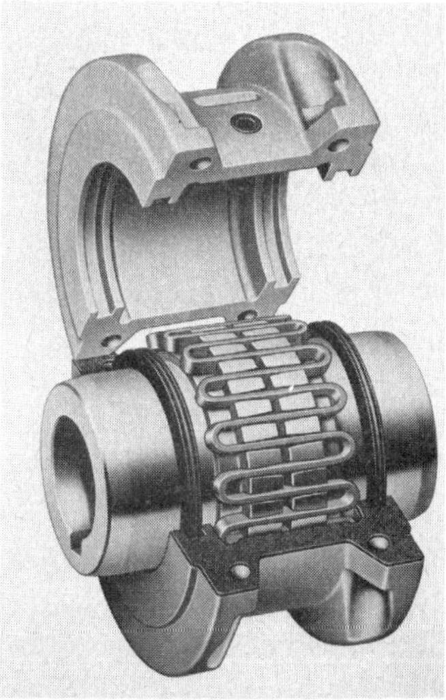

Figure 3-13 Steel-grid coupling. *(Reproduced by permission of The Falk Corporation.)*

1. Because couplings rely on the centrifugal effect to force the lubricant between the sliding surfaces, heavy greases are not good choices. NLGI No. 1 greases are the best compromise between good lubrication and adequate sealing.
2. Because the wear of the coupling decreases as the viscosity of the base oil of a grease increases, one should select a grease blended with an oil having a viscosity of no less than 900 SSU (Saybolt second(s) universal) at 100°F. This information can be obtained from the grease manufacturer.
3. Because greases will separate into oil and soap when subjected to centrifugal forces for a long time and because the soap used in greases is not a lubricant, one should select a grease having very little soap—preferably less than 8 percent of the total weight.

Couplings should be lubricated every 6 months, and before new grease is pumped the coupling should be opened and cleansed of the old lubricant.

Flexing-Element Couplings

These types of couplings accommodate misalignment through flexing of one or more of their components. This flexing can in time cause the failure of the element which must then be replaced. It is obvious that the less misalignment the coupling must accommodate, the less flexing the elements must suffer, and the coupling can provide longer trouble-free service.

Depending on the material used for the flexing element, couplings can be divided into two types: (1) metallic-element and (2) elastomer-element. Metallic-element couplings can accommodate only angular misalignment at each flexing point. To accommodate parallel (offset) misalignment, a coupling needs two flexing elements. As shown in Fig. 3-10, the greater the distance between the elements, the more offset the coupling can accommodate. Elastomer-element couplings can accommodate offset with only one element. They are designed for close-coupled machines; however, when used with a special centering bushing, they can be used for long shaft separations.

Metallic-Element Couplings

The flexible element is not a single piece; rather it is a pack of many thin stamped disks, usually made of stainless steel (Fig. 3-14). Coupling sizes vary from miniature to large. With few exceptions, they are not used at high speeds. The multiple-disk pack offers the advantage of a redundant system, and the coupling can operate even after one or more of the disks fail. However when the disks are replaced, the pack should be replaced as a whole rather than only the broken disks. One drawback of the metal-disk couplings is that they tolerate very little error in the axial spacing of the machines. On the other hand, this disadvantage becomes an advantage when a limited end-float coupling is required, as is the case with sleeve-bearing motors that rely on their magnetic centering and have no thrust bearings.

Figure 3-14 Disk-pack coupling. *(Reproduced by permission of Rexnord Corp.)*

Elastomer-Element Couplings

There are quite a few designs using elastomer elements: some using rubber with or without plies and some using plastics. Each model has its own advantages and disadvantages. Many times, availability in one particular area determines which couplings will be used. Only the most popular types will be discussed here.

The Rubber Tire. The rubber-tire element (Fig. 3-15) is clamped at each hub, and it is split axially to permit the element's replacement without moving the connected machines.

The Rubber Doughnut. The rubber-doughnut element (Fig. 3-16) is bolted with radial fasteners to the hubs, and in the process it is also precompressed so that it never

Figure 3-15 Rubber-tire coupling. *(Reproduced by permission of Reliance Electric Co.)*

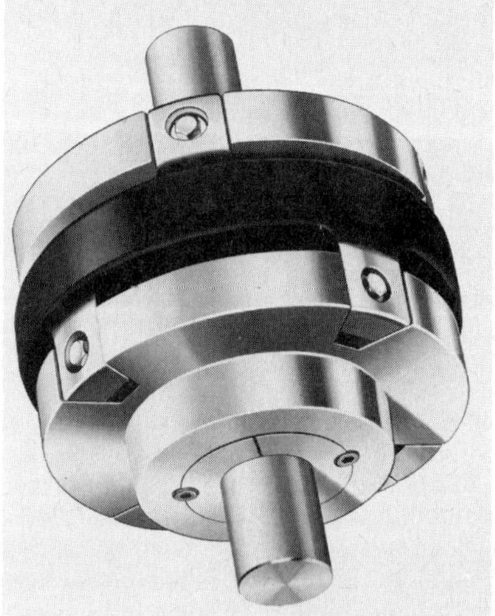

Figure 3-16 Rubber-doughnut coupling.

works in tension. It is split axially at one of the inserts to facilitate installation without the need to disturb the connected machines.

The Splined Element. The splined element (Fig. 3-17) slides inside the hubs axially, and it is either rubber or plastic. To replace the element, one of the hubs must be pushed back axially. For close-coupled shafts, the element is split axially so that the machines do not have to be moved at installation of the element.

The Jaw. This coupling (Fig. 3-18) is also known as the spider coupling because of the shape of the elastomer element. This type of coupling is perhaps the simplest, but it has the following disadvantages: (1) it can accommodate little misalignment and (2) it can usually transmit less than 100 hp (74.6 kW) and, similarly to the splined-element coupling, one of the hubs has to be moved axially in order to replace the element.

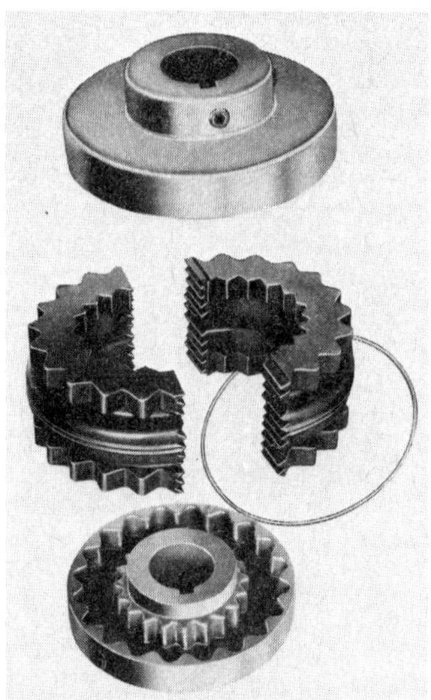

Figure 3-17 Splined-element coupling. *(Reproduced by permission of T. B. Wood's Sons Co.).*

Figure 3-18 Jaw coupling. *(Reproduced by permission of Lovejoy, Inc.)*

COUPLING SELECTION

Usually couplings are supplied as part of any new equipment. Instead of having to select a new coupling, one is faced only with the need to replace an old one, or part of an old one. Assuming that the equipment manufacturer selected the right coupling type and size, couplings generate few problems. There are cases, however, when either the coupling does not live up to expectations or when a new piece of equipment is purchased without a driver and a coupling must be selected. The process is not simple because there is no application for which only one type of coupling would work. An excellent article on this subject can be found in Ref. 11. The best approach is to let an application engineer from a coupling manufacturer make the selection. Today most manufacturers make more than one type of coupling and can objectively recommend the best one for the application.

Choosing a coupling of the correct size is very important. To do this one must know not only the required power and speed, but also the severity of service the coupling must accommodate. A correction factor, or service factor, must be applied.

Coupling manufacturers rate their couplings in horsepower per 100 r/min. For instance, if a pump requires 50 hp (37.3 kW) at 1750 r/min, it needs a coupling that can handle 2.86 hp (2.13 kW) at 100 r/min. This is correct only if the pump is centrifugal and is driven by an electric motor. In this case the service factor is 1. If we have a double-acting reciprocating pump driven by an internal-combustion engine, we have to use a service factor of 2.0 + 1.0 = 3.0 for a gear coupling and 2.0 + 0.5 = 2.5 for an elastomer coupling, according to one manufacturer. As a result, we must choose a gear coupling that can handle 8.58 hp (6.4 kW) at 100 r/min or an elastomer coupling that can handle

7.15 hp (5.3 kW) at 100 r/min. It seems that we can choose a smaller coupling if we choose the elastomer type. However, the elastomer coupling will be about 8¾ in (222 mm) in diameter, while the gear coupling will be only half that size! If size is not important, price can be the next selection criterion. But the price of the coupling alone is not a good guide; one should consider the total cost including maintenance, replacement parts, lost production, etc.

Although couplings represent a small percentage of the total cost of a piece of machinery, they can cause as much, if not more, trouble than the rest of the equipment if they are not properly selected. Buying an inadequate size or type of coupling will never be economical in the long run.

REFERENCES

1. Dreymala, J.: "Try Dial Indicators for Close Alignment of Coupling Connected Machinery," *Power,* June 1971, pp. 96–98.
2. Jackson, C.: "Techniques for Alignment of Rotating Equipment," *Hydrocarbon Processing,* January 1976, pp. 81–85.
3. *Compressor Handbook,* Gulf Publishing Co., Houston, Texas, 1979.
4. Campbell, A. J.: "Optical Alignment Saves Equipment Downtime," *Oil and Gas Journal,* November 1975, pp. 54–56.
5. Murray, M. G.: "Minimum Movement Machinery Alignment," *Hydrocarbon Processing,* January 1979, pp. 112–114.
6. "Reverse Indicator Method of Alignment," Hughes & Associates, Houston, Texas, 1974.
7. "M905 Series, Dyn Align Bars," Dymac, San Diego, California, 1974.
8. "Acculign Gauges," Boyce Engineering, Houston, Texas.
9. Calistrat, M. M.: "Flexible Coupling Installation," *Proceedings of the National Conference on Power Transmission, 1981,* Illinois Institute of Technology, Chicago.
10. Wright, J.: "Which Shaft Coupling is Best—Lubricated or Non-lubricated?," *Hydrocarbon Processing,* April 1975, pp. 191–193.
11. Wright, J.: "Which Flexible Coupling?," *Power Transmission & Bearing Handbook,* 1971/1972, Industrial Publishing Co., Cleveland.

PART 3

Chain Drives

by

Thomas J. Rost, P.E.
Senior Product Engineer
Industrial Chain Division
PT Components, Inc.
Indianapolis, Indiana

INTRODUCTION

Chain drives are one of the most efficient methods used to transmit mechanical power between two or more rotating shafts that cannot be directly coupled. A chain drive consists of a series of links assembled together and two or more sprockets (Fig. 3-19). The positive-engagement sprockets are keyed to the rotating shafts between which power is transmitted. Roller chain, engineering steel chain, and silent chain are the three types of chain used in industrial drive applications. The benefits of a chain drive compared to a belt drive or gear drive are:

1. Shaft center distances are relatively unrestricted.
2. Chains are easily installed.
3. Chain drives do not slip or creep, resulting in overall high efficiency.
4. The load in a chain drive is distributed over a number of sprocket teeth simultaneously.
5. Chain drives operate in adverse environmental conditions.

This presentation will deal only with chain-drive applications. Additional information on this subject and other chain applications, along with sprocket-selection procedures, can be obtained from manufacturers' catalogs and the sources listed in the reference section. The reader is cautioned that all calculations are examples to aid in understanding concepts. All design work should be done with the cooperation of a manufacturer.

DRIVE-SELECTION PROCEDURE

The proper steps for selecting a chain drive are listed below:

1. Determine and record all applicable operating factors, such as
   a. Source and type of power

Figure 3-19 A roller chain drive operates multiple rolls in a heat-treating furnace.

- **b.** Size and speed of driving shaft
- **c.** Size and speed of driven shaft
- **d.** Approximate center distances between shafts
- **e.** Relative position of shafts
- **f.** Type of driven equipment
- **g.** Operating conditions
- **h.** space limitations

2. Establish the service factor from the actual operating conditions. The service factor is a factor by which the transmitted horsepower is multiplied to compensate for drive conditions. The composite or final service factor is the product of the separate service factors multiplied together (see Table 3-3).
3. Calculate the factored horsepower value by multiplying the horsepower to be transmitted by the final service factor.
4. Make a trial chain selection based on the factored horsepower and revolutions per minute of the small sprocket (Fig. 3-20).
5. Determine the number of teeth in the small sprocket from the speed-horsepower charts.
6. Check the small sprocket for bore capacity, number of teeth, and availability.
7. Divide the speed of the faster-turning shaft by the speed of the slower shaft to determine the drive ratio.
8. Calculate the chain length and exact sprocket centers.
9. Determine the chain sag. The chain sag should be approximately 2 percent of the distance between shaft centers at the initial installation.
10. Determine the method of lubrication.

DRIVE-SELECTION EXAMPLE

Application

Select a roller chain drive for the following conditions:

Source of power	Electric motor
Horsepower to be transmitted	10 hp (7.5 kW)
Size of driving shaft	2.438 in (62 mm) diameter

Speed of driving shaft	100 r/min
Drive equipment	Uniformly fed coal elevator for power plant
Size of driven shaft	2.938 in (75 mm) diameter
Speed of driven shaft	42 r/min
Approximate center distance	24.00 in (610 mm)
Relative position of shafts	On same horizontal plane
Space limitations	None

Solution

Service Factor. The service factor listed in Table 3-3 for a uniformly fed elevator driven by a gear motor is 1.0.

Equivalent Horsepower. The equivalent horsepower is 10 × 1.0 = 10 hp.

Trial Chain. From Fig. 3-20 note that the intersection of the 100-r/min vertical line and the 10-hp single-strand horizontal line falls in the area for No. 100 chain. Thus, the trial chain is No. 100 single strand.

Small Sprocket. In Table 3-4 for No. 100 roller chain, the 100-r/min column lists 10.3 hp which corresponds closely to the equivalent horsepower of 10 required for this application. This rating is for single-strand chain when used with a 17-tooth sprocket.

Check the Small Sprocket. As shown in the rating table, the maximum bore of a 17-tooth No. 100 sprocket is larger than the 2.438 in (62 mm) bore required; therefore, the selection is satisfactory.

Drive Ratio. The drive ratio equals

$$\frac{100 \text{ r/min}}{42 \text{ r/min}} = 2.38{:}1$$

Number of Teeth in Large Sprocket. The number of teeth in the large sprocket equals 2.38 × 17 = 40.4 teeth. Use a 40-tooth sprocket.

Center-Distance and Chain-Length Computations. To calculate the sprocket centers and chain length for a given drive, these symbols are used for the formulas in the following text:

e = desired sprocket centers, in (given as 24.00 in)
E = exact sprocket centers, in
g = pitch diameter of small sprocket, in (for a 17-tooth, No. 100 sprocket the pitch diameter = 6.803 in)
G = pitch diameter of large sprocket, in (for a 40-tooth, No. 100 sprocket the pitch diameter = 15.932 in)
N = actual length of chain, pitches
P = chain pitch, in (the No. 100 = 1.250 in pitch)
t = number of teeth in small sprocket (17 teeth)
T = number of teeth in large sprocket (40 teeth)

Calculate factor A using the formula

$$A = \frac{G - g}{2e} = \frac{15.932 - 6.803}{(2)(24.00)} = 0.19019$$

Refer to Table 3-5 and select factors B, C, and D corresponding to value A or the next higher value. Since A = 0.19019, select the next higher listed value of 0.19081. Corresponding factors for B, C, and D are 1.9633, 0.4389, and 0.5611, respectively. The number of pitches in the chain equals the sum of the pitches between sprockets and the pitches around the sprockets, or

$$\text{N} = \frac{Be}{P} + Ct + Dt = \frac{(1.9633)(24.00)}{1.250} + (0.4389)(17) + (0.5611)(40)$$
$$= 67.601 = 68 \text{ pitches}$$

TABLE 3-3 Service Factors for Various Operating Conditions

Driven equipment	Service Factors		
	Input power		
	Internal combustion engine with hydraulic drive	Electric motor or turbine	Internal combustion engine with mechanical drive
Agitators, liquid stock	1.0	1.0	1.2
Beaters	1.2	1.3	1.4
Blowers, centrifugal	1.0	1.0	1.2
Boat propellers	1.4	1.5	1.7
Compressors			
centrifugal	1.2	1.3	1.4
reciprocating, 3 or more cylinders	1.2	1.3	1.4
reciprocating, singular, 2 cylinders	1.4	1.5	1.7
Conveyors			
uniformly loaded or fed	1.0	1.0	1.2
not uniformly loaded or fed	1.2	1.3	1.4
reciprocating	1.4	1.5	1.7
Cookers, cereal	1.0	1.0	1.2
Crushers	1.4	1.5	1.7
Elevators, bucket			
uniformly loaded or fed	1.0	1.0	1.2
not uniformly loaded or fed	1.2	1.3	1.4
Fans, centrifugal	1.0	1.0	1.2

Feeders			
rotary table	1.0	1.0	1.2
apron, belt, screw, rotary vane	1.2	1.3	1.4
reciprocating	1.4	1.5	1.7
Generators	1.0	1.0	1.2
Grinders	1.2	1.3	1.4
Hoists	1.2	1.3	1.4
Kettles, brew	1.0	1.0	1.2
Kilns and dryers, rotary	1.2	1.3	1.4
Lineshafts			
light or normal service	1.0	1.0	1.2
heavy service	1.2	1.3	1.4
Machinery			
uniform load, nonreversing	1.0	1.0	1.2
moderate pulsating load, nonreversing	1.2	1.3	1.4
severe impact or variable load, reversing	1.4	1.5	1.7
Mills			
ball, pebble and tube	1.2	1.3	1.4
hammer, rolling	1.4	1.5	1.7
Pumps			
centrifugal	1.0	1.0	1.2
reciprocating, 3 or more cylinders	1.2	1.3	1.4
Screens, rotary, uniformly fed	1.2	1.3	1.4
Basis for service factors: Uniform load	1.0	1.0	1.2
Moderate shock load	1.2	1.3	1.4
Heavy shock load	1.4	1.5	1.7

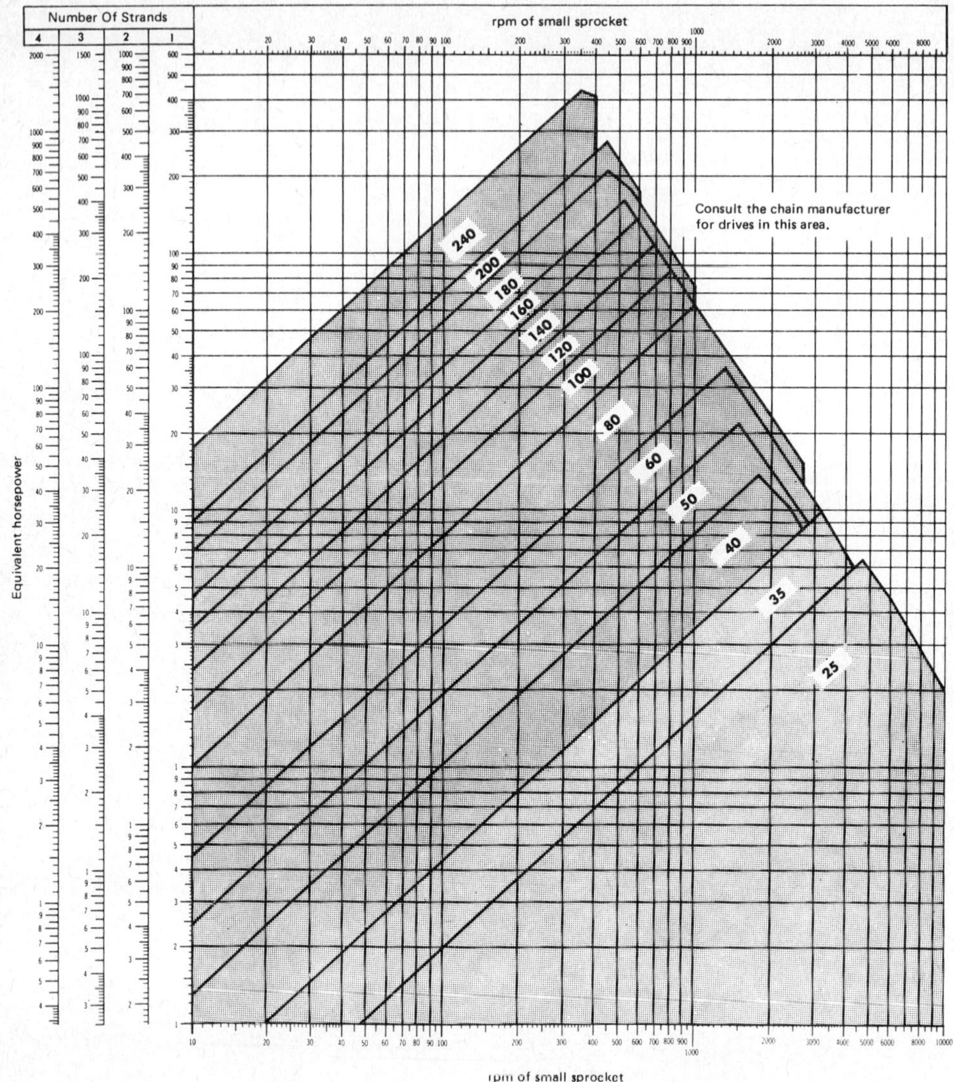

Figure 3-20 Trial selection chart for ANSI standard roller chains.[1]

Select an even whole number nearest to the calculated number of pitches. The exact sprocket center is found by the following formula:

$$E = \frac{(N - Ct - Dt)P}{B} = \frac{68 - (0.4389)(17) - (0.5611)(40)}{1.9633} 1.250$$
$$= 24.254 \text{ in}$$

Lubrication. The No. 100 rating table specified type B bath or disk lubrication. The drive selected for this application consists of:

17-tooth No. 100 driving sprocket

40-tooth No. 100 driven sprocket

Oil-retaining casing for oil-bath lubrication

TABLE 3-4 Speed-Horsepower Ratings for No. 100 Roller Chain

Number of teeth in small sprocket	Maximum bore, inches	Horsepower for single strand chain▲																			
		RPM of small sprocket																			
		100	500	900	1200	1800	2500	3000	3500	4000	4500	5000	5500	6000	6500	7000	7500	8000	8500	9000	10000
11	.313	0.05	0.23	0.39	0.50	0.73	0.98	1.15	1.32	1.38	1.16	0.99	0.86	0.75	0.67	0.60	0.54	0.49	0.45	0.41	0.35
12	.375	0.06	0.25	0.43	0.55	0.80	1.07	1.26	1.45	1.57	1.32	1.12	0.97	0.86	0.76	0.68	0.61	0.56	0.51	0.47	0.40
13	.438	0.06	0.27	0.47	0.60	0.87	1.17	1.38	1.58	1.77	1.49	1.27	1.10	0.96	0.86	0.77	0.69	0.63	0.57	0.53	0.45
14	.563	0.07	0.30	0.50	0.65	0.94	1.27	1.49	1.71	1.93	1.66	1.42	1.23	1.08	0.96	0.86	0.77	0.70	0.64	0.59	0.50
15	.563	0.08	0.32	0.54	0.70	1.01	1.36	1.61	1.85	2.08	1.84	1.57	1.36	1.20	1.06	0.95	0.86	0.78	0.71	0.65	0.56
16	.563	0.08	0.34	0.58	0.76	1.09	1.46	1.72	1.98	2.23	2.03	1.73	1.50	1.32	1.17	1.05	0.94	0.86	0.78	0.72	0.61
17	.625	0.09	0.37	0.62	0.81	1.16	1.56	1.84	2.11	2.38	2.22	1.90	1.64	1.44	1.28	1.14	1.03	0.94	0.86	0.79	0.67
18	.750	0.09	0.39	0.66	0.86	1.24	1.66	1.96	2.25	2.53	2.42	2.07	1.79	1.57	1.39	1.25	1.12	1.02	0.93	0.86	0.73
19	.813	0.10	0.41	0.70	0.91	1.31	1.76	2.07	2.38	2.69	2.62	2.24	1.94	1.70	1.51	1.35	1.22	1.11	1.01	0.93	0.79
20	.875	0.10	0.44	0.74	0.96	1.38	1.86	2.19	2.52	2.84	2.83	2.42	2.10	1.84	1.63	1.46	1.32	1.20	1.09	1.00	0.86
21	.875	0.11	0.46	0.78	1.01	1.46	1.96	2.31	2.66	2.99	3.05	2.60	2.26	1.98	1.76	1.57	1.42	1.29	1.17	1.08	0.92
22	.938	0.11	0.48	0.82	1.07	1.53	2.06	2.43	2.79	3.15	3.27	2.79	2.42	2.12	1.88	1.69	1.52	1.38	1.26	1.16	0.99
23	1.000	0.12	0.51	0.86	1.12	1.61	2.16	2.55	2.93	3.30	3.50	2.98	2.59	2.27	2.01	1.80	1.62	1.47	1.35	1.24	1.06
24	1.063	0.13	0.53	0.90	1.17	1.69	2.27	2.67	3.07	3.45	3.73	3.18	2.76	2.42	2.15	1.92	1.73	1.57	1.44	1.32	1.12
25	1.188	0.13	0.56	0.94	1.22	1.76	2.37	2.79	3.21	3.61	3.96	3.38	2.93	2.57	2.28	2.04	1.84	1.67	1.53	1.40	1.20
28	1.250	0.15	0.63	1.07	1.38	1.99	2.68	3.15	3.62	4.09	4.54	4.01	3.47	3.05	2.70	2.42	2.18	1.98	1.81	1.66	1.42
30	1.313	0.16	0.68	1.15	1.49	2.15	2.88	3.40	3.90	4.40	4.89	4.45	3.85	3.38	3.00	2.68	2.42	2.20	2.01	1.84	1.57
32	1.500	0.17	0.73	1.23	1.60	2.30	3.09	3.64	4.18	4.72	5.25	4.90	4.25	3.73	3.30	2.96	2.67	2.42	2.21	2.03	1.73
35	1.688	0.19	0.80	1.36	1.76	2.53	3.41	4.01	4.61	5.20	5.78	5.60	4.86	4.26	3.78	3.38	3.05	2.77	2.53	2.32	1.98
40	1.875	0.22	0.92	1.57	2.03	2.93	3.93	4.64	5.32	6.00	6.68	6.85	5.93	5.21	4.62	4.13	3.73	3.38	3.09	2.83	2.42
Lubrication type■		A					B									C					

▲ Ratings are based on a service factor of 1. For a complete list of service factors, refer to Table 3-3.
The ratings tabled above apply directly to lubricated, single strand, standard roller chains.

■ Type A: Manual or drip (Maximum chain speed 500 FPM)
Type B: Bath or disk (Maximum chain speed 3500 FPM)
Type C: Forced (pump)

TABLE 3-5 Factors for Chain-Length and Center-Distance Computations

A	B	C	D	A	B	C	D
.00000	2.0000	.5000	.5000	.19937	1.9598	.4361	.5639
.00436	2.0000	.4986	.5014	.20364	1.9581	.4347	.5653
.00873	1.9999	.4972	.5028	.20791	1.9563	.4333	.5667
.01309	1.9998	.4958	.5042	.21218	1.9545	.4319	.5681
.01745	1.9997	.4944	.5056	.21644	1.9526	.4306	.5694
.02181	1.9995	.4931	.5069	.22070	1.9507	.4292	.5708
.02618	1.9993	.4917	.5083	.22495	1.9487	.4278	.5722
.03054	1.9991	.4903	.5097	.22920	1.9468	.4264	.5736
.03490	1.9988	.4889	.5111	.23345	1.9447	.4250	.5750
.03926	1.9985	.4875	.5125	.23769	1.9427	.4236	.5764
.04362	1.9981	.4861	.5139	.24192	1.9406	.4222	.5778
.04798	1.9977	.4847	.5153	.24615	1.9385	.4208	.5792
.05234	1.9973	.4833	.5167	.25038	1.9363	.4194	.5806
.05669	1.9968	.4819	.5181	.25460	1.9341	.4181	.5819
.06105	1.9963	.4806	.5194	.25882	1.9319	.4167	.5833
.06540	1.9957	.4792	.5208	.26303	1.9296	.4153	.5847
.06976	1.9951	.4778	.5222	.26724	1.9273	.4139	.5861
.07411	1.9945	.4764	.5236	.27144	1.9249	.4125	.5875
.07846	1.9938	.4750	.5250	.27564	1.9225	.4111	.5889
.08281	1.9931	.4736	.5264	.27983	1.9201	.4097	.5903
.08716	1.9924	.4722	.5278	.28402	1.9176	.4083	.5917
.09150	1.9916	.4708	.5292	.28820	1.9151	.4069	.5931
.09585	1.9908	.4694	.5306	.29237	1.9126	.4056	.5944
.10019	1.9899	.4681	.5319	.29654	1.9100	.4042	.5958
.10453	1.9890	.4667	.5333	.30071	1.9074	.4028	.5972
.10887	1.9881	.4653	.5347	.30486	1.9048	.4014	.5986
.11320	1.9871	.4639	.5361	.30902	1.9021	.4000	.6000
.11754	1.9861	.4625	.5375	.31316	1.8994	.3986	.6014
.12187	1.9851	.4611	.5389	.31730	1.8966	.3972	.6028
.12620	1.9840	.4597	.5403	.32144	1.8939	.3958	.6042
.13053	1.9829	.4583	.5417	.32557	1.8910	.3944	.6056
.13485	1.9817	.4569	.5431	.32969	1.8882	.3931	.6069
.13917	1.9805	.4556	.5444	.33381	1.8853	.3917	.6083
.14349	1.9793	.4542	.5458	.33792	1.8824	.3903	.6097
.14781	1.9780	.4528	.5472	.34202	1.8794	.3889	.6111
.15212	1.9767	.4514	.5486	.34612	1.8764	.3875	.6125
.15643	1.9754	.4500	.5500	.35021	1.8733	.3861	.6139
.16074	1.9740	.4486	.5514	.35429	1.8703	.3847	.6153
.16505	1.9726	.4472	.5528	.35837	1.8672	.3833	.6167
.16935	1.9711	.4458	.5542	.36244	1.8640	.3819	.6181
.17365	1.9696	.4444	.5556	.36650	1.8608	.3806	.6194
.17794	1.9681	.4431	.5569	.37056	1.8576	.3792	.6208
.18224	1.9665	.4417	.5583	.37461	1.8544	.3778	.6222
.18652	1.9649	.4403	.5597	.37865	1.8511	.3764	.6236
.19081	1.9633	.4389	.5611	.38268	1.8478	.3750	.6250
.19509	1.9616	.4375	.5625	.38671	1.8444	.3736	.6264

TABLE 3-5 *(continued)*

A	B	C	D	A	B	C	D
.39073	1.8410	.3722	.6278	.56641	1.6483	.3083	.6917
.39474	1.8376	.3708	.6292	.57000	1.6433	.3069	.6931
.39875	1.8341	.3694	.6306	.57358	1.6383	.3056	.6944
.40275	1.8306	.3681	.6319	.57715	1.6333	.3042	.6958
.40674	1.8271	.3667	.6333	.58070	1.6282	.3028	.6972
.41072	1.8235	.3653	.6347	.58425	1.6231	.3014	.6986
.41469	1.8199	.3639	.6361	.58779	1.6180	.3000	.7000
.41866	1.8163	.3625	.6375	.59131	1.6129	.2986	.7014
.42262	1.8126	.3611	.6389	.59482	1.6077	.2972	.7028
.42657	1.8089	.3597	.6403	.59832	1.6025	.2958	.7042
.43051	1.8052	.3583	.6417	.60182	1.5973	.2944	.7056
.43445	1.8014	.3569	.6431	.60529	1.5920	.2931	.7069
.43837	1.7976	.3556	.6444	.60876	1.5867	.2917	.7083
.44229	1.7937	.3542	.6458	.61222	1.5814	.2903	.7097
.44620	1.7899	.3528	.6472	.61566	1.5760	.2889	.7111
.45010	1.7860	.3514	.6486	.61909	1.5706	.2875	.7125
.45399	1.7820	.3500	.6500	.62251	1.5652	.2861	.7139
.45787	1.7780	.3486	.6514	.62592	1.5598	.2847	.7153
.46175	1.7740	.3472	.6528	.62932	1.5543	.2833	.7167
.46561	1.7700	.3458	.6542	.63271	1.5488	.2819	.7181
.46947	1.7659	.3444	.6556	.63608	1.5432	.2806	.7194
.47332	1.7618	.3431	.6569	.63944	1.5377	.2792	.7208
.47716	1.7576	.3417	.6583	.64279	1.5321	.2778	.7222
.48099	1.7535	.3403	.6597	.64612	.15265	.2764	.7236
.48481	1.7492	.3389	.6611	.64945	1.5208	.2750	.7250
.48862	1.7450	.3375	.6625	.65276	1.5151	.2736	.7264
.49242	1.7407	.3361	.6639	.65606	1.5094	.2722	.7278
.49622	1.7364	.3347	.6653	.65935	1.5037	.2708	.7292
.50000	1.7321	.3333	.6667	.66262	1.4979	.2694	.7306
.50377	1.7277	.3319	.6681	.66588	1.4921	.2681	.7319
.50754	1.7233	.3306	.6694	.66913	1.4863	.2667	.7333
.51129	1.7188	.3292	.6708	.67237	1.4804	.2653	.7347
.51504	1.7143	.3278	.6722	.67559	1.4746	.2639	.7361
.51877	1.7098	.3264	.6736	.67880	1.4686	.2625	.7375
.52250	1.7053	.3250	.6750	.68200	1.4627	.2611	.7389
.52621	1.7007	.3236	.6764	.68518	1.4567	.2597	.7403
.52992	1.6961	.3222	.6778	.68835	1.4507	.2583	.7417
.53361	1.6915	.3208	.6792	.69151	1.4447	.2569	.7431
.53730	1.6868	.3194	.6806	.69466	1.4387	.2556	.7444
.54097	1.6821	.3181	.6819	.69779	1.4326	.2542	.7458
.54464	1.6773	.3167	.6833	.70091	1.4265	.2528	.7472
.54829	1.6726	.3153	.6847	.70401	1.4204	.2514	.7486
.55194	1.6678	.3139	.6861	.70711	1.4142	.2500	.7500
.55557	1.6629	.3125	.6875				
.55919	1.6581	.3111	.6889				
.56280	1.6532	.3097	.6903				

ROLLER CHAIN DRIVES

Roller chain drives are used in a wide range of power-transmission applications for all basic industries such as food processing, materials handling, textiles, and machine tools.

Fourteen standard sizes of single- and multiple-width roller chain listed in ANSI B29.1[1] are available for service (Figs. 3-21 and 3-22). Table 3-6 shows the chain number,

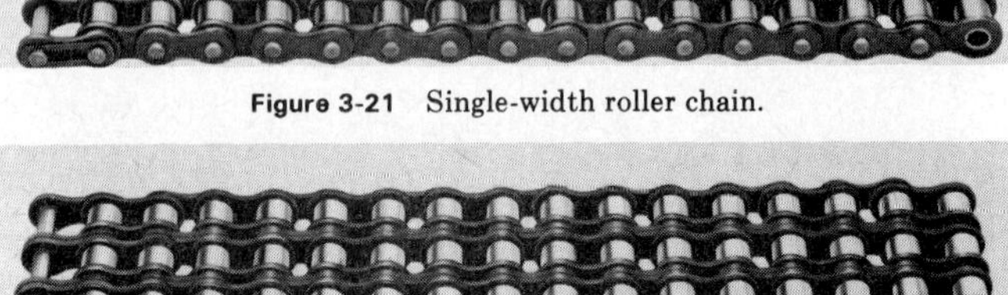

Figure 3-21 Single-width roller chain.

Figure 3-22 Multiple-width roller chain.

the pitch size, and the average ultimate strength for single-width chain. The average ultimate strength for multiple-width roller chain is equal to the average ultimate strength of the single-width chain times the number of strands in the multiple-width chain. Speed and horsepower ratings are the prime considerations in selecting a chain drive. The ratings are normally listed for the smaller sprocket, regardless of whether it is the drive or driven member. Chain manufacturers should be consulted when special conditions such as composite duty cycles, idlers, or more than two sprockets are involved in the drive cycle.

The speed and horsepower ratings are based upon approximately 15,000 h of service life at full-load operation and a service factor of 1.0. Operating conditions which establish the service factor are shown at the bottom of Table 3-3. Strand factors for multiple-width chains are given in Table 3-7. Note that the strand factors are not equal to the number of widths in a multiple-width chain.

Specialty roller chains have been developed for particular applications. For example,

TABLE 3-6 Standard Single-Width Roller Chain Size and Strength

	Chain pitch		Average ultimate strength	
Chain number	in	mm	lb	kg
25	0.250	6.35	875	400
35	0.375	9.52	2,100	950
40	0.500	12.70	3,700	1,680
41	0.500	12.70	2,000	900
50	0.625	15.88	6,100	2,770
60	0.750	19.05	8,500	3,850
80	1.000	25.40	14,500	6,580
100	1.250	31.75	24,000	10,900
120	1.500	38.10	34,000	15,400
140	1.750	44.45	46,000	20,900
160	2.000	50.80	58,000	26,300
180	2.250	57.15	80,000	36,300
200	2.500	63.50	95,000	43,100
240	3.000	76.20	130,000	59,000

TABLE 3-7 Multiple-Strand Factors for Multiple-Width Chains

Number of strands	Multiple-strand factor
2	1.7
3	2.5
4	3.3
5	4.1
6	5.0
7 or more	Consult manufacturer

double-pitch chain is an economical choice for slower-speed drives on relatively long centers. Heavy-series roller chain is used when conditions demand additional capacity to withstand occasional shock loads. Flexible-joint-type chain has been designed for smaller-horsepower drives where shafts must operate out of normal alignment. Lubricated-joint chains have oil-impregnated bushings to provide cleaner and longer operating life where restricted lubrication or absence of external lubrication is essential. Standard roller chain made of stainless steel is recommended for applications where high resistance to corrosive attack is required.

ENGINEERING STEEL CHAIN DRIVES

Engineering steel chain drives are especially suited for heavy-duty applications. The normally offset sidebar chain (Fig. 3-23) can handle speeds up to 1000 ft/min (305 m/min) and power requirements as high as 500 hp. These chains are commonly used in elevator drives, conveyor drives, drum drives, and applications with poor operating conditions.

The eight sizes of engineering steel chain available are listed in ANSI Standard B29.10[1]. Table 3-8 shows the chain number, the pitch size, and the average ultimate strength for the eight chains. Speed and horsepower ratings are the prime considerations in selecting a chain drive. Normally, the ratings are listed for the smaller sprocket, regardless of whether it is the drive or driven member. Chain manufacturers should be consulted for proper drive selections when special conditions are encountered.[2]

Figure 3-23 Engineering steel drive chain.

TABLE 3-8 Engineering Steel–Chain Size and Strength

Chain number	Chain pitch		Average ultimate strength	
	in	mm	lb	kg
2010	2.500	63.5	62,500	28,300
2512	3.067	77.9	90,000	40,800
2814	3.500	88.9	122,000	55,3000
3315	4.073	103.4	141,000	63,900
3618	4.500	114.3	193,000	87,500
4020	5.000	127.0	250,000	113,400
4824	6.000	152.4	360,000	163,300
5628	7.000	117.8	490,000	222,200

The speed and horsepower ratings are based upon approximately 15,000 h of service life at full-load operation and a service factor of 1.0. Operating conditions which establish the service factor are shown at the bottom of Table 3-3.

Two additional types of engineering chain are also available for use in special applications. Welded steel chain is used in some slow-speed drives subject to shock loads and other difficult operating conditions. Cast pintle chain is used for drives handling light loads at slow speeds.

SILENT-CHAIN DRIVES

Silent chain is a drive chain constructed of sidebars, pins, and bushings with no rollers (Fig. 3-24). The sidebars are designed to mesh with sprocket teeth in a gear-type engagement.

Figure 3-24 Silent drive chain.

Silent drive chain is selected for high-speed–high-load applications and smooth, quiet operations for many industrial services such as electric generating plants, automotive test stands, machine tools, and ventilating systems.

There is a wide range of silent chain-link configurations available from various manufacturers.[2] For this reason, silent chains are not normally interchangeable on different manufacturers' sprockets. However, the "SC" series of silent chains shown in ANSI Standard B29.2[1] is interchangeable on sprockets between manufacturers.

DRIVE ARRANGEMENTS

Illustrated in Fig. 3-25 are the drive arrangements recommended for optimum drive life. The preferred direction of rotation is indicated, although arrangements *A*, *B*, and *C* will

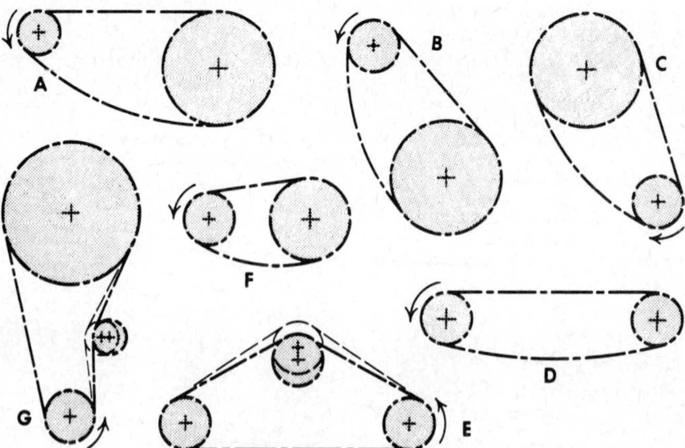

Figure 3-25 Preferred drive arrangements.

operate satisfactorily in either direction. Consult the chain manufacturer for approval on other drive arrangements.[3]

CHAIN TIGHTENERS

Chain tighteners are used to obtain a desired tension between shaft centers. Where centers approach the vertical relative position or there is pulsating operation, it is important to provide a means for adjusting the chain tension in order to prolong chain life. A chain tightener, sometimes referred to as an idler sprocket, should be located so as to operate against the slack or return side of the chain. Pictured below are illustrations of chain tighteners employed in chain drives (Fig. 3-26).

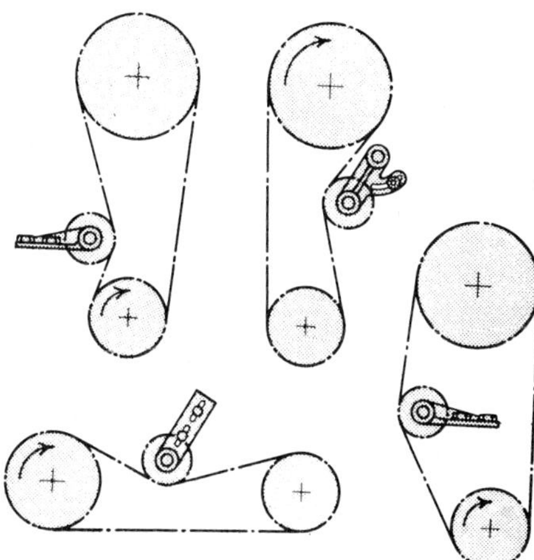

Figure 3-26 Chain tighteners used to maintain tension.

LUBRICATION

Lubrication is the most important factor in maintaining high chain efficiency and providing a long service life.

The primary purpose of chain lubrication is to provide a clean film of oil at all load-carrying points where relative motion occurs. The lubrication method recommended in the speed-horsepower selection charts should be used to ensure proper lubrication at all times for a drive chain selection.

Several methods of lubrication have been developed to fit a particular range of horsepower, chain speed, and relative position of shafts. Manual or drip lubrication is used for open running drives which operate in a nonabrasive atmosphere. These methods should be confined to low-horsepower drives with a chain speed under 600 ft/min (183 m/min). Bath lubrication is the simplest automatic method of lubricating encased chain drives and is highly satisfactory for low or moderate speeds. Disk lubrication is used for moderately high-speed drive arrangements unsuitable for oil-bath lubrication. Forced lubrication or oil-pump lubrication is recommended for large-horsepower drives, heavily loaded drives, high-speed drives, or where oil-bath or oil-disk lubrication cannot be used (Fig. 3-27).

Chains can be lubricated with any neutral grade of straight mineral oil in the 20 to 50 viscosity range, depending on the temperature. For difficult operating conditions, such as high temperature or an abrasive atmosphere, consult a lubricant manufacturer for proper lubricant selection.

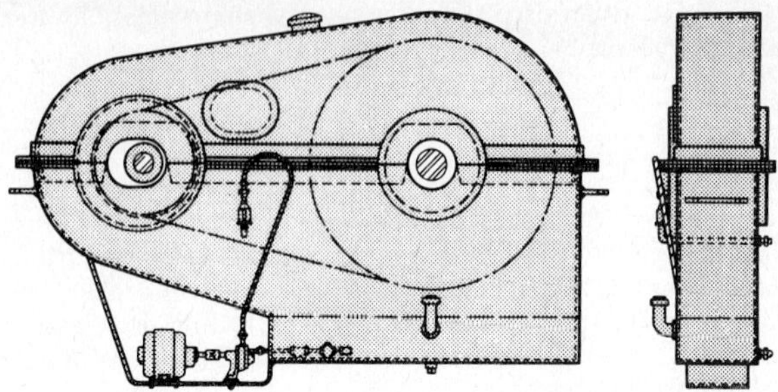

Figure 3-27 Forced lubrication. An oil pump is used to provide a continuous spray of oil to the inside of the lower span of chain.

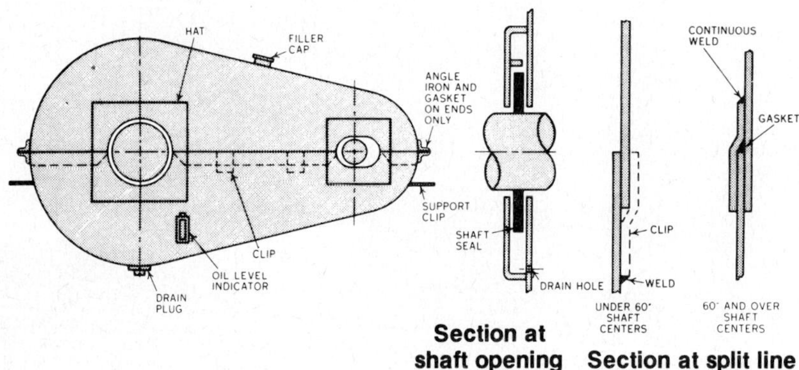

Figure 3-28 Oil-bath or oil-disk casing.

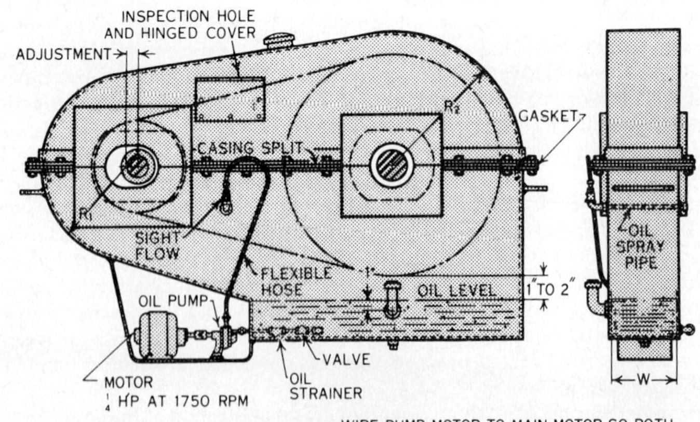

Figure 3-29 Casing for forced lubrication.

CASINGS

Casings are important considerations in providing adequate lubrication to prolong the efficiency and service life of a chain drive. Casings contain the lubricant, exclude foreign materials from the lubricant, and function as a safety guard. There are four basic types of casings used with chain drives.

The oil-bath or oil-disk casing is used with the bath and disk methods of lubrication, with variations from the standard casing available depending on the chain speed and shaft center distances (Fig. 3-28). The second type of casing is used in outdoor applications when water, dust, and other contaminants are present, when extra protection against oil leakage is needed, or when forced lubrication is used (Fig. 3-29). The weathertight casing is the third type of construction, and it is used when additional precaution against contamination is desired. Guard-type casings are used primarily as a safety device rather than a lubrication accessory and are employed when other types of casings are not compulsory.

MAINTENANCE

A good maintenance program for a chain drive should include proper installation, alignment, and periodic inspections. After an installation inspection, an initial inspection should come after 100 h of operation; a second inspection should come after 500 h, followed by regular intervals thereafter. The following is recommended as part of the maintenance routine:

1. **Check Sprocket and Shaft Alignment** Wear on the inner surface of the roller link sidebars and/or wear on both sides of the sprocket teeth indicate drive misalignment.
2. **Check Sprocket Tooth Wear** Hook-shaped teeth are an indication of a worn sprocket.
3. **Check Chain Tension** Incorrect chain tension can cause improper sprocket action.
4. **Check the Oil Level** The oil level should be inspected when the drive is idle, allowing sufficient time for the lubricant to accumulate in the sump.
5. **Check Oil Flow** Forced lubrication should be frequently inspected. Sight flow gauges must be mounted above the spray line to allow visual examination to ensure proper lubrication.
6. **Change Oil** Oil should be changed initially after the first 500 h of operation. Thereafter, the oil should be changed after 2500 h of operation.
7. **Inspect the Chain Parts** Look for visual indications of any improper operation such as unusual wear.

REFERENCES

1. American National Standards Institute, Standards (use most recent issue).
ANSI B29.1, "Precision Power Transmission Roller Chains, Attachments and Sprockets."
ANSI B29.2, "Inverted Tooth (Silent) Chains and Sprocket Teeth."
ANSI B29.10, "Heavy Duty Offset Sidebar Power Transmission Roller Chains and Sprocket Teeth."
2. American Chain Association Publications
Design Manual for Roller and Silent Chain Drives; Applications Handbook for Engineering Steel Chains.
3. Chain Manufacturers' Catalogs.

PART 4

Flexible Shafts for the Transmission of Rotary Motion

prepared by

Thomas C. Roberts
Product Engineering Manager
S. S. White Industrial Products Division
Pennwalt Corporation

INTRODUCTION

A rotary-motion flexible shaft can be a useful alternative to other types of power transmission when the engineer is faced with (1) hard-to-align shafts, (2) relative motions between shafts, (3) one or more angles in the power-transmission path, (4) power transmission at angles other than 90°, or (5) the need to absorb the shock of sudden stops and starts and reduce vibration.

Shaft Alignment

When mounting a motor to drive a pump, a gearbox, or any device with self-contained bearings, a conventional coupling can absorb only small amounts of misalignment (typically less than 2° angularity and 0.005 in parallel misalignment). The large tolerance for misalignment inherent in the flexible shaft allows the production engineer to do away with time-consuming and expensive realignment, requiring dial indicators, every time a motor is removed and replaced.

Relative Motion Between Shafts

Use in printing machinery and paper-coating equipment, or whenever power must be transmitted from a stationary motor to a moving carriage, are natural applications for flexible shafting. The shafting will absorb the relative motion between the driving and the driven member without resorting to constant-velocity universal joints.

More than One Angle in Power-Transmission Path

When the power must be transmitted through several angles or a complex path to the driven member, a flexible shaft is much easier to apply than a series of universal gearboxes. When using gears, an angle other than 90° will require two gearboxes with their attendant alignment and expense problems.

Power Transmission at Angles Other than 90°

Flexible shafting can be used for small angles and angles other than the standard 90° available with gearboxes. As a side benefit, the flexible shafting does not have to be mounted as a gearbox needs to be.

Shock and Vibration Absorption

The flexible shaft will act as a heavily damped spring in the system to absorb the shock of sudden machinery stops and starts. It will also tend to smooth out vibration in the drive system. This heavy internal damping can cause a temperature rise in the shaft if it is used with a continuously pulsating driver, such as a single-cylinder engine. The shaft should be oversized in applications like this to reduce the heating.

The engineer can apply flexible shafting in many applications where solid shafting and/or gearboxes would provide an expensive system that is difficult to align and maintain. It is suggested that the shafting manufacturer be consulted as early as possible in the design process to allow use of predesigned shafting whenever possible. Consultation in the early design stage will eliminate use of nonstandard sizes, which, in turn, will result in lower final cost and shorter delivery cycles.

WHAT IS A ROTARY-MOTION FLEXIBLE SHAFT?

A rotary-motion flexible shaft (RMFS) is a relatively simple and trouble-free device. The manufacturer will take one central wire mandrel, wrap several successive layers of wire concentrically around this mandrel, and square the ends or attach suitable fittings to make a flexible shaft. Depending on the number of wires in each layer and their diameter, the manufacturer can design either (1) a stiff, flexible shaft which exhibits very little backlash in either direction, making it ideal as a remote-control shaft, or (2) a more flexible shaft which is ideal for power transmission in one direction, sometimes at speeds as high as 20,000 r/min. Available flexible-shaft diameters range from 0.050 to 1.0 in (12.5 to 25 mm) and power shafts are capable of transmitting up to 50 hp (Fig. 3-30).

Critical Properties

Typically, the plant engineer is not expected to design a flexible shaft; that can be left to the engineering department of the flexible-shaft manufacturer. However, the engineer must know what is expected of the flexible shaft and must provide the manufacturer with certain basic data so that a suitable selection or design can be provided. Among the factors that are critical are: (1) type of shaft, remote control or power; (2) manual or dynamic operation; (3) maximum torque to be transmitted; (4) horsepower of driving unit; (5) peak allowable revolutions per minute; (6) direction(s) of rotation; (7) total permissible angular deflection; (8) cycling time or duty-cycle rest periods available; (9) minimum bend radius; (10) length of shaft; (11) ambient temperature; (12) unusual service conditions. Note that not all these data will apply in all instances, but the more background the manufacturer can be given, the more suitable and economical will be the flexible-shaft design.

In many instances, a predesigned shaft, or RMFS coupling, will solve the plant engineer's power transmission problem. However, if the requirements are unusual, a special shaft may have to be designed. Obviously, whenever possible, predesigned shafts should be selected because they are the less costly solution.

Casings and Fittings

In many applications, a bare flexible shaft as described above, equipped with suitable fittings, is perfectly adequate for the task at hand. In other cases, however, the flexible shaft should be provided with special fittings and a casing.

End fittings are installed on both the casing and the shaft core itself. The casing fittings keep the shaft assembly fixed in place; neither these fittings nor the casing rotate.

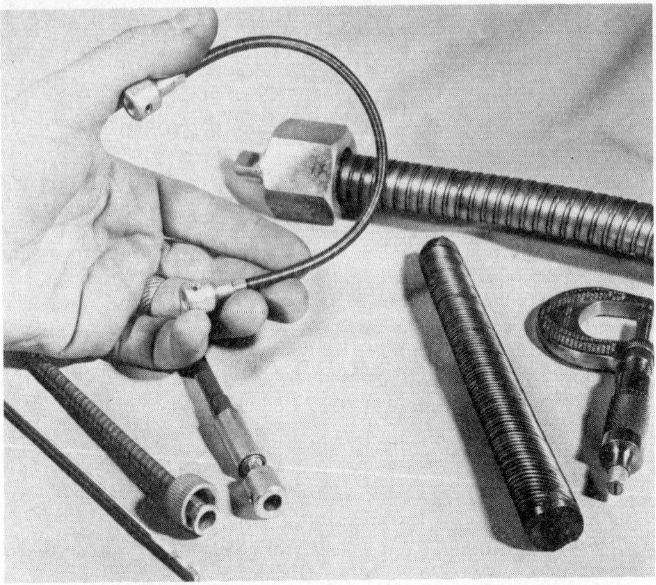

Figure 3-30 Standard flexible-shaft assemblies come in a wide range of sizes and configurations, with several types of end fittings; some of these are shown here.

Casing fittings may be of several types, among them a loose male (or female) threaded coupling nut, or quick disconnect. With *shaft*-end fittings, even more choices are available: integrally formed drive square, fork or tang fitting, male or female spline, hollow square, panel mounting, or set screw. The choice of fittings depends to a large extent on the application, how accessible the shaft is, and how often it is expected to be serviced. If a shaft is to rotate at high speeds for prolonged periods and has to be lubricated regularly to compensate for this severe service, it is recommended that the assembly be designed for easy withdrawal of the shaft from the casing. In that event, one end of the shaft is usually swaged square, because this offers a drive that does not permanently lock the shaft into the casing at both ends.

A casing provides the following benefits: (1) It keeps the flexible shaft properly lubricated for longer periods of time. (2) It offers intermediate points of attachment, when shaft length exceeds 18 in. (3) It keeps plant personnel from coming into contact with a high-speed rotating member. (4) It protects the flexible shaft against hostile environments.

Casings can be classified into two types: metallic and covered. Metallic casings are of either two- or four-wire construction. They are neat, strong, durable, and bendable. Although they will retain grease, they are not oil- or watertight.

Covered casings are more expensive because their construction is more elaborate. Following an inner liner that may be either metallic or plastic, the casing is reinforced with layers of steel braid, and finished off with a final outer sheath of rubber or plastic. This type of casing permits the flexible shaft to operate under water, hydraulic fluid, or other liquid, so long as the fluid does not attack the elastomeric outer covering.

Shafts Thrive on Speed

Power shafts function most efficiently when operated at high speed. The reason stems from basic engineering principles: Horsepower transmitted is directly proportional to speed.

$$\text{hp} = kTN,$$

where T is torque, N is speed, and k is a constant to reconcile the units of torque and peripheral speed being used in the equation

Typically, if torque is expressed in foot pounds and speed in revolutions per minute, hp = 0.00019 × ft·lb × r/min. For torque expressed in inch pounds, hp = 0.000016 × in·lb × r/min. Therefore, for a given horsepower requirement, the higher the shaft speed, the less the torque on the shaft, therefore, the smaller the shaft needed to accomplish the task. This factor should be kept in mind during the early stages of design, because quite often the insertion of a suitable speed-reduction device at the correct location in the drive train may serve to reduce the overall cost of the assembly.

A power shaft is to rotate at the highest permissible speed; it is, therefore, important to consider this in the design. If a speed reducer is necessary, insert it *at the correct end of the shaft.* Thus, if motor speed is 5000 r/min and final operating speed is to be 500 r/min, the speed reducer should be inserted at the output, *not* at the motor end. Conversely, if a speed*up* is desired, the speed increaser should be inserted at the motor end (Fig. 3-31).

Shaft Radius of Curvature

Although rotary-motion flexible shafts are almost snakelike in their ability to reach into inaccessible areas, there is a limit to how much they may be bent without damaging them. This limit is called the *n*on*d*amaging *r*adius (NDR). In addition, there is a larger radius which is the *m*inimum *o*perating *r*adius (MOR). Each shaft construction (diameter, number of wires per layer, and number of layers) leads to a different MOR, a figure that is always available from the manufacturer.

NDR is that minimum radius beyond which the shaft would be permanently damaged. Such a condition generally would only occur during handling or installation. The NDR is a smaller radius than the MOR. It is determined by test.

MOR is the zero torque point on the torque capacity curve for any particular shaft. (It is a kind of misnomer because in normal circumstances a shaft would not be run in the MOR because technically it could not carry any load.) The MOR is established by calculations based on the diameter and material of the mandrel wire.

It should come as no surprise that the torque load a flexible shaft can comfortably manage is an inverse function of the MOR. The more a flexible shaft is curved, the higher the internal friction, the greater the heat generated, and the smaller the permissible continuous load. Here again, these figures are known for each construction and should be rigidly adhered to if overload or early failure of the shaft is to be avoided. Table 3-9 shows typical values for a range of high-tensile-steel power shafts, including such basic parameters as diameter, radius of curvature vs. torque capacity, torsional breaking load, and weight per foot.

In using Table 3-9, it must be remembered that the torque-carrying capacities are shown for shafts rotating in the direction that tightens up the outer layer. If the same shaft is rotated so as to unwind the outer layer—clockwise for a right-lay flexible shaft or counterclockwise for a left-lay shaft—its torque capability is reduced by up to 50 percent. See Fig. 3-32 for an example of right- and left-hand shaft application.

Shaft Life

As a rule of thumb, a power shaft running under no load at the MOR can be expected to have a useful life in excess of 10^8 revolutions. Such factors as service temperature, number of shaft bends, torque output, speed, shock loading, and operating radius can reduce shaft life.

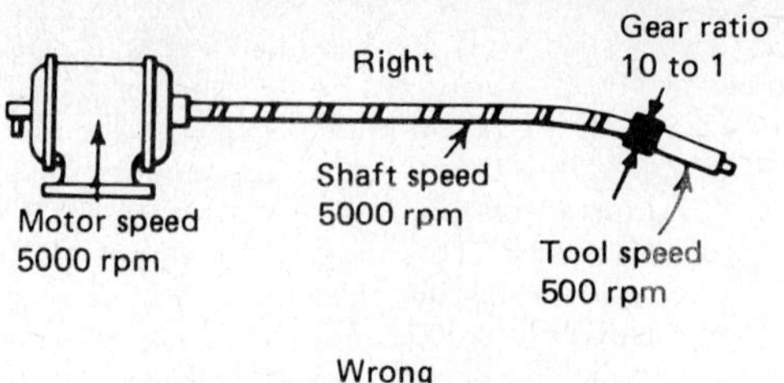

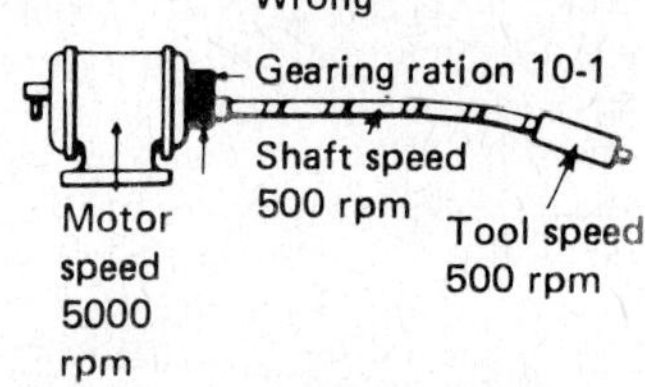

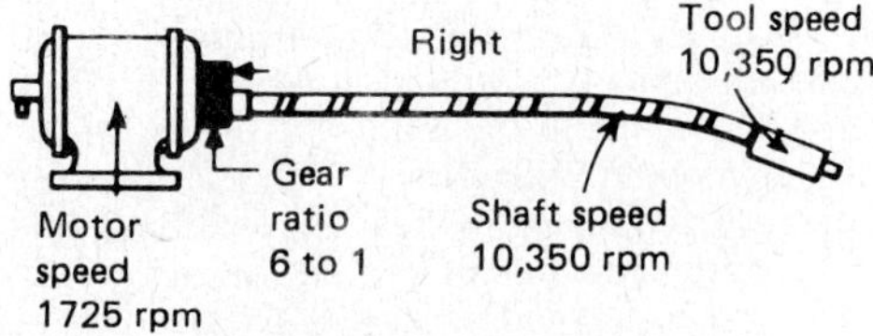

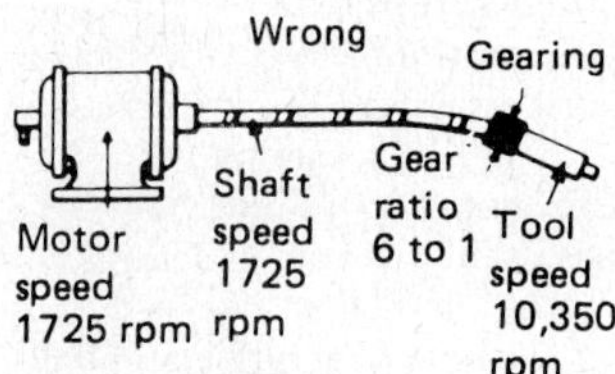

Figure 3-31 Changing tool speed with gearing.

A remote-control shaft, under manual operation, will last many years. When remote-control shafts are used for dynamic operations, it is very important that the duty cycle not exceed 5 min ON and 2 min OFF.

SHAFT SELECTION

To select the proper shaft, we need to determine the following items:

1. Torque requirement
2. Shaft radius of curvature
3. Length of path between driving and driven systems
4. Operating speed
5. Acceptable torsional deflection or backlash

TABLE 3-9 Power-Drive Flexible Shafts*,†

Shaft number‡	Shaft diameter, in	Min operating radius, in	Dynamic torque capacity winding direction, lb·in input								Torsional breaking load for straight shafts winding dir, lb·in§	Approx weight of shafting, lbs. per 100 ft
			Radius of curvature, in									
			25	20	15	12	10	8	6	4		
050–9	0.050	2.0					0.26	0.24	0.22	0.16	0.9	0.5
068–9	0.068	2.5					0.60	0.55	0.46	0.30	2.6	0.9
098–9	0.098	2.5				1.9	1.8	1.7	1.4	0.91	8.5	2.0
130–9	0.130	3.0			3.8	3.6	3.4	3.1	2.4	1.7	15.0	3.3
150–9	0.150	4.0			5.0	4.7	4.4	3.9	3.1	1.4	24	4.5
187–9	0.187	4.0		13.5	12.6	11.8	11.0	9.8	7.8	4.0	55	7.2
250–9	0.250	4.0	25	24	22	21	19	16	12		100	12.4
312–9	0.312	4.5	49	46	43	40	37	32	24		200	19.6
375–9	0.375	4.5	67	63	58	53	48	40	26		280	28.5
437–9	0.437	5.0	100	95	85	77	68	54	30		440	38.1
500–9	0.500	5.5	133	125	113	102	90	70	40		550	51.8
625–9	0.625	6.5	230	210	190	160	130	95			1100	79.0
750–9	0.750	9.0	405	365	300	235	170				2000	114.0

*The data shown in the shaft selection tables are suitable for most applications.

†The most popular sizes of shafts are listed. Other sizes can be made on special order if the volume warrants.

‡Shafts can be supplied in either "left-lay" or "right-lay." When ordering, the letter "L" or "R" should be inserted in the shaft number to indicate the lay (i.e., 150L9 is a 0.150 shaft, "left-lay").

§Shaft will break or helix under these loads. For routine tests, do not specify more than approximately 50 percent of these figures, otherwise the flexible shaft may be seriously damaged.

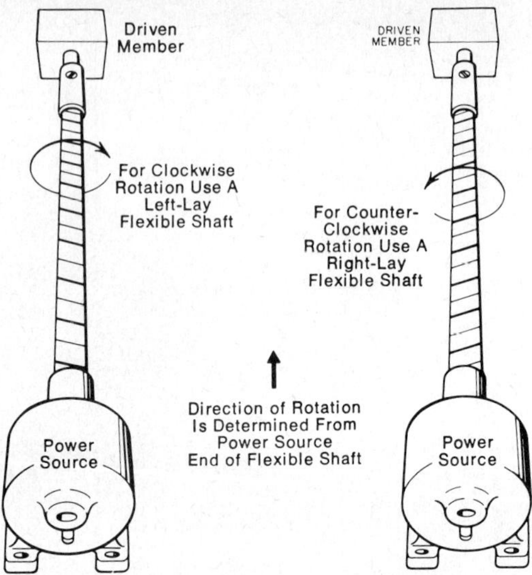

Figure 3-32 How to determine proper shaft lay.

Determining the torque requirement on a shaft may not be easy. Remote-control shaft requirements can be determined by attaching a lever of known length to the device to be turned and pulling the end of the lever with a spring scale. Use the highest torque as the required torque. Measuring torque on a power shaft is more difficult because the requirement is dynamic. The best way is to instrument load with a torque cell and measure the torque under operating conditions. Component manufacturers usually have these data available. A shaft efficiency of about 90 percent should be attainable for most applications; therefore, increasing the output torque by about 10 percent leads to the input torque.

The shaft path should then be examined. A drawing should be made which shows the path in a true view. The MOR should be determined from this drawing. A full-scale prototype can be used. This length of the shaft can also be determined at this time.

For power shafts, the operating speed is generally given. For remote-control shafts, in which the speed is essentially zero, the amount of torsional deflection or dead band acceptable between the turning device and the turned device is the important consideration.

Sample Problem

A power shaft is driven by a motor to drive a fan. The fan requires 2 hp at 1728 r/min. A layout of the installation shows that a 6-ft shaft with one bend of 12-in radius and one bend of 8-in radius are required. How big a shaft is required and what will the approximate life be in hours?

Solution

$$\text{Torque} = \frac{\text{hp} \times 63{,}025}{\text{r/min}}$$

$$= \frac{2.0 \times 63{,}025}{1728} = 72.9\ \text{in} \cdot \text{lb}$$

When assumed efficiency e is 90 percent,

$$e = \frac{\text{torque out}}{\text{torque in}}$$

so

$$\text{Torque in} = \frac{\text{torque out}}{e} = \frac{72.9}{0.90} = 81 \text{ in}\cdot\text{lb}$$

Entering Table 3-9 at the smallest radius of curvature (8 in in this case) we determine that a ⅝-in shaft is required.

The shaft selected would be a 6-ft-long shaft of ⅝-in diam. See Fig. 3-31 to determine if a left- or right-hand-lay shaft is required.

This diameter shaft is more expensive than a smaller shaft. For this reason, consider increasing each bend to at least a 25-in radius, which will allow use of a 7⁄16-in shaft at considerable cost saving.

$$\text{Shaft life} = \frac{10^8 \text{ r}}{N \text{ r/min} \times 60 \text{ min/h}}$$

$$= \frac{10^8}{1728 \times 60} = 964.5 \text{ h}$$

Exact life is dependent on the service and environment of the shaft.

DOS AND DON'TS

Used intelligently, flexible shafts are the product designer's friend. They are more flexible than universal joints; they are more versatile than gear systems because they are totally unaffected by the exact angle or offset necessary; finally, flexible shafts offer an inherent shock-absorption capability and ease of installation and maintenance unmatched by other forms of rotary-motion transmission.

However, flexible shafts do have to be treated carefully if they are to provide the service life built into them. They cannot be bent completely out of shape and must be used at radii equal to or greater than the MOR specified by the manufacturer. They cannot serve as steps of a ladder. They should be secured approximately every 18 in to prevent the possibility of helixing. They must be lightly lubricated for preventive maintenance at regular intervals. Replacement must match the original design because a control shaft is not interchangeable with a power shaft. The type of service expected must be clearly specified. Two differently built shafts, even of the same diameter and length, are not interchangeable. A remote-control shaft is intended for low-speed continuous or high-speed intermittent operation in either direction with minimum backlash. On the other hand, a power shaft is intended for high-speed continuous operation, but in only one direction. Above all, flexible shafts must be designed for the task they are to perform. For this reason, early consultation between the user's engineering department and the shaft manufacturer is highly recommended.

Follow these simple, basic suggestions, and the design flexibility of a rotary-motion flexible shaft will be applied to utmost advantage.

TYPICAL APPLICATIONS

Flexible-shaft couplings, in lengths of 6 in or less, are ideal for connecting a driven member to its power source when a small amount of angular, parallel, or combined misalignment must be accommodated. Not only does such a motor-type coupling readily compensate for the misalignment, but it will also handle a certain amount of shock loading and attenuate vibration.

Longer segments of flexible shaft, from 1 to 100 ft (0.3 to 30.5 m), are ideal for transmitting rotary motion to very distant locations, either for remote control of valves and other apparatus or for power transmission.

Flexible-shaft applications in industry range far and wide, so that it would be quite a task to try to list them all. However, a few typical cases may help to demonstrate how rotary-motion problems have been avoided or solved with a flexible shaft:

An operating device on a swinging oven door has to rotate freely, no matter how much the angle changes as the door opens and closes.

Two rotating devices have to be synchronized through a central gearbox, even though one of the devices moves laterally while the other stays fixed in place.

A crane operator has to know which way the cable is unreeling, even when neither reel nor load is visible from the cab.

Condenser or other heat-exchange tubing has to be cleaned out, 35 ft (10.7 m) from the header end.

Several door-actuating devices have to be moved simultaneously, all by one driving motor and each one at a different angle to it.

A small, buried control has to be readily turnable from "outside," a valve has to be closed from a remote location, a potentiometer has to be adjusted in an "unreachable" spot, and a switch has to be operated from far away.

The above examples serve to illustrate the wide variety of problems that flexible shafts can cope with. This type of drive should always be investigated by the plant engineer as a possible alternative to conventional drives in many applications.

chapter 1-4

Fluid Seals

prepared by

Parker-Hannifin Corp.
Seal Group
Lexington, Kentucky

Authors

Chris S. Louskos
Seal Group Staff

John B. Painter
Gasket Division

Ralph E. Peterson
Packing Division

Richard G. Ramsdell
Seal Group Staff

John B. Scannell
Packing Division

INTRODUCTION

Packings and *seals* are devices or materials designed to create and/or maintain a fluid-pressure differential across the interface or gap between two relatively movable and/or separable components of a fluid system. (*Fluids*, in this context, include liquids and gases, with or without entrained solids.)

Included within this category of packings and gaskets are static seals, reciprocating dynamic seals, rotary dynamic seals, and flexural sealing devices, involving an almost unlimited variety of sizes and configurations and a broad range of common and exotic materials and material combinations.

Because of the complexity of the subject, only a brief overview is possible in this handbook. Although the information provided may help in solving relatively simple problems, a competent supplier should be consulted for assistance with most sealing applications.

MATERIALS

Table 4-1 lists the most commonly used seal and packing materials and gives a few pertinent facts about each. All of the facts given for any material do not necessarily apply to all members of the group. They merely suggest the range of uses of materials within that group. For instance, elastomers are represented as having a temperature range of −178 to +500°F (−115 to 260°C), but only a few specific compounds will serve very long at either of these extremes and no compound will function at both. Similarly, asbestos is said to have good resistance to strong acids and bases, but only the blue asbestos has good resistance to strong basic solutions.

STATIC SEALS

Gaskets

Gaskets are seals placed between two static faces (Fig. 4-1). They are made of deformable materials which, when clamped, will flow into surface imperfections of the mating surfaces to prevent fluids from escaping through the joint. For relatively low pressures and temperatures, nonmetallic gaskets may be used. For more severe applications, metallic gaskets are required.

As a rough guide to this selection, it is common practice to multiply the operating pressure, in pounds per square inch, by the operating temperature, in degrees Fahrenheit. If this value exceeds 250,000, the use of metallics is indicated.* In any case, however, nonmetallic gaskets should not be used at temperatures above 850°F (450°C) or at pressures above 1200 lb/in^2 (8.3 MPa). Generally pressure, temperature, and fluid determine the gasket material, while dimensional and mechanical features of the joint determine the gasket type.

*In SI units, find the value of $[P(1.8C + 32)]$, where P is the pressure in megapascals and C is temperature in degrees Celsius. If the result exceeds 1720, the use of metallics is indicated.

TABLE 4-1 Materials Commonly Used in Seals

Materials and notes	Temperature range, °F (°C)	Commonly sealed fluids	Types of seals in which the material is commonly used
Aluminum	−300 to +800 (−185 to +430)	Water, weak acids except acetic, steam, air, oxygen, dry bromine and chlorine, aliphatic and aromatic fluids, acetone, alcohols, petroleum oils, hot sulfur-bearing gases	Compression packing (as foil), gaskets (solid or as jacketing over asbestos, rubber, or other filler)
Asbestos, white (chrysotile) and blue (crocidolite); fire-resistant fibrous minerals: May be woven into fabrics, braided, or pressed into solid form; often impregnated with elastomer or TFE (tetrafluoroethylene), or reinforced with metal	−300 to +1000 (−185 to +540)	Water, strong acids and bases, steam, air, chlorine, alcohols, petroleum fluids	Compression packing, gaskets (as sheet or with metals in corrugated, jacketed, or spiral-wound types)
Brass	−300 to +500) (−185 to +260)	Water, mild acids and bases, oxygen, steam, dry bromine and chlorine, aliphatic and aromatic fluids, acetone, alcohols, petroleum oils except sulfur-containing oils	Diaphragms, gaskets (as sheet, corrugated, or spiral-wound types)
Bronze	−300 to +500 (−185 to +260)	Similar to brass	Gaskets (with asbestos or TFE in spiral-wound types)
Cast iron	−50 to +1000 (−45 to +540)	Similar to iron	Mechanical seals, piston rings; not usually a gasket material
Ceramics: Excellent resistance to oxidation at elevated temperatures, but quite abrasive unless given a very smooth finish	−300 to >1500 (−185 to >815)	Water, strong acids and bases, steam, air, chlorine, alcohols, petroleum fluids; chemical resistance similar to glass	Mechanical seals, filler material for metallic types of gaskets
Copper	−300 to +600 (−185 to +315)	Water, mild acids and bases, oxygen, steam, dry bromine and chlorine, aliphatic and aromatic fluids, acetone, alcohols, petroleum oils except sulfur-containing oils; not recommended for use above 600°F (315°C) unless oxygen-free	Compression packing (foil), gaskets (solid, as jacketing over filler or with filler in spiral-wound-type)
Cork: Cork particles bonded together in sheet form with resin or an elastomer	−22 to +300 (−30 to +150)	Mild acids and bases, water, coolants, petroleum oils	Gaskets
Elastomers: See Table 4-2 for details	−178 to +500 (−115 to +260)	Wide range of fluids; see Table 4-2	Cup and hat packings, diaphragms, gaskets, O rings, oil seals, PolyPak, T seals, U cups, V packings, backup rings, wipers; reinforced with fabric in some of these products

TABLE 4-1 Materials Commonly Used in Seals (*Continued*)

Materials and notes	Temperature range, °F (°C)	Commonly sealed fluids	Types of seals in which the material is commonly used
Glass fibers	——	Added to nylon, TFE, etc., to improve resistance to wear, extrusion, creep, etc.	——
Graphite: Excellent chemical resistance, but oxidizes at elevated temperatures; good lubricant; porosity and oxidation effect can be modified by impregnating	−450 to +1500 (−270 to +815)	Water, all concentrations of acids and bases if not strongly oxidizing, nitric acid to only 100°F if concentration over 25%, steam, air to 960°F, aliphatic and aromatic fluids, acetone, alcohols, petroleum oils	Compression packing (in fiber form), gaskets, mechanical seals; in powdered form, added to elastomers and plastics to reduce friction
Hastelloy®: Union Carbide trademark for a series of high-strength nickel-base corrosion-resistant alloys	−300 to +2000 (−185 to +1100)	Noted for their superior corrosion resistance; individual applications should be investigated; boiling acids and bases, salts, chlorine, hypochlorites, and sea-water	Gaskets (solid, as jacketing over filler or with filler in spiral-wound type), mechanical seals
Inconel®: International Nickel trademark for nickel-chromium alloys	−300 to +2000 (−185 to +1100)	Noted for high-temperature strength and corrosion resistance; individual applications should be investigated; resists chloride-ion-stress corrosion cracking, organic acids in food products, alkaline sulfur compounds, ammonia, dry gases, steam, air, carbon dioxide; resists progressive oxidation to 2000°F (1100°C), sulfur atmospheres	Gaskets (solid, as jacketing over filler or with filler in spiral-wound type), mechanical seals
Iron: Soft and low-carbon steel	−50 to +1000 (−45 to +540)	Widely used on an economical heavy-cross-section gasket; sulfuric acid at high concentrations, hydrochloric not satisfactory at any concentration, most alkalies, air, water, steam, oxygen, acetone, acetylene	Gaskets (solid as jacketing over filler or with filler in spiral-wound type) mechanical seals
Lead: Soft metal, melts above 500°F (260°C)	−300 to +212 (−185 to +100)	Water, dry bromine and chlorine, aliphatic and aromatic fluids, acetone, alcohols, petroleum oils	Compression packing (as foil or insert in channel-type packing); gaskets (solid or jacketing over filler)
Leather: Porosity can be controlled or virtually eliminated with waxes, oils, etc., that also extend its temperature range and the types of fluids it can withstand	−70 to +212 (−55 to +100)	Water, weak acids and bases, air, aliphatic and aromatic fluids, alcohols, petroleum oils	Cup and hat packings, gaskets, oil seals, U cups, V packings

TABLE 4-1 Materials Commonly Used in Seals (*Continued*)

Materials and notes	Temperature range, °F (°C)	Commonly sealed fluids	Types of seals in which the material is commonly used
Molybdenum disulfide (MoS_2) in a silvery powder form	——	——	May be mixed with most seal materials to reduce friction without causing corrosion
Monel®: International Nickel trademark for group of alloys primarily of nickel and copper plus small amounts of other ingredients	−300 to +1500 (−185 to +815)	Noted for high-temperature properties and corrosion resistance; resists chloride-ion-stress corrosion cracking; most acids (including hydrofluoric) and alkalies; not satisfactory with strong oxidizing acids; fresh and sea-water, air, dry gases, neutral and alkaline salts	Gaskets (solid, as jacketing over asbestos, or with asbestos in spiral-wound type), mechanical seals
Nickel	−300 to +1400 (−185 to +760)	Not as all-around-resistant as Monel; resists chloride-ion-stress corrosion cracking; fresh and seawater, alkalies, natural and alkaline salts; not satisfactory with strong, hot, sulfurous and oxidizing acids	Gaskets, as jacketing over asbestos
Nylon (polyamide)	−65 to +300 (−55 to +150)	Air, hydraulic fluids	Fabric, as reinforcement for U cups, V rings, diaphragms; solid, for backup rings
Paper: Due to porosity it is impregnated except where needed merely to exclude dust and dirt	to +300 (to +150)	Aliphatic and aromatic fluids, petroleum oils	Gaskets
Plastics: See nylon, polyurethane, TFE			
Polyamides: See nylon			
Polyurethane: A very tough, wear-, extrusion-, and abrasion-resistant group of materials spanning the range from elastomers to plastics; good resistance to petroleum fluids, air, and aging	−65 to +200 (−54 to +130)	Petroleum-base hydraulic fluids, especially in high-pressure heavy-duty systems; not satisfactory with hot water	Cup and hat packings, O rings, PolyPak, U cups, V packings, adapters, backup rings, scrapers
Rubber: See elastomers (Table 4-2)			
Silver	−300 to +1200 (−185 to +650)	Food and drug industry; acetic acid and acetic anhydride, carbon tetrachloride, wet chlorine, formaldehyde, formic acid, hydrofluoric acid over 65%, magnesium chloride, oxalic acid	Gaskets, plating for spring-type metal seals

TABLE 4-1 Materials Commonly Used in Seals (*Continued*)

Materials and notes	Temperature range, °F (°C)	Commonly sealed fluids	Types of seals in which the material is commonly used
Stainless steel	−300 to +1600 (−185 to +870)	Resistant to multitude of corrosive media depending on operating conditions and alloy selection; water, air, acids, bases, gases, alcohols, petroleum fluids,	Mechanical seals, gaskets (solid, as jacketing over asbestos in spiral-wound type)
Steel: Low-carbon	−50 to +1000 (−45 to +540)	Same as iron	Gaskets, mechanical seals
TFE (also PTFE): (poly) tetrafluoroethylene or Teflon® Du Pont; a plastic having excellent chemical resistance and low friction; softens above 500°F (260°C); creeps under stress	−300 to +500 (−185 to +260)	Water, all concentrations of acids and bases, steam, air, oxygen, bromine, chlorine, aliphatic and aromatic fluids, acetone, alcohols, petroleum oils	Compression packing, cup and hat packings, diaphragms, O rings, piston rings, spring-actuated U cups, tape, backup rings, cap strips, coating for spring-type metal seals, impregnant for asbestos, rubber, etc., mechanical seals
Titanium	−300 to +2000 (−185 to +1100)	Nitric acid, except fuming, other oxidizing acids, mixed acids, wet chlorine, phosphoric acid to 30%, chlorine compounds, hydrogen sulfide, fresh and sea-water, salt solutions, alkaline solutions, organic acids	Gaskets
Vegetable fibers, such as cotton, flax, hemp, jute, and ramie: Usually impregnated with neoprene or other rubber to reduce porosity, while the fibers reinforce the rubber and reduce swelling in some fluids; exposed fibers hold fluid, aiding lubrication	−20 to +200 (−30 to +95)	Water, ammonia, aliphatic and aromatic fluids, petroleum oils	Compression packing, gaskets
Wool felt	−100 to +160 (−75 to +70)	Water, ammonia, aliphatic and aromatic fluids, petroleum oils	Gaskets, impregnated with an elastomer or plain, for dust seals

Nonmetallic Gaskets

Nonmetallic gaskets in general are more economical than metallic. They are also softer, thereby sealing at a lower seating stress. These gaskets are made from rubber, both synthetic and natural, paper, plant fibers, cork, cork combined with rubber, asbestos with rubber or other binders, compressed asbestos sheet packing, PTFE (polytetrafluoroethylene or Teflon®), and carbon sheet.

An almost limitless variety of nonmetallic gasket materials is produced by combinations of the above materials. This allows the designer to select the most economical combination to provide the sealing capability, strength, and durability required to fit the operating conditions of a given system.

Metallic and Combination Gaskets

Corrugated Gaskets. Corrugated gaskets are made from thin metal which is corrugated with concentric waves (Fig. 4-2). These are essentially line-contact seals with multiple corrugations providing a labyrinth effect. They are used in special shapes, with complicated hole and corrugation patterns, as engine head and manifold gaskets, and also used in lightweight fuel and hydraulic systems. This type of gasket, with asbestos cord or other filler, is used extensively in large, lightly bolted hot-gas duct systems. With this construction, large, oddly-shaped gaskets can be fabricated in pieces and assembled on the flange.

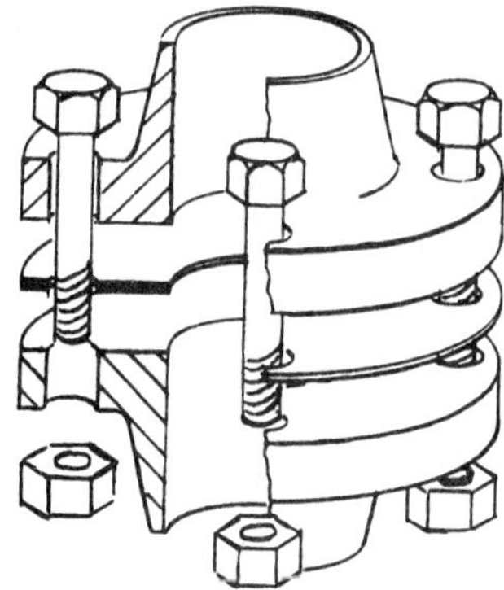

Figure 4-1 Flat gasket between two flanges.

Figure 4-2 Corrugated gasket with filler. *(Reproduced by permission of Parker Gasket Division, North Brunswick, N.J.)*

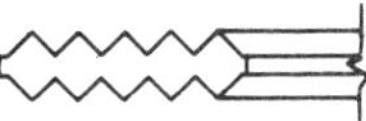

Figure 4-3 Flat-metal gasket with grooves. *(Reproduced by permission of Parker Gasket Division, North Brunswick, N.J.)*

Flat Metal Gaskets. Flat metal gaskets are washer-shaped and are relatively thin. The face width is at least 1.5 times the thickness (Fig. 4-3). They can be used with flat surfaces as cut, or grooves may be machined in the surfaces to reduce the contact area. These reduced-area types have less friction, and are therefore useful in screwed attrition joints.

All types seal by brute compressive force flowing the gasket into the flange contact surfaces, and finishes are therefore important. Nevertheless, this can be an economical gasket. Some uses include valve bonnets, ammonia fittings, heat exchangers, and tongue-and-groove joints.

Spiral-Wound Gaskets. Spiral-wound gaskets are made by spirally winding a V-shaped metal strip with a soft filler such as asbestos (Fig. 4-4). These gaskets have good resilience and sealability. This type is suited to assemblies subject to extremes in joint relaxation, temperature or pressure cycling, and shock or vibration. They are available in a wide variety of metals and filler materials and are produced in circular or moderately noncircular shapes. The inner and outer metal plies of the gasket must be under compression. Preferred flange surface finish is 125 to 250 rms.

Metal-Jacketed Gaskets. Metal-jacketed gaskets are made with a soft compressible filler partially or wholly enclosed in a metal jacket (Fig. 4-5). The entire inner lap must be under compression since this is the primary seal. They are used for circular and noncircular applications including heat exchangers, valves, pumps, compressors, and boilers. In some instances the double-jacketed type is used as rib work in a spirally wound outer

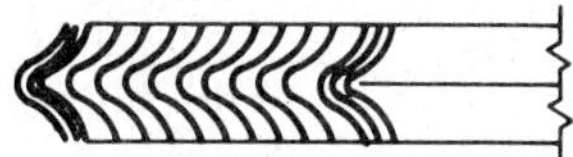

Figure 4-4 Spiral-wound gasket with asbestos filler. *(Reproduced by permission of Parker Gasket Division, North Brunswick, N.J.)*

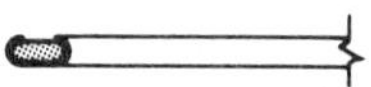

Figure 4-5 Metal-jacketed gasket with filler partially enclosed. *(Reproduced by permission of Parker Gasket Division, North Brunswick, N.J.)*

gasket for heat exchangers. They require 20 to 30 percent compression and are not normally used for joints requiring close maintenance of compressed thickness. When temperatures exceed 900°F (480°C), metallic fillers can be used.

Heavy-Cross-Section Gaskets. The heavy-cross-section gaskets are widely used in the petroleum and processing industries. These gaskets are designed for use in flanges that are specially machined to accept them. Among this group are specialized cross sections including oval, octagonal, lens, Bridgeman, and delta (Fig. 4-6). They are used in high-pressure and high-temperature services including oil-field drilling and production equipment, pressure vessels, valve bonnets, and piping systems. Some of these gaskets are pressure-actuated.

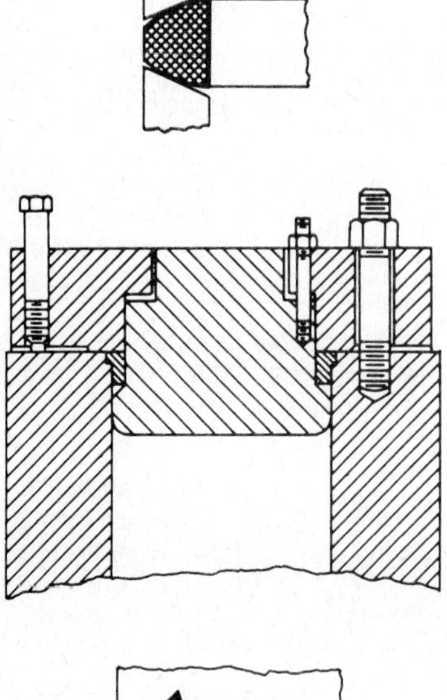

Figure 4-6 Heavy-cross-section gaskets; from the top down: oval gasket *(Reproduced by permission of Parker Gasket Division, North Brunswick, N.J.)*; octagonal gasket *(Reproduced by permission of Parker Gasket Division, North Brunswick, N.J.)*; lens gasket *(Reproduced by permission from Machine Design, September 13, 1973)*; Bridgeman joint *(Reproduced by permission from Petroleum Refinery, May 1956)*; delta gasket *(Reproduced by permission from Machine Design, September 13, 1973)*.

Round-Cross-Section Solid-Metal Gaskets. These gaskets are usually made from wire formed to size and welded (Fig. 4-7). They seal by line contact with high local gasket stresses at low flange loadings. Flange faces are usually grooved or otherwise faced to accurately locate the gasket during assembly. This type is useful in vacuum or other light bolted flange systems where efficient seals are needed.

Light-Cross-Section Gaskets. This type can be used in elastomeric O-ring-type installations sealing extremes in vacuum, high-temperature or -pressure, cryogenic, or nuclear applications (Fig. 4-8). They require very little load for initial

Figure 4-7 Round-cross-section metal gasket. *(Reproduced by permission of Parker Gasket Division, North Brunswick, N.J.)*

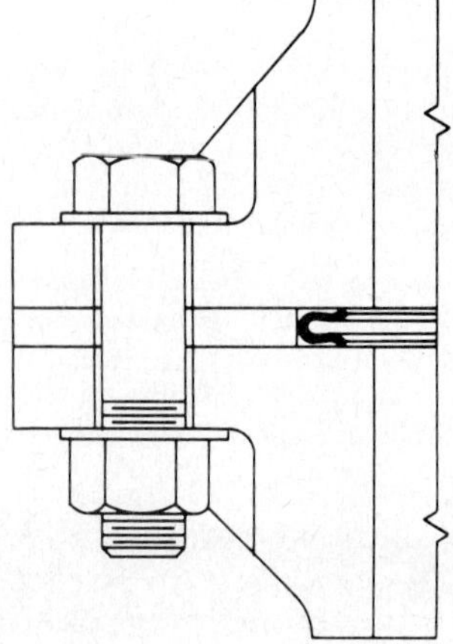

Figure 4-8 Light-cross-section metal gasket. *(Reproduced by permission of Parker O-Seal Division, Culver City, California)*

sealing. Pressure of the system acts on the specially designed seal bellows area, increasing tightness. Flange finish can be 32 to 100 rms for coated seals and 4 to 32 for uncoated.

Installation

Installation of any gasket is a very important factor in achieving a sealed joint. Flange surfaces must be clean and free from nicks, scratches, weld splatter, and any other foreign material. They must be flat and free from warpage. The clamping force or bolt load must be applied in an even stepwise manner to assure uncocked flanges and a uniformly applied load. *It is the joint that leaks. It is the gasket that seals.*

O Rings

The O ring is a versatile, compact, inexpensive sealing device (Fig. 4-9). It is made in a variety of elastomers and may be used as a face seal, a radial squeeze seal, a tube-fitting seal, or occasionally in other configurations.

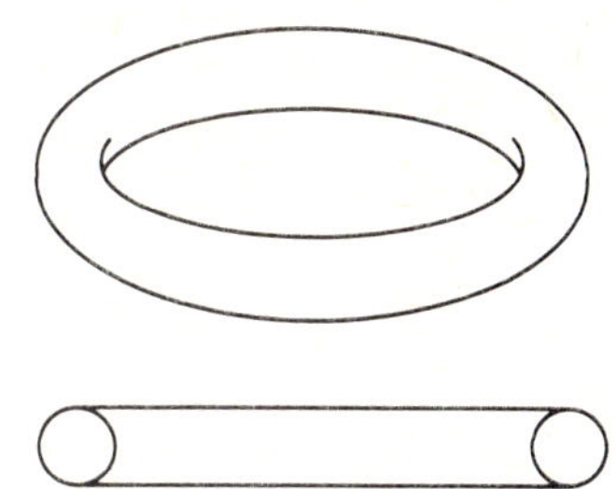

Figure 4-9 O ring. *(Reproduced by permission from Parker O-Ring Handbook No. ORD5700, page A1-1, Parker O-Ring Division, Lexington, Kentucky.)*

Materials

O-ring elastomers are selected for their ability to withstand the conditions of particular sealing applications. The most commonly used O-ring-seal elastomers, together with some other elastomers that are occasionally discussed, are described in Table 4-2.

Applications

O rings in use are confined in a cavity, or "gland," in which the cross section is squeezed in one direction to provide the lines of contact with mating surfaces that create the sealing function. The combination of an O ring and its gland makes up the O-ring seal.

In using an O ring as a seal, the amount of squeeze on the cross section may be critical. With too much squeeze, there may be assembly problems, or the O ring may be damaged at assembly. With too little, the seal is likely to leak at low temperatures.

Though the O ring is squeezed in one direction, it must be free to bulge at right angles to the direction of squeeze, and have extra space to allow for both normal volume swell in contact with the system fluid and for thermal expansion (Fig. 4-10). In some applications, it is acceptable to provide little excess void if the fluid causes little or no swelling, if the surrounding structure is so rugged that it can contain the high pressures caused by a confined rubber that would normally swell, or if the O ring is so completely confined that the fluid can contact only a minuscule portion of its surface, thus effectively preventing swell.

Standard gland dimensions established by O-ring manufacturers generally allow for a moderate amount of swelling.

1. Radial squeeze static O-ring seals, whether of the rod (male gland, Fig. 4-11) or piston (female gland, Fig. 4-12) type, usually have a moderate amount of squeeze that is sufficient to assure good low-temperature performance but not enough to cause undue assembly problems.
2. Face-type rectangular grooves for O-ring seals may provide a heavier squeeze, which makes for a more reliable sealing function (Fig. 4-13). The extra squeeze is acceptable in this design because the cross section is generally deformed by a straight axial pull exerted by flange or cover bolts, so that assembly is not a problem. A variation is the dovetail groove, designed to capture the O ring in a face-type assembly so that the ring cannot drop out when the cover is removed (Fig. 4-14). These are particularly

TABLE 4-2 Elastomers Commonly Used for Seals

No. and designations*	Temperature range, °F (°C)	Properties and uses
1. Butadiene (P, C) BR (A)	−80 to +225 (−62 to +110)	Very good low-temperature flexibility, good resistance to automotive-brake-fluid, water, low-molecular-weight alcohols; preferred because of better high-temperature resistance; often blended with other polymers; primary uses are tires and vibration mounts for low temperature
2. Butyl (P) isobutene-isobutene-isoprene (C) IIR (A)	−75 to +225 (−60 to +110)	Used for vacuum and gases because of its low permeability rate; in other sealing applications, generally superseded by ethylene propylene because of the superior heat resistance and recovery of ethylene propylene; types of seal: diaphragms, gaskets, O rings
3. Epichlorohydrin (P) Hydrin® (Goodrich) 3a. Polychloromethyl Oxirane (C) CO (A)	−40 to +325 (−40 to +163)	Low permeability rate, good resistance to weathering and to oils and other hydrocarbons; usefulness limited because it tends to corrode metals and has poor recovery; types of seal: diaphragms, gaskets
3b. Ethylene oxide (oxirane) + chloromethyl oxirane (C) ECO (A)	−50 to +325 (−45 to +163)	
4. Ethylene propylene (P) 4a. Ethylene propylene copolymer (C) EPM (A) 4b. Ethylene propylene terpolymer (C) EPDM (A)	−70/60 to +250/400 (−57/−50 to +120/205)	Very good resistance to water, steam, weak acids and bases, Skydrols® (Monsanto), and other phosphate esters; swells severely in petroleum products and other hydrocarbons; types of seal: diaphragms, gaskets, O rings, oil seals, T seals, U cups, V packings
5. Fluorocarbon elastomer (P, C) Fluorel (P) (® 3M) Viton (P) (® Du Pont) FKM (A) formerly FPM (A)	−40/0 to +400 (−40/−20 to +200)	Good resistance to wide range of fluids including hydrocarbons and air; hardens in high-temperature water; special, very expensive type (Du Pont Viton GLT) good down to −40°F, otherwise low-temperature limit for seals is 0 to −15°F; excellent high-temperature resistance; types of seal: diaphragms, gaskets, O rings, oil seals, T seals, U cups
6. Fluorosilicone (P, C) FVMQ (A)	−100 to +350/400 (−73 to +175/205)	Has resistance to a wide range of fluids, including hydrocarbons; an expensive elastomer generally restricted to static seals owing to lack of toughness; types of seal: gaskets, O rings
7. Natural Rubber (P) natural polyisoprene (C) NR (A)	−30 to +225 (−34 to +110)	Has excellent recovery, but seldom used for seals because of poor resistance to aging and to most fluids
8. Neoprene (P) chloroprene (P, C) CR (A)	−70/−40 to +250/300 (−55/40 to +120/150)	An "in-between" elastomer, having fair resistance to both hydrocarbons and to oxidation and aging; one of the earliest synthetics; primary seal usage is in refrigerants and silicate esters; types of seal: braided packing, diaphragms, gaskets, O rings, U cups

TABLE 4-2 Elastomers Commonly Used for Seals (*Continued*)

No. and designations*	Temperature range, °F (°C)	Properties and uses
9. Nitrile (P) Buna-N (P) Nitrile butadiene (C) NBR (A)	−70/−20 to +180/275 (−55/−30 to +80/135)	The most commonly used elastomer for seals owing to its resistance to petroleum fluids (fuels, hydraulic and lubricating oils), its useful temperature range, and its low cost; types of seal: braided packing, cup and hat packings, diaphragms, gaskets, O rings, oil seals, sealants, T seals, U cups, V packings
10. Perfluoroelastomer (P, C) Kalrez® (Du Pont) (P)	−30 to +500 (−34 to +260)	Resists a very wide range of fluids and extra-high temperatures; use limited owing to extremely high cost.
11. Phosphazene (P) PNF ® (Firestone) (P) Phosphonitrilic fluoroelastomer (C)	−85 to +350 (−65 to +177)	Similar to fluorosilicone in its fluid resistance and temperature range but tougher, making it suitable for dynamic use; as this is written, it is too new and too expensive to have gained wide acceptance as a seal material, but this may change in the next few years
12. Polyacrylate (P, C) ACM (A)	0 to +350 (−18 to +177)	Resists hydrocarbons, ozone, sunlight, but water resistance, mechanical properties, and recovery are poor; often used to seal automatic transmission and power-steering fluids; types of seal: gaskets, O rings
13. Polysulfide (P, C) Thiokol (P) EOT (A)	−75 to +225 (−60 to +107)	Resists many solvents that no other elastomer can handle; has good resistance to oxygen and ozone and good low-temperature flexibility; heat resistance, strength, and recovery poor; types of seal: cup and hat packing, diaphragms, gaskets, O rings
14. Polyurethane (P) 14a. Polyester polyurethane (C) AU (A) 14b. Polyether polyurethane (C) EU (A)	−65 to +200 (−54 to +93)	Very tough, high-tensile-strength elastomer with excellent resistance to abrasion and wear; good resistance to petroleum-type hydraulic fluids and other hydrocarbons and to ozone and aging; poor resistance to water, acids, high temperatures; poor recovery; types of seal: cup and hat packings, O rings, U cups, V seals; also used for backup rings and wipers
15. SBR (P, A) GR-S (P) Buna-S (P) styrene butadiene (C)	−70 to +225 (−57 to 107)	Most commonly used elastomer today because most automobile tires are made from it; very limited use in seals owing to its poor resistance to hydrocarbons, ozone, sunlight, and aging
16. Silicone (P) Silastic® (Dow Corning) (P) 16a. Phenyl vinyl methyl silicone (C) PVMQ (A)	−75/60 to 400/500 (−60/−50 to +205/260)	Has wide temperature range with good resistance to oxygen, ozone, high-temperature-air aging; excellent recovery, but is not very tough; has poor resistance to most fluids, and is quite permeable to gases; in seal use, is confined almost entirely to static applications; types of seal: diaphragms, gaskets, O rings, sealants
16b. Vinyl methyl silicone (C) VMQ (A)	−175 to +400 (−115 to +205)	

*P = popular term, T = trade name, C = Chemical name, A = designation per ASTM D1418.

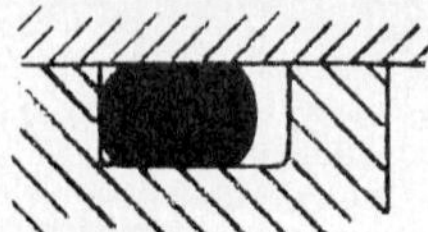

Figure 4-10 Assembled O ring showing squeezed section and extra void.

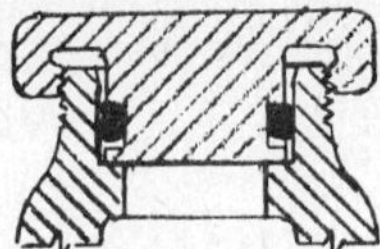

Figure 4-11 Static O-ring seal—male gland. *(Modified with permission from Parker O-Ring Handbook ORD5700, Figure A5-1, Parker O-Ring Division, Lexington, Kentucky.)*

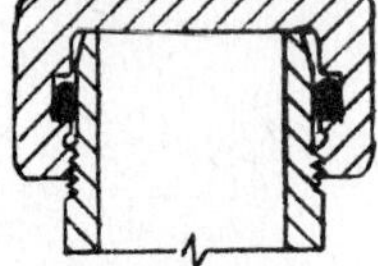

Figure 4-12 Static O-ring seal—female gland. *(Modified with permission from Parker O-Ring Handbook ORD5700, Figure A5-1, Parker O-Ring Division, Lexington, Kentucky.)*

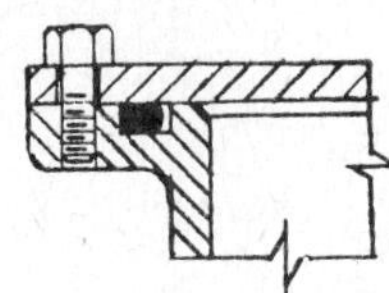

Figure 4-13 Static-face-type O-ring seal.

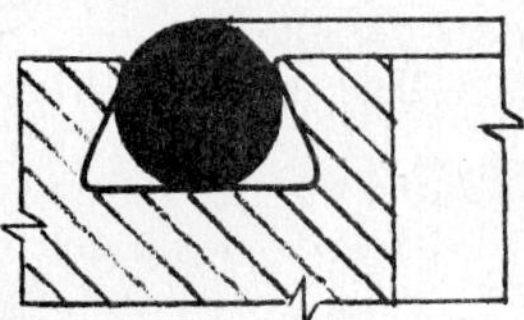

Figure 4-14 O ring in a dovetail groove. *(Modified with permission from Parker O-Ring Handbook ORD5700, Parker O-Ring Division, Lexington, Kentucky.)*

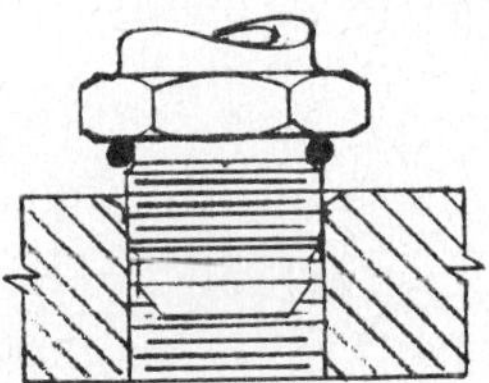

Figure 4-15 MS33656-tube-fitting end with O ring and MS33649 boss.

vulnerable to modification. If deviation from the standard is desired, it must be studied carefully to assure that a maximum-tolerance O ring will not overfill it and a minimum-tolerance ring will still be retained.

3. Tube fittings are sealed with a separate series of O-ring sizes established for the purpose. In this type of application, the gland is essentially triangular. There are two standards in the United States for the O-ring cavity, or "boss." The older is described in Military Standard MS33649 (Fig. 4-15). The new improved style is per MS16142 (Fig. 4-16).

Figure 4-16 Improved tube-fitting boss, SAE J514 and MS16142.

Sealants and Tapes

Sealants are liquids or pastes that are applied between mating surfaces to prevent leakage. Some are formulated to harden after being applied to a joint so that they can contain high pressures. Others remain semiliquid for long periods of time so that joints can be opened easily after prolonged use. Sealants of this latter type are well known for threaded pipe joints. Tapes are also used for this and similar applications. In selecting a sealant, as for seals of other types, a material must be found that will function through the full anticipated temperature range and will not be adversely affected by the fluids and pressures that will be encountered.

RECIPROCATING SEALS

Introduction

The effectiveness of a reciprocating piston or rod seal made of an elastomer or a deformable thermoplastic material depends on the three variables of seal design and the effect of many factors on these variables. The primary factors influencing reciprocating seal design are the fluid to be sealed, the temperatures and pressures to be encountered, the length and speed of stroke, the surface finishes, the amount of clearance, the type of bearings, the space available for the seal, and the desired performance.

Seal Design

The three variables of seal design are the overall shape, the lip configuration, and the properties of the material.

Seal Shape

If permeability is disregarded, the only requirement for an effective seal is an unbroken line of contact between the sealing element and the mating surfaces. In a dynamic seal, this ideal situation can only be approached. Preferably, the seal surface rides on a thin film of the system fluid that provides lubrication and retards wear (Fig. 4-17).

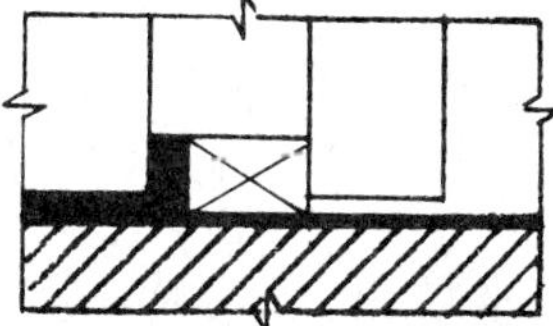

Figure 4-17 Seal riding on thin film of fluid.

Figure 4-18 shows a number of typical seal shapes ranging from pure lip seals on the left to pure compression or pure squeeze types on the right.

When there is little or no system pressure, and assuming that all the seal designs are made in the same material, the only force available to cause a pure lip-type seal to conform with the mating surface is the light load required to bend the lip. A pure compression type, however, generates a much higher wiping or sealing load because a relatively incompressible material must be deformed.

Examining these seal shapes, it will be seen that with low fluid pressure, a pure lip-type seal should have little friction and wear, but also it cannot be expected to seal as positively as the pure compression types. The pure compression types, however, will produce more friction and will wear more rapidly. Similarly, the intermediate shapes can be expected to produce intermediate results.

Lip Configuration

A refinement of the basic seal shape analysis takes into account the shape of the sealing lip. Figure 4-19 shows the five common lip shapes as they appear in the free state and as they appear in the installed state. Also shown is a force vector distribution curve associated with each lip shape. High unit loading improves sealability and film breaking. Therefore seals, especially squeeze seals, with the lip shape shown in No. 4, a back-beveled lip, would be extremely dry seals and would be excellent seals for wiping oil film. If lower friction (and lower sealability) is desired, a lip shape such as No. 2 would be selected because of the low force vectors associated with this particular shape.

Properties of Seal Material

Stability. Long strokes, too slow or too rapid speed, poor lubrication, eccentricity due to inaccurate machining or eccentric loads, and a number of other factors tend to cause a seal to twist in its gland, resulting in premature seal failure. The ability of a seal to resist this twisting action is called *stability*.

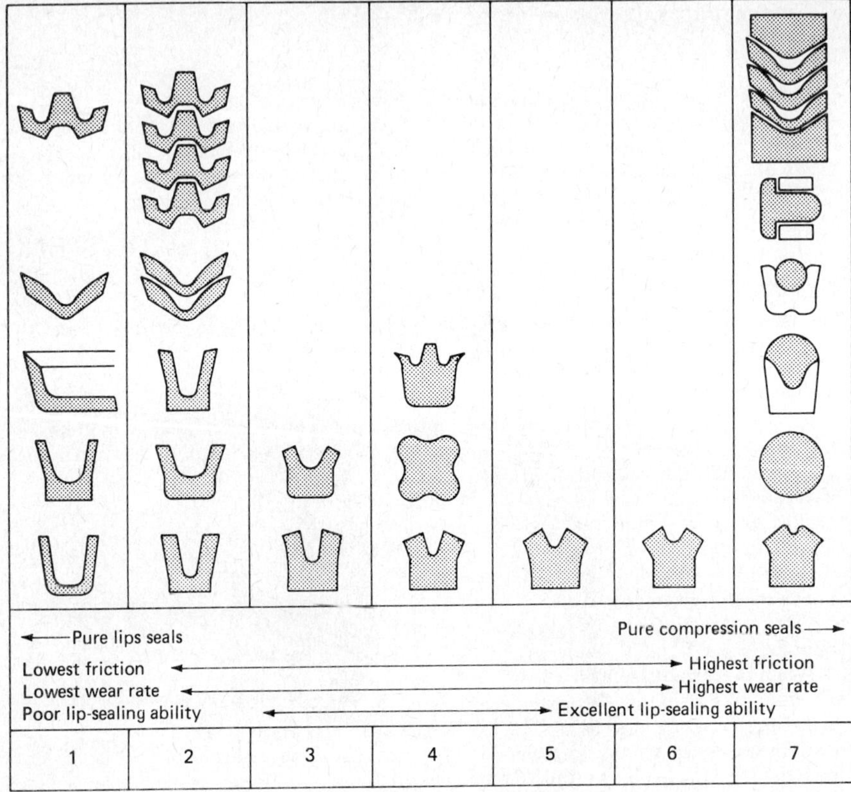

Figure 4-18 Basic shape chart. *(Reproduced by permission of Parker Packing Division, Salt Lake City.)*

Seal stability generally depends on two factors: the basic shape of the seal and the amount of gland fill (Fig. 4-20). A square seal is more stable than a round seal, and a rectangular seal is more stable than a square seal. Usually, the more completely a seal fills the gland, the greater will be its stability. For most seals, however, it is important that the gland not become overfilled since this will cause excessive friction, wear, and extrusion of the seal. When a seal becomes unstable, leakage usually results and the seal is often damaged as well.

In addition to the shape factors mentioned above, a stiffer, higher-modulus material will resist twisting much better than a softer, more flexible material.

Fluid Compatibility. When the basic seal shape has been determined and the lip shape has been established, the seal material must be selected. One of the first considerations should be fluid compatibility. A fluid is considered incompatible with a compound if it causes enough physical-property changes to reduce the sealing function or shorten the working life of the compound. Many fluids will cause seals to swell enough to produce excessive friction and wear. Other fluids may cause physical breakdown of the basic polymer used in the seal, whereas another class of fluids may harden the material so that it no longer has sufficient resilience to conform and establish a sealing line.

Memory and Resilience. The "memory" of a seal material is another important property. It is defined as the ability to return to the original shape on removal of a deforming force, and it is measured by use of the compression set test. Resilience is sim-

Contact type number	Free shape of contact	Low-pressure sealing ability	Ability to clean surface without trapping particles	Hydroplaning tendency	Deformed lip shape and force vector distribution
1		High	Low	Medium	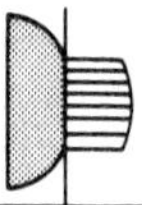
2		Low	Very high	High	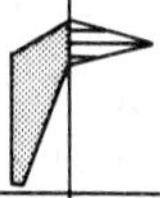
3		High	High	Medium	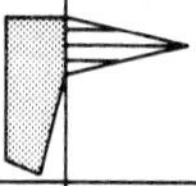
4		Very high	Low	Low	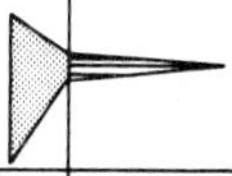
5		Medium/ high	High	Medium/ high	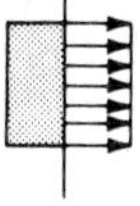

Notes:

(a) Highest unit loading will give highest sealability.
(b) Concentration of vector forces best for wiping.
(c) Lubrication will pass low unit loading patterns.

Figure 4-19 Lip shapes and unit loading chart. *(Reproduced by permission of Parker Packing Division, Salt Lake City.)*

ilar, but implies quick recovery, enabling the seal material to follow rapid variation in the surface passing under the sealing line. With rapid response, very little fluid will be lost on each cycle. The resilience of elastomers is determined with a Bashore resiliometer.

Temperature Range. Closely associated with fluid compatibility is the heat resistance of seal compounds. Heat resistance is a time-temperature function. The higher the temperature experienced by a seal, the shorter its life will be. Some problems stem from the fact that most seal materials expand about 10 times as much as do the metallic materials that contain them. The more serious problems, however, are related to resilience and hardness. High temperatures tend to reduce the resilience and the hardness of seal materials, though many of these materials will eventually harden if the high temperature persists for an extended time. The loss of memory at elevated temperatures, however, means that the seal will assume the shape of the gland. As the temperature falls, the deformed seal then shrinks more than its surroundings and pulls away from the mating surfaces, resulting in leakage.

Seal materials also lose resilience at low temperatures, though the resilience is regained on warming. Nevertheless, as the temperature drops below the point where resilience is lost, the seal shrinks away from the contacting surfaces and leaks until it is warmed again.

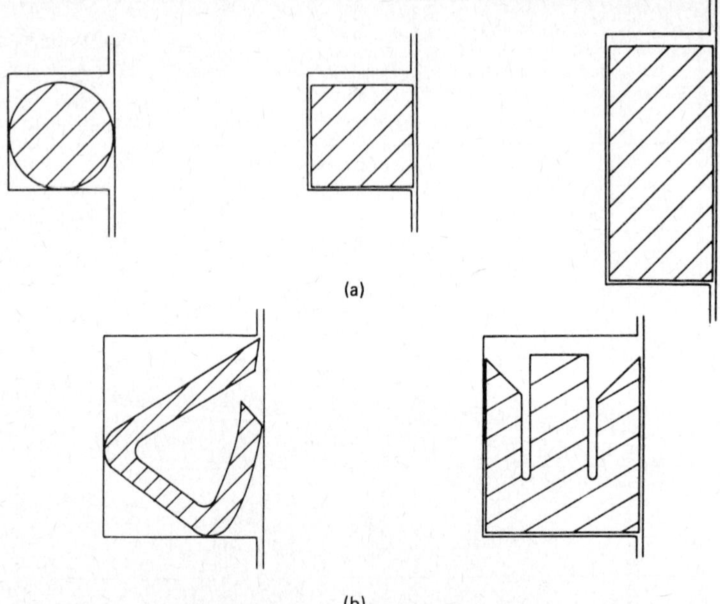

Figure 4-20 Stability guide: *(a)* basic shape; *(b)* gland fill. The square shape is better than the round and the rectangular is better than the square. The higher the gland fill, the more stable the seal. Overfill can cause high friction and heat, and subsequent seal damage. Stiffer seal materials also improve sealability *(Reproduced by permission of Parker Packing Division, Salt Lake City.)*

The temperatures at which these two effects become a problem vary with the seal material, and in many cases the fluid being sealed poses more restrictive limits on the temperature range than does the seal material.

Abrasion Resistance. The life of a seal can be directly related to its abrasion resistance. This property can be measured by various types of abrader wheels or pads. However, these tests generally give only relative abrasion resistance between polymers and do not give good results for predicting seal life in a particular application. Polyurethane and carboxylated nitrile are known for their excellent abrasion resistance, with polyurethanes being the leader.

Extrusion Resistance. Extrusion resistance is another property which must be considered in maximizing seal life. Extrusion resistance is a function of the seal compound and the size of the extrusion gap between the piston bore and the piston, or between the rod and the rod throat. It is also inversely proportional to temperature, pressure, and the frictional drag on the seal.

Extrusion of the seal is one primary cause of seal leakage. This type of leakage can be massive in that a seal can be progressively torn away until a major gap develops at the lip.

Extrusion can be controlled by following one or more of the directions listed below.

1. Use a backup ring.
2. Select an extrusion-resistant compound for the seal itself. (High-tensile-strength modulus and, sometimes, hardness will correlate with increased extrusion resistance.)
3. Reduce the clearance gap between the two dynamic surfaces.

4. Maintain concentricity between the cylinder and bore and between the rod and its housing. (This requires the use of bearings within the cylinder or nonmetallic wear bands on the piston or rod to avoid side loading the seals. Seals are not designed to carry side loads.)

Installation

The elongation of a seal compound is important if stretch-in applications are being considered. The forces required to install a seal should also be taken into account if large cross sections are being considered, even in rod applications, where the seal is folded, rather than stretched, in assembly. The 100 percent modulus of a seal compound is a good guide to ease of installation. Figures 4-21 and 4-22 are also helpful.

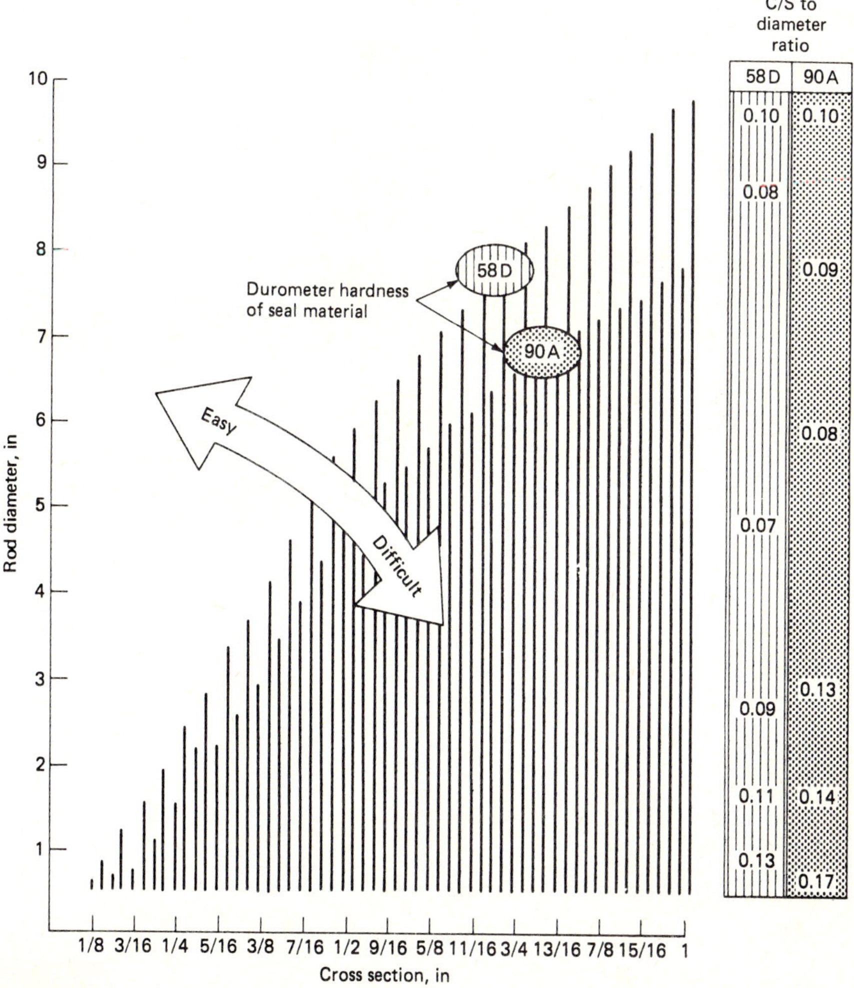

Figure 4-21 Rod-seal installation guide. For thermoplastic materials only: Rigid materials require split seals or separable gland. *(Reproduced by permission of Parker Packing Division, Salt Lake City.)*

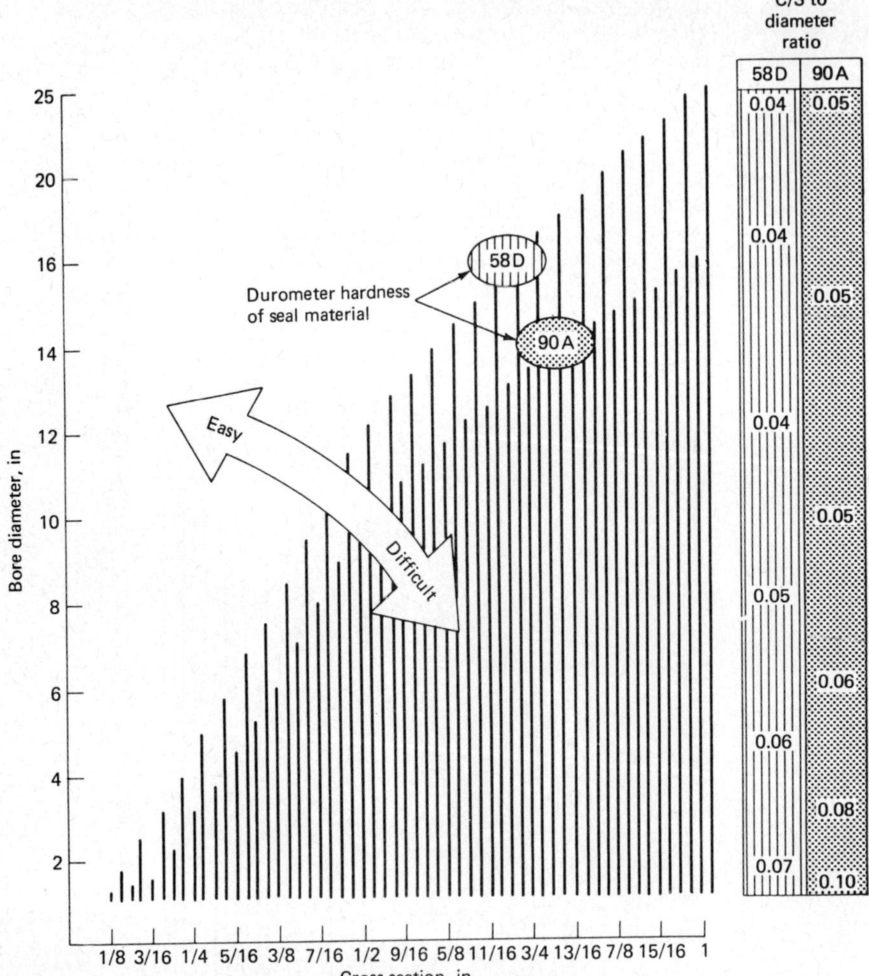

Figure 4-22 Piston-seal installation guide. For thermoplastic materials only: Rigid materials require split seals or split piston. *(Reproduced by permission from Parker Packing Division, Salt Lake City.)*

Backups (Auxiliary Devices)

The most common technique for preventing extrusion is to use auxiliary devices or backups. Backups are added to seals to improve performance. Not only will they prevent extrusion, but many times they will improve the stability of a seal package and its seal ability.

There are two primary types of backups in use, the positively actuated backup and the non-positively actuated backup (Fig. 4-23).

Non-positively actuated backups depend on axial forces which cause lateral deformation of the backup material to close the extrusion gap. Because lateral deformation of the material is required for this type of backup, the materials are limited to those having a fairly high degree of plastic or elastic deformation.

Positively actuated backups use radial forces to close the gap. Since this type of

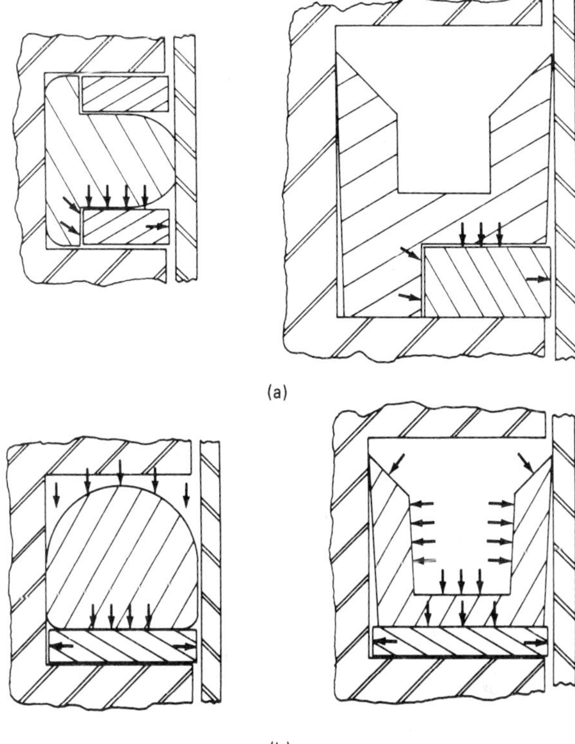

Figure 4-23 Extrusion resistance: (*a*) Positively actuated backups. (*b*) non-positively actuated backups. *(Reproduced by permission of Parker Packing Division, Salt Lake City.)*

backup does not require lateral deformation, a wide variety of hard, stiff materials, including metals, can be used. This type of backup is usually split to allow for radial movement.

Exclusion Devices (Wipers and Scrapers)

Although exclusion devices are not seals, they play an important part in extending seal life and minimizing leakage. Whenever a reciprocating rod is exposed to dust, ice, snow, mud, or other abrasive materials, it is important that a wiper be included in the system to clean the rod before it can carry these foreign materials in through the seal, abrading it and contaminating the system.

There are many types and styles of exclusion devices available, and one can be found that is appropriate for almost any environment.

Metal Considerations

Seal leakage can also be a function of face-roughness values. Tests and experience have shown that, for dynamic surfaces, surface roughness should be in the range of 10 to 20 rms. With rougher finishes, friction and wear increase, instability increases, and the potential for leakage increases. Tests and experience also show that finishes can become *too smooth*! Dynamic surfaces below 10 rms have also been shown to increase friction and wear and may shorten seal life (Fig. 4-24). It can be seen that the optimum dynamic

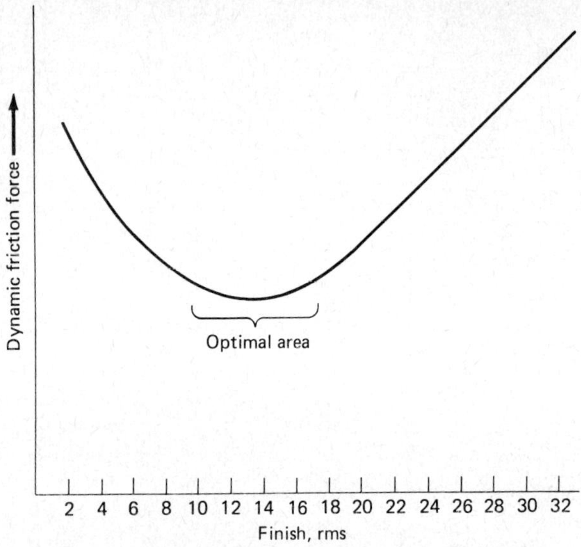

Figure 4-24 Friction-vs.-surface-roughness curve. *(Reproduced by permission of Parker Packing Division, Salt Lake City.)*

surface finish lies between 10 and 20 rms. (Static surface finish can range from 16 to 32 rms and still maintain a good seal line.)

It is important that all sharp corners be removed from the gland area or any area the seal passes during installation. If the seal is damaged by the metal prior to reaching its designed location, leakage can be an immediate result. Seal installation should be designed as carefully as the seal gland. Lead-in tapers should be used, avoiding threaded areas during installation. Other precautions like this will help assure reduced leakage and better sealing applications. Installation tools are also of great value. Gland dimensions should be adhered to carefully, as recommended by the seal supplier. If deviations must be used, the seal supplier should be consulted and the modification tested extensively prior to production.

Notes

1. When in doubt contact the seal supplier for guidance.
2. Test all new seal designs under conditions as close as possible to the conditions of use. (The vendor can provide guidance but cannot actually test all seals under all the conditions encountered in the field.)
3. If a premature failure occurs, save the failed seal and the mating parts to assist in a failure analysis.

Rigid Packings

Rigid packings, generally of cast iron or reinforced TFE (tetrafluoroethylene), are most commonly used on gas compressors. These range in form from the simple bevel or step-joint piston ring to rather complex spring-assisted segmental styles (Fig. 4-25). The need for "clear" (unlubricated) compressed gas or air systems has accentuated the use of the reinforced TFE style.

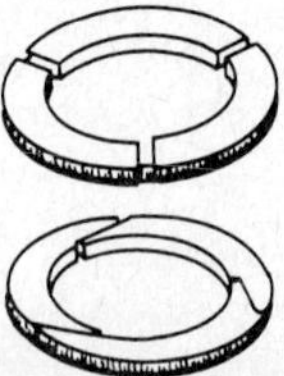

Figure 4-25 Segmental metal rings. *(From: W. Staniar, "Plant Engineering Handbook," 2d ed., Fig. 27-23d, p. 27-22. Copyright © 1959, McGraw-Hill Book Co., Inc. Reproduced by permission of the publisher.)*

ROTARY SEALS

Introduction

Rotary sealing devices may be subdivided into four categories:

1. Clearance seals (controlled clearance, labyrinth, wind-back)
2. Mechanical face seals (mechanical seals, face seals)
3. Compression packings
4. Lip seals (oil seals)

Although most types are commonly used alone, severe conditions often require two or more types in a single application.

Rotary packings and seals present a specific group of problems to the seal designer. Typically rubbing speeds are high, and the seal-rubbing interface is confined to a very limited area. The problem of dissipating the resultant heat differentiates rotary applications from reciprocating applications where the heat is incurred over a relatively large area of a rod or a cylinder bore.

Factors such as fluid, temperature, pressure, interface velocity, allowable leakage, necessary life, space, and cost will determine which type of rotary seal is best suited for any individual application. No attempt can be made in this limited discussion to provide all the information needed to make such a selection. Rather, the following section is intended to acquaint the reader with the basic types, principles, and terminology

Clearance Seals

Clearance seals, as a group, are intended to reduce leakage while avoiding actual contact between the moving parts.

Labyrinth Seals

A simple labyrinth is shown in Fig. 4-26. Numerous variations of this type are feasible, involving stepped diameters, barrier fluid areas, drains, etc. Since clearances must be kept small and contact generally avoided, substantial design problems result when wobble, end play, and thermal expansion are taken into account.

Labyrinths have, essentially, no pressure or speed limitations because of their noncontacting nature. Similarly, no temperature limits are imposed other than those of the materials of construction. Since these materials are not required to be deformable, and no rubbing contact is involved, material choices for high temperatures normally present no problem.

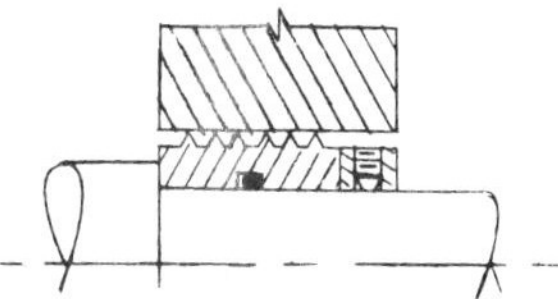

Figure 4-26 Labyrinth seal.

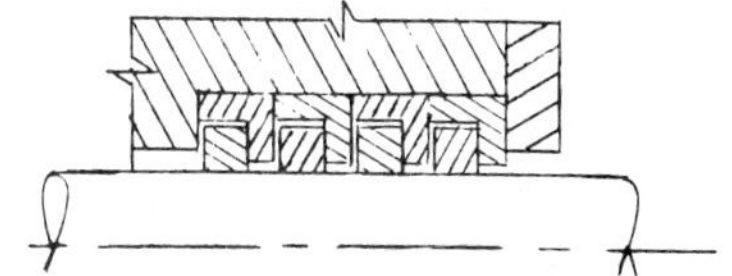

Figure 4-27 Multiple-unit floating bushing seal.

Bushing Seals

Bushings, most simply, are sleeves with a small clearance around a shaft, leakage being limited by laminar- and turbulent-flow considerations in the small annulus thus created. A noncontact situation is desired, and the need for small clearance, when viewed in the light of shaft runout, angularity, vibration, thermal expansion, etc., generally results in a floating-sleeve design. Figure 4-27 represents a multiple-unit floating-sleeve or bushing design, each bushing being free to center itself diametrally with reference to the shaft. As in other rotary-seal types, designs will often allow for drains, buffer or barrier fluids, etc.

Wind-Back Seals

The tendency of fluid in close proximity to a moving surface (as in a small annulus) to move with the surface is made use of in a variety of self-pumping or *wind-back* applications. Figure 4-28 shows a sample representational wind-back, where the tendency of the fluid to flow to the left is balanced by the pumping force of the helical land.

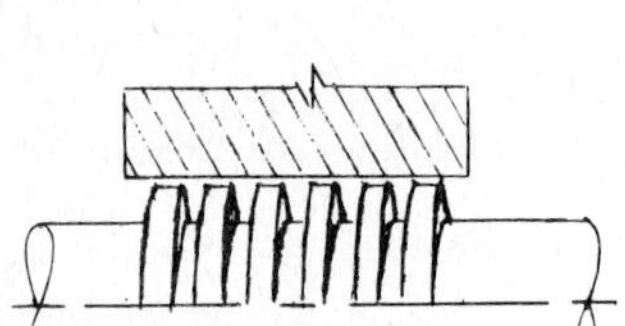

Figure 4-28 Wind-back seal.

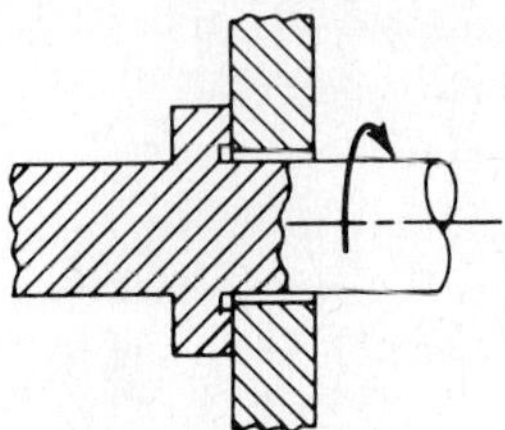

Figure 4-29 Simplified face seal. *(Reproduced by permission from "Sealing Is Our Business," Fig. 5-15. Copyright © January 1975, Parker Seal Group, Lexington, Kentucky.)*

Mechanical Face Seals

A face seal in its simplest conceptual form would be as illustrated in Fig. 4-29. As may be observed, the sealing interface is on a plane at right angles to the rotational axis. Thus a certain amount of shaft runout could be accommodated without distorting the relationship between the two parts at the sealing interface.

This conceptual design, however, could not compensate for wear at the seal interface, nor could it accommodate axial motion.

In Fig. 4-30 the design is modified to provide replaceable rubbing portions, and to provide axial spring force to accommodate wear and small amounts of axial motion. A replaceable seat *A* has been inserted in the housing, a replaceable head *B* replaces the rotating shoulder, a spring *C* bearing against a collar *D* keeps the seal interface in contact, while secondary seals *E* and *F* seal the leak paths thus created. Secondary seal *E* is essentially static, whereas *F* will experience motion as end play and wear are accommodated. Figure 4-30 is thus a simple mechanical face seal. Seal types discussed here may be considered to be modifications or expansions of this basic configuration.

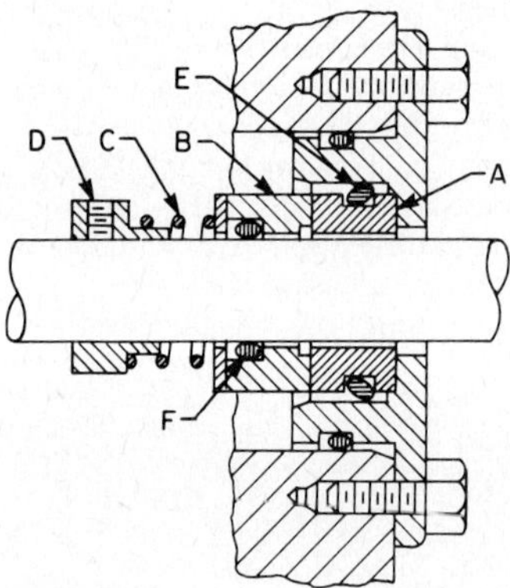

Figure 4-30 Typical mechanical face seal. *(Reproduced by permission from "Sealing Is Our Business," Fig. 5-18. Copyright © January 1975, Parker Seal Group, Lexington, Kentucky.)*

Balance Seals

Figure 4-30 shows the contained fluid pressure against the left radial surface of the head *B* as it tends to force the head to the right, thus increasing the interface pressure established by the spring in proportion to the fluid pressure contained. This is shown in simplified form in Fig. 4-31. If the design is modified by use of a stepped shaft, as in Fig. 4-32, one will observe that the annular area on the left end of the head is reduced, proportionately reducing the fluid-pressure induced component of the interface pressure. In industry terminology, Fig. 4-32 is referred to as "balanced" and Fig. 4-31 as "unbalanced." Balanced seals are normally specified for higher pressures or for the sealing of "light" liquids.

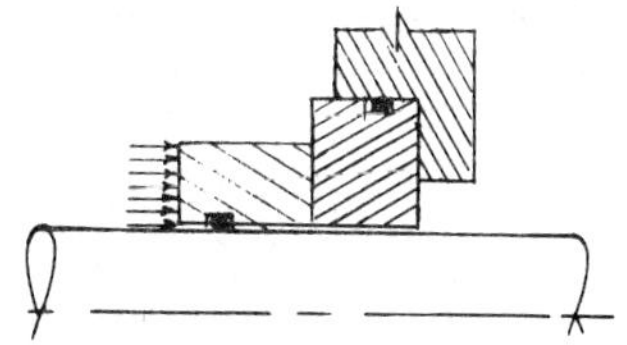

Figure 4-31 Unbalanced face-seal head.

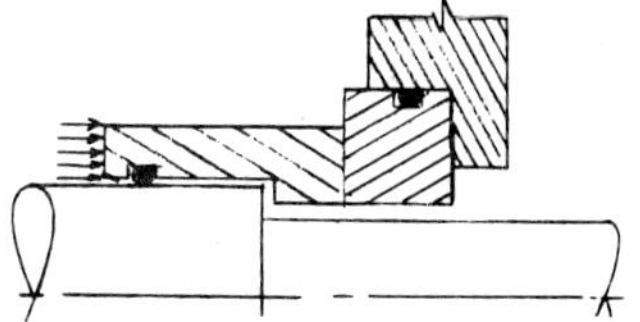

Figure 4-32 Balanced face-seal head.

Rotating Seat. Seals may be designed and installed with the seat stationary (as in the preceding examples) or rotating (Fig. 4-33).

Outside Installation. A further variation is shown later in Fig. 4-35. Figures 4-30 and 4-33 show the seal head on the fluid side of the seat (an inside installation), and Fig. 4-34 shows an outside installation.

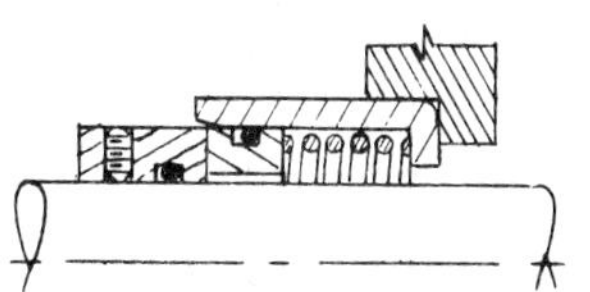

Figure 4-33 Mechanical seal with rotating seat.

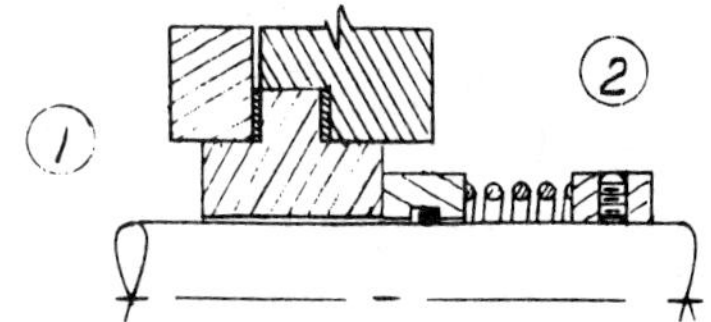

Figure 4-34 Seal head on outside of seat: (1) fluid; (2) atmosphere.

Clamped Seat. Figure 4-34 shows a further variation with the seat clamped, as opposed to the relatively free O-ring mounting shown in the preceding sketches. The secondary seal at the seat is a flat gasket.

Double Seals

It is often necessary or desirable to circulate a secondary fluid in a chamber formed by a double (Fig. 4-35) or tandem (Fig. 4-36) seal arrangement. This secondary fluid could

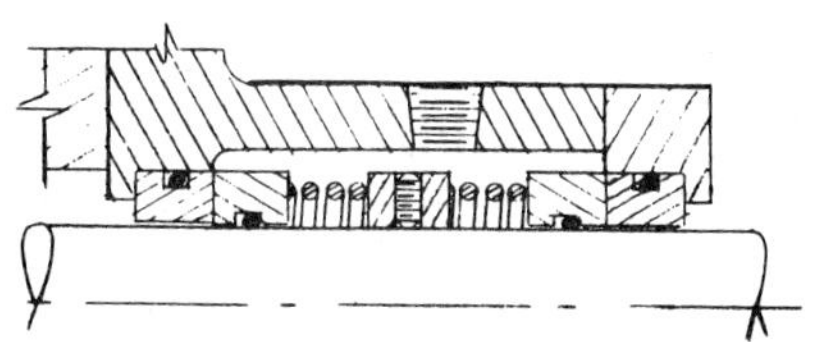

Figure 4-35 Double mechanical seal.

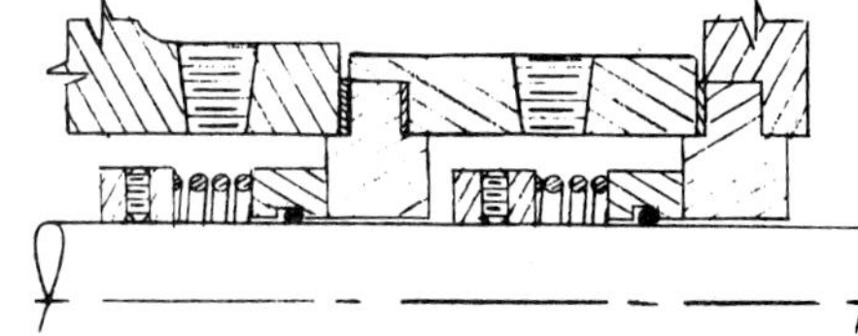

Figure 4-36 Tandem-type mechanical seal.

be at a higher or lower pressure than the primary fluid. Such an arrangement may be used for a wide variety of purposes, ranging from safety (in the case of a toxic or flammable primary fluid) to simple redundancy. Problems with fluids containing abrasives, or those which tend to crystallize on cooling or on contact with the atmosphere, are commonly handled in this fashion. This method may also be utilized to provide interface lubrication when sealing gases.

An alternative form of rotary seal (a throttle bushing, rings of soft packing, an oil seal, etc.) may be employed either inboard or outboard of the primary mechanical face seal for similar purposes.

For simplicity all of the examples shown depict a single spring. However, a number of small, evenly spaced springs are often employed to provide a more uniform load around the circumference of the head.

Seals with Bellows. The motion of the secondary seal associated with the head tends, in most designs, to "fret" or wear the adjacent sealed surface. In order to avoid this, recourse is often had to bellows designs, of elastomer or TFE, in which the motion in question is absorbed by deformation of the bellows (Fig. 4-37).

A further modification (Fig. 4-38) uses a welded metal bellows which performs a dual function as a motion-absorbing seal and as the actuating spring.

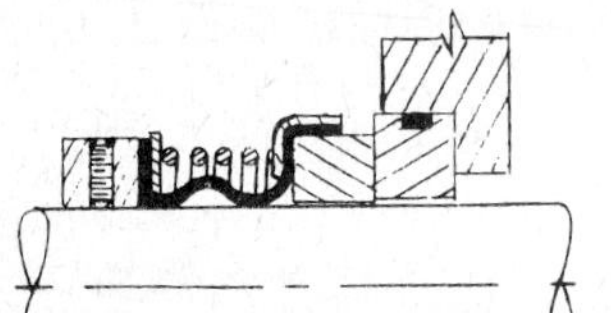

Figure 4-37 Mechanical seal with elastomeric or TFE bellows.

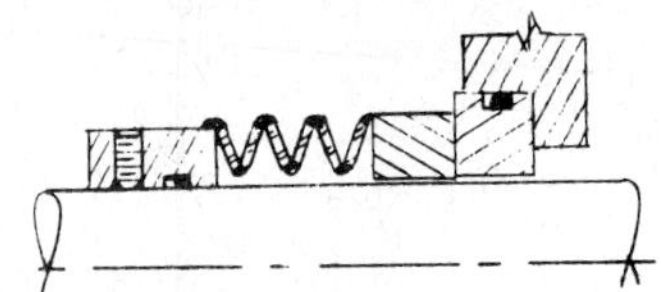

Figure 4-38 Mechanical seal with metal bellows.

The Secondary Seal.

In the preceding paragraphs, O rings, gaskets, and bellows have been mentioned as secondary seals. Actually, a wide variety of secondary seals are used in mechanical face-seal assemblies. The secondary seal at the seat is generally static, leading most often to use of a simple O-ring gasket. The secondary seal associated with the head, however, is subjected to various degrees of small-magnitude motion as the head moves in response to axial shaft motion and face wear. Thus, a wider variety of wedges, X rings, V rings, coated O rings, and other, more sophisticated, seal forms will be noted.

Compression Packings

Compression packings, also called soft packings, braided packings, and rope packings, are among the oldest packings known to industry. Such packings characteristically depend on axial pressure provided mechanically by a gland follower to radially distort the rings into intimate contact with the sealed surfaces. A broad range of materials and combinations of materials and impregnants or saturants, ranging from flax, cotton, and asbestos to the latest exotics, are used in compression-packing manufacture.

Types of Construction

Cross Braid. Strands of yarn cross the surface diagonally, interlocking to form a dense, yet flexible, structure that cannot unravel in service. Lubrication of individual strands in the braiding process provides a more uniform distribution of antifriction materials and yarn density for increased life of the packing.

Square Braid. Strands of yarn cross over and under other strands in the same running direction and produce square cross sections. Other names for this construction are square, plaited, or flax braid. This type of construction is normally specified for high-speed rotary service at low pressure.

Braid over Braid. This type is formed by one or more layers of braided yarns covering a core of braided, twisted, or homogeneous materials and producing an initially round construction. It may be calendered square or rectangular into a dense, highly lubricated packing for low-speed and high-pressure applications.

Twisted. Strands of yarn are twisted in the same running direction to the required round size. This type is recommended only for low-speed, low-pressure utility use when the packing space is small or when the available cross section is larger than the required packing space. The individual strands are easily removed to fit emergency requirements.

Laminated. Coils, spirals, or rings are cut from molded slabs of fabric and rubber materials. This construction is normally used for the liquid end of reciprocating rams or pump pistons.

Wrapped. Rubber-impregnated fabric materials are rolled or calendered into a square or rectangular cross section alone or over a homogeneous core for high resilience. Typical use is in applications with high lateral movement in the equipment.

Plastic. Lubricated plastic packings contain oil, fiber materials, graphite, grease, or mica. They are usually square and within a cotton jacket for handling or within an asbestos jacket to prevent extrusion. They are normally used to seal gases or in applications in which the packing must supply part or all of the lubrication.

Metallic. Metal packings of lead, copper, aluminum foil, or ribbon are spirally wrapped, folded, or twisted into a square or rectangular cross section. Normal use is in high-pressure, high-temperature service and corrosive fluids.

Cutting and Installing Packing Rings

Pumps and Agitators.

1. Remove old packing from the stuffing box. Clean the box and shaft thoroughly and examine the shaft for wear or scoring. Replace the shaft or sleeve if wear is excessive. Check the bearing by moving the shaft up and down.
2. Use the right size of coil packing to be sure the packing will fit and can compensate for shaft wear, if any. To determine the correct packing size, measure the diameter of the shaft and then the inside diameter of the stuffing box. Subtract the shaft diameter from the inside-diameter measurement and divide by 2 for the required size.
3. Cut the packing into individual rings. *Never wind packing in a continuous length into a stuffing box.* For pumps and agitators cut rings with a butt (square) joint (Fig. 4-39). The best way to cut packing rings is to cut them on a mandrel of the same diameter as the shaft in the stuffing-box area. Hold the coiled packing tightly and firmly on the mandrel but do not stretch it excessively. Cut the ring and try it in the stuffing box to make certain that it fills the packing space properly, with no gap in the joint at the outside diameter of the ring.

Figure 4-39 Butt joint. *(Reproduced by permission of Parker Packing Division, Carson City, Nevada.)*

4. Install one ring at a time. Make sure it is clean and has not picked up any dirt in handling before installing it. If clean oil is available, lubricate each ring thoroughly; the shaft and the inside of the stuffing box would also benefit from lubrication. Joints of successive rings should be staggered and be kept at least 90° apart. Each ring should be firmly seated with a tamping tool. When enough rings have been individually seated to allow the nose of the follower to reach them, the individual tamping should be supplemented by the follower. Never depend entirely on the follower to seat a set of rings properly; this practice will jam the last rings installed but leave the front rings loose in the box. The result is excessive and rapid wear of rear rings, erratic packing performance, or sometimes, twisting and tearing of the front rings which are loose in the stuffing box.
5. After last ring is installed, take up bolts finger tight. Start the pump, and take up bolts until leakage is decreased to no more than 10 drops per minute. Stopping leakage entirely at this point will cause the packing to burn up.
6. Allow the packing to leak freely when starting up a newly packed pump. It will take about one working day to break in a set of packing to a point where the leakage is stabilized at a uniform acceptable rate.
7. If at all possible, provide, through a lantern ring (Fig. 4-40), means of lubricating the

shaft and packing by supplying grease, oil, water, or the liquid handled in the pump. Make sure the lantern ring, as installed, is slightly behind the lubricant fitting so it will move under the fitting as follower pressure is applied.

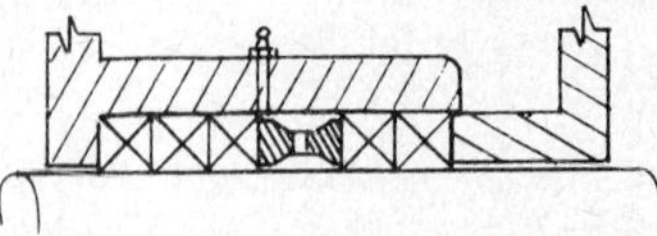

Figure 4-40 Packing set with lantern ring in a stuffing box. *(Modified from Parker Compression Packing Catalog, No. PPD3901-A, by permission of Parker Packing Division, Carson City, Nev.)*

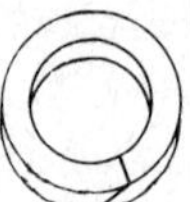

Figure 4-41 Skive joint. *(Reproduced by permission of Parker Packing Division, Carson City, Nev.)*

Valves and Expansion Joints.

1. Carefully perform all the operations outlined under steps 1, 2, 3, and 4 as given for pumps and agitators. Rings used on valves and expansion joints should be cut with a skive joint (45°). See Fig. 4-41.
2. Bring the follower down on the packing to the point where heavy resistance to wrenching is felt. During this time turn the valve stem back and forth to determine ease of turning. Do not wrench to the point at which the stem will not turn.
3. After the valve has been on the line a day or so, even if no leakage exists, the follower should be tightened slightly. If it is leaking at all, tighten the follower until there is no leakage.

Lip-Type Rotary Seals

Lip-type rotary seals are often called *oil seals*, though they are by no means restricted to sealing oils, or *rotary-shaft seals*. Elements common to this class of seals are a flexible sealing lip that rides on the surface of the rotating shaft and a formed metal cup that holds the sealing element and is pressed into the seal housing. Generally there is also a garter spring that helps maintain contact of the lip (Fig. 4-42). The sealing element may be elastomeric, leather, or plastic.

The primary use of oil seals is in low-pressure applications, generally below 8 lb/in^2. Surface speeds may be quite high, ranging up to 4000 ft/min. Occasionally there are special designs that can tolerate much higher pressures, but at high pressure the surface speed must be slow.

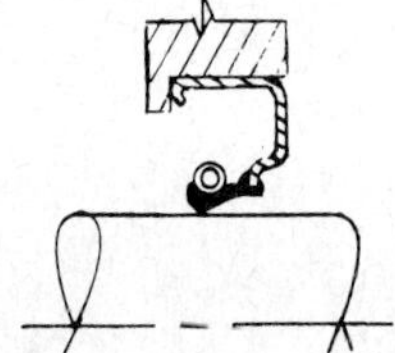

Figure 4-42 Typical oil seal with garter spring.

Oil seals are well suited to low-pressure applications because the narrow contact band of the sealing lip keeps friction and its associated heat low while permitting ready dissipation of the heat that is generated. The sealing-element material may be varied to resist special fluids and unusual temperature extremes.

Dirt-Lip Design

One of the most common variations on the basic design is a second lip pointing outward for use where abrasive dust could damage the primary sealing lip or where external fluid splash could enter and contaminate the internal medium (Fig. 4-43). The secondary or *dirt lip* is not spring-loaded, but relies on the resilience of the material to maintain contact with the shaft.

Nose-Seal Gasket

Another modification includes a nose-seal gasket to prevent static leakage around the drawn cup (Fig. 4-44).

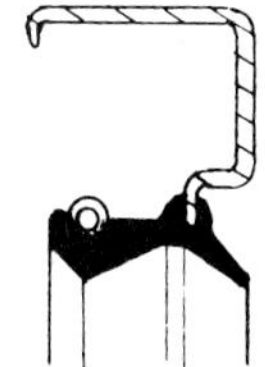

Figure 4-43 Oil seal with dirt lip. *(Reproduced by permission of Parker O-Seal Division, Culver City, California.)*

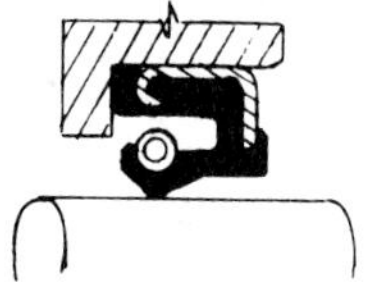

Figure 4-44 Oil seal with nose-seal gasket. *(Modified with permission from "Sealing Is Our Business," Copyright © 1975, Parker Seal Group, Lexington, Kentucky.)*

Cup Designs

Besides variations in the sealing element, the metal cup may be made in any number of styles. For instance, a pry-out flange may be incorporated to make the seal accessible from either the outside (Fig. 4-45) or from the fluid side. Often an inner case, which protects the spring and the sealing lip and makes the assembly more rigid, is provided (Fig. 4-46).

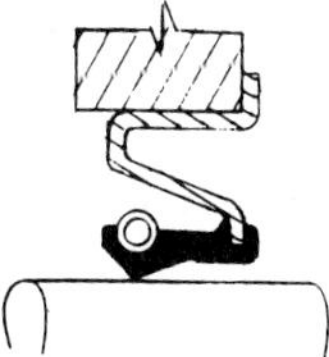

Figure 4-45 Oil seal with pry-out flange. *(Modified with permission from "Sealing Is Our Business," Copyright © 1975, Parker Seal Group, Lexington, Kentucky.)*

Figure 4-46 Oil seal with inner case. *(Modified with permission from "Sealing Is Our Business," Copyright © 1975, Parker Seal Group, Lexington, Kentucky.)*

FLEXURAL SEALING DEVICES

Diaphragms, bellows, and expansion joints are variations of a class of sealing devices that are attached to one or more components of a system, absorbing the relative motion between such components or absorbing fluid displacement by deformation of the sealing device itself.

The simplest of these devices is the flat diaphragm (Fig. 4-47). These are generally cut from reinforced or unreinforced elastomeric sheet, although metal, leather, and plastics find use in some situations. The range of motion which can be accommodated by the

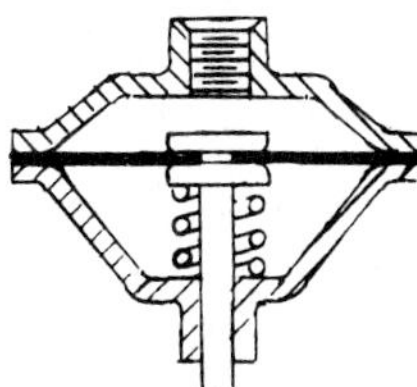

Figure 4-47 Flat diaphragm. *(From sketch by J. B. Scannell.)*

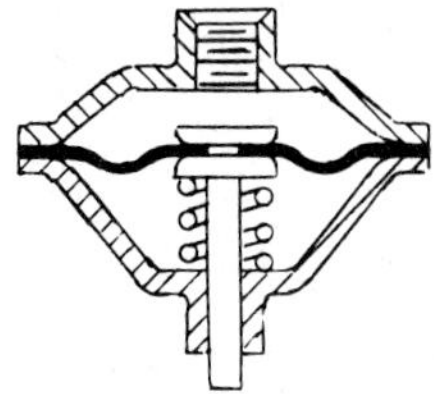

Figure 4-48 Diaphragm with convolution. *(From sketch by J. B. Scannell.)*

flat diaphragm is limited, and a molded or formed convolution is often added to accommodate additional travel (Fig. 4-48).

The *tubular diaphragm*, or bellows, accommodates large axial motion for a given diameter (Fig. 4-49), as does the rolling diaphragm (Fig. 4-50).

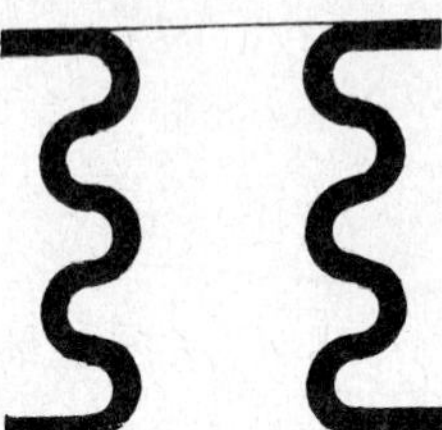

Figure 4-49 Bellows. *(From sketch by J. B. Scannell.)*

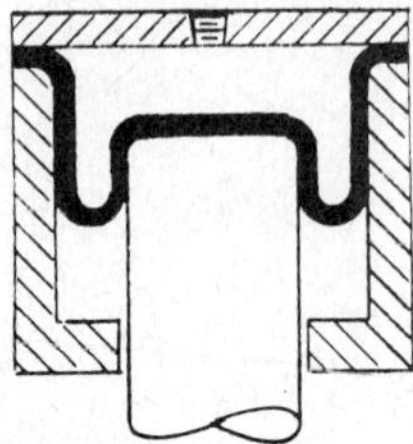

Figure 4-50 Rolling diaphragm. *(From sketch by J. B. Scannell.)*

A heavy-duty form of bellows is the *expansion joint* (Fig. 4-51), most commonly fabricated of rubber with fabric and metal reinforcement, but also of TFE or with a TFE liner. These devices absorb vibration and relative movement between sections of piping or duct systems.

Although a degree of standardization exists among these products, most are produced to order for specific applications, and competent suppliers should be consulted for detailed information.

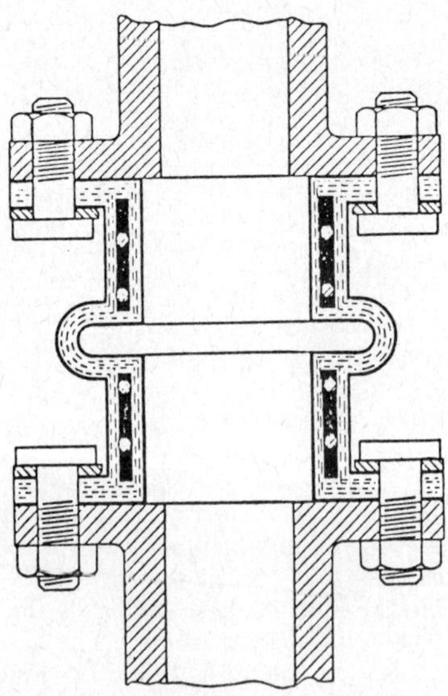

Figure 4-51 Expansion joint. *(From: W. Staniar, "Plant Engineering Handbook," 2d ed., Fig. 27-43, p. 27-52. Copyright © 1959, McGraw-Hill Book Co., Inc. Reproduced by permission of the publisher.)*

INFORMATION FOR SEAL DESIGN

When assistance is needed in seal design, the consultant will need specific details about the application. The following comments may be helpful in collecting the necessary pieces of information.

Data Needed for All Seal Designs

1. Temperature

- **a.** Maximum operating temperature
- **b.** Time at maximum temperature. The time at maximum temperature is particularly important when the maximum is too high for normal seal materials. If the time is sufficiently short and will be experienced once or only a few times, an inexpensive material may be suitable since degradation due to high temperature, though generally irreversible, is time-dependent.
- **c.** Minimum operating temperature. If the minimum operating temperature is very low, it will have a great influence on the selection of a seal material. In general,

however, low-temperature effects are reversible. If a seal does not operate at low temperature, it will generally function normally when warmed up. A seal is considered to be operating, in this sense, if it must contain a fluid, even though the device in which it is installed may not operate at the low temperature.

   d. Normal operating temperature

2. **Pressure**
   a. Maximum pressure
   b. Operating pressure
   c. Minimum pressure (or vacuum level for vacuum seals).
3. **Medium** (to be sealed) Length of time of seal contact with the medium.
4. **Configuration** The seal consultant needs to know the arrangement, materials, and sizes of surfaces in the vicinity of the seal. In some cases, a few dimensions will suffice. Often it is necessary to provide dimensional drawings or sketches of the pertinent parts. Surface roughness values of mating parts are also important to seal design.

Additional Data for Reciprocating Seals

1. Length of stroke
2. Stroke rate
3. Mode of operation (i.e. actuator, operated by the fluid pressure, or pump, generating fluid pressure).
4. Duty cycle
5. Eccentricity (There should be bearings to minimize the effects of side loading. How much eccentricity do they permit?)

section 2

Materials Handling

Prepared by
K. W. Tunnell Company, Inc.
King of Prussia, Pennsylvania

chapter 2-1

Planning Materials Handling

MATERIALS-HANDLING DEFINITIONS

Materials-handling technology includes hardware and systems which can be categorized as follows:

- Containerization
- Fixed-path handling
- Mobile handling
- Warehousing

Containerization. This classification covers the broad spectrum of confinement methods that are used for storage through all phases of the manufacturing, or process, cycle. The materials-handling engineer employs the unit-size principle to optimize the quantity, size, and weight of the load to be handled or moved and is able to specify the best container after considering material and other production-system parameters. Pallets, skids, tote boxes, wire-mesh containers, covering a wide range of sizes and materials, are included within the category.

Fixed-Path Handling. This classification applies to movement and storage of unit loads of material with an intermittent or a continuous flow over a fixed path from one point to another. Fixed-path-handling equipment is secured to, and is considered part of, the facility. Once installed it is more difficult to modify or replace; therefore a considerable amount of planning and interfacing with other functions has to be considered. This equipment, if installed above the floor surface, can effectively utilize what would otherwise be dead space. Chutes, conveyors, elevators, bridge cranes, palletizing equipment, and robots are examples of fixed-path-handling equipment.

Mobile Handling. This classification inclues all handling systems that move material over various paths within a manufacturing or processing cycle. Handling equipment allows a considerable degree of flexibility in moving material in an intermittent flow but requires special facility requirements, such as aisle sizes, clearances, door openings, and running and maneuvering surfaces. Equipment in this category consumes more energy per unit load moved than most other systems and generally requires trained personnel for operation. Equipment in this category ranges from simple two-wheeled hand trucks to specially designed over-the-road vehicles and also includes skid trucks, floor trucks, powered walkie lift trucks, powered lift trucks, and mobile hydraulic cranes.

Warehousing. This classification of materials handling considers the systems, equipment practices, and requirements dedicated to the following operations within the manufacturing, processing, or distribution cycle:

- Receiving
- Storage of raw, in-process, and finished materials
- Movement in and out of storage
- Order picking and accumulation
- Containerization for shipping
- Loading and shipping

This area of materials handling involves a wider range of planning and analysis. Consider some of the following factors:

1. Location of activity
2. Sizing and physical characteristics relating to product size, type, and volume
3. Number of stockkeeping units
4. Storage equipment
5. Selection of materials-movement methods
6. Packaging methods for shipping

The range of solutions covers the full spectrum, from a single warehouse that shares movement equipment with other parts of the operation to a self-contained, specially equipped, fully automated warehouse.

INTRODUCTION

Materials handling deals with the movement of materials from receiving, through fabrication, to the shipment of the finished product. More broadly, handling and distribution are considered to be one overall system. This viewpoint gives consideration to all handling activities involved in movement of materials from all sources of supply through the various plant and central warehouse operations to the customer distribution network.

The activities related to the flow of the materials are either viewed as separate activities or treated as one element in an integrated handling system. Not everyone agrees that materials-handling activities should be viewed as an overall system. Progressive plant engineering personnel, however, recognize that materials handling represents the efficient integration of workers, materials-control systems, and equipment, as well as the movement of materials. Materials-handling applications must take into account operating costs and time-phased material flow.

Inefficient materials handling and storage increase product cost, delay product delivery, and consume excessive square feet of plant and warehouse space. Studies have indicated that actual materials-handling cost runs between 20 and 50 percent of the total product cost *even though it does not add any value to the finished product.* In addition, from 80 to 95 percent of the total overall time devoted to processing a customer order from fabrication to shipment is devoted to materials handling and storage. Product manufacturing time, therefore, is only a small percentage of the overall process time.

A properly designed and integrated materials-handling system provides tremendous cost-saving opportunities and customer-service improvement potential. The correct selection of a handling method can reduce handling costs per unit by as much as 200 percent. Improvement in storage, such as high-rise storage applications, can reduce unit storage space cost by 20 to 40 percent. Work-in-process inventory can be reduced 30 to 50 percent through compressed cycle times. The reduced cycle times will also result in shorter customer delivery cycles.

The significant impact of materials-handling costs on total product cost has resulted in a great deal of attention and substantial resources being directed toward discovering more efficient methods to reduce handling costs. This effort is expected to receive even greater concentration in the future. The following trends are beginning:

- Most manufacturing managers now recognize that materials handling is a prime area of cost-reduction opportunities.
- Most companies will either establish a full department or assign an individual to the responsibilities of analyzing and solving materials-handling problems.
- Materials handling and storage will increasingly be treated as an integral part of the manufacturing and processing cycle.
- Techniques which analyze problems and evaluate alternatives for materials flow and storage requirements in quantitative terms will increasingly replace intuitive solutions.
- Computer-based technology will employ more powerful and sophisticated techniques such as queueing theory, simulation facilities, and flow-planning techniques to consider and select optimum solutions from a wide range of variables and materials-handling options.
- Handling systems will become more automated by employing computer-controlled systems, robot loading and unloading, driverless vehicles, and automatic storage and retrieval systems. These automation principles will be applied in receiving, manufacturing operations, warehousing, and shipping.

It should be recognized that materials handling is an extremely broad-based subject which more often deals with the application of equipment and mechanical devices than fundamental engineering principles or basic physical laws. At least in part, it requires the application of subjective and experienced judgment and has even been described (with some justification) as more of an art than a science. Based on this fact and the trends in materials handling already discussed, this chapter outlines the methodology for solving materials-handling problems as well as the classification of hardware.

However, it is suggested that plant engineers involved continually with materials-handling projects should be familiar with additional sources of current information not only from reference books but also from specialized trade periodicals, professional societies, and trade associations. (See the reference section at the end of this chapter.)

SOLVING MATERIALS-HANDLING PROBLEMS

Materials-handling activities and installations vary in complexity from operation to operation. The plant engineer can solve most materials-handling problems by keeping two points in mind: one is to thoroughly understand the materials-handling principles; the other is to recognize that a materials-handling system is composed of a series of interrelated handling activities. The plant engineer should therefore apply materials-handling principles to improve each separate materials-handling activity and then interrelate the handling activities by applying flow-planning principles.

Principles of Materials Handling

The collective experience and knowledge of many materials-handling experts has been organized into a framework of generalized principles. The principles are basic and can be used universally. They include:

1. Integrate as many handling activities, such as receiving, storage, production, and inspection, as is practical into a coordinated system.
2. Arrange operation sequence and equipment layout so as to optimize materials flow.
3. Simplify handling by reducing, eliminating, or combining unnecessary movement and/or equipment.
4. Use gravity to move material wherever practical.
5. Make optimum use of the building cube.
6. Increase the quantity, size, and weight of the load handled.
7. Use mechanized or automated handling equipment whenever it can be economically or safely justified.
8. Select handling equipment on the basis of lowest overall cost when considering the material to be handled, the move to be made, and the methods to be utilized.
9. Standardize methods as well as types and sizes of handling equipment.
10. Use methods and equipment that perform a variety of tasks and applications.
11. Plan preventive maintenance, and schedule regular repairs on all handling equipment.
12. Determine the effectiveness of handling performance in terms of expense per unit handled.
13. Move materials in as direct a path as possible, minimizing backtracking.
14. Deliver materials directly to work areas whenever practical, and plan the minimum of material in the area.
15. Move the greatest weight or bulk the shortest distance.
16. Provide alternative plans in case of a breakdown.

Steps in Solving Handling Problems

The general methods that are used for solving other operational problems are applicable in the materials-handling area. The factors that must be considered relate to how to most efficiently move certain volumes and types of materials by a particular method. The steps involved in systematically solving these problems consist of:

- Problem identification
- Problem analysis and quantification
- Selection and evaluation of alternatives
- Project justification

Problem Identification

Identification of materials-handling problems includes determining the impact of interfacing activities such as production control, manufacturing, vendors, shipping, and receiving. The buildup of material in front of a machine or a truck dock may not be a problem of too little storage but rather one of lot sizing or inefficient truck-loading systems. Most importantly, a costly handling route between two distant machines may not be caused by the handling mechanism but by the location of the equipment itself.

Problem Analysis and Quantification

Qualitative and quantitative answers are obtained through use of industrial engineering techniques such as flow diagrams, flowcharts, from-to charts, and activity-relationship charts. For the detailed application of these manual techniques see Refs. 1 to 4.

Computer-aided techniques are useful when large amounts of data are involved.[5]

Selection and Evaluation of Alternatives

In the selection of alternatives three general types of criteria are involved.

Movement. Movement involves the study of routes in terms of the combination of handling equipment and containers jointly rather than on an individual product basis. Under this criterion, distances from and to locations, outside travel, and frequencies would be minimized.

Criteria Which Cannot be Directly Costed. These involve criteria such as:

1. **Performance** The potentials for relocation of equipment at future time periods, as related to the handling equipment, and for material design changes, as they relate to the containers themselves.
2. **Delicacy** The nature of the part and its requirements for special handling, dunnage, or containers—particularly to avoid damage.
3. **Interfaces** Production control and manufacturing departments, vendors, shipping, receiving, and intersite movement requirements.
4. **Uniformity** The need to standardize or at least unitize containers to provide uniform handling characteristics. This provides for the use of idealized container sizing and packing techniques, including proper dunnage for irregular loads. On the other hand, it requires the special handling and designs for special irregular-sized parts.

Cost-Effectiveness. This involves the analysis in concept of all standard operating costs, equipment life, maintenance and spare usage for equipment, and intermediate storage.

Generally, the analysis will be dominated by the cost-effectiveness of the alternatives involving cost components as follows.

- **Capital Investment Costs.** One-time charges incurred at the time of equipment procurement that include:
  1. Equipment cost, including freight
  2. Installation costs
  3. Special maintenance requirements
  4. Special power and fuel facilities
  5. Rearrangement and alteration of facilities to accommodate equipment
  6. Engineering support
  7. Supplies
- **Fixed Costs.** Determined or assigned to a system, a piece of equipment, or an activity on a time-period basis; include:
  1. Depreciation
  2. Taxes
  3. Insurance
  4. Supervision
- **Variable Costs.** Can be considered the cost of performing an operation or activity. In the case of equipment, it is the cost of using the equipment. The following items are included within this component.
  1. Equipment-operating personnel or personnel manually performing a materials-handling task

TABLE 1-1 Annual Operating Cost

Cost component	Present manual method, $	Proposed method, $ Conveyor	Fork lift, electric	Fork lift, propane
1. Capital equip. investment				
Equipment cost		6,000	25,000	27,000
Freight		500	800	800
Installation		8,000	—	—
Fuel-power facilities		—	4,000	—
Alterations to facility		10,000	—	—
Engineering support		1,000	—	—
Supplies		—	—	—
Total capital investment		25,500	29,800	27,800
2. Fixed cost				
Depreciation		5,100	6,560	6,160
Taxes		750	3,500	4,000
Insurance		200	1,000	1,000
Supervision		—	1,400	1,400
Total fixed cost		6,050	12,460	12,560
3. Variable cost				
Operators-loaders	(5)* 50,000	(2)* 20,000	(1)* 12,000	(1)* 12,000
Power-fuel	—	1,200	2,300	800
Lubrication	—	20	150	150
Maintenance	—	—	1,500	2,000
Total variable cost	50,000	21,220	15,950	14,950
4. Indirect cost				
Space occupied	—	1,500	—	—
Effect on taxes	—	1,000	—	—
Changes in prod. rate	—	—	—	—
Downtime	—	200	—	—
Total indirect	—	2,700	—	—
Total operating cost (2 + 3 + 4)	50,000	29,970	28,410	27,510
Annual cost savings		20,030	21,590	22,490

*Number of units employed.

2. Power and fuel costs
3. Lubricants
4. Maintenance labor supplies

- **Indirect Costs.** Affected in other areas of company operation as a result of changing a method, adding new equipment, or changing the materials-handling system, and may consist of:
  1. Space occupied
  2. Effect on taxes
  3. Values of repair parts
  4. Changes in production rate and quality of product
  5. Downtime

  A summary of such an analysis is contained in Table 1-1.

JUSTIFICATION OF MATERIALS-HANDLING PROJECTS

The most common methods of determining the profitability of materials-handling investment are payoff period, return on investment (ROI), and discounted cash flow.

The *payoff-period* method indicates the amount of time that new equipment or a system will take to produce the savings to recover the capital investment. The investment is divided by the annual savings to give the time (in years) needed to break even. In Table 1-1:

$$\text{Payoff period} = \frac{\text{total capital investment}}{\text{annual cost savings}}$$

This method is a good risk indicator and measure which can be useful to indicate the projects that would be worth considering for closer study, but the actual profitability of new equipment depends on how much useful life is left after the payoff period. Some caution is therefore advised if the payoff period is to be the sole determinant for equipment justification, because cheaper equipment having a low useful life will always appear to be the best investment opportunity.

Simple ROI is another gross indicator that can be used to set priorities for capital investments. Here again, the effect of useful equipment life is not considered, so this method should not be used for determining the profitability of the proposed equipment:

$$\text{ROI} = \frac{\text{annual cost savings}}{\text{total capital investment}}$$

The ROI method that considers the effect of useful equipment life is

$$\text{ROI} = \frac{\text{annual savings} - \text{capital investment/useful equipment life}}{\text{capital investment}}$$

The discounted-cash-flow method of determining ROI indicates in a more realistic manner the equipment cost and return on investment by considering:

1. Savings and cost over equipment life period.
2. Net cash flow of the savings and depreciation.
3. Present worth of each year's cash flow. A factor is used to reduce the cash flow for each year to the amount of cash that would be required today to earn a desired rate of interest.
4. The effect of taxes on the rate of return.

The ROI is calculated as follows (Table 1-2):

1. Determine the cost savings for each year of the equipment useful life.

TABLE 1-2 Example of Calculating ROI* by Cash-Flow Analysis

Factors	Amounts
Total capital investment from annual operating cost	\$30,800
Equipment life	5 years
Depreciation straight line	5 years, 20%/year
Savings per year	\$22,490

Cash flow						Trial 1 @ 45%		Trial 2 @ 50%	
Yr	Cost savings	Taxes	Savings after taxes	Depreciation	Net cash flow	Factor	Present worth	Factor	Present worth
1	22,490	11,245	11,245	6,160	17,405	0.690	12,009.45	0.667	11,609.14
2	22,490	11,245	11,245	6,160	17,405	0.476	8,284.78	0.444	7,727.82
3	22,490	11,245	11,245	6,160	17,405	0.328	5,708.84	0.296	5,151.88
4	22,490	11,245	11,245	6,160	17,405	0.226	3,933.53	0.198	3,446.19
5	22,490	11,245	11,245	6,160	17,405	0.156	2,715.18	0.132	2,297.46
					78,715		32,651.78		30,232.49

*Interpolating

$$\frac{32{,}651.78 - 30{,}800}{32{,}651.78 - 30{,}232.49} \times 5 = \frac{1{,}851.78}{2{,}419.29} \times 5 = 3.83$$

Add 3.83% + 45% = 48.83% Return on Investment

TABLE 1-3 Present-Worth Values

	Interest, percent											
Years	6	8	10	12	15	20	25	30	35	40	45	50
1	0.943	0.926	0.909	0.893	0.870	0.833	0.800	0.769	0.741	0.714	0.690	0.667
2	0.890	0.857	0.826	0.797	0.756	0.694	0.640	0.592	0.549	0.510	0.476	0.444
3	0.840	0.794	0.751	0.712	0.658	0.579	0.512	0.455	0.406	0.364	0.328	0.296
4	0.792	0.735	0.683	0.636	0.572	0.482	0.410	0.350	0.301	0.260	0.226	0.198
5	0.747	0.681	0.621	0.568	0.497	0.402	0.328	0.269	0.223	0.186	0.156	0.132
6	0.705	0.630	0.564	0.507	0.432	0.335	0.262	0.207	0.165	0.133	0.108	0.088
7	0.665	0.583	0.513	0.452	0.376	0.279	0.210	0.159	0.122	0.095	0.074	0.058
8	0.627	0.540	0.466	0.404	0.327	0.323	0.168	0.123	0.091	0.068	0.051	0.039
9	0.592	0.500	0.424	0.361	0.284	0.194	0.134	0.094	0.067	0.048	0.035	0.026
10	0.558	0.463	0.386	0.322	0.247	0.162	0.107	0.072	0.050	0.035	0.024	0.017
11	0.527	0.429	0.350	0.288	0.215	0.135	0.086	0.056	0.037	0.025	0.017	0.012
12	0.497	0.397	0.319	0.257	0.187	0.112	0.069	0.043	0.027	0.018	0.012	0.008
13	0.469	0.368	0.290	0.229	0.162	0.094	0.055	0.033	0.020	0.013	0.008	0.005
14	0.442	0.340	0.263	0.205	0.141	0.078	0.044	0.025	0.015	0.009	0.006	0.003
15	0.417	0.315	0.239	0.183	0.123	0.065	0.035	0.020	0.011	0.006	0.004	0.002
16	0.394	0.292	0.218	0.163	0.107	0.054	0.028	0.015	0.008	0.005	0.003	0.002
17	0.371	0.270	0.198	0.146	0.093	0.045	0.022	0.012	0.006	0.003	0.002	0.001
18	0.350	0.250	0.180	0.130	0.081	0.038	0.018	0.009	0.004	0.002	0.001	0.001
19	0.330	0.232	0.164	0.116	0.070	0.031	0.014	0.007	0.003	0.002	0.001	0.000
20	0.312	0.214	0.149	0.104	0.061	0.026	0.012	0.005	0.002	0.001	0.001	0.000
	1.030	1.039	1.049	1.059	1.073	1.097	1.120	1.143	1.166	1.189	1.211	1.233

2. Deduct the estimated percentage for taxes for each year of equipment life.
3. Add depreciation for each year of the depreciation period.
4. Determine net cash flow, which is the algebraic total of items 1, 2, and 3 above.
5. Consult present-worth value table (Table 1-3) and select an interest value for the first trial.
6. Multiply the net cash flow by the factor selected in step 5.
7. If the present-worth cash flow is higher than the capital investment, select the present-worth factor for the higher percentage; if lower, select present-worth factor for the lower percentage.
8. Continue the discounted-cash-flow trials until the total discounted cash flow equals the capital investment cost. Interpolation will generally be necessary to determine the exact percent of ROI.
9. The trial present-worth calculations that equal the net cash flow total are those that are used to determine the ROI.

REFERENCES AND BIBLIOGRAPHY

1. Sims, E. Ralph, Jr.: *Planning and Managing Material Flow,* Industrial Education Institute, Boston, 1968.
2. Apple, James M.: *Material Handling Systems Design,* Ronald, New York, 1971.
3. Muther, Richard, and Knut Haganas: *Systematic Handling Analysis,* Management & Industrial Research Publications, 1969.
4. Muther, Richard: *Systematic Layout Planning,* Industrial Education Institute, Boston, 1961.
5. Tompkins, J. A.: "Computer-Aided Plant Layout," *Modern Material Handling,* 7 part series, May–September 1978.
6. Merkle, W.: "Dock Planning Guide," *Material Handling Engineering,* August 1980.
7. Bolz, Harold A., et al (ed.): *Materials Handling Handbook,* Wiley-Interscience, New York, 1958.

chapter 2-2

Containerization

INTRODUCTION

One of the basic principles of materials handling is that materials should be converted wherever possible to unit loads to avoid manual handling. A unit load is defined as a standard container package containing one or more items that can be handled in a standard way. The *unit-load principle* suggests that the larger the load to be handled or moved, the lower the overall handling cost. To meet this objective, materials-handling systems must be designed to handle the materials-handling volume within the constraints imposed by load size as well as the material properties involved in the production or process cycle. The decisions regarding size, shape, and configuration of the unit load should also take into account compatibility.

Some *guidelines for the specification of unit load sizes* leading to the design of containerization methods and hardware to transport and store materials include:

1. Use the same pallet or container throughout the system, or at least standardize on a limited number of containers wherever possible.
2. Plan to use raw material or parts directly out of the original container.
3. Use stackable containers to permit stacking without racks.
4. Consider collapsible containers to save space and freight costs, if they are to be used also as returnable shipping containers.
5. Use nesting.
6. Be sure that the size selected fits efficiently into standard trailers and/or railcars if containers are to be used for shipping.
7. Design or select containers suitable for mechanical handling.
8. Plan containers to accommodate a wide range of products and parts.

9. Design containers to fit into building geometry.
10. Design containers that do not require special orientation to accomplish movement.
11. Use the lightest-weight material possible.
12. Consider the use of expendable materials
13. Use containers through which contents can be identified.
14. Keep the design simple and inexpensive.

CONTAINERIZATION HARDWARE

Containerization hardware can generally be grouped into five main categories:

- Pallets
- Containers
- Tote boxes and bins
- Dunnage
- Outer securement

Standard Pallets

Pallets are used mainly as supports, carrier surfaces, or storing structures for unit loads.

The most commonly used material is *wood*, and pallets are available in a number of different hardwood and softwood varieties (Table 2-1). The type of wood, like any other material that is specified, should depend on load capacity, load requirements, durability, and the handling and storage environment. In general, softwood pallets are lighter and suitable for shipping pallets, while hardwood pallets are stronger, have a longer life, and are less susceptible to the wear and tear associated with interplant movement. Local, indigenous woods should be specified wherever possible to minimize costs.

Principles of Pallet Construction

Nomenclature, Design, Style, and Size. The principal pallet parts and the most commonly used construction features are indicated in Fig. 2-1. By convention, the length of the pallet is the first-stated dimension, the dimensions are always stated in inches, and the width is the dimension that is parallel to the top of the deck boards.

Types. Wood pallets fall into three general groups:

1. **Expendable** (one-way pallets) Cost is the major factor and the design and construction must meet the requirements for this purpose.
2. **Special-Purpose** Design and construction must meet the special requirements for the product or material to be moved or stored.
3. **General-Purpose** Uses standard design and features which enable the pallet to be used in a wide range of applications and also to be replaced and exchanged easily.

Pallet Configuration. This is specified by a combination of design, style, and construction features. The National Wooden Pallet and Container Association has established the descriptions of each parameter.

Typical pallet configurations are shown in Fig. 2-2. Pallets are available in a wide range of sizes; however, the most popular size is the 48 × 40 in pallet which accounts for over 27 percent of all pallets produced.

There is movement within some industries, particularly the food and grocery industries, to standardize pallet sizes. It has been determined that size standardization, among other obvious benefits, could also increase the use of pallet pools or exchange programs, which would have cost advantages throughout the distribution cycle.

TABLE 2-1 Strength Properties of Commercial Woods Employed for Pallets (Figures shown are for 12 percent moisture content.)

Species	Static bending fiber stress at proportional limit, lb/in²*	Compression perpendicular to grain, lb/in²*	General properties
Group IV			
Oak, red	8,400	1,260	Heaviest hardwood species; greatest nail-holding power and beam strength; best shock-resisting capacity; greatest tendency to split at nails; difficult to dry
Oak, white	7,900	1,410	
Maple, sugar	9,500	1,810	
Beech	8,700	1,250	
Birch	10,100	1,250	
Hickory, true	10,900	2,310	
Ash, white	8,900	1,510	
Pecan	9,100	2,040	
Group III			
Ash, black	7,200	940	More inclined to split when nailed; greater nail-holding and shock-resisting power, beam strength, and easier to dry than group IV
Gum, black	7,300	1,150	
Maple, silver	6,200	910	
Gum, red	8,100	860	
Sycamore	6,400	860	
Tupelo	7,200	1,070	
Elm, white	7,600	850	
Group II			
Douglas fir	7,400	950	Relatively free from splitting when nailed; moderate nail-holding power and shock-resisting capacity; lightweight, easy to work, holds shape well, and easy to dry
Hemlock (W)	6,800	080	
Larch (Tamarack)	8,000	990	
N.C. pine	7,700	1,000	
Southern yellow pine (longleaf)	9,300	1,190	
Group I			
Aspen	5,600	460	
Cottonwood	5,700	470	
Redwood	6,900	860	
Spruce	6,700	710	
Sugar pine	5,700	590	
Ponderosa pine	6,300	740	
White fir	6,300	610	
White pine (N)	6,300	550	
White pine (W)	6,200	540	
Yellow poplar	6,100	580	

*Multiply by 6900 for newtons per square meter.

Design. The most common designs of wood pallets are:

1. **Two-Way Pallets** Permit the entry of forklift or hand pallet trucks from two sides only and in opposite directions
2. **Four-way pallets** Permit entry on all four sides
   a. **Notched Stringer Design** Has four-way entry *only* with forklift trucks, and two-way entry with hand pallet trucks
   b. **Block Design** Has four-way entry with both forklift and hand pallet trucks.

Style. There are two styles of wood pallets, and they are (Fig. 2-2):

1. **Single-Face Pallet** Has only one deck as the top surface;
2. **Double-Face Pallet** Has both top and bottom decks and comes in two different designs, viz.:
   a. **Reversible** Has identical top and bottom decks, and goods may be stacked on either deck

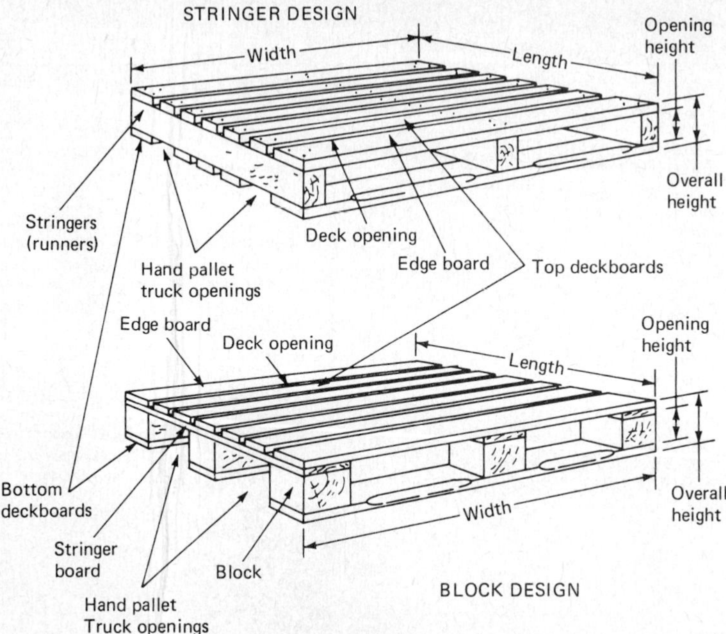

Figure 2-1 Principal parts of wooden pallets. *(National Wooden Pallet and Container Association.)*

b. Nonreversible Top and bottom decks have different configurations, and substitute goods may be stacked only on the top deck

Constructions. Wood pallet constructions are as follows:

1. **Flush Stringer** A pallet in which the outside stringers or blocks are flush with the ends of the deckboards
2. **Single Wing** A pallet in which the outside stringers are set inboard of the top deck, while the stringers are flush with the ends of the bottom deckboards
3. **Double Wing.** A pallet in which the outside stringers are set inboard of both top and bottom deckboards to accommodate bar slings or other devices for handling pallets

Maintenance and Repair

Procedures should also be established within the system to identify worn pallets that require repair or disposal. To accomplish this effectively, the acquisition date should be marked on the pallet and older pallets should be inspected periodically to detect wear. The following are guidelines for repair operations:

1. Never repair a pallet a second time.
2. Never repair more than three deckboards or one stringer on a given pallet. If the *average* replacement is more than 1½ boards per pallet, repair is uneconomical.
3. Productivity should average 100 repaired pallets per worker per 8-h shift for those on the repair line—forklift support and supervisory personnel excluded.
4. Cost of repair should not exceed half the price of a new, similar pallet.

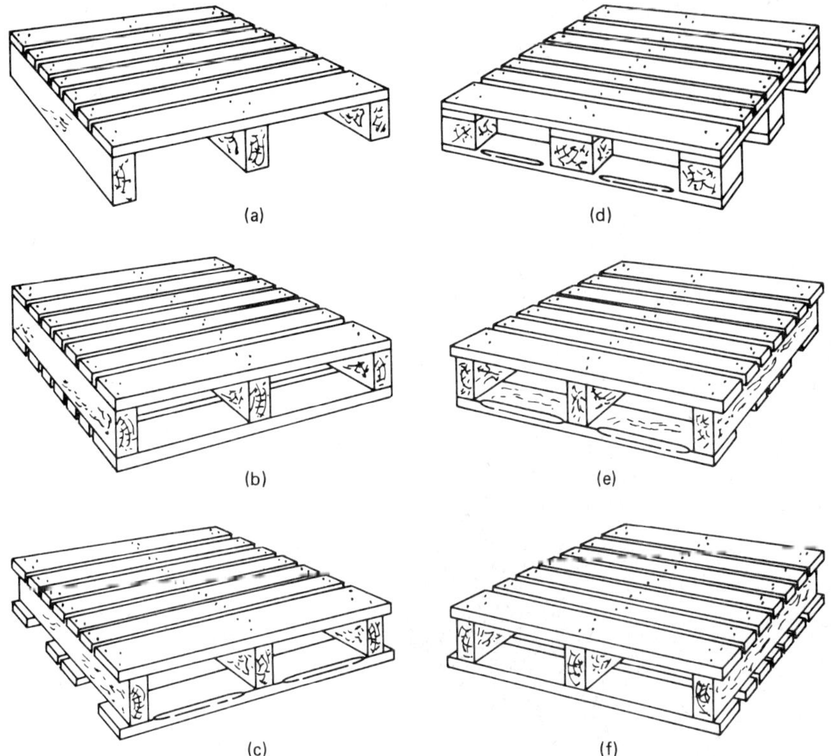

Figure 2-2 Typical pallet configurations. (*a*) Single-face; (*b*) double-face, reversible; (*c*) double-wing, double-face, nonreversible; (*d*) double-face, nonreversible; (*e*) single-wing, double-face, nonreversible; (*f*) double-wing, double-face, reversible. *(National Wooden Pallet Association.)*

Pallets for Use with Forklifts

Expendable Wood Pallets. These pallets are used to support a unit load for one-way and one-time use. Pallets of this type must be specified with the capacity to carry unit load but do not require the durability of reusable types. The single-face style (Fig. 2-2) is primarily used for this purpose. Plywood deck surfaces are frequently used for expendable pallets.

Metal Pallets. These pallets can be made of corrugated steel, expanded metal, steel wire, aluminum, and combinations of metal and wood. Metal pallets are more expensive than wood pallets and are used mainly for movement of materials inside the plant where additional strength and life is required.

Corrugated-Metal Pallet Bases. These pallets (Fig. 2-3) are often integrated into the design of corrugated-steel containers with a number of other features. This permits wide versatility in parts handling and storage in the plant. The style variations available are similar to those of their wood pallet counterparts to permit movement in both two- and four-way entry bases by forklift trucks and pallet hand trucks.

All-Steel, Single-Face Pallets. Supported on three runners, this type is designed to handle heavy loads and containers. Recessed side channels bound in flanges can be incorporated in the design to permit safe movement by hand. Their double-faced, reversible design eliminates sharp edges, and thus prevents damage to bagged materials.

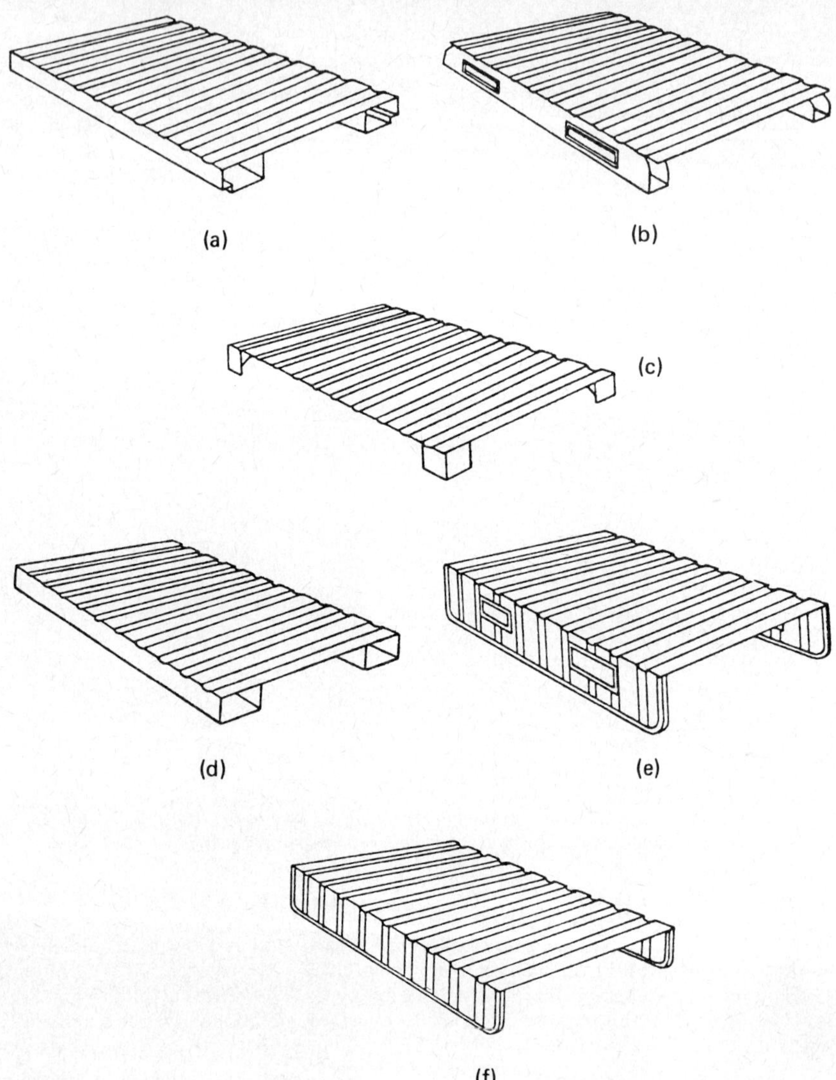

Figure 2-3 Corrugated-steel pallet container bases. (*a*) Two-way entry pallet style, (*b*) four-way entry pallet style, (*c*) box or angle-style pallet, (*d*) two-way box runner, (*e*) skid or pallet truck, (*f*) standard pallet with runners.

One-Piece, Formed-Metal Pallets. These pallets have a built-in nesting feature that permits a number of empty pallets to be stored conveniently. They are useful where pallet storage space is scarce.

Wire-Mesh Pallets. These pallets use galvanized or painted steel or aluminum deck sections with formed, corrugated support structures and are used where durability and light weight are required. The wire-mesh pallet, like the corrugated-metal pallets, are often incorporated as bases in wire-mesh containers.

Cardboard Pallets. These pallets (Fig. 2-4) are useful for light unit loads that are less than 1500 lb (700 kg) and for stacked loads that are less than 1000 lb (450 kg) per pallet leg. Because of their low cost, they are ideal as expendable pallets and can be expanded on a modular basis to become shipping containers.

Plastic Pallets. These pallets are more expensive than wood ones and in some cases are more expensive than metal ones. The main uses of plastic pallets are in the food or pharmaceutical industries where a high standard of cleanliness is required.

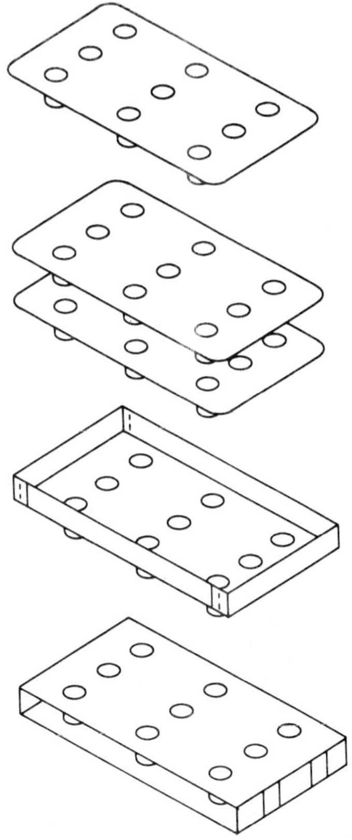

Figure 2-4 Cardboard expendable pallet. *(Menosha Corporation.)*

Other Types of Pallets

In addition to using forklift trucks or hand-operated equipment for major movements in the plant, there are other handling requirements for movement of materials within and between manufacturing operations. Generally there is very little standardization in this area, since the carrier surfaces have to be specified to be compatible with equipment or product.

Slip-Sheet Systems. Such systems enable a unit load to be handled and moved without being supported on a pallet type of platform. Slip sheets (Fig. 2-5) are made of heavy corrugated paperboard, plastic, or kraft fiber composition and function as the base surface for the unit load. Special equipment or *push-pull* forklift truck attachments are required to move and handle loads unitized by this method. The cost benefits of slip-sheet systems are obvious: in addition to lower initial cost, storage space requirements are 1/100 of the cube required for empty pallets and shipping costs are less than for comparable loads using wood pallets.

Slave Pallets. These are used for assembling unit loads before transfer to other containers, for moving odd-shaped loads on conveyors, for serving as accumulating and transfer platforms for automated computer-controlled storage and retrieval systems, and for supporting unit-load containers that are not designed for use on conveyors.

Slave pallets are normally plywood-sheet surfaces, but if interim storage is required in racks where the pallet-edge surfaces are supported by shelf angles, pallet specifications become more critical from the standpoint of supporting loads and safety. As a general rule, to achieve maximum strength and stiffness, face grain should be across supports. Design criteria include pallet size, total uniform load, permissible deflection, clear span, and uniform load in pounds per square foot (newtons per square meter). Selection of the proper grade and thickness of plywood can then be determined. Information regarding recommended maximum uniform loads and deflections is available from the American Plywood Association.

Air Pallets. This type uses a bed of air to support a unit load and enables large loads to be moved and maneuvered. Air pallets or air-film equipment can be used to convey

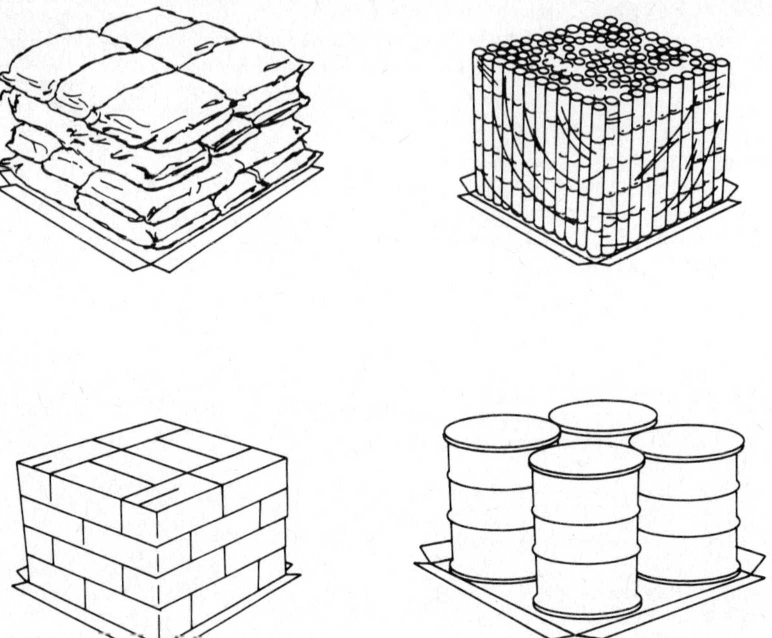

Figure 2-5 Typical uses of slip-sheet system. *(Little Giant Products.)*

parts, rotate work stock, and move palletized loads in and out of buildings as well as trucks, railcars, and other conveyances.

Portable Stacking Racks

These racks are used for storage of palletized loads that cannot be stacked on each other. Pallet stacking frames (Fig. 2-6) are used to confine and protect irregularly shaped, fragile, or nonuniform loads during in-process or temporary storage. The pallet itself is the base unit and rests on the frame of the pallet beneath it. The second kind of portable stacking rack (Fig. 2-7) consists of a base unit and removable post and end frames. Pallets are stored on the base units and the base units, when stacked, nest in the end frame.

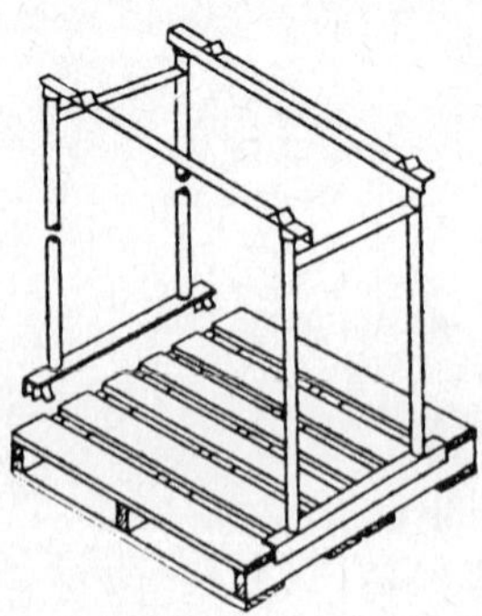

Figure 2-6 Pallet stacking frame using pallet as base unit.

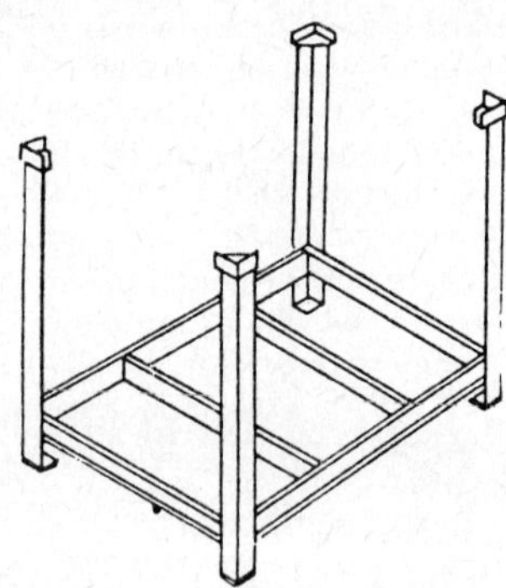

Figure 2-7 Pallet stacking frame with base unit.

Pallet Loading Patterns

A pallet pattern is an arrangement of units on a pallet and ideally is the most effective way of loading a pallet with the least loss of cube. There are a number of ways to select the optimum pattern, ranging from trial and error to the use of computer models. No matter what technique is used, the following factors must be taken into consideration:

1. **Size of Material** There may be several ways, one way, or no way to place a given size material onto a given pallet.
2. **Weight of Material** In the case of very heavy material, fewer layers will be stacked on a pallet. To a certain extent the number of layers will depend on the strength of the containers, if any are used.
3. **Size of Unit Load** Taken as a whole, the length, width, and especially the height of the load must be considered.
4. **Loss of Space within Unit Load** Some patterns have too many large gaps between units. This kind of piling is particularly bad when paper pallets are used because the weight should be distributed evenly and the units should brace each other.
5. **Compactness** Some patterns do not tie together well; they will not interlock.
6. **Methods of Binding Products in Patterns** If the units of a load are glued together, one kind of pattern may be ideal; with strapping, another type may be the best; and if no fastening at all is used, some combination of stacking may be the most suitable method to interlock and hold the load together.

Some general rules that should be followed in establishing pallet patterns are:

1. Interlocking unit loads should be used when possible to make the most effective use of the cube and to provide load stability.
2. Overhang, where unit loads extend beyond the edge of a pallet, should be avoided or minimized to a point where container damage or load stability is not affected. The added dimensions caused by this condition should not exceed the width or length of the shipping conveyance or reduce the utilization of the conveyance.
3. Underhang, where unit loads do not fill the deck surface and where there are large voids, should be avoided.
4. Utilize the basic pallet patterns (Fig. 2-8) effectively. Use block patterns for containers of equal width and length. This type of pattern is the least stable and may require bonding and fastening if considerable movement is involved.
5. Brick, row, and pinwheel patterns are used for containers of unequal length or width. All three patterns result in the interlocking that stabilizes a load.

Pallet patterns have been developed empirically for materials that have rectangular dimensions. The U.S. Navy Research and Development Facility has developed such a pattern.[1]

Container capacities range from 500 to 6000 lb (230 to 2700 kg), and standard base sizes cover the range of sizes of wood pallets, including 40 × 48 in (1.4 × 1.7 m). The 44 × 54 × 40 in (1.6 × 2 × 1.4 m) size is ideal for use when making shipments by rail or truck because the container dimensions are exact multiples of trailer and railcar dimensions and therefore can use the cube of these conveyances fully.

Metal Containers

Three types of metal containers are in general use: wire mesh, noncorrugated steel, and corrugated steel. Current development trends tend toward increasing versatility by incorporating features such as stacking and dumping capability and pallet-type bases to allow and facilitate movement.

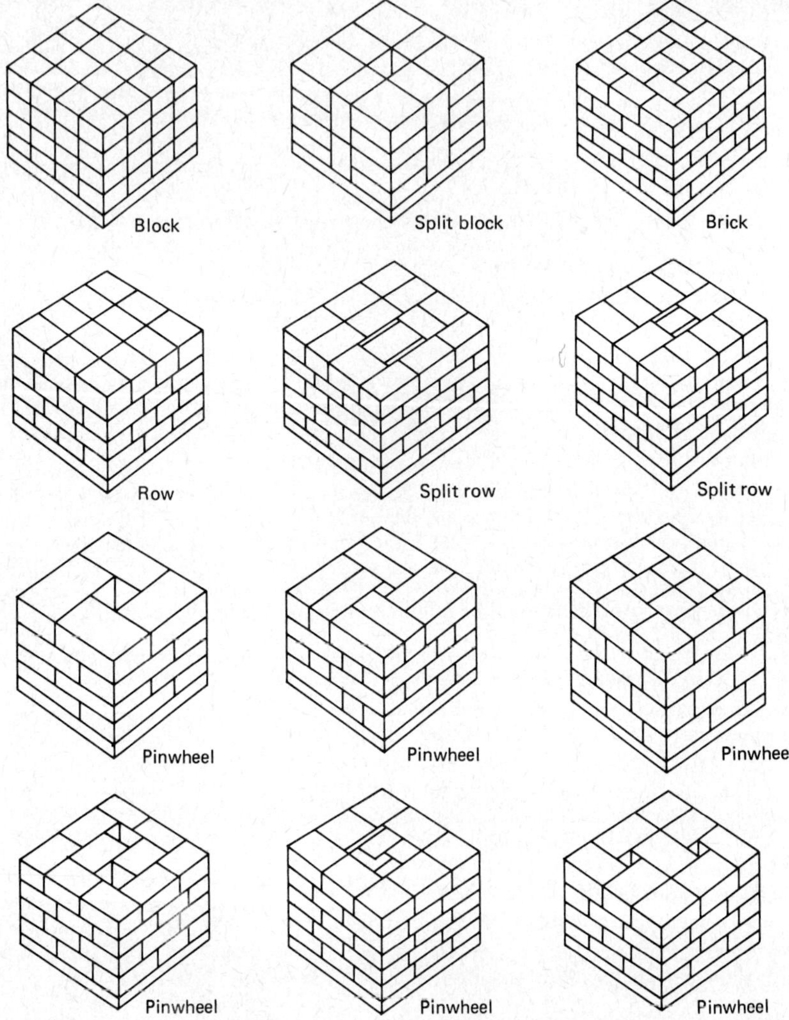

Figure 2-8 Typical pallet patterns. Paper or fiberboard binders are used between layers if necessary.

Welded-Wire-Mesh Containers

These containers (Fig. 2-9) are fabricated from welded wire for containment of materials. Additional structural sections are added for additional strength, and optional features can be included for specific uses and applications. The kinds of material and product that can be handled is limited only by size that would fall through the wire mesh and total container volume. Mesh openings from ½ × ½ in (1.3 × 1.3 cm) to 4 × 4 in (10 × 10 cm) are available to accommodate a wide range of product or material sizes.

Advantages that are generally cited in using this type of container are:

- Lightweight as compared with other metal containers
- Allows visibility of product for easy and quick identification

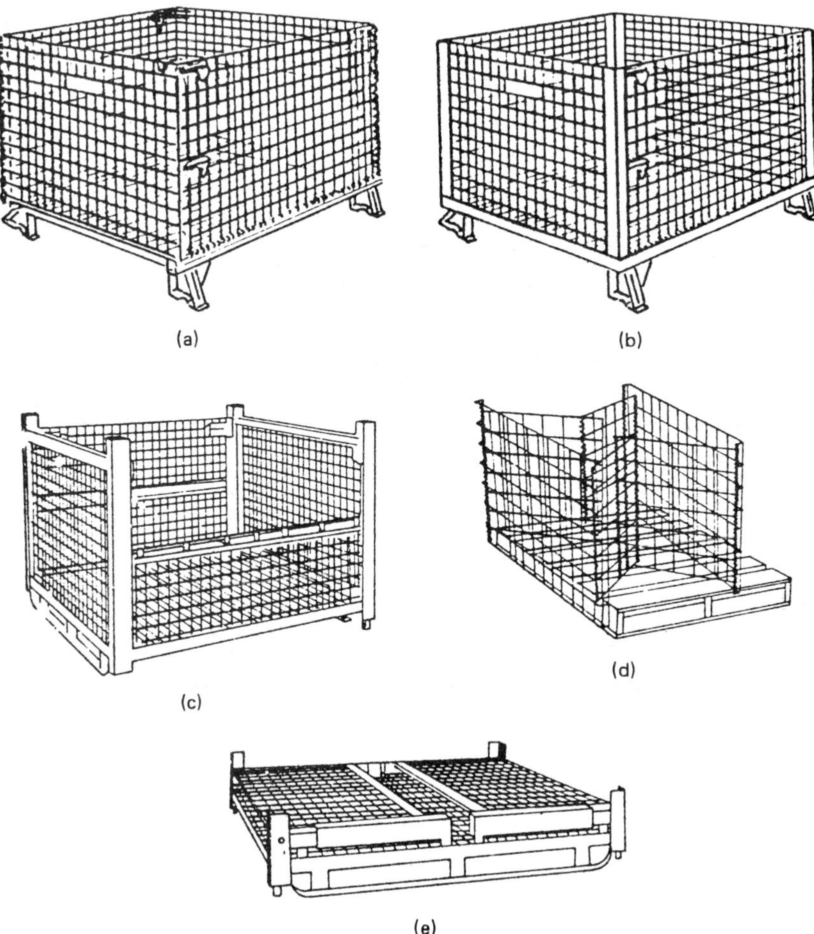

Figure 2-9 Kinds of welded-wire-mesh containers. (*a*) Collapsible wire container, (*b*) rigid wire container, (*c*) heavy-duty rigid, (*d*) wrap-around, (*e*) folded-down collapsible rigid.

- Self-cleaning by shedding debris
- Material can sometimes be processed in the container in such processes as degreasing, cleaning, and air drying.

Wire-mesh container stacking is accomplished in two ways: (1) interlocking through the corner posts and (2) the arch (saddle) of the upper container rests on the rim of the container below.

Wrap-Around Collapsible Frame. This forms the most basic type of wire-mesh container by using a standard pallet as a base.

Collapsible Container. This type of container is constructed to permit the sides to fold down when not in use. The obvious advantage of this design, saving space, must be weighed against its shorter life and smaller payload capacity.

Rigid Wire Container. This type is designed with additional vertical structural members that increase the strength of the containers, resulting in longer life and greater capacity than the collapsible types can provide. A heavy-duty rigid container has been developed to handle heavier loads in more strenuous environments. Additional horizontal bracing has been added to achieve the capability.

Collapsible Rigid Container. This is a heavy-duty version of the collapsible container. It is nearly equal in strength to the heavy-duty rigid type, but the higher original cost must be compared with savings on return freight costs.

Corrugated-Steel Containers

These are probably the strongest type of metal container available, since corrugation permits a longer surface of material to be used for a given size of a container than any other method of construction, and it is fabricated from hot-welded steels 0.105 to 0.179 in (2.7 to 4.6 mm) thick. This type of container has progressed from the one-type-only gondola bin to configurations that are almost virtually materials handling systems in themselves. They can be considered as consisting of three basic *modules:* base, lifting and stacking aids, and container box, with options for specific applications.

Container Bases

Container bases are used to facilitate surface movement and can be essentially categorized as metal pallets since there are provisions for two- and four-way entry with forklift trucks and other materials handling equipment. The general features of these bases were covered in the pallet section of this chapter.

Container Sections

Container sections can include a number of options (Fig. 2-10) to provide hopper dumping or accessibility, drop-bottom dumping, and end gates. Air- or hydraulic-actuated equipment and tilt stands are used in conjunction with, and as means to extend the versatility of, containers in assembly or machining stations. Container sections come in a wide range of heights; the deeper containers are used for dumping applications, and the shallower sizes are more suitable for manual handling of parts.

Several lifting and stacking aids can be designed or are available as part of the container system. Overhead movement can be accomplished by the addition of overhead crane lugs, chain-sling lugs, and bar-lift or sling-lift notches in the base. Stacking is accommodated by corner stacking angles, self-aligning hairpin brackets, and stacking lugs.

Industry Specifications

Specifications issued by the Industrial Metal Containers Product section of the Material Handling Institute specify the materials, design, construction, testing procedures, environmental considerations, utilization, and safety practices regarding the use of metal containers. The safety guidelines do not recommend the use of wire-mesh containers for applications where parts and material protrude or for dumping purposes. A load and capacity rating plate is used to indicate the load capacity of each container and the number of containers that can be safely stacked on top of each other.

Wood Containers

Wood containers are of three basic types.

Types

1. Bins with bases and closed sides and ends
2. Boxes with bases, closed sides and ends, and a top
3. Crates with open or slatted sides and ends

Wood containers with standard pallet bases are finding increasing use in applications where mechanized handling and storage are required for products ranging from solid

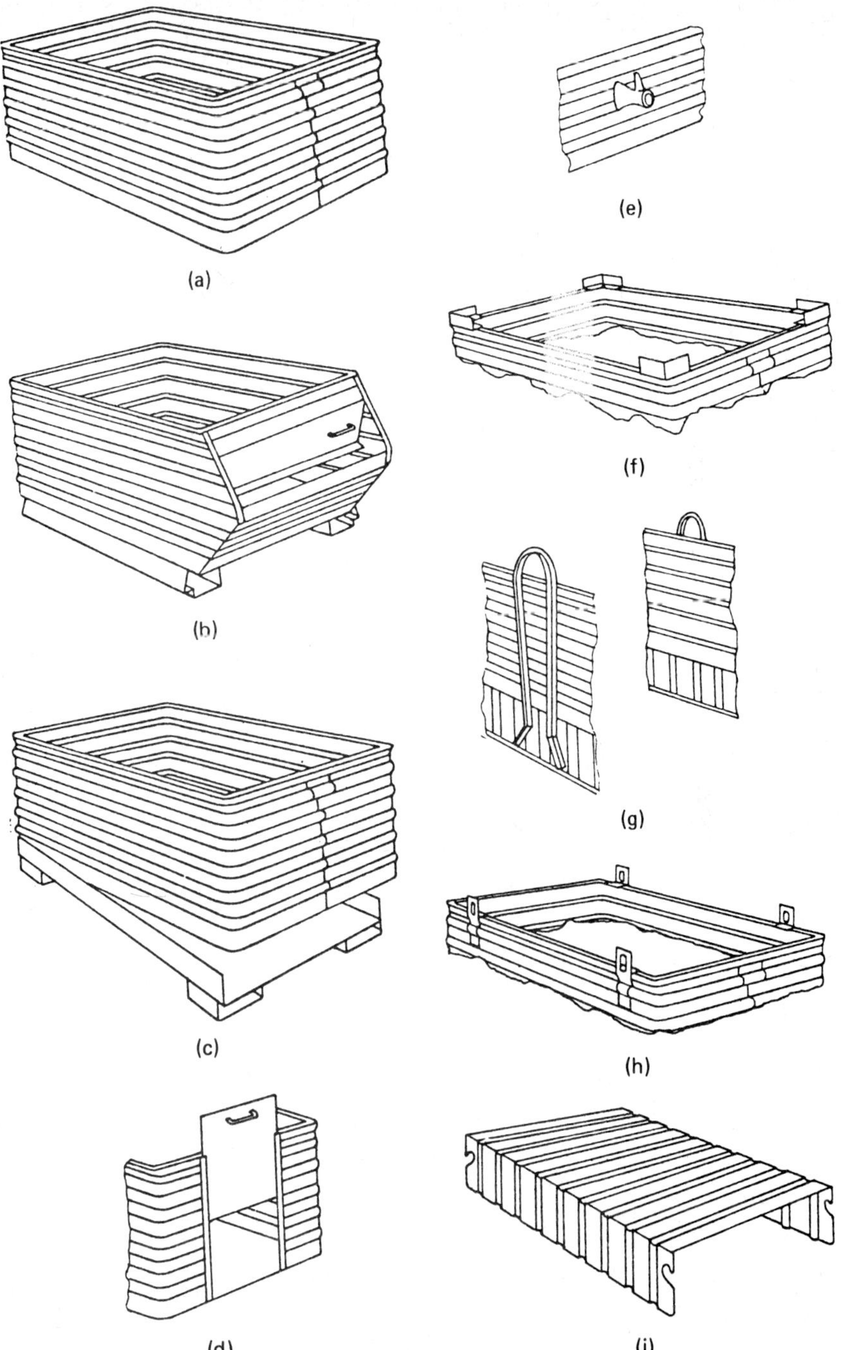

Figure 2-10 Corrugated-metal-container styles and lifting and stacking aids. (*a*) Basic corrugated box, (*b*) hopper-front pallet base container, (*c*) drop-bottom container, (*d*) end gate option, (*e*) overhead crane lugs, (*f*) corner-stacking lugs, (*g*) chain sling and stacking lugs, (*h*) bar lift or sling sling lift notches.

materials of irregular shapes and sizes to granular materials. Pallet containers are now being used in the agricultural area, where fruits and vegetables are loaded into pallet containers when picked and then washed, transported, and placed in supermarkets in the same container.

The pallet container design may include collapsible sides, a feature which saves space when the container is not in use or when it is being returned empty.

Construction

The four major types of end-panel and side-panel construction in wood containers are:

1. **Solid** Usually of plywood, this type of construction provides great strength as well as a smooth interior surface.
2. **Vertical-Slat** Used for deep containers and requires more bracing than the solid or horizontal-slat construction to prevent rocking.
3. **Horizontal-Slat** Used for shallower containers and requires less bracing than containers with vertical-slat construction.
4. **Wirebound** Provides the economy of lightweight construction with the added strength of being wire-bound (Fig. 2-11). The main advantages of this construction is a high-strength container with low tare weight.

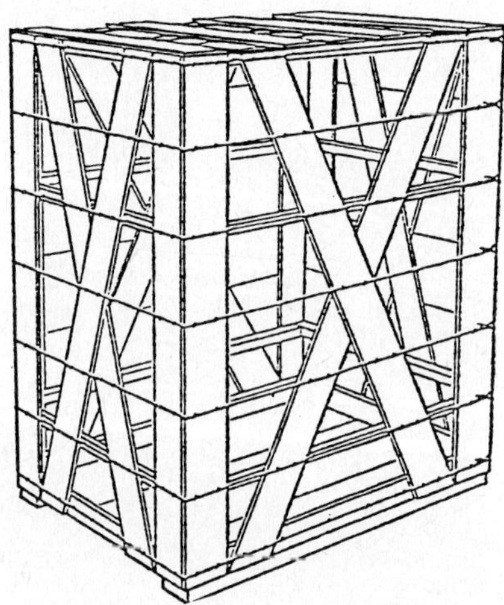

Figure 2-11 Typical wirebound container.

Selection of Wood

In the selection of the wood species to be used for wood containers, the following factors should be considered:

1. Intended usage and life requirements of the container
2. Pressure to be exerted on the bottom and sides by the goods
3. Permissible bulging of sides and edges
4. Degree of impact resistance required
5. Type of handling equipment used

6. Degree of weather and water resistance required for exposure to weather or cleaning operations

Corrugated-Cardboard Containers

Corrugated-cardboard containers offer a wide range of economical solutions for packaging or materials handling problems. There are literally hundreds of unique designs of corrugated containers. Because of the relatively low cost, it is feasible to "tailor-make" a configuration that is an optimum design for a particular product and situation. This container may be one of, or a modification of five common types.

Types

1. **Regular Slotted Container** The most commonly used style. All flaps are of equal length, and other flaps meet when closed. Contents of the box are protected by one thickness of corrugation on the side and two thicknesses on top. If additional top and bottom protection is required, both outer top and bottom flaps can be designed to overlap.
2. **Telescope Box** A two-piece box designed so one part fits into the other.
3. **Five-Panel Folder** A corrugated flat sheet, scored into five panels, which folds into a four-sided tube-type container that is closed by end flaps.
4. **One-Piece Folder or Book Fold** Used for shipping books, catalogs, wearing apparel.
5. **Gaylord** A corrugated container that has a base, generally consisting of two or more wood runners that allow the container to be moved by fork lift equipment.

Corrugated Material and Construction

The basic corrugated material is referred to as "double-face" or single-wall, consisting of outer facing, corrugated medium, and inner facing, joined by adhesive. Corrugated material is also available in single-face, double-wall, and triple-wall designs.

Types of Board. These include Fourdrinier kraft linerboard, the highest-quality and -cost material made from virgin pulpwood, and cylinder linerboard which is generally made from a combination of reclaimed fibers and virgin pulp. The weight of a linerboard required for the contents of a corrugated container is specified in the Uniform Freight Classification and Motor Truck Classification.

Corrugating Medium. Straw, reclaimed fibers, or woods can be used as a corrugating medium. The combination of corrugated-medium thickness, weight, and flute configuration determines the strength and moisture-lockout properties of the container. Corrugating mediums are specified by thickness and weight per 1000 ft^2 (MSF).

Flute Configuration. This specifies the number of corrugations per lineal foot. The three most common flutes used are:

1. **A Flute** The highest flute with the least number of corrugations (36 per linear foot, 0.1875-in high) (12 percm, 4.2 mm). When used in an upright position, A flute has the best stacking strength and greater capacity to absorb shock in the direction of the thickness.
2. **B Flute** Has the greatest number of corrugations per foot with the lowest flute height and is stiffer and less shock-absorbent than A flutes, but it has greater crush resistance to loads placed in the direction of thickness.
3. **C Flute** A compromise between A and B flutes.

Methods of Fastening Joints. The accepted methods are taping, stitching, or gluing. Common carriers publish detailed regulations governing the method to be used in regard to content type, weight, and other factors.

Corrugated boards may be specially treated to provide additional properties to the container, such as a coating and lamination, to:

- Retard slippage
- Inhibit mold
- Retain temperature
- Increase water or moisture resistance

Tote Boxes and Bins

Tote boxes are used for unit loads of smaller parts that can be moved manually through the operation or can be stacked in a larger container to become part of a unit load. Tote boxes are available mainly in metal and plastic and are also fabricated from other materials such as wood, cardboard, fiberboard, and Plexiglas.

Plastic Tote Boxes

Plastic tote boxes have many applications where small and light parts are handled and in environments where protection from corrosive chemicals or a high degree of cleanliness is required. Plastic containers are easily cleaned without harm or deterioration to the material. Molding capabilities permit desirable features to be incorporated as an

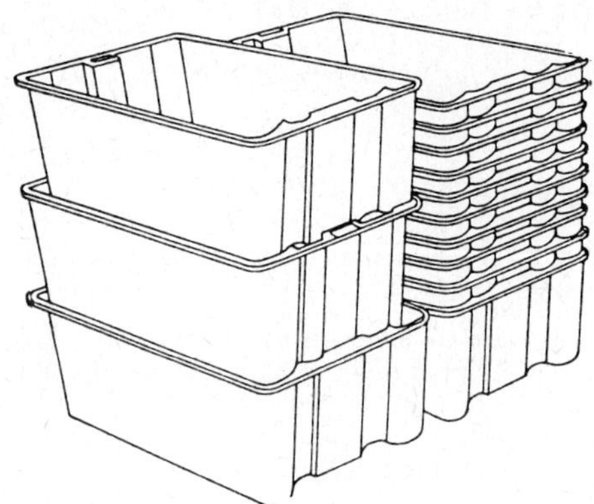

Figure 2-12 Typical tote boxes.

integral part of the container, permitting a number of provisions for nesting and stacking. A typical tote box is shown in Fig. 2-12. There are three major types:

Straight-Nesting. This refers to tote boxes that can be nested when not used. This type requires the use of lids or covers if stacking is required but results in maximum product protection, since one box will not fall into another. The design of straight-nesting boxes is characterized by tapered sides which reduce cube utilization and should not be used for storing on shelves.

Straight Stacking. This method of stacking boxes is ideal, because of minimum tapered sides, for shelf storage and use as an inner container of maximum cube utilization. Since the sides and ends support the loads, greater weights can be stacked than when using the straight-nesting variety.

Combination of Stack and Nest. This feature is available in some boxes and can be altered by the orientation of boxes in relation to each other.

The most common plastic materials and their major properties are indicated in Table 2-2.

TABLE 2-2 Plastic Materials Used in Tote Boxes

Material	General properties
ABS (acrylonitrile-butadiene-styrene)	High impact absorption, good compressive strength; more expensive than other thermoplastics
High-density polyethylene	Good to excellent stiffness, excellent temperature range, −40 to 150°F (−71 to 51°C); commonly used for food applications with USDA and FDA approval
High-impact polypropylene	More durable than polyethylene, not as stiff; tendency to crack at temperatures below 0°F (−32°C)
High-impact polystyrene	Extremely stiff, excellent compressive load strength, good temperature range; tendency to crack easily under high impact; readily attacked by solvents and oils
FRP (fiberglass-reinforced polyester)	Exceptional compressive load strength; can be heat-, fire-, and wear-resistant

DUNNAGE AND OUTER SECUREMENT OF CONTAINERS AND UNIT LOADS

Dunnage

Dunnage refers to inner package containment methods or material that is used to protect the contents of a container from damage. This is done in one of two ways, either by preventing the movement of the contents or by providing a cushioning medium to absorb shocks.

Plastic and other petroleum-base materials are used for dunnage because of their light weight and low bulk density.

Dunnage Materials

Polystyrene is used to cushion package contents in three general forms:

1. Loose foam strands are used to fill the air space in the package and provide a cushioning barrier around the contents.
2. Polystyrene can be molded to the general form of the part in the container and actually becomes an inner case for the part.
3. Polystyrene is also used for corner forms to strengthen corrugated containers.

Bubblepack is two thin sheets of polyethylene with air entrapped within the "bubble" sections when laminated together. The contents of a container may be wrapped in bubblepack for protection during shipment or movement.

Corrugated board can be easily formed into many shapes to protect, support, and cushion products. The trend is to reduce the number of inserts by combining features into a single interior form, reducing the cost and inventory expense of packaging materials. Typical inner packings, portions, and sheets used for this purpose are shown in Fig. 2-13.

Outer Securements

Container closure can be achieved by sealing flaps with glue and tape by stapling, and by strapping with plastic or steel bands.

The glue and tape sealing method lends itself to automation by use of case-sealing equipment which automatically dispenses glue or tape close to the flaps. Stapling can also be automated by passing containers through, under, and between staple heads.

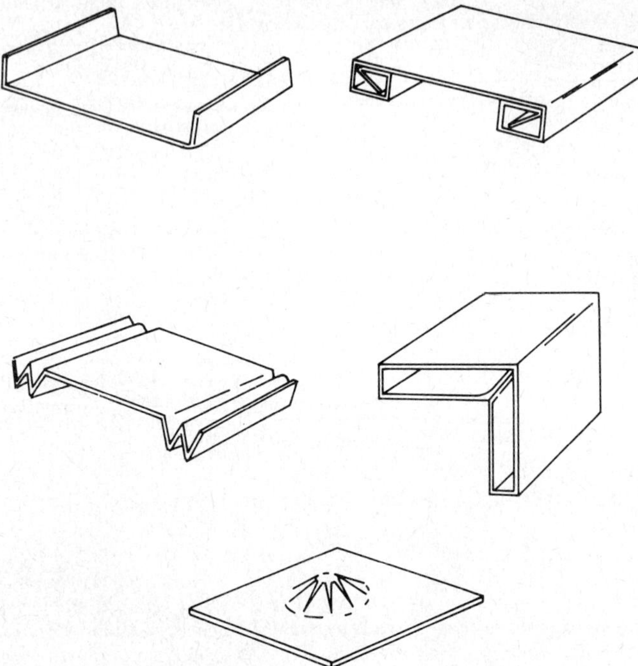

Figure 2-13 Typical inner packings, portions, and sheets.

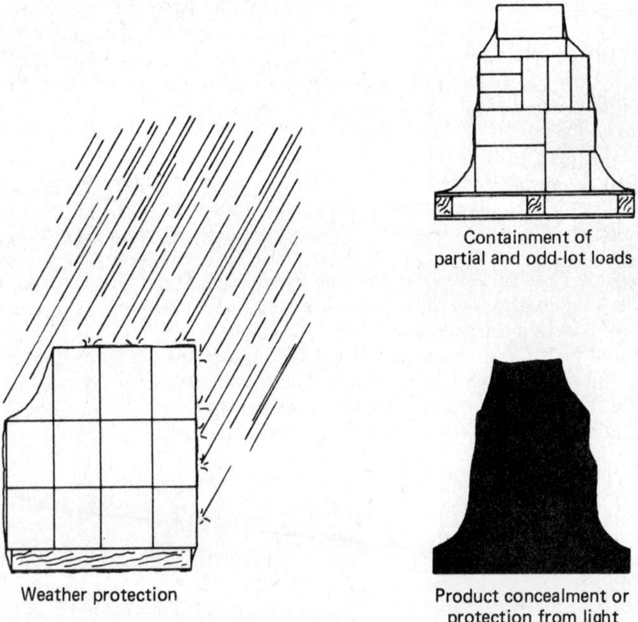

Figure 2-14 Other benefits from use of shrink wrap. *(Black Body Corp.)*

Strapping

Strapping can be used to secure not only containers but stand-alone unitized loads and palletized unit loads as well. There are two major strapping classifications: steel and plastic. Steel strapping is dimensionally stable under all but extreme conditions and plastic strapping is more resilient.

Common cold-rolled-steel strapping, the least stretchable of all steel strapping, is good for strapping lightweight packages or pallet loads that are not subject to high impact or shock. Heavy-duty steel strapping, both hot- and cold-rolled steel, absorbs high impacts without breaking; however, it stretches under great stress and requires staples to keep it in place.

Plastic strapping continues to stretch under tensioned loads and therefore should not be used where continuous loads are present. However, their resilience keeps them tight on a package or load that shrinks or settles.

Steel strapping may be applied by manually operated tensioning tools which notch or crimp steels to retain the strapping. Plastic strapping can also be applied both manually and automatically in the same manner as steel strapping. Friction-welding and heat-sealing systems are also available.

Stretch Wrap

Stretch wrap is one method of confining unitized or pallet loads by wrapping a film of polyethylene material around the load. The stretch film is wound under light tension by rotating the load on a platform on horizontal- and spiral-type stretch film equipment. Horizontal-type equipment uses a full sheet wrap, while spiral-type equipment bands the load by dispensing several overlapping layers. Vertical wrapping equipment rotates the film around the load and is generally used when the load is not on a pallet.

Shrink Wrap

Shrink wrap uses a polystyrene film that can be a bag or flat sheet that is put over or around the unit load and then placed in an oven. Bag design should be selected to achieve the desired protection required or to optimize the holding power for the load configuration. While in the oven, the film reaches a temperature of 240°F (116°C). During the cooling cycle, the film molecules try to return to their original compact orientation, shrinking the film tightly around the load.

Three types of ovens are used for this purpose, depending on the volume handled. Closet types can normally handle up to 20 loads per hour; bell-type ovens are for higher volume operations, up to 20 loads per hour; and tunnel or feedthrough ovens are generally specified for automated lines and handle more than 75 loads per hour.

In addition to securing products in a unit load, there are extra benefits in using shrink wrap (Fig. 2-14). Irregular, hard-to-pack loads (like those in order-picking warehouses) can be secured with shrink wrap. Products can be protected from damage caused by weather, dirt, or moisture. Clear film allows for easy viewing for identification and inventorying of product. Opaque film can be used to conceal or protect product from light.

REFERENCES AND BIBLIOGRAPHY

1. "Storage and Materials Handling," Departments of the Army, Navy, Air Force, and U.S. Marine Corps, Washington, D.C., 1955.
2. "Unit Load Stretch Wrapping," *Modern Materials Handling,* October 1979.
3. Schumf, G., "Slip Sheets," *Material Handling Engineering,* July 1980.

chapter 2-3

Fixed-Path Equipment

INTRODUCTION

Conveyors, cranes, and hoists are generally considered fixed-path materials-handling equipment, since they often become a fixed part of the physical plant. Once in place, a considerable amount of time, disruption, and cost is needed to change the arrangement of the equipment. It is therefore very important to plan the installation of these pieces of equipment very carefully.

A complete materials-handling system can include a wide variety of fixed-path equipment for unit loads and bulk materials handling, as well as mobile handling equipment and storage racks. This further complicates the planning process, since the fixed-path equipment not only has to satisfy the requirements of the fixed-path handling but must also be compatible with the overall flow of the total handling system.

There are many considerations involved in planning fixed-path equipment installations; many of these considerations are unique to a specific type or class of equipment, but general areas that must be addressed in the planning and exploration stage are:

- **Flexibility of the system.** Must a wide range of unit-load sizes or bulk material be handled or conveyed?
- **State of Materials To Be Handled.** Is it in a unit load or bulk form?
- **Weight, Dimensions, and Physical Properties of the Material Being Handled or Moved.** Is it fragile, light, firm, or does it have other properties that require special attention?
- **Loading and Unloading Methods.** Is it handled manually or received from or delivered to other equipment such as lift trucks, palletizers, or packaging equipment?
- **Capacity of Equipment.** Does the conveying speed match the speed or capacities of the equipment it is being interfaced with? Is there sufficient capacity or length to accumulate material when required?

- **Supporting-System Requirements.** Is the material to be sorted, accumulated, weighed, or further processed while being handled or conveyed?
- **Environmental Conditions.** Must provisions be made for dust, high or low temperature, high humidity, or other ambient conditions in the plant or outside?
- **Safety.** What special precautions must be taken to protect operating personnel or personnel working near the equipment? What provisions must be made to comply with regulatory requirements?
- **Maintenance.**
- **Facility Restrictions.** Are overhead heights or floor loading capacities adequate for supporting and accommodating equipment? Is there sufficient plant area? Will the fixed-path system impede access to equipment and the flow of personnel and other materials within the plant?
- **Horizontal or Vertical Distances To Be Covered.** What hardware is required to negotiate inclines and declines throughout the system?
- **Power and Energy Requirements.**

There are many types and varieties of fixed-path equipment. Each of the major classifications of this type of equipment are discussed and described here, but no effort will be made to list all of the items that are contained in each classification. The classifications that are covered include:

- Conveyors
- Sorting, consolidating, and diverting devices
- Hoists and cranes
- Guided vehicles
- Robots

CONVEYORS

Conveyors are gravity or power devices commonly used to move uniform loads continuously from point to point over fixed paths. The primary function of the conveyor is to move materials when the loads are uniform, and the routes do not vary. The movement rate and direction is usually fixed although the system can be designed to bypass cross traffic. The major types of conveyor and related devices are chutes and wheel and roller conveyors.

Chutes

Chutes are the simplest fixed-path devices that use gravity to convey bulk or unit loads down declines. Straight and spiral types are available. The spiral chute (Fig. 3-1) is a continuous trough over which bulk materials or discrete objects are guided in a helical path.

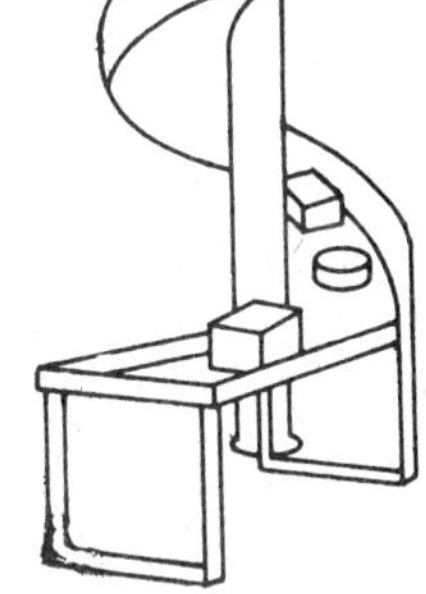

Figure 3-1 Spiral chute.

Wheel and Roller Conveyors

These depend on both gravity and power to move materials. Objects of various shapes can be handled by changing the cross section of the rolling surface or by aligning the objects in the conveyor framework. These conveyors are generally used to move materials horizontally.

Considerations for Chutes and Wheel and Roller Conveyors

The following sections discuss points that must be considered in specifying and designing both of these classes of conveyors.

Load Characteristics. These include maximum and minimum sizes of loads and the shapes and carrying surfaces of all units. The suitability of a load configuration to be handled on roller or wheel conveyor is important. Unsupported packages such as bags (Fig. 3-2) are not recommended for this type of equipment.

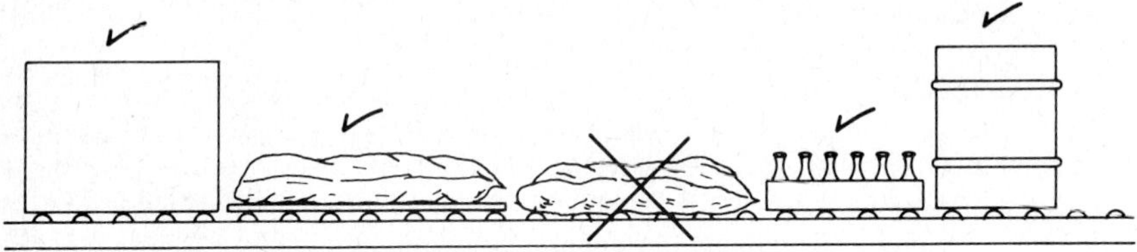

Figure 3-2 Can it be handled on a roller conveyor? *(Litton UHS.)*

Operating Conditions. These include the size and weight of the conveying surfaces, environmental conditions, and loading and unloading methods. These considerations determine the type and capacity of frame, roller, or wheel material and sizes, and the type of bearings that should be used.

Roller or Wheel Spacing and Pattern. This is determined by the size of the minimum package or unit load. (See Fig. 3-3.) To determine roller centers, divide the minimum load length by three. Wheel pattern should be specified to provide a minimum of five wheels under the package. Other guidelines include:

1. A minimum of three rollers under a hard bottom surface
2. A minimum of four rollers under a flexible bottom surface

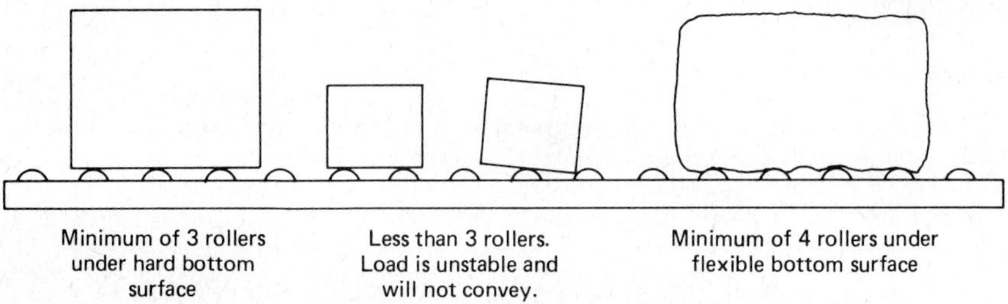

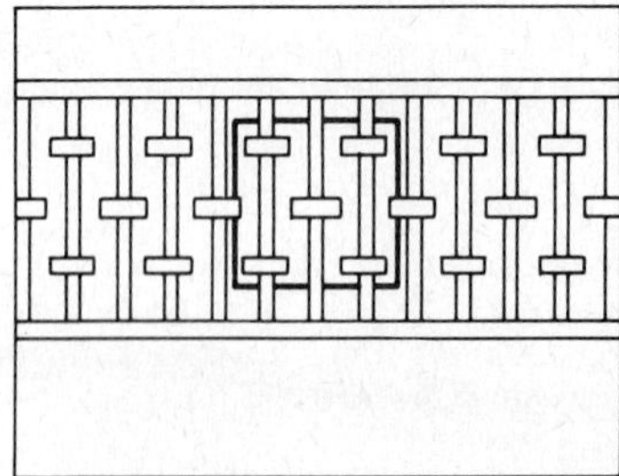

Figure 3-3 Roller and wheel spacing guidelines. *(Litton UHS.)*

Roller and wheel capacity is determined by dividing the weight of the heaviest load to be handled by the minimum number of rollers or wheels under the carrying surface of the load. If special requirements must be considered, such as drop, shock, or side loads, then a roller with a higher load rating will have to be considered.

Conveyor Width; Wheel and Roller Setting. The width of conveyor is determined by the back-to-back frame dimension required to ensure sufficient clearance to carry the load around a 90° curve. The minimum clearance is dependent on the roller setting. If rollers are set high, the conveyor can handle loads up to 1.25 times the width of the conveyor. If the rollers are set low, a minimum of 1 in (2.5 cm) between frame and load must be allowed on each side. Package skew should also be considered in determining this dimension. Specification of curve sections can be calculated graphically from the chart in Fig. 3-4. The design of curve sections depends on the size and shapes of loads. Tracking of packages is important, especially when there are a number of turns involved and the skewing effect becomes cumulative. This effect can be minimized by using tapered rollers (Fig. 3-5) or a double-roller differential section.

Bearing Selection. This is dependent upon conveyor operating conditions. Plain ball bearings are used for indoor use where severe environmental conditions are not present. Dust-tight bearings, which are designed to run dry, are ideal in dusty atmospheres. Greased bearings require more force to turn and their use should be kept to a minimum on gravity-type conveyors.

Conveyor Pitch. The pitch that is required cannot be easily determined because it is dependent on a combination of many factors such as:

- Weight and stability of unit load
- Smoothness of bottom of load
- Firmness of load surface
- Length of rollers
- Number of rollers under load
- Length of runs
- Types of bearings

The suggested pitch for a number of types of unit loads is shown in Table 3-1 and should be used as a general guide for determining pitch.

Conveyor Support and Frame Capacity. Conveyor supports can be one of three types: permanent-floor, ceiling-hung, or portable. Supporting points (Fig. 3-6) should be located to handle the load equally. The design load is the weight of the conveyor section plus the maximum unit load for that section of the conveyor.

Special Considerations for Wheel Conveyors. Wheel conveyors are used for light-duty applications and have several advantages over roller conveyors for lightweight unit loads. Gravity wheel conveyors consist of a series of wheels which can be one of many different styles and materials, mounted on common axles and supported between two frames. They are generally less expensive and lighter in weight, making them ideal for portable applications. Light unit loads travel better on wheel conveyors because less pitch is needed and less force is required to start the wheels in motion (see Table 3-1). Another advantage inherent in the use of multicontact conveying surfaces is that the wheels provide a turning action which enables the package to maintain its original position.

Frames that support the wheel axles are either steel or aluminum. Low weight and corrosion resistance are properties that make aluminum attractive, but steel should be used where conditions require the conveyor to be more rugged.

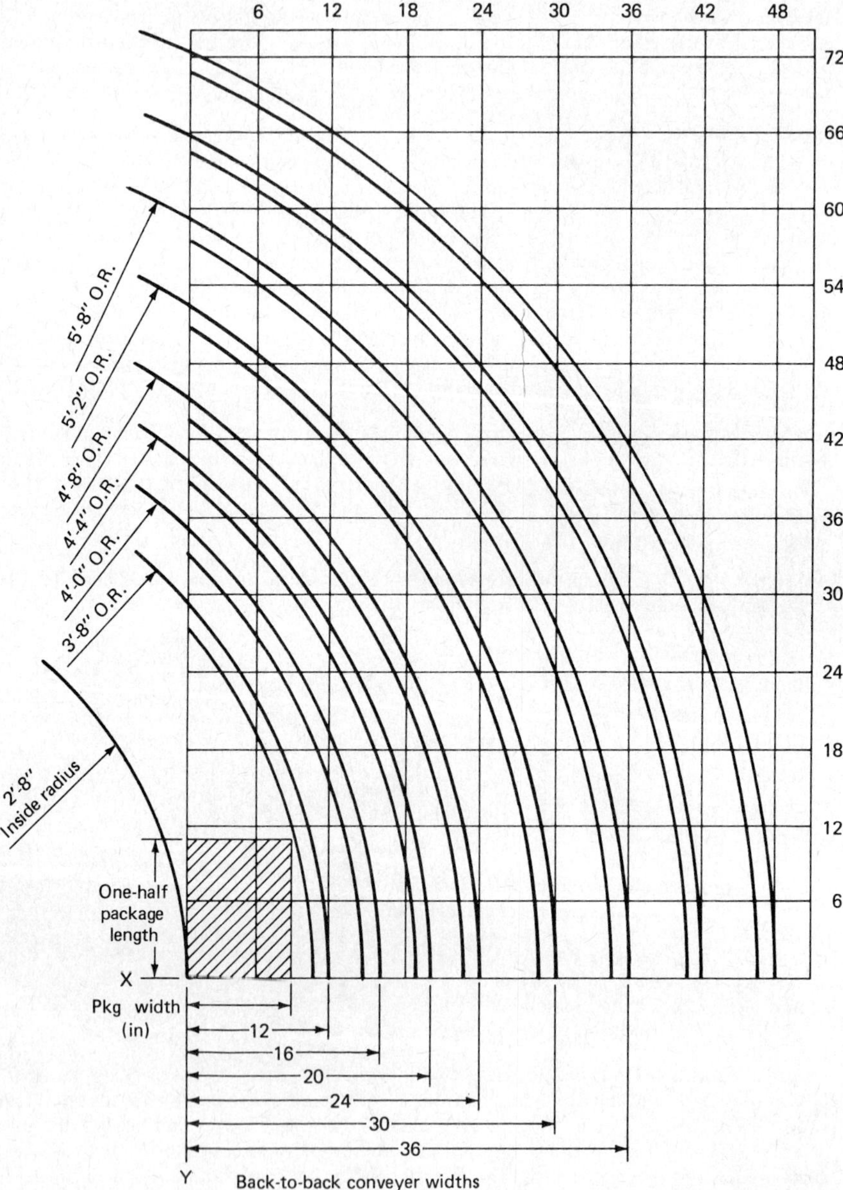

Figure 3-4 Curve-radius selection.

Types of Wheels. There is a wide variety of both metal and plastic wheels now available, including:

- **Steel Wheels with Ball Bearings.** The most rugged and commonly used wheels, these are used where long life is a requirement. Life potential of this wheel is 10 times that of aluminum. Steel wheels can be covered with *neoprene tires* and are used to

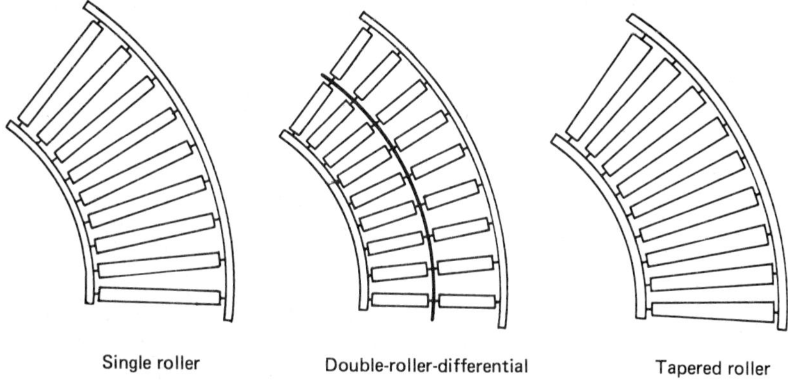

Figure 3-5 Roller-conveyor curve sections. *(Litton UHS.)*

reduce shock, prevent slippage, increase traction, prevent marring fragile surfaces, and reduce noise.

- **Aluminum Wheels with Steel or Plastic Ball Bearings.** Used in applications where weight is a factor, particularly for portable conveyors. Plastic ball bearings should be used in corrosive atmospheres.
- **Nylon Wheels.** Used where resistance to salt, water, and chemicals is required, as well as in applications where conveyors are cleaned frequently. Nylon wheels also do not mar or mark containers.
- **Polypropylene Plastic Wheels.** Have many properties which make them ideal for a wide range of applications. This material is highly impervious to a wide range of corrosive materials and is temperature-resistant from 230 to −30°F (110 to −34°C). The wheels do not absorb moisture and can be steam-cleaned repeatedly.
- **Hysteretic Wheels.** Metal wheels surrounded by a tire formed by elastomeric material, these are used for line storage of heavy loads. The purpose is to absorb the energy of the initial load impact and to control the movement of the load to a safe speed.

Lubrication. Metal wheels can be greased, oiled, or dry. Nylon and plastic wheels are always furnished dry. Oiled or dry bearings should be used where high temperatures can cause thinning and leakage of grease. Dry bearings are recommended where temperatures are below 0°F (−18°C).

Special Types of Wheel Conveyors. These have been designed for handling specific products or for special industries. Samples of these are shown in Fig. 3-7.

Powered Conveyors. These are designed for continuous control of products on level surfaces, through inclines and declines, and around curves. Many of the same points that must be considered for gravity conveyors are applicable to powered conveyors. Powered roller and belt conveyors are the major powered types used to move unit loads.

Powered Roller Conveyors. These are used mainly to accumulate loads because power-drive disengagement can be accomplished very simply when the forward motion of the unit load is stopped. Generally, power-drive disengagement is triggered when the unit load meets an obstruction, creating an opposite reaction which causes the carrying roller bushing to move up an angular slot, thus relieving the pressure and contact between the drive belt and rollers.

Powered roller conveyors are either chain- or belt-driven. Chain-driven units are used in heavy-duty applications and where oil or contaminants would have an adverse effect on the belting. The belt-drive-powered trains are designed for accumulation where the

TABLE 3-1 Suggested Pitch for Gravity Conveyor

Container being conveyed	Approx. cont. wt., lb	Gravity roller conv. plain and dust-tight brgs.			Gravity roller conv. pres. lubricated brgs.			Pi gravity wheel conv. or live rail conv.		
		Pitch per ft, pitch per 10′ 0″ sec, and pitch per 90° curve								
		1″/ft	10′ 0″	90°	1″/ft	10′0″	90°	1″/ft	10′0″	90°
Cartons (fiber smooth bottom)	10	5/8	6 1/4	5	—	—	—	3/8	3 3/4	4
Cartons (fiber smooth bottom)	20	1/2	5	5	—	—	—	1/4	2 1/2	3 1/2
Cartons (fiber smooth bottom)	45	3/8	3 3/4	3 1/2	3/4	7 1/2	6	3/16	1 7/8	3
Cartons (fiber smooth bottom)	100	1/4	2 1/2	3	5/8	6 1/4	5	1/4–3/16	—	3
Fiber beverage carton empty	5	—	—	—	—	—	—	1 1/2	15	10
Fiber beverage carton empty bottles	35	1/2	5	4–5	3/4	7 1/2	6	3/8	3 3/4	4
Fiber beverage carton filled bottles	45	3/8	3 3/4	3 1/2	5/8	6 1/4	5	1/4	2 1/2	3 1/2
Wood cases or boxes	5	5/8	6 1/4	5	—	—	—	3/8	3 3/4	5
Wood cases or boxes	10	3/8	3 3/4	4	3/4	7 1/2	6	1/4	2 1/2	5
Wood cases or boxes	25	1/4	2 1/2	4	5/8	6 1/4	5	1/4	2 1/2	4
Wood cases or boxes	50	3/16	1 7/8	3	1/2	5	4	1/4	2 1/2	3 1/2
Wood cases or boxes	100	3/16	1 7/8	3	3/8	3 3/4	3 1/2	1/8	1 1/4	2 1/2
Steel drums	20	5/8	6 1/4	5 1/2	3/4	7 1/2	6	—	—	—
Steel drums*	55	7/16–1/2	4 1/4–5	5	5/8	6 1/4	5	—	—	—
Steel drums	120	3/8	3 3/4	4	1/2	6	4 1/2	—	—	—
Steel drums	250	1/4	2 1/2	3	3/8	3 3/4	3 1/2	—	—	—
Steel drums*	550	1/4–3/16	2 1/2–1 7/8	3	1/4	2 1/2	3	—	—	—

Wood barrels	100	½	5	5	⅝	6¼	6	—	—	—
Wood barrels	400	⅜	3¾	4	½	5	5	—	—	—
Metal and fiberglass tote boxes	10	½	5	5	—	—	—	⅜	3¾	4
Metal and fiberglass tote boxes	25	⅜	3¾	4	¾	7½	6	¼	2½	3½
Metal and fiberglass tote boxes	50	¼	2½	3	⅝	6¼	5	3/16	1⅞	3
Milk cans empty	25	⅝	6¼	6	¾	7½	6½	—	—	—
Milk cans full	105	⅜	3¾	4	½	5	5	—	—	—
Milk crates empty	10	¾	7½	6	1	10	7½	—	—	—
Milk crates empty bottles	45	½	5	5	⅝	6¼	5	—	—	—
Milk crates full bottles	60	⅜	3¾	4	½	5	5	—	—	—
Wood pallets smooth runner†	350	⅜	3¾	—	7/16	4¼	—	—	—	—
Wood pallets smooth runner†	750	5/16	3⅛	—	⅜	3¾	—	—	—	—
Wood pallets formica base	350	¼	2½	—	5/16	3½	—	—	—	—
Wood pallets formica base†	750	⅛	1¼	—	3/16	2½	—	—	—	—
Wood pallets pine plywood†	350	⅜	3¾	—	7/16	4¼	—	—	—	—
Wood pallets pine plywood†	750	5/16	3⅛	—	⅜	3¾	—	—	—	—
Wood beverage case empty	6	13/16	8⅜	6¾	—	—	—	¼	2½	3½
Wood beverage case empty bottles	30	½	5	5	¾	7½	6	⅛	1¼	2½
Wood beverage case full bottles	40	7/16	4¼	4	½	5	5	⅛	1¼	2½
Multiwall bags, firm	50	—	—	—	—	—	—	⅝	6¼	6
Multiwall bags, firm	100	—	—	—	—	—	—	½	5	5

*Varies with new or reconditioned drums

†Depends on the way it is nailed and if banded or not. Bad nailing or banding will considerably increase the pitch and make a planned slope impractical.

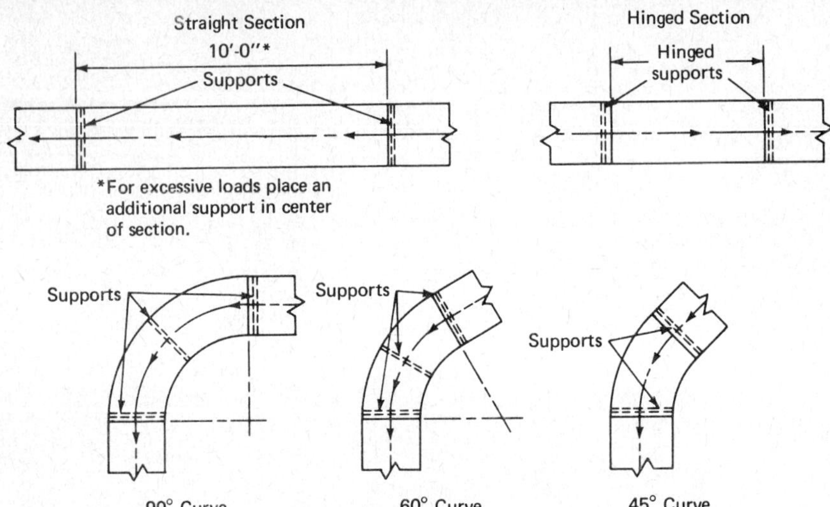

Figure 3-6 Suggested support locations.

belt-to-roller pressure is very light or for transportation sections where belt-to-roller pressure is increased by center takeup rollers and the use of high friction drive belts.

Powered roller conveyors are not used for inclines greater than 5° since the contact force between the unit load and roller surface is not sufficient to overcome the gravity force due to a low coefficient of friction. This type of conveyor is not normally used over straight runs because of the higher cost as compared with belt conveyors.

Belt Conveyors

Belt conveyors consist of an endless moving belt which carries materials within a supporting frame. The belt can be made from a variety of materials and may or may not be

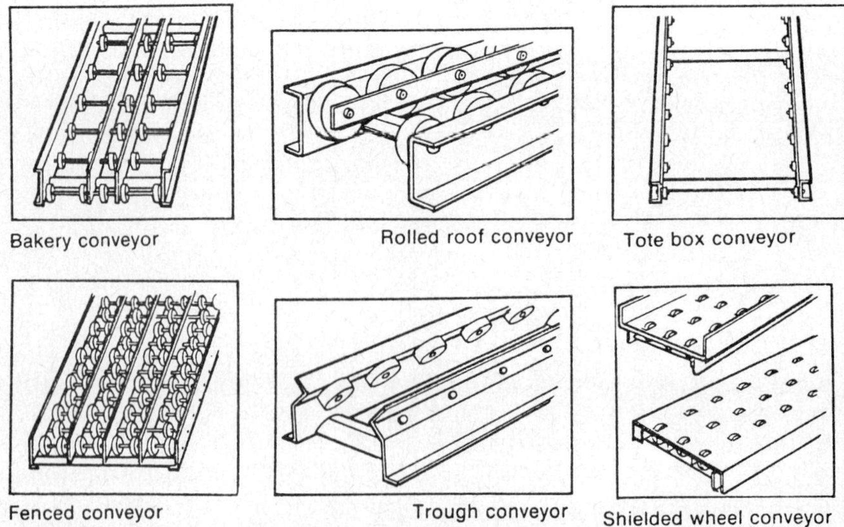

Figure 3-7 Special types of wheel conveyors.

equipped with cleats or other grabbing devices. The belt may be supported by a solid-slider type bed of wood or metal or by rollers.

Conveyor manufacturers suggest friction surface belts on inclines up to 13°, and rough-top rubber belts should be used for inclines up to 25°. In applications where a steeper incline is required, heavily textured, nubbed or cleated surface belts can be used. Also, special requirements for belt surface material should be considered in applications where chemical or oil resistance is needed or for mandated cleanliness requirements.

Belt Conveyor Parameters. The parameters that must be defined before equipment is specified are belt speed, length, maximum load on belt at one time, tension loads, power requirements, and support and mounting hardware. Belt speed should be specified to be compatible with process equipment and other materials-handling hardware. Belt length should be adequate to accumulate the maximum expected product capacity. The maximum load on the belt at any one time can be calculated from the following formula:

$$P = \frac{K}{S \times 60 \text{ min}} \times C \text{ (ft)}$$

where

P = maximum product weight
K = load per hour, lb
S = speed of belt, ft/min
C = center-to-center distance

For example, if

Load per hour in pounds = 10,000 lb
Speed of belt = 50 ft/min
Center-to-center distance = 36 ft

Then, the maximum product weight is

$$P = \frac{10{,}000}{50 \times 60} \times 36$$
$$= 1200 \text{ lb}$$

Considerations for Bulk-Materials Belt Conveyors. These considerations are similar to those for all conveyors; however, the properties of the materials to be moved affect the parameters of and the specification of the conveyor. The use of belt conveyors for bulk materials is limited by the characteristics of materials, some of which are:

- Stickiness, which may prevent materials from completely discharging from the conveyor or interfering with the power-train components.
- Temperatures that exceed 150°F (71°C) would cause deterioration or damage to most belt materials.
- Chemical reactions of conveyed materials with belt material. Some oils, chemicals, fats, and acids can damage belts.
- Large lump size becomes a factor and generally requires the system to be oversized for the amount of weight being moved.

Weight and friction are the common factors that determine the amount of incline that is possible for unit-load and bulk-materials-handling belt conveyors. Bulk-materials belt conveyors must also include materials characteristics such as size consistency, shape of lumps, moisture content, angle of repose, and flowability. The maximum angle of incline for various bulk materials is shown in Table 3-2. The ideal combination of belt width and speed (Table 3-3) is determined by the characteristics of the materials handled.

Metal-Belt Conveyors. Similar in design to standard belt conveyors, these differ in that their surface is a belt of woven or solid metal. The materials include carbon steel,

TABLE 3-2 Maximum Angle of Incline

Material carried	Maximum angle of incline, deg*	Material carried	Maximum angle of incline, deg*
Alumina, dry, free-flowing, ⅛″ lumps	10 to 12	Grain	8–16
Beans, whole	8	Ore (see stone)	15–20
Coal, anthracite	16	Packages	15–25
Coal, bituminous, sized, lumps over 4 in	15	Pellets, depending on size, bed of material and concentricity (taconite, fertilizer, etc.)	5–15
Coal, bituminous, sized,† lumps 4 in and under	16	Rock (see stone)	15–20
Coal, bituminous, unsized†	18	Sand, very free-flowing¶	15
Coal, bituminous, fines, free-flowing‡	20	Sand, sluggish (moist)§	20
Coal, bituminous, fines, sluggish§	22	Sand, tempered foundry	24
Coke, sized	17	Stone, sized, lumps over 4 in	15
Coke, unsized	18	Stone, sized, lumps 4 in and under, over ⅜ in	16
Coke, fines and breeze	20	Stone, unsized, lumps over 4 in	16
Earth, free-flowing‡	20	Stone unsized, lumps 4 in and under, over ⅜ in	18
Earth, sluggish§	22	Stone, fines ⅜ in and under	20
Gravel, sized, washed	12	Wood chips	27
Gravel, sized, unwashed	15		
Gravel, unsized	18		

*For ascending conveyors when uniformly loaded and with constant feed.

‡Angle of repose 30 to 45°.

†See second footnote (†) to Table 3-3 for definitions of "sized" and "unsized" as used in Materials Carried columns of Table 3-2.

§Angle of repose greater than 45°.

¶Very wet or very dry, with angle of repose less than 30°.

TABLE 3-3 Recommended Belt Speed as Determined by Material Handled

Material		Recommended belt speed, ft/min*												
		Belt width, in												
Characteristics	Example	14	16	18	20	24	30	36	42	48	54	60	72	84
Maximum size lumps, sized or unsized														
Mildly abrasive	Coal, earth	300	300	400	400	450	500	550	600	600	650	650	650	650
Very abrasive, not sharp	Bank gravel	300	300	400	400	450	500	550	550	600	600	600	600	600
Very abrasive, sharp and jagged	Stone, ore	250	250	300	350	400	450	500	500	550	550	550	550	550
Half max. lumps, sized or unsized														
Mildly abrasive	Coal, earth	300	300	400	400	500	600	650	700	700	700	700	700	700
Very abrasive	Slag, coke, ore, stone, cullet	300	300	400	400	500	600	650	650	650	650	650	650	650
Flakes	Wood chips, bark, pulp	400	450	450	500	600	700	800	800	800	800	800	800	800
Granular ⅛″ to ½″ lumps	Grain, coal, cottonseed, sand	400	450	450	500	600	700	800	800	800	800	800	800	800
Fines														
Light, fluffy, dry, dusty	Soda ash, pulverized coal							220–250 ft/min						
Heavy	Cement, flue dust							250–300 ft/min						
Fragile, where degradation is harmful	Coke, coal												200–250 ft/min	
	Soap chips							150–200 ft/min						

*Normal for belts traveling horizontally on ball- or roller-bearing idlers. For picking belts, speed is usually 50 to 100 ft/min. Belts with discharge plows should not travel faster than 200 ft/min. For tripers, recommended speed is 300 to 400 ft/min. Trippers for higher-speed applications can be furnished. A speed of at least 300 ft/min should also be maintained for proper discharge when using 35 to 45° idlers, and also for materials tending to cling to belt.

†*Unsized* means a uniform mixture of material in which not more than 10 percent are lumps ranging from maximum size to ½ maximum size, at least 15 percent are fines or lumps smaller than 1/10 maximum, and the remaining 75 percent are lumps of any size smaller than ½ maximum. *Sized* means a uniform mixture in which not more than 20 percent are lumps ranging from maximum size to ½ maximum size, and the remaining 80 percent are lumps no larger than ½ maximum size and no smaller than 1/10 maximum size.

galvanized steel, chromium stainless steels, and other metals or special alloys that are required for a specific application and environment. Wire belts are also available for use where processing temperatures vary from 320 to 2500°F (160 to 1416°C). Wire-belt conveyors are used primarily to move product or unit loads through processes that include liquid or chemical treatment, heat treatment, or kiln firing operations. Wire belts can be cleaned or sterilized while in motion. Mesh openings in the belt permit circulation of water, gases, heat, and cooling air. Typical uses for metal-belt conveyors include operations such as spray-washing glass containers, moving baked goods to ovens, conveying cathode-ray tubes through various processes, and moving hot forgings from automatic die casting equipment.

Tracking of a wire-mesh belt is a problem since the belt is formed by a number of different sections joined together and a wide range of temperatures used in processing operations causes expansion and contraction of the belt material. The conveyor specification must frequently include one of the following features to compensate for these conditions and assure straight belt tracking.

- Multitooth sprocket belt drive
- Belt aligners, consisting of pulleys or rollers mounted on the supporting frame
- Self-tracking belt that has V-shaped wires on the underside that run through grooved driver drums

Surface Chain Conveyors

Surface chain conveyors (Fig. 3-8) include sliding chain, pusher bar, slat, and tow types and car-type trolley conveyors.

Sliding Chain Conveyors. These are the simplest type since they use the chain itself to convey packages down two sliding tracks. The conveyors are used to handle heavier loads than can belt conveyors, such as loaded pallets or unitized loads, but they have the same incline limitations as belt and powered-roller conveyors.

Pusher Bar Conveyors. These are able to convey loads up steeper inclines (to 45°) because the load is pushed by a car connected to the chain drives and this arrangement moves the load along a metal bed or trough. Pusher bar conveyors are generally used for floor-to-floor movement in multistory warehouses or facilities.

Slat Conveyors. These employ an endless chain to drive a conveyor surface of nonoverlapping, noninterlocking slats made of wood or metal. Slat conveyors can be used as moving work tables and to move heavy unit loads and are ideal for applications where the conveyor surface must be flush with a work station or with a floor surface. In the latter application, the installation will permit industrial trucks to cross over or be carried on the slat surface. Slat conveyors can be operated on inclines or declines, the angle of which is limited by the friction between slat surface and the load. Cleats may be added to support loads where steeper inclines are required.

Tow Conveyor. A tow conveyor uses an endless chain either supported from an overhead track or running in a track below the floor surface to tow trucks, dollies, or carts. The under-the-floor tow conveyor system is the type most commonly found in warehouses, and it is versatile because it can be looped around storage areas and moved down aisles and can be moved on spur sections for loading and unloading operations and empty car storage. The track recessed in the floor allows the use of the floor surface for other equipment, but the track and chain drive system cannot easily be relocated once it is installed. Carts and trucks used in a tow conveyor system can range from the ordinary pallet truck fitted with tow pins to engage with the chain drive to special carts or trucks designed for a specific application.

Car-Type Trolley Conveyors. These employ an endless chain to pull a series of small cars or trolleys which carry the material to be moved. These often have fixtures to be used in assembly lines or contain molds for use in foundry processing operations.

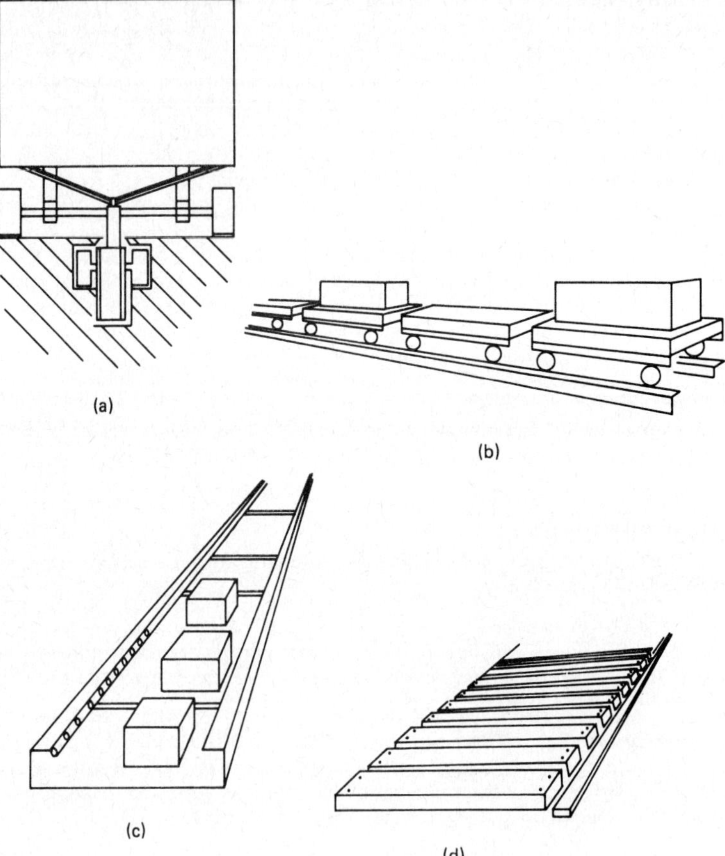

Figure 3-8 Types of conveyors. (*a*) In-floor towline, (*b*) trolley conveyor, (*c*) pusher bar conveyor, (*d*) slat conveyor.

Overhead Conveyors

Overhead conveyors include both trolley conveyors and power and free types of equipment. These conveyors are supported and function within a trolley track driven by a chain power drive to move parts or product. The path of the conveyor can be straight, inclined, declined, and around corners; it can make optimum use of building geometry and follow the general work flowpath within the limitations of building constraints and equipment design parameters. Conveyors can be supported independently or attached to existing beams and trusses, depending on the load factors involved.

In order to determine equipment design parameters, the following procedure should be followed.

1. From process flowcharts, determine all operations to be serviced by the conveyor.
2. Determine the path of the conveyor on a scaled plant layout (Fig. 3-9), showing all obstructions such as columns, walls, machinery, and work aisles.
3. Develop a vertical elevation view to determine incline and decline dimensions (Fig. 3-10). At this point a three-dimensional view could be prepared to give a multiplan view of the proposed installation.

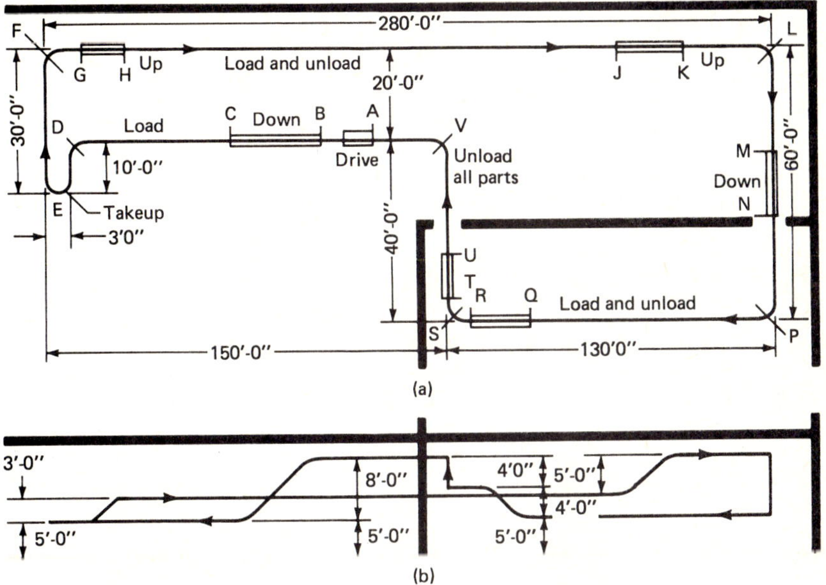

Figure 3-9 Plan and vertical elevation views of trolley conveyor system.

4. Determine the movement rate, unit load size, spacing, and carrier design.
5. Modify turn radii to provide clearances that allow for desired clearance (Fig. 3-10) on turns.
6. Modify loading spacing to provide clearance on inclines and declines. As inclines and declines get steeper, load spacing has to be increased to provide a constant clearance or separation between loads. Table 3-4 shows the load spacing required for a given separation at various incline angles.
7. Redraw the conveyor path and vertical elevation views using new radii and incline information.
8. Compute the chain pull, which is the total weight of the chain, trolleys, and other

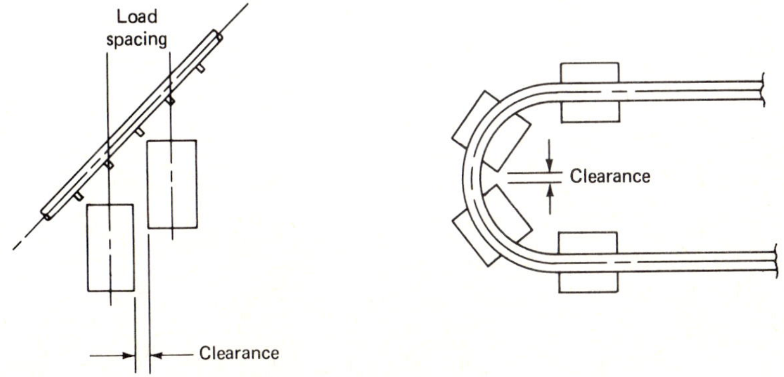

Figure 3-10 Load-spacing considerations for overhead conveyors.

TABLE 3-4 Load Clearance on Inclined Track for Overhead Conveyors

Load spacing, in	Incline angle, deg											
	5	10	15	20	25	30	35	40	45	50	55	60
	Horizontal centers, in											
12	12	11⅞	11⅝	11¼	10⅞	10⅜	9⅞	9¼	8½	7¾	6⅞	6
16	15⅞	15¾	15½	15⅛	14½	13⅞	13⅛	12¼	11⅜	10⅜	9¼	8
18	18	17¾	17⅜	17	16⅜	15⅝	14¾	13⅞	12¾	11⅝	10⅜	9
24	24	23⅝	23¼	22⅝	21¾	20⅞	19¾	18⅜	17	15½	13¾	12
30	29⅞	29⅝	29	28¼	27¼	26	24⅝	23	21¼	19⅜	17¼	15
32	31⅞	31½	31	30⅛	29	27¾	26¼	24½	22⅝	20⅝	18⅜	16
36	35⅞	35½	34¾	33⅞	32⅝	31¼	29½	27⅝	25½	23⅛	20⅝	18
40	39⅞	39⅜	38⅝	37⅝	36¼	34⅝	32¾	30⅝	28¼	25¾	23	20
42	41⅞	41⅜	40⅝	39½	38⅛	36⅜	34⅜	32¼	29¾	27	24⅛	21
48	47⅞	47¼	46⅜	45⅛	43½	41⅝	39⅜	36¾	34	30⅞	27⅝	24
54	53⅞	53¼	52¼	50¾	49	46¾	44¼	41⅜	38¼	34¾	31	27
56	55⅞	55⅛	54⅛	52¾	50¾	48½	45⅞	42⅞	39⅝	36	32⅛	28
60	59¾	59⅛	58	56⅜	54⅜	52	49⅛	46	42½	38⅝	34½	30
64	63¾	63	61⅞	60⅛	58	55½	52½	49	45¼	41⅛	36¾	32
72	71¾	70⅞	69⅝	67¾	65¼	62⅜	59	55¼	51	46¼	41⅜	36
80	79¾	78⅞	77¼	75¼	72½	69⅜	65½	61⅜	56⅝	51½	45⅞	40

components, as well as the weight of the carriers and load. For example, for a given system the tentative chain pull is calculated as follows:

$$\text{Total tentative chain pull} = 700 \times 60.0 \times 0.03 = 1260$$

where 700 = conveyor length, ft
0.03 = coefficient of friction, %
60.0 = 10.0 lb/ft (chain and trolleys) + 12.5 lb/ft (carriers) + 37.5 lb/ft (line load)

For this initial calculation, inclines and declines are assumed to be level sections if the number of declines balances out the number of inclines; however, for each additional incline, the weight has to be added to determine the total chain pull. If, in our example, a vertical incline that raises a load 8 ft is required, then the additional chain pull is

$$37.5 \text{ lb} \times 8\text{-ft lift} = 300 \text{ lb}$$

The total chain pull then becomes 1260 + 300 lb = 1560 lb.

9. Select tentative conveyor size based on trolley load and chain pull.

10. Select vertical curve radii.

11. Determine power requirements and drive locations. This requires a point-to-point calculation of chain pull around the complete path of the conveyor, which is shown in Fig. 3-9. The following three formulas are used to compute point-to-point chain pull.

a. Pull for each straight horizontal run:

$$P_H = XWL$$

where
X = 0.02 for standard ball-bearing trolleys
W = total moving weight, lb/ft (empty or loaded, as the case may be)
L = length of straight run, ft

b. Pull for each traction wheel or roller turn:

$$P_T = YP$$

where
Y = 0.02 for traction wheel or roller turn
P = pull at turn, lb

c. Pull for each vertical curve:

$$P_V = XWS + ZP + HW(1 + Z)$$

where
X = 0.02 for standard ball-bearing trolleys
W = total moving weight, lb/ft
S = horizontal span of vertical curve, ft
H = total change of level of conveyor, ft (plus, when conveyor is traveling up the curve; minus, when conveyor is traveling down the curve).
Z = 0.03 for 30° incline; 0.045 for 45° incline; 0.06 for 60° incline; 0.09 for 90° incline
P = pull at start of curve, lb

Drive horsepower may be calculated from the following formula:

$$\text{Drive hp} = \frac{\text{drive capacity (lb)} \times \text{maximum speed}}{33{,}000 \times 0.6}$$

12. Design conveyor supports and superstructures.

13. Design guards which are required by federal, state, and other codes under high trolley runs, particularly over aisles and work areas. Guard panels are normally made from woven or welded wire mesh with structural angles and channels to suit the size and weight of the material being handled.

Power and Free Conveyors. These consist of two separate trolley systems: one moves and is powered by a chain drive; the other has a track under the powered track that accommodates a free-moving trolley containing a carrier or fixture from which a load is suspended (Fig. 3-11). In the powered mode, the powered trolley system is engaged with the free trolley through contact of a pusher dog on the powered system to a retractable dog on the free system. Disengagement is accomplished by contact with another load or by actuating the dog acutator. The main advantage of this system is that one carrier can be stopped whenever and wherever desired without interrupting the whole system. The versatility of the power and free conveyor can be utilized in a process or production line where operations do not take the same amount of time to complete, or where units have to be accumulated for off-line operations, such as for repair. This type of conveyor is used in many diverse industries in applications which include engine and transmission assembly, butchering operations in meat-packing plants, and nonindustrial environments such as hospital distribution of medical supplies and patient meals.

The same design criteria apply to power and free conveyors as to other chain-driven trolley systems.

Bulk-Materials Vertical Conveyors

Bulk-materials vertical conveyors (Fig. 3-12) are generally used to lift bulk materials up to silos, hoppers, or other storage containers from which the material may be dispensed into a mixing, packing, truck-loading operation, or directly to a process. Some of the industries that use this equipment include glass, agricultural fertilizer, and powdered chemicals.

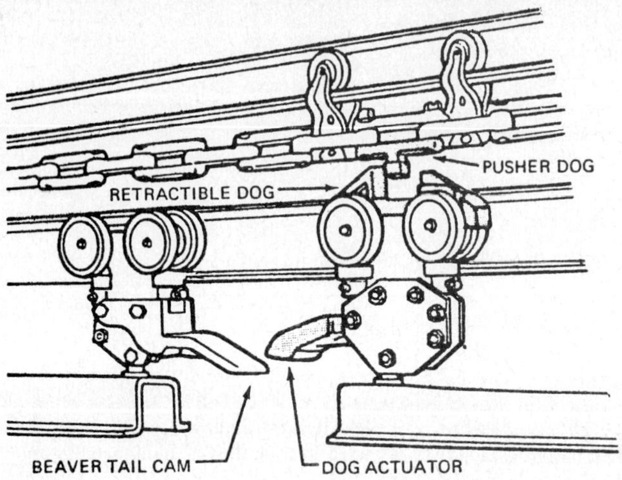

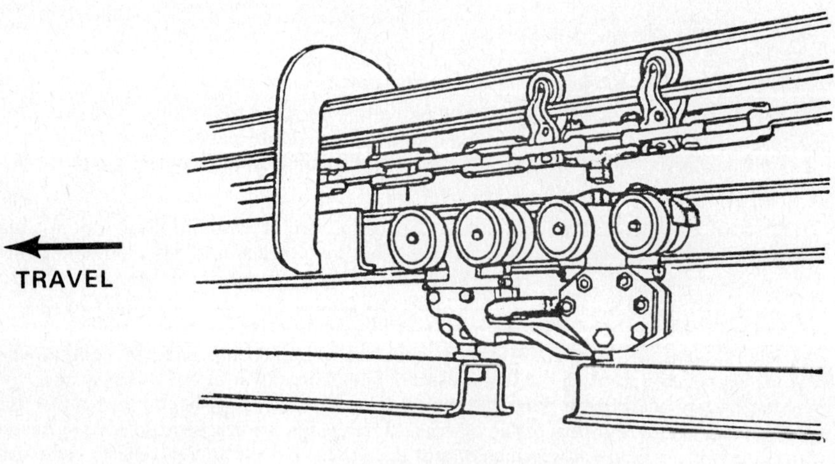

Figure 3-11 Power and free trolley.

Skip Hoists. These are used to lift bulk materials handled in batches to very high points. A bucket which carries the material moves vertically in guides and is raised and lowered by a hoist-operated cable.

Gravity Discharge Conveyor-Elevators. These carry material in both horizontal and vertical paths. The buckets are rigidly mounted on two strands of chain running in tracks. Material is loaded into a bucket at the base of the equipment by feeding material into a lower trough, and discharge is effected when the bucket position changes in the horizontal run.

Bulk-Flos. These lift material by the use of flights attached to a chain drive which is

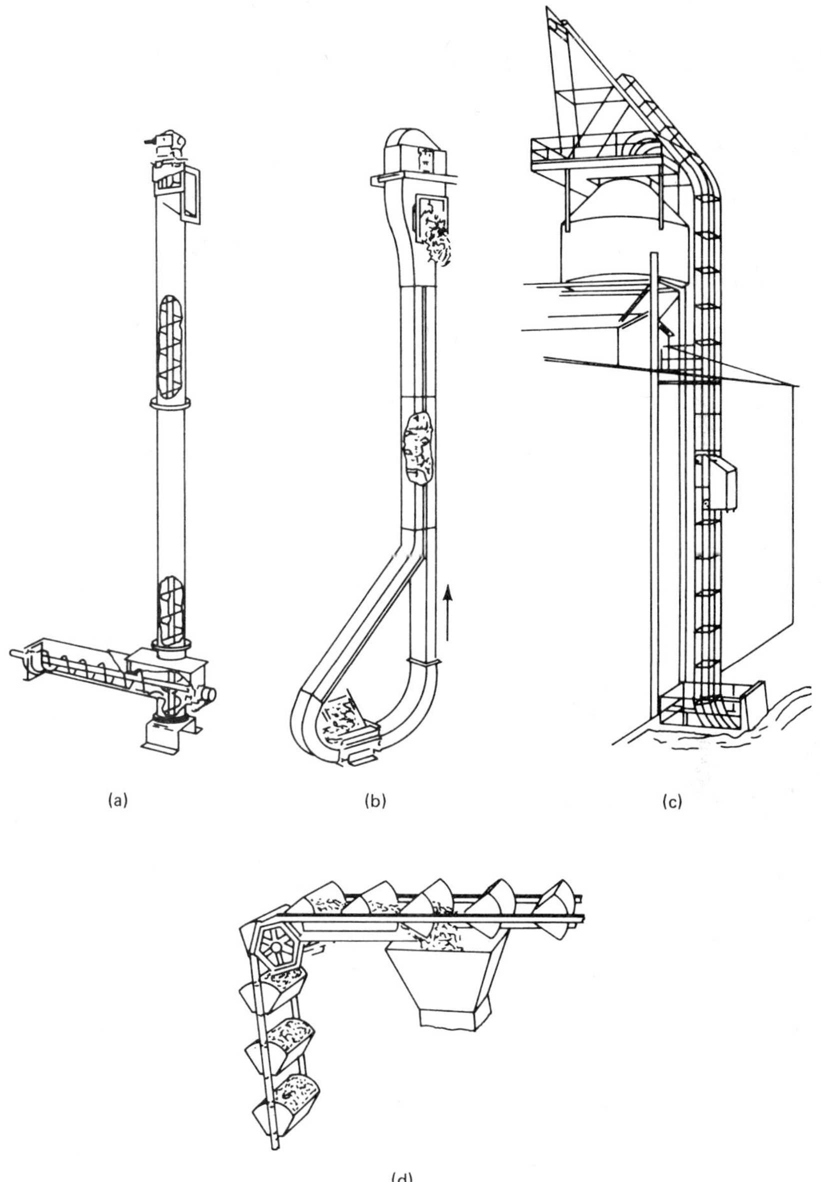

Figure 3-12 Bulk material vertical conveyors.

contained in a dust-tight casing. Bulk-Flos are self-feeding and -discharging and lend themselves to continuous bulk-material processes.

Rotor Lifts. These are similar to screw conveyors but are mounted vertically to effect the lifting of bulk materials and are contained in a dust- and weatherproof casing. Screw feeders or conveyors are generally used to deliver material to rotor lifts.

Other Specialty Conveyors

There are innumerable variations on standard conveying systems, some of which are unique to individual industries. Six common examples are described below.

Screw Conveyor. This conveyor (Fig. 3-13) consists of a screw rotating in a stationary trough and the material moving along its length by rotation of the screw. This type of conveyor serves a dual purpose since it can also be used to perform processes such as blending and mixing of material while the material is being moved. The conveyor is generally enclosed to prevent dust or fumes from escaping and allow the conveyor to be cooled or heated. Loading or discharging can be located at any point along a conveyor.

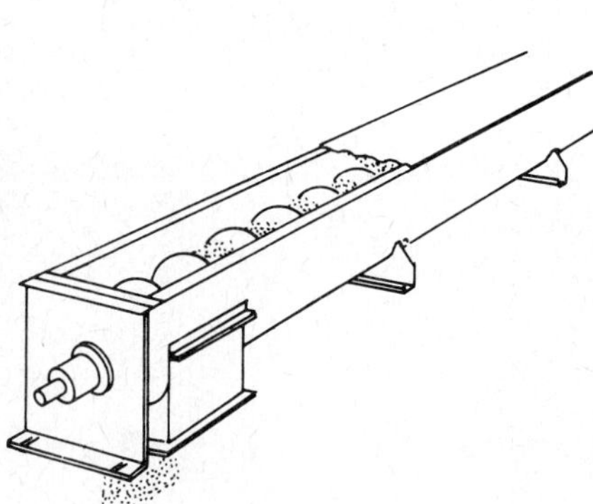

Figure 3-13 Screw conveyor.

Figure 3-14 Spiral track conveyor.

Spiral Track Conveyors. These conveyors (Fig. 3-14) consist of a continuous spiral track with a power drive which turns the track, moving anything which is hung on it. It has wide application in the garment industry. It is generally used for items weighing less than 10 lb (5 kg). Interlocking nylon wafers can permit turns to be made in any direction on a radius of 18 in (46 cm).

Oscillating and Vibrating Conveyors. These use the natural frequency vibration of a trough to provide a conveying action to move material. Oscillating conveyors use a mechanically driven power train to move a trough carrying material against spring supports which provide a fast return and downward stroke, causing the trough to vibrate and convey the material. Vibrating conveyors utilize some form of magnetic pulsation to create this vibration motion. Wider variations of frequency are possible by simple control for vibrating conveyors, enabling speed changes compensating for material differences.

Application of both types of conveyors is growing in a number of different industries for uses such as: conveying light food products such as cereal in the food industry; moving, cooling, and breaking up lumps of casting sand in foundries; quenching and removal of glass cullet in water-filled troughs in the glass industry; removing ferrous from nonferrous materials in separation systems; and feeding small parts into automatic packaging or assembly equipment.

Flight Conveyors. These conveyors (Fig. 3-15) use scraper plates to push nonabrasive bulk material through a trough which can be horizontal or inclined.

Apron Conveyors. These conveyors (Fig. 3-16) use a series of interlocking apron pans supported in a stationary frame for conveying materials that are heavy, abrasive and lumpy, such as ore, stone, industrial refuse, and waste materials.

Pneumatic Tubes. These use a pressure or vacuum system to move materials or a container at relatively high speed. The major application is that of an internal mail carrier, although it can also be used to move certain types of high-volume fine particulate.

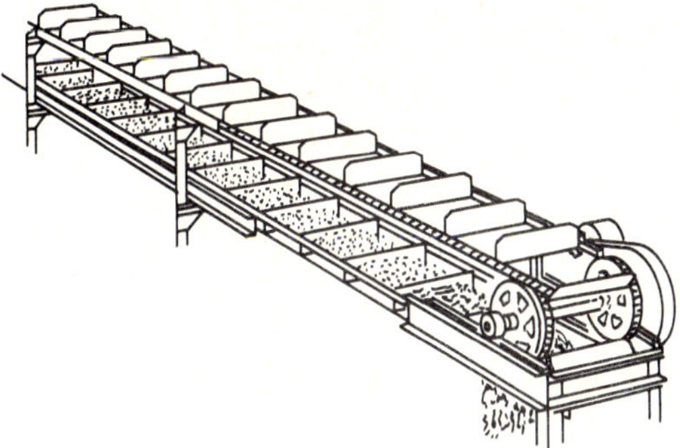

Figure 3-15 Flight conveyor.

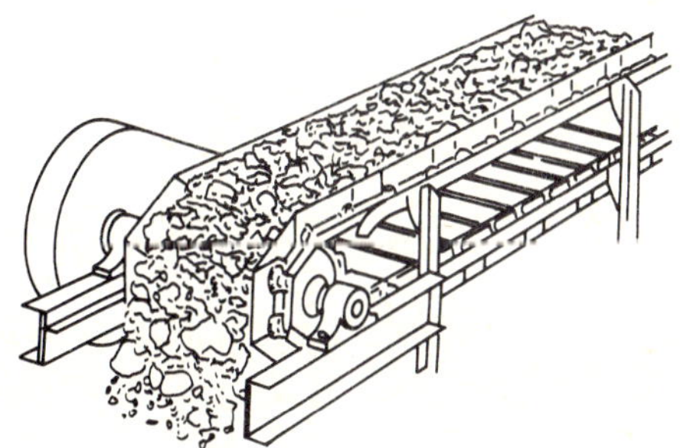

Figure 3-16 Apron conveyor.

SORTING, CONSOLIDATING, AND DIVERTING DEVICES

A materials handling system must frequently have the ability at some point to identify, sort, and divert parts, products, or unit loads. Peripheral accessories and equipment do this, ranging from simple mechanical diverters to sophisticated optical recognition reading devices, which can actually read and identify alphanumeric characters and sort 20,000 items per hour and which are used mainly for check and mail handling. Whatever the complexity of the system, three basic elements must be considered: identification of the item to be sorted or consolidated, recognition of the item, and the command to activate the mechanisms to divert the item.

Simple Mechanical Sorting

Simple mechanical sorting utilizes inherent differences such as size, shape, weight, or other physical differences to identify or recognize items; it generally is contact sorting in which an item must make contact with a channel or feeler guides or discerns physical differences and contacts a cam or other simple mechanism to activate a diverter chute or other diverting device.

Diverting Mechanisms

Diverting mechanisms can be grouped into devices that deflect, push off, drive off, or tilt; many variations are included within each group (Fig. 3-17).

Electromechanical Sorting

Electromechanical sorting uses noncontacting identification devices that can sense both inherent differences and applied differences. These are identified on the load or package

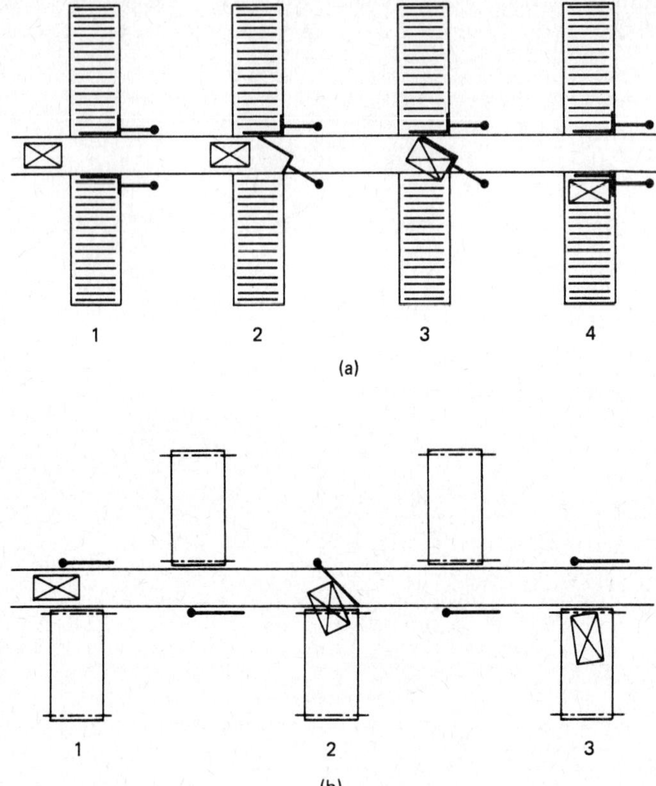

Figure 3-17 Diverting mechanisms.

by a code that can be discriminated by a sensing or scanning device and that triggers a diverting mechanism.

Photosensors. These are the most commonly used sensing and scanning devices. A photoelectric control consists of a light source, photoreceiver, amplifier, and output. A beam of light from the light source activates the photosensitive elements of the photoreceiver which produces an electric signal, which drives a relay to activate the diverting mechanism. When used as a sensor, photoelectric controls can be used to:

- Sense the presence or absence of containers or products on a conveyor
- Detect over- and undersize products
- Sort products by size
- Count items

Photoelectric controls can also be arranged in an array or ganged in a manner that a code on a container can be read. Each sensor that goes into the scanning device will detect a specific mark or blank space from the code which when scanned produces binary information into a logic function of a controller. This in turn supplies a signal for the diverting mechanism. Typical codes are ladderlike in format, and this allows a scanning device to read the code vertically.

The use of code scanning in a materials-handling system must consider the following criteria:

- Required information content
- Method of code application
- Range and depth of field requirements
- Nature and speed of product flow
- Value and volume of products

The state of the art in scanning technology now includes laser scanners and fiber optics. These are used as optical recognition devices which can actually identify alphanumeric characters. The charge-coupled device (CCD) is a self-contained device that scans a package and is able to detect many different levels of light intensity. It can produce an accurate representation of the object being scanned, enabling the CCD to be used as an observation or inspection device.

Palletizers

Palletizers receive individual packages, cases, or bags from a conveyor and automatically arrange them on a pallet in a predetermined pattern with the required number of tiers. Each tier need not have the same predetermined pattern. Generally, units to be stacked are received on a control belt at the entrance to the machine. At this point the unit will be counted and oriented depending on the patterns required. As each row is completed, a pusher moves the cases onto an apron. When the tier is completed, the apron is withdrawn, depositing the tier on the pallet or tier below. The operation is repeated until the pallet load is completed, when it is discharged and replaced with an empty pallet.

Large volumes of standard units are required, and it is estimated that palletizers become economical when approximately 900 units per hour require palletizing. The larger palletizers can handle in excess of 6000 cases per hour of certain products.

Depalletizers

Depalletizers are highly specialized pieces of equipment which automatically depalletize cartons and cases. Automatic squaring mechanisms permit the handling of loose pallet loads. Depalletizers generally operate in the range of 3500 cases per hour and are seen primarily in beverage distribution.

HOISTS AND CRANES

Hoists and cranes are materials-handling equipment used to move varying loads intermittently within a fixed area. The loads vary in size and weight and are not uniform. Most of the materials movement is devoted to raising and lowering loads, although some units are so constructed as to permit them to travel laterally over a specific area. The types of hoists, cranes, and attachments are listed below.

Hand and Powered Hoists

Hand and powered hoists (Fig. 3-18) are the most basic and economical lifting equipment which enables an operator to move a large load, up to 50 tons, vertically by using some kind of mechanical advantage.

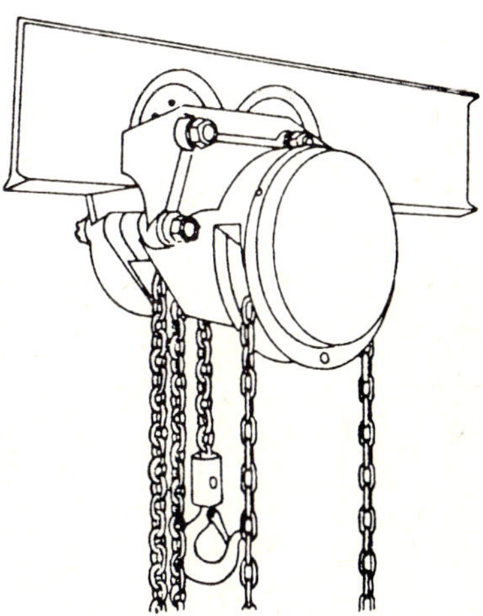

Figure 3-18 Hoist.

Jib Cranes

Jib cranes (Fig. 3-19) consist of a hoist that is mounted on a boom track. The hoist mechanism can be moved laterally in the track and the boom can be turned in an arc limited by the building restrictions or the mounting arrangement of the boom. Jib cranes are classified into basic groups of bracket jib, cantilever jib, and pillar jib. Load capacities range from small manually operated cranes to loading towers that exceed 300 tons.

Bridge Cranes

Bridge cranes consist of a hoist mounted on a guider bridge which is supported by two trucks on each end and rides on runways supported by building members. Top-running bridges, where end trucks ride on top of runway tracks are able to support a total bridge and load weight of hundreds of tons, but underhung or bottom-running bridges, where the trucks are suspended from the lower flanges of the runway track, normally are used for loads less than 20 tons. Bridge cranes can be operated manually or powered or in the cases of very large cranes can be operated by remote control (Fig. 3-19).

Gantry Cranes

The gantry crane is very similar to a bridge crane except it is supported by self-contained vertical support members that travel in tracks on the floor surface and it is generally used where overhead runways are not feasible due to building restrictions. The gantry crane system also has the advantage of being usable in outdoor operations without the construction of an expensive supporting structure (Fig. 3-19).

Stacker Cranes

The stacker crane consists of a rigid mast suspended from an overhead bridge that travels laterally. A platform or a set of forks moves up and down on slider bars to lift and lower loads. The stacker crane is most commonly used to place or retrieve loads to and from racks from both sides of an aisle. In automatic storage and retrieval systems, the stacker crane is computer-controlled. The computer has the rack location of each item stored in the memory and is able to command the load-carrying platform to a specific location for storage and retrieval of a load.

Lifters

A lifter (Fig. 3-20) is an attachment suspended from the load hook of a hoist or crane that permits a load to be handled more easily or quickly than possible with a hook, and many load configurations cannot be handled with a hook. In many cases, lifters are designed for a specific application, but there are many standard types that are available for a wide range of applications. Lifters are categorized by the method in which the load is carried.

Supporting Lifters. These carry the load on the surface of the lifter, on bearing surfaces of cradles, or hooks and slings attached to lifters.

Clamping Lifters. These hold the load by surface friction or by squeezing load.

Surface-Attaching Lifters. These consist of both magnetic or vacuum types. Magnetic lifters can use either a permanent magnet that requires a strip-off device to release the load or an *on-off magnet* that can be activated by applying a voltage. Vacuum pads can be used to lift loads with nonporous and smooth surfaces and are commonly used to handle glass and aluminum.

Manipulating Lifters. These move the load through one or more axes for operations such as positioning or dumping.

GUIDED (DRIVERLESS) VEHICLES

Guided (driverless) vehicles move material over fixed paths but do not require the use of an operator or a mechanical drive train located below the floor surface or an overhead

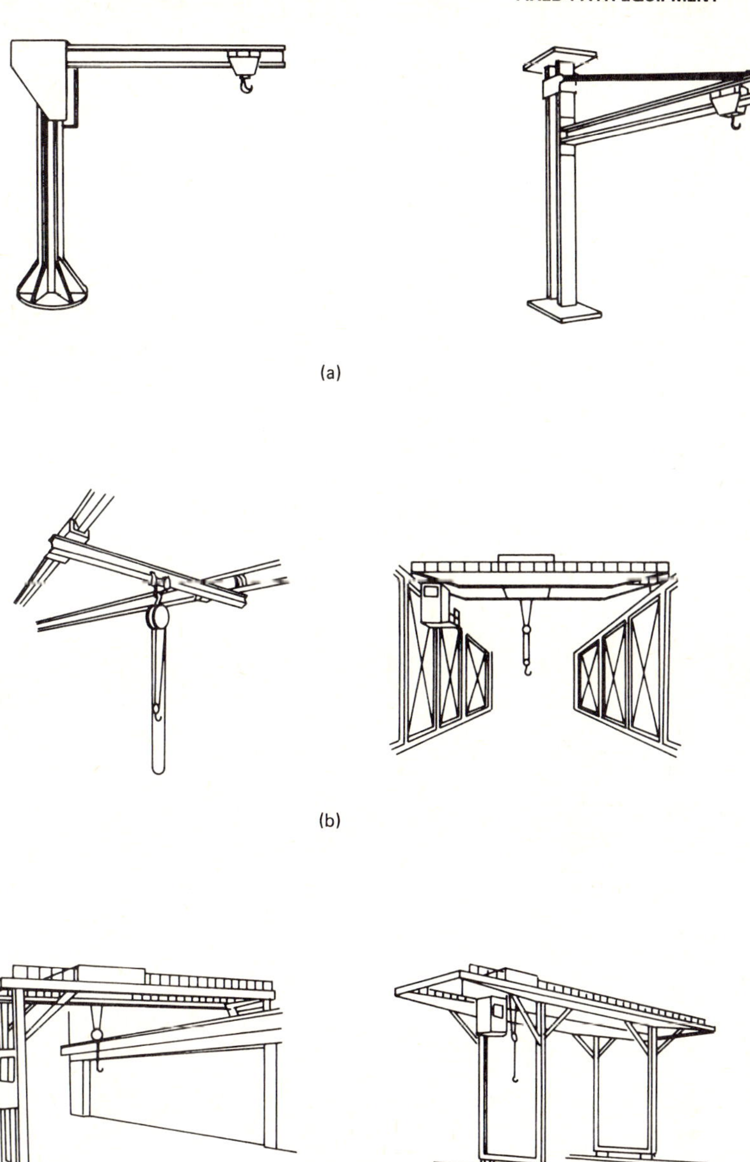

Figure 3-19 Types of cranes. (*a*) Jib crane, (*b*) bridge crane, (*c*) gantry crane.

towline. They are useful when a variety of materials must be moved over long distances to and from a variety of fixed destinations. There are three identifiable types of vehicles: first, the driverless tractor (Fig. 3-21) which hauls trailers or cartloads of material; second, the individual unit-load or pallet mover (Fig. 3-22); and third, the multishelved self-contained vehicle. The last type is used primarily to move mail in office buildings or for food and supply deliveries in hospitals.

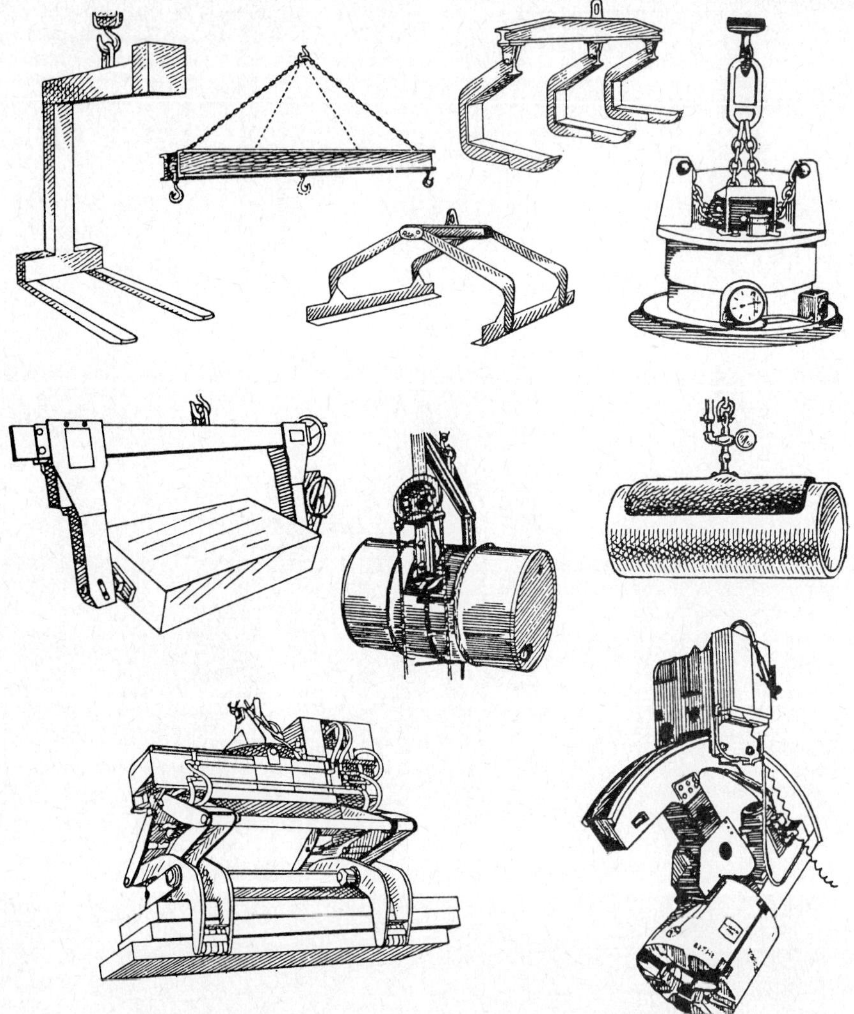

Figure 3-20 Types of lifters. *(Reproduced with permission from Material Handling Engineering Handbook and Directory, 1977/1978,* published by *Material Handling Engineering,* Cleveland.)

Guidance and Control Systems

Guidance and control systems are similar for all three systems. Two systems are used: optical, where the unit follows a line taped or painted on the floor surface; or magnetic, where a thin wire is set in a shallow channel sealed over in the floor. This latter system is less flexible and more costly to control but is not subject to obliteration or wear, which can be a problem in certain factory environments.

The driverless tractor, being unable to reverse on its own trailers, generally requires a closed-loop system. However, multiple-loop systems can be used. Unit-load movers are generally reversible and can operate on a spur.

The programming information which determines the paths and stops can be preset on the tractor programmer or can be controlled from a central dispatching point. These systems generally have the logic to allow the tractor to take the shortest route to the destination without traveling through the entire loop. Radio-control-transmitters are often used to reposition the train within a loading station, eliminating unnecessary walking in operations such as order picking or loading the train at the receiving dock.

Loading and Unloading

Although all vehicles can be loaded and unloaded with operator assistance, both tractors and unit-load movers can have automatic load and unload features. The tractor-trailer arrangement can have an automatic uncoupling option. More common are options whereby the trailers have rollers on the carrying surface and the loading-unloading stations where a pusher can be used to move the load. Similar systems can be used for unit-load movers, sometimes using powered roller systems. More common is the lifting device established in Fig. 3-22. This has particular potential in manufacturing operations where materials can be brought directly into the work station.

Routes are dependent on surface conditions. Cracked and broken slab can cause discontinuity in the tape or wire guides. Inclines and declines within a plant must be considered, in which case an acceleration or deceleration feature must be specified for the equipment. External routes linked with automatic door control, internal traffic lights, and automatic ramps to cover rail lines have been used. However, external use of this equipment is not widespread, and external surfaces must be prepared very carefully, especially in regions where snow and ice are involved.

Safety

Driverless tractors are available with many more safety options than any other automatic conveying system and include such features as encounter detection, sonic detectors, and optical detectors, which will all shut the tractor down if an object is detected in the path. Additional safety devices include a strobe light, siren, and panic buttons which can override all other controls. Using warning signs and placing mirrors at corners and blind spots are good preventive measures and so is keeping the tractor speed below 5 mi/h.

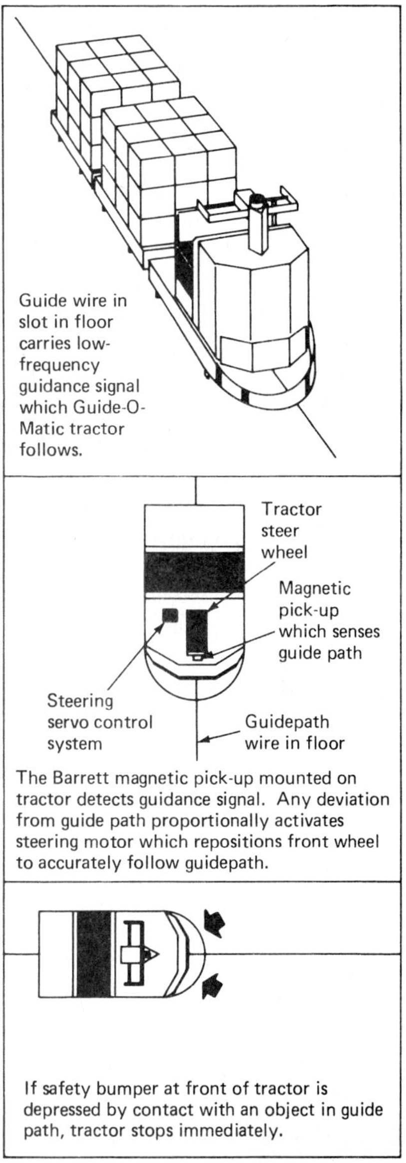

Figure 3-21 Typical features of a driverless tractor system.

ROBOTS

Robots are programmable machines capable of automatically moving individual parts or objects over precise paths in space.[1] A robot can also be programmable so that it is able

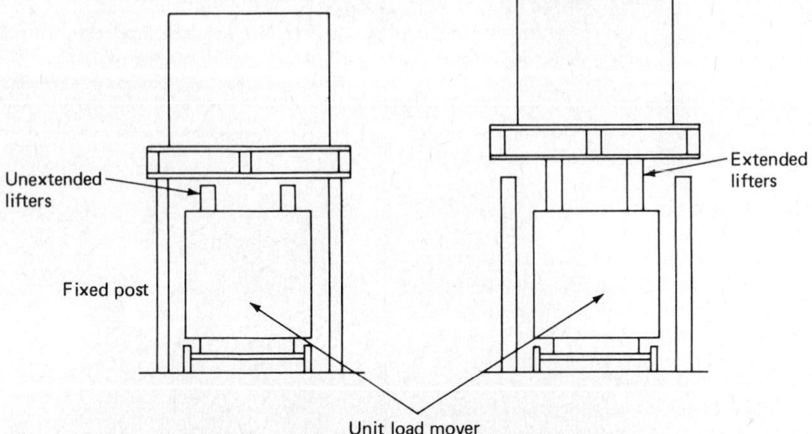

Figure 3-22 Individual unit load or pallet mover.

to move parts through different paths, capable of performing repetitive motions, able to duplicate the movements of the human arm by moving parts through four axes in space.

Applications

Present applications related to materials handling include machine loading and unloading, conveyor transfer, and pallet loading. The most practical applications for materials handling will be those areas that require repetitious manual operations, particularly those involving the interface between workers and machines. Robots are also ideal for these types of operations in poor working environments, such as those where heat, cold, fumes, or radiation exposure is present. Painting and welding are typical major potential application areas.

Design Components

Robots (Fig. 3-23) are available with a wide range of capabilities and in various design configurations. The major components include a manipulator which actually performs an operation and moves parts, a controller that stores data and directs the movements of the manipulator, and the energy source to power the robot.

A sophisticated robot with six axes of motion can perform many of the same movements as the shoulder, elbow, and wrist. Simpler, less expensive units with two degrees of freedom, called *put-and-place units,* are typically used for machine loading and should become widely used in the materials handling field in the next decade.

Manipulator. The handling of objects by the manipulator is facilitated by the use of tools that give the robot "hand" capability. The general categories for this purpose are either grippers or surface-lift devices.

Mechanical Grippers. These grippers (Fig. 3-24) are usually movable fingerlike levers paired to work in opposition to each other. They can be thought of as mechanical equivalents of the thumb and forefinger.

Surface-Lift Devices. These can include simple forklift attachments, vacuum pickups (Fig. 3-25), hooks, or magnetic devices.

Controller. The controller initiates the motions of the manipulator through a sequence at the desired points and stops the motion when required. The controller can be programmed by adjustment of mechanical cams, stops, and limit switches on the simpler types of put-and-place robots. The more sophisticated robots can be "taught" a sequence of movements by an operator. In the teaching mode, the programmer manually moves the manipulator through the motions of the operation, and the coordinates of the path are stored in the controller memory.

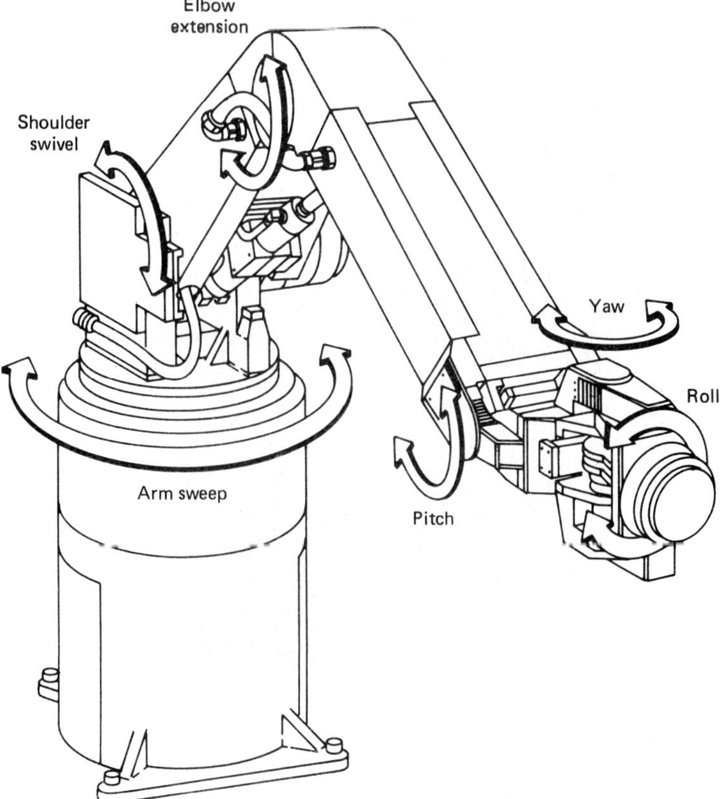

Figure 3-23 Robot with six axes of motion.

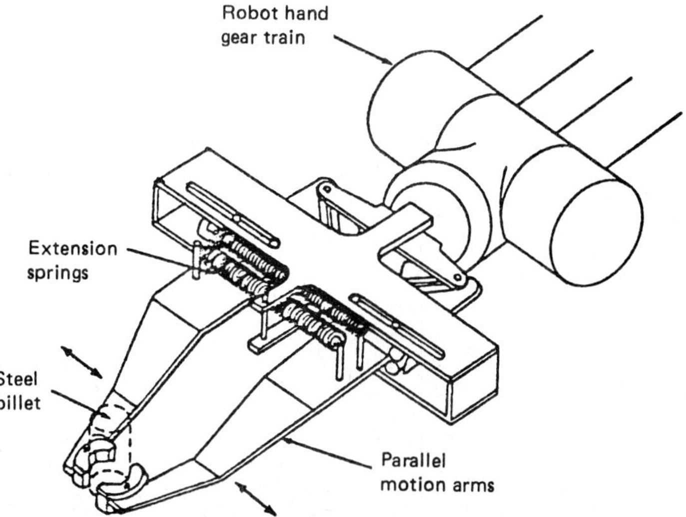

Figure 3-24 Robot grippers equipped with spring-loaded fingers.

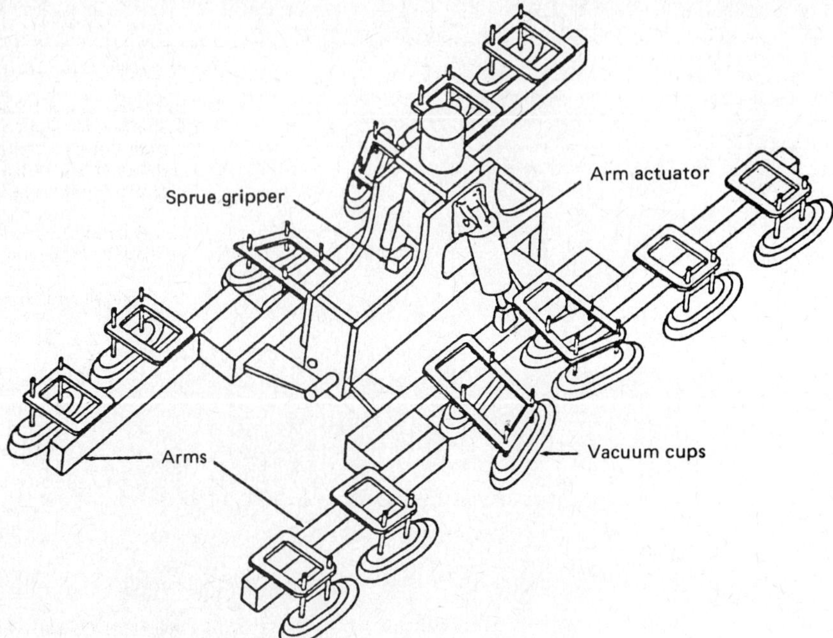

Figure 3-25 Vacuum pickup device for robot.

Energy Sources. Nonservo, or *pick-and-place* robots operate through activation of a hydraulic or pneumatic system and are the simplest, lowest-cost units. They have limited flexibility in terms of program capability and positioning capability but are highly reliable. In the operation of this type of robot, as the sequence is indexed, the manipulator members move until the present limit of travel is reached. Since there are only two positions for each axis to assume, programming can be done by adjusting the end stops for each axis to establish the operation sequence.

Servo-type robots use servo motors or valves to move the manipulator members and can be further classified into either point-to-point or continuous-path types. *Point-to-point* servo robots are programmed or taught by feeding them manipulator-position data at discrete points and, in performing a task, they will internally select a path to that point. *Continuous-path* servo robots are programmed or taught to follow a precise path and are used for operations where movement is important, particularly in spray painting.

Future Developments

The technology of robots will be expanded in the future to include the capability to discriminate differences in objects by optical- or mechanical-sensing devices which would send a feedback signal to the controller which will make a decision to initiate a movement command to the manipulator. Further future developments include speech recognition for robot programming and three-dimensional optical-sensing devices. Also, while robots now in operation are generally large, floor-mounted units, future robots will also include table-mounted units able to assist in small subassembly and final assembly operations.

Planning Considerations for the Use of Robots

There are four points that must be considered when evaluating the feasibility of using a robot in materials handling. They are rate of handling, weight of the object, orientation of the object, and number of different items to be handled.

Rate of Handling. Robots are not high-speed handling equipment. If the handling rate is greater than 15 items per minute, another approach should be considered.

Weight of Object. The weight-handling capacity of robots is presently 500 to 2000 lb, depending on the type of robot. The heavier the load, the lower the handling rate.

Orientation of the Object. Position of the object is important and should be consistent. A primary limitation of current robots is the precise orientation required of parts to be picked up by the robot and, hence, a feeding or positioning mechanism to the robot itself is often required.

Number of Items to be Handled. Setup time for product changes can be reduced by quick changeover grippers and automatic program selecting capability. In cases where dissimilar parts are handled in the same operation, a multipurpose gripper or "hand" should be used, along with a sensing device that can command the robot to switch to a preset program.

REFERENCE AND BIBLIOGRAPHY

1. Tanner, W. R.: *Industrial Robots,* Vol. 1 and 2, Society of Manufacturing Engineers, Dearborn, Mich., 1979.
2. *Automated Storage/Retrieval Systems Planning Guide,* Clark Handling Systems, 525 W. 26th St., Battle Creek, MI 49016.
3. *Automated Storage/Retrieval Systems Justifications,* Clark Handling Systems 525 W. 26th St., Battle Creek, MI 49016.
4. Industrial Robots, *Modern Material Handling,* April 1980.

chapter 2-4

Mobile Materials-Handling Equipment

INTRODUCTION

The group of equipment that is described as mobile materials-handling equipment is made up of machines that essentially depend on a self-contained power source for movement and are independent in their movement route. The equipment, being self-contained material movers, provides a flexible, relatively inexpensive transportation link between plant activities. This broadly classified group of equipment includes devices and equipment from the simplest two-wheeled hand truck to highly sophisticated movers controlled by computer-based systems.

Within the mobile materials-handling equipment group there is a wide array of general-purpose and specialized material movers. Basically, there are two broad categories of mobile equipment. The powered equipment depends on a built-in power source for its operation. The unpowered device relies on a detachable prime mover, either a piece of powered equipment, or in many cases, the equipment operator. The least complex equipment provides transportation between two points without positioning or lifting capabilities. Other units lift or roughly position the load being transported as well as move the material. The multiple-axis movers transport the load; they also have a position capability along two or more axes to accomplish loading and unloading.

Generically, mobile materials-handling equipment falls into five groups, each of which will be discussed in this chapter:

1. Floor trucks and operator-powered movers
2. Powered lift trucks

3. Burden carriers
4. Tractors and tractor trains
5. Mobile industrial cranes

APPLICATION CONSIDERATIONS

Equipment Utilization and Selection

From available records, it appears that mobile equipment often has a low level of utilization. Powered equipment is often employed well beyond its economic life, generating penalty costs in spare-parts inventories, maintenance, and productivity.* Eleven thousand (11,000) engine hours, or approximately 5 years, has been calculated to be the average economic life of a powered vehicle. Other general considerations in establishing equipment requirements include:

- Unit-load condition and size and center of load
- Terrain, environment, and aisle width in the movement area
- Length, type, and frequency of moves
- Positioning requirements of load(s)
- Hazards inherent in movement area
- Operating economies and maintenance ease
- Maintenance and spares
- Standardization of equipment
- Critical nature of operation(s) serviced

Factors in Wheel Selection and Use

Solid Wheels. These are made in semi-steel, forged steel, or molded plastic, hard rubber, and composite materials. They should be limited to small diameters and low-speed movement and should not be used to transmit power. They have low resistance to roll, but a short life span when overloaded or subjected to rough floor conditions. They will cause load vibration because of a lack of cushioning.

Rubber-Cushioned Tired Wheels. These consist of a metal wheel having a machined diameter onto which a rubber tire is pressed or molded. It has the lightest load-carrying capacity of those used on mobile equipment. Minimal power is required to move material, since rolling friction is minimized.

Oil-Resistant Tired Wheels. The tires are made of special oil-resistant rubber compounds which will resist the degrading effects of oil on rubber.

High-Traction Tired Wheels. The tires are made of rubber impregnated with abrasive or other materials to give additional traction on ice or in wet conditions.

Low-Power Tired Wheels. The tires are fabricated from rubber compounds that offer minimum roll resistance and have lower power requirements, causing less drain on battery-operated equipment.

Nonmarking Tired Wheels. The tires use a rubber compound filler other than carbon to avoid floor marking and contamination.

Conductive Tired Wheels. The tires avoid the chance of static sparking in hazardous or explosive environments by maintaining vehicle-to-floor conductivity.

*"When to Replace Your Lift Truck," *Material Handling Engineering,* July 1980.

Laminated Tired Wheels. The tires for these wheels are made up of sections of pneumatic tire carcasses threaded onto a steel band. Such tires are extremely tough, with a harsh ride. They are well suited to littered environments, such as scrap yards, and trash handling.

Polyurethane Tired Wheels. Though more expensive than rubber, these wheels have a significantly higher load-carrying capacity and are less susceptible to cuts than most rubber and rubber-compound wheels. Wheel hardness of polyurethane tires results in a harsher ride and increased plant floor damage.

Inflatable Tired Wheels. These wheels have vulcanized, reinforced rubber tires similar to automotive tires. The tires are both tube and tubeless. They generally carry a lower load rating for their size than solid-tire wheels. Their use will provide greater load cushioning, higher speed capability, easier maintenance, and less floor damage.

Factors in Internal-Combustion-Engine Selection and Use

Internal-Combustion Engines. These are used in outdoor applications, in well-vented interiors, in nonhazardous environments, and where noise is not a factor.

Industrial Engine. Typically, this heavier engine is designed to operate in a lower rpm range than an automobile engine. It can be expected to give about 10,000 h of useful life before overhaul. At an equivalent operating speed of 20 mi/h in an automobile, this would equate to 200,000 mi.

Automotive Engine. This is of lighter construction than the industrial engine and, because of the quantities in which it is produced, is of relatively lower cost. It generally operates most efficiently in a higher rpm range than the industrial engine and can be expected to give about 7000 h of useful life prior to overhaul. This life is equivalent to about 140,000 mi of automobile travel. An advantage of this type of engine is the availability of replacement parts through automotive supply firms.

Air-Cooled Engine. This is restricted to lighter-duty applications where weight, size, and initial cost are the prime concerns. The absence of a separate cooling system is a distinct advantage, although this engines life expectancy is a relatively short 1500 to 2000 h of operation.

Diesel Engine. Typically, this type is installed in large pieces of equipment where the additional size and cost is not significant. However, because of recent improvements in engine design, diesel engines are becoming more prominent in smaller trucks. This is largely due to the reduced need for periodic maintenance, greater fuel economy per hour of operation, and longer expected life—up to 20,000 h.

Factors in Battery-Powered-Vehicle Selection and Use

Battery-Electric Equipment. This is mechanically simpler in design than engine-driven equipment. Typically, the high-torque dc electric-drive motor is coupled directly to the drive axle through a constant-mesh drive train. An electronic SCR speed-control device regulates the motor's revolutions per minute through operator foot control. Direction is reversed electrically with a delay interlock to avoid reversing motor direction while in motion.

Storage Battery. These must be replenished frequently either by recharging or by exchanging them for fully charged batteries. Batteries used in a given piece of equipment should provide ample power to operate effectively for an 8-h day as determined by their ampere-hour (Ah) rates. The Ah rating, to some degree, limits the effective operating range of battery-operated equipment and requires that routine schedules for replenishment are followed. Also, because of the weight of a large storage battery, equipment application is sometimes adversely limited.

Advantages of Battery Vehicles. The advantages are low fume emission and heat contamination, quietness and cleanliness, and generally lower maintenance requirements.

Types of Batteries. The two primary types of batteries used are lead-acid and nickel-iron-alkaline. A lead-acid battery will provide 2.0 to 2.3 V per cell, while the nickel-iron-alkaline battery will provide 1.2 V per cell. Voltages used for modern battery-powered mobile equipment are 12, 24, 36, 48, and 72, with some higher voltages used in larger equipment.

Advantages. The advantages of the lead-acid battery are a lower initial cost, high ampere-hour capacity, and low resistance to self-discharge. The nickel-iron-alkaline battery is desirable because of its longer life expectancy, resistance to physical damage, noncorrosive electrolyte (KOH), and more rapid and less critical recharge rates.

Recharging Times. These are adjusted for different batteries by dividing the Ah rating of the battery by the 8-h Ah rating of the charger and multiplying by 8. For example, a battery having a 600-Ah rating and a 450-Ah charger will require

$$(600 \div 450) \times 8 = 10.64 \text{ h}$$

FLOOR TRUCKS AND OPERATOR-POWERED MOVERS

This type of equipment is the most fundamental materials-handling aid available. The basic simplicity permits easy adaptation for single-purpose application. Standard catalogs indicate the wide variety available, often designed for specific industries. However, custom design may be specified with very little, if any, cost penalty.

Generally, floor trucks are described as follows.

Two-Wheeled Hand Trucks

Two-wheeled hand trucks (Fig. 4-1) are essentially levers on two wheels. The axle connecting the wheels serves as the fulcrum of the lever and carries up to 80 percent of the total load moved. The two-wheeled cart is normally used for short nonrepetitive moves of smaller loads over smooth floors. Carts are generally 48 to 64 in (1.2 to 1.6 m) high,

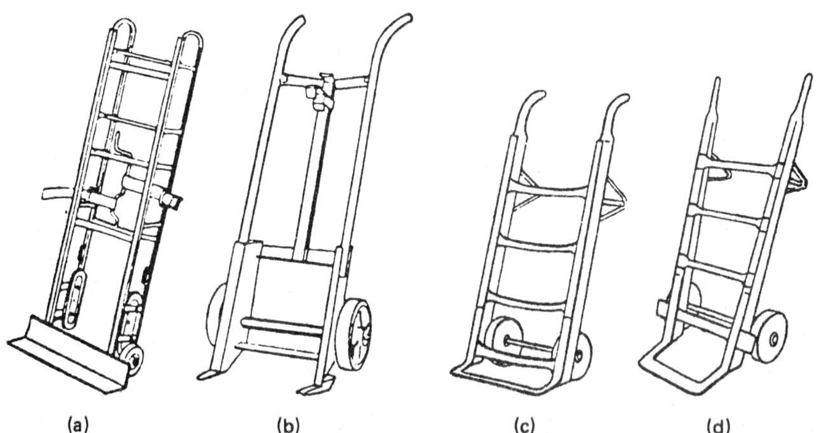

Figure 4-1 Two-wheeled hand trucks. (*a*) Appliance type, (*b*) drum and barrel mover, (*c*) general type with Western handle, (*d*) general type with Eastern handle. (Reproduced with permission from *Material Handling Engineering Handbook and Directory, 1979–1980*, published by *Material Handling Engineering* magazine, Cleveland, Ohio.)

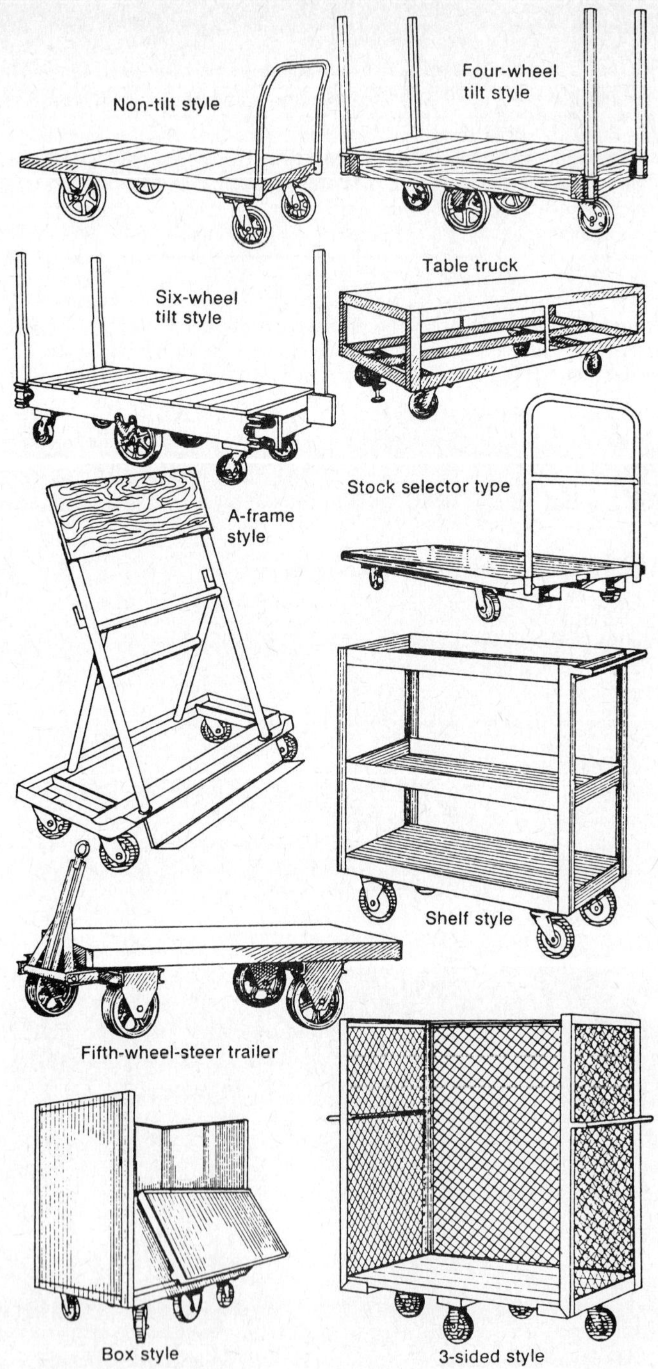

Figure 4-2 Factory trucks and wheel arrangement patterns. (Reproduced with permission from *Material Handling Engineering Handbook and Directory, 1979–1980*, published by *Material Handling Engineering* magazine, Cleveland, Ohio.)

and are designed to carry a variety of materials in bags, barrels, bales, boxes, and bins. Typical accessories include height extension, stair climbers, safety brake, spread clamps, and straps.

Dollies

Dollies are smaller-wheeled platforms upon which a load is placed for short distance and intermittent moves. Typically, dollies are fitted with caster-type wheels and are either pulled or pushed by an operator.

Factory Trucks

Factory trucks (Fig. 4-2) are wheeled platforms or containers either moved by an operator or towed by detachable power units. There is a wide variety of devices in this group and an even wider variety of uses for materials movement and as mobile storage.

The hand factory truck is hand-powered, guided by the direction of the moving force, and closely related to the dolly. Several wheel-arrangement patterns are available with tradeoffs between maneuverability and stability.

The towed factory truck is connected to the prime mover by a tow bar which provides the steering direction. Both two-wheel and four-wheel steering are available on towed factory trucks. Two-wheeled steering is generally the least expensive and most commonplace. Because of the steering geometry involved, each truck will follow a turn of shorter radius than the preceding vehicle. As several of these units are connected in trains, the continual tightening of turns requires more space for maneuvering.

The four-wheel-steered truck, with properly adjusted steering, is capable of following the same path as the vehicle in front of it. Where long trains are economically justified and desirable, the four-wheel-steered devices may be used to minimize commitment of valuable manufacturing space to aisles.

The Semilive Skid

The semilive skid is a rectangular platform or box having two wheels on one end and two fixed supports on the other. The end having the fixed supports is also fitted with a heavy pickup pin to which a two-wheeled jack is attached. The jack and handle are used as the lifting device and tiller, allowing the skid to be maneuvered by the operator.

Hydraulic-Lift Trucks

Hydraulic-lift trucks (Fig. 4-3) are used for moves at the workplace and occasional moves over short distances. They generally range in capacity from 2500 to 4500 lb. The operator uses a jacklike manually operated hydraulic system to elevate a loaded pallet sufficiently off the floor to move it. Some units use an electrically driven hydraulic system to lift the load, often above the maximum 5 in of the manually operated system. Hydraulic-lift trucks generally use forks for lifting pallets or platforms for special containers and for moving and positioning heavy loads such as dies.

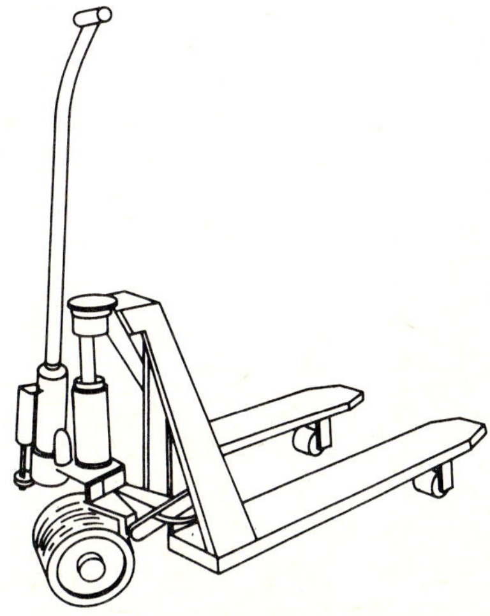

Figure 4-3 Hydraulic-lift truck.

POWERED-LIFT TRUCKS

This equipment group represents what is probably the largest and most varied of equipment for materials handling. The powered-lift truck owes its popularity to its versatility, being able to easily pick up a unit load, transport it quickly in a variety of environments, and then position the load vertically at almost any point within the capability of the equipment. Depending on the volumes involved, they become less economical for moves over 300 ft (90 m) since rated speeds are generally between 5 and 10 mi/h (7 and 14 km/h). Powered lift trucks are usually fitted with lifting forks to carry a unit load, although a wide variety of special load-carrying attachments can be used in place of forks. Power for lift trucks is either by internal-combustion engine or battery electricity.

The various pieces of equipment in this group can be operated over a variety of terrains, depending upon the design and, specifically, the wheel and tire combination used. Load-carrying capacities from 1000 to over 40,000 lb (450 to 18,000 kg) are common. Large vehicles are available with capacities in excess of 100,000 lb (45,000 kg). The very large vehicles are generally used outside, particularly for the moving and stacking of shipping containers.

Establishing aisle widths and their relation to fork-truck selection are critical when significant storage areas are involved. Clearly, the narrower the aisles, the more rows of storage. Equipment manufacturers have been ingenious in designing specialty trucks to operate in narrow aisles. It should be noted that manufacturers specify equipment turning circles and, thus, aisles will require space to aid in fork-truck maneuverability. Specialty trucks designed to operate in narrow aisles permit better space utilization, but tend to trade off some aspect of performance, a factor to be considered in specifying specialty as opposed to general-purpose equipment.

Truck capacity is generally calculated as follows (see Fig. 4-4):

A = distance, in, from center of front axle to heel of fork
B = distance, in, from heel of fork to center of load
C = distance ($A + B$) from center of front axle to center of load
D = length, in, of load on fork
W = weight of load, lb

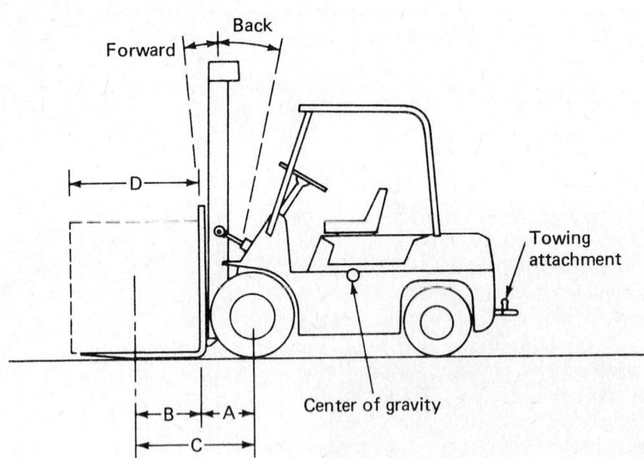

Figure 4-4 Rated truck capacity and counterbalanced truck.

1. Inch·Pound Rating

$$\text{Inch}\cdot\text{pound rating} = W \times C$$

2. Maximum Load Length for Given Load

$$C = \frac{\text{inch}\cdot\text{pound rating}}{W}$$

3. Maximum Load for Given Load Length

$$W = \frac{\text{inch·pound rating}}{C}$$

A specific example is given to illustrate the actual calculations.

1. Truck is rated 3000 lb (W) at 20 in [3000-lb load which has a center 20 in (B) from heel of fork].
2. Distance from center of axle to heel of fork is 10 in (A).
3. Pallet load to be handled is 2000 lb:

$$C = A + B = 10 + 20 = 30 \text{ in}$$
$$\text{Inch-pound rating} = W \times C = 3000 \times 30 = 90{,}000 \text{ in·lb}$$
$$C = \frac{\text{inch·pound rating}}{W} = \frac{90{,}000}{2000} = 45 \text{ in}$$
$$B = C - A = 45 - 10 = 35 \text{ in}$$
$$D = 2 \times B = 2 \times 35 = 70 \text{ in allowable load length}$$

4. When selecting attachments, refer to the truck manufacturer to determine the amount of negative effect the attachment has on the truck's useful load-carrying capacity.

Aisle widths are generally established as follows:

A = aisle width
TR = turning radius of truck
L = load length
C = aisle clearance (total on both sides)
AX = distance from rear corner of load to centerline of axle:

$$A = TR + L + C + AX$$

The several varieties of powered-lift trucks are described below.

Counterbalanced Trucks

The counterbalanced trucks (Fig. 4-4) use their large, carefully positioned weight mass to offset (counterbalance) the moved load mass. These trucks are generally equipped with a tilting mast which will "tilt" the lifting mechanism rearward from the vertical lifting position and further counterbalance the load during movement. The load is positioned fully in front of the truck so that the truck structure does not interfere with adjacent stacks of material. This minimizes the aisle widths that are required.

Straddle Trucks

The straddle trucks (Fig. 4-5) differ from the counterbalanced type in that they do not depend on weight mass to counteract the weight of the load being handled. Instead, the straddle forklift positions the two main load-carrying wheels at or forward of the material load center. The truck is extremely stable as a result of this arrangement.

The straddle design is more compact and of lighter weight than the counterbalanced type. It is necessary, when negotiating loads into or out of racks, that either the straddle truck be equipped with an extending fork mechanism (pantograph) or the racks be positioned or constructed to allow the forward wheels of the truck to enter them.

Side-Loading Trucks

Side-loading trucks (Fig. 4-6) are a unique combination of a straddle-lift truck and a narrow-aisle truck. They are used where there are narrow aisles, where rapid transportation is called for, and where long narrow loads such as pipe and bar stock are handled. Side-loading trucks do not have to be turned to engage or place loads.

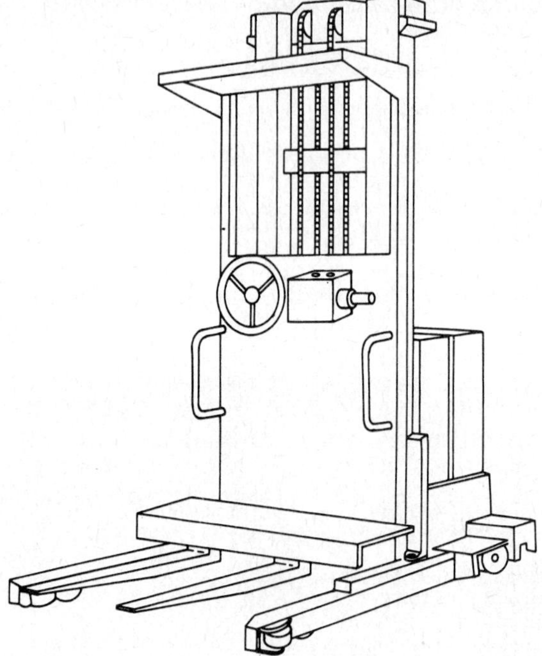

Figure 4-5 Straddle truck.

Nonrider Lift Trucks

Nonrider lift trucks (Fig. 4-7) are those where the operator walks along with the truck, directing the operation through a control unit attached to the truck. These units have basically the same features found in larger counterbalanced and straddle trucks. They are used for lifting and stacking light loads and moving these loads short distances.

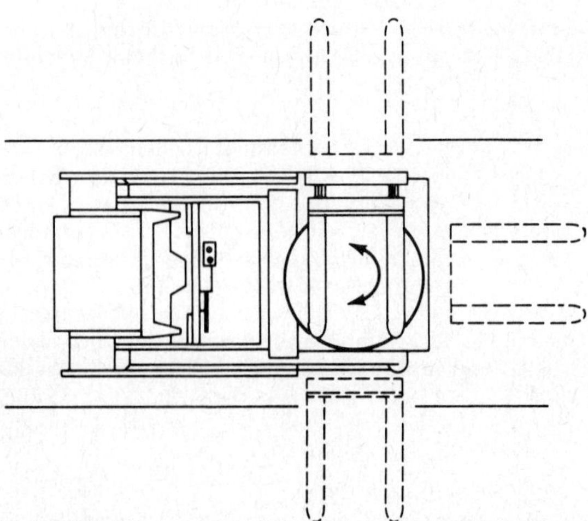

Figure 4-6 Side-loading truck.

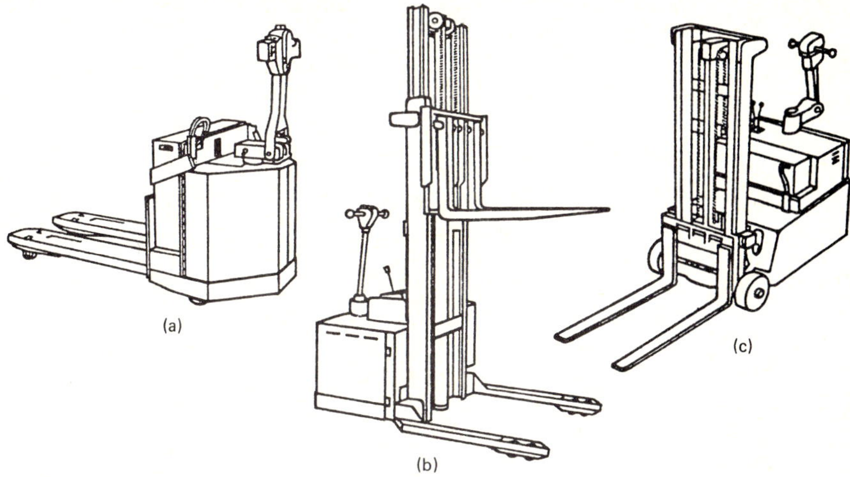

Figure 4-7 Nonrider lift trucks. (Reproduced with permission from *Material Handling Engineering Handbook and Directory, 1979–1980*, published by *Material Handling Engineering* magazine, Cleveland, Ohio.)

Straddle Carriers

Straddle carriers (Fig. 4-8) are large-capacity, highly maneuverable, powered lift trucks. To load and unload, the vehicle is driven over the unit load(s). The actual loading and unloading is extremely fast, although precise positioning of loads requires other methods. Unit loads can be transported at rates approaching highway speeds.

Order-Picker Trucks

Order-picker trucks have an elevated platform forward of the mast from which the truck and the platform can be operated. Typically, the trucks are used for picking partial loads in narrow aisles to heights of 24 ft, allowing for significant labor and space saving.

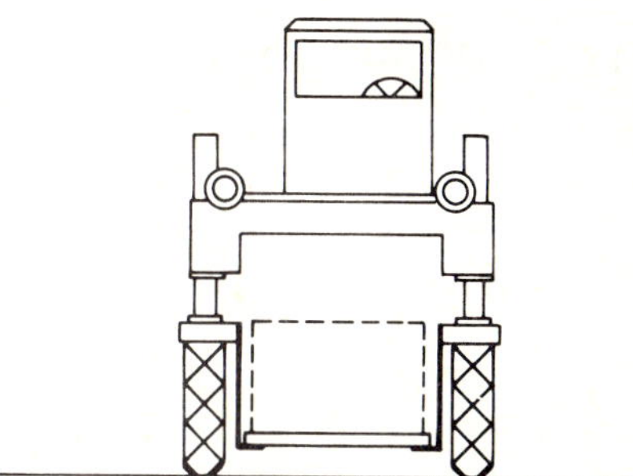

Figure 4-8 Straddle carrier.

Materials-Handling Attachments

The most widely used attachments are the forks themselves. They can be set at various widths and generally range between 30 and 60 in (80 and 160 cm) in length. The forks should be at least two-thirds the length of the maximum load to be lifted.

Standard two-stage uprights provide a lift height of approximately 18 ft (5.5 m), and three- and four-stage uprights provide heights to 20 ft (6 m). Certain specialty vehicles are designed to operate above 20 ft (6 m). The difference in fork height and total extended height is generally 4 ft (1.2 m), reflecting the height of the backrest. For low buildings, *free-lift* trucks should be specified. This feature permits the forks to be raised to lift loads to nearly half the total lift height without extending the uprights.

Frequently, a forklift truck will be fitted with an attachment or combination of attachments which allows the vehicle to perform special handling functions or simply allows it to operate more efficiently in a given situation. In some cases, these attachments replace the conventional forks for handling products which the forks cannot. In other instances, the attachments are used to augment the original fork function by giving the load-carrying forks additional motions.

When selecting attachments, it is always wise to consult with the truck manufacturer since attachments have a negative effect on a truck's useful load-carrying capability. When attachments are installed, the truck's information plate must be restamped, indicating the new effective truck capacity as required by OSHA 1910.178(4).

Attachments usually limit a forklift truck to a specialized function and, to some extent, limit its overall in-plant versatility. Some of the more simply designed attachments mount on the fork attachment rails and require only a few minutes to install or remove. The more complicated attachments, particularly those requiring hydraulic connections, should be considered as permanent conversions.

The following is a list and a brief description of some of the more common attachments (Fig. 4-9):

1. **Ram** A single projection, mounted in place of the forks for carrying coiled materials which can be easily entered horizontally. Rams have a variety of lengths and diameters to handle a variety of products, from steel coils to rolled carpet.

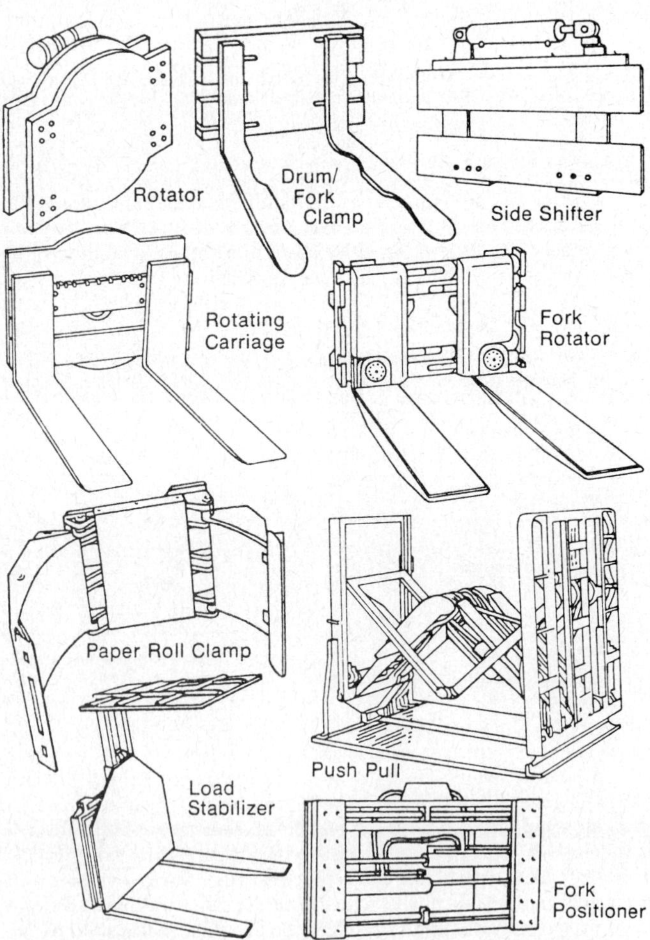

Figure 4-9 Common materials handling attachments. (Reproduced with permission from *Material Handling Engineering* 1979–1980, published by *Materials Handling Engineering* magazine, Cleveland, Ohio.)

2. **Barrel Attachment** Used to grasp the top seam of a steel drum and transport it in the vertical position.
3. **Concrete-Block Fork** This and the similar brick fork are designed specifically for handling stacks of masonry products without pallets.
4. **Paper-Roll Clamp** Specifically designed to carefully grasp and transport rolled materials in the vertical position. It is frequently combined with a rotator which allows the roll to be carried horizontally in the case of loosely rolled or easily damaged materials.
5. **Push-Pull** Uses a polished platen instead of forks to carry the load. Its purpose is to position loads in dense environments without the use of a pallet. In place of a pallet, a thin slip sheet is used under the load. This sheet is grasped by hydraulic clamps and pulled into the truck platen for loading and pushed off again into its next position.
6. **Bale Clamp** Used to grasp and carry baled materials and depends on hydraulic pressure to grasp the bales from the sides.
7. **Scoop** Used to handle loose or granular bulk material and consists of a metal bucket mounted in place of the fork, with a dumping capability usually provided. Tilting the bucket for loading and transport is accomplished by tilting the lift truck's mast forward and back.
8. **Squeeze Clamp** Used to grasp the sides of boxed products in a manner similar to the bale clamp, except that the grasping arms are smooth and deliver an even pressure to the carton to avoid damaging its contents. This device eliminates the need for pallets. It requires, however, additional side clearance on each side of the material moved to accommodate the clamps.
9. **Top-Handling Lift** Used to handle folded cartons by hooking into the folded lip of the top carton. The most important advantage is that extremely high storage density can be accomplished since only minimum side clearances are required without the use of either pallets or slip sheets.
10. **Side Shifter** Used with almost any type of attachment as well as forks, the side shifter allows loads to be positioned accurately from right to left without relocating the truck. Its major function is to speed the positioning of loads and to minimize rack space between loads. The side shifter will also reduce wear on the truck itself by reducing repositioning.
11. **Adjustable Forks** Where a variety of pallet and load sizes are encountered, adjustable forks are used. While most fork arrangements are manually adjustable, the mechanically adjustable forks allow the operator to accomplish the operation while remaining in the driver's seat.
12. **Load Stabilizers** To assure that loosely arranged and unstable loads are firmly contained during transit, various load stabilizers are available. Such a device is essentially a vertical clamp which exerts a downward pressure on a load and thus holds it in position while it is being moved.
13. **Clamping Forks** Similar in design to the fork positioner, clamping forks can be used to pick up loads in a conventional manner or may be used to clamp loads between the forks. This device is quite commonly used with special notched forks for transporting drums.
14. **Rotator** The use of a rotator allows a load to be rotated through 360 degrees, generally for dumping. The rotator is used with unit-load devices that fully enclose the forks and thus remain attached to the fork during rotation. They are also used with various clamping devices when rotation is required.
15. **Extended-Reach Forks** Commonly used with straddle forklifts to enable the truck to reach a load in the racks while the forward wheels remain outside of the rack space. The attachment also allows racked materials to be reached when two-deep storage is used. The reaching mechanism consists of a hydraulically operated pantograph system between the truck mast and the forks.

BURDEN CARRIERS

In the manufacturing process where sufficient volumes are involved, conveyor systems are often used to move materials from point to point. When smaller volumes or several moves of varying density are involved, however, a fixed-platform device is often used. These fixed-platform vehicles depend on an auxiliary loading and unloading method and are not tied to a specific unit-load module. Such devices are called burden carriers.

Burden carriers come in a wide variety of sizes and shapes. They are available in two basic types (Fig. 4-10). One is the walkie (nonriding) type and the other is the riding type. Both are available with battery-electric and internal-combustion power sources. They are usually limited in load-carrying capacity. High loads are generally handled by other types of handling equipment.

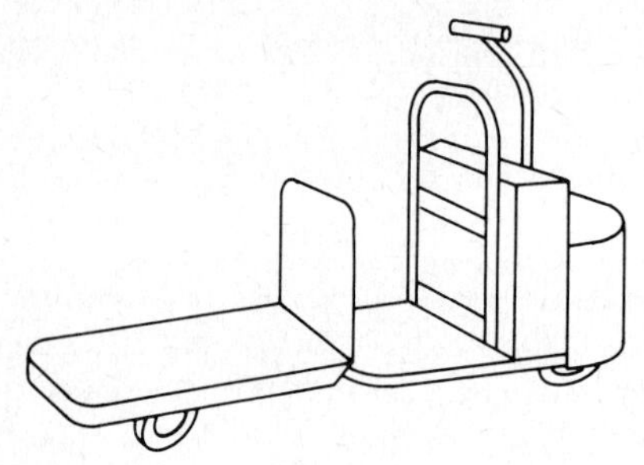

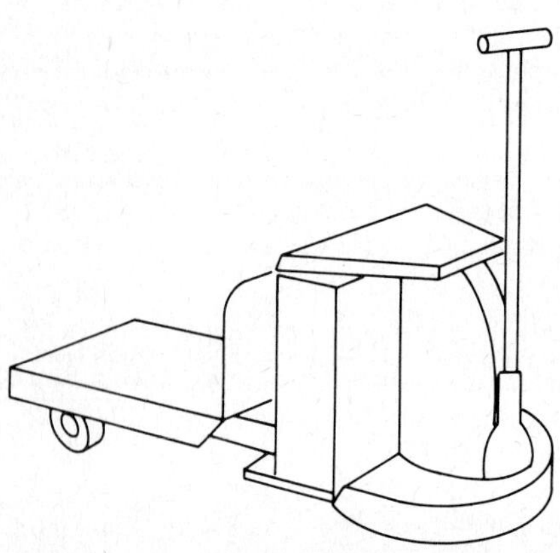

Figure 4-10 Types of burden carriers.

Walkie Burden Carrier

The walkie burden carrier is typically a three-point suspension hauler using battery-electric power, although some units are available that are powered by a small air-cooled engine. They are similar in design to the previously discussed walkie lift trucks, except that they have a fixed platform. Load ranges of 1000 to 3000 lb are available and application is limited to noncontained loads. Loading is generally done by hand or, in the case of heavier loads, by hoists and cranes.

Rider Burden Carrier

The rider-type burden carrier is often tailored to a variety of special applications such as personnel carriers, fire trucks, and portable maintenance shops. In its simplest form, the truck provides a driver's seat and a flat load-carrying bed. In this configuration it serves most commonly as a miscellaneous hauler to deliver supplies and materials in-house for distances of more than 300 ft (90 m). The power source for the rider-type truck is fairly evenly divided between air-cooled gasoline engines and battery-electricity. Both three-point and four-point suspensions are common, with an operating suspension system being incorporated in many larger units, along with pneumatic tires. These vehicles are able to negotiate rougher terrains and may attain speeds of up to 20 mi/h (30 km/h).

TRACTORS AND TRACTOR TRAINS

The term tractor (Fig. 4-11) refers to a detachable power source supplying locomotion to one or a group of load-bearing vehicles not having on-board power. The tractor is a steerable mover which is directed by an operator. They are generally classified according to their drawbar pulling rating (DPR) into small, medium, and large sizes.

On all grades above 5 percent, the individual manufacturer should be consulted since a variety of other factors which must be considered vary with individual tractor designs.

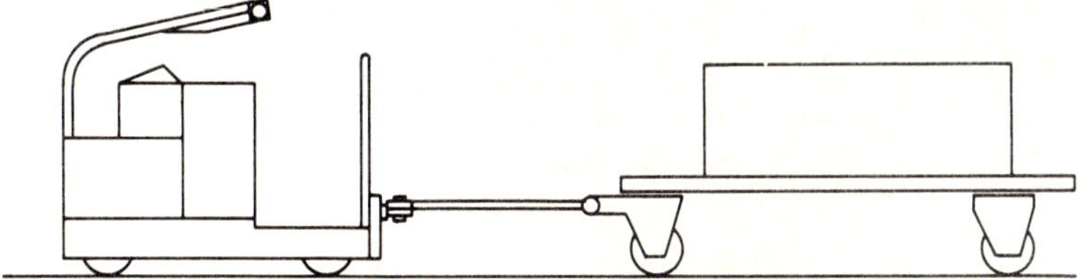

Figure 4-11 Tractor used in industrial applications.

Minimum safety criteria for industrial tractors are covered in OSHA Standard Section 1910.178 (Powered Industrial Trucks) and should be referred to when equipment is being selected.

The main application for these vehicles is the movement of goods in volume over distances too long to be economically moved by fork trucks—approximately 300 ft (90 m). Since the tractor trains are not self-loading, a system of tractor loading stations and surplus tractors and trailers is required, involving constant hitching and unhitching of tractors and trailers. An alternative is to use a forklift as a tractor with the operator of the fork truck loading the trailers.

Apart from fork trucks, five types of tractors are used for most industrial applications.

Highway Tractors

Highway tractors are typically used in over-the-road applications and are relatively specialized to serve this purpose. They do, however, find application in large manufacturing complexes for the movement of materials between remote locations where warranted by the speed and density of materials flow. This type of tractor is also frequently used in factory shipping yards for the positioning of both loaded and unloaded semitrailers.

Walkie Tractors

Walkie tractors are the smaller variety of industrial tractors. These tractors are battery-electric with motive power, braking, and steering being provided by a single wheel or a close-coupled pair of wheels. The drive mechanism is tiller-controlled through hand controls, as in other walkie equipment, and dead-man controls are provided. Two other wheels are provided for stability at the rear of the unit. A variety of coupling devices are available for attaching the tractor to trailers and semilive skids.

Walkie-Rider Tractors

The walkie-rider tractor is essentially a larger version of the walkie tractor. The major differences are that in a walkie-rider tractor a platform is provided for the operator to stand on during operation and two travel speeds are provided. A slow speed comparable to the operator's walking pace, on the order of 3 mi/h (5 km/h) and a higher speed of roughly 7 mi/h (11 km/h) is common in this type of equipment. Owing to higher operating speeds, these tractors have wider operating ranges. Because of the longer range of these tractors, larger-capacity batteries are used and, therefore, the units are heavier and larger than pure walkie tractors.

Rider Tractors

Rider tractors are available in both stand-up and sit-down configurations. The stand-up variety is more compact and generally applicable to more congested situations. The sit-down tractor is generally larger and is used where higher speed and longer distances, up to ½ mi (0.8 km) or more, are encountered. Battery-electric and internal-combustion engines are used as power sources in both versions; however, battery-electric power is more prevalent in the stand-up models, and the internal-combusion engine is the frequent choice in sit-down tractors.

Specialty Tractors

Specialty tractors are usually confined to very heavy load applications and are often built as an integral part of the load carrier itself. Two more common applications of these specialty tractors are large bulk-handling carriers for molten metals and granular materials and for spotting of railway cars.

MOBILE INDUSTRIAL CRANES

Mobile industrial cranes (Fig. 4-12) serve a variety of plant and production-related materials-handling functions. They are especially adaptable to loads of large or unusual

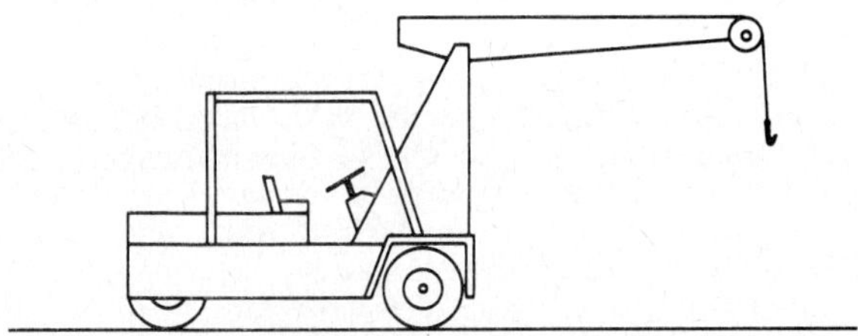

Figure 4-12 Mobile industrial crane.

size and where careful placement is required. In some applications, they are used only to position a given load, while in other applications they are used as both prime mover and positioner.

Mobile cranes differ from other plant hoisting equipment in that they operate independently of any supporting structure. The primary advantage of a crane is its ability to reach into places not normally accessible by other types of materials-handling equipment. With the exception of straddle cranes, the industrial crane depends on a boom for its reach and lift capability. It is the positioning of the boom that ultimately determines where a load will be placed and how large a load can be safely lifted.

The following text discusses the types of mobile cranes in use.

Portable Hand-Powered Crane

The portable hand-powered crane is similar in design to a small manual-lift truck, except that the load-carrying forks have been replaced by a boom and hook. This equipment is commonly used to move and position work pieces into and out of process equipment where volumes do not warrant a permanently installed hoisting system. It is also frequently found in maintenance and repair shops to assist in the disassembly and reassembly of in-plant equipment. Lifting is accomplished either through a hand winch and cable system or a manually operated hydraulic system. Typical lifting capacities are limited to 2000 lb (900 kg) or less.

Stevedore Crane

The stevedore crane is a nonswinging crane which requires that the hook be positioned by maneuvering the entire vehicle. This limits its use to relatively unobstructed areas. The boom may be extended outward by the operator to reach the load and returned back to a position closer to the vehicle for transport. The crane is a relatively fast vehicle which is used to pick up a load and transport it to a final destination.

The front, load-carrying wheels are also the powered wheels, with steering being achieved by the trailing wheels. Both three- and four-point suspensions are used, and the crane is often used to tow factory trucks while also loading and unloading them. Typical load capacities range from 2 to 4 tons (1800 to 3600 kg).

Swing-Boom Crane

The swing-boom crane is a larger-capacity crane than the stevedore crane and is used more for positioning loads than for transportation. The boom structure is constructed so that it can be rotated by the operator through 180°. Outriggers are provided for stability. They are usually powered by diesel or spark-ignition engines, with battery-electric power also being available.

Full Revolving Cranes

Full revolving cranes are capable of swinging a load through a full 360° and are generally the largest of mobile cranes. Their use is normally one of positioning loads as opposed to transportation. This type of crane will often be mounted on a truck-type chassis for rapid movement between jobsites. Power is provided by diesel or spark-ignition engines through direct hydraulic torque converters.

Load-lifting capacities at the boom's most upright position can be as high as 100 tons (90,000 kg) with reaches in excess of 100 ft (30 m) possible. Power is provided by diesel or spark-ignition engines through direct hydraulic torque converters. Industrial applications are almost totally limited to construction activities and maintenance of large structures.

Straddle Crane

The straddle crane has no boom but has a wheel-mounted framework on which are mounted two hoists. These hoists are capable of moving within the limits of the framework for precise load positioning. The straddle crane is related to the straddle carrier. It is a highly versatile crane, finding application as both positioner and mover of materials. In addition, its design is such that it is extremely stable and can move at relatively high speeds.

The framework consists of four vertical columns mounted above the vehicle's high flotation wheels, supporting two horizontal crane rails which carry the traveling hoists. Load-carrying capacity ranges from approximately 10 to 60 tons (9000 to 55,000 kg) per hoist for an aggregate capacity of approximately 20 to 120 tons (18,000 to 110,000 kg).

A variety of power systems are used, all of which are engine-driven. Hydraulic systems are those most commonly used for transport, hoisting, and positioning. However, one manufacturer employs an engine-driven electrical power plant to operate the crane's functions through electric motors.

The straddle crane is capable of operating in high-density areas and is highly maneuverable since all four of its wheels can be turned and powered independently. Common applications are in steel storage yards, loading and handling of shipping containers, commercial concrete castings, truck and car loading, and boatyards. Special load-handling devices are easily adapted to this crane, increasing its versatility.

chapter 2-5

Warehousing and Storage

INTRODUCTION

In the overall materials-handling system, warehousing provides the facilities, equipment, personnel, and techniques required to receive, store, and ship raw materials, goods in process, and finished goods. Storage facilities, equipment, and techniques vary widely depending on the nature of the material to be handled. Characteristics of materials, such as size, weight, durability, shelf life, and order lot size, are factors in designing a warehousing system and in solving warehousing problems.

Economics is also of great importance in the design of warehousing systems. Storage and retrieval costs are incurred, but add no value to the product. Thus, the investment in storage and handling equipment and in floor space must be based on minimizing unit storage and handling costs.

Other factors to be considered in designing warehousing systems include control of inventory size and location, provisions for quality inspection, provisions for order picking and packing, staging for receiving and shipping, appropriate numbers of shipping and receiving docks, and maintenance of records:

WAREHOUSING ACTIVITIES

Warehousing activities vary according to material amounts and characteristics. However, the activities associated with warehousing generally include the following procedures:

1. Unload inbound vehicles.
2. Accumulate received material in a staging area.
3. Examine the quantity and quality and assign a storage location.

4. Transport the material to the storage area.
5. Place the material in the assigned storage location.
6. Retrieve the material from storage and place it in an order-picking line, if a picking line is used.
7. Fill orders, if applicable.
8. Sort and pack, if applicable.
9. Accumulate for shipping.
10. Load and check outbound vehicles.

WAREHOUSING ADMINISTRATIVE CONTROL

Associated with the physical handling and storage of materials is an administrative control system. The administrative control system provides for:

1. Acknowledging receipt of goods for accounting purposes
2. Verifying the quality and quantity of received goods
3. Updating the inventory records to reflect receipts
4. Locating all goods in storage
5. Updating the inventory records to reflect shipments
6. Notification to the accounting function of shipments for billing purposes

Many administrative control systems are computerized and/or automated. The cost-effectiveness of such systems over manual systems depends on such factors as:

1. The number of line items in storage
2. The number of customers served
3. The volume of goods shipped

Generally, computerization and automation are cost-effective for industries and distribution centers having many line items in storage, many customers, and a large volume of goods shipped. Distributors of grocery, health, and beauty-aid products often have computerized and automated systems.

TYPES OF MATERIALS

Materials to be stored may be broadly classified as bulk materials or packaged goods. Bulk materials such as fuels, chemicals, minerals, and grain are stored in specialized storage facilities and transported in pipes, screw conveyors, power shovels, etc. In the many industries which handle and store bulk goods, each accomplishes these tasks with very specialized equipment and techniques. This discussion will be limited to warehousing packaged goods. The reader should consult specific publications that apply to bulk materials-handling industries for particulars in bulk handling.

Within the packaged-goods classification, materials are subdivided into categories according to their state of completion in the manufacturing process. Categories include raw materials, goods in process, and finished goods.

Raw Materials. These vary widely in characteristics, depending on the industry. A few examples are raw foods and ingredients for food processors, thousands of small parts for electronics assemblers, engines and motors for manufacturers of vehicles, and wood and finishes for furniture manufacturers. Raw materials are the goods on which the manufacturing process will operate to produce salable products. Indeed, the finished goods of one manufacturer often become the raw materials of another.

Goods in Process. This refers to goods which have completed some but not all of the manufacturing process. Typically, a manufacturing process involves several operations utilizing different equipment, skills, and materials. Goods in process are stored while awaiting the next manufacturing operation. They are often stored along the manufacturing process rather than in the warehouse proper.

Finished Goods. These goods are those which have completed the manufacturing process and are stored in inventory to fill customer orders. Finished goods may be further subdivided into reserve and order-picking stock. Customer orders are filled from order-picking stock while the picking stock is replenished from reserve stock.

The amount of raw materials, goods in process, and finished goods to be handled and stored varies considerably from industry to industry. Industries having large inventories of raw materials usually are converters of bulk materials such as paper and steel. Manufacturers of highly complex equipment such as computers and automobiles require a significant amount of raw-materials storage for parts as well.

Industries having significant needs for goods in process handling and storage are those whose manufacturing process is not automated. Machine-shop and electronic-assembly operations are examples.

Finished-goods handling and storage capacity are a function of manufacturing volume and product bulk. Industries having high-volume and high-bulk output generally require a considerable handling capacity for finished goods. The paper conversion and bottling industries are examples.

CONSIDERATIONS IN WAREHOUSE PLANNING

The objective of warehouse planning is to provide space and equipment to hold and preserve goods until they are used or shipped in the most cost-effective manner. The efficient accomplishment of warehousing activities listed in Chap. 2-1 is dependent on thorough planning. The following sections discuss these considerations as a guide to the warehouse planner.

Type and Number of Materials

The type and number of materials to be stored and handled form the basis for warehouse planning. The physical characteristics of the material, to a great extent, determine materials-storage and -handling methods. Physical factors include dimensions, weight, shape, and durability. As a first step in warehouse planning, all materials to be stored must be identified and their physical characteristics listed.

The quantity of each material item to be stored must be established. The planner may require assistance from sales management for finished-goods inventory levels and manufacturing management for establishing levels of raw materials and goods in process. In establishing inventory levels, seasonality, changes in product mix, and expected turnover rate become factors.

With the inventory level of each item of stored material established, a storage unit is selected. A storage unit is the least number of an item which is stored as one unit. Examples include a single crated refrigerator, a pallet containing 20 cases of canned goods, and a bundle of pipe. The storage unit is usually selected according to the physical characteristics of the material, the available handling and storage equipment, the quantity, and the manner in which the material is received or shipped.

A storage unit may be larger than a shipping unit or a manufacturing unit. In this case, order-picking facilities are provided for items used or shipped in lots smaller than a storage unit. The service level of storage in an order-picking operation must be established as well.

Factors affecting order-picking stock levels include minimum order quantity, volume, and the physical characteristics discussed earlier. Sheet-metal screws, for example, might be in 3-months supply, while cased canned goods might be in only 8-h supply.

Storage Equipment

Storage-equipment selection follows the establishment of the reserve and order-filling inventory storage units and levels.

In the case of selecting equipment for an existing building, the constraints of the building itself must be taken into consideration. Storage equipment must be compatible with floor loading capacity, clear height beneath sprinklers and structural steel, column spacing, and location of shipping and receiving docks, etc.

The characteristics of the storage unit, pallet, drum, bundle, etc., largely determine the type of storage equipment required. The inventory levels to be maintained determine the number of pieces of storage equipment. Materials characteristics and the volume of materials movement generally are deciding factors in selecting materials-handling equipment. Materials-handling equipment is discussed in Chaps. 2-3 and 2-4.

Storage equipment usually consists of general-purpose or specialty storage racks of varying height, depth, and load capacity. However, the warehouse floor may serve as all or part of the required equipment. Storage units such as pallets of cased canned goods, which have the rigidity and stability to support loads placed on top of them, are normally stored on the floor in stacks. Rolls of paper and coils of steel are frequently stacked on end. Storage units which have rigidity and are many in number lend themselves to floor-stacking techniques.

Heavy or bulky storage units which lack rigidity or which are few in number are generally better stored in storage racks. Storage units which are small, such as wristwatches or thumbtacks, are suitable for storage in shelving and bins. Containers used in conjunction with shelving or by themselves are discussed in Chap. 2-2.

Some types of available storage equipment are described below. Custom-designed special equipment is offered by many storage-equipment manufacturers.

Pallet Frames

Pallet frames (Fig. 5-1) are useful where materials lack the rigidity or stability to be stacked on the floor and where there are a large number of storage units in inventory. The pallet frame attaches to the pallet and extends above the material. The frame acts as a structure on which another pallet is stacked. Pallets so stacked are often placed several stacks deep and thus conserve floor space as compared with pallet racks which require aisle access. The frames are removable for pallet loads not requiring support.

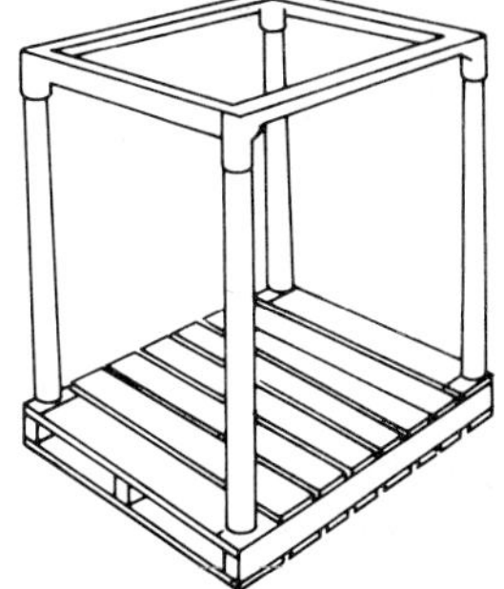

Figure 5-1 Pallet frame.

Pallet Racks

Pallet racks are the most commonly used storage aids and are available in many configurations adapted to particular materials characteristics and turnover rates. Pallet racks, for the purpose of this discussion, are classified into five groups.

One-Deep Pallet Racks. These are used when many items with small inventory quantities must be stored for ready accessibility. They may be configured to accommodate containers and other unit loads in addition to pallets. They are also used for order picking when it is most economical to pick directly from storage units.

One-deep pallets racks consist of vertical upright frames connected by horizontal crossbeams on which pallets and containers are placed one deep. Uprights are available in various heights and depths, and crossbeams are available in various lengths to accommodate most storage unit sizes. The load-carrying capacity for the upright and beam combinations is established by the manufacturers.

The normal storage height for this type of rack is from 20 to 24 ft (6 to 7 m) from the floor to the top of the top load. Lifting operations tend to be inefficient at greater heights because it becomes too difficult for the lift operator to accurately place the storage unit. Specialized equipment for heights greater than 24 ft (7 m) is available, however.

The horizontal crossbeams are adjustable so that the vertical height of the rack may be divided into as many storage levels as desired. The individual storage-opening height is tailored to suit the height of the storage unit. Clearance of 4 to 6 in (10 to 15 cm) from the top of the load to the bottom of the crossbeam above it is usually provided. In establishing the maximum height of the top load, the height of the fire-protection sprinklers must be considered. Most fire-protection codes and fire-insurance underwriters require a minimum clearance of 18 in (45 cm) between the top storage unit and the sprinklers. The warehouse planner should consult the local applicable fire code and the insurance underwriter.

The horizontal width of the storage opening is determined by two factors. These are the maximum weight and the maximum width of the loads to be stored in the opening. It should be noted that the load width may be larger than the pallet width because of load overhang. Normally, 4 in (10 cm) is provided horizontally between loads and between loads and uprights. Typically, two pallet loads are placed side by side in one opening. When the horizontal dimension of the opening has been determined, the rack manufacturer's catalog is consulted to select a compatible crossbeam length and weight capacity.

At this point in warehouse planning, the planner should calculate the floor load resulting from the fully loaded pallet rack. The floor loading will become a design parameter for new construction. In the case of an existing facility, the floor will be confirmed as adequate or the rack arrangement will be shown to be unfeasible.

The number of pallet racks required is determined by dividing the maximum number of storage units by the number of those units contained in one rack.

Two-Deep Pallet Racks. These racks (Fig. 5-2) are similar in design to the one-deep pallet rack except that two pallets, one behind the other, are stored in each position.

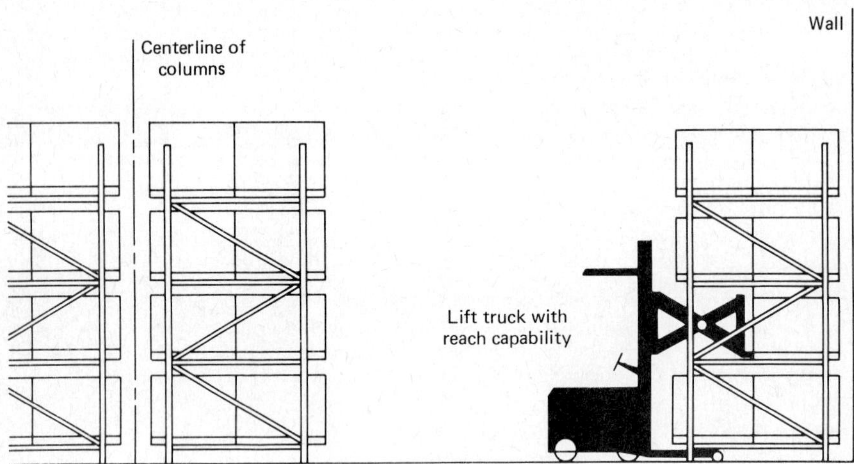

Figure 5-2 Two-deep pallet racks.

Two-deep racks are used when there is insufficient floor space to accommodate the required number of one-deep racks. Two one-deep racks are normally placed side by side and require aisle access from each side of the two-rack combination. The two-deep rack requires access from only one side but stores the same amount of material as the two one-deep racks placed side by side. The ratio of aisle to storage is thus reduced by using two-deep racks.

Two-deep racks have some costs associated with them, however. Lift equipment must be fitted with extended reach capability in order to position loads in the rear storage position. The efficiency of storing and retrieving the load in the rear storage position is less than with single-position loading and unloading. Two-deep racks are often more expensive than their two, one-deep, side-by-side counterparts. Storage units are sometimes damaged when positioned in or retrieved from the rear position.

The manner of selecting the height, depth, width, and number of two-deep racks is similar to that for one-deep racks.

Drive-In or Drive-Through Pallet Racks. These racks (Fig. 5-3) are designed to provide storage several pallets deep. The racks consist of vertical uprights which are

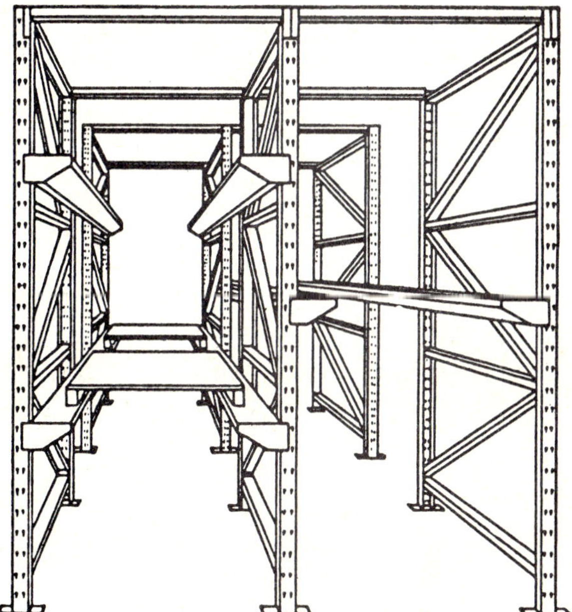

Figure 5-3 Drive-in or drive-through rack. (Reproduced with permission from *Material Handling Handbook and Directory, 1977/1978,* published by *Material Handling Engineering* magazine, Cleveland, Ohio.)

braced across the top of the rack. Angle-iron ledges are welded or bolted to the insides of the uprights to support the pallets. This arrangement allows the lift vehicle to enter the rack to place or retrieve a pallet.

Drive-in or -through pallet racks are used where floor space is limited and where there are many storage units of a particular item to be accommodated. Palletized items shipped or delivered by the entire truckload would be candidates for storage in this type of rack. The total number of positions of storage in one storage aisle in the rack could be designed to contain one truckload.

There are several limitations of drive-in or -through racks. Pallets stored in the rear or middle of a storage aisle cannot be retrieved until those in front are removed. This feature limits first-in–first-out inventory control except by loading or unloading entire storage-aisle lots one at a time. When entire storage-aisle lots are so treated, empty pallet positions are created if less than the entire aisle is filled or emptied. Storage efficiency is thereby reduced.

The lift operator must move and lift the storage unit in very confined spaces. Damage to the goods as a result of close tolerances is more frequent in this rack than in others.

It is also clear that the lift operator's efficiency is reduced by the requirement of driving into the rack in confined spaces.

Since the ledges that support the pallets are fixed, storage units of uniform dimensions are required. Storage units having overhang on the side of the pallet generally do not lend themselves to this type of storage. Finally, the lift vehicle may be no wider than the distance between the ledges. In selecting drive-in or -through racks, the planner should keep in mind that the usual maximum height is again 20 to 24 ft (6 to 7 m) and that drive-in racks usually are no more than six pallets deep. Due to their depth and the relative lack of bracing of their own, higher drive-in and -through racks generally require bracing to the building structure.

Gravity-Flow Racks. These racks (Fig. 5-4) are constructed to contain several pallets in depth and to support the pallets on inclined roller conveyors. Pallets are loaded on the high side of the roller conveyor and removed from the low side. As a pallet is removed from storage, the pallets behind it roll down toward the retrieval opening.

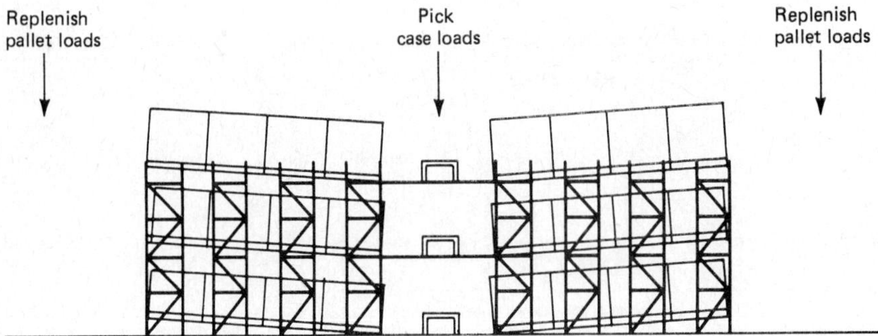

Figure 5-4 Gravity-flow racks.

Gravity-flow pallet racks and smaller versions for individual cartons and containers are commonly employed in order-picking operations. A continuous supply of an item is presented to the order picker without replenishment interference. Gravity-flow racks are also useful in maintaining first-in–first-out inventory control.

Gravity-flow pallet racks generally do not exceed six pallets in depth because of the high cost. However, the depth of the rack may be designed to contain a particular time period's supply. This configuration is appropriate when continuous replenishment is not employed.

The height of gravity-flow racks seldom exceeds 24 ft (7 m). The height of the rack is often limited to that conveniently reached by the order picker. In some cases two-level picking on the inside by personnel on foot is replenished by lift vehicles from the back side. See Fig. 5-4.

The height and width of storage or picking openings in a gravity-flow system are usually fixed with little or no adjustment conveniently possible. Instead, storage units are arranged to fit the gravity-flow configuration.

Gravity-flow racks may not be suited to very unstable storage units due to the shock of impact of movement and sudden stops on the inclined roller conveyor. Such difficulties are overcome by placing the unstable items in suitable containers.

Logic-Flow Racks. These are, in principle, designed to accomplish the same functions as gravity-flow racks. Instead of gravity providing the motive force to move full storage units to the picking opening, a powered conveyor does so. In most applications the order picker operates the powered conveyor with start-stop control. This arrangement eliminates the shock of impact experienced in gravity conveyors. Unstable and very delicate storage units are handled and stored in this manner.

In general, due to its high cost, the logic-flow rack is employed only for very specific, small storage situations. For the most part, these systems are manufactured from custom designs.

Bins and Shelving

Bins and shelving are widely used for the storage of goods in small lot sizes as raw materials, goods in process, and as finished goods, particularly in order-picking applications. They are available in many sizes, strengths, and degrees of closure. Indeed, pallet racks, previously discussed, may easily be converted to shelving.

In selecting shelf and bin storage, the planner determines for each storage item an appropriate shelf opening and depth or bin-drawer size. Shelves, bins, and drawers may be fitted with dividers to contain more than one item. The degree of protection from dust, light, theft, etc., determines the degree of closure required. Shelving and bin closures are available from totally open to totally enclosed and individually locked. Where many items require the same degree of protection, the shelving-bin system may be enclosed in a protective enclosure such as a clean room or refrigerated room.

Shelf, bin, and drawer arrangements can be obtained as separate units or in customized combinations. Customized combinations are more costly but may be justified in situations where stock may be advantageously stored in some picking order. Order-picking efficiency is maximized by reducing search and travel time on the part of the order picker.

Automatic Storage and Retrieval Systems

Automatic storage and retrieval systems (Fig. 5-5) are employed to achieve highly dense storage and very efficient placement and retrieval of materials. Many of the other activities listed under "Warehousing Activities" at the beginning of this chapter may also be mechanized and partially or fully automated as well.

The mechanization and automation of warehousing activities require a high capital investment and a very comprehensive feasibility study to justify the investment. The success of mechanized and automated warehousing also requires the *complete commitment by management* to support the planning, design, procurement, installation, and *especially* debugging. The time period from planning to start-up is often in excess of 3 years.

Mechanized and automated warehousing systems may be considered by the planner if some or all of the following conditions exist:

- Many varieties of items in storage
- High-volume storage items
- High turnover in general
- Highly seasonal storage items
- High cost of land and floor space
- High labor costs
- Need for rapid customer service
- Random storage desirable
- Storage units uniform in size

Automatic storage and retrieval systems, whether automated or not, achieve their high density by storing goods at greater heights than in conventional racks. *High cube* warehousing from 20 to 100 ft (6 to 30 m) is in use. At heights above 20 ft (6 m) the system may become the structure of the building to which walls and the roof are attached.

The materials-handling equipment, normally stacker cranes, travels on rails between the storage units and is guided by rails at the top of the storage units. The stacker cranes are capable of servicing the storage units on either side. Very narrow aisles result, further increasing density.

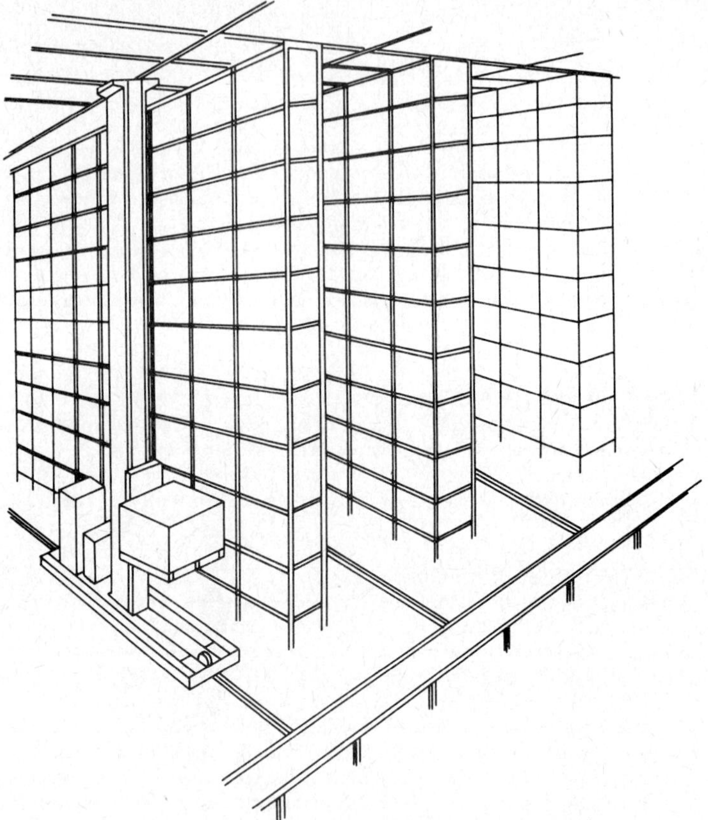

Figure 5-5 Automated storage and retrieval system.

The operator of the stacker crane travels with it both horizontally and vertically. In semiautomatic installations, the operator selects the bay and level on a keyboard. The stacker crane positions itself and the operator at the storage opening for placement or retrieval of a load. In a fully automatic system, a single operator controls the movements of several stacker cranes from a console, usually with the aid of a computer.

Goods to be placed in storage are often delivered to the stacker crane by conveyor; goods to be removed are delivered to the user by conveyor, too. Conveyors for these functions may be semi- or fully automatic as well.

The degree of mechanization and automatic control of warehousing varies from user to user and from manufacturer to manufacturer. The planner should consider engaging consultants and equipment manufacturers in the planning process. Most manufacturers provide planning guides to assist in identifying requirements. See Refs. 2 and 3, page 2-59, for more information. Prior to or during the identification of requirements for a mechanized-automated system, the planner should also determine requirements for a comparable conventional warehouse system. The capital investment and the operating cost of each system are than analyzed for economic justification.

Typically the automated system will require a higher initial capital investment but incur lower annual operating costs than a conventional system. The automated system would be economically justified if its payback period and return on investment are satisfactory to management. There also may be tax implications in choosing an automated storage and retrieval system. Where the racking structure supports the building walls and roof, the structure may be considered equipment. Equipment may be depreciated at a faster rate than buildings. Other factors influencing the decision to mechanize and automate include:

- Competitive advantage in servicing customers
- Image
- Reliability and the need for backup systems
- Degree of sophistication of current warehousing personnel
- Degree to which the market will change
- Time to become operational
- Availability of capital

Storage and retrieval for small lots can be accomplished with mechanized and automated arrangements. Mechanization usually provides for transporting bins of material to a stationary order picker. Carousels are frequently used for this purpose. See Fig. 5-6.

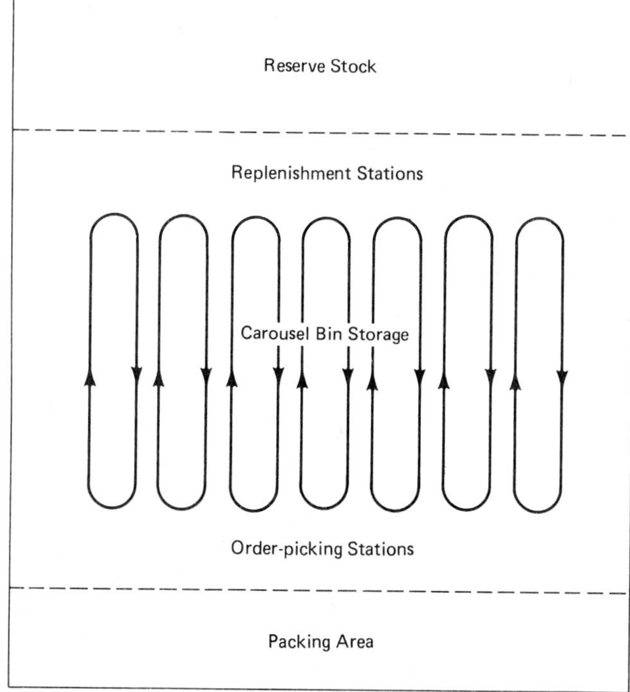

Figure 5-6 Schematic plan of automated bin storage.

The operator may select the picking face to be transported by an on-off control panel in a semiautomatic configuration. Frequently, the picking list is prepared by computer in the order of storage. In an automated system, the computer automatically positions the next picking face as the order picker indicates completion of the last pick or that the item is out of stock. Automation may also include provisions for automatic packing-list preparation, billing, and reordering.

Whether manual or mechanized, shelving and bin storage may be arranged in multi-levels by the use of mezzanines. See Fig. 5-7. This configuration is appropriate when high bay space is available and high storage density is required. In general, high-volume and high-weight items are stored in lower levels, and lighter, slow-moving items in upper levels.

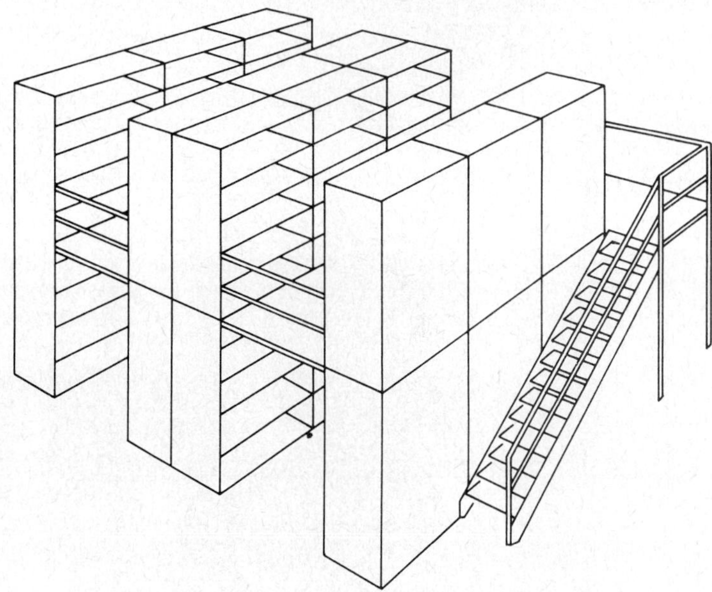

Figure 5-7 Multilevel shelving.

Shipping and Receiving Docks

Shipping and receiving docks are an important factor in warehouse planning. Of prime importance is the number of docks to be provided and how the docks are to be equipped.

In warehousing situations involving very high inbound and outbound volume, the principles of queuing theory are applied to assess peak dock activity. The peak dock activity, using alternative numbers of docks, may then be simulated using computer methods to determine the best combination of receiving and shipping docks. Suppliers of dock equipment will often assist in the planning.

In practice, warehouses having moderate to low inbound-outbound traffic are analyzed from historic or expected data. The peak rate of deliveries and shipments is determined and expressed in vehicles per hour. The rate at which vehicles are loaded and unloaded is also determined and expressed in vehicles per hour per dock. Dividing the vehicle rate of arrival by the rate at which the vehicles are loaded or unloaded results in the number of docks required.

Example

$$\text{Peak rate of truck deliveries} = 10 \text{ trucks per hour}$$

$$\text{Rate at which trucks are unloaded} = 2 \text{ trucks per hour per dock}$$

$$\text{No. receiving docks required} = \frac{\text{rate of truck arrivals}}{\text{rate trucks are unloaded}} = \frac{10 \text{ trucks per hour}}{2 \text{ trucks per hour per dock}} = 5 \text{ docks}$$

The type of dock and the equipment provided at the dock are dependent upon the type of material to be handled, the need for specialized loading/unloading equipment, and the need for security or weather protection.

Types of Docks

Types of docks to be considered include the following.

Indoor Docks. These are designed to accommodate the delivery vehicle under the roof of the warehouse building. Accommodation for entire tractor-trailers, trailers only, small delivery trucks, and railcars are common. These arrangements are appropriate where there is need to contain heated or cooled air, where security is important, and where materials-handling systems such as bridge cranes load or unload the material.

Flush Docks. These are constructed flush with the warehouse floor and with the outer wall of the building. This type of dock is generally built at a height above the outside grade level to accommodate the height of the vehicles to be serviced. Where the terrain is flat, ramps down to the docks are usually provided. This type of dock is normally provided with dock seals for weather protection. Also typical are dock leveler installations to accommodate minor differences in the heights of vehicle load beds.

Open Docks. These are less expensive than those previously discussed. However, they offer the least protection from weather and pilferage. Open docks are appropriate in warmer climates when handling goods which are not weather-sensitive or pilferable. When installed, a canopy over the dock area is usually provided.

Sawtooth Docks. These are arranged at an angle to the face of the building wall. This configuration is applicable where maneuvering room for shipping and delivery vehicles is limited. The sawtooth arrangement may be fully enclosed, covered, or open, as described above.

Dock Equipment

This must accommodate the materials to be handled, the vehicle to be serviced, the loading and unloading equipment, and the need for weather and security protection. Some standard equipment includes the following.

Dock Doors. These should be of the overhead type, counterbalanced for ease of operation. The doors should ride vertically along the wall of the building. Doors with tracks curving inward, such as a household garage door, could be damaged by materials-handling equipment. Doors are sized to accommodate the loaded materials-handling equipment passing through them and/or the size and configuration of the delivery-shipping vehicle. Options available for dock doors include the material from which the door is made, insulation, windows, and mechanized door-opening or -closing operators.

Dock Levelers and Dock Boards. These provide a bridge between the delivery or shipping vehicle and the dock (Fig. 5-8). They also serve to accommodate differences in height between vehicles and the dock. Dock levelers are typically installed to be flush with the floor of the dock when retracted. Models which attach to the outside of the dock as a retrofit are also available. The levelers are activated or retracted by spring pressure or hydraulics. Dock boards are reinforced steel plates which are manually lifted into place to form the bridge.

Factors used in selecting the appropriate dock leveler or board include: the weight of the heaviest load and materials-handling vehicle combination to cross it, the distance which the unit must span for bridging, and the combined width of the load and materials-handling vehicle. Manufacturers of this equipment offer a wide variety of weight capacities and sizes.

Weather-Protection Equipment. This equipment (Fig. 5-9) consists of devices to seal or cover the loading-dock area. Truck-dock seals used in conjunction with flush docks are popular. They have flexible construction and are often inflated. The arriving truck backs into the seal surrounding the door to effect the seal. Also available are hood-type seals which are mechanically activated. Made of flexible material, the hood extends out from the dock and conforms to the shape of the truck or railcar.

Other devices in common use are inside docks (discussed above) and canopies extend-

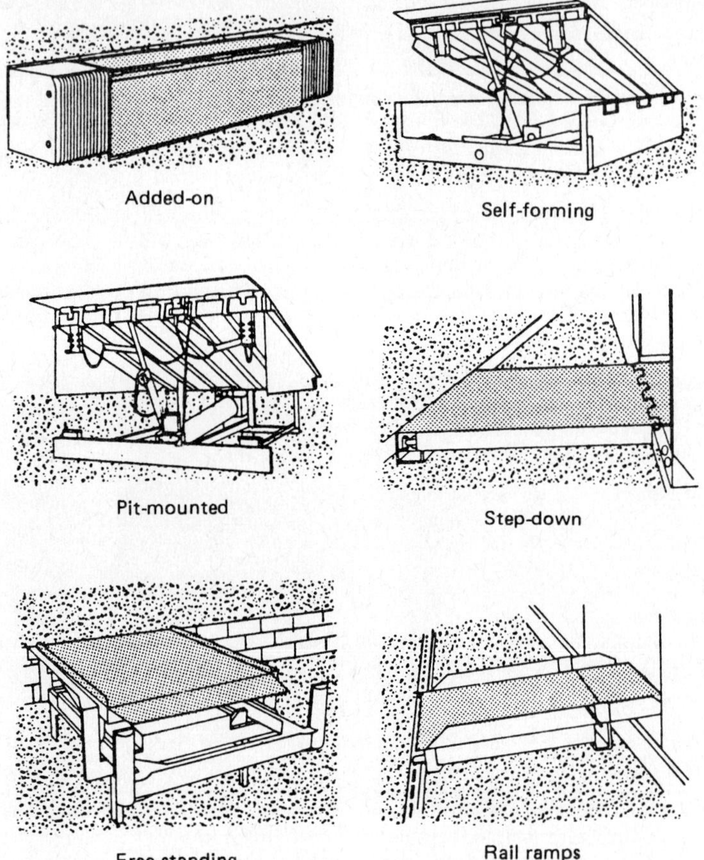

Figure 5-8 Types of dock levelers and dock boards. (Reproduced with permission from *Material Handling Engineering Handbook and Directory, 1977/1978,* published by *Material Handling Engineering* magazine, Cleveland, Ohio.)

ing over the outside dock area for rain protection. Where high activity means that dock doors are seldom closed, weather curtains are indicated. The curtains consist of strips of clear flexible material covering the door opening. Materials-handling equipment can drive through the curtain. When there is no traffic through the curtain, the strips act, to some extent, to seal the door from weather.

Dock Lighting

Dock lighting is required for nighttime operations. Floodlights are typical for lighting outside driveways, rails, and maneuvering areas to facilitate spotting delivery and shipping vehicles. Lighting for open docks is required to facilitate loading- and unloading-vehicle movement. When dock seals or hoods are used, lighting may be required for the inside of the truck or railcar. Portable or adjustable fixed lighting for these purposes is available. See Chap. 3-5, "Lighting."*

Warehouse Layout

The warehouse layout is the final and perhaps the most important step in the planning process. Prior to undertaking the layout, the planner establishes the activities to be com-

*See Rosaler and Rice: *Standard Handbook of Plant Engineering,* 1983.

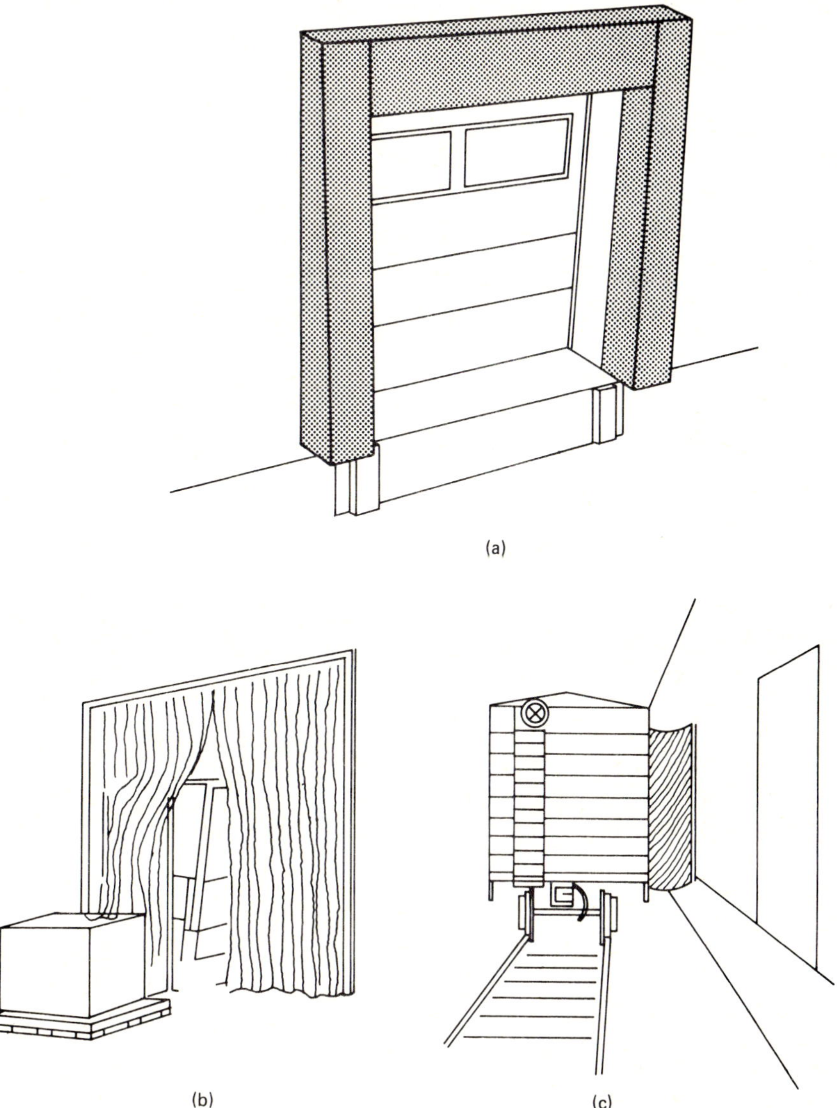

Figure 5-9 Types of dock weather protection. (*a*) One-deep pallet rack, (*b*) storage floor plan.

pleted and the type and number of materials to be stored and handled, storage and handling equipment, and docks. (See "Warehousing Activities" at the beginning of this chapter.) The warehouse layout should be planned to provide the space and arrangement that makes the best use of

- Storage cubes
- Efficiency of the flow of materials from activity to activity
- Effective communications between activities

Because thousands of combinations of types, sizes, weights, and volumes of materials have been observed, specific warehouse layout characteristics cannot be described in this book. However, general principles for warehouse design are discussed below.

Location in Storage

The location in storage (Fig. 5-10) of particular items is of importance. The following points should be considered.

- Items having a high turnover should be located near the user. The user may be a manufacturing operation, the shipping docks, or a quality-inspection area.
- Items having a high turnover should be stored and retrieved in the most convenient level vertically—slow movers high and fast movers low.
- Heavy and/or difficult-to-move items should be stored low.
- Where few items but large volumes of commodities are characteristic, individual loads of an item should be stored together in semidedicated areas.
- Where many items, but few of each, are encountered, random storage should be considered. A locator system, perhaps computerized, may be necessary.

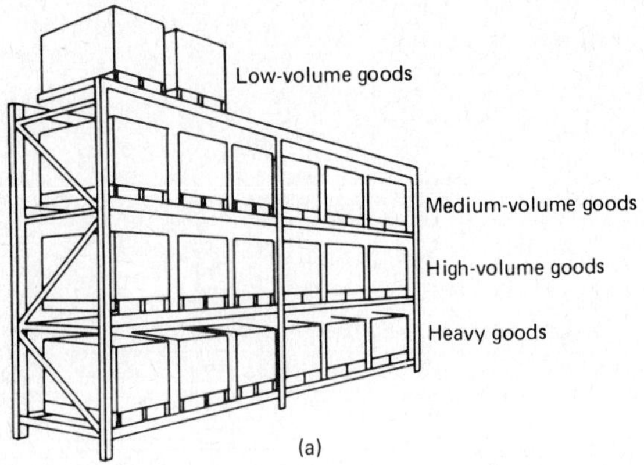

(a)

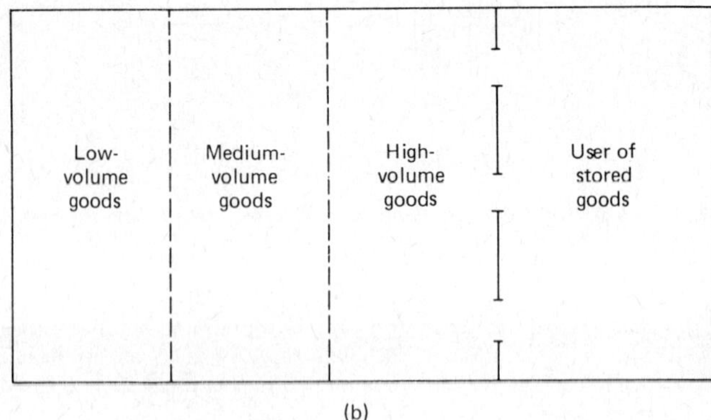

(b)

Figure 5-10 Location in storage according to volume.

- The nature of some storage items may require them to be stored in dedicated space. Some examples include hazardous materials, items of high value, and perishable goods.

Aisles

- Minimum aisle width is determined by the loaded maneuvering characteristics of materials-handling equipment. Determination of minimum aisle width is discussed in Chap. 2-4.
- Aisle width may be reduced by imposing one-way traffic.
- Aisles should open from the supplying area and open to the user area for maximum efficiency.
- Aisles should not be located next to walls as only one storage face is presented.

Storage Equipment Location and Arrangement

- The arrangement may be effected by column spacing in existing buildings or may determine the column spacing in new facilities. Normally, one-deep, two-deep, and drive-in racks, as well as shelving, are placed end on end, back-side along column lines. This arrangement eliminates column interference with aisles. See Fig. 5-2.
- One-deep, two-deep, and drive-in racks, as well as shelving, are most effectively placed back-to-back in open floor areas. This arrangement minimizes access aisle requirements. In the case of racks, space between them must be provided for pallet overhang. Frequently, the width of a line of columns provides this space.
- Storage racks, except gravity-flow and logic-flow, in addition to shelving, are efficiently placed along walls with openings for doors and fire-protection equipment. See Fig. 5-2.
- The height of the storage equipment is limited to that which provides no less than 18 in (45 cm) of clearance beneath fire-protection sprinklers. Local codes or fire underwriters may require a different clearance.

Docks

- Receiving and shipping docks are usually located to accommodate the flow of materials in the manufacturing process. The most common manufacturing flow patterns are straight-through and U-shaped. See Fig. 5-11.
- In the straight-through processes, raw materials are received and stored at the beginning of the manufacturing process. Finished goods appear at the end of the process and are stored and shipped from that location. Receiving and shipping areas which are so separated generally require more personnel and docks than an equivalent U-shaped arrangement.
- In U-shaped process flow, raw materials arrive at the same side of the building as that from which the finished goods are shipped. Shipping and receiving docks may be separated by no more than an imaginary line. This arrangement may offer economies in lower personnel requirements and in the number of docks since receiving and shipping personnel and equipment may be interchanged when necessary. The U-shaped process flow is also advantageous if high bay storage for both raw materials and finished goods is required.
- Spacing between docks is established to minimize interference between the docks during operations.
- Areas adjacent to docks for staging off-loaded material or material awaiting shipment are normally required.
- Provisions for enclosed space near docks may include administrative offices, personnel comfort facilities, and quality-inspection areas.

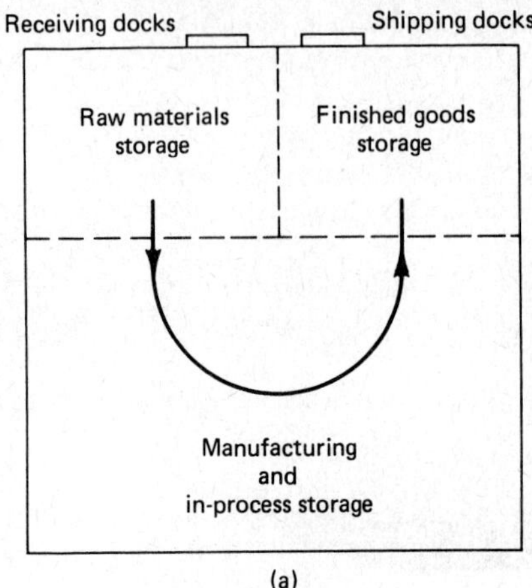

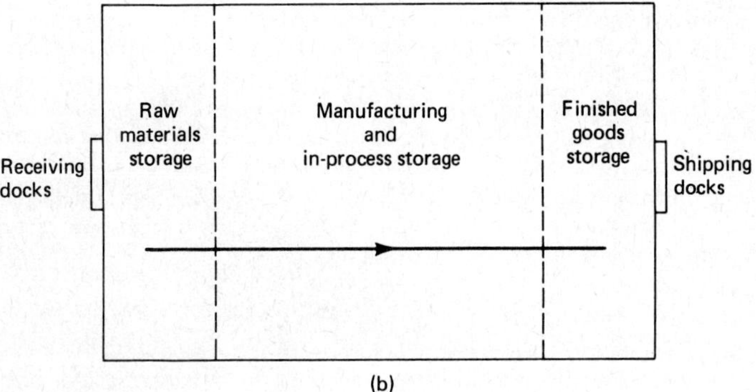

Figure 5-11 Materials flow patterns. (*a*) U-shaped materials flow, (*b*) straight-through materials flow.

Building Characteristics

- Storage height may be limited not only by fire codes but also by local zoning restrictions.
- Building services such as piping and space heaters should be placed in aisles to avoid interfering with storage equipment and to be accessible for maintenance.
- Lighting is normally designed to aid materials-handling equipment operators and order-picking personnel in locating and identifying stored items. Architectural and engineering firms normally assist the planner in determining the number and location of lighting fixtures.
- The type and number of units of fire-protection equipment are governed by fire code and fire underwriter requirements. Generally, the flammability and the amount of material stored determines the requirements.
- Floors may be enhanced to increase durability and housekeeping qualities. Coatings are available to increase surface hardness and wearability and to reduce dust. Typically, 3- to 4-in (8- to 10-cm) lines are painted on the floor to mark traffic and storage aisles.

section 3

Hydraulic and Pneumatic Systems

chapter 3-1

Hydraulic Systems

by

Ken S. Satija, Ph.D., P.E.
United Engineers & Constructors, Inc.
Philadelphia, Pennsylvania

INTRODUCTION

A system comprises a group of devices forming a network. A hydraulic system involves the devices and networks that operate by the pressure of liquids, whereas a pneumatic system works with gases. The most predominant liquid in hydraulic systems is water, the second is oil. In pneumatic systems, air is the most common fluid with natural gas and steam also in frequent use. Water, oil, and/or air systems and components will be mainly discussed in these chapters.

A typical system includes a pressurizer source (a high elevation tank or a pump), a conveyance system (channels, pipes, valves, etc.), and a sink. Thus, in a fluid (liquid or gas) system, a fluid is conveyed from one point with a certain energy to another point. Because of the resistance in the system, some pressure energy is lost as heat energy in the network.

It is necessary to know the fundamentals of fluid mechanics in order to understand, design, and maintain a system. The first part of this chapter is devoted to just that; the remainder discusses typical system components and some examples of actual plant systems.

BASIC FLUID MECHANICS

Pressure

Pressure is the most important term in fluid mechanics. The movement of the fluid and the forces on the equipment due to the fluid are basically due to the pressure in the fluid. By definition, pressure is the force per unit area of a surface with which the fluid is in contact, and always acts normal to the surface considered:

$$p = \frac{F}{A} = wH \tag{1}$$

where

p = pressure intensity, lb/ft² (kg/m²)
F = total force acting on a surface, lb (kg)
A = area of the surface, ft² (m²)
w = weight density of the fluid, lb/ft³ (kg/m³)
H = head = height of free surface of the fluid above a certain point on the surface, ft (m)

TABLE 1-1 Density of Common Fluids (at Atmospheric Pressure)

Fluid	Temperature, °F	°C	Specific density, lb/ft³
Air	0	−18	0.0862
	60	16	0.0763
	100	38	0.0709
	200	93	0.0601
Ammonia	60	16	0.0455
Carbon dioxide	60	16	0.117
Gasoline	60	16	42.0
Helium	60	16	0.0106
Hydrogen	60	16	0.00531
Mercury	60	16	847.0
Methane (natural gas)	60	16	0.0424
Oxygen	60	16	0.0844
Nitrogen	60	16	0.0739
Sulfur dioxide	60	16	0.173
Water (fresh)	32	0	62.4
	70	21	62.3
	180	83	60.6
	212	100	59.8
Water (salt, sea)	60	16	64.0

The last part of Eq. (1) really emanates from the principles of hydrostatics, in which the only energy available comes from the static head of the column of fluid. Thus the hydrostatic pressure at a point is equal to the pressure created by the height of the fluid column above that point. Since the density of gases is usually very small compared with liquids (see Table 1-1), the hydrostatic pressure of gases is usually neglected in practical engineering unless a very high level of precision is desired.

A standard atmosphere (the atmospheric pressure) is taken as 14.7 psia (pounds per square inch absolute), or 34 ft of a column of water, or 760 mmHg (29.92 inHg). *Absolute* is used with reference to zero pressure. A pressure gauge normally measures and displays pressure *relative to the atmosphere.* Thus a gauge reading of zero is actually a pressure exactly equal to prevailing atmospheric pressure. In fluid mechanics, unless specifically mentioned otherwise, gauge pressures are understood. Thus a pressure given as 30 lb/in^2 automatically means 30 psig (pounds per square inch gauge). However, it must be kept in mind that consistent units must be used while working out problems with the help of the equations such as Eq. (1).

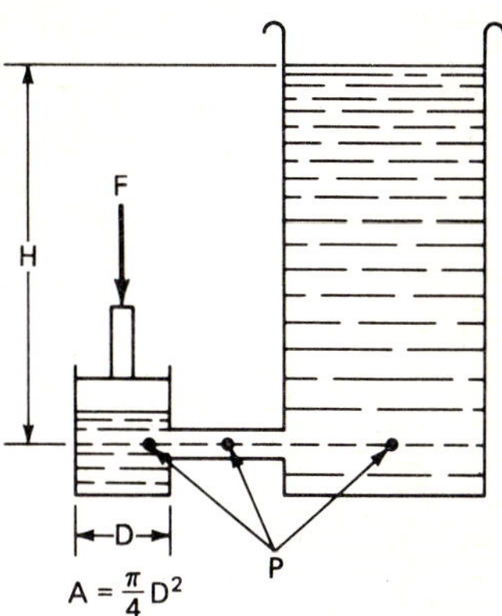

Figure 1-1 Relation of pressure to head.

The term *pressure head* with reference to Eq. (1) can easily be understood by referring to Fig. 1-1, which also depicts that hydrostatic pressure is transmitted equally in all directions at a point, the so-called *Pascal's law*. If a point *P* is surrounded by fluid on all sides, the hydrostatic forces are balanced, and the net force at that point is zero; that is, the point stays in equilibrium. Otherwise, there will be a net force trying to move that point.

Density

The density or the specific weight w of a substance is the weight of a unit volume of the substance. For water, w = 62.4 lb/ft^3 (999.6 kg/m^3) is commonly used (1 g/cm^3 = 1000 kg/m^3). Table 1-1 lists specific weight of several common fluids.

$$\text{Mass density} = \rho = \frac{\text{mass}}{\text{unit volume}} = \frac{w}{\mathbf{g}} \tag{2}$$

where

$\mathbf{g}$ = acceleration due to gravity, = 32.2 ft/s^2 (9.81 m/s^2)

G_f = relative density

= specific gravity

$$= \frac{\text{weight of the substance in air}}{\text{weight of an equal volume of water}}$$

$$= \frac{\text{weight of the substance in air}}{\text{loss of weight of the substance in water}}$$

$$= \frac{\text{weight of the substance in air}}{\text{buoyant force on the substance if immersed in water}}$$

$$= \frac{\text{density of the substance}}{\text{density of water}}$$

This is Archimedes' principle.

Viscosity

The resistance of a fluid to a shearing force is determined by viscosity. If the force tends to move a fluid particle with a velocity of V ft/s (m/s), then dynamic or absolute viscosity is

$$\mu = \frac{\tau}{dV/dy} \quad \text{lb}\cdot\text{s/ft}^2 \quad (\text{poise} = \text{P} = \text{dyn}\cdot\text{s/cm}^2) \tag{3}$$

where dV/dy is the velocity gradient and distance y is measured normal to V.

The kinematic viscosity is

$$\nu = \frac{\mu}{\rho} \quad \text{ft}^2/\text{s} \quad (\text{stokes} = \text{cm}^2/\text{s} = \text{St}) \tag{4}$$

and τ is the shear stress in fluid, lb/ft² (kg/cm²). Viscosity values for common fluids are shown in Table 1-2.

TABLE 1-2 Viscosity of Common Fluids

Fluid	Temperature °F	(°C)	Kinematic viscosity, ft²/s
Air	0	(−18)	1.26×10^{-4}
	60	(16)	1.46×10^{-4}
	100	(38)	1.80×10^{-4}
	200	(94)	2.40×10^{-4}
Water (fresh)	32	(0)	1.93×10^{-5}
	70	(21)	1.05×10^{-5}
	180	(82)	0.385×10^{-5}
	212	(100)	0.319×10^{-5}
Water (salt, sea)	60	(16)	0.319×10^{-5}

Hydrostatic Forces

Consider an inclined plane surface (with area A) of a tank containing a liquid as shown in Fig. 1-2. The center of gravity of the plane is at point cg. Point O is the point of intersection of the plane (extended if necessary) and the surface of liquid. Other nomenclature is clear from the figure. Then the total hydrostatic force F (always normal to a surface) on the inclined surface and its point of action P are given by

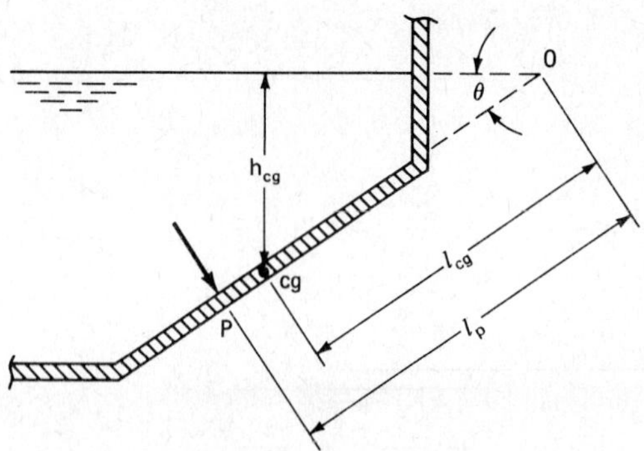

Figure 1-2 Hydrostatic forces.

$$F = wh_{cg}A \qquad \text{lb (kg)} \tag{5}$$

$$l_p = l_{cg} + \frac{I_{cg}}{Al_{cg}} = \frac{I_0}{Al_{cg}} \qquad \text{ft (m)} \tag{6}$$

where

I_{cg} = second moment of the area (moment of inertia) about the center of gravity of the plane (as usually given in mechanics books)

I_0 = moment of inertia about point O

Continuity Equation

For no other inflow or outflow from a subsystem, the mass flow rate at upstream and downstream end are equal; i.e.,

$$M_1 = M_2 \qquad (Q_1 = Q_2 \text{ if } \rho \text{ is constant}) \tag{7}$$

or

$$\rho_1 A_1 V_1 = \rho_2 A_2 V_2 \tag{8}$$

If a branch comes into a point on the subsystem, the total outflow from the point is equal to the total inflow to the point, i.e.,

$$M_1 + M_2 = M_3 \tag{9}$$

Energy Equation

Consider flow in a closed conduit of nonuniform circular cross section (Fig. 1-3). Also consider two cross sections 1 and 2 of this conduit with the following quantities applicable to them:

p_1 = pressure at cross section 1, lb/ft² (kg/m²)
p_2 = pressure at cross section 2, lb/ft (kg/m²)
v_1 = velocity at cross section 1, ft/s (m/s)
v_2 = velocity at cross section 2, ft/s (m/s)
z_1 = height above datum of cross section 1, ft (m)
z_2 = height above datum of cross section 2, ft (m)

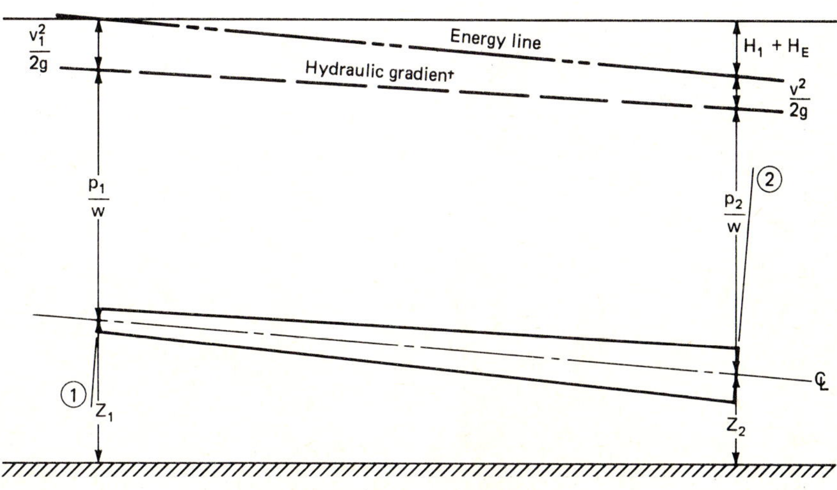

Figure 1-3 Energy equation.

All these quantities are referenced to the centerline of the conduit. Then, in terms of head, the energy equation is

$$\frac{p_1}{w} + \frac{V_1^2}{2g} + z_1 = \frac{p_2}{w} + \frac{V_2^2}{2g} + z_2 + H_1 + H_e \tag{10}$$

where

H_1 = head losses in the flow, such as friction, ft (m)

H_e = head energy extracted from the system (negative if energy was added), ft (m)

If H_1, and H_e are zero, as for an ideal flow, the remaining equation (10) is also called the Bernoulli theorem.

Each side of Eq. (10) is equal to the total energy in the system. The total energy must be constant, so the plot of both sides of Eq. (10) falls on a horizontal line, representing the total energy in the system. If H_1 and H_e are excluded from the equation, then we get a net *energy line.* If the velocity head is also excluded, then we get the *hydraulic gradient line,* representing the gradient of the hydrostatic pressure.

Momentum Principles

According to Newton's second law

$$\begin{aligned} \text{Momentum force} &= \text{rate of change of momentum in a direction} \\ &= \text{initial momentum} - \text{final momentum in that direction} \qquad (11) \\ &= \frac{wq}{g} V_1 - \frac{wq}{g} V_2 \qquad \text{lb (kg)} \end{aligned}$$

Therefore, the total force on an element will be the sum of the pressure force, pA, defined in Eq. (1) and the momentum force given by Eq. (11), both considered in the same direction. For example, on a 90° elbow (Fig. 1-4), the forces must be considered separately along the x and y axes, and then the resultant computed:

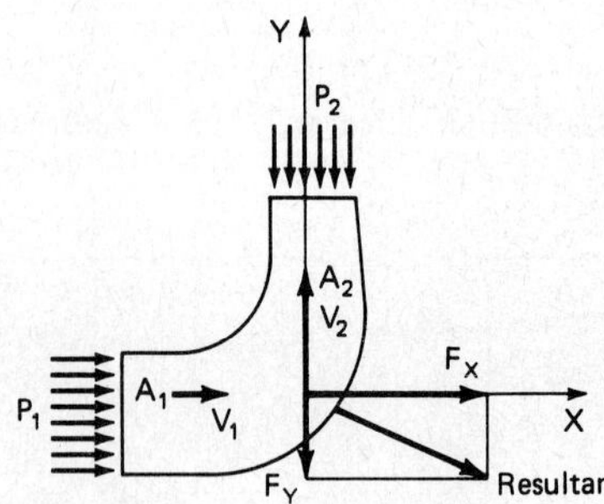

Figure 1-4 Forces on a bend.

$$F_x = p_1 A_1 + \frac{wq}{g} V_l - \frac{wq}{g} 0 \tag{12}$$

(Final velocity in the x direction is zero.)

$$F_y = -p_2 A_2 + \frac{wq}{g} 0 - \frac{wq}{g} V_2 \tag{13}$$

(Positive y is upward; initial momentum is zero; pressure always acts normal to the surface and toward the surface of the element considered.)

Head Losses

Frictional Flow in Closed Pipes

A pipe flow is considered closed and pressurized if the water surface is not exposed to atmospheric pressure. Under actual conditions, there is always a resistance to flow which causes a pressure head loss. (Basically, this loss of energy is converted to heat and is usually dissipated.)

The most commonly used equation for head loss due to friction, $H_{L,f}$, is the Darcy-Weisbach equation

$$H_{L,f} = \frac{fL}{D}\frac{V^2}{2g} \qquad \text{ft (m)} \tag{14}$$

where

f = friction coefficient
D = diameter of pipe, ft (m)
L = length of pipe, ft (m)

The friction coefficient f depends upon the material and roughness of the pipe and the Reynolds number (N_R) of flow:

$$N_R = \frac{VD}{\nu} = \frac{V\,4R}{\nu} \tag{15}$$

where

R = hydraulic mean radius of flow = A/P = $D/4$ for pipe, ft (m)
P = wetted perimeter of flow boundary, ft (m)

Figure 1-5 is a chart for the determination of f for a certain value of N_R and relative roughness. The lowest curve is for any pipe called "smooth," i.e., with negligible roughness.

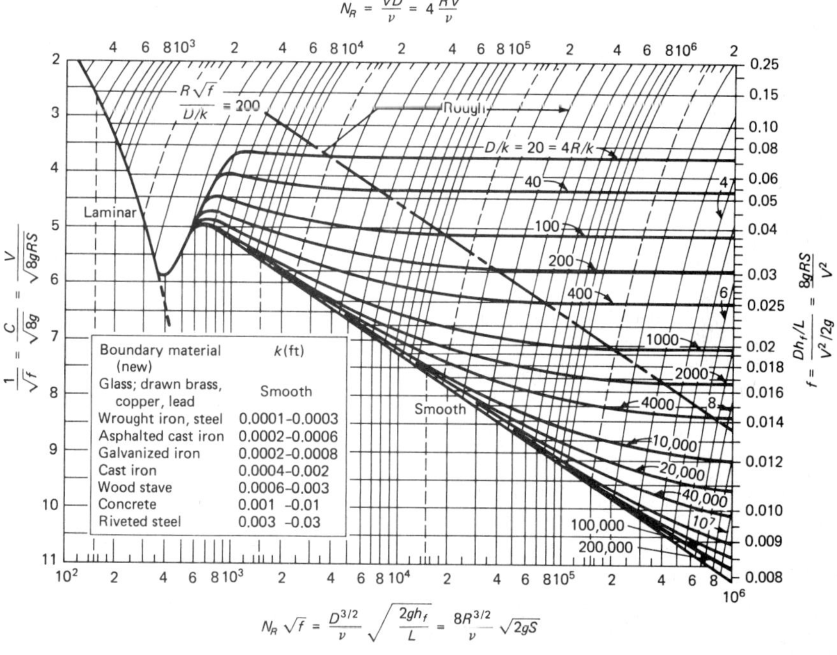

Figure 1-5 Friction in pipes and conduits.

Sometimes the Hazen-Williams formula is used for computing frictional flow and is given by

$$V = 1.318 C_1 R^{0.63} S^{0.54} \qquad \text{feet per second} \tag{16}$$

where

C_1 = Hazen-Williams coefficient (see Table 1-3)
S = slope of the hydraulic gradient

$$= \frac{H_{L_f}}{L} \tag{17}$$

TABLE 1-3 Some Values of the Hazen-Williams Coefficient C_1 (U.S. Customary Units)

Extremely smooth and straight pipes	140
New, smooth cast-iron pipes	130
Average cast-iron, new riveted steel pipes	110
Vitrified sewer pipes	110
Cast-iron pipes, some years in service	100
Cast-iron pipes, in bad condition	80

Other Losses

There are also other losses in a flow besides the frictional pressure drop; losses occur at every transition, control, bend, valve, etc. These are usually given as a function of the velocity head $V^2/2\mathbf{g}$. Table 1-4 provides the values of the coefficient K with which $V^2/2\mathbf{g}$ (or its variations, as indicated) is multiplied to get the head losses due to fittings on a piping system. (Sometimes head loss is computed in terms of equivalent length of pipe.)

Flow in Open Channels

If the water surface in a conduit (as in a channel) is open to atmospheric pressure, it is called *open-channel flow*. In this instance, the surface pressure is the same all along the flow, i.e., atmospheric. That is why, instead of total energy, *specific energy E* is employed in open-channel flow:

$$E = y + \frac{V^2}{2g} \qquad \text{ft (m)} \tag{18}$$

where y = depth of water in open-channel flow, ft (m)

The plot of Eq. (18) of E vs. y (for a constant Q) will yield a curve of the type shown in Fig. 1-6.

For any particular energy value, there are two depths y_A and y_B, except at one particular point where there is only one depth (the *critical depth* y_c), corresponding to a

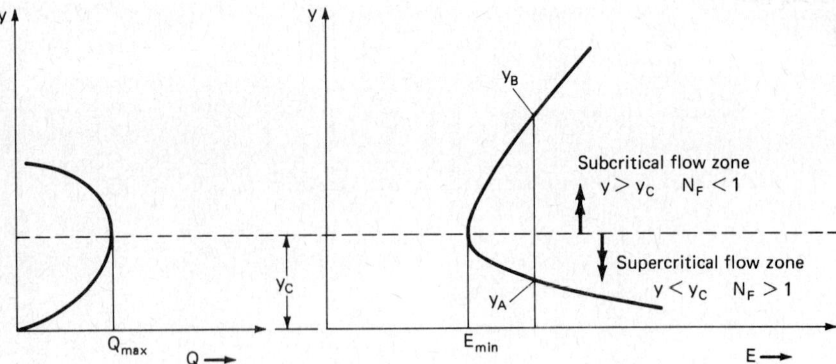

Figure 1-6 Critical depth.

minimum specific energy E_{min}. It is also known that if discharge vs. depth is plotted, the maximum discharge point on this curve corresponds to the critical depth. This is indicated at the left end of Fig. 1-6. It is also known that the Froude number N_F of flow is equal to one for a flow with critical depth $N_{F,c}$ = 1), where

$$N_F = \frac{V}{\sqrt{\mathbf{g}y}} \tag{19}$$

TABLE 1-4A Typical Loss of Head Coefficients (Subscript 1 = Upstream and Subscript 2 = Downstream)

Item		Average lost head
1. From tank to pipe (entrance loss)		
Flush connection		$0.50 \frac{V_2^2}{2g}$
Projecting connection		$1.00 \frac{V_2^2}{2g}$
Rounded connection		$0.05 \frac{V_2^2}{2g}$
2. From pipe to tank (exit loss)		$1.00 \frac{V_1^2}{2g}$
3. Sudden enlargement		$\frac{(V_1 - V_2)^2}{2g}$
4. Gradual enlargement		$K \frac{(V_1 - V_2)^2}{2g}$
5. Sudden contraction		$K_c \frac{V_2^2}{2g}$
6. Elbows, fittings, valves*		$K \frac{V_2^2}{2g}$
Some typical values of K are:		
45° bend	0.35 to 0.45	
90° bend	0.50 to 0.75	
Tees	1.50 to 2.00	
Gate valves (full open)	about 0.25	
Check valves (open)	about 3.0	
Ball valve (open)	about 0.1	
Butterfly valve (open)	about 1.0	
Globe valve (open)	about 2.0	
Plug valve (open)	about 0.5	

*In the one-quarter-open position, a valve will have a K value several times that in the full-open position.

TABLE 1-4B Values of K*—Contractions and Enlargements

Sudden contraction*		Gradual enlargement for total angles of cone*						
d_1/d_2	K_c	4°	10°	15°	20°	30°	50°	60°
1.2	0.08	0.02	0.04	0.09	0.16	0.25	0.35	0.37
1.4	0.17	0.03	0.06	0.12	0.23	0.36	0.50	0.53
1.6	0.26	0.03	0.07	0.14	0.26	0.42	0.57	0.61
1.8	0.34	0.04	0.07	0.15	0.28	0.44	0.61	0.65
2.0	0.37	0.04	0.07	0.16	0.29	0.46	0.63	0.68
2.5	0.41	0.04	0.08	0.16	0.30	0.48	0.65	0.70
3.0	0.43	0.04	0.08	0.16	0.31	0.48	0.66	0.71
4.0	0.45	0.04	0.08	0.16	0.31	0.49	0.67	0.72
5.0	0.46	0.04	0.08	0.16	0.31	0.50	0.67	0.72

*Values from H. W. King and E. F. Brater, *Handbook of Hydraulics*, McGraw-Hill, New York, 1976.

The critical depth for a rectangular channel can be calculated by

$$y_c = \left(\frac{q^2}{\mathbf{g}}\right)^{1/3} \tag{20}$$

where

$$q = \frac{Q}{B} = \text{discharge per unit width} \tag{21}$$

Also

$$y_c = \frac{2}{3} E_{\min} = \frac{V_c^2}{\mathbf{g}} \tag{22}$$

where

$$V_c = \text{velocity corresponding to critical depth} = \frac{q}{y_c} \tag{23}$$

Equation (20) can also be rewritten as

$$q_{\max} = \sqrt{\mathbf{g} y_c^3} \tag{24}$$

For a *trapezoidal* channel, the equations are

$$\frac{Q^2}{g} = \frac{A_c^3}{\mathrm{B}} \quad \text{or} \quad \frac{V_c^2}{\mathbf{g}} = \frac{A_c}{B}$$

where

$$B = \text{top width of the channel} \qquad \text{ft (m)} \tag{25}$$

A channel flowing with a depth greater than the critical depth has a *subcritical* flow. If normal depth of flow is smaller than the critical depth, it is called *supercritical* flow.

The *normal flow* (uniform flow, where depth is not changing with distance, and the water surface is parallel to the bed) is calculated by using the Manning or the Chezy formula with the hydraulic gradient slope S equal to the bed slope S_0 (see Table 1-5).

TABLE 1-5 A Few Average Values of n for Use in Kutter's and Manning's Formulas

Type of open channel	n
Smooth cement lining, best planed timber	0.010
Planed timber, new wood-stave flumes, lined cast iron	0.012
Good vitrified sewer pipe, good brickwork, average concrete pipe, unplaned timber, smooth metal flumes	0.013
Average clay sewer pipe and cast-iron pipe, average cement lining	0.015
Earth canals, straight and well maintained	0.023
Dredged earth canals, average condition	0.027
Canals cut in rock	0.040
Rivers in good condition	0.030

The Manning formula is

$$V = \frac{1.486}{n} R^{2/3} S^{1/2} \qquad \text{(feet per second)} \tag{26}$$

where

S = hydraulic slope
S_0 = bed slope = vertical drop per length of channel
n = Manning roughness coefficient (Table 1-5)

The Chezy formula is

$$V = C\sqrt{RS} \tag{27}$$

where

$$C = \sqrt{\frac{8g}{f}} \qquad \text{(Darcy-Weisbach } f) \tag{28}$$

Also

$$C = \frac{1.486}{n} R^{1/6} \qquad \text{(Manning } n) \tag{29}$$

Kutter gave a more complex value:

$$C = \frac{41.65 + (0.00281/S) + (1.811/n)}{1 + (n/\sqrt{R})[41.65 + (0.00281/S)]} \tag{30}$$

If the depth of flow at a particular cross section of a channel is supercritical ($y < y_c$) and if the conditions downstream of that section can allow subcritical flow ($y > y_c$), for example, a normal flow depth downstream, then the flow will jump from the smaller depth to the larger depth. This jump is called the *hydraulic jump* and is associated with

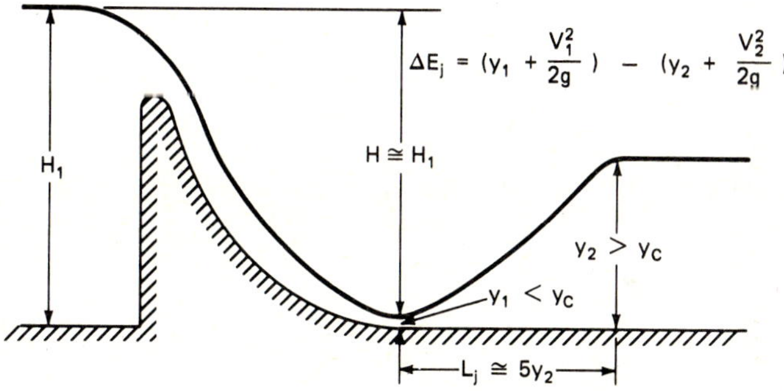

Figure 1-7 Hydraulic jump.

a lot of turbulence, and therefore loss of energy. Consider the flow downstream of a fall (say from a dam). The velocity downstream of the fall (Fig. 1-7) is such that almost all the potential energy has changed to kinetic energy; i.e.,

$$V_1 = \sqrt{2gH} \tag{31}$$

Then

$$y_1 = \frac{q}{v_1} \tag{31a}$$

and the depth after the jump y_2 is given by

$$y_2 = \frac{y_1}{2}\left(\sqrt{1 + 8N_{f1}^2} - 1\right) \tag{32}$$

The most economical section of a rectangular channel is obtained when

$$R = \frac{y}{2} \tag{33}$$

This means that there will be the least-wetted perimeter and therefore the least frictional energy losses and hence the maximum flow.

Nonuniform Flow

When the water surface is not parallel to the bed, the depth of flow varies with the distance (depth is not equal to the normal depth mentioned before), and this flow is termed nonuniform flow. The water flowing at all the flow restrictions, transitions, falls, controls, spillways, etc., takes on a curved surface profile. The calculations of such surface profiles are made in small steps for better accuracy, or may be done in a single step for less accuracy with the equation

$$\frac{dy}{dL} = \frac{S_0 - S}{1 - \dfrac{V^2}{gy}} \tag{34}$$

Another form of this equation is

$$\Delta L = \frac{E_1 - E_2}{S - S_0} \tag{35}$$

Here, S is the normal slope computed from the Chezy or the Manning formula. E_1 and E_2 are the specific energies at the two ends of a reach of length ΔL (under consideration).

Measurements

The static pressure, velocity, and mass rate of flow are the usual quantities required to be measured in fluid mechanics. The *pressure* is usually measured by a U-tube manometer containing a liquid of known density (Fig. 1-8). One leg of the U tube is connected to the point where the pressure is to be measured. The other leg is open to the atmosphere. The vertical distance between the liquid surfaces in the two legs is a measure of the pressure head in feet of that liquid. It can be converted into psig by multiplying by the density of the liquid [see Eq. (1)]. As an alternative, use the pressure of a liquid immiscible with the first one. The pressure of a liquid can also be measured by just inserting a standpipe (piezometer) at the point where pressure is to be measured. The

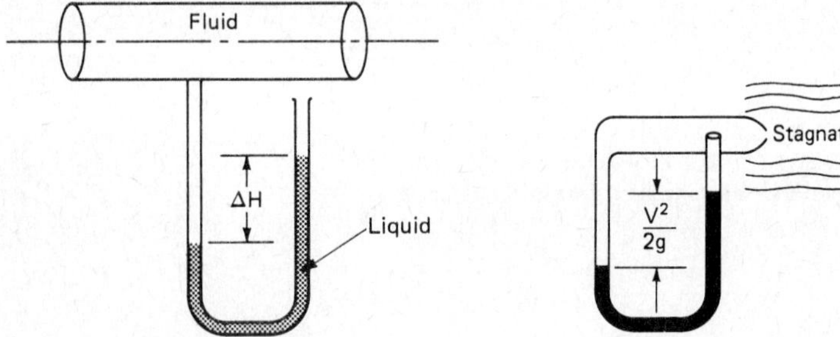

Figure 1-8 U-tube manometer.

Figure 1-9 Pitot tube.

pressure in a fluid is also measured by using a diaphragm pressure gauge in which the scale is calibrated such that the displacement of the diaphragm within the system transmitted through mechanical, pneumatic, or hydraulic linkages directly gives a pressure reading on the dial.

The velocity of flow is usually obtained by measuring the differential pressure head ΔH between two points of flow (usually at a transition) and using proper formulation, such as

$$V = C_v\sqrt{2g\,\Delta H} \tag{36}$$

where

ΔH = differential head, ft (m), of fluid flowing (not gauge fluid)

C_v = coefficient of velocity

The Pitot tube (Fig. 1-9) was specifically developed to measure velocity and is a special form of U tube in which the stagnation leg measures $p_0/w + V^2/2\mathbf{g}$ and the other leg gets signal $p_0/2w$. Therefore, the differential head is equal to $V^2/2\mathbf{g}$, thus giving velocity. The quantity or mass *rate of flow* is most accurately measured by collecting in a weighing or graduated tank for a definite time. However, it is usually calculated by measuring the velocity of flow and multiplying it by a suitable area of cross section.

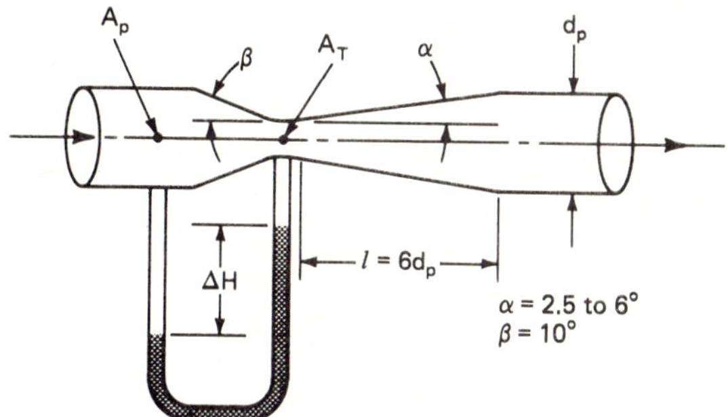

Figure 1-10 Venturimeter.

This principle has been further extended by using special transition elements or fittings for which the coefficient of discharge is already known from laboratory measurements. For example, a venturimeter (Fig. 1-10) has the following formula:

$$Q = A_T C_d \frac{\sqrt{2\mathbf{g}\,\Delta H}}{1 - (A_T/A_p)^2} \tag{37}$$

where

Q = quantity of flow, ft³/s (m³/s) (multiply by density to get mass rate)
C_d = coefficient of discharge of the meter, 0.95 to 0.99
A_T = area of cross section at throat, ft² (m²)
A_p = area of cross section of pipe, ft² (m²)
ΔH = differential head, ft (m) of fluid flowing (not the gauge liquid)

Orifices and nozzles have also been used with the same basic principle of acquiring velocity and then converting to discharge. However, the jet of flow in such a case is contracted at the discharge, "vena contracta" (Fig. 1-11), giving

$$C_c = \text{coefficient of contraction} = 0.65 \text{ usually}$$
$$= \frac{A_{jet}}{A_o} \tag{38}$$

where

A_{jet} = area of vena contracta, ft² (m²)
A_o = area of orifice nozzle, ft² (m²)

Then

$$Q = VA_{jet} \tag{39}$$

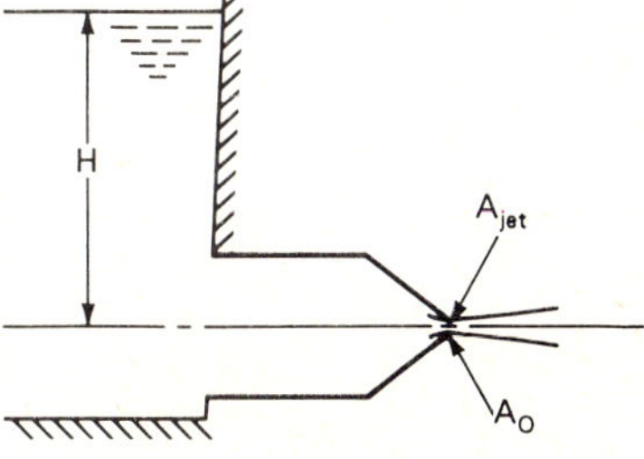

Figure 1-11 Vena contracta.

V is calculated from the static head H on the nozzle. Direct flow-reading, magnetic, ultrasonic, and other meters have also been used recently.

The flow in open channels is usually measured by utilizing *weirs* (Fig. 1-12). For a rectangular, sharp-crested weir, Francis gave the following relation:

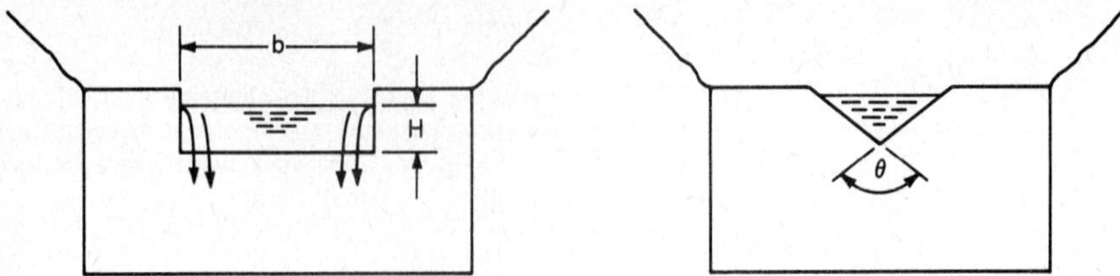

Figure 1-12 Weirs.

$$Q = C\left(b\frac{NH}{10}\right)\left[\left(H+\frac{V_0^2}{2g}\right)^{3/2} - \left(\frac{V_0^2}{2g}\right)^{3/2}\right] \tag{40}$$

where

Q = quantity of liquid flow, ft^3/s (m^3/s)
b = length of rectangular weir, ft (m)
H = head above weir crest, taken far upstream, ft (m)
V_0 = velocity of approach flow, taken far upstream, ft/s (m/s)
N = number of end contractions
= 2, if weir length < channel width at both ends
= 1, if weir length < channel width at one end
= 0, if weir length = channel width
C = 3.33 (U.S. customary units) = 1.84 (SI units)

However, Bazin gave the following relation for rectangular weir:

$$Q = \left(0.405 + \frac{c}{H}\right)\sqrt{2g}\, bH^{3/2} \tag{41}$$

where c = 0.00984 (U.S. customary units) = 0.03 (SI units)

If the end contractions and approach velocity are neglected, a simplified formula can be used. For a triangular weir (V notch), the formula is

$$Q = \begin{cases} \frac{8}{15} c \tan\frac{\theta}{2} \sqrt{2g}\, H^{5/2} & \text{(U.S. customary units)} \\ 2.36c \tan\frac{\theta}{2} H^{5/2} & \text{(SI units)} \end{cases} \tag{41a}$$

where

c = a coefficient of contraction = 0.6
H = head above the lowest point on the notch, taken far upstream, ft (m)
θ = total angle subtended by the notch

Measurements by Using Electrical Data

Motor Input-Output. The flow-rate pump head can also be determined by measurement of voltage-current at the pump motor (or turbine-generator). If the current, voltage, and power factor of the pump motor are measured in the control room, the bus-voltage drop should be properly accounted for.[14*]

For a three-phase motor, the motor output brake horsepower (bhp) is given by

$$\text{bhp} = \sqrt{3}\,\frac{VI\cos\theta}{746}\,\eta_m \tag{42a}$$

where

V = terminal voltage, V
I = current, A

*Reference for Chapters 3-1 and 3-2 are combined following Chapter 3-2.

$\cos\theta$ = power factor
η_m = motor efficiency
746 W = 1 hp

For getting flow rate, the following equation for pump water horsepower (whp) is used:

$$\text{whp} = \begin{cases} \dfrac{\gamma Q H}{550} & \text{(U.S. customary units)} \\ \dfrac{\gamma Q H}{75} & \text{(SI units)} \end{cases} \tag{42b}$$

where

γ = density of water being pumped, lb/ft^3 (kg/m^3)
Q = rate of flow, ft^3/s (m^3/s)
H = TDH of pump, ft (m) of water
550 ft·lb/s = 1 hp(U.S. customary units)
75 kg·m/s = 1 hp (SI units)

Another relation required is for pump bhp:

$$\text{bhp} = \frac{\text{whp}}{\eta_p} \tag{42c}$$

where η_p is the efficiency of the pump. Using Eqs. (3), (4), and (5),

$$Q = \begin{cases} \dfrac{\sqrt{3}VI\cos\theta\ \eta_m\eta_p}{746}\dfrac{550}{\gamma H} & \text{(U.S. customary units)} \\ C\dfrac{VI\cos\theta\ \eta_m\eta_p}{\gamma H} & \text{(SI units)} \end{cases} \tag{42d}$$

where C = 1.275 for Q, ft^3/s.

η_m and η_p are involved as unknowns. They have to be obtained from manufacturers' curves by trial and error, by assuming a Q, finding efficiencies from those curves, and using those efficiencies to get a new Q.

Note—For a turbine, the only difference is that the inverse of $\eta_m\eta_p$ is used.

1-3. PRINCIPAL COMPONENTS OF FLUID SYSTEMS

Following is a list of the components most frequently used in fluid systems:

Accumulators
Actuators
Compressors, vacuum pumps, and blowers
Couplings, quick-connect, hydraulic
Couplings, quick-connect, pneumatic
Cylinders
Demineralizers
Dryers
Expansion joints
Filters, air-line
Filters, hydraulic
Fittings, tube and port, hydraulic
Fittings, tube and port, pneumatic
Fluids
Gauges, pressure and flow, hydraulic
Gauges, pressure and flow, pneumatic
Heat exchangers
Hose, hose fittings, and hose assemblies, hydraulic
Hose, hose fittings, and hose assemblies, pneumatic
Hydrostatic drives
Intensifiers
Interface devices
Joints, rotating and swivel, hydraulic and pneumatic
Logic devices, air
Lubricators
Manifolds
Measuring devices and meters
Motors, air rotary

Motors, electrohydraulic stepping
Motors, hydraulic rotary
Motors, low-speed–high-torque
Pipe
Pulsation dampers
Pumps
Reservoirs and accessories
Restrictive orifices
Seals, scrapers, backup rings, and boots
Shock absorbers
Steam traps
Strainers
Switches, flow
Switches, level
Switches, limit and proximity
Switches, pressure and temperature, hydraulic
Switches, pressure and temperature, pneumatic
Transducers, pressure
Tubing
Valves, directional control
Valves, electrohydraulic servo
Valves, flow control
Valves, pressure control
Valves, level control

Various materials are used in these components. It is important to be aware of their compatibility with fluids used in industry. Table 1-6 provides such data.

Some of the major components of fluid systems will be discussed here in detail. Consult other sections of this handbook as well.

PUMPS

A pump is the most essential component of a hydraulic system in the absence of another source of pressure. The most common other source of pressure is a reservoir kept at high elevation (or potential energy). However, water at high elevation automatically exerts a static pressure and does not require much more explanation. Since pumps have diverse characteristics according to their speed and size and require extended discussion for a thorough understanding, only the most essential details will be discussed here. The Hydraulic Institute[7] classification of pumps given in Fig. 1-13 is useful.

Types of Pumps

Positive-Displacement-Type Pumps

These pumps (reciprocating plunger, rotary-gear, screw, piston, vane, etc.) create pressure by a direct-contact push to the liquid. They are employed generally for medium-pressure, low-flow applications, such as a metering pump to displace a known volume of fluid in a specified time. Large-size positive-displacement pumps usually become uneconomical because of space requirements.

Rotary-Gear Pumps

A rotary-gear pump, shown in Fig. 1-14, depends for its action on a pair of gears closely fitted in a housing. Fluid enters the tooth spaces usually on the inlet side and is carried along the casing to the outlet side.

Centrifugal Pumps

The pumps more commonly used in plant applications are of the centrifugal type, in which the rotary motion of the impeller imparts centrifugal force to the liquid, thereby raising the pressure. The selection of the impeller of a particular centrifugal pump (Fig. 1-15) is based upon the specific speed N_s as defined below:

$$N_s = \frac{N\sqrt{Q}}{H^{3/4}} \tag{43}$$

Text continues on p. 3-24

TABLE 1-6 Suitability of Various Metals and Alloys for Various Fluids*,†

Corrosive	Steel‡ C.I. D.I.	Brz.	316SS	CA-20	CD4MCu	Mon	Ni	H-B	H-C	Ti	Zi
Acetaldehyde, 70°F	B	A	A	A	A	A	A		A	A	A
Acetic acid, 70°F	X	A	A	A	A	B	B	A	A	A	A
Acetic acid, <50%, to boiling	X	B	A	A	B	B	B	C	A	A	A
Acetic acid, >50%, to boiling	X	X	B	A	C	B	B	X	A	A	A
Acetone, to boiling	A	A	A	A	A	A	A	A	A	A	A
Aluminum chloride, <10%, 70°F	X	B	C	B	C	B	C	A		B	A
Aluminum chloride, >10%, 70°F	X	X	C	B	C	C	X	A		B	A
Aluminum chloride, <10%, to boiling	X	X	X	C	X	X	X	A		X	A
Aluminum chloride, >10%, to boiling	X	X	X	X	X	X	X	A	X	X	A
Aluminum sulfate, 70°F.	X	B	A	A	A	B	B	B	B	A	A
Aluminum sulfate, <10%, to boiling	X	B	B	A	B	X	X	A	A	A	A
Aluminum sulfate, >10%, to boiling	X	C	C	B	C	X	X	B	B	C	B
Ammonium chloride, 70°F	X	X	B	B	B	B	B		A	A	A
Ammonium chloride, <10%, to boiling	X	X	B	B	C	B	B	B	A	A	A
Ammonium chloride, >10%, to boiling	X	X	X	C	X	C	C		C	C	C
Ammonium fluosilicate, 70°F	X	X	C	B	C	X	X		C	X	X
Ammonium sulfate, <40%, to boiling	X	X	B	B	C	B	B	X	B	A	A
Arsenic acid, to 225°F	X	X	C	B	C	X	X				
Barium chloride, 70°F, <30%	X	B	C	B	C	B	B	B	B	B	B
Barium chloride, <5%, to boiling	X	B	C	B	C	B	B	B	B	A	A
Barium chloride, >5%, to boiling	X	C	X	C	X	C	C	C	C	C	C
Barium hydroxide, 70°F	B	X	A	A	A	B	A	B	B	A	A
Barium nitrate, to boiling	C	X	B	B	B		B	B		B	B
Barium sulfide, 70°F	C	X	B	B	B	X	X			A	A
Benzoic acid	X	C	B	B	B	B	B	A	A	A	A
Boric acid, to boiling	X	C	B	B	B	C	C	A	A	B	B
Boron trichloride, 70°F, dry	B	B	B	B	B	B	B	B	B		
Boron trifluoride, 70°F, 10%, dry	B	B	B	A	B	A	A		A		
Brine (acid), 70°F	X	X	X	X	X				B	B	
Bromine (dry), 70°F	X	X	X	X	X	X	C	B	B	X	X
Bromine (wet), 70°F	X	X	X	X	X	X	C		B	X	X
Calcium bisulfite, 70°F	X	X	B	B	B	X	X		B	A	A
Calcium bisulfite, to hot	X	X	C	B	C	X	X		C	A	A

TABLE 1-6 Suitability of Various Metals and Alloys for Various Fluids*,† (*Continued*)

Corrosive	Steel‡ C.I. D.I.	Brz.	316SS	CA-20	CD4MCu	Mon	Ni	H-B	H-C	Ti	Zi
Calcium chloride, 70°F	B	C	B	B	B	B	B	A	A	A	A
Calcium chloride, <5%, to boiling	C	C	B	B	B	A	A	A	A	A	A
Calcium chloride, >5%, to boiling	X	C	C	B	C	C	C	A	A	B	B
Calcium hydroxide, 70°F	B	B	B	B	B	B	B		A	A	
Calcium hydroxide, <30%, to boiling	C	B	B	B	B	B	B		A	A	
Calcium hydroxide, >30%, to boiling	X	X	C	C	C	C	C		B	A	
Calcium hypochlorite, <2%, 70°F	X	X	X	C	X	X	X		A	A	A
Calcium hypochlorite, >2%, 70°F	X	X	X	C	X	X	X		B	A	B
Carbolic acid, 70°F (phenol)	C	B	A	A	A	A	A	A	A	A	A
Carbon bisulfide, 70°F	B	B	A	A	A	B	B			A	
Carbonic acid, 70°F	B	C	A	A	A	C	B	A	A	A	A
Carbon tetrachloride, dry to boiling	B	B	A	A	A	A	A	B	B	A	A
Chloric acid, 70°F	X	X	X	B	C	X	X	X	C		
Chlorinated water, 70°F	C	C	B	B	B				A	A	A
Chloroacetic acid, 70°F	X		X	X						A	B
Chlorosulfonic acid, 70°F	X	X	X	C	X	X	X	A	A	B	X
Chromic acid, <30%	X	X	C	B	C	X	X		B	A	A
Citric acid	X	C	A	A	A	C	C	A	A	A	A
Copper nitrate, to 175°F	X	X	B	B	B	X	X	X	X	B	
Copper sulfate, to boiling	X	C	C	B	C	X	X		A	A	A
Cresylic acid	C	C	B	B	B	C	C	B	B		
Cupric chloride	X	C	X	X	X	C	X		C	B	X
Cyanohydrin, 70°F	C		B	B	B						
Dichloroethane	C	B	B	B	B	C	B	B	B	A	B
Diethylene glycol, 70°F	A	B	A	A	A	B	B	B	B	A	A
Dinitrochlorobenzene, 70°F, dry	C	B	A	A	A	A	A	A	A	A	A
Ethanolamine, 70°F	B	X	B	B	B	C	X			A	A
Ethers, 70°F	B	B	B	A	A	B	B	B	B	A	A
Ethyl alcohol, to boiling	A	A	A	A	A	A	A	A	A	A	A
Ethyl cellulose, 70°F	A	B	B	B	B	B	B	B	B	A	A
Ethyl chloride, 70°F	C	B	B	A	B	B	B	B	B	A	A
Ethyl mercaptan, 70°F	C	X	B	A	B			B	B		
Ethyl sulfate, 70°F	C	B	B	A	B	B					
Ethylene chlorohydrin, 70°F	C	B	B	B	B	B	B	B	B	A	A
Ethylene dichloride, 70°F	C	B	B	B	B	B	B	B	C	A	A

Ethylene glycol, 70°F	B	B	B	B	B	B	B	A	A	A	A
Ethylene oxide, 70°F	C	X	B	B	B	B	B	A	A	A	A
Ferric chloride, <5%, 70°F	X	X	X	X	X	X	X	X	A	A	B
Ferric chloride, >5%, 70°F	X	X	X	X	X	X	X	X	B	B	X
Ferric nitrate, 70°F	X	X	B	A	B	X	X		B		
Ferric sulfate, 70°F	X	X	C	B	C	C	C		B	B	B
Ferrous sulfate, 70°F	X	C	C	B	C	C	C	B	B	A	A
Formaldehyde, to boiling	B	B	A	A	A	B	B	B	B	A	A
Formic acid, to 212°F	X	C	X	A	B	C	C	A	A	C	A
Freon, 70°F	A	A	A	A	A	A	A	A	A	A	A
Hydrochloric acid, <1%, 70°F	X	X	C	B	C	B	B	B	A	B	A
Hydrochloric acid, 1–20%, 70°F	X	X	X	X	X	X	X	C	B	X	A
Hydrochloric acid, >20%, 70°F	X	X	X	X		X	X	C	B	X	B
Hydrochloric acid, <½%, 175°F	X	X	C	C	C	X	X	B	A	X	A
Hydrochloric acid, ½–2%, 175°F	X	X	X		X	X	X	B	B	X	A
Hydrocyanic acid, 70°F	X	X	C	B	C	C	C	C	C		
Hydrogen peroxide, <30% <150°F	C	X	B	B	B	B	B	B	B	A	A
Hydrofluoric acid, <20%, 70°F	X	B	X	B	C	C	C	C	B	X	X
Hydrofluoric acid, >20%, 50°F	X	C	X	C	X	C	C	C	B	X	X
Hydrofluoric acid, to boiling	X	X	X	X	X	C	X		C	X	X
Hydrofluosilicic acid, 70°F	X		C	B	C				B		
Lactic acid, <50%, 70°F	X	B	A	A	A	X	C	B	B	A	A
Lactic acid, >50%, 70°F	X	B	B	B	B	C	C	B	B	A	A
Lactic acid, <5%, to boiling	X	X	C	B	C	X	X	B	B	A	A
Lime slurries, 70°F	B	B	B	B	A	B	B	B	B	B	B
Magnesium chloride, 70°F	C	C	B	A	B	C	C	A	A	A	A
Magnesium chloride, <5%, to boiling	X	C	C	B	C	C	C	A	A	A	A
Magnesium chloride, >5%, to boiling	X	C	X	C	X	C	C	B	B	B	B
Magnesium hydroxide, 70°F	B	A	B	B	A	B	A	B	B	A	
Magnesium sulfate	C	C	B	A	B	B	B	C	C	B	B
Maleic acid	C	C	B	B	B	C	C	B	B	A	
Mercaptans	A	X	A	A	A	X	X				
Mercuric chloride, <2%, 70°F	X	X	X	X	X	X	C		B	A	A
Mercurous nitrate, 70°F	C	X	B	B	B	C			C		
Methyl alcohol, 70°F	A	A	A	A	A	A	A	A	A	A	A
Naphthalene sulfonic acid, 70°F	X	C	B	B	B	C	C	B	B		
Naphthalenic acid, to hot	C	C	B	B	B	C	C	B	B		
Nickel chloride, 70°F	X	X	C	B	C	C	X	A		B	B
Nickel sulfate	X	C	B	B	B	C	C		B		A

TABLE 1-6 Suitability of Various Metals and Alloys for Various Fluids*·† (*Continued*)

Corrosive	Steel‡ C.I. D.I.	Brz.	316SS	CA-20	CD4MCu	Mon	Ni	H-B	H-C	Ti	Zi
Nitric acid	X	X	B	B	B	X	X			B	B
Nitrobenzene, 70°F	A	C	A	A	A	B	B	B	B	A	
Nitroethane, 70°F	A	A	A	A	A	A	A	A	A	A	A
Nitropropane, 70°F	A	A	A	A	A	A	A	A	A	A	A
Nitrous acid, 70°F	X	X	X	C	X	X	X				
Nitrous oxide, 70°F	C	C	C	C	C	X	X		C		
Oleic acid	C	C	B	B	B	C	C	C	C	C	C
Oleum, 70°F	B	X	B	B	B	X	X	B	B	B	
Oxalic acid	X	C	C	B	C	C	C	B	B	X	A
Palmitic acid	B	B	B	A	B	B	B				
Phenol (see carbolic acid)											
Phosgene, 70°F	C	C	B	B	B	C	C	B	B		
Phosphoric acid, <10%, 70°F	X	C	A	A	A	C	C	A	A	A	A
Phosphoric acid, >10–70%, 70°F	X	C	A	A	A	C	C	B	C	B	B
Phosphoric acid, <20%, 175°F	X	C	B	B	B	C	C	A	A	C	B
Phosphoric acid, >20%, 175°F <85%	X	C	C	B	C	C	C	B	C	C	C
Phosphoric acid, >10%, boil, <85%	X	C	X	C	C	C	C	C	C	C	C
Phthalic acid, 70°F	C	B	B	A	B	B	B	B	B	A	A
Phthalic anhydride, 70°F	B	C	A	A	A	A	A	A	A		
Picric acid, 70°F	X	X	C	B	C	C	X		B		
Potassium carbonate	B	B	A	A	A	B	B	B	B	A	A
Potassium chlorate	B	C	A	A	A	C	C		B	A	A
Potassium chloride, 70°F	C	C	B	A	B	B	B	B	B	A	A
Potassium cyanide, 70°F	B	X	B	B	B	C	C	B	B		
Potassium dichromate	B	B	A	A	A	B	B		B	A	A
Potassium ferricyanide	C	B	B	B	B	B	B	B	B	A	A
Potassium ferrocyanide, 70°F	X	B	B	B	B	B	B	B	B		B
Potassium hydroxide, 70°F	C	C	B	A	B	A	A	B	C	B	A
Potassium hypochlorite	X	C	C	B	C	X	X		B	A	
Potassium iodide, 70°F	C	B	B	B	B	B	B	B	B	A	A
Potassium permanganate	B	B	B	B	B	C	B		B		
Potassium phosphate	C	C	B	B	B					B	B
Seawater, 70°F	C	B	B	A	B	A	A	A	A	A	A
Sodium bisulfate, 70°F	X	C	C	B	C	C	C	B	B	B	A
Sodium bromide, 70°F	B	C	B	B	B	B	B	B	B		

Sodium carbonate	B	B	B	A	B	B	B	B	B	A	A
Sodium chloride, 70°F	C	B	B	B	B	A	A	B	B	A	A
Sodium cyanide	B	X	B	B	B	X	X			B	
Sodium dichromate	B	X	B	B	B					B	
Sodium ethylate	B	A	A	A	A	A	A				
Sodium fluoride	C	C	B	B	B	B	B	C	C	B	B
Sodium hydroxide, 70°F	B	B	B	A	B	A	A	A	A	A	A
Sodium hypochlorite	X	X	C	C	C	X	X		B	A	B
Sodium lactate, 70°F	B	C	C	C	C	C		C	C		
Stannic chloride, <5%, 70°F	X	C	X	C	X	C	C	B	B	A	A
Stannic chloride, >5%, 70°F	X	X	X	X	X	X	X	B	C	B	B
Sulfite liquors, to 175°F	X	C	B	B	B	C	C		B	A	
Sulfur (molten)	B	X	A	A	A	C	C	C	A	A	
Sulfur dioxide (spray), 70°F	C	C	B	B	B	C	C		B	C	
Sulfuric acid, <2%, 70°F	X	C	B	A	B	C	C	A	A	B	A
Sulfuric acid, 2–40%, 70°F	X	C	C	B	C	C	C	A	A	X	A
Sulfuric acid, 40%, <90%, 70°F	X	X	X	B	X	X	X	A	A	X	C
Sulfuric acid, 93–98%, 70°F	B	X	B	B	B	X	X	B	B	X	C
Sulfuric acid, <10%, 175°F	X	C	X	B	X	X	X	A	C	X	B
Sulfuric acid, 10–60% & >80%, 175°F	X	X	X	B	X	X	X	B	C	X	C
Sulfuric acid, 60–80%, 175°F	X	X	X	X	X	X	X	B	C	X	C
Sulfuric acid, <¾%, boiling	X	X	C	B	C	X	X	B	B	X	B
Sulfuric acid, ¾–40%, boiling	X	X	X	C	X	X	X	B	C	X	B
Sulfuric acid, 40–65% & >85%, boil	X	X	X	X	X	X	X	X	X	X	X
Sulfuric acid, 65–85%, boiling	X	X	X	X	X	X	X	X	X	X	X
Sulfurous acid, 70°F	X	C	C	B	C	X	X	B	B	A	B
Titanium tetrachloride, 70°F	C		C	B	C	C			C		
Trichlorethylene, to boiling	B	C	B	B	B	B	B	B	B	A	A
Urea, 70°F	C	C	B	B	B	C	C	C	C	B	B
Vinyl acetate	B	B	B	B	B				B		
Vinyl chloride	B	C	B	B	B	C	C	C	B	A	
Water, to boiling	B	A	A	A	A	A	A	A	A	A	A
Zinc chloride	C	C	B	A	B	B	B	B		A	A
Zinc cyanide, 70°F	X	B	B	B	B	B	B	B	B	B	B
Zinc sulfate	X	C	A	A	A	C	C	C	C	A	

* *Source:* Goulds Pumps, Inc., Seneca Falls, N.Y.

†Ni, Nickel, ASTM A 296 Gr. CZ-100; H-Bm Hastelloy alloy-B, ASTM A 494; H-C, Hastelloy alloy-C, ASTM A 494; Ti, Titanium unalloyed, ASTM B 367 Gr. C-1; Zi, Zirconium.

‡Code: A, fully satisfactory; B, useful resistance; C, limited use; X, unsuitable.

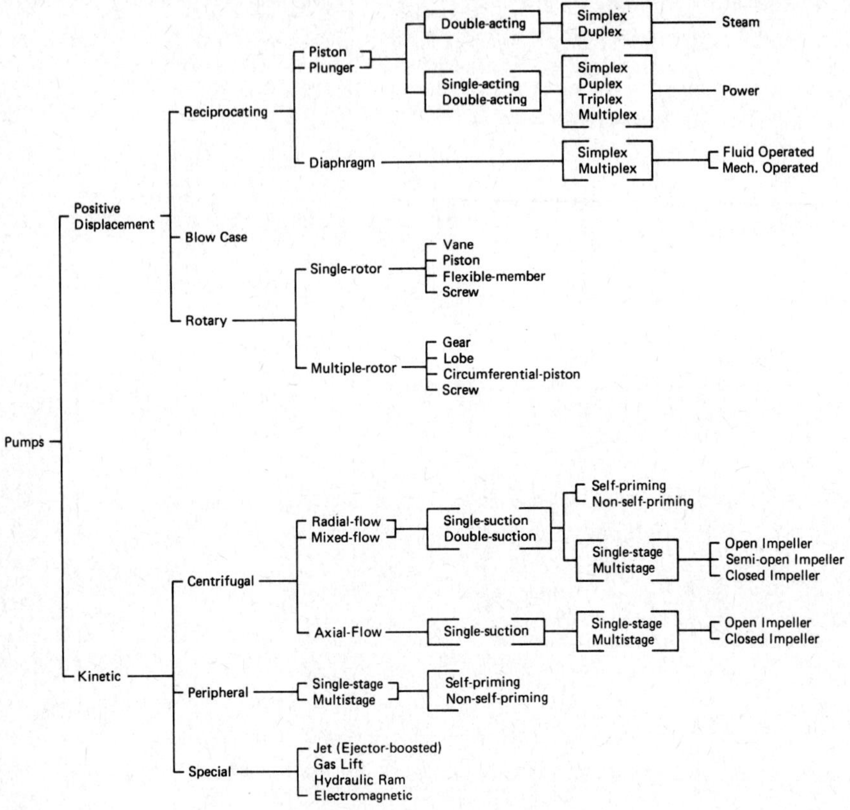

Figure 1-13 Classification of pumps.

where

N = pump rotation speed, r/min

Q = nominal rate of flow to be pumped, gal/min (m^3/s) (should be halved for double suction pump)

H = total pressure head to be created by pump, ft (m)

In the specific speed range of 1000 to 6000 r/min double-suction impellers can also be used if axial thrust needs to be balanced.

Pump Terms

Total Dynamic Head

The total dynamic head (TDH) of the pump needs to be determined. With reference to Fig. 1-16,

$$H = \text{TDH} = h_s^* + h_d + h_L \tag{44}$$

Also

$$\text{H} = h_{g,1} - h_{g,2} + \frac{V_2^2}{2\mathbf{g}} - \frac{V_1^2}{2\mathbf{g}} \tag{45}$$

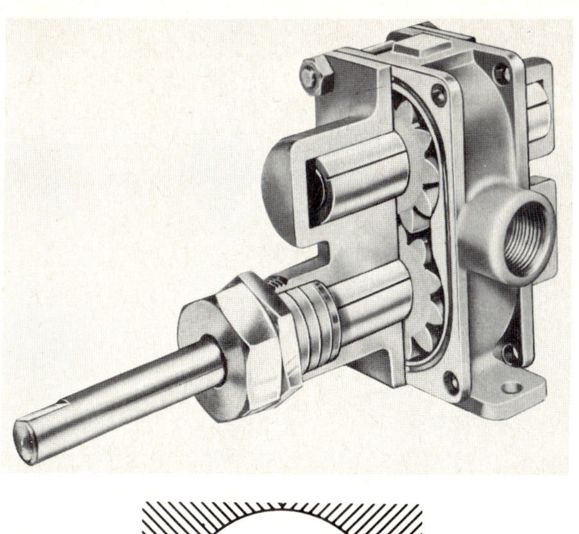

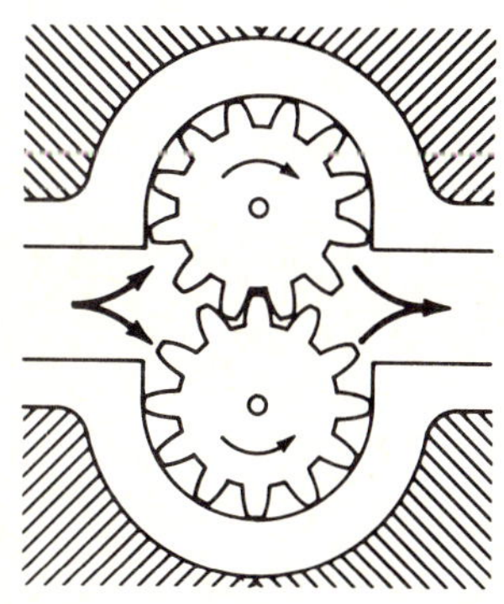

Figure 1-14 Rotary gear pump. *(ECO Pump Co.)*

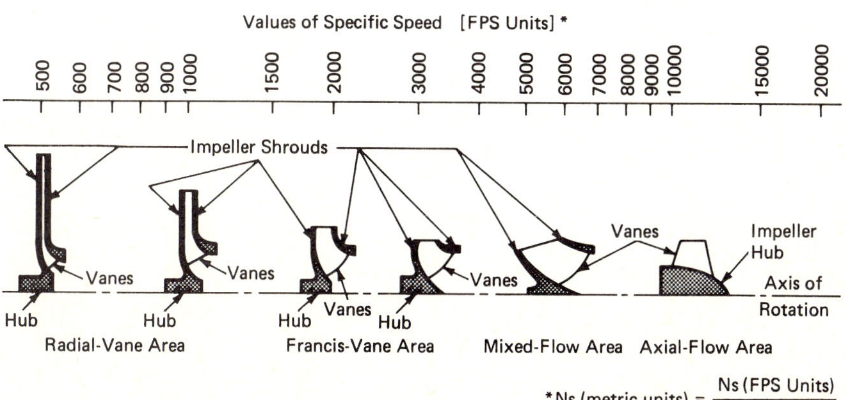

$$*N_s \text{ (metric units)} = \frac{N_s \text{ (FPS Units)}}{47.25}$$

Figure 1-15 Selection of impeller design based on specific speed.

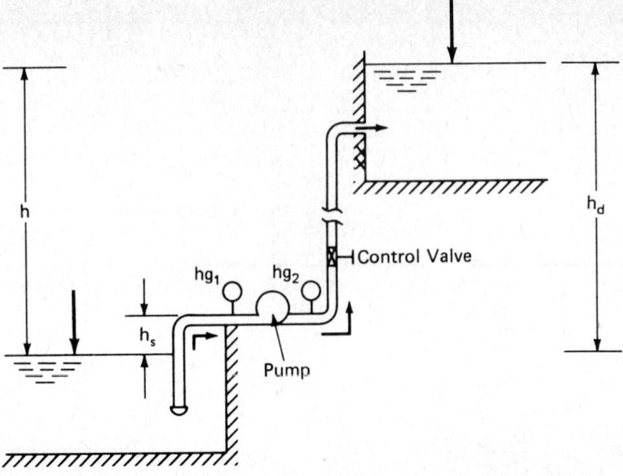

Figure 1-16 Definition sketch for pump heads.

TABLE 1-7 Centrifugal Pump Steam Turbine Specification Sheet

Client	Job no.
Plant location	Date
Inquiry no.	Equip. no.
Pump service	No. required

Operating Conditions		Specifications*	
Fluid pumped		Pump type	Cat. no.
		Stages	r/min
Pumping temp., °F		Suct. size, in.	Location
Spec. grav. @ P.T.		Noz. series	Flange facing
Visc. @ P. T., cP		Disch. size, in.	Location
Vap. press. @ P.T.		Noz. series	Flange facing
Solids % & comp.		Hydraulic hp	Efficiency, %
Suct. head, psia	ft	bhp	Max. hp
Disch. head, psia	ft	NPSH required by pump, ft fluid	
TDH, ft, normal	Design	Pump materials & details*	
Pump cap., gal/min, normal	Design	Case	Max. imp.
NPSH available, ft, fluid		Case Th'kn.	Corr. all.
Nonoverloading		Shaft Dia.	Keyway
Rotation (facing coupling end)		Shaft sleeves	Brinell
Motor by		Impeller Type	O.D.
hp V Ph. Freq.		Impeller wear ring	
Frame St. torque		Case wear ring	
Mfr. & type		Radial brg.	Mfg. no.
Enclosure Class Group		Thrust brg.	Mfg. no.

where

h_s^* = static suction head of the pump, ft (m)
h_d = static delivery discharge head of the pump, ft (m)
h_L = head losses in the piping, including entrance, friction, bends, fittings, exit, etc., ft (m)
$h_{g,1}^*$ = pressure (converted to head) indicated by suction gauge (vacuum gauge), ft (m). The asterisk indicates $+Ve$ if pump is higher than suction water surface and $-Ve$ if pump is lower than suction water surface.
$h_{g,2}$ = pressure (converted to head) indicated by discharge gauge, ft (m)
H = total static head of the pump, ft (m)

Net Positive Suction Head

The net positive suction head (NPSH) is another term that needs to be determined for the cavitation-free design and performance of a pump. It is the net absolute static pressure above the vapor pressure of the liquid in the pump:

$$\text{NPSH} = P_b - h_s^* - h_{f,s} - P_v \tag{46}$$

where

P_b = barometric pressure (absolute), ft (m)
P_v = vapor pressure of the liquid being pumped, ft (m)
$h_{f,s}$ = head losses in suction line up to pump, ft (m)

Operating Conditions			Specifications*			
Turbine by			Packing		No. rings	
Mfr.	Model		Seal cage		Seal fluid	
Type			Packing gland			
Rated hp			Mech. seal		Mfr.	
Inlet steam, psia	temp., °F		Coupling			
Exh. steam, psia	Temp., °F	Quality	Base plate			
No. hr steam fl	3/4 L	1/2 L	Drip (lip) (pan)			
Turbine performance curves, details, cuts, and			Wt. pump, lb.			
dimension sheets, etc., must be submitted by bidder			Wt. base plate, lb.			
Driver mounted by			Wt. driver, lb.			
Remarks			Total wt., lb.			
			Case design temp. °F			
			Case design press., psig			
			Case hydrostatic test, psig			
			Pump performance curves, details, cuts, dimension			
			sheets, etc., must be submitted by bidder			
Bidder in proposal to complete these spaces and others						
left vacant. Successful bidder shall supply:						
Copies of certified dimension drawings						
Copies of performance curves for fluid pumped						
Copies of spare parts list						
Copies of operating instructions			No.	Date	Revision	Appr.

The available NPSH (NPSHA) at a pump application (in a pump house) should be greater than the NPSH required (NPSHR) by the pump for an efficient (cavitation-free) operation. If NPSHA is less than NPSHR, it means that vapor is likely to form inside the pump. This vapor can accumulate and form bubbles which, when burst, can produce cavities (nonsmooth surfaces on impeller vanes).

Suction Specific Speed

Here another term comes into the picture, the *suction specific speed* S_s:

$$S_s = \frac{N\sqrt{Q}}{(\text{NPSHR})^{3/4}} \tag{47}$$

The Hydraulic Institute[7] recommends S_s between 7480 and 10,690 r/min. See p. 3-22 for definitions of N and Q.

Data Specifications

A useful blank form for a specification sheet of pump data is given in Table 1-7. All data should be filled in and sent to vendors for further completion and bid preparation.

Pump Characteristic Curves

Pump characteristic parameters that are a measure of the pump performance are discharge, head, horsepower, efficiency, and NPSH. These characteristics vary with the pump type, size, speed, etc. Some typical characteristics are shown in Fig. 1-17.

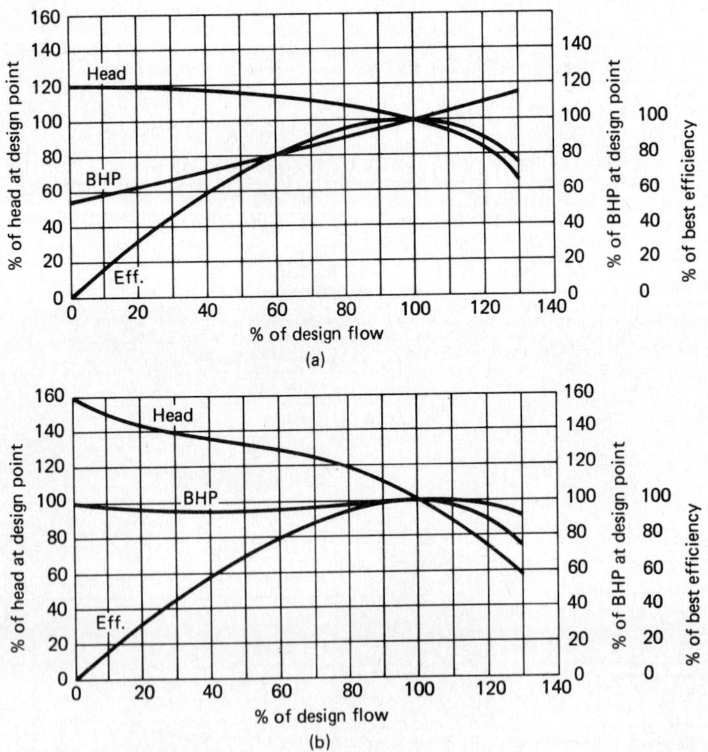

Figure 1-17 Pump characteristic curves: (*a*) radial-flow pump, (*b*) mixed-flow pump. *(Goulds Pumps, Inc.)*

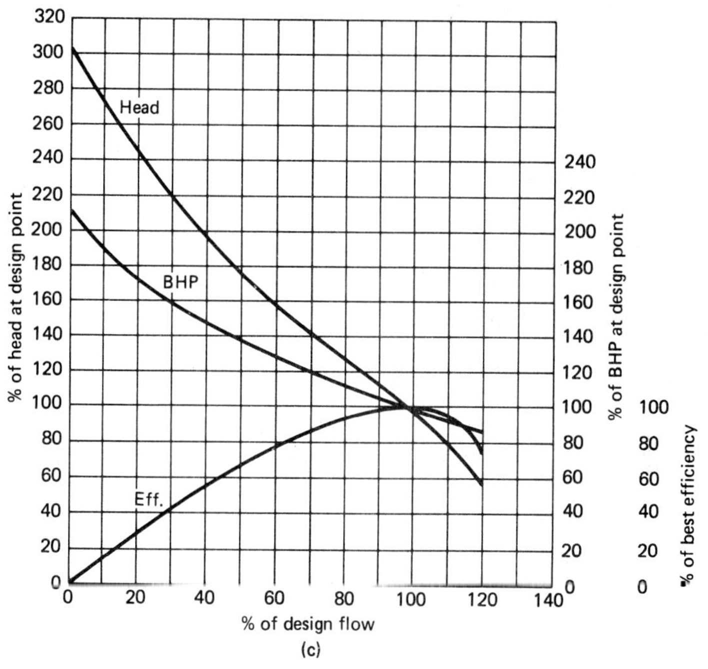

(c)

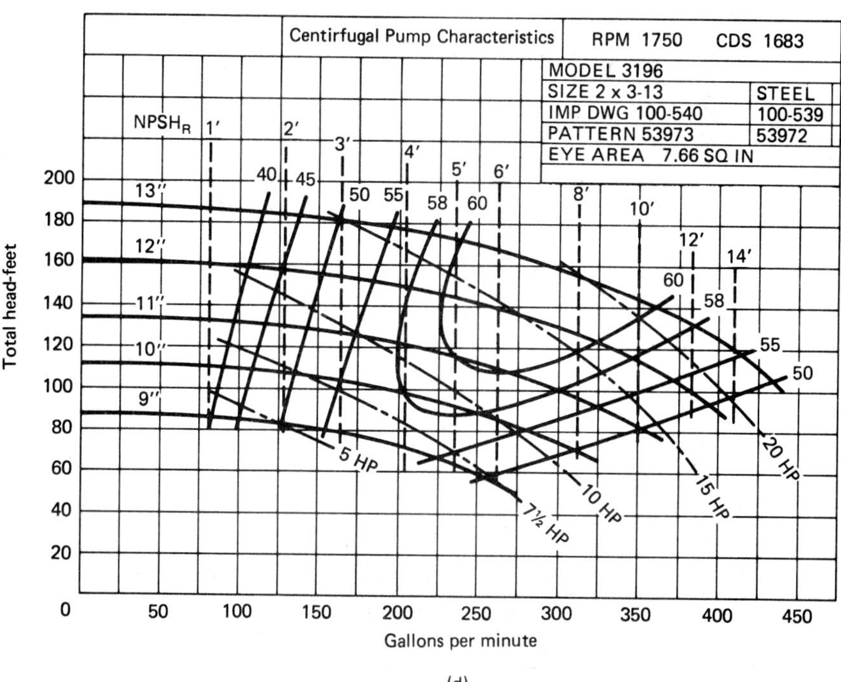

(d)

Figure 1-17 (*Continued*) (*c*) axial-flow pump, (*d*) composite performance curve. (*Goulds Pumps, Inc.*)

The normal operating point of the pump should be selected somewhere on the sloping portion of the curve, not too close to shutoff or run-out conditions. The manufacturer usually recommends a pump that supplies within +10 percent of design flow or +5 percent of design TDH.

Figures 1-18 and 1-19 are typical curves developed by manufacturers to aid in pump selection.

Pump Construction and Maintenance

Figures 1-20 and 1-21 show cross sections through some typical pumps. To control leakage, mechanical seals (discussed elsewhere in this book) serve best. The American National Standards Institute is often referred to for a proper design of pumps. ANSI B73.1 is a standard for horizontal centrifugal pumps.

Proper design of the pumphouse and pump-suction bays, inlet conditions, etc., is very important for proper performance of the pumps.[8]

Basic dimensions of the pumps required are first evolved on the basis of the Hydraulic Institute publication, "Standards for Centrifugal Pumps"; however, the final dimensions on the construction drawings have to be based on the pump manufacturers' recommendations, as their requirements are evolved on the basis of their own model tests. A matter of great importance is to determine the pump TDH and the NPSH which in turn reflect the overall dimensions of the suction intakes. Shape of the intake, depth, and clearances are among the important critical factors that affect the performance of large circulating-water pumps.

Hydraulic performance of a large-capacity pump is also affected by the formation of submerged or surface vortices in the sump area, which in practice are not visually observable because of the cover on the sump necessary to mount the pumping equipment and motors. Therefore, it is imperative to give careful consideration to the sump dimensions, preferably by the way of model experiments (discussed later). Moreover, the size of the pump house (including intake canal, cooling-tower basin for closed system, etc.) should be enough to store (or bypass) any transient quantities caused by sudden pump failures or operation of valves. Another important aspect of the pump house is to provide a device for measuring the actual flow and discharge pressure of each pump.

Preliminary dimensions of a pump house can be arrived at by using the guidelines given in Ref. 7 or a vendor's engineering data book.

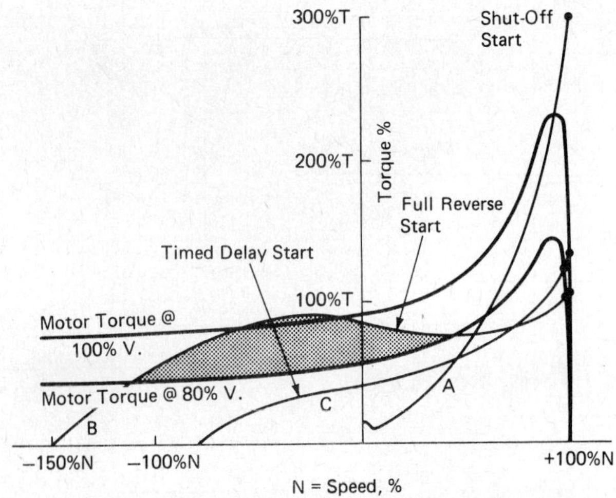

Figure 1-18 Start-up and shutdown of pumps. *(Allis-Chalmers.)*

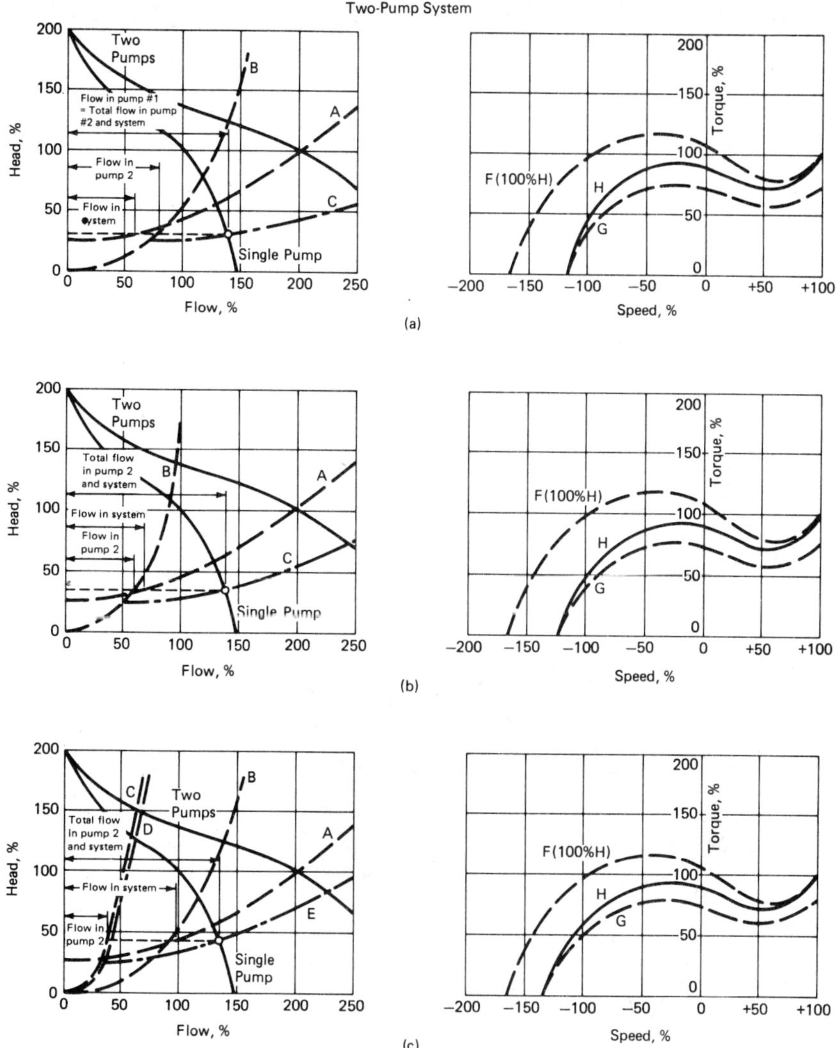

Figure 1-19 Starting large custom pumps against reverse flow. (*a*) Friction curves with pump characteristics are used to determine complete speed-torque curve. Pump 1 is operating and pump 2 is being started against reverse flow with valve open. (*b*) Starting at zero speed with valve open and reverse flow in pump, the speed-torque curve is identical. Because anti-reverse device prevents reverse rotation, maximum torque requirements are less. (*c*) Interlocking of valve operator and motor starter requires consideration of friction-head characteristic of the valve. The figure gives head vs. flow curves for single-pump operation and for two pumps in parallel operation, when both are running in the forward direction. The transient phase of starting is shown by dashed/dotted lines. The corresponding torque requirements are shown on the right. *(Allis-Chalmers.)*

Maintenance

Basically, pump maintenance requires proper lubrication according to the manufacturer's recommendations. Stuffing boxes and packings must be inspected and maintained as necessary. If proper setup, operation, and maintenance are not performed, the defects shown in Table 1-8 will be evident.[9] They should be corrected as soon as discovered.

TABLE 1-8 Causes and Symptoms of Malfunctions of Pumps

Causes	Symptoms									
	The pump fails to deliver liquid	Capacity is inadequate	Head is inadequate	The pump cuts out after starting	Power required by pump is too high	Excessive leakage of stuffing box	Packing must be renewed too frequently	The pump vibrates or is noisy	Bearings heat up	Pump runs heavily or seizes
Pump and suction line are not sufficiently primed with the liquid handled	●	●		●				●		●
Excessive suction lift	●	●		●				●		
Insufficient margin between suction lift and vapor pressure	●	●						●		●
Liquid contains gas		●	●	●						
Air pocket in suction line	●	●		●						
Air leak in suction line		●		●						
Air leaks into pump along stuffing box		●		●						
Foot valve too small		●								
Foot valve partly clogged		●						●		
Foot valve and suction line not fully submerged	●	●		●				●		
Connection of water seal on suction stuffing box blocked				●			●			
Lantern ring in stuffing box incorrectly fitted				●		●	●			
Speed too low	●	●	●							
Speed too high					●					
Incorrect direction of rotation	●		●							
Total manometric head of system greater than manometric head of the pump	●	●	●		●					
Total manometric head of system lower than manometric head of the pump					●					
Specific gravity of liquid handled not what it was originally supposed to be					●					

TABLE 1-8 Causes and Symptoms of Malfunctions of Pumps (*Continued*)

Causes	Symptoms: The pump fails to deliver liquid	Capacity is inadequate	Head is inadequate	The pump cuts out after starting	Power required by pump is too high	Excessive leakage of stuffing box	Packing must be renewed too frequently	The pump vibrates or is noisy	Bearings heat up	Pump runs heavily or seizes
Viscosity of liquid handled not what it was originally supposed to be		●	●		●					
Pump operating at too low a capacity								●		●
Parallel connection unsuitable for the specific operating conditions	●	●	●							●
Line, impeller, or pump casing clogged	●	●			●			●		
Pump set incorrectly aligned					●	●	●	●	●	●
Foundation not level								●		
Pump shaft warped					●	●	●	●	●	
A rotating part runs against a stationary part, e.g., impeller runs against wearing rings					●			●	●	●
Bearing(s) defective								●	●	●
Wearing rings worn		●	●							
Impeller damaged		●	●					●		
Pump shaft or shaft sleeve worn locally at the stuffing box						●	●			
Stuffing box incorrectly packed					●	●	●			
Type of packing used unsuitable for liquid handled						●	●			
Impeller out of balance						●	●	●	●	●
Failure to apply water cooling when handling hot liquids						●	●			
Clearance between pump shaft and bore of pump casing at the bottom of the stuffing box too great							●			

TABLE 1-8 Causes and Symptoms of Malfunctions of Pumps (*Continued*)

Causes	Symptoms: The pump fails to deliver liquid	Capacity is inadequate	Head is inadequate	The pump cuts out after starting	Power required by pump is too high	Excessive leakage of stuffing box	Packing must be renewed too frequently	The pump vibrates or is noisy	Bearings heat up	Pump runs heavily or seizes
Liquid for water seal contains impurities							●			
Gland overtightened							●			
Axial fit of complete pump shaft with impeller incorrect								●	●	●
Insufficient or excessive lubrication									●	
Lubricant contains impurities									●	●
Bearings incorrectly fitted									●	●

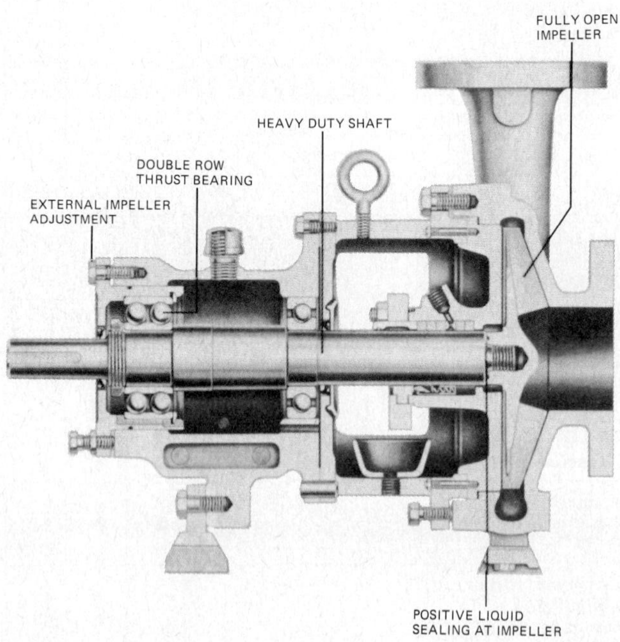

Figure 1-20 Cross-sectional view of a horizontal centrifugal pump. *(Goulds Pumps, Inc.)*

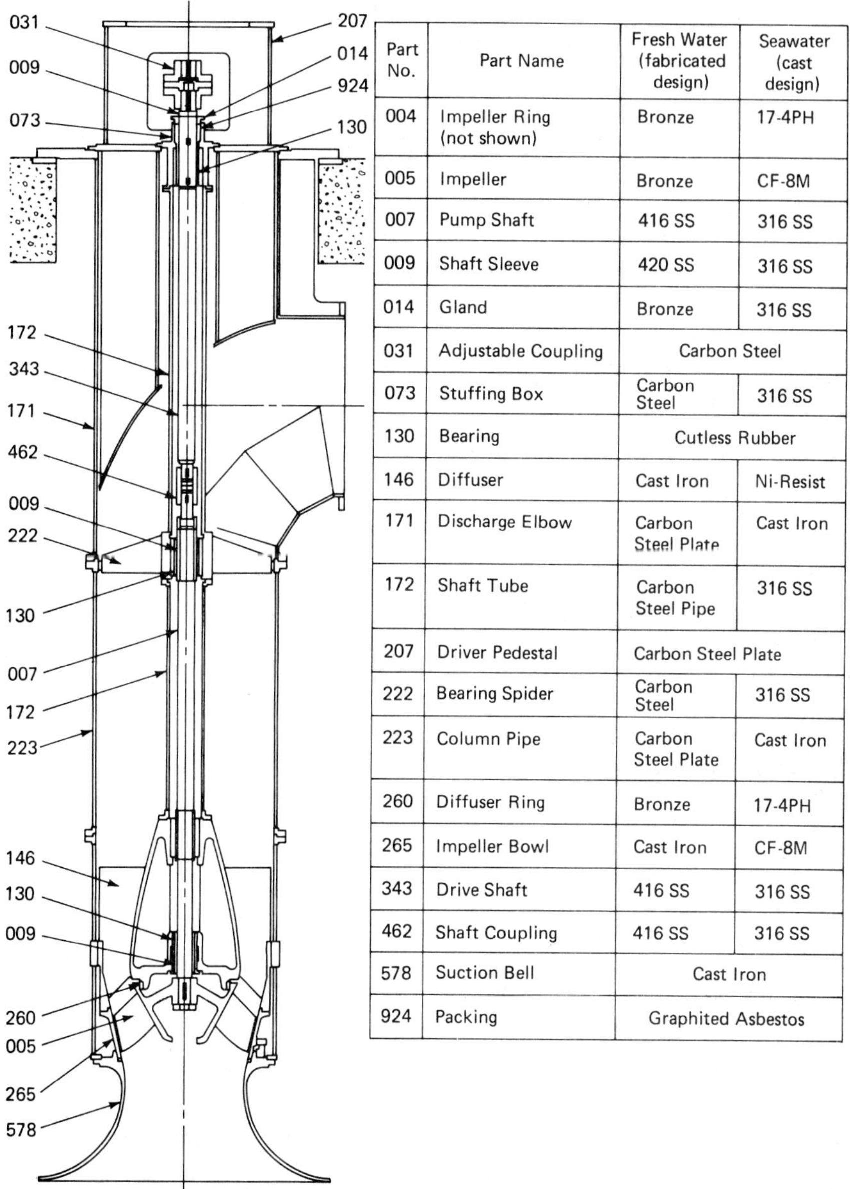

Part No.	Part Name	Fresh Water (fabricated design)	Seawater (cast design)
004	Impeller Ring (not shown)	Bronze	17-4PH
005	Impeller	Bronze	CF-8M
007	Pump Shaft	416 SS	316 SS
009	Shaft Sleeve	420 SS	316 SS
014	Gland	Bronze	316 SS
031	Adjustable Coupling	Carbon Steel	
073	Stuffing Box	Carbon Steel	316 SS
130	Bearing	Cutless Rubber	
146	Diffuser	Cast Iron	Ni-Resist
171	Discharge Elbow	Carbon Steel Plate	Cast Iron
172	Shaft Tube	Carbon Steel Pipe	316 SS
207	Driver Pedestal	Carbon Steel Plate	
222	Bearing Spider	Carbon Steel	316 SS
223	Column Pipe	Carbon Steel Plate	Cast Iron
260	Diffuser Ring	Bronze	17-4PH
265	Impeller Bowl	Cast Iron	CF-8M
343	Drive Shaft	416 SS	316 SS
462	Shaft Coupling	416 SS	316 SS
578	Suction Bell	Cast Iron	
924	Packing	Graphited Asbestos	

Figure 1-21 Vertical mixed-flow column pump. *(Allis-Chalmers.)*

1-5. VALVES

A valve is used to regulate the flow in a pipe. A hydraulic system may contain several valves, some of which may be just for full open-close positions, while other valves regulate the quantity of flow as desired. The latter are usually categorized as control valves. Control valves can be either manually or automatically operated depending upon the desired amount of flow, according to pressure, temperature, level, etc. (See Chap. 5-2,

"Automatic Controls.") Instrumentation is usually provided in the system to measure and sense the quantities. In automatic operation of the valve, the sensors send signals directly to the air-electric or other operating mediums that supply force to the automatic-controlling mechanisms. The valves are available as flanged, screwed, or welded inlet-outlet ports. The standards normally referenced for valves are ANSI-B16.10 and ASME Boiler and Pressure Vessel Code VIII, III. See also Chap. 4-7, "Valves."

Valve Pressure Drop

The valve loss coefficient K was mentioned previously (p. 3-10). Another way to represent the pressure drop through the valves (especially for control valves) is called the *flow coefficient* C_v. It is defined as the number of gallons per minute of water which will pass through a given flow restriction with a pressure drop of 1 lb/in^2. (The actual fluid does not necessarily have to be water. There are formulas available for calculating C_v for other fluids also.)

The relationship between C_v and K is easy to derive from definitions. For liquid service,

$$C_v = Q\sqrt{\frac{G_f}{\Delta P}} \tag{48}$$

$$\Delta P = K\frac{V^2}{2} \tag{48a}$$

where

Q = liquid flow rate, gal/min = $448.8AV\ldots$
G_f = specific gravity of liquid, for water = 1
ΔP = pressure drop at the valve for flow Q, lb/in^2
V = flow velocity 1 ft/sp

For critical and flashing flows, when the pressure drop cannot be obtained from Eq. (48*a*), the equation is

$$\Delta P = C_f^2\,\Delta P_s \tag{49}$$

$$\Delta P_s = P_l - P_v \tag{50}$$

where

C_f = critical flow factor = 0.6 to 0.9 depending upon valve type
P_1 = pressure upstream of the valve
P_v = vapor pressure of liquid

For gas and vapor flow, the above formulas are modified to suit the flow units.[10] Chapter 4-7, "Valves," describes the more important valves in plant use.

OTHER COMPONENTS OF A SYSTEM

Besides pumps and valves, system piping must be considered. This includes associated fittings such as bends, reducers, branches, elbows, and T connections. (See Chaps. 4-2 and 4-3.) Standard pipe dimensions, i.e., nominal diameter, inside diameter, thickness, etc., for various schedule pipes, are available from literature.[6,15] Table 1-9 gives the important dimensions and burst pressures for some of the sizes and classes. The velocity of flow in a pipe should normally range from 8 to 15 ft/s (2.44 to 4.57 m/s) in order to keep the pressure drop reasonably low for a given pump size and also to deliver an optimum amount of fluid.

Other components are often a part of hydraulic systems. These include heat exchangers, filters, strainers, demineralizers, spray nozzles, steam traps, expansion joints, restrictive orifices, and measuring devices (already discussed).

Graphic Symbols for Fluid Power Systems

According to the American National Standards Institute definition, fluid power systems are those that transmit and control power through the use of a pressurized fluid within

TABLE 1-9 Important Pipe Dimensions and Burst Pressures (U.S. Customary Units)

		Schedule 40 (standard)		Schedule 80 (extra strong)		Schedule 160		Double (extra strong)	
Nominal pipe size, in	Outside diameter of pipe, in	Inside diameter of pipe, in	Burst pressure, lb/in²	Inside diameter of pipe, in	Burst pressure, lb/in²	Inside diameter of pipe, in	Burst pressure, lb/in²	Inside diameter of pipe, in	Burst pressure, lb/in²
¼	0.540	0.364	16,000	0.302	22,000	—	—	—	—
⅜	0.675	0.493	13,500	0.423	19,000	—	—	—	—
½	0.840	0.622	13,200	0.546	17,500	0.466	21,000	0.252	35,000
¾	1.050	0.824	11,000	0.742	15,000	0.614	21,000	0.434	30,000
1	1.315	1.049	10,000	0.957	13,600	0.815	19,000	0.599	27,000
1¼	1.660	1.380	8,400	1.278	11,500	1.160	15,000	0.896	23,000
1½	1.900	1.610	7,600	1.500	10,500	1.338	14,800	1.100	21,000
2	2.375	2.067	6,500	1.939	9,100	1.689	14,500	1.503	19,000
2½	2.875	2.469	7,000	2.323	9,600	2.125	13,000	1.771	18,000

an enclosed circuit. ANSI Y32.10 gives graphic symbols commonly used for drawing circuit diagrams of fluid power systems. They show connections, flow paths, and functions of components represented. Some of the important symbols are given here in Fig. 1-22.

Figure 1-23 is an illustration of a power system and its symbolic representation. This example illustrates a pump that transfers liquid from reservoir through a check valve and a four-way control valve to hydraulic cylinders, the other ends of which are connected to another open reservoir containing liquid.

A TYPICAL HYDRAULIC SYSTEM

In order to illustrate the use of hydraulic components, one hydraulic system will be discussed here in detail. It is the circulating-condenser, cooling-water system, one of the very important systems for a thermal power plant.[8] It is important because of the large quantity of water required which in turn necessitates large open channels, pipes, valves, etc., in the system.

Hydraulics

A typical circulating-water system for a power plant is shown schematically in the plan view in Fig. 1-24. Large quantities of cooling water are withdrawn from a river by pumps provided in the intake canal. Trash racks and traveling screens are usually provided in the intake canal in order to keep the unwanted debris from entering the pumps and condensers. The discharge from the condensers can be thrown off-site into a large body of water (e.g., the ocean) for final heat dissipation. A weir is usually provided downstream of the condenser to avoid the impact on the condensers from the downstream water-level fluctuations (e.g., due to tides). In order to dissipate the extra energy available from the fall over the weir and to accommodate a hydraulic jump, a stilling basin is usually provided downstream of the weir. If there is another waterway (say river) to be crossed downstream before approaching the ocean, pipes can pass under that waterway. Finally, under modern environmental practice, the hot water cannot be discharged into the ocean too near the shore. Therefore, water is pumped from the discharge canal to give offshore final discharge submerged for good heat dissipation. Valves are provided in the piping to control flow. No further discussion of heat dissipation is required here because the purpose of this chapter is not just to understand one system, but how to engineer a fluid system in general. A comprehensive knowledge of circulating water systems, if required, can be obtained from Ref. 8.

The first step in design is to plot a developed view of the system to scale, with all the hydraulic components (mentioned previously) indicated at least schematically. All the horizontal and vertical distances to these components should be plotted and the piping drawn properly. *Developed view* indicates plotting in two dimensions only. If there is a third dimension, it should be rotated horizontally to bring it to the second dimension at the point where it occurs. Thus all the dimensions get accounted for. The vertical scale can be distorted (different from horizontal scale) in order to be able to see the elevations of each point clearly. All the components, transitions, bends, etc., should be properly accounted for.

For the circulating water system mentioned above, a developed picture is shown in Fig. 1-25, with some assumed dimensions. The next step is to plot a hydraulic gradient line for this system. In order to do this, all the head losses in the open channels, piping, and appurtenances have to be calculated and pump TDH established for the known flow rate, as mentioned in previous sections. In Fig. 1-25, the losses for valves, bends, etc., and the friction loss in the pipeline drop from the upstream end to the downstream end of the pipe as shown on the hydraulic gradient line.

The calculations of head losses in pipe fittings, valves, etc., and determination of pump TDH can best be accomplished by using a tabular form, e.g., Table 1-10. This keeps the number of computation sheets to a minimum, and the important factors are not omitted. It should be noted that the head-loss data for certain components, e.g., heat

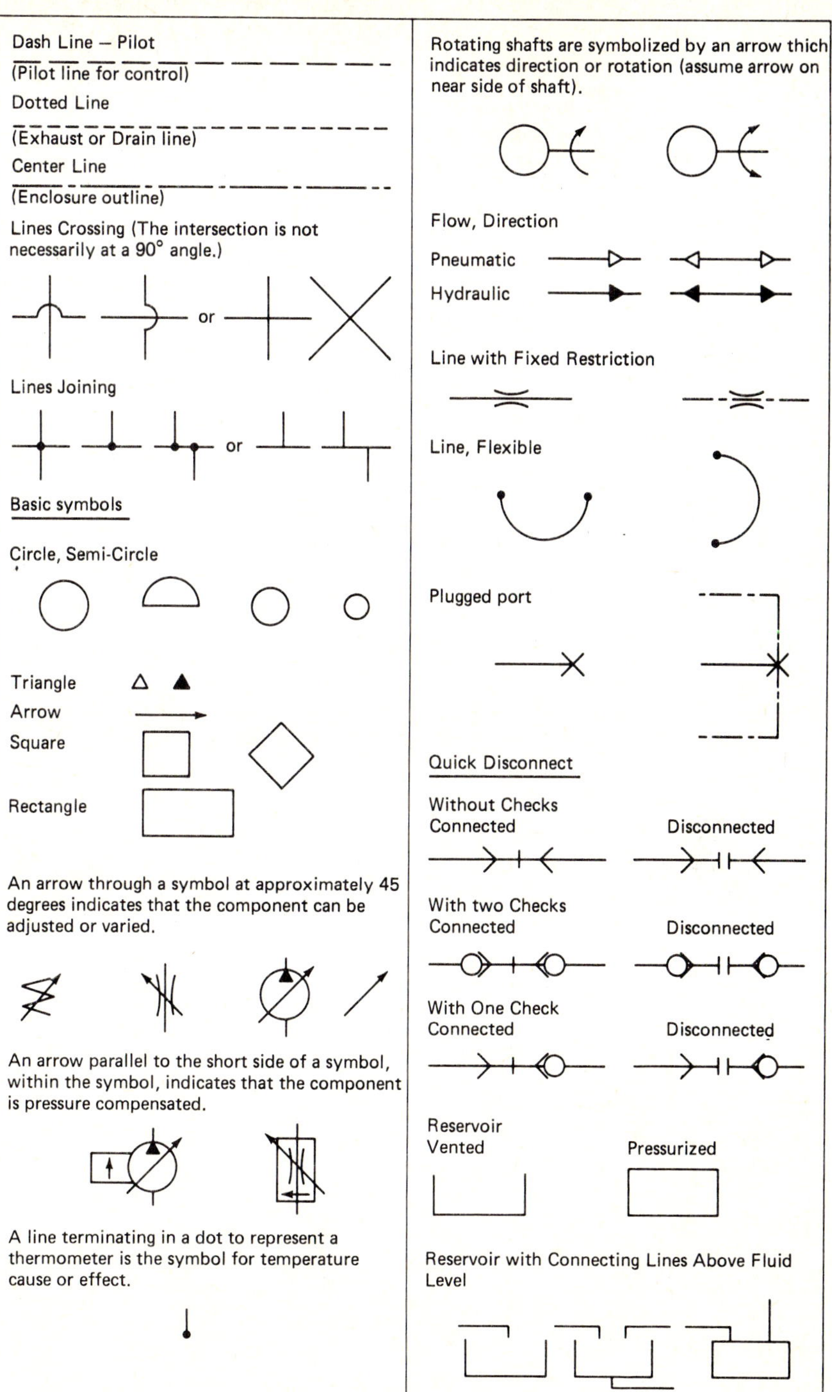

Figure 1-22 Graphic symbols.

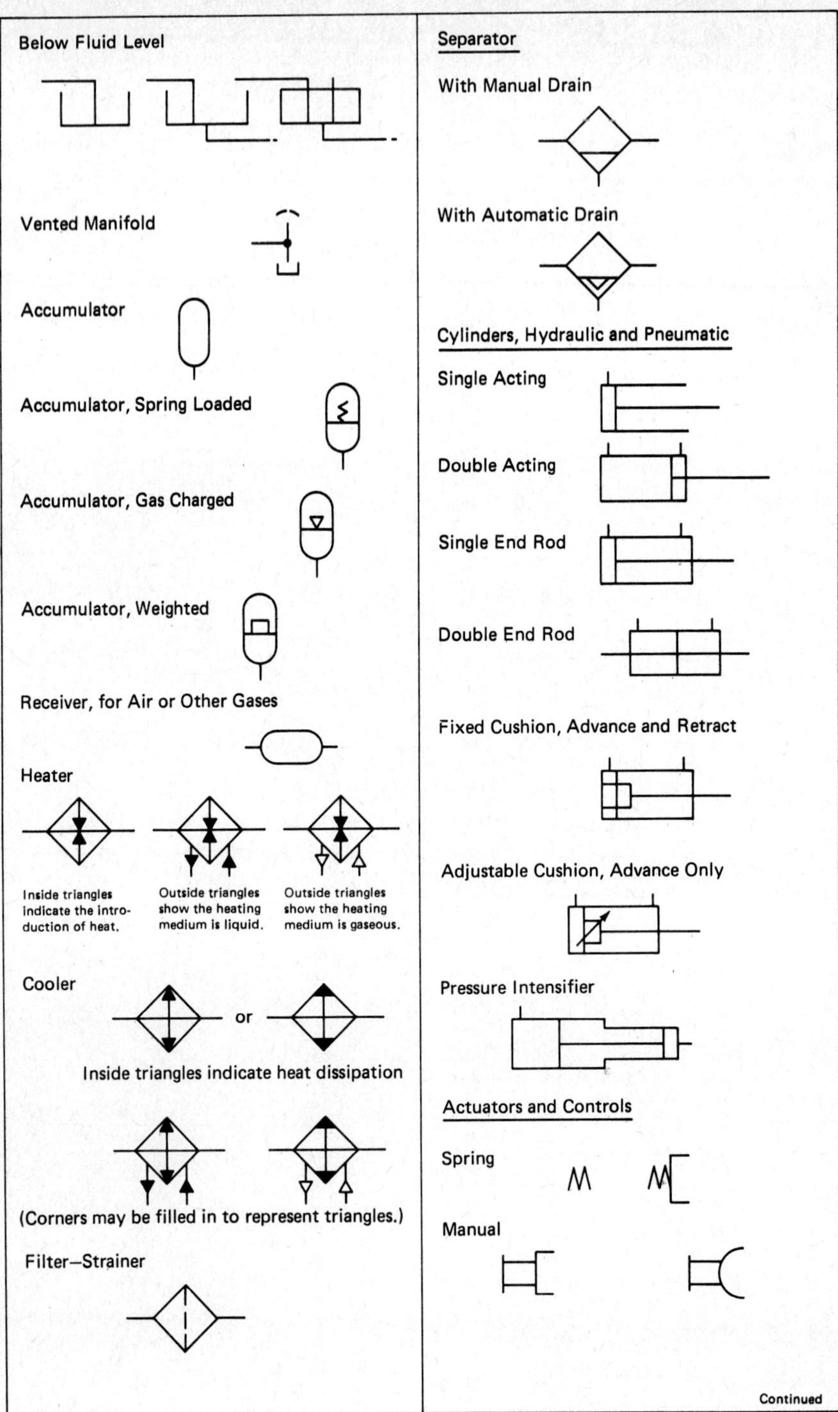

Figure 1-22 *(continued)* **Graphic symbols.**

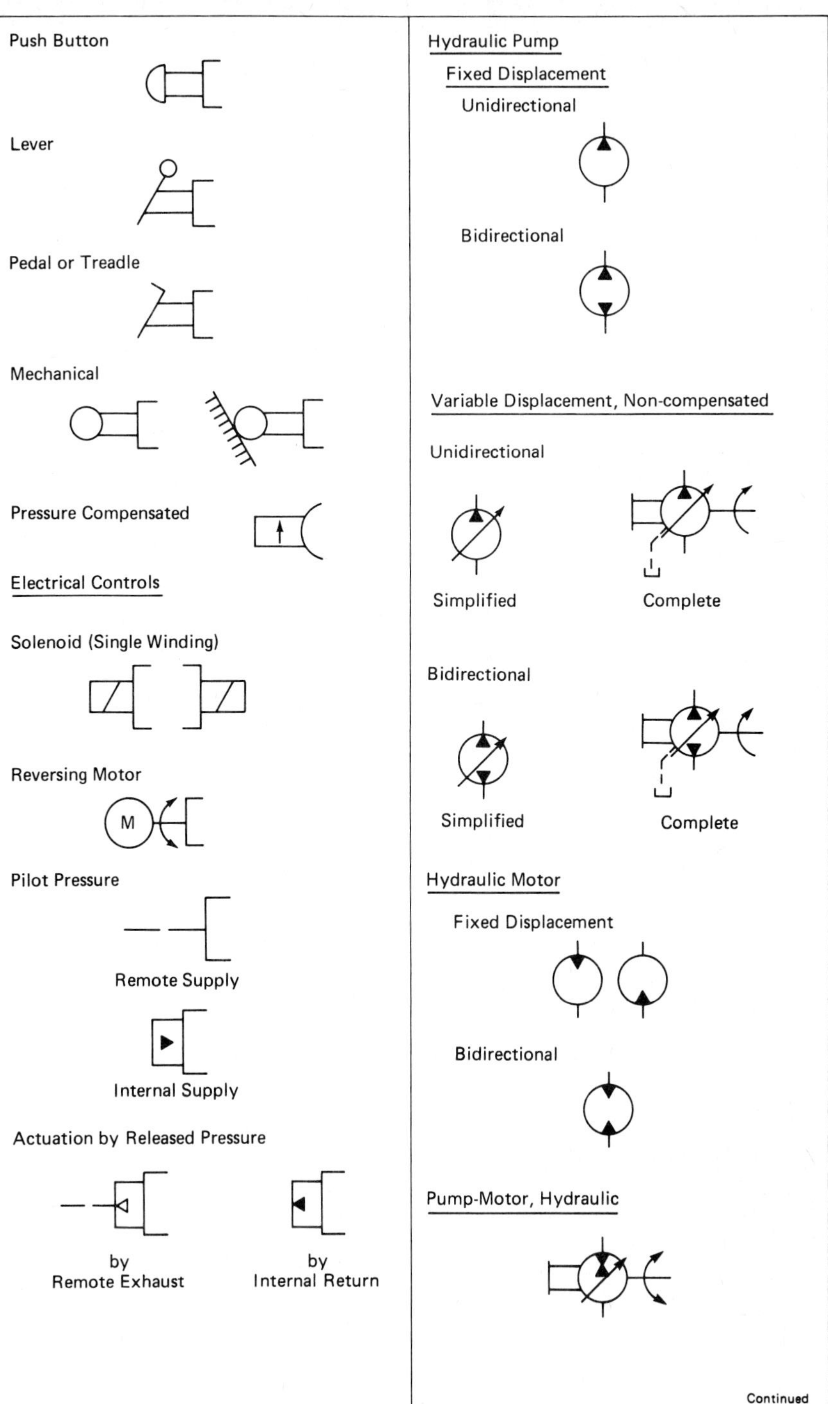

Figure 1-22 *(Continued)* Graphic symbols.

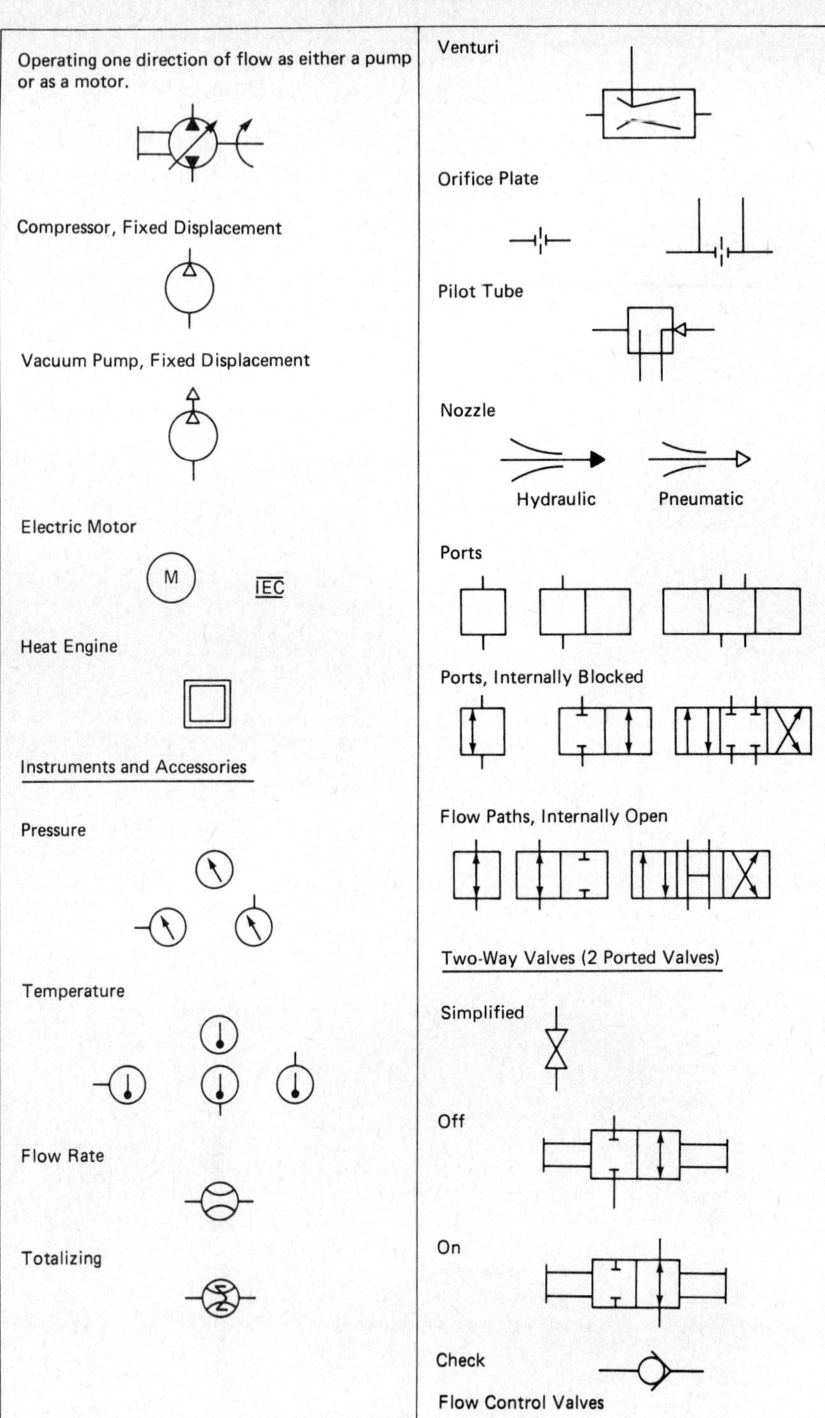

Figure 1-22 *(Continued)* Graphic symbols.

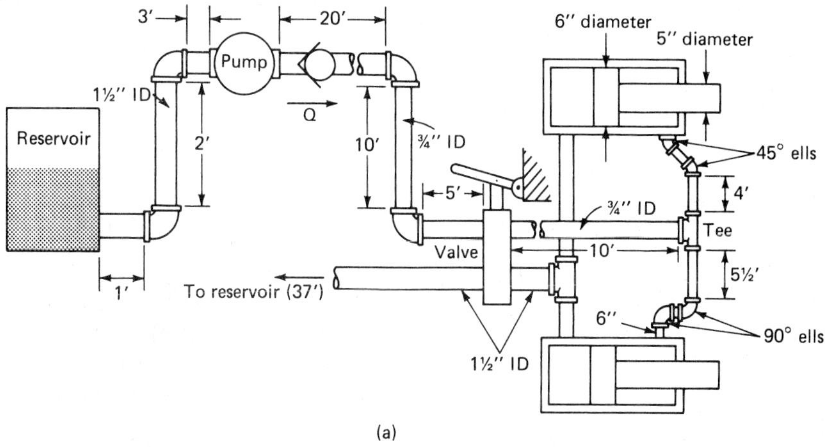

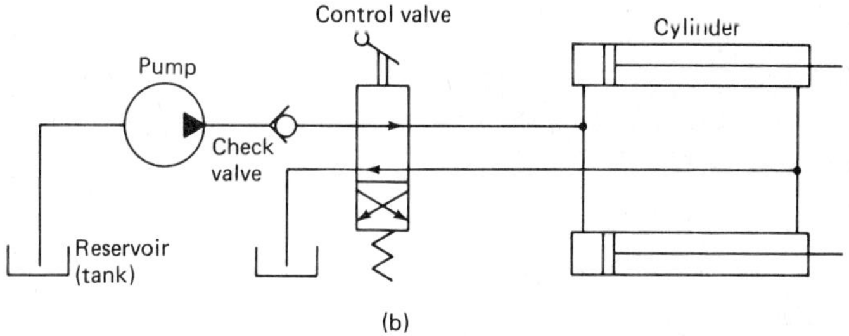

Figure 1-23 Illustration of power systems symbols.

exchanger (condenser in this typical example), orifice plates, control valves, filters, etc., should be obtained from the catalogs of the manufacturers of specific components.

The TDH of the pump should be conservatively estimated so as not to fall short of pressure in the system, especially as the system gets older and scale deposits form, thereby increasing the pressure drops. If a little extra pressure is available from the pump in the beginning, the runout of the pump (excessive flow) can be controlled by throttling the flow at the control valves in the system. Therefore, it is usually good to have some throttling-type valves in the system.

The TDH of the pump and performance of a system can be determined or verified in the field on the large-size systems because laboratory facilities are usually not large enough to handle full-scale flows. Field testing can be achieved by following one of the several measurement techniques outlined in a previous section. More sophisticated instrumentation and techniques are also available these days, such as, annubar, current meter, etc., which are discussed in Ref. 14. More sophisticated flowmeters, e.g., magnetic-acoustic, are also available in the market.

Hydraulic transients and water-hammer aspects for varied system-pump-valve operations should be studied.

Process Instrumentation

The engineering of systems, such as the circulating water system discussed above, of course involves other aspects besides hydraulics. The most important, from a systems

TABLE 1-10 Pump Calculation Sheet (U.S. Customary Units)

Project ________ Design gal/min ____ Design TDH ____ Pump no. ________

Pump service ________ Date ____ Sch. & pipe matl. ________ Viscosity @ min P.T. ________ cP(μ)

Pump no. ________ Fluid pumped ________ Spec. grav. @ min P.T. ________ (SG)

Lb/h ________ Temp., max ____ Min ____ F Kinematic viscosity ________ cS(z)

Normal gal/min ________ From ____ To ________ Vapor press. @ max P.T. ________ mmHg(VP)

Design gal/min ________

Reference drawings

		Suction line				Discharge line		
Line no. & schedule		___	___	___		___	___	___
Nom. line size, inches I.D. in; ft D/d	d	___	___	___	d	___	___	___
Velocity, ft/s	V	___	___	___	V	___	___	___
Frict. loss, ft fluid/100 ft	or f	___	___	___	F	___	___	___
Reynolds number $V^2/2g$	Re	___	___	___	Re	___	___	___
Lineal feet of pipe, or k$		___	___	___		___	___	___
Gate valves	___ @	___ = ___	___ = ___	___ = ___	___ @	___ = ___	___ = ___	___ = ___
Globe valves	___ @	___ = ___	___ = ___	___ = ___	___ @	___ = ___	___ = ___	___ = ___
Ells	___ @	___ = ___	___ = ___	___ = ___	___ @	___ = ___	___ = ___	___ = ___
Tees, thru branch	___ @	___ = ___	___ = ___	___ = ___	___ @	___ = ___	___ = ___	___ = ___
Tees, thru run	___ @	___ = ___	___ = ___	___ = ___	___ @	___ = ___	___ = ___	___ = ___
Ells 45° or other L	___ @	___ = ___	___ = ___	___ = ___	___ @	___ = ___	___ = ___	___ = ___
Butterfly/diaphragm	___ @	___ = ___	___ = ___	___ = ___	___ @	___ = ___	___ = ___	___ = ___
Check valves	___ @	___ = ___	___ = ___	___ = ___	___ @	___ = ___	___ = ___	___ = ___
Total equivalent length	EL				EL			
Line frict. (F × EL/100)1.15	F_s	− ___	− ___	− ___	F_d	+ ___	+ ___	− ___
Vessel pres. psig × 2.31/SG	P_s**	+ ___	+ ___	+ ___	P_d**	+ ___	+ ___	+ ___
Liquid elev. above pump, ft	h_s*	+ ___	+ ___	+ ___	h_d	+ ___	+ ___	+ ___
Tank contr. or enl. loss	h_c	− ___	− ___	− ___	h_e	+ ___	+ ___	+ ___
Orifice pl., lb/in² × 2.31/SG						+ ___	+ ___	+ ___
Contr. valve lb/in² × 2.31/SG						+ ___	+ ___	+ ___
Exchanger, lb/in² × 2.31/SG						+ ___	+ ___	+ ___
Filter (dirty)						+ ___	+ ___	+ ___
						+ ___	+ ___	+ ___
Alg. sum = suct. or disch. head	Hs				Hd			
Total dynamic head, Hd–Hs	TDH	___ Ft(	" Suct.,	" Disch.)				

NPSH available:				
Abs pressure head	P_s† + 34/SG	+ ___	+ ___	+ ___
Static head	h_s*	+ ___	+ ___	+ ___
Friction head	$h_c + F_s$	− ___	− ___	− ___
Vap. pres. head	0.045 × VP/SG	− ___	− ___	− ___
Alg. sum = NPSH ft fluid		___	___	___

Formulas

1 lb/in² = (2.31/SG) = ________ ft. fluid

$V = 0.408\ \text{gal/min} \cdot d^2$

$d = \sqrt{\dfrac{\text{gal/min}}{2.45v}}$

$z = \mu/SG$

$h_c = V^2/128.8$

$h_e = V^2/64.4$

$Re = \dfrac{3162 \times \text{gal/min} \times SG}{d \times \mu}$

Notes typ. calc., etc.

*Minus for suction lift.

†Minus for pressure less than atmospheric. For elevations above sea level, decrease 34 by 1.1 ft for each 1000 ft of altitude above sea level.

$-K can be given in terms of f. K for pipe = fl/d

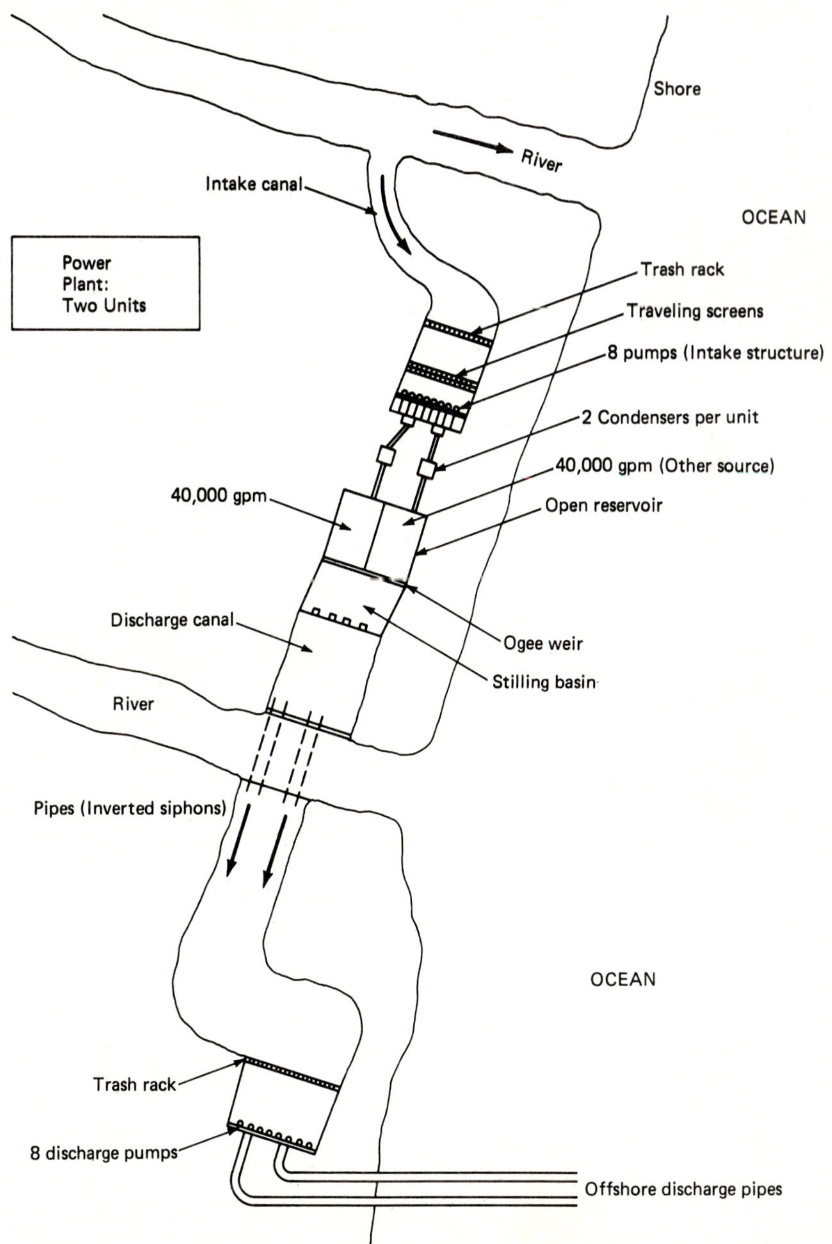

Figure 1-24 Cooling water system for a power plant.

point of view, is to establish the various modes of operation and logic of the system components, e.g., pumps, control valves, etc. For this purpose, process instrumentation and control and logic diagrams are developed with reference to the system descriptions. More details on these are shown in Chapter 3-2, "Pneumatic Power Systems," but is applicable to both.

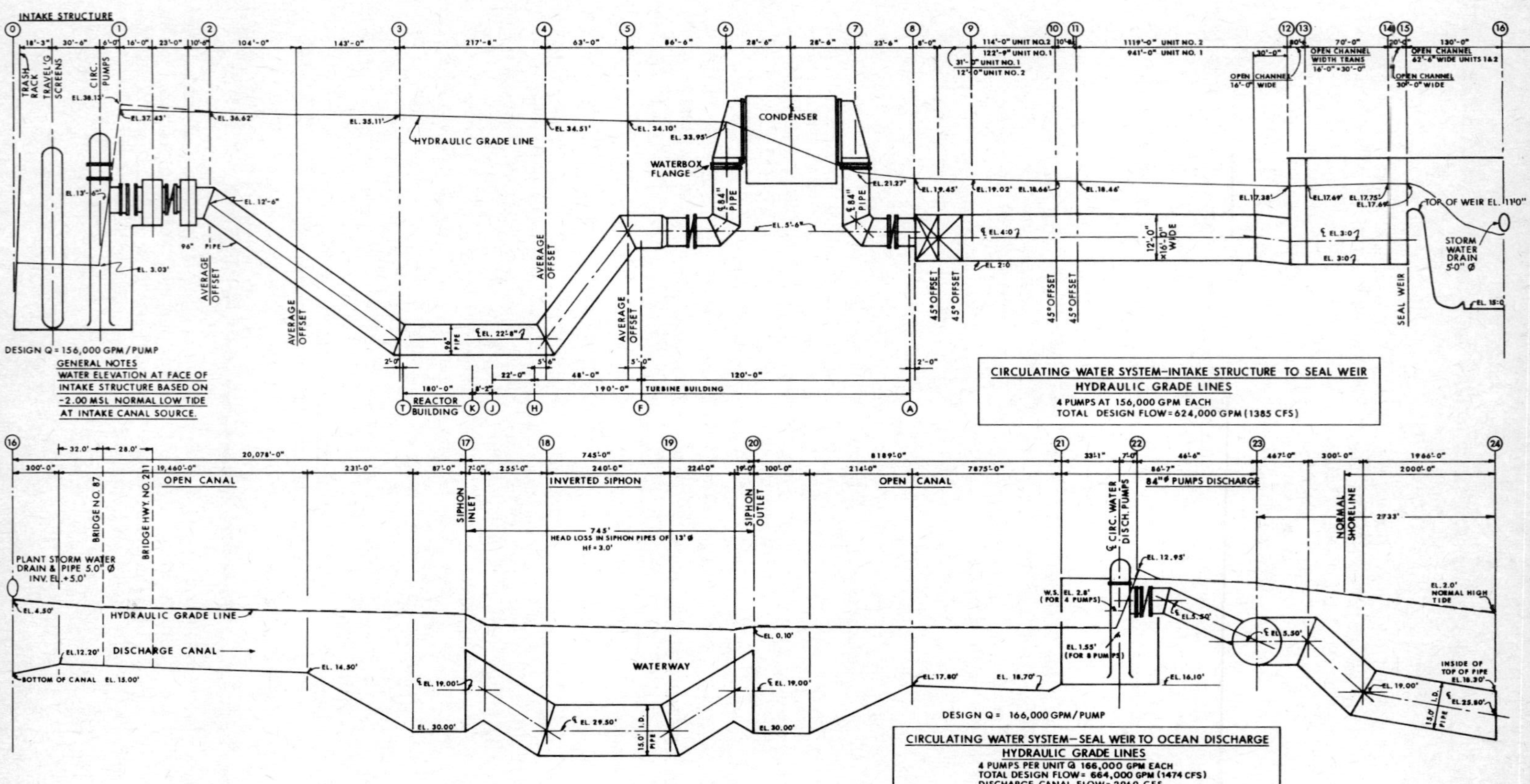

Figure 1-25 Hydraulic gradient of a circulating water system.

Other Aspects

Other important aspects include preparation of detailed technical specifications for the various types of equipment in the system. Besides the process aspects of the equipment, the materials and safety aspects should also be accounted for. A table of recommended materials and metals for various applications has already been discussed. It is recommended that the manufacture of the equipment should be closely monitored for quality assurance reasons and proper shop and field testing should be conducted *before* commercial operation.

Of course, maintenance of the systems and components is also a very important aspect. Only equipment properly inspected and maintained through lubrication and replacement of parts at periodic intervals will fulfill its life expectancy. Some checkpoints similar to those given in the discussion on pumps should be developed for all systems and components.

MODEL EXPERIMENTS

The science of hydraulic engineering depends on practical experience obtained from prototypes, or models. For the design of any hydraulic system, e.g., circulating water system, it is advisable to conduct model studies in order to determine the flow, pressure, and velocity distributions. In recent years, hydraulic-model studies have been conducted on the pump houses, channels, pipes, intake structures, diffusers, surrounding fluid fields, etc.

Intake, Discharge, and Pump House

Pump-house (and channels) hydraulics is unusual in that the flow pattern in the approach to the pump suction must be uniform, tranquil, and nonvortexing; it has a flow velocity of about 1 ft/s (0.305 m/s). Failing to conform to these criteria can result in inefficient performance of the system. Model studies on pump houses are conducted in order to ensure these criteria by providing enough length, width, transition, splitters, baffles, rounding of corners, etc., in the approach channel. Such model studies are usually conducted at a geometric scale of 1:8 to 1:15. Kinematic and dynamic similarities are achieved by further scaling according to the laws of Froude, Reynolds, etc. One important consideration is to see that supercritical flow and hydraulic jump (limiting the amount of flow and creating turbulence) do not exist in the approach channel. Sometimes flags or strings are attached at the bottom of the sump below the suction bell in order to see that all the flags are pointing radially and uniformly all around the bell. Lighted transparent (glass) window sections are provided at the important points, and dye injection tests are conducted in order to visualize the flow patterns.

Velocities are usually measured by miniature current meters, and pressures (or depth of water) are generally measured by piezometers. The pump bells are made geometrically similar, but the pumps themselves are usually not modeled in these tests. Instead, a siphon arrangement usually serves for the correct amount of simulated flow. The models for pumps and their internal hydraulics of vanes are generally tested by pump manufacturers only in order to determine their pump characteristics for the prototype.

A general question comes up, "Why it should be necessary to test a sump pump and not just utilize the existing experience or literature for design?" The answer is that most layouts and flow quantities are typical. If a pump house has been working efficiently for a long time and another similar one (geometrically and dynamically) is required, there is no need to test it. However, if a design is provided based on existing literature, e.g., Hydraulic Institute standards, usually the dimensions are very large and, therefore, result in expensive construction. Usually, a lot of construction quantities and costs can be saved by reducing the size of pump house to its optimum by a model test, the cost of which is usually only a fraction of the savings. Moreover, a design based on literature or experience may not guarantee a certain desirable flow pattern without conducting a test.

Hydrothermal Diffusion Problems

Another field in which extensive test work on models has been conducted recently is hydrothermal diffusion at the point of hot-water discharge into a natural body of water. The importance of this field has been increased due to the recent public concern for environmental protection. It is generally considered that hot-water discharge is harmful to ecological balance. Therefore, a careful study in the area of hot-water discharge should be conducted in order to determine the extent of potential harm and to reduce it to allowable limits. Physical model studies have proved useful to determine the isotherm patterns in the vicinity of discharges. However, there is a general concern that some of the hot water may be returning from the far field. This type of model study will need very large models and has been tried in some cases with only a limited success.

For the near field studies, physical models have generally been constructed to a scale of 1:80 to 1:130. The densimetric Froude number is generally used for similarity. This is also done for the bottom topography of the body of water. Most of the time, the purpose of testing the model is to determine what dimensions, shape, and size of discharge structure (diffuser with ports) will give the most efficient mixing in the least surface area resulting in maximum temperature isotherms. Temperature distributions in the surrounding area are measured by a series of thermostats. Data accumulation and processing is done side by side with the experiment by using computers. Several arrangements are tried, and the best ones are selected with their corresponding merits and demerits. Dye tests are also conducted to observe the flow patterns. It is important to model enough area around the diffuser. The temperature in the model room has to be properly controlled. The tidal effects, currents, and surface tension should be properly controlled or modeled.

Other Model Tests

Another field in which model studies are helpful is determining flow patterns (velocity distributions) near intake and discharge structures in order to determine biota intake and sedimentation-erosion quantities. Also, some model studies have been conducted for flow into and out of the individual intakes and diffuser ports, respectively. Some model studies have also been conducted to determine the hydrodynamic forces on immersed structures.

HYDRAULIC FORCE AND TORQUE

What we have discussed so far in hydraulic systems is largely related to what is involved in transmitting the fluid from one point to another. However, sometimes the purpose is not just to deliver a bulk quantity but also to deliver pressure to activate other components. The pressure is ultimately utilized either as a force or a torque.

Some simple uses of hydraulic pressure as a force become evident in applications such as a hydraulic press (which operates basically on Pascal's law, discussed previously), a hydraulic-brake system on an automobile, or a hydraulic crane for lifting objects. Some indirect uses, by passing through actuators and valves, become evident on applications such as the hydraulic milling machine, shaper or surface grinder, or lathe. For rotary action, hydraulic couplings and torque converters are used.

Pressure Accumulator

This is a device to accumulate or store liquid under pressure delivered by the pump when it is not required by the machine. The pressure can be later supplied to the machine when needed.

Various industrial presses require separate pumping units to furnish liquid at the desired pressure. Such pumps are known as *press pumps.* Normally, the pressure generated by these pumps ranges from 50 to 150 kg/cm^2 and is uniform throughout the supply period. However, the demand for liquid and its required pressure is variable. At some intervals the machine may not be doing any work at all, and to deal with such operating conditions an arrangement to receive and store the pressurized liquid being constantly supplied by the pump is necessary. The device must be able to deliver the liquid back to the machine on demand. In some cases it may be even desirable to retrieve the stored liquid at a pressure higher than that provided by the pump itself. All this is done by the pressure or hydraulic accumulator.

Hydraulic Accumulator

The hydraulic accumulator (refer Fig. 1-26) consists of a cylinder and a plunger generally known as a *ram*. One side of the cylinder is connected to the press pump and other to

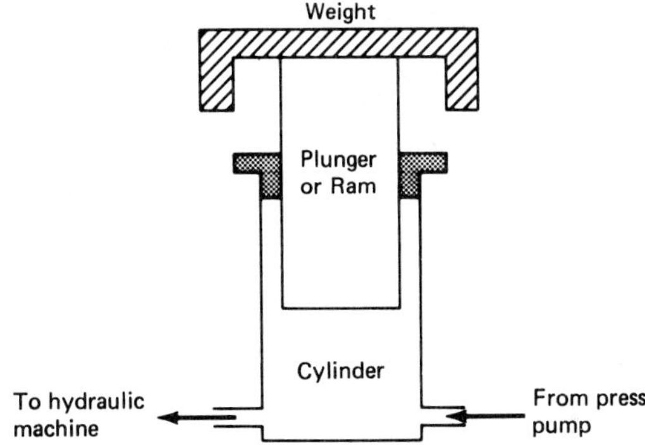

Figure 1-26 Hydraulic accumulator.

the hydraulic machine. Either the cylinder or the ram may be fixed. Generally the cylinder is fixed and the ram moves up and down to accommodate a variable quantity of liquid inside by the cylinder.

Accumulators may be *dead-load* or *variable-load* type. In the former, dead weights are employed to press the plunger in, while the latter employs steam pressure. The main advantage of the steam-pressure type is that the pressure may be varied at will, but it is handicapped in many applications by the need of a boiler to supply steam. However, it can be used on ships if steam is readily available.

The accumulator also serves the purpose of a pressure regulator. A suitable arrangement can be easily designed to switch on a pump motor after a predetermined travel of the ram.

Capacity of Accumulator

This is the maximum amount of energy stored by an accumulator. The storage capacity is equal to the potential energy of the lifted ram together with its weight.

Let

d = diameter of ram
s = stroke or lift of ram
p = intensity of pressure of water supplied

The total moving weight or weight of the ram is

$$W = \frac{\pi}{4} d^2 p \tag{51}$$

The work done in lifting the ram, or capacity of the accumulator, is

$$Ws = \frac{\pi}{4} d^2 ps \tag{52}$$

The volume of the accumulator is $(\pi/4)d^2s$, and

$$\text{Capacity of ram} = p \times \text{volume} \tag{53}$$

The capacity of the accumulator and that of the ram are same.

Pressure Intensifier

The pressure intensifier, sometimes known as *differential accumulator,* is a device to multiply the pressure supplied by the pump to suit the requirements of a high-pressure machine. Often a fluid pressure machine requires a high pressure at a particular stage in its operation. It can be easily provided by the intensifier.

Normally, a simple intensifier (refer Fig. 1-27) consists of two coaxial rams or pistons moving in cylinders as shown in Fig. 1-24. Low-pressure liquid is admitted to the ram or piston of large cross-sectional area, which then transmits force to a small ram or piston

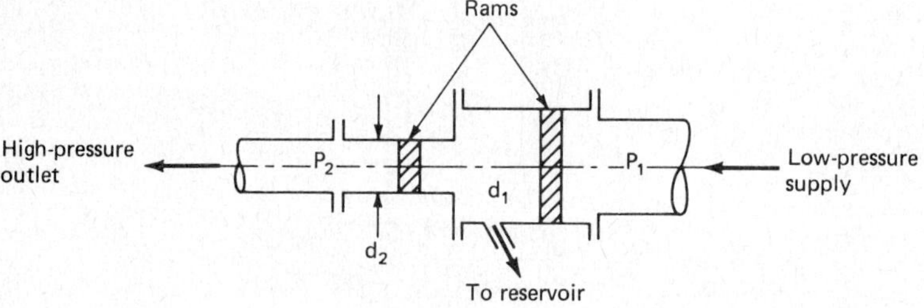

Figure 1-27 Pressure intensifier.

by a rod connecting the two. Since the piston on the left-hand side has a smaller cross-sectional area, the pressure of the liquid coming out will be high. The volume between the two pistons must be vented.

Let d_1 and d_2 be the diameters of the two rams and p_1 and p_2 the respective pressure of the liquid inside them. Then, if the ram moves slowly, forces

$$p_1 = p_2$$

or

$$p_1 \frac{\pi}{4} d_1^2 = p_2 \frac{\pi}{4} d_2^2$$

or

$$p_2 = p_1 \frac{d_1^2}{d_2^2} \tag{54}$$

Figure 1-28 shows a modified form of intensifier which consists of two coaxial rams inside a cylinder. The cylinder and the outer ram are fixed. *A*, *B*, and *C* are the valves provided for fixed cylinder, sliding ram, and fixed ram, respectively. To start with, the hollow sliding ram is filled with liquid; valves *A* and *C* are open. A liquid having a low pressure p_1 enters from the mains through *A* and forces the sliding ram upwards, so that the liquid inside the sliding ram goes out through valve *C* at a greater pressure p_2. When the sliding ram has reached its top position, valve *B* opens while valves *A* and *C* close. The liquid, now trapped between the fixed cylinder and sliding ram, enters inside the hollow sliding ram which then comes back to its starting position. This completes one cycle of the intensifier. Sometimes, compressed air is supplied to the larger cylinder in place of low-pressure hydraulic supply, in which case the intensifier is known as a hydropneumatic accumulator or intensifier.

Steam can also be supplied to the larger cylinder in place of low-pressure hydraulic supply or compressed air. It is then called a *steam intensifier.*

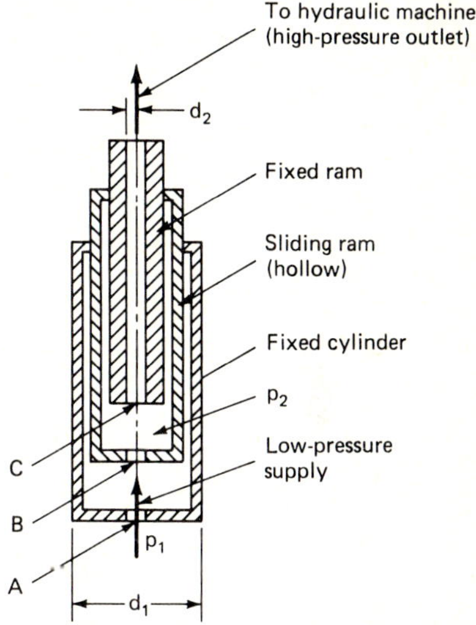

Figure 1-28 Modified form of intensifier.

Fluid or Hydraulic Coupling

The fluid coupling consists of a radial pump impeller keyed to a driving shaft *A* (refer to Fig. 1-29) and a reaction (radial-type) turbine runner keyed to driven shaft *B*. There is no mechanical connection between the driving and driven shafts. The impeller and turbine runner together form a casing completely filled with oil, the fluid with which shafts *A* and *B* are to be coupled. The fluid is usually conventional lubricating oil. If the shaft *A* is made to revolve slowly, the oil, due to a forced vortex, will flow out from the impeller and will strike the turbine runner

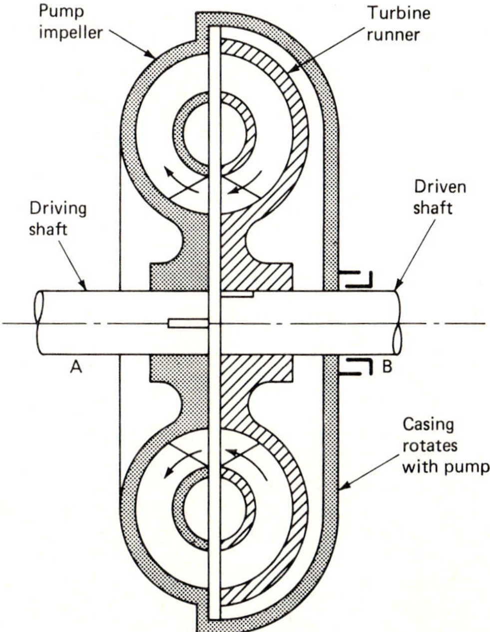

Figure 1-29 Fluid or hydraulic coupling.

blades. After sufficient head has been built up by increasing the speed of A, the fluid will drive the turbine runner and thus set the shaft B in motion. When at full speed, the two shafts rotate at almost the same rate; in practice, owing to slip, the driven shaft speed is typically about 2 percent less. Thus, the efficiency of the typical coupling would be 98 percent. If both driver and follower were to rotate at the same speed, no circulation of oil could take place. It is the difference between centrifugal forces set up in the driver and the follower which causes oil circulation. The necessary reduction in the speed of the driven shaft thus maintains the continuous flow of oil from impeller to the turbine runner. The blades of impeller and runner are generally a straight radial type.

REFERENCES

References are listed at the end of Chap. 3-2.

chapter 3-2

Pneumatic Systems

by
Ken S. Satija, Ph.D., P.E.
United Engineers & Constructors, Inc.
Philadelphia, Pennsylvania

INTRODUCTION

Pneumatic systems involve gases (usually compressed air, including negative gauge pressures for a vacuum system). The most important component of a pneumatic system, therefore, is a compressor (or blower, or fan). Other components include filters, aftercoolers, moisture separators, air tanks, air dryers, piping, valves, etc., the basic functions of which can be deduced just from their names. Some of these components have already been discussed as parts of hydraulic systems. A compressor usually supplies pressurized gas into a line. A pressure regulator is provided on every branch where gas is to be used at a specific pressure.

The fundamentals of airflow are similar to those of the flow of water which was discussed previously in great detail. The methods of calculating the pressure drops in order to compute the pressure requirements of a compressor are similar to those with water, with a few minor exceptions due to the compressibility of air. First, the density, viscosity, etc., of air at the operating temperature must be considered.

Second, the flow computations with air are made in terms of cubic feet per minute. In dealing with equations for gases, absolute temperature and pressure are used. Absolute temperature is the temperature above absolute zero, which is equivalent to $-460°F$ ($-273°C$). When the Fahrenheit temperature is converted to absolute, it is sometimes referred to as degrees Rankine (°R), and °C converted to absolute is kelvins (K). Thus

$$T(\text{K}) = 273 + T(°\text{C})$$

$$T(°R) = 460 + T(°F)$$

Similarly,

$$P \text{ absolute (lb/in}^2) = P \text{ gauge (lb/in}^2) + 14.7$$

$$P \text{ absolute (kg/m}^2) = P \text{ gauge (kg/m}^2) + 10{,}335$$

The pressure drop for flow through pipes is calculated usually to start with airflow conditions at 100 psig and 60°F. Then the pressure drop is converted to the actual flow temperature t and pressure p conditions by the following equation:

$$\frac{460 + t}{520}\,\frac{100 + 14.7}{p + 14.7} = \frac{\Delta p}{\Delta p_{100}} \quad \text{(U.S. customary units)} \tag{1}$$

where

Δp_{100} = pressure drop for flow of air at 100 psi, 60°F
Δp = pressure drop for actual air

It should be kept in mind that this pressure drop is for a flow volume, actual cubic feet per minute (acfm), at p, t conditions. The corresponding standard cubic feet per minute (scfm) of air, that is, at 14.7 psia and 60°F, can be computed from the following equation:

$$\frac{520}{460 + t}\,\frac{14.7 + p}{14.7} = \frac{\text{scfm}}{\text{acfm}} \quad \text{(general gas)} \tag{2}$$

Reference 6 gives tables for specific numbers corresponding to the above equations, in order to simplify calculations.

Tables[6] are also available for computing the pressure drop at different flow rates. These tables are usually specific p and t values only, but one can easily proceed with calculating pressure drop Δp using the Harris formula:

$$p = \frac{cL}{R}\,\frac{Q^2}{D^5} \tag{3}$$

where

c = a coefficient of pressure drop = $(0.1025/D^{0.33})$ (U.S. customary units)
L = length of pipe, ft
R = compression ratio = $(p + 14.7)/14.7$
Q = flow rate, ft^3/s of free air
D = pipe diameter, in

FANS

In plant applications, fans are used for circulating air in rooms, as well as delivering air for plant systems.

Forward-curved-fan characteristics are given in Fig. 2-1. The horsepower curve is continuously increasing with flow rate. Such fans are used in low-pressure applications.[17]

Backward-curved fans have a horsepower curve that rises slightly in the beginning and then drops. This is the type most commonly used in large air-handling systems.

Vane-axial fans that are mostly used in air-conditioning systems have a dip in the horsepower curve.

All types of fans mentioned above have an unstable surge zone in the low-flow zone, as the fan is started or stopped. This zone is considered unsuitable for normal fan operation. Similarly, a steep portion of the pressure curve towards the end, where pressure drops fast with a small change in flow, is unsuitable (and not very predictable) for normal operation. Therefore the system-demand curve should fall within the selection area. The best zone of selection on the curve is where there is a uniform gradual well-defined variation of pressure with flow (one pressure for one flow and not varying too fast).

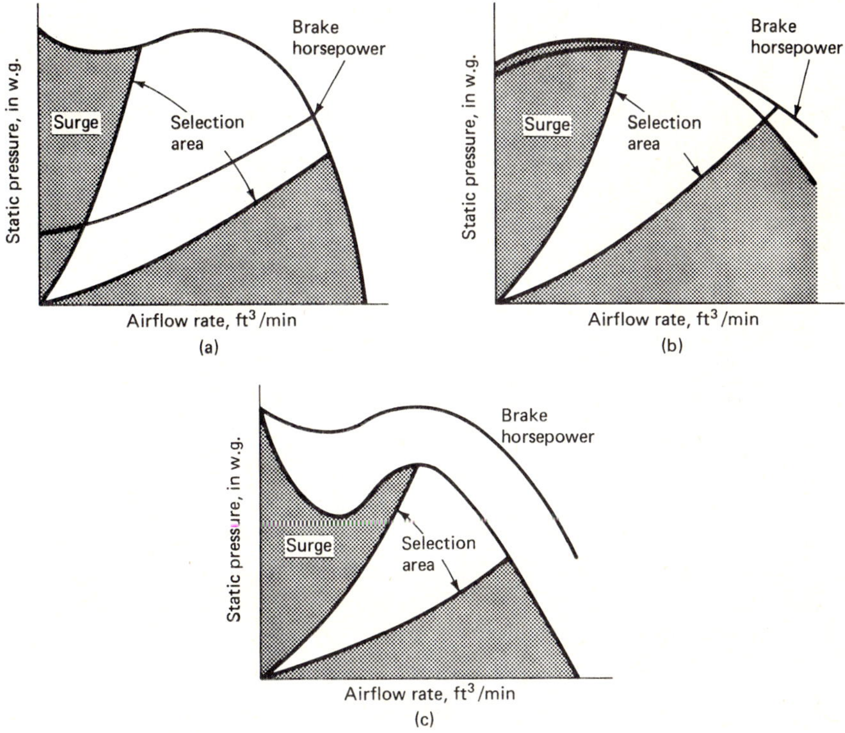

Figure 2-1 Fan characteristics: (*a*) forward-curved vanes, (*b*) backward-curved vanes, (*c*) axial vanes.

Small fans usually have a bigger zone to modulate airflow but are usually less efficient.

COMPRESSORS

Compressors are just like pumps in basic construction as well as function. The basic function of a compressor is to pressurize gas just as a pump pressurizes liquid. Like pumps, compressors have positive displacement (reciprocating, rotary screw, sliding vane) as well as turbo types. Metal diaphragm compressors[16] are used for compressing gases where leakage to the environment could be hazardous or wasteful. Reciprocating compressors operate on the principle of a piston sliding in a cylinder. A turbocompressor is shown in Fig. 2-2.

The horsepower of a centrifugal compressor is given by the following equation:

$$\text{hp} = \begin{cases} \dfrac{WH_{ad}}{550\eta_{ad}} & \text{(U.S. customary units)} \\ \dfrac{WH_{ad}}{75\eta_{ad}} & \text{(SI units)} \end{cases} \tag{4}$$

where

W = weight of flow

η_{ad} = adiabatic efficiency of the compressible flow and all other efficiencies, including mechanical, disk friction, motor, etc.

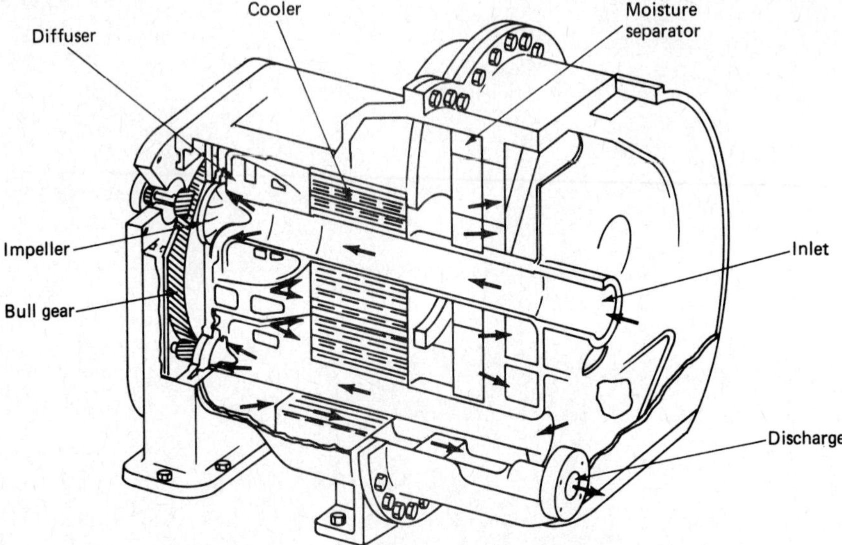

Figure 2-2 Centrifugal compressor. *(Ingersoll-Rand Corp.)*

H_{ad} = adiabatic head of the compressible flow including all the pressure losses in the system and the final pressure required for use.

H_{ad} can also be calculated from the following equation:

$$H_{ad} = \frac{1545\ T_1}{\overline{m}\sigma}(R_c - 1) \tag{5}$$

where

$\sigma = k_1/k$

k = coefficient of adiabatic expansion

= 1.4 for air

$\overline{m}$ = molar weight

R_c = compression ratio = $\dfrac{\text{after compression abs. press.}}{\text{before compression abs. press.}}$

T_1 = absolute temperature at suction

There are two basic types of turbocompressors, axial and centrifugal. Input energy is of course provided by an engine or motor. In the axial compressor the air is forced to move parallel to the centerline of the propeller. Stationary vanes divert the flow to the succeeding row of rotating vanes.

Centrifugal compressors are more common. A high-speed impeller forces airflow in a radial direction inside the compressor, thereby causing a pressure rise. The air at the periphery is then ducted to the successive stages.

Pulseless, high volumetric flow rates with oil-free discharge are the basic attractions of centrifugal compressors. Typical performance charts of compressors are shown in Fig. 2-3, for backward-leaning vs. radial vanes. Notice the general loss of capacity and higher power requirements of the radial vanes. The heat is rejected by the compressor to the air (therefore, there are cooling water requirements). The horsepower required by the compressor for a certain scfm capacity and the number of stages are important parameters for a final selection.

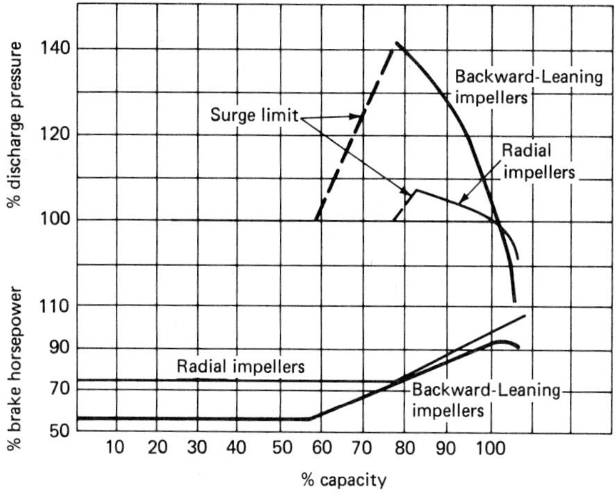

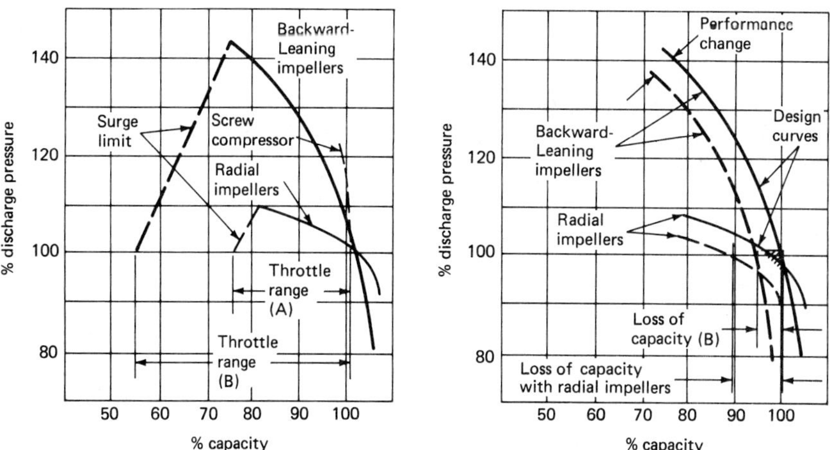

Figure 2-3 Typical performance curves of centrifugal compressors with backward-leaning impellers (A) vs. compressors with radial impellers (B). *(Ingersoll-Rand Corp.)*

More stages are often desirable. API Standards 617 and 618 are commonly used in connection with compressors.

TYPICAL PNEUMATIC SYSTEMS

A typical air-supply system diagram along with the primary controls and final uses is shown in Fig. 2-4. Free air is sucked through a filter into the compressor and discharged into a service air receiver (after passing through an aftercooler to remove the heat from the compressor and moisture separator). The air receiver is kept under pressure and also serves as a surge chamber. The relief valve protects against overpressures. From the receiver, the air can either be directly used or dried and chilled before use, as in instrument-air applications.

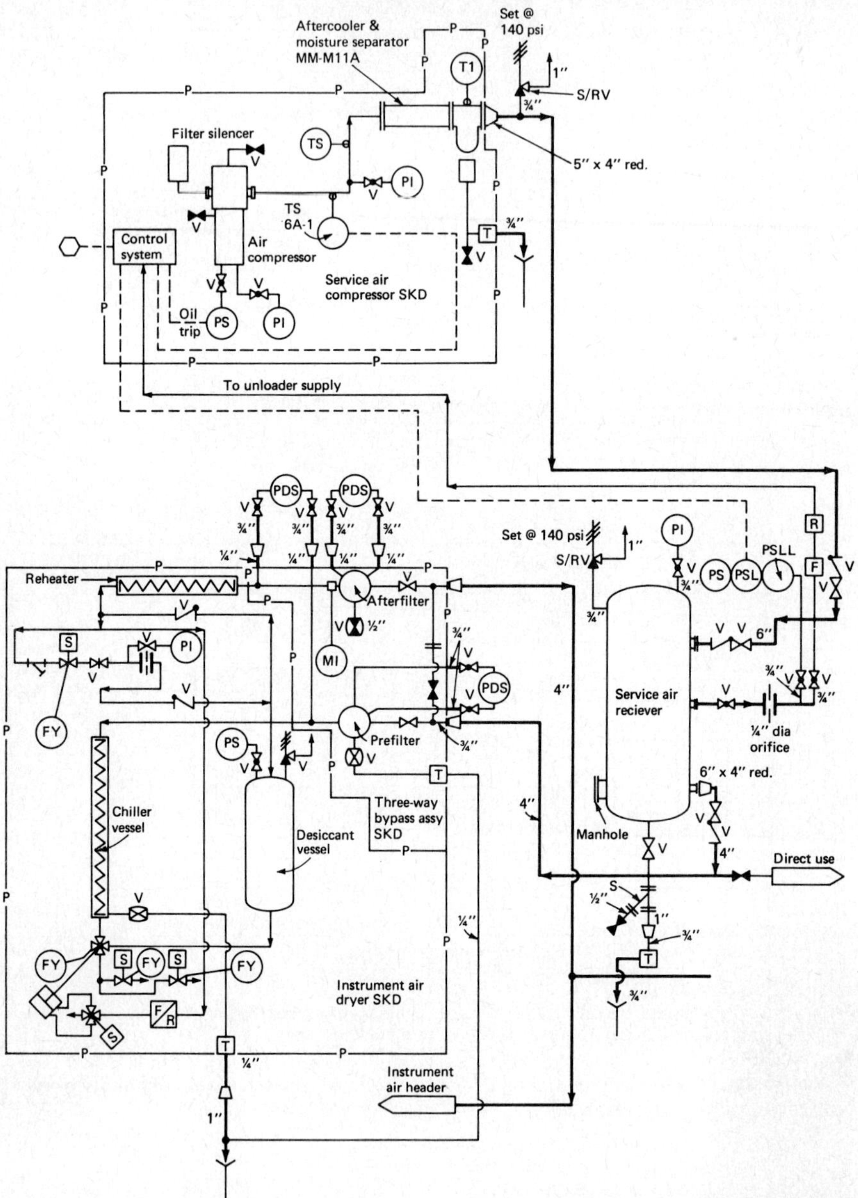

Figure 2-4 Typical pneumatic system. Key: PI = pressure indicator, T1 = temperature indicator, PS = pressure switch, TS = temperature switch, T = moisture trap, S/RV = safety/relief valve, V = valve, PDS = pressure differential switch, S = solenoid, FI = flow indicator, PSL = pressure switch low, PSLL = pressure switch low-low.

Another typical pneumatic system would include a vacuum pump (really a "negative" air compressor) to pump out air *from* a system as in applications to a condenser that functions best under vacuum, or in a negative-pressure air-vent system. The vacuum pumps function in a way similar to vacuum cleaners. One important component in such a system is a valve that shuts itself and the vacuum system upon sensing a particular water level in a container at the top of the water box of a condenser. The valve reopens and the vacuum system starts when air is again sensed inside the water box. There are other arrangements that require air to enter the system when a vacuum is sensed in order to avoid either pipe collapse or system cavitation due to vapor pressures forming inside. One such air and vacuum valve is shown in Fig. 2-5, with the corresponding performance graph shown in Fig. 2-6. Such a valve has a large orifice through which a great amount of air escapes when the system is being filled with liquid. Once the system is filled, the fluid lifts the float in the valve, closes the orifice, and stays closed until the system is drained.

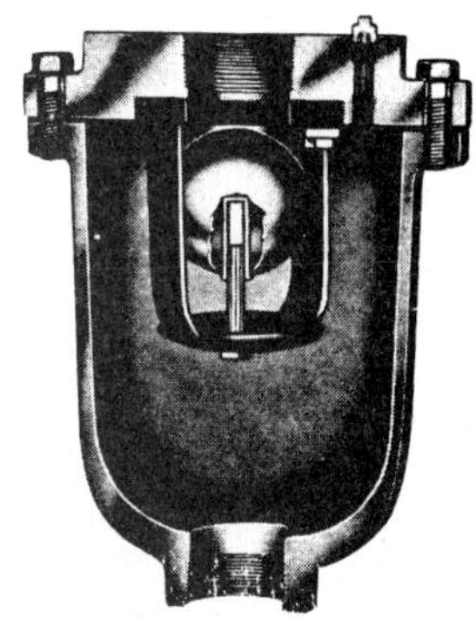

Figure 2-5 Air and vacuum valve. *(APCO Valve & Primer Corp.)*

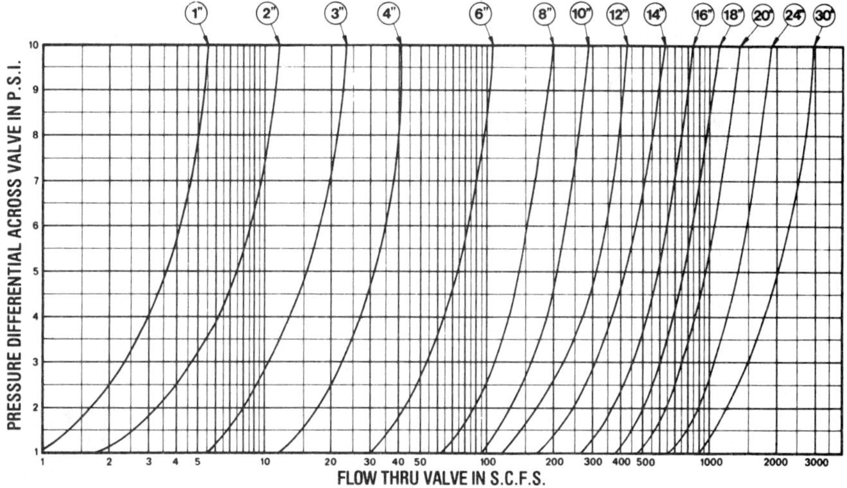

Figure 2-6 Performance graph for air and vacuum valve. *(APCO Value & Primer Corp.)*

REFERENCES AND BIBLIOGRAPHY

1. Miller, Donald S.: *Internal Flow: A Guide to Losses in Pipe and Duct Systems,* British Hydrodynamic Research Association, Cranfield-Bedford, England, 1971.
2. Giles, Donald V.: *Fluid Mechanics & Hydraulics,* Schaum Outline Series, McGraw-Hill, New York, 1962.
3. King, H. W., and E. F. Brater: *Handbook of Hydraulics,* McGraw-Hill, New York, 1976.
4. Rouse, Hunter: *Elementary Mechanics of Fluids,* Wiley, New York, 1946.
5. Simon, Andrew L.: *Practical Hydraulics,* Wiley, New York, 1976.

6. *Flow of Fluids Through Valves, Fittings, and Pipe,* Crane Company, New York, 1976.
7. "Standard for Centrifugal, Rotary & Reciprocating Pumps," Hydraulic Institute, Cleveland, Ohio, 13th ed., 1975.
8. Satija, K. S., and N. M. Shah: *On Circulating Water for Power Plants,* Proceedings Second World Congress, International Water Resources Association, India, 1975.
9. Lal, Jagdish: *Hydraulic Machines,* Metropolitan Book Company Pvt. Ltd., Delhi, India, 1975.
10. *Masoneilan Handbook for Control Valve Sizing,* 6th ed., Masoneilan International, Inc., Norwood, Mass., 1977.
11. *Power Engineers' Valve Manual,* Technical Publishing Co., Burlington, Ill., 1962.
12. *Standard of Tubular Exchanger Manufacturers Association (TEMA),* 6th ed., White Plains, N. Y., 1978.
13. Henke, R. W.: *Introduction to Fluid Mechanics,* Addison-Wesley, Reading, Mass., 1972.
14. Satija, K. S.: "Prototype Testing of Circulating Water Systems," paper presented at the *ASCE Hydraulics Division Conference,* Seattle, Wash., 1975.
15. Wylie, E. B., and V. L. Streeter: *Hydraulic Transients,* McGraw-Hill, New York, 1969.
16. Lingston, E. H.: "Metal Diaphragm Compressors," *Plant Engineering,* April 19, 1979.
17. Patterson, N. R.: "Variable Air Volume Systems", *Plant Engineering,* February 7, 1980.
18. Stewart, H. L.: "Plant Engineer's Fluid Power Handbook," *Plant Engineering,* June 1977 to December 1979.

section 4

Piping and Valving

chapter 4-1

Piping System Design

by

D. G. Wilson
Vice President and General Manager,
Facility Plans, Engineering and Construction
United States Steel Corp
Pittsburgh, Pennsylvania

INTRODUCTION

In industrial plants where fluids are transferred from one place to another, it has been found that the application of piping is extremely important. Therefore, a knowledge of how this piping is designed is critical to the plant engineer, for both the engineer's understanding of plant functions and his or her ability to make plant modifications to improve efficiency of operations and optimize costs.

Fundamental Considerations

Many elements make up a successful piping installation, and a conscientious effort should be made by the designer and the installer to achieve a trouble-free, maintainable system.

Proper and adequate system planning is considered the first step. With pipeline diameters established and equipment and end-point locations determined, an accurate sketch or drawing should be prepared. Lines should be routed to minimize length of pipe runs, yet allow for adequate flexibility for thermal expansion. Piping subject to water hammer from sudden valve closures must be provided with suitable anchors or pulsation-damping devices.

Lines for conveying condensable vapor or gases, such as steam, should be sloped for drainage, with provision for liquid removal by manual or automatic blowdown. Field investigations to determine existing site conditions may be necessary to confirm proposed line routings. Piping materials should be selected to be compatible with the fluid. Maximum pressures and temperatures of fluid within the line should be used to determine thicknesses of pipe and fitting wall. Bare carbon steel piping and supports should be painted with a suitable enamel or epoxy paint to minimize atmospheric corrosion. Consideration should be given to cathodic protection and/or coating for underground lines.

The method of fabrication of a piping system should be selected on the basis of maintainability, initial cost, service life, and overall end use. Smaller-diameter, low-pressure piping is frequently installed with threaded joints. Hazardous or flammable fluids may necessitate the use of welded fittings to minimize the potential for leakage. Periodic inspection of piping materials, weld joints, fitting makeup, etc., should be undertaken during the installation phase. Welding qualifications, tests, and procedures may be obtained from the American Welding Society or the American Society of Mechanical Engineers and used as guidelines for fabrication.

After fabrication and erection, the piping system should be pressure tested. Pressure testing should be performed with water and limited to a hydrostatic test pressure of 1.5 times the design pressure. When greater sensitivity for detecting leakage is desired, halide leak testers or helium mass spectrometers may be employed.

PRESSURE DROP AND SIZE DETERMINATION

System Definition

Before starting calculations, the system being studied must be defined. The following information should be gathered or developed.

1. A flow diagram of the process or plant section
2. Physical and mechanical properties of the system and fluids
3. Pumping requirements or power input to the fluids

Fluid Flow Equation

The basic equation for fluid flow can be derived from a consideration of thermodynamics and the interrelation of various forms of energy.

For 1 lb (kg) of an incompressible fluid, flowing between two points in a horizontal circular pipe under isothermal steady conditions and turbulent flow, with no work done on the system, the expression takes the form of the familiar Fanning or Darcy equation:

$$F = \left(2f\frac{L}{D}\right)\frac{V^2}{g_c} = \left(2f\frac{L}{D}\right)\frac{G^2}{g_c P^2} \qquad (1)$$

or

$$p = \left(2f\frac{L}{D}\right)\frac{V^2 P}{g_c} = \left(2f\frac{L}{D}\right)\frac{G^2}{g_c P}$$

where

F = friction loss, ft·lbf/lb of fluid flowing = (N·m/kg)*
V = average linear velocity, ft/s (m/s)
L = length of pipe, ft (0.305 m)
f = friction factor, dimensionless
= factor, 32.17 ft·lb/(s²)(lbf) [kg/m/(s²)(N)]
D = pipe diameter, ft (m)
G = VP = mass velocity, lb/(s)(ft²) (cross section) [kg/(s)(m²)]
P = fluid density, lb/ft³ (kg/m³)
p = absolute pressure, lbf/ft² (Pa)

Equation (1) can be used to find values for pressure drop or head loss, size of the pipeline, and the rate of fluid flowing. Given any two unknowns, the third can be found. The friction factor f has been shown to be dependent on the dimensionless Reynolds number, $N_{Re} = DVP/u = DG/u$, where u = fluid viscosity, lb/(ft)(h) [kg/(m)(h)].

A correlation developed by Moody,[1] relating the friction factor, pipe roughness, and N_{Re} is shown in Fig. 1-1 where E/D is the ratio of the surface roughness and pipe diameter and is dimensionless.

A useful chart for solving Eq. (1) has been developed by Generaux[2] and is shown in Fig. 1-2. Convenient nomographs and calculation methods for fluid flow parameters have also been developed by Crane Company[3] for both liquids and gases.

The Fanning equation (1) can also be used for determining the unknown parameters for gases, provided that the overall system pressure drop does not exceed 10 percent of the absolute internal pressure.

Miscellaneous Friction Losses

Friction losses due to sudden expansion and contraction and losses in valves and fittings can be included in Eq. (1). The friction head F can be split into a summation of terms covering the different losses.

The equations presented below for evaluating these frictional effects utilize the "velocity head" method.[4]

Friction Loss from Sudden Expansion of Cross Section

$$F_e = \frac{K_e(V_1 - V_2)^2}{2g_c} \tag{2a}$$

and

$$K_e = 1 - \frac{S_1}{S_2} \tag{2b}$$

Friction Loss from Sudden Contraction of Cross Section

$$F_c = K_c \frac{V_1^2}{2g_c} \tag{3a}$$

and

$$K_c = 0.4\left(1 - \frac{S_1}{S_2}\right) \tag{3b}$$

*The second set of units for each symbol denotes corresponding International System (SI) units.

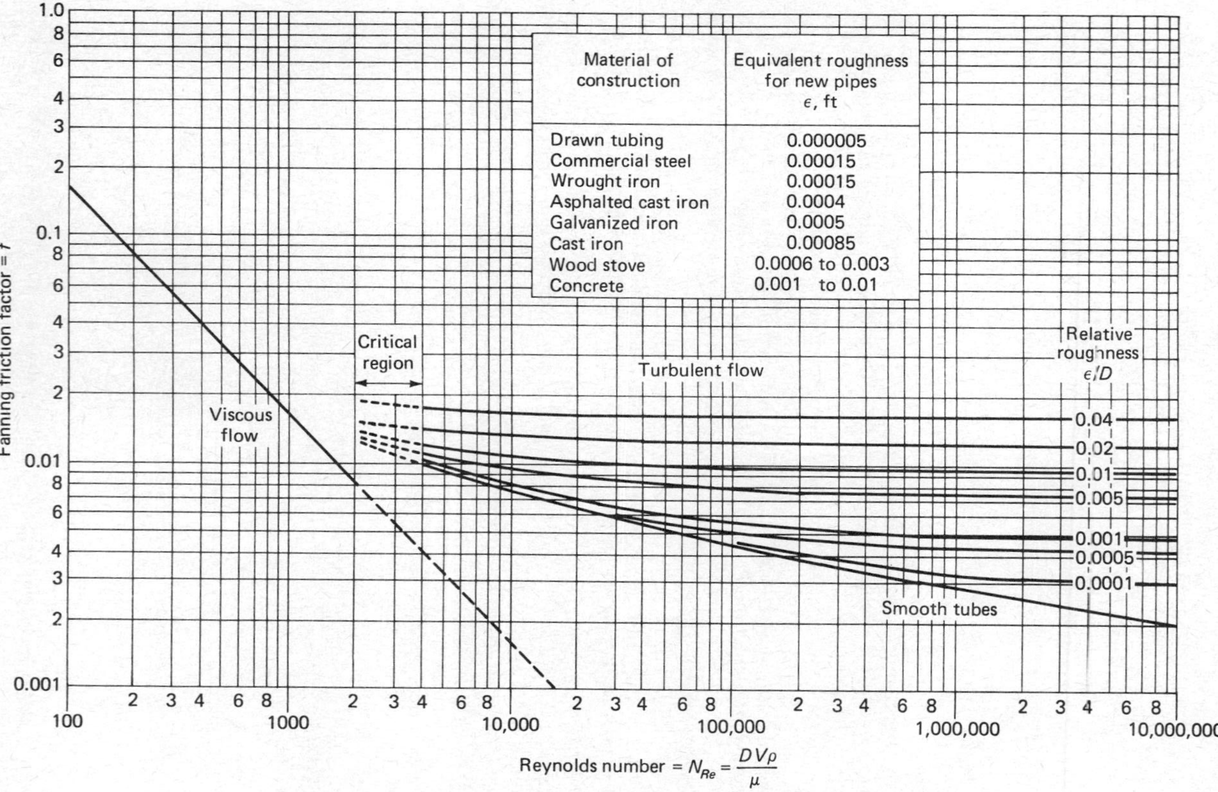

Figure 1-1 Fanning friction factors for long straight pipes.

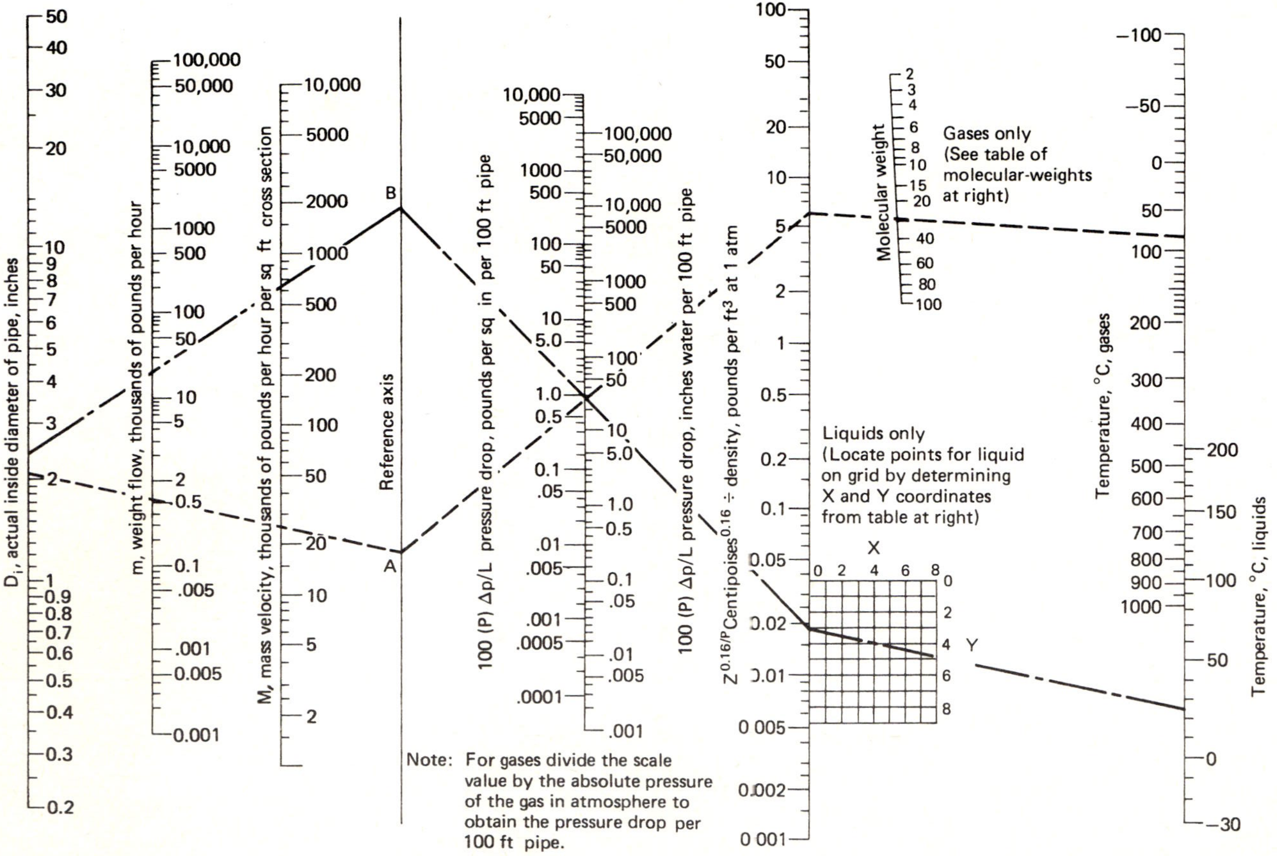

D_i, actual inside diameter of pipe, inches

m, weight flow, thousands of pounds per hour

M, mass velocity, thousands of pounds per hour per sq ft cross section

Reference axis

100 (P) $\Delta p/L$ pressure drop, pounds per sq in per 100 ft pipe

100 (P) $\Delta p/L$ pressure drop, inches water per 100 ft pipe

$Z^{0.16/P}$Centipoises$^{0.16}$ ÷ density, pounds per ft^3 at 1 atm

Molecular weight

Gases only (See table of molecular-weights at right)

Liquids only (Locate points for liquid on grid by determining X and Y coordinates from table at right)

Temperature, °C, gases

Temperature, °C, liquids

Note: For gases divide the scale value by the absolute pressure of the gas in atmosphere to obtain the pressure drop per 100 ft pipe.

Figure 1-2 Pipe flowchart for turbulent flow.

TABLE 1-1 Additional Frictional Loss for Turbulent Flow Through Fittings and Valves[k]

Type of fitting or valve	Additional friction loss, equivalent no. of velocity heads, K_f
45° ell, standard[a,b,e,g,i]	0.35
45° ell, long radius[b]	0.2
90° ell, standard[a,b,d,g,i,m]	0.75
Long radius[a,b,e,g]	0.45
Square or miter[m]	1.3
180° bend close return[a,b,g]	1.5
Tee, std., along run, branch blanked off[g]	0.4
Used as ell, entering run[d,h]	1.0
Used as ell, entering branch[b,d,h]	1.0
Branching flow[f,h,l]	1[o]
Coupling[b,g]	0.04
Union[g]	0.04
Gate valve,[a,g,j] open	0.17
¾ open[p]	0.9
½ open[p]	4.5
¼ open[p]	24.0
Diaphragm valve,[n] open	2.3
¾ open[p]	2.6
½ open[p]	4.3
¼ open[p]	21.0
Globe valve,[g,j] bevel seat, open	6.0
½ open[p]	9.5
Composition seat, open	6.0
½ open[p]	8.5
Plug disk, open	9.0
¾ open[p]	13.0
½ open[p]	36.0
¼ open[p]	112.0
Angle valve,[a,g] open	2.0
Y or blowoff valve,[a,j] open	3.0
Plug cock[c] 0 = 5°	0.05
10°	0.29
20°	1.56
40°	17.3
60°	206.0
Butterfly valve,[c] 0 = 5°	0.24
10°	0.52
20°	1.54
40°	10.8
60°	118.0
Check valve,[a,g,j] swing	2.0[q]
Disk	10.0[q]
Ball	70.0[q]
Foot valve[g]	15.0
Water meter[m] disk	7.0[r]
Piston	15.0[r]
Rotary (star-shaped disk)	10.0[r]
Turbine-wheel	6.0[r]

[a] "Flow of Fluids through Valves, Fittings, and Pipe," Tech. Paper 410, Crane Co., 1969.

[b] Freeman, *Experiments upon the Flow of Water in Pipes and Pipe Fittings*, American Society of Mechanical Engineers, New York, 1941.

[c] Gibson, *Hydraulics and Its Applications*, 5th ed., Constable, London, 1952, p. 250.

[d] Giesecke and Badgett, *Heating/Piping/Air Conditioning*, **4**(6):443–447 (1932).

[e] Giesecke, *Journal of the American Society of Heating and Ventilation Engineers*, **32**:461 (1926).

[f] Gilman, *Heating/Piping/Air Conditioning*, **27**(4):141–147, (1955).

[g] *Pipe Friction Manual*, 3d ed., Hydraulic Institute, New York, 1961.

[h] Hoopes, Isakoff, Clarke, and Drew, *Chemical Engineering Progr.*, **44**:691–696 (1948).

[i] Ito, *Journal of Basic Engineering*, **82**:131–143 (1960).

Friction Loss from Valves and Fittings

$$F_f = K_f \frac{V_1^2}{2g_c} \tag{4}$$

where

F_e, F_c, F_f = friction loss, ft·lbf/lb [(m)(N)/kg]

$V_1, V_2,$ = volumetric average velocity for the upstream and downstream points, ft/s (m/s)

S_1, S_2 = upstream and downstream pipe cross-sectional areas, ft^2 (m^2)

K_f = valve and fitting component factor, dimensionless (see Table 1-1)

Economics

While the equations discussed above permit determination of the pipe diameter, the selection of a particular pipe size will depend upon a balance between the cost of materials, power consumption, and maintenance. Hence, when selecting pipe sizes, the overall costs of the system must be considered. One method of approach is that taken by Peters and Timmerhaus[5] for establishing optimum pipe diameters.

Small Computers

Book type computers are fast and reduce the need for engineers to access large computers for solving problems.

One such unit, the Hewlett-Packard HP 67/97, is capable of storing programs on magnetic strips for reuse without having to reprogram it each time a program is used. The availability of a large library reduces most engineering calculations to routine operations which can be completed quickly.

Benenati[6] has developed a program for pressure-drop calculations that uses the Hewlett-Packard 67/97 computer for solving the previously discussed Fanning equation (1).

PIPE STRESSES

To determine pressure-retaining ability and mechanical strength in piping components, a thorough knowledge of strength of materials is required. All standard piping materials in common use today have been developed with this theoretical background and have

---

[j]Lansford, *Loss of Heat in Flow of Fluids through Various Types of 1½ M. Valves,* Univ. Illinois Eng. Expt. Sta. Bull., ser. 340, 1943.

[k]Lapple, *Chemical Engineering,* **56**(5):96–104 (1949), general survey reference.

[l]McNown, *Proceedings of the American Society of Civil Engineers,* **79**(separate 258):1–22 (1953) discussion, *ibid.* **80**(separate 396):19–45 (1954).

[m]Schoder and Dawson, *Hydraulics,* 2d ed., McGraw-Hill, New York, 1934, p. 2130

[n]Streeter, *Product Engineering,* **18**(7):89–91 (1947).

[o]This is pressure drop (including friction loss) between run and branch, based on velocity in the main stream before branching. Actual value depends on the flow split, ranging from 0.5 to 1.3 if main stream enters run and from 0.7 to 1.5 if main stream enters branch.

[p]The fraction open is directly proportional to stem travel or turns of hand wheel. Flow direction through some types of valves has a small effect on pressure drop (see Freeman, op. cit.). For practical purposes this effect may be neglected.

[q]Values apply only when check valve is fully open, which is generally the case for velocities more than 3 ft/s for water.

[r]Values should be regarded as approximate because there is much variation in equipment of the same type from different manufacturers.

Source Perry, R. H. and C. H. Chilton (eds), *Chemical Engineers Handbook,* 5th ed., McGraw-Hill, New York, 1973, p.5–36. Reproduced with permission.

been service-tested to achieve the predicted performance. In addition, all piping systems or replacement piping components are now being specified to conform with an applicable code requirement. Most municipalities require compliance with building codes such as those issued by the American National Standards Institute for building permits to be issued.

Circumferential Stress

Neglecting external loading effects, thermal expansion, or corrosion allowance, the expression used to determine the pipe wall thickness from the internal pressure takes the form

$$t = \frac{pD}{2S}$$

where

t = wall thickness, in (m)
D = pipe diameter, in (m)
S = hoop stress, lb/in^2 (Pa)
p = internal pressure, lb/in^2 (Pa)

Usually the pipe diameter and internal pressure have been determined by flow requirements, and it is necessary to find the pipe wall thickness. The allowable hoop stress S is a function of the pipe material selected and the temperature to which the material will be exposed (i.e., normally the pipe contents' temperature).

Problem

A 12-in-diam carbon steel pipe is to be used to convey water at 60°F (15.4°C) and a pressure of 400 psig (2758 kPa). The pipe wall thickness must be determined.

Solution

By reference to handbook data it is found that the minimum yield strength corresponds to 30,000 lb/in^2 (20 700 MPa) for carbon steel and a 3:1 safety factor is considered adequate; therefore

$$t = \frac{400 \text{ lb/in}^2 \times 12 \text{ in}}{2 \times 10{,}000 \text{ lb/in}^2}$$

$$= 0.24 \text{ in } (0.61 \text{ m})$$

Some pressure piping codes require the use of a more rigorous determination of wall thickness using empirical factors for manufacturing tolerances and finite allowable stress-temperature ranges. As presented in ANSI B31.1, "Power Piping Code," the equation takes the form

$$t_{in} = \frac{PD_o}{2(S_E + Py)} + A$$

where

t_{in} = required wall thickness, in (m)
P = internal pressure, lb/in^2 (Pa)
D_o = outside diameter of pipe, in (m)
S_E* = allowable stress of material, lb/in^2 (Pa), temperature-dependent
A* = additional thickness, in (m) to compensate for corrosion, threading, etc.
y* = temperature coefficient

*Values determined by reference to code appendixes.

Combined Stresses

In multiple-plane systems, it often becomes necessary to examine the torsional stresses resulting from thermal expansion. When these and bending stresses are combined, a resultant fiber stress of higher intensity may develop. Piping codes often require the calculated combined stress to fall within an allowable stress range for hot and cold conditions.

Combined stress may be calculated by use of the formula

$$S = \tfrac{1}{2}[S_L + S_C + \sqrt{4S_t^2 + (S_L - S_C)^2}]$$

where

S = total bending stress
S_L = maximum combined stress
S_C = hoop stress
S_t = torsional or shear stress

Bending Stresses

Bending stresses are developed in piping systems by loads resulting from thermal expansion or weight of the pipe.

In high-temperature piping systems, it is required that a thorough analysis be made of maximum bending stresses resulting from thermal expansion. Often, it is necessary to reconfigure pipeline geometry and to provide specially designed spring hanger supports to maintain allowable pipe stresses.

The bending stress resulting from piping weight effects may be calculated by use of the following equation:

$$S = \frac{WL^2}{\text{SM}}$$

where

S = bending stress, lb/in^2 (Pa)
W = weight, lb/in (kg/m)
L = distance between supports, in (m)
SM = section modulus of the pipe cross section, in^3 (m^3)

For a circular cross section the section modulus can be determined by

$$\text{SM} = \frac{\pi}{32}\,\frac{D_o^4 - D_2^4}{D_o}$$

where

D_o = outside diameter, in (m)
D_2 = inside diameter, in (m)

Table 1-2 gives the recommended pressure-temperature ratings for ASTM A 53 grade A carbon steel pipe.

Pipe Deflection

Consideration should also be given to pipe deflection between supports; this is often the governing parameter in determining supporting spans. Deflections in horizontal pipelines should be maintained between 0.1 and 0.5 in (2.5 and 12.5 mm) at their midpoint. Additional supports should be provided at concentrated load areas such as valves, flanges, or direction changes. The following equation may be used to determine the midspan deflection of a horizontal pipeline with free ends:

$$\Delta = \frac{5WL^4}{384EI}$$

TABLE 1-2 Pressure-Temperature Ratings of Pipe, Seamless Carbon Steel to ASTM A 53 Grade A

Pipe size, in	Schedule number	Wall thickness, in	Temperature and stress: to 650°F, 12,000 lb/in² stress — Working pressure, lb/in²	Temperature and stress: 750°F, 10,700 lb/in² stress — Working pressure, lb/in²
½	40	0.109	882	787
	80	0.147	1947	1736
1	40	0.133	960	856
	80	0.179	1777	1585
2	40	0.154	724	646
	80	0.218	1329	1185
3	40	0.216	875	780
	80	0.220	1422	1218
4	40	0.237	1146	1022
	80	0.337	1660	1481
6	40	0.280	914	815
	80	0.432	1435	1279
8	20	0.250	622	555
	40	0.322	806	718
	80	0.500	1270	1133
10	20	0.250	497	443
	40	0.365	729	650
	80	0.594	1208	1077
12	20	0.250	418	373
	40	0.406	683	609
	80	0.688	1178	1050

where

Δ = deflection, in (m)
W = weight, lb/in (kg/m)
L = length of span, in (m)
E = modulus of elasticity
I = moment of inertia

When calculations are not performed, Table 1-3 may be used to determine recommended spans. Recommendations are based upon deflections of 0.1 in and combined bending and shear stress of 1500 lb/in² using Schedule 40 pipe.

TABLE 1-3

Pipe Size, in	1	2	3	4	6	8	12	16	20	24
Water service, ft	7	10	12	14	17	19	23	27	30	32
Air service, ft	9	13	15	7	21	24	30	35	39	42

STANDARDS AND SPECIFICATIONS

"Power Piping," ANSI B31.1. = 1977
"Fuel Gas Piping," ANSI B31.2. = 1968
"Chemical Plant and Petroleum Refinery Piping," ANSI B31.3. = 1976
"Refrigeration Piping," ANSI B31.5. = 1974
"Gas Transmission and Distribution Piping systems," ANSI B31.8. = 1975
"Building Services Piping," ANSI B31.9.
"Thickness Design of Cast-Iron Pipe," ANSI/AWWA C101-67 (1977) [A21.1]
"Thickness Design of Ductile-Iron Pipe," ANSI/AWWA C150-76 [A21.50.]
"Pipe Hangers and Supports, Materials and Design," MSS Standard SP-58.
"Pipe Hangers and Supports—Selection and Application," MSS Standard SP-69.
"Steel Water Pipe 6″ and Larger," AWWA Standard C-200-75.
"Recommended Practice for the Design and Installation of Pressure Relieving Systems in Refineries," API Standard RP 520; "Guide for Pressure Relief and Depressuring Systems," API Standard RP 521.
Hydraulic Institute Standards.

BIBLIOGRAPHY

Perry, R. H. and C. H. Chilton (eds.): *Chemical Engineers' Handbook,* 5th ed. McGraw-Hill, New York, 1973.
King, R. C.: *Piping Handbook,* 5th ed., McGraw-Hill, New York, 1967.
Clark, L., and R. Davidson: *Manual for Process Engineering Calculations,* 2d ed., McGraw-Hill, New York, 1962.
"Flow of Fluids through Valves, Fittings and Pipe", Tech. Paper 410, Crane Company, New York, N.Y., 1976.
Shanley, F. R.: *Strength of Materials,* McGraw-Hill, New York, 1957.
Piping Design & Engineering, Grinell Co., (DIV.ITT) Providence R.I. 3d ed., 1971.
Baumeister, Theodore, Eugene A. Avallone, and Theodore Baumeister III (eds.): *Marks' Standard Handbook for Mechanical Engineers,* 8th ed. McGraw-Hill, New York, 1978.

REFERENCES

1. *Transactions of the American Society of Mechanical Engineers,* **66**:671–684 (1944).
2. Generaux: *Chemical and Metallurgical Engineering,* **44**:241 (1937).
3. Tech. Paper 410, Crane Company, 1976.
4. McCabe and Smith: *Unit Operations of Chemical Engineering,* 3d ed., McGraw-Hill, New York, 1967, p. 108.
5. Peters and Timmerhaus: *Plant Design and Economics,* McGraw-Hill, New York, 1968.
6. Benenati, R. F.: "Solving Engineering Problems on Programmable Pocket Calculators," *Chemical Engineering,* February 1977.

chapter 4-2

Metallic Piping and Fittings

by

D. G. Wilson

Vice President and General Manager,
Facility Plans, Engineering and Construction
United States Steel Corp.
Pittsburgh, Pennsylvania

INTRODUCTION

Piping and tubing are the most commonly used construction materials today. The primary purpose of conveying liquid, gas, air, and slurry is augmented by piping's use as a support member as well as in the manufacture of products such as rollers, cylinders, conduit, recreation equipment, partitions, etc.

Orders for piping and tubing are, in addition to specifying standard weight designations and/or wall thickness, accompanied with a specification such as those published by ASTM, API, WWP, ASME, AWWA, or ANSI, which establishes the minimum standards required of the product. Standard pipe is generally covered under specifications published by ASTM.

Classifications for pipes are standard weight (Std), extra strong (XS), and double extra strong (XXS). In pipe sizes ⅛ to 10 in nominal, ANSI Schedule 40 thicknesses are identical to standard-weight pipe. Schedule 80 (⅛ to 8 in nominal) is identical to extra-strong pipe, and Schedule 160 falls between extra-strong and double extra-strong pipe. It should be noted that wall thickness does not apply to the outside diameter (OD) of the pipe, but only affects the inside diameter (ID), i.e., the thicker the wall the smaller the ID. Refer to Table 2-1.

DESCRIPTION, CLASSIFICATION, AND APPLICATIONS

Standard-weight pipe is used for low-pressure applications for gas and water or general plumbing. Extra-strong pipe with its heavier wall is for medium-pressure applications, whereas double extra-strong pipe is for high-pressure applications.

Pipe is manufactured from steel, cast iron, wrought iron, brass, and copper. Tubing is manufactured from steel, copper, stainless steel, and aluminum.

Pipe, as manufactured, is generally classified as seamless, continuous weld, electric weld, and double submerged-arc weld pipe. Tubing is classified as seamless or welded. The type of pipe to be used will depend upon the service condition, internal pressure, temperature, life expectancy, and corrosion. When specifying pipe, always remember to observe the applicable federal, state, local, and industrial codes.

Several Applications

Seamless Pipe

Seamless types are manufactured by piercing a solid billet and are available in sizes ⅛ to 26 in. Because of the seamless method of manufacturing, this type is widely used in industry for air, gas, steam, water, and oil piping.

Continuous Weld Pipe

This type is manufactured by butt welding a continuous metal sheet which passes through forming rolls, is welded, and then passes through a stretch reduction mill; it is available in sizes to 4 in and is used for gas, water, air, steam, structural applications, sprinkler systems, electric raceways, fencing, etc. It generally is less expensive than seamless pipe.

Electric Weld Pipe

This is manufactured by forming a strip from flat steel using high-frequency welding and sizing methods. It is available in sizes 4 to 20 in and is used in the steel, oil, and natural gas industries as well as for pipe piling and for slurry lines.

Double Submerged-Arc Weld Pipe

This type is manufactured from a prepared flat steel plate, is formed on presses, and is either mechanically or hydrostatically expanded: It is welded inside and outside and generally comes in sizes ranging from 20 to 48 in. It is used in construction, oil, natural gas, and water applications.

Pipe sizes larger than 48 in are true fabrication, rolled, and welded.

Tubing

There exists in the industry an intermingling of the terms "tubing" and "pipe." Round tubing is used for conveying fluids, tubing with other shapes for fabrication purposes.

TABLE 2-1 Pipe Dimensions and Properties

	Wall thickness		Dimensions			Weights		Areas				Properties		
								Surface		Cross-sectional				
Nom. pipe size, in	Iron pipe size	Sch. no.	Outside diam, in	Inside diam, in	Wall thkn., in	Plain end pipe, lb/ft	Water in pipe, lb/ft	Outside, ft²/ft	Inside, ft²/ft	Flow, in²	Metal, in²	Moment of inertia, in⁴	Section modulus, in³	Radius of gyration, in
½	Std	40	0.840	0.622	0.109	0.850	0.132	0.220	0.1637	0.3040	0.2503	0.0171	0.0407	0.2613
	XS	80	0.840	0.546	0.147	1.087	0.101	0.220	0.1433	0.2340	0.3200	0.0201	0.0478	0.2505
	XXS		0.840	0.252	0.294	1.714	0.022	0.220	0.0660	0.0499	0.5043	0.0242	0.0577	0.2192
¾	Std	40	1.050	0.824	0.113	1.130	0.230	0.275	0.2168	0.5330	0.3326	0.0370	0.0705	0.3337
	XS	80	1.050	0.742	0.154	1.473	0.187	0.275	0.1948	0.4330	0.4335	0.0448	0.0853	0.3214
	XXS		1.050	0.434	0.308	2.440	0.063	0.275	0.1137	0.1479	0.7180	0.0579	0.1103	0.2840
1	Std	40	1.315	1.049	0.133	1.678	0.374	0.344	0.2740	0.8640	0.4939	0.0873	0.1328	0.4205
	XS	80	1.315	0.957	0.179	2.171	0.311	0.344	0.2520	0.7190	0.6388	0.1056	0.1606	0.4066
	XXS		1.315	0.599	0.358	3.659	0.122	0.344	0.1570	0.2818	1.0760	0.1405	0.2136	0.3613
1½	Std	40	1.900	1.610	0.145	2.717	0.882	0.497	0.4213	2.0361	0.8001	0.3099	0.3262	0.6226
	XS	80	1.900	1.500	0.200	3.631	0.765	0.497	0.3927	1.7672	1.0689	0.3912	0.4118	0.6052
	XXS		1.900	1.100	0.400	6.408	0.412	0.497	0.2903	0.9502	0.8859	0.5678	0.5977	0.5489
2	Std	40	2.375	2.067	0.154	3.65	1.45	0.622	0.540	3.355	1.075	0.666	0.561	0.787
	XS	80	2.375	1.939	0.218	5.02	1.28	0.622	0.507	2.953	1.477	0.868	0.731	0.766
	XXS		2.375	1.503	0.436	9.03	0.77	0.622	0.393	1.774	2.656	1.312	1.104	0.703
2½	Std	40	2.875	2.469	0.203	5.79	2.07	0.753	0.646	4.788	1.704	1.530	1.064	0.947
	XS	80	2.875	2.323	0.276	7.66	1.83	0.753	0.610	4.238	2.254	1.924	1.339	0.924
	XXS		2.875	1.771	0.552	13.70	1.07	0.753	0.463	2.464	4.028	2.871	1.997	0.844
3	Std	40	3.500	3.068	0.216	7.58	3.20	0.916	0.802	7.393	2.228	3.017	1.724	1.164
	XS	80	3.500	2.900	0.300	10.25	2.86	0.916	0.761	6.605	3.016	3.892	2.225	1.136
	XXS		3.500	2.300	0.600	18.58	1.80	0.916	0.601	4.155	5.466	5.993	3.424	1.047

4	Std	40	4.500	4.026	0.237	10.79	5.51	1.178	1.055	12.730	3.174	7.231	3.214	1.510
	XS	80	4.500	3.826	0.337	14.98	4.98	1.178	1.002	11.497	4.407	9.610	4.271	1.477
	XXS		4.500	3.152	0.674	27.54	3.38	1.178	0.826	7.803	8.101	15.284	6.793	1.374
6	Std	40	6.625	6.065	0.280	18.97	12.5	1.73	1.59	28.90	5.58	28.14	8.50	2.24
	XS	80	6.625	5.761	0.432	28.57	11.3	1.73	1.51	26.07	8.40	40.49	12.22	2.19
	XXS		6.625	4.897	0.864	53.16	8.1	1.73	1.28	18.83	15.64	66.33	20.02	2.06
8	Std	40	8.625	7.981	0.322	28.55	21.6	2.26	2.09	50.03	8.40	72.49	16.81	2.94
	XS	80	8.625	7.625	0.500	43.39	19.8	2.26	2.01	45.67	12.76	105.70	24.51	2.88
	XXS		8.625	6.875	0.875	72.42	16.1	2.26	1.80	37.13	21.30	161.98	37.56	2.76
10	Std	40	10.750	10.020	0.365	40.48	34.1	2.81	2.62	78.85	11.91	160.71	29.90	3.67
	XS	60,80S	10.750	9.750	0.500	54.74	32.3	2.81	2.55	74.66	16.10	211.94	39.43	3.63
	XXS	140	10.750	8.750	1.000	104.13	26.1	2.81	2.29	60.13	30.63	367.81	68.43	3.46
12	Std	40	12.750	11.938	0.406	53.6	48.5	3.34	3.13	111.9	15.74	300.3	47.1	4.37
	XS	80	12.750	11.374	0.688	88.6	44.0	3.34	2.98	101.6	26.07	475.7	74.6	4.27
	XXS	120	12.750	10.750	1.000	125.5	39.3	3.34	2.81	90.8	36.91	641.7	100.7	4.17
14	Std	30	14.000	13.250	0.375	54.6	59.7	3.67	3.47	137.9	16.05	372.8	53.2	4.82
	XS		14.000	13.000	0.500	72.1	57.4	3.67	3.40	132.7	21.21	483.8	69.1	4.78
16	Std	30	16.000	15.250	0.375	63	79.1	4.19	4.00	182.6	18.41	562	70.3	5.53
	XS	40	16.000	15.000	0.500	83	76.5	4.19	3.93	176.7	24.35	732	91.5	5.48
18	Std		18.000	17.250	0.375	71	101.2	4.71	4.51	233.7	20.76	807	89.6	6.23
	XS		18.000	17.000	0.500	93	98.2	4.71	4.45	227.0	27.49	1053	117.0	6.19
20	Std	20	20.000	19.250	0.375	79	126.0	5.24	5.04	291.1	23.12	1113	111.3	6.94
	XS	30	20.000	19.000	0.500	104	122.8	5.24	4.97	283.5	30.63	1457	145.7	6.90
22	Std	20	22.000	21.250	0.375	87	153.7	5.76	5.56	354.7	25.48	1490	135.4	7.65
	XS	30	22.000	21.000	0.500	115	150.2	5.76	5.50	346.4	33.77	1953	177.5	7.61
24	Std	20	24.000	23.250	0.375	95	183.9	6.28	6.09	424.6	27.83	1942	161.9	8.35
	XS		24.000	23.000	0.500	125	180.0	6.28	6.02	416.0	36.90	2550	213.0	8.31

Both shapes are seamless and welded, ferrous and nonferrous, and come in sizes ranging to 10 in. However, for this discussion tubing can be differentiated from piping when used with compression, bite, swaged, or flared-tube fittings to convey liquids, gases, or air. Tubing can be further defined as that product which can be easily bent or formed, generally because it has a thinner wall. It is beyond the scope of this chapter to adequately convey all the information available. It is therefore suggested that when the end use is known (boiler tubes, heat-exchanger tubes, hydraulic, refining tubes, etc.), the ASME, ANSI, and SAE standards be used.

FITTINGS AND COUPLINGS

Standard fittings are available in the form of 45 and 90° elbows, tees, reducers, unions, crosses, Y branches, couplings, and caps that have been screwed together, socket-welded, flanged, butt-welded, or soldered.

Materials used in fabrication are forged steel, cast steel, cast iron, malleable iron, galvanized steel, and brass.

Because of the difficulty of assembly, screwed fittings are limited to smaller sizes and are generally classified as 125, 150, and 250 lb; they are suitable for service with low-pressure steam, gas, air, and water, and are supplied in brass and malleable and cast iron. Forged-steel fittings are for higher pressure ratings to 6000 lb/in^2. Although available in small sizes, flanged and butt-welded fittings are generally for larger pipes.

Tubing is most frequently supplied in sizes 2 in and smaller for use with flared, bite-type, soldered, and swaged tube fittings.

AVAILABLE SIZES AND FORMS

The ANSI (formerly ASA) standards cover the dimensional and physical properties of fittings. Because there are large numbers of fittings, information about type and material should be sought in standards and manufacturers' handbooks, such as those published by Crane, Grinnell, Tube Turns Division of National Cylinder Gas, and Midwest Piping. Tables 2-3 to 2-10 (pages 4-22 to 4-29) are a sampling of those available.

INSTALLATION AND MAINTENANCE

The joining of metallic pipe and fittings to form a complete piping system may be accomplished in several ways; the following methods are most popular.

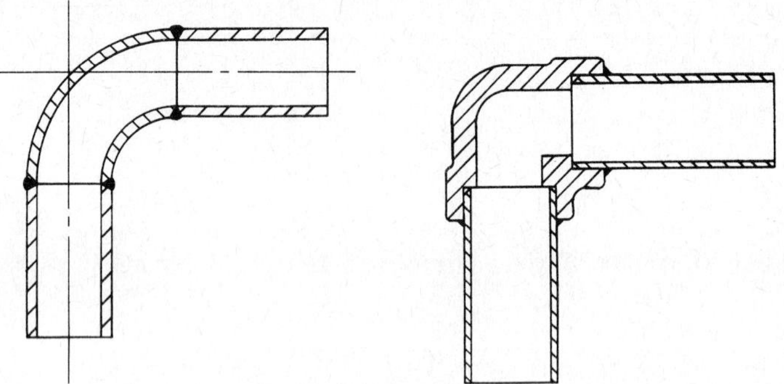

Figure 2-1 (*a*) Butt-welded elbow; (*b*) socket-welded elbow.

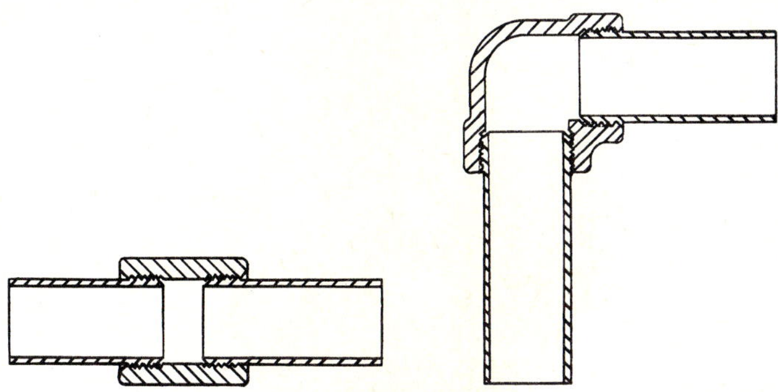

Figure 2-2 (*a*) Threaded coupling; (*b*) threaded elbow.

Welding

This technique requires the fusion of two metal surfaces to form one. Arc welding uses heat produced by an electric arc to coalesce the metals. Filler metals may also be used as fluxing agents. Gas welding uses a gas flame to produce the required fusion temperatures. See Fig. 2-1.

Brazing

Most commonly used in nonferrous domestic piping systems, brazing is a process whereby a filler metal, having a melting point above 800°F (425°C) but lower than that of the base metal, is heated to accomplish the joining. A capillary action occurs, drawing the filler metal between the closely fitted base-metal surfaces.

Threading

This method is customarily used in smaller-diameter, lower-pressure piping systems. Straight- and tapered-thread designs are available. Threaded fittings are made in a variety of material, including malleable iron, forged steel, bronze, etc. Pressure ratings are provided by the American National Standards Institute (ANSI) for threaded fittings and appear stamped on the body of each fitting produced. Fabrication of pipe is performed by die-cutting threads onto the pipe ends to mate with those of the fitting. See Fig. 2-2.

Flanging

Flanged joints are used in systems requiring frequent disassembly or to facilitate maintenance at equipment connections. Successful application of flanges relies upon achieving the correct mating surfaces of the flange faces and the seating element of the gasket into a leak-tight joint. Pipe mating the ends of flanges may be threaded or suitable for butt-welding, slip-on attachment, or socket-welding.

Utmost importance is placed upon gasket selection and uniformity of bolt tension. See Table 2-2. Gaskets must be selected for fluid compatibility, temperature, and pressure conditions. Bearing surfaces of nuts must have a smooth machine finish, and threads should be properly lubricated at assembly. Final tensioning of bolts or studs should be performed with a torque wrench. See Fig. 2-3.

Tubing and Tube Connectors

Tubing is in widespread use throughout the industrial market; its applications include fluid power, instrumentation, and general-service piping.

TABLE 2-2 Recommended Bolt Torques (Based upon Carbon Steel with 30 × 10^6 lb/in² Electric Modules)

Stud diameter, in	Threads per in	Torque, lb·ft
⅜	10	107
⅝	11	89
⅞	9	162
1	8	244
1⅛	8	322
1⅛	8	410
1⅜	8	510
1½	8	615

The use of smaller diameters and thinner wall sections allows tubing to be easily bent into a variety of configurations. This increased flexibility and workability often results in a cost savings by minimizing fitting requirements.

Tube fittings are primarily designed for use with extruded tubing (see Fig. 2-4) and are not used for standard pipe joining. Fittings are designed so that the interlocking threaded unions and ferrules maintain a pressure-tight seal. It is necessary to properly tension locking nuts, debur cut tube ends, and avoid crimped or scored tube surfaces.

STANDARDS AND SPECIFICATIONS

ASTM Ferrous Material Specifications

Pipe—Seamless for High-Temperature Service

A 106 Carbon Steel
A 335 Ferritic Alloy Steel
A 376 Austenitic Central Station Service
A 405 Ferritic Alloy, Special Heat-Treated

Pipe—Seamless and Welded

A 53 Carbon Steel
A 120 Black and Zinc-Coated (Ordinary Use)
A 312 Austenitic Stainless Steel
A 333 Carbon Steel (Low-Temperature Service)
A 530 General Requirements for Specialized Carbon- and Alloy-Steel Pipe

Pipe—Welded

A 134 Arc-Welded Steel Plate 16 in and over
A 135 Electric-Resistance-Welded Steel
A 139 Arc-Welded Steel 4 in and over
A 155 Arc-Welded Steel for High-Temperature Service
A 211 Spiral-Welded Steel or Iron
A 252 Specification for Welded and Seamless Pipe Piles
A 358 Arc-Welded Chrome-Nickel Alloy High-Temperature Service
A 523 Specification for Plain-End Seamless and Electric-Resistance-Welded Steel Pipe for High-Pressure Pipe-Type Cable Circuits
A 589 Specifications for Seamless and Welded Carbon Steel Water Well Pipe

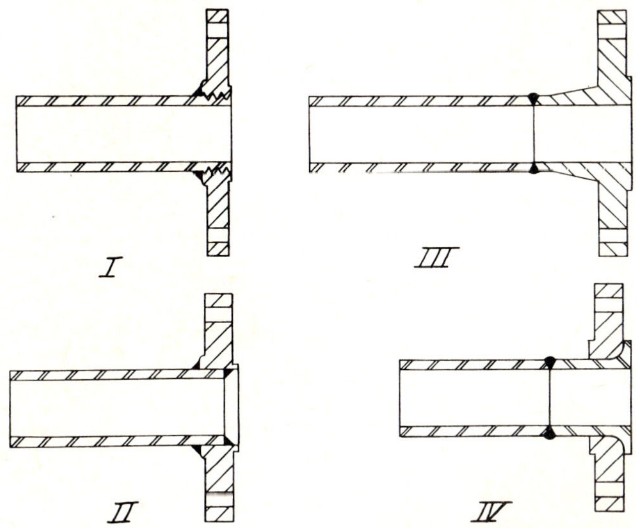

Figure 2-3 Standard flange-to-pipe configurations: I, screwed and back welded; II, slip-on welding flange; III, welding neck, butt-welded to pipe; IV, lap joint, stub-end butt-welded to pipe.

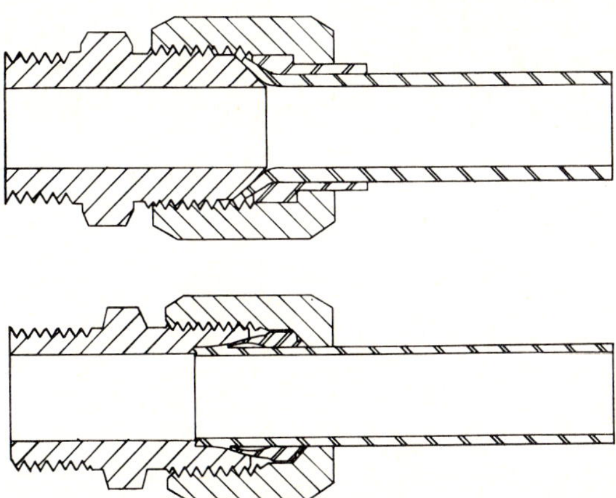
Figure 2-4 Compression fittings to join tubing to threaded unions.

TABLE 2-3 Dimensions of Long Radius, 90° Butt-Welding Elbows (Standard Weight: ANSI B16.9-1978, A 234)

Nominal pipe size*	OD	ID	Wall thickness	Center to face	Pipe sched. numbers	Approx wt, lb
2½	2.875	2.469	0.203	3¾	40	2.92
3	3.500	3.068	0.216	4½	40	4.58
3½	4.000	3.548	0.226	5¼	40	6.43
4	4.500	4.026	0.237	6	40	8.70
5	5.563	5.047	0.258	7½	40	14.7
6	6.625	6.065	0.280	9	40	22.9
8	8.625	7.981	0.322	12	40	46.0
10	10.750	10.020	0.365	15	40	81.5
12	12.750	12.000	0.375	18	St†	119
14	14.000	13.250	0.375	21	30	154
16	16.000	15.250	0.375	24	30	201
18	18.000	17.250	0.375	27	St†	256
20	20.000	19.250	0.375	30	20	317
22	22.000	21.250	0.375	33	St†	385
24	24.000	23.250	0.375	36	20	458
26	26.000	25.250	0.375	39	St†	539
30	30.000	29.250	0.375	45	St†	720
34	34.000	33.250	0.375	51	St†	926
36	36.000	35.250	0.375	54	St†	1040
42	42.000	41.250	0.375	63	St†	1420

*All dimensions are in inches except the last column.
†Standard weight.
Source: Tube Turns Division, National Cylinder Gas Co.

TABLE 2-4 Dimensions of Concentric and Eccentric Butt-Welding Reducers (Standard Weight: ANSI B16.9-1978, ASTM A 234)*

Nominal pipe size	Length	Approx wt, lb	Nominal pipe size	Length	Approx wt, lb	Nominal pipe size	Length	Approx wt, lb
2½ × 1	3½	1.30	8 × 6	6	13.4	22 × 20	20	157
2½ × 1¼	3½	1.47	8 × 6	6	13.9	24 × 16	20	160
2½ × 1½	3½	1.51	10 × 4	7	21.1	24 × 18	20	163
2½ × 2	3½	1.60	10 × 5	7	21.8	24 × 20	20	167
3 × 1¼	3½	1.70	10 × 6	7	22.3	26 × 18	24	200
3 × 1½	3½	1.89	10 × 8	7	23.2	26 × 20	24	200
3 × 2	3½	2.00	12 × 5	8	30.5	26 × 22	24	200
3 × 2½	3½	2.16	12 × 6	8	31.1	26 × 24	24	200
3½ × 1¼	4	2.35	12 × 8	8	32.1	30 × 20	24	220
3½ × 1½	4	2.52	12 × 10	8	33.4	30 × 24	24	220
3½ × 2	4	2.71	14 × 6	13	55.8	30 × 26	24	220
3½ × 2½	4	2.96	14 × 8	13	57.2	30 × 28	24	220
3½ × 3	4	3.05	14 × 10	13	60.4			
4 × 1½	4	2.73	14 × 12	13	63.4			Conc. Ecc.
4 × 2	4	3.17	16 × 8	14	70.2	34 × 24	24	270 229
4 × 2½	4	3.34	16 × 10	14	72.9	34 × 26	24	270 237
4 × 3	4	3.50	16 × 12	14	75.6	34 × 30	24	270 253
4 × 3½	4	3.61	16 × 14	14	77.5	34 × 32	24	270 261
5 × 2	5	5.05	18 × 10	15	86.9	36 × 24	24	340 237
5 × 2½	5	5.52	18 × 12	15	89.2	36 × 26	24	340 245
5 × 3	5	5.73	18 × 14	15	90.9	36 × 30	24	340 261
5 × 3½	5	5.86	18 × 16	15	94.0	36 × 32	24	340 269
5 × 4	5	5.99	20 × 12	20	134	36 × 34	24	340 277
6 × 2½	5½	7.61	20 × 14	20	135	42 × 24	24	260
6 × 3	5½	8.00	20 × 16	20	138	42 × 26	24	270
6 × 3½	5½	8.14	20 × 18	20	142	42 × 30	24	285
6 × 4	5½	8.19	22 × 14	20	148	42 × 32	24	295
6 × 5	5½	8.65	22 × 16	20	151	42 × 34	24	300
8 × 3½	6	12.8	22 × 18	20	154	42 × 36	24	310
8 × 4	6	13.1						

*All pipe sizes and lengths are in inches
Source: Tube Turns Division, National Cylinder Gas Co.

TABLE 2-5 Dimensions and Weights of Steel Pipe—Plain End—by Iron Pipe Size Weight Class

Sizes, in		Standard (Std)		Extra strong (XS)		Double extra strong (XXS)	
Nom. ID*	OD	Wall, in	Wt per ft, lb	Wall, in	Wt per ft, lb	Wall, in	Wt per ft, lb
⅛	0.405	0.068	0.24	0.095	0.31		
¼	0.540	0.088	0.42	0.119	0.54		
⅜	0.675	0.091	0.57	0.126	0.74		
½	0.840	0.109	0.85	0.147	1.09	0.294	1.71
¾	1.050	0.113	1.13	0.154	1.47	0.308	2.44
1	1.315	0.133	1.68	0.179	2.17	0.358	3.66
1¼	1.660	0.140	2.27	0.191	3.00	0.382	5.21
1½	1.900	0.145	2.72	0.200	3.63	0.400	6.41
2	2.375	0.154	3.65	0.218	5.02	0.436	9.03
2½	2.875	0.203	5.79	0.276	7.66	0.552	13.69
3	3.500	0.216	7.58	0.300	10.25	0.600	18.58
3½	4.000	0.226	9.11	0.318	12.50	0.636	22.85
4	4.500	0.237	10.79	0.337	14.98	0.674	27.54
5	5.563	0.258	14.62	0.375	20.78	0.750	38.55
6	6.625	0.280	18.97	0.432	28.57	0.864	53.16
8	8.625	0.322	28.55	0.500	43.39	0.875	72.42
10	10.750	0.365	40.48	0.500	54.74	1.000	104.13
12	12.750	0.375	49.56	0.500	65.42	1.000	125.49
14	14.000	0.375	54.57	0.500	72.09		
16	16.000	0.375	62.58	0.500	82.77		
18	18.000	0.375	70.59	0.500	93.45		
20	20.000	0.375	78.60	0.500	104.13		
22	22.000	0.375	86.61	0.500	114.81		
24	24.000	0.375	94.62	0.500	125.49		
26	26.000	0.375	102.63	0.500	136.17		
28	28.000	0.375	110.64	0.500	146.85		
30	30.000	0.375	118.65	0.500	157.53		
32	32.000	0.375	126.66	0.500	168.21		
34	34.000	0.375	134.67	0.500	178.89		
36	36.000	0.375	142.68	0.500	189.57		

*Although pipe 2 in and larger will be identified by OD former nominal sizes are shown.
Source: "Wrought Steel and Wrought Iron Pipe," ANSI B36.10-1970.

TABLE 2-6 Dimensions of Long Radius 45° Butt-Welding Elbows (Standard Weight: ANSI B16.9-1978, ASTM A 234)

Nominal pipe size*	OD	ID	Wall thickness	Center to face	Radius	Pipe schedule numbers	Approx wt, lb
2½	2.875	2.469	0.203	1¾	3¾	40	1.64
3	3.500	3.068	0.216	2	4½	40	2.43
3½	4.000	3.548	0.226	2¼	5¼	40	3.29
4	4.500	4.026	0.237	2½	6	40	4.31
5	5.563	5.047	0.258	3⅛	7½	40	7.30
6	6.625	6.065	0.280	3¾	9	40	11.3
8	8.625	7.981	0.322	5	12	40	22.8
10	10.750	10.020	0.365	6¼	15	40	40.4
12	12.750	12.000	0.375	7½	18	St†	59.5
14	14.000	13.250	0.375	8¾	21	30	76.5
16	16.000	15.250	0.375	10	24	30	100
18	18.000	17.250	0.375	11¼	27	St†	128
20	20.000	19.250	0.375	12½	30	20	158
22	22.000	21.250	0.375	13½	33	St†	192
24	24.000	23.250	0.375	15	36	20	229
26	26.000	25.250	0.375	16	39	St†	269
30	30.000	29.250	0.375	18½	45	St†	358
34	34.000	33.250	0.375	21	51	St†	463
36	36.000	35.250	0.375	22¼	54	St†	518
42	42.000	41.250	0.375	26	63	St†	707

*All dimensions are in inches except the last column.
†Standard weight
Source: Tube Turns Division, National Cylinder Gas Co.

TABLE 2-7 ANSI Schedule Number Wall Thicknesses for Plain End Steel Pipe*

Sizes, in		ANSI Schedule number wall thicknesses									
Nom. ID	OD	10	20	30	40	60	80	100	120	140	160
1/8	0.405				0.68		0.095				
1/4	0.540				0.88		0.119				
3/8	0.675				0.091		0.126				
1/2	0.840				0.109		0.147				0.188
3/4	1.050				0.113		0.154				0.219
1	1.315				0.133		0.179				0.250
1¼	1.660				0.140		0.191				0.250
1½	1.900				0.145		0.200				0.281
2	2.375				0.154		0.218				0.344
2½	2.875				0.203		0.276				0.375
3	3.500				0.216		0.300				0.438
3½	4.000				0.226		0.318				
4	4.500				0.237		0.337		0.438		0.531
5	5.563				0.258		0.375		0.500		0.625
6	6.625				0.280		0.432		0.562		0.719
8	8.625		0.250	0.277	0.322	0.406	0.500	0.594	0.719	0.812	0.906
10	10.750		0.250	0.307	0.365	0.500	0.594	0.719	0.844	1.000	1.125
12	12.750		0.250	0.330	0.406	0.562	0.688	0.844	1.000	1.125	1.312
14	14.000	0.210	0.312	0.375		0.594	0.750	0.938	1.094	1.250	1.406
16	16.000	0.250	0.312	0.375	0.500	0.656	0.44	1.031	1.219	1.438	1.594
18	18.000	0.250	0.312	0.438	0.562	0.750	0.938	1.156	1.375	1.562	1.781
20	20.000	0.250	0.375	0.500	0.594	0.812	1.031	1.281	1.500	1.750	1.969
22	22.000	0.250	0.375	0.500		0.875	1.125	1.375	1.625	1.875	2.125
24	24.000	0.250	0.375	0.562	0.688	0.969	1.219	1.531	1.812	2.062	2.344
26	26.00	0.312	0.500								
28	28.000	0.312	0.500	0.625							
30	30.000	0.312	0.500	0.625							
32	32.000	0.312	0.500	0.625	0.638						
34	34.000	0.344	0.500	0.625	0.638						
36	36.000	0.312	0.500	0.625	0.750						

*The weight of steel pipe is expressed in pounds per foot. All weights of carbon and alloy steel pipe are calculated on the basis of a cubic inch of steel weighing 0.2833 lb. Two useful formulas relating to pipe weights are

$$\text{Outside diameter} \times \text{wall thickness} \times 10.68 = \text{pounds per foot}$$

$$\text{Pounds per foot} \times 2.64 = \text{tons per linear mile}$$

Source: "Wrought Steel and Wrought Iron Pipe," ANSI B36.10-1970.

TABLE 2-8 Pressure-Temperature Ratings of Pipe (Seamless Carbon Steel to ASTM A 53 Grade A)

Pipe size, in	Schedule number	Wall thickness, in	Working pressure, lb/in²: −20 to 650°F 12,000 lb/in² stress	Working pressure, lb/in²: 750° 10,700 lb/in² stress
½	40	0.109	882	787
	80	0.147	1947	1736
1	40	0.133	960	856
	80	0.179	1777	1585
2	40	0.154	724	646
	80	0.218	1329	1185
3	40	0.216	875	780
	80	0.220	1422	1218
4	40	0.237	1146	1022
	80	0.337	1660	1481
6	40	0.280	914	815
	80	0.432	1435	1279
8	20	0.250	622	555
	40	0.322	806	718
	80	0.500	1270	1133
10	20	0.250	497	443
	40	0.365	729	650
	80	0.594	1208	1077
12	20	0.250	418	373
	40	0.406	683	609
	80	0.688	1178	1050

TABLE 2-9 **Dimensions of American 150-lb Standard Malleable-Iron Threaded Fittings (Straight Sizes)***

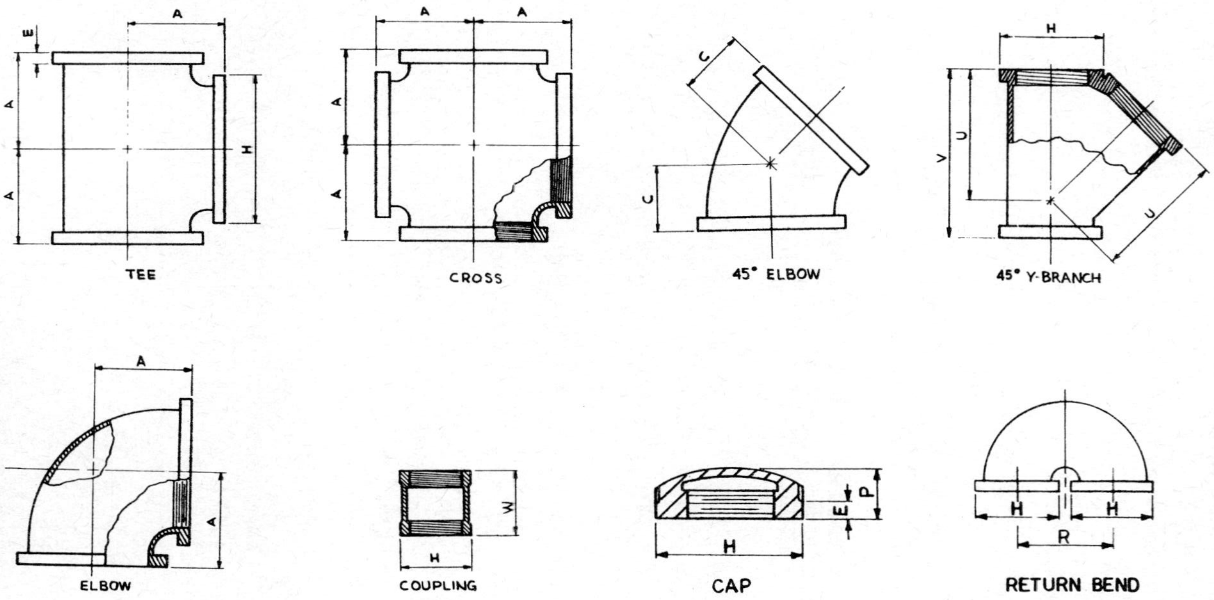
A
E
H
TEE
A
A
CROSS
C
45° ELBOW
H
U
V
C
45° Y-BRANCH
A
ELBOW
W
H
COUPLING
P
E
H
CAP
H
H
R
RETURN BEND

Size	Dimension								R		
	A	H	E	C	V	U	W	P	Closed	Medium	Open
⅛	0.69	0.693	0.200	—	—	—	—	0.96			
¼	0.81	0.844	0.215	0.73	—	—	1.06				
⅜	0.95	1.015	0.230	0.80	1.93	1.43	1.16				
½	1.12	1.197	0.249	0.88	2.32	1.71	1.34	0.87	1.000	1.25	1.50
¾	1.31	1.458	0.273	0.98	2.77	2.05	1.52	0.97	1.250	1.50	2.00
1	1.50	1.771	0.302	1.12	3.28	2.43	1.67	1.16	1.500	1.875	2.50
1¼	1.75	2.153	0.341	1.29	3.94	2.92	1.93	1.28	1.750	2.25	3.00
1½	1.94	2.427	0.368	1.43	4.38	3.28	2.15	1.33	2.188	2.50	3.50
2	2.25	2.963	0.422	1.68	5.17	3.93	2.53	1.45	2.625	3.00	4.00
2½	2.70	3.589	0.478	1.95	6.25	4.73	2.88	1.70	—	—	4.50
3	3.08	4.285	0.548	2.17	7.26	5.55	3.18	1.80	—	—	5.00
3½	3.42	4.843	0.604	2.39	—	—	3.43	1.90			
4	3.79	5.401	0.661	2.61	8.98	6.97	3.69	2.08			
5	4.50	6.583	0.780	3.05	—	—	—	2.32			
6	5.13	7.767	0.900	3.46	—	—	—	2.55			

*The complete standard (ANSI B16.3-1977) covers also reducing couplings, elbows, tees, crosses, and service or street elbows and tees. All dimensions in inches.

TABLE 2-10 Dimensions of Americn 125-and 250-lb Standard* Cast-Iron Threaded Fittings (Straight Sizes)(ANSI B16.4-1977)

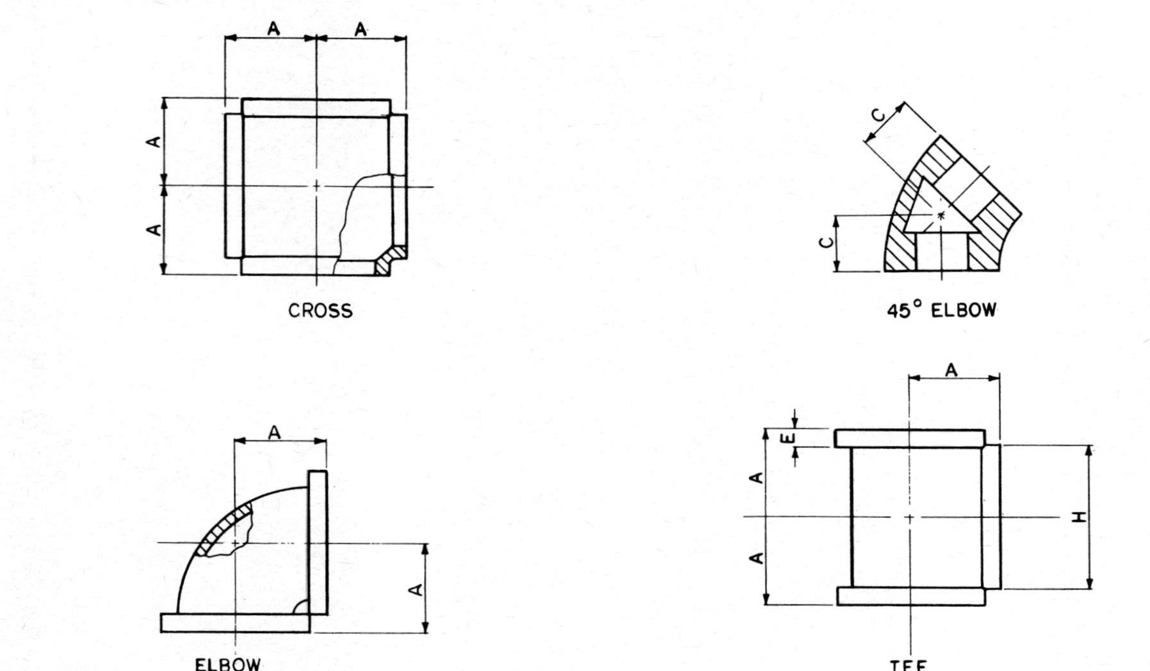

Size	125-lb fittings				250-lb fittings			
	A	H	E	C	A	H	E	C
¼	0.81	0.93	0.38	0.73	0.94	1.17	0.49	0.81
⅜	0.95	1.12	0.44	0.80	1.06	1.36	0.55	0.88
½	1.12	1.34	0.50	0.88	1.25	1.59	0.60	1.00
¾	1.13	1.63	0.56	0.98	1.44	1.88	0.68	1.13
1	1.50	1.95	0.62	1.12	1.63	2.24	0.76	1.31
1¼	1.75	2.39	0.69	1.29	1.94	2.73	0.88	1.50
1½	1.94	2.68	0.75	1.43	2.13	3.07	0.97	1.69
2	2.25	3.28	0.84	1.68	2.50	3.74	1.12	2.00
2½	2.70	3.86	0.94	1.95	2.94	4.60	1.30	2.25
3	3.08	4.62	1.00	2.17	3.38	5.36	1.40	2.50
3½	3.42	5.20	1.06	2.39	3.75	5.98	1.40	2.63
4	3.79	5.79	1.12	2.61	4.13	6.61	1.57	2.81
5	4.50	7.05	1.18	3.05	4.88	7.92	1.74	3.19
6	5.13	8.28	1.28	3.46	5.63	9.24	1.91	3.50
8	6.56	10.63	1.47	4.28	7.00	11.73	2.24	4.31
10	8.08	13.12	1.68	5.16	8.63	14.37	2.58	5.19
12	9.50	15.47	1.88	5.97	10.00	16.84	2.91	6.00

*The 125-lb standard covers also reducing elbows and tees. The 250-lb standard covers only the straight sizes.

All dimensions in inches.

†This applies to elbows and tee only.

Pipe—Forged and Bored High-Temperature Service

A 369 Ferritic Alloy Steel
A 430 Austenitic Stainless Steel

Pipe—Centrifugally Cast High-Temperature Service

A 426 Ferritic Alloys
A 451 Austenitic Alloys
A 452 Austenitic Cold-Wrought Alloys

Tube—Seamless

A 179 Low-Carbon Steel for Heat Exchanger and Condensers
A 192 Carbon Steel for Boilers at High-Temperature Service
A 199 Cold-Drawn Intermediate Alloy-Steel
A 210 Medium-Carbon Steel for Boilers
A 213 Ferritic and Austenitic Alloys for Boilers

Tube—Seamless and Welded

A 268 Ferritic Stainless Steel for General Use
A 269 Austenitic Stainless Steel Tubing for General Service
A 450 General Requirements for Carbon, Ferritic, and Austenitic Alloy

Tube—Welded

A 178 Arc-Welded Carbon Steel Boiler Tubes
A 214 Electric-Resistance-Welded Carbon Steel
A 226 Carbon Steel, High-Pressure Service
A 249 Austenitic Stainless Steel
A 254 Copper-Brazed Steel

Plates for Pressure Vessels

A 240 Stainless, Chromium and Chrome-Nickel
A 285 Low- and Intermediate-Temperature-Service Carbon Steel
A 299 Carbon-Manganese-Silicon Steel
A 387 Chromium-Molybdenum Alloy Steel
A 515 Carbon Steel for Intermediate- and High-Temperature Service
A 516 Carbon Steel for Moderate- and Low-Temperature Service

Welding Fittings—Factory-Made

A 234 Wrought Carbon and Ferritic Alloy Steel
A 403 Austenitic Steel

Castings

A 47 Malleable Iron
A 48 Gray Iron
A 126 Gray Iron for Valves, Flanges, and Pipe Fittings

A 197 Cupola, Malleable-Iron
A 389 Ferritic Alloy, Special Heat-Treated
A 395 Ductile Iron for Pressure Parts

ASTM Nonferrous Material Specifications

Pipe—Seamless

B 42 Copper (Standard Sizes)
B 43 Red Brass (Standard Sizes)
B 302 Threadless Copper

Pipe and Tube—Seamless

B 161 Nickel
B 165 Nickel-Copper Alloy
B 167 Nickel-Chromium-Iron Alloy
B 241 Aluminum Alloy Extruded
B 251 General Requirements for Wrought Copper and Copper Alloy
B 315 Copper Silicon Alloy
B 423 Nickel-Iron-Chromium-Molybdenum-Copper Alloy

Tube—Copper Seamless

B 68 Bright Annealed
B 75 For General Engineering Purpose
B 88 For General Plumbing Purpose
B 111 Condenser Tubes and Ferrule Stock
B 280 For Air-Conditioning and Refrigeration Field Service

Tube—Aluminum Seamless

B 210 Drawn
B 234 Drawn, for Condenser and Heat Exchangers

Tube—Welded

B 547 Aluminum-Alloy, Formed and Arc-Welded

Welding Fittings—Factory-Made

B 361 Wrought Aluminum and Aluminum-Alloy

American National Standards (ANSI)

A21.1-1977, Thickness Design of—Cast-Iron Pipe
A21.6-1975, Cast-Iron Pipe Centrifugally Cast in Metal Molds, for Water or Other Liquids
A21.10-1977, Gray-Iron and Ductile-Iron Fittings, 3 in through 48 in, for Water and other Liquids
A21.50-1976, Thickness Design of Ductile-Iron Pipe

A21.51-1976	Ductile-Iron Pipe, Centrifugally Cast in Metal Molds or Sand-Lined Molds for Water or Other Liquids
B1.1-1974	Unified Inch Screw Threads
B2.1-1968	Pipe Threads (except Dryseal)
B1.20.3-1976	Dryseal Pipe Threads (Inch)
B16.1-1975	Cast Iron Pipe Flanges and Flanged Fittings, Class 25, 125, 250, and 800
B16.3-1977	Malleable-Iron Threaded Fittings, Class 150 and 300
B16.4-1977	Cast Iron Threaded Fittings, Class 125 and 250
B16.5-1977	Steel Pipe Flanges and Flanged Fittings, Including Ratings for Class 150, 300, 400, 600, 900, 1500, and 2500
B16.9-1978	Factory-Made Wrought Steel Buttwelding Fittings
B16.11-1973	Forged Steel Fittings, Socket-Welding and Threaded
B16.14-1976	Ferrous Pipe Plugs, Bushings, and Locknuts with Pipe Threads
B16.15-1978	Cast Bronze Threaded Fittings, Class 125 and 150
B16.18-1978	Cast Bronze Alloy Solder-Joint Pressure Fittings
B16.22-1973	Wrought Copper and Bronze Solder-Joint Pressure Fittings
B16.24-1979	Bronze Pipe Flanges and Flanged Fittings, 150 and 300
B16.25-1979	Buttwelding Ends
B16.28-1978	Wrought Steel Buttwelding Short Radius Elbows and Returns
B16.34-1977	Steel Valves , Flanged and Butt-Welding End
B31.3-1976	Chemical Plant and Petroleum Refinery Piping
B31.4-1974	Liquid Petroleum Transportation Piping Systems
B31.8-1975	Gas Transmission and Distribution Piping Systems
B36.10-1979	Welded and Seamless Wrought Steel Pipe
B36.19-1976	Stainless Steel Pipe

MSS Standard Practices

SP-6	Finishes—On Flanges, Valves, and Fittings
SP-9	Spot-Facing for Bronze, Iron, and Steel Flanges
SP-25	Marking for Valves, Fittings, Flanges, and Unions
SP-43	Wrought Stainless Steel Butt-Welding Fittings
SP-45	Bypass and Drain Connection
SP-48	Steel Butt-Welding Fittings, 26 in and larger
SP-51	Corrosion-Resistant Cast Flanges and Flanged Fittings—150 lb
SP-58	Pipe Hangers and Supports—Materials and Design
SP-69	Pipe Hangers and Supports—Selection and Application
SP-75	High-Strength Wrought Welding Fittings
SP-79	Socket-Welding Reducer Inserts

API Specifications

5L	Line Pipe

ASME Codes

ASME Boiler and Pressure Vessel Code

AWWA Standards

C-101	Computation of Strength and Thickness of Cast Iron Pipe
C-106	Cast Iron Pipe Centrifugally Cast in Metal Molds
C-108	Cast Iron Pipe Centrifugally Cast in Sand-Lined Molds
C-110	Cast Iron Fittings, 2 through 48 in
C-111	Rubber-Gasketed Joint for Cast Iron Pipe and Fittings
C-112	2 and 2¼-in Cast Iron Pipe, Centrifugally Cast
C-150	Thickness Design of Ductile-Iron Pipe
C-151	Ductile-Iron Pipe, Centrifugally Cast in Metal or Sand-Lined Molds
C-201	Fabricated Electrically Welded Steel Pipe
C-208	Dimensions for Steel Water Pipe Fittings
C-600	Installation of Cast Iron Water Mains

Federal Specifications

WW-P-421	Pipe, Cast, Gray, and Ductile Iron, Pressure (for Water and Other Liquids)

SOURCES OF LISTED SPECIFICATIONS AND STANDARDS

API	American Petroleum Institute 1801 K Street, N.W. Washington, DC 20006
ANSI	American National Standards Institute 1430 Broadway New York, NY 10018
ASME	American Society of Mechanical Engineers, Inc. 345 East 47th Street New York, NY 10017
ASTM	American Society for Testing and Materials 1916 Race Street Philadelphia, PA 19103
AWS	American Welding Society, Inc. 2501 N.W. 7th Street Miami, FL 33125
AWWA	American Water Works Association 6666 W. Quincy Avenue Denver, CO 80235
MSS	Manufacturers Standardization Society of the Valve and Fittings Industry 1815 North Fort Myer Drive Arlington, VA 22209
PFI	Pipe Fabrication Institute 1326 Freeport Road Pittsburgh, PA 15238
SAE	Society of Automotive Engineers 400 Commonwealth Drive Warrendale, PA 15096

Federal Specifications

Superintendent of Documents
United States Government Printing Office
Washington, DC 20402

BIBLIOGRAPHY

King, R. C.: *Piping Handbook,* 5th ed., McGraw-Hill, New York, 1967.
Piping Design and Engineering, 3d ed., ITT Grinnell Co., Providence, R.I. 1971.
Rase, H. F.: *Piping Design for Process Plants,* Wiley, New York, 1963.
Welding Fittings and Forged Flanges, Crane Co., New York, 1967.

chapter 4-3

Plastics Piping

by
Stanley A. Mruk
Technical Director, Plastics Pipe Institute
A Divison of the Society of the Plastics Industry
New York, New York

INTRODUCTION

Outstanding chemical resistance is a principal reason for the growing use of plastics piping in practically every phase of U.S. industry including the pharmaceutical, chemical, food, paper and pulp, electronics, oil and gas production, water and waste treatment, mining, power generation, steel production and metal-refining industries. With the general recognition of its other features (e.g., it is easy to work with, durable, and economically advantageous), plastics piping has also become widely accepted for a broad range of applications for other reasons than just its chemical inertness. Current major uses include water mains and services, gas distribution, storm and sanitary sewers, plumbing (drain, waste, and vent piping and hot- and cold-water piping), electrical conduit, power and communications ducts, chilled-water piping, and well casing.

The diversity of plastic pipe—reflecting the many available materials, wall constructions, diameters, and techniques for joining—is so great and the attendant technology is evolving at such a rate that it is difficult to present in a concise reference all the pertinent information available. This chapter, therefore, offers only basic and general information. In applying plastic pipe, the reader is advised to ensure that all appropriate code requirements and government safety regulations are complied with. Sources and references for additional information are provided for this purpose at the end of the chapter.

DESCRIPTION, CLASSIFICATION, AND TYPICAL USES

"Plastic" pipe is as indefinite a term as "metal" pipe. As with metal products, plastic pipes are made from a variety of materials. Plastics used for pipe exhibit a wide range of properties and characteristics. The variabilities in properties of plastics are derived not only from the chemical composition of the basic synthetic resin, or polymer, but are also largely determined by the kind and amount of additiyes, the nature of reinforcement, and the process of manufacture. For example, it is possible to formulate mixtures of polyvinyl chloride (PVC) resins plus appropriate additives that range from a clear, soft, and pliable product (such as that used for laboratory tubing and upholstery) to a rigid and strong product (such as for pressure pipe). Another example is the construction of reinforced thermosetting resin pipe (RTRP) by using glass-fiber winding techniques that can adjust to the desired value the ratio of the circumferential (or hoop) strength to the axial strength.

Plastics are divided into two basic groups, thermosetting and thermoplastic, both of which are used in the manufacture of plastic pipe. Thermoplastics, as the name implies, soften upon the application of heat and reharden upon cooling: they can be formed and reformed repeatedly. This characteristic permits them to be easily extruded or molded into a wide variety of useful shapes, including pipe and fittings. Because thermoplastics are shaped, by a die or mold, while in a "molten" state their properties are essentially isotropic (i.e., independent of direction). In some processes and under some conditions, some anisotropy (i.e., direction dependence) may result.

Thermosetting plastics, on the other hand, form permanent shapes when cured by the application of heat or a "curing" chemical. Once shaped and cured during the manufacturing process, they cannot be reformed by heating. The excellent adhesion properties of thermosetting resins permits their utilization in composite structures by which strength and stiffness can be greatly enhanced through the use of reinforcements and fillers. The greater strength and higher temperature limits of these composite structures permit RTRP to handle fluids at temperatures and pressures beyond the limits for thermoplastic pipe. Piping systems now available are capable of operating at temperatures in excess of 300°F (148°C). All important commercial constructions of thermosetting pipe utilize some form of reinforcement and/or filler. By orienting the reinforcement, RTR pipes can be given directionally dependent properties.

Thermoplastic Pipe

Most thermoplastic pipes and fittings are made from materials containing no reinforcements, although fillers are occasionally used. Pipe is manufactured by the extrusion process, whereby molten material is continuously forced through a die that shapes the product. After being formed by the die, the soft pipe is simultaneously sized and hardened by cooling it with water. Fittings and valves are usually produced by the injection molding process, in which molten plastic is forced under pressure into a closed metal mold.

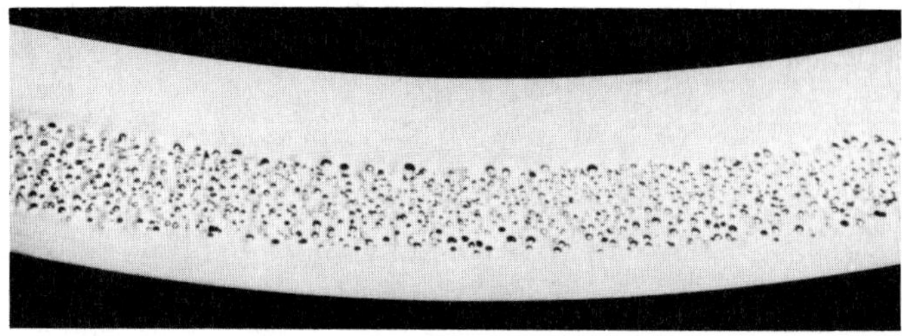

Figure 3-1 ABS foam-core construction. (Borg-Warner Chemicals.)

After cooling, the mold is opened and the finished part is removed. Some items, especially larger-sized fittings for which there is insufficient demand to justify construction of injection-molding tooling, are fabricated from pipe sections, or sheets, by utilizing thermal or solvent cementing fusion techniques. To compensate for the lower strength, the fitting may either be made from a heavier wall stock or reinforced with a fiberglass-resin overwrap. The engineer designing a pressure-rated system should make sure that the pressure ratings of the selected fittings are adequate.

There is some thermoplastic pipe made of a foam-core construction (for example, ASTM* F 628) in which the pipe wall consists of thin inner and outer solid skins sandwiching a high-density foam (Fig. 3-1). The primary benefit of such construction is improved ring and longitudinal (beam) stiffness in relation to the material used. Because the foam-wall structure results in some loss of strength, applications for foam-core pipe are in nonpressure uses, such as for above- and below-ground drainage piping, which can take advantage of the more material-efficient ring and beam stiffness.

For buried nonpressure applications, a composite pipe (ASTM D 2680) is produced that consists of two concentric tubes that are integrally braced with a truss webbing. The resultant openings between the concentric tubes are filled with a lightweight concrete. This construction increases both the ring and the beam stiffness. Composite pipe is used only for nonpressure buried applications such as sewerage and drainage.

Several other processes for improving the radial (i.e., ring) stiffness of thermoplastic pipe for buried applications have in common the formation of some type of rib reinforcement. A well-established technique is forming corrugations in the pipe wall. Corrugated polyethylene pipe (ASTM F 405) in sizes from 2 to 12 in (5 to 30 cm) is widely used for building foundations, land, highway, and agricultural drainage, and communications ducts. Ribbed pipe also is commercially made by the continuous spiral winding of the plastic over a mandrel of a specially shaped profile. Adjacent layers of this profile are fused to each other to form a cylinder that is smooth on the inside and has ribbed reinforcements on the outside (Fig. 3-2). The smooth inside diameter is preferable for many applications, such as sewerage, because it creates no flow disturbances. Pipes with ribbed construction are available in PVC and polyethylene (PE). PE pipes, which are made with hollow ribs to minimize material usage, are available in sizes from 18 to 96 in (45 cm to 1.8 m) in diameter.

The distinctive characteristics of the principal thermoplastic piping materials are discussed in the following pages.

Polyvinyl Chloride

Polyvinyl chloride (PVC) piping is made only from compounds containing no plasticizers and minimal quantities of other ingredients. To differentiate these materials from flexible, or plasticized PVCs (from which are made such items as upholstery, luggage, and laboratory tubing) they have been labeled rigid PVCs in the United States and unplasticized PVC (RuPVC) in Europe. Rigid PVCs used in piping range from type I to type III, as identified by an older classification system that is still much in use. In this system,

*American Society for Testing and Materials.

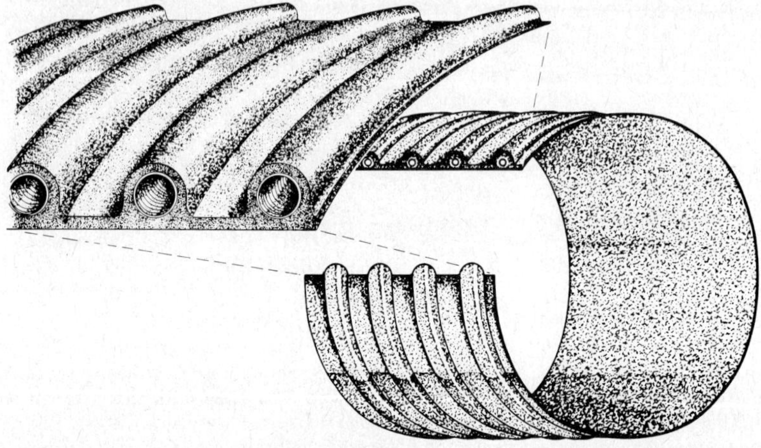

Figure 3-2 Typical profile of hollow ribbed polyethylene pipe. (*Spiral Engineered Systems.*)

the type designations are supplemented by grade designations (e.g., grade 1 or 2) which further define the material's properties. Type I materials, from which most pressure and nonpressure pipe is made, have been formulated to provide optimum strength as well as chemical and temperature resistance. Type II materials are those formulated with modifiers that improve impact strength but that also somewhat reduce, depending on modifier type and quantity, the aforementioned properties of Type I materials. There is little call for Type II pipe, as the impact strength of the stronger Type I pipe is more than adequate for most uses. Type III materials contain some inert fillers which tend to increase stiffness concomitant with some lowering of both tensile and impact strength and chemical resistance. Some nonpressure PVC piping, such as that used for conduit, sewerage, and drainage, is made from Type III PVCs.

The currently used classification system for rigid PVC materials for piping and other applications is described in ASTM D 1784, "Standard Specification for Rigid Polyvinyl Chloride and Chlorinated Polyvinyl Chloride Compounds." This specification categorizes rigid PVC materials by numbered cells that designate value ranges for the following properties: impact resistance (toughness), tensile strength, modulus of elasticity (rigidity), deflection temperature (temperature resistance), and chemical resistance. The following table cross-references the designations of the principal PVC materials from the older to the newer classification system.

By cell classification system of ASTM D1784 minimum cell class	*By older system*	
	Type and grade	*Designation*
12454-B	Type I, Grade 1	PVC 11
12454-C	Type I, Grade 2	PVC 12
14333-D	Type II, Grade 1	PVC 21
13233	Type III, Grade 1	PVC 31

Because (as expanded in the discussion on properties) short-term properties of plastic materials are not a reliable predictor of long-term capabilities, those PVC materials that have been formulated for long-term pressure applications are also designated by their categorized maximum recommended hydrostatic design stress (RHDS) for water at 73.4°F (23°C) as determined from long-term pressure testing. The most commonly used designation system for PVC pressure-piping materials is based on the above older designation system with two added digits that identify, in hundreds of pounds per square inch, the maximum recommended design stress.* For example: PVC 1120 is a type I, grade 1 PVC (minimum cell class 12454-B) with a maximum recommended HDS of 2000 lb/in^2 (13.8 MPa) for water at 73.4°F (23°C); PVC 2110 is a type 2, grade 1 PVC (minimum cell class 14333-D) with an RHDS of 1000 lb/in^2 (6.9 MPa). Other PVC material designations available for pressure pipe are listed later in Table 3-4. Most pressure-rated PVC pipe is made from PVC 1120 materials.

The combination of good long-term strength with higher stiffness explains why PVC has become the principal plastic pipe material for both pressure and nonpressure applications. Major uses include: water mains; water services; irrigation; drain, waste, and vent (DWV) pipes; sewerage and drainage; well casing; electric conduit; and power and communications ducts. A much broader range of fittings, valves, and appurtenances of all types is available in PVC than in any other plastic.

Chlorinated Polyvinyl Chloride

As implied by its name, chlorinated polyvinyl chloride (CPVC) is a chemical modification of PVC. CPVC has properties very similar to PVC but the extra chlorine in its structure extends its temperature limitation by about 50°F (28°C), to nearly 200°F (93°C) for pressure uses and about 210°F (99°C) for nonpressure applications. ASTM D 1784, the rigid PVC materials specification, also covers CPVC which it classifies as Class 23477-B. By the older designation system, it is known as type IV, grade 1 PVC. CPVCs for pressure pipe are designated CPVC 4120 (i.e., type IV, grade 1 CPVC with a maximum recommended hydrostatic design stress of 2000 lb/in^2 (13.8 MPa) for water at 73.4°F (23°C). At 180°F (82°C) the maximum recommended hydrostatic design stress* for CPVC is 500 lb/in^2 (3.4 MPa).

Principal applications for CPVC are for hot or cold water piping and for many industrial uses which take advantage of its higher temperature capabilities and superior chemical resistance.

Polyethylene

Polyethylene (PE) is the best-known member of the polyolefin group—plastics that are formed by the polymerization of straight-chain hydrocarbons—known as olefins—that include ethylene, propylene, and butylene. Polyethylene plastics are tough and flexible even at subfreezing temperatures. They are generally formulated with only an antioxidant (for protection during processing) and some pigment (usually carbon black) or other agent designed to screen out ultraviolet radiation in sunlight which over long-term exposure could be damaging to the natural-color polymer.

ASTM D 1248, the PE molding and extrusion materials specification, classifies these materials into three types depending on the density of the natural resin. Type I consists of lower-density materials which are relatively soft and flexible and have low heat resistance. Type II PEs are of medium density, slightly harder, more rigid and more resistant to elevated temperatures; they also have better tensile strength. Type III materials show maximum hardness, rigidity, tensile strength, and resistance to the effects of increasing temperature. Pipe is made almost exclusively from type II and type III PEs. ASTM D 1248 also provides for grade designations to further classify PEs according to other physical characteristics. PE piping materials for pressure piping are classified by a designation system that combines the type and grade coding with that for the maximum RHDS* for water at 73.4°F (23°C). PEs utilized for pressure piping, and so designated, are listed in Table 3-4 (p. 10-49).

The more recently issued standard, ASTM D 3350, classifies PE piping materials according to broader physical property criteria, including long-term strength, and it is expected to become the primary PE piping material standard.

*Since the maximum recommended design stress is for continuous water pressure at 73°F, it is up to the designer to determine the extent, if any, by which this stress should be reduced to account for any departure from these conditions and the need for a suitable margin of safety against other considerations. See the discussion on design.

Outstanding toughness and relatively low flexural modulus, which permits coiling of smaller-diameter pipe, are large factors in PE's prominence in gas-distribution and water-service piping. Other features, such as heat fusibility, good abrasion resistance, and availability in large diameters [up to 96 in (2.5 m)], account for the use of PE piping for chemical transfer lines, slurry transport, sewage force mains, intake and outfall lines, power ducts, and renewal (by insertion into the old pipe) of deteriorated sewer, gas, water, and other pipes.

Polybutylene

Polybutylene (PB) is a unique polyolefin. Its stiffness resembles that of low-density PE, yet its strength is higher than that of high-density PE. However, its most significant feature is that it retains its strength better with increasing temperature. Its upper temperature limit is higher than that for any PE: nearly 200°F (93°C) for pressure uses and somewhat higher for nonpressure applications.

PB piping materials are covered by ASTM D 2581. Materials for pressure applications are designated PB 2110; the last two digits signify a maximum recommended hydrostatic design stress* of 1000 lb/in^2 (6.9 MPa) for water at 73.4°F (23°C). At 180°F (83°C) this design stress is 500 lb/in^2 (3.4 MPa).

Major applications of PB pipe tend to take advantage of its improved temperature resistance and its toughness. They include hot or cold water piping and industrial uses such as hot effluent lines. Because of its excellent abrasion resistance, PB pipe is also used for slurry lines.

Polypropylene

Polypropylene (PP) is a polyolefin similar in properties to Type III PE but is slightly lighter in weight, more rigid, and more temperature-resistant. PPs are classified by ASTM D 2146 into two types: Type I covers homopolymers that generally have the greatest rigidity and strength but offer only moderate impact resistance; Type II covers copolymers of propylene and ethylene, or other olefins, which are less rigid and strong but have much improved toughness, particularly at lower temperatures. As indicated by Table 3-4 there are some PPs classified for use in pressure pipe.

Although on the basis of its short-term properties PP shows better resistance to temperature, this material is inherently somewhat more sensitive to thermal aging than PE. To overcome this sensitivity, specially formulated grades containing appropriate heat-stabilizer systems have been developed. For pressure uses, the adequacy of long-term heat stability is evaluated by means of long-term pressure tests of the PP formulation, conducted at various higher temperatures. A PP material containing flame-retardant additives is available for drainage-piping applications.

Polypropylene piping is used for chemical waste and drainage and various other industrial uses that take advantage of excellent chemical resistance, good rigidity and strength, and higher-temperature operating limits. One feature that accounts for its use in laboratory and industrial drainage is its superior resistance to many organic solvents.

Acrylonitrile-Butadiene-Styrene

Acrylonitrile-butadiene-styrene (ABS) is a family of materials formed from three different monomers (chemical building blocks), acrylonitrile, butadiene, and styrene. The proportions of the components and the way in which they are combined can be varied to produce a wide range of properties. Acrylonitrile contributes rigidity, strength, hardness, chemical resistance, and heat resistance; butadiene contributes toughness; and styrene contributes gloss, rigidity, and ease of processing.

*Since the maximum recommended design stress is for continuous water pressure at 73°F, it is up to the designer to determine the extent, if any, by which this stress should be reduced to account for any departure from these conditions and the need for a suitable margin of safety against other considerations. See the discussion on design.

ASTM D 1788 classifies ABS plastics into numbered cells that designate value ranges for each of three properties: impact strength (toughness), tensile stress at yield (strength), and deflection temperature under load. ABS pipe materials are categorized into types and grades in accordance with established minimum cell requirements for each type and grade. Like the other major thermoplastics, ABS materials for pressure pipe are designated by a coding that identifies both short-term properties and long-term strength. For example, ABS 1316 is a type I, grade 3 (minimum cell classification 3-5-5 per D 1788) material with a maximum RHDS* of 1600 lb/in^2 (11.2 MPa) for water at 73.4°F (23°C). Other ABS pressure-pipe materials are listed in Table 3-4 (p. 10-49).

An advantageous combination of toughness with good strength and stiffness largely accounts for the most common uses for ABS piping, i.e., for drain, waste, and vent (DWV) applications as well as for sewers, well casings, and communications ducts. One especially tough formulation of ABS is utilized to manufacture piping for compressed-air service. Most other thermoplastic materials are not recommended for above-ground compressed-air service because, should they fail, the pipe failure mechanism would sometimes produce flying fragments which could be injurious.

Polyvinylidine Fluoride

Polyvinylidine fluoride (PVDF) is a fluoroplastic, i.e., its chemical composition includes the element fluorine. Fluoroplastics are distinguished by their exceptional chemical and solvent resistance, excellent durability, and broad working-temperature range. To these features PVDF adds good strength and toughness and radiation resistance (which explains its use for conveying radioactive materials). The ASTM Specification covering PVDF molding and extrusion materials is D 3222. A standard for PVDF pressure piping is under development at ASTM. Standard recommended hydrostatic design stresses for PVDF have not yet been established, and an effort to develop these values is underway. Manufacturers of PVDF piping systems should be consulted for recommended design-stress values.

PVDF piping is utilized primarily for chemical processing and other industrial applications that are beyond the reach of the more common thermoplastics. These include the handling of active materials such as chlorine and bromine and piping materials that subject it to higher temperatures. Because of its immunity to radiation, PVDF piping is also used in reprocessing nuclear wastes.

Other Thermoplastics

Some thermoplastics of lesser current commercial importance are either used for special reasons or were more popular some time ago. Among these is *chlorinated polyether,* a tough rigid material of outstanding chemical resistance and useful to about 220°F (104°C). PVDF has largely displaced it as a specialty material. *Cellulose acetate butyrate* (CAB) was extensively used in petroleum production for conveying saltwater and crude oil as well as for natural gas distribution. However, because of its relatively low strength and moderate resistance to temperature and chemicals, it has been largely displaced by other materials. Nylon, a tough, strong, and heat-resistant material that is also very resistant to aromatic and chlorinated solvents, has some special uses. One of these is for coiled small-diameter tubing for compressed-air service. Nylon insert (with barbs) fittings are also made for use with polyethylene pipe.

Polyacetal, a strong, hard plastic with relatively good temperature resistance, is being increasingly used for various hot- or cold-water plumbing components such as fittings, valves, and faucets.

*Since the maximum recommended design stress is for continuous water pressure at 73°F, it is up to the designer to determine the extent, if any, by which this stress should be reduced to account for any departure from these conditions and the need for a suitable margin of safety against other considerations. See the discussion on design.

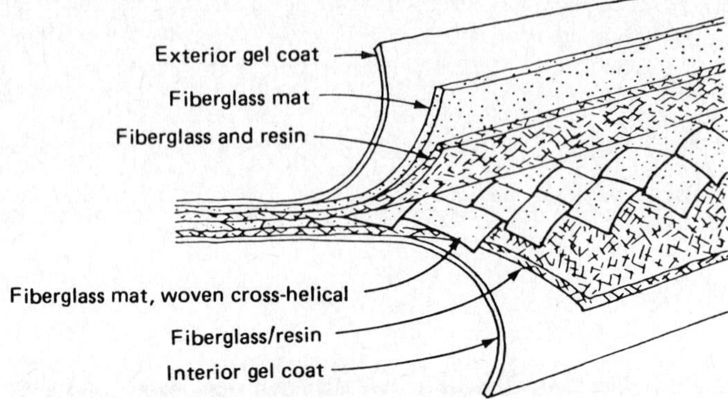

Figure 3-3 Fiberglass-reinforced pipe.

Reinforced Thermosetting Resin Pipe

Reinforced thermosetting resin pipe (RTRP) is a composite largely consisting of a reinforcement imbedded in, or surrounded by, cured thermosetting resin. Included in its composition may be granular or platelet fillers, thixotropic agents, pigments, or dyes. The most frequently used reinforcement is fiberglass, in any one or a combination of the following forms: continuous filament, chopped fibers, and mats (Fig. 3-3). While reinforcements such as asbestos or other mineral fibers are sometimes used, fiberglass-reinforced pipe (FRP) is by far the most popular. One form of FRP, called reinforced plastic mortar pipe (RPMP), consists of a composite of layers of thermosetting resin–sand aggregate mixtures that are sandwiched by layers of resin-fiberglass reinforcements. In another construction, the sand is replaced by glass microspheres. The high content of reinforcements in RTRP, which may run from 25 to 75 percent of the total pipe weight, and the specific design of the composite wall construction are the major determinants of the ultimate mechanical properties of the pipe. The resin, although also influencing these properties somewhat, is the binder that holds the composite structure together, and it supplies the basic source of temperature and chemical resistance. Glass fibers, as well as many other reinforcements, do not have high resistance to chemical attack. For enhanced chemical and/or abrasion resistance, RTRP construction may include a liner consisting of plastic (thermosetting or thermoplastic), ceramic, or other material. The outer surface of the pipe—especially that of the larger diameter sizes—may also be made "resin rich" to better resist weathering, handling, and spills. Reinforced thermosetting resin pipe is available in a variety of resins, wall constructions, and liners with diameters ranging from 1 in (2.5 cm) to more than 16 ft (5 m). Stock and specially fabricated fittings are readily available.

Filament Winding

Pipe is produced by machine-winding, under controlled tension and in predetermined patterns, of glass reinforcement—which may consist of continuous-filament strands, woven roving, or roving tape—onto the outside of a mandrel. The reinforcement may be saturated with a liquid resin or pre-impregnated with partially cured resin. After the pipe has been formed and cured, the mandrel is removed. The dimension of the inside diameter is set by the mandrel; that of the outside diameter by the thickness of the wall.

A high glass content and a precise machine-controlled fiber orientation which permits control of the ratio of circumferential to axial strength makes this pipe more economically suited for certain uses including higher-pressure applications. Since for a given pressure rating the axial strength of filament-wound pipe tends to be lower than that for pressure-molded pipe, it may have to be offset by more frequent support spacing. The use of automatic machines results in pipes of closer dimensional tolerance and in mechanical properties that are very consistent. To provide a stronger barrier against corrosion of the fiberglass filaments, a resin-rich liner, 0.02 to 0.1 in (0.5 to 2.5 mm) thick, is usually deposited on the inside of the pipe. In one process a thick, highly chemically resistant liner is produced by overwrapping a PVC, CPVC, or PVDF pipe (the only materials used at this time) with resins that can bond tightly to the thermoplastic.

The matched bell-and-spigot socket and the butt-and-strap methods are those primarily used to join filament-wound pipe.

Centrifugal Casting

In this process, resin and reinforcements (such as chopped fibers) are applied to the inside of a rotating mold. After the pipe has cured by action of heat or a catalyst, it is removed from the mold. The outside diameter of the pipe is set by the mold dimensions: the inside diameter is determined by the amount of material introduced into the mold. This type of pipe is almost fully machine-made and it provides very consistent mechanical properties and the closest tolerances. Because its glass fiber content is lower than that of filament-wound pipe it offers generally better chemical resistance. For additional resistance it may be made with a corrosion-resistant liner. The strength properties of centrifugally cast pipe are less direction-dependent than those of filament-wound pipe. Fittings are normally of the socket type for bell and spigot joining. The joint may be overwrapped if added strength is desired.

The smaller-diameter fittings for RTRP are usually produced by a molding (compression or injection) process that is similar to that used with thermoplastics. Fittings may also be made by the filament-winding or contact-molding process whereby the part is first fabricated on a steel or fiberglass form, then cured and removed. Centrifugal casting is also sometimes employed. By all these procedures (except filament winding) random orientation of the reinforcement is obtained.

Resins Used for Piping

Brief descriptions of the more important resins used for piping are given in the following text.

Epoxies

Epoxy resins are strong and have good resistance to solvents, salts, caustics, and dilute acids. Epoxies are cross-linked by curing agents which become an integral part of the polymer and affect the thermal, chemical, and physical properties of the polymer. For instance, the maximum service temperature of epoxy pressure pipe cured with anhydrides is 180°F (83°C), and it has little resistance to caustics; that cured with aromatic amines can be used at temperatures above 225°F (107°C), and it has good caustic resistance.

The major use of epoxy pipe is in oil fields where its resistance to corrosion and paraffin buildup makes it preferable to steel pipe for crude collection and saltwater injection lines. Other uses are in the chemical process industry, in heating and air conditioning, in food processing, for gasoline and solvents, and in mining applications (including abrasive slurry transport, communications ducts, and power conduits).

Polyesters

Although typically not as strong as the epoxies, polyesters offer good resistance to mineral acids, bleaching solutions, and salts. The most commonly used polyester resins for pipe are isophthalic polyesters and bisphenol A fumarate polyesters. Isophthalics have poorer resistance to caustics and oxidizers. Bisphenol A fumarates have improved resistance to these materials and are widely used in paper mills for bleach lines.

Isophthalic resin pipe is used in waste-treatment and power plants in services where corrosive conditions are not severe. Maximum operating-temperature limits for pressure pipe vary, depending upon the specific material, but are generally below 200°F (93°C).

Vinyl Esters

These resins include chemical features of both epoxies and polyesters. Vinyl ester resins offer better chemical resistance, somewhat higher temperature limits, and better solvent resistance than ordinary polyesters but generally do not compare to epoxies in these properties. Vinyl ester resins are preferred over polyesters because they are more chemical-resistant than the isophthalics and less brittle than the bisphenol A fumarates. Typical services are in fertilizer plants (acid lines), chlorine plants (chlorine-saturated brine lines), and paper mills (caustic and black-liquor lines).

Furans

Furan resins offer very good chemical, solvent, and temperature resistance—up to about 300°F (150°C). Because they extend the limitations of the other resins, they are often selected for use in the processing industries in place of exotic metal piping.

Desirable Qualities of FRP Resins

The performance criteria that usually determine the choice of one or more FRP resins over another include chemical resistance, mechanical strength, heat resistance, and (for the manufacturer) processability. A qualitative summary of FRP resin performance is presented in Table 3-1. For the final choice of the appropriate FRP resin(s) and liner combination of a given service (chemical environment, weathering exposure, abrasion resistance, etc.) more detailed information, including case history data, should be obtained.

PROPERTIES

Compared with other materials, pipes made of plastic have less strength and rigidity and are more temperature-sensitive. However, they offer these essential properties sufficiently to satisfy the performance requirements of most industrial piping applications. Moreover, they have good to excellent strength-to-weight ratios; are durable and easy to install and maintain; and have outstanding chemical resistance. The thermoplastics are lower than thermosets in strength, rigidity, and maximum operating temperature. However, their chemical resistance tends to be superior. Thermosets, in contrast, are capable of handling corrosive fluids at pressures and temperatures well beyond the service limits of most thermoplastic pipe.

Physical and Mechanical

Typical physical, mechanical, and thermal properties of the major thermoplastic piping materials are presented in Table 3-2. Those for thermosets are given in Table 3-3. The actual values for any pipe will vary according to the specific material(s) used and, in the case of composite products, such as the thermosets, will also depend on the specific wall construction.

Because the properties of plastics are influenced by time of loading, temperature, and environment, data-sheet values for mechanical properties such as those presented in Tables 3-2 and 3-3 are not satisfactory for design purposes. The stress-strain, and strain-stress, responses of plastics reflect their viscoelastic nature. The viscous, or fluidlike, component tends to damp or slow down the response between strain and stress. For example, if a load is continuously applied on a plastic material, it creates an instantaneous initial deformation that then increases at a decreasing rate. This further deformation response is known as *creep*. If the load is removed at any time, there is a partial immediate recovery of the deformation followed by a gradual creep recovery. If on the other hand the plastic is deformed (i.e., strained) to a given value that is then maintained, the initial load (stress) created by the deformation slowly decreases at a decreasing rate. This is known as the *stress-relaxation response*. The ratio of the actual values

TABLE 3-1 Qualitative Summary of FRP Resin Performance*

Resin	Chemical resistance			Other properties		
	Acids	Bases	Solvents	Processability	Strength	Heat resistance
Polyester resins	Fair–good	Poor	Fair	Good	Fair–good	Fair–good
Isophthalic acid based	——	O	+	+	+	——
Het acid based	——	O	+	+	+	+
BPA/fumarates	+	——	——	+	+	——
Vinyl ester terminated polyesters	——	——	——	+	+	+
Vinyl ester resins	Good	Fair–good	Fair–good	Good	V. good	Good–v. good
BPA/ECH epoxy derived						
$n = 0$	+	——	+	+	++	+
$n = 2$	+	+	——	+	++	——
Phenolic-Novolac epoxy derived	+	——	+	——	+	+
Epoxy resins	Fair	Good	V. good	Fair	V. good	Good–v. good
Aliphatic amine cured	——	——	+	——	+	+
Aromatic amine cured	——	+	++	——	++	++
Anhydride cured	——	O	+	+	+	+
Lewis acid cured	+	+	+	——	+	++
Furan resins	Fair	Good	V. good	Poor	Good	V. good
Furfuryl alcohol derived	——	+	++	O	+	++

*++ = Very Good, + = Good, —— = Fair, and O = Poor.

Source: Based on table from M. B. Launikitis, "Chemically Resistant FRP Resins," *Proceedings of 1977 Plastics Seminar*, National Association of Corrosion Engineers, Houston, Texas.

TABLE 3-2 Typical Physical Properties of Major Thermoplastic Piping Materials*†

	ASTM	ABS		PVC			PE				
	Test no.	I	II	I	II	CPVC	II	III	PB	PP	PVDF
Specific gravity	D 792	1.04	1.08	1.40	1.36	1.54	0.94	0.95	0.92	0.92	1.76
Tensile strength, lb/in² (× 10³)	D 638	4.5	7.0	8.0	7.0	8.0	2.4	3.2	4.2	5.0	7.0
Tensile modulus, lb/in² (× 10⁶)	D 638	0.3	0.3	0.41	0.36	0.42	0.12	0.13	0.06	0.2	0.22
Impact strength, Izod, ft·lb/in notch	D 256	6	4	1	6	1.5	>10	>10	>10	2	3.8
Coeff. of linear expansion, in/(in)(°F)(× 10^{-5})	D 696	5.5	6.0	3.0	5.0	3.5	9.0	9.0	7.2	4.3	7.0
Thermal conductivity, (Btu)(in)/(h)(ft²)(°F)	C 177	1.35	1.35	1.1	1.3	1.0	2.9	3.2	1.5	1.2	1.5
Specific heat, Btu/(lb)(°F)	—	0.32	0.34	0.25	0.23	0.20	0.54	0.55	0.45	0.45	0.29
Approx operating limit‡											
°F, nonpressure	—	180	180	150	130	210	130	160	210	200	300
°F, pressure	—	160	160	140	120	200	120	140	200	150	280

*The properties of a piping material may vary from one commercial material to another. The pipe manufacturer should be consulted for specific properties.

†Consult Table 3-4 for values of long-term strength at 73°F.

‡Exact operating limit may vary from each particular commercial plastic material (consult manufacturer). Effects of environment should also be considered.

TABLE 3-3 Typical Physical Properties of Glass-Fiber-Reinforced Thermosetting Resin Pipe*†

Property at 75°F	Test no.	Polyester	Filament wound epoxy	Filament wound vinyl ester	Filament wound reinforced plastics mortar
Specific gravity	D 792	1.6	1.9	1.9	1.7 to 2.2
Tensile strength, lb/in² (× 10³)	D 2105				
Hoop direction		35	20 to 50	20 to 50	15 to 60
Axial direction		25	2 to 10	2 to 10	3 to 20
Tensile modulus, lb/in² (× 10⁶)	D 2105				
Hoop direction		1.4	1.5 to 4	1.5 to 4	1.4 to 3.6
Axial direction		1.4	0.8 to 2.0	0.8 to 2.0	0.8 to 1.8
Coeff. of linear expansion, in/(in)(°F)(× 10^{-5})	D 696				
Hoop direction		1.3	0.5 to 0.7	0.5 to 0.8	1.0
Axial direction		1.3	1.0 to 1.8	1.0 to 1.8	1.5
Thermal conductivity, (Btu)(in)/(h)(ft²)(°F)	C 177	0.9	1.5 to 4.2	1.5 to 4.2	1.3
Approximate temperature limit‡					
°F, nonpressure		240	300	250	200
°F, pressure		200	240	220	140

*The properties of a piping material may vary from one commercial material to another. The pipe manufacturer should be consulted for specific properties.

†Consult Table 3-4 for values of long-term strength at 73°F.

‡Exact operating limit may vary from each particular commercial plastic material (consult manufacturer). Effects of environment should also be considered.

of stress to strain for a specific time under continuous stressing, or straining, is commonly referred to as the *effective creep modulus,* or the *effective stress-relaxation modulus.* In the case of thermoplastics this effective modulus is significantly influenced by time. For continuous loading of 20 years' duration, it can be from one-quarter to one-third the value of the short-term modulus. In the case of reinforced thermosets the viscous response is of lower order and the long-term effective modulus tends to be at least three-quarters of the short-term values. Most pipe manufacturers are prepared to provide values of effective moduli for specific materials and loading conditions.

The effective strength of plastics is influenced by time, temperature, and environment. For example, the breaking point for a thermoplastic material under short-term tensile testing is reached only after considerable material deformation has taken place, at least 10 percent, and in some cases over 100 percent (the ultimate elongation for reinforced thermosets is lower than for thermoplastics). Under long-term continuous loading material failure (or an unacceptable level of material damage) will occur at much lower deformations than in tensile testing. With thermoplastics the strain levels at failure can be as low as 3 percent, and with some reinforced thermosets they may be below 0.5 percent. Material damage, and not creep or excessive deformation, represents the durability limit for plastics subject to long-term loading. These durability limits, are time-, temperature-, and environment-dependent.

Durability under Continuous Loading

To establish its longer-term hydrostatic strength, plastics pipe is tested in accordance with ASTM D 1598, "Time to Failure of Plastic Pipe Under Constant Internal Pressure." After obtaining a sufficient number of stress vs. time-to-fail points that must span a testing time from 10 to 10,000 h, the data are extrapolated to determine the estimated average 100,000 h strength. The extrapolating procedures are those of ASTM method D 2837, "Obtaining Hydrostatic Design Basis for Thermoplastic Materials," or procedure B of ASTM D 2992, "Obtaining Hydrostatic Design Basis for Reinforced Thermosetting Resin Pipe and Fittings." The extrapolated long-term hydrostatic strength (LTHS) so determined is then rounded off into the appropriate hydrostatic design basis (HDB). The HDBs are design-stress categories represented by a series of preferred numbers (i.e., 2000, 2500, 3150, 4000, 5000, etc.) that ascend in steps of 25 percent. The following relationship, known as the ISO (International Standards Organization) plastics pipe hoop stress equation, is used to relate the test pressure and pipe dimensions to the resultant circumferential stress in the pipe wall:

$$S = \frac{p}{2}\frac{D_m}{t} \tag{1a}$$

where

S = hoop stress, lb/in² (N/m²)
p = internal hydraulic pressure, lb/in² (N/m²)
D_m = mean pipe diameter, in (cm)
t = minimum pipe wall thickness, in (cm)

Once the HDB is established and then reduced to a hydrostatic design stress (HDS) by the application of an appropriate service (or design) factor, the same formula, rewritten as follows, may be used to compute the pipe pressure rating:

$$p = 2\text{HDS}\frac{t}{D_m} \tag{1b}$$

where

HDS = hydrostatic design stress, lb/in² (N/m²)
= HDB × SF
SF = service factor

The selected service factor considers two general groups of conditions: The first encompasses the normal variations in material and pipe manufacture and the second those of the pipe's application and use (environment, hazards, life expectancy). The general practice has been to use service factors of not more than 0.5 for static water pressure at 73°F (23°C). Smaller factors are used for more demanding conditions.

The HDBs for static water pressure at 73°F (23°C) for the major thermoplastic pipe materials are presented in Table 3-4. As indicated by this table, the ASTM material designations for pressure thermoplastics code the material according to its short-term

TABLE 3-4 Hydrostatic Design Basis* (Strength Categories) for Thermoplastic Pipe Compounds Determined with Water at 73.4°F (23°C)

Material designation†	HDB lb/in² (MPa), 73.4°F (23°C)‡
PE 1404	800 (5.5)
PE 2305	1000 (6.9)
PE 2306	1250 (8.6)
PE 3306	1250 (8.6)
PE 3406	1250 (8.6)
PE 3408	1600 (11.2)
PVC 1120	4000 (27.6)
PVC 1220	4000 (27.6)
PVC 2110	2000 (13.8)
PVC 2112	2500 (17.2)
PVC 2116	3150 (21.7)
PVC 2120	4000 (27.6)
CPVC 4120	4000 (27.6)
ABS 1208	1600 (11.2)
ABS 1210	2000 (13.8)
ABS 1316	3150 (21.7)
ABS 2112	2500 (17.2)
CAB MH08	1600 (11.2)
CAB SO04	800 (5.5)
PB 2110	2000 (13.8)
PP 1110	2000 (13.8)
PP 1208	1600 (11.2)
PP 2105	1000 (6.9)

*Per ASTM D 2837.

†The last two digits code the maximum recommended hydrostatic design stress (RHDS), expressed in hundreds of pounds per square inch. RHDS = HDB × 0.5, where 0.5 is the generally accepted maximum value for the design factor. Lower values than 0.5 may be justified by certain operating and safety considerations. The first two digits code the material according to short-term properties.

‡Since thermoplastics, even though of the same ASTM designation, may be affected differently by increasing temperature, HDBs at higher temperatures must be established for each specific commercial product. A number of such products have HDBs for temperatures as high as 180°F (82°C). Consult the most current Plastics Pipe Institute (355 Lexington Ave, New York, NY 10017) Technical Report TR-4 for latest listing of HDBs of commercial pipe compounds.

properties and its maximum RHDS. The RHDS is determined by applying a factor of 0.5 to the material's HDB for water at 73°F (23°C). Values of HDB for other environments and temperatures will be different, even for materials of the same ASTM designation. For example, not all PE 3306s will show the same long-term strength at 120°F (49°C). For other than the standard HDB rating conditions the pipe manufacturer should be consulted for data or recommendations on the specific pipe material.

In the case of thermosets, because of their composite construction, their HDB is determined not only by the properties of the materials used but also by the specific wall construction and manufacturing process. Typical HDBs for some machine-made RTRPs for water service at 73°F (23°C) are given in Table 3-5. The pipe manufacturer should be consulted for HDBs for the specific pipe construction under consideration.

Durability under Cyclic Loading

The higher shorter-term strength of plastics permits them to easily tolerate relatively infrequent short-lived excursions in pressure beyond those established for static pressure conditions. However, under repeated cyclic stressing, as in pressure surging, plastics tend to fatigue and their long-term strength is reduced. The fatigue sensitivity varies from plastic to plastic, and for each material it is dependent on various factors including the amplitude and the frequency of the cyclic stressing. The reduction in limiting strength of thermoplastics may range from slight to as much as one-half the value under static conditions. For services with only moderate and infrequent surging, as is the case in many water piping systems, selecting the pipe pressure rating solely on the basis of the maximum static pressure has been demonstrated to be effective. The American Water Works Association Standard for PVC Water Piping (AWWA C-900) establishes pipe pressure classes with a built-in consideration of cyclic pressure.

TABLE 3-5 Hydrostatic Design Basis for Machine-Made Reinforced Thermosetting Resin Pipe

ASTM pipe specification*	Classification per ASTM D 2310†				HDB, lb/in²‡	
	Type	Grade	Class	Designation	Static	Cyclic
D 2517 (gas pressure pipe)	Filament-wound	Epoxy, glass-fiber-reinforced	No liner	RTRP-11AD RTRP-11AW	— 16,000	5,000 —
D 2996	Filament-wound	Epoxy, glass-fiber-reinforced	No liner	RTRP-11AD RTRP-11AW	— 16,000	5,000 —
	Filament-wound	Epoxy, glass-fiber-reinforced	Reinforced epoxy liner	RTRP-11FE RTRP-11FD	— —	6,300 5,000
	Filament-wound	Polyester, glass-fiber-reinforced	Reinforced polyester liner	RTRP-12EC RTRP-12ED RTRP-12EU	— — 12,500	4,000 5,000 —
	Filament-wound	Polyester, glass-fiber-reinforced	No liner	RTRP-12AD RTRP-12AU	— 12,500	5,000 —
D 2997	Centrifugally cast	Polyester, glass-fiber-reinforced	Polyester liner	RTRP-22BT RTRP-22BU	10,000 12,500	— —
	Centrifugally cast	Epoxy, glass-fiber-reinforced	Epoxy liner	RTRP-21CT RTRP-21CU	10,000 12,500	— —

*See Table 3-8 (p. 4-57) for the abbreviated title of the pipe standard.

†"Standard Classification for Machine-Made Reinforced Thermosetting Resin Pipe," ASTM D 2310. (The first three symbols following RTRP designate type, grade, and class; the last symbol codes the HDB.)

‡Multiply by 6894 to convert to pasculs.

Reinforced thermosetting plastic pipes are somewhat more sensitive to fatigue than thermoplastic pipes. Accordingly, they are often pressure-rated for surge service on the basis of their HDB as determined by Procedure A of the aforementioned ASTM D 2992 from test data obtained in accordance with ASTM D 2143, "Test for Cyclic Pressure Strength of Reinforced Thermosetting Plastic Pipe." Typical HDB values for cyclic pressure service for certain RTRPs are also shown in Table 3-5.

Durability under Continuous Straining

Stress relaxation gradually reduces the initial stress level that is first generated when a plastic material is strained to a given level and then maintained there. As a consequence of this reduction in stress, plastics can tolerate somewhat larger strains under constant-strain (i.e., stress-relaxation) conditions than the ultimate strain that ensues in a constant-load condition (under which there is no stress relaxation). At present there is no unanimously preferred method for establishing limiting strain values for conditions of stress relaxation. Appendix X2 to ASTM D 3262, "Standard Specification for Reinforced Plastic Mortar Sewer Pipe," includes a method, not officially part of the standard, by which one may evaluate the strain limits of the subject pipe in an acid environment. The strain limits for RTRPs tend to be significantly lower than those for thermoplastics piping. In fact, under conditions of continuous straining, limiting strain is not a design constraint for most thermoplastics. When dealing with situations such as in pipe bending or in deflection of pipe under earth loading (in which limiting strain is a possible consideration), recommended design values for maximum permissible strain should be obtained for the specific material and end-use conditions.

Durability under Cyclic Straining

Both ultimate strength and allowable strain may be reduced by fatigue. Guidance should be sought from the pipe manufacturer.

Other Mechanical Properties

In addition to the test for long-term strength, a number of special tests have also been established for evaluating other relevant properties of plastic pipe:

ASTM D 2444	"Impact Resistance of Thermoplastic Pipe and Fittings by Means of a Tup (Falling Weight)"
ASTM D 2105	"Longitudinal Tensile Properties of Reinforced Thermosetting Pipe and Tube"
ASTM D 2925	"Measuring Beam Deflection of Reinforced Thermosetting Pipe Under Full Bore Flow"
ASTM D 2412	"External Loading Properties of Plastic Pipe by Parallel-Plate Loading"
ASTM D 2924	"External Pressure Resistance of Reinforced Thermosetting Resin Pipe"

Flow Properties

Plastic pipes offer minimal resistance to fluid flow. Tests indicate that they may be characterized as hydraulically smooth conduits according to the well-established rational flow formulas. The Hazen-Williams equation, a frequently used engineering approximation formula, yields good correlations for water flow when a C_H factor of 150 to 155 is used. For calculating water flows in open channels, or nonfull flow in conduits, Manning's equation is often used. The Manning n factor for plastics for clean water is approximately 0.0085 to 0.0095. With sewerage, values of n = 0.010 to 0.012 are used since they provide margin for flow-disturbing influences such as sedimentation and slime growth.

Since plastics do not corrode, their original flows do not deteriorate with time. Plastics are also easier to clean than other materials.

Chemical Resistance

Plastics do not corrode in the sense that metals do. Being nonconductors they are immune to galvanic or electrochemical effects. If they are affected by an environment, it is generally through direct chemical attack, solvation, or strain corrosion. In addition to differences in chemical resistance due to the nature of the plastic and pipe wall construction, the extent of resistance may also depend on time and temperature of contact and, in some cases, on the presence of an externally applied stress.

In direct chemical attack, the molecules of either the polymer or the reinforcement are altered, and the alteration leads to a gradual deterioration of properties. Such attack can be brought about by strong oxidizing and reducing agents and by ultraviolet and other radiation. Thermoplastics as a group tend to be inherently more resistant than thermosets to chemical attack. Good protection against ultraviolet radiation (which might be required with the more ultraviolet-sensitive pipes that are to be used in continuous exposure to the weather) is afforded by incorporation into the formulation of an opacifier such as a finely divided carbon black, titanium dioxide, or some other opaque pigment.

Solvation is the absorption of an organic solvent by the plastic. Its effect may range from a slight swelling and softening, with minor effects on properties, to a complete solution. The solvent cementing of ABS and PVC pipe is based on solvation. By the use of selective solvents that evaporate after their task is completed, solvent cementing makes it possible to create a monolithic joint that retains the properties of the base material. Thermosets, because of their cross-linked chemical structure, tend to have superior resistance to solvation.

In strain corrosion, damage will occur only under the combined action of strain (i.e., stress) and environment. In the case of thermoplastics this form of attack is called environmental stress cracking. The mechanism, although not fully understood, is essentially the development and ensuing slow growth and propagation of cracks by the combined action of stress and a sensitizing agent. Stress-cracking agents tend to be materials such as detergents and alcohols that have a surface-wetting tendency. Stress cracking may be controlled by selecting stress-crack-resistant grades of material or by creating designs that ensure that the stress, or strain, is below the threshold value necessary to set this mechanism in action.

In the case of thermosetting materials, the formation of crazes or microcracks exposes the glass fiber, or other reinforcement, to possible chemical attack. To protect against this, a tough resilient liner is used, and the stress, or strain, level is limited to established safe values which will preclude the formation of crazing under the specific exposure conditions.

Table 3-6 presents a broad guide to the chemical resistance of the major thermoplastic piping materials. Table 3-1 includes a general guide to FRP resin performance. Final selection of material and pipe wall construction for chemical or corrosive service should follow consultation with more detailed chemical-resistance information. Because ultimate resistance is affected by stress, time, and temperature, it may not be reliably predicted by shorter-term "soak" tests. Successful previews in similar service are the best guide. Lacking these, new applications would best be evaluated by actual service testing. An advantage of service testing is that it is sure to include some minor (but often overlooked) contaminant which could influence the final result.

In the case of RTRP piping, the liner is considered the first line of defense and therefore a critical factor in corrosive service selection. Special resin liners, which may be deposited in extra thicknesses, are available for extra protection. Pipes with thermoplastic liners are also available.

Inertness to Potable Water and Other Fluids

Nearly all of the base materials from which plastic pipes are made are inert to potable water. It is possible, however, that through the addition of ingredients such as stabilizers,

TABLE 3-6 Thermoplastic Piping Materials: Chemical-Resistance Guide for Ambient Temperatures*

Attacking chemicals	ABS	PVC		CPVC	PE	PB	PP	PVDF
		I	II					
Inorganic compounds								
Acids, dilute	G	G	L	G	G	G	G	G
Acids, concentrated 80%	L	L	L	G	L	L	L	G
Acids, oxidizing	L	P	P	L	P	P	P	G
Alkalies, dilute	G	G	G	G	G	G	G	G
Alkalies, concentrated 80%	L	G	L	G	G	G	G	G
Gases, acid (HCl and HF), dry	L	L	L	L	G	G	G	G
Gases, acid (HCl and HF), wet	L	G	L	G	G	G	G	G
Gases, ammonia, dry	L	G	L	G	G	G	G	G
Gases, halogens, dry	L	L	L	L	L	L	P	G
Gases, sulfur gases, dry	P	G	L	G	G	L	P	G
Salts, acidic	G	G	G	G	G	G	G	G
Salts, basic	G	G	G	G	G	G	G	G
Salts, neutral	G	G	G	G	G	G	G	G
Salts, oxidizing	L	L	L	L	G	G	G	G
Organic compounds								
Acids	G	G	G	L	G	G	G	G
Acid anhydrides	L	L	L	P	L	L	L	L
Alcohols, glycols	L	G	L	G	L†	G	G	G
Esters, ethers, ketones	P	P	P	P	L	L	L	L
Hydrocarbons, aliphatic	L	L	L	G	L	L	L	G
Hydrocarbons, aromatic	P	P	P	L	P	P	P	G
Hydrocarbons, halogenated	L	L	L	L	P	P	P	L
Natural gas (fuel)	G	G	G	G	G	G	G	G
Mineral oil	G†	G	G	G	L†	G	G	G
Oils, animal and vegetable	G†	G	G	G	L†	G	G	G
Synthetic gas (fuel)	L	L	L	L	L	L	L	G

*G, good; P, poor, L, limited knowledge: determination requires precise knowledge of individual conditions.

†Stress-crack-resistant grade should be used.

catalysts, modifiers, and pigments, the final formulation may render a pipe inadequate for potable water in terms of its effect on the toxicological safety and the taste and odor quality of the water. To safeguard against this, most potable water piping standards require that the pipe be evaluated for this purpose by a laboratory recognized by the public health profession. A very commonly used laboratory is the National Sanitation Foundation (NSF), Ann Arbor, Michigan. NSF evaluates, and lists as acceptable, those plastic pipes that satisfy the requirements of their standards for potable water service.

Because they do not contaminate fluids with metallic ions, plastic pipes are often utilized in the transport of pure materials, including de-ionized water. For food service there are pipes available that have been made from materials approved by the Food and Drug Administration.

COMMON METHODS FOR JOINING PLASTIC PIPES

Plastic pipe may be joined by a variety of methods (see Table 3-7), the choice of which is influenced in some cases by the properties of the basic material. For example: the polyolefins (PE, PB, and PP) may not be solvent-cemented; the vinyls (PVC and CPVC) and ABS heat-fuse with relative difficulty; and the thermosets may not be either heat-fused or solvent-cemented. Of the available choices, the one that is selected will depend upon pipe performance, installation, and maintenance requirements as well as availability of fitting and joining equipment. Heat fusion (of PE, PP, PB, and PVDF), solvent-

TABLE 3-7 Techniques for Joining Plastic Pipe

Method of joining	Thermoplastic pipe ABS	PVC	CPVC	PE	PP	PB	PVDF	RTR pipe
Adhesive	—	—	—	—	—	—	—	o
Solvent cements	o	o	o	—	—	—	—	—
Heat fusion	—	—	—	o	o	o	o	—
Threading*	o	o	o	o	o	—	o	o
Flanged connectors†	o	o	o	o	o	o	o	o
Grooved joints‡	o	o	o	o	o	—	o	o
Mechanical compression§¶	o	o	o	o	o	o	o	o
Elastomeric seal¶	o	o	o	o	o	o	o	o
Flaring	—	—	—	o	—	o	—	—
Insert	—	—	—	—	—	o	—	—

*Molded thread adapters are available for attachment on the pipe by another technique. Threads may not be cut in thermosetting pipe. Some thermosetting threaded connections may be adhesive-bonded for extra strength. For threading, the wall thickness of thermoplastic pipe should be not less than Schedule 80.

†Flanged adapters are applied on pipe by heat fusion, solvent-cementing, or threading.

‡Minimum wall thicknesses are prescribed depending on the pipe material.

§With thinner-walled pipe, stiffening inserts must be used.

¶Many designs of elastomeric seal and compression fittings provide no thrust restraint and therefore may be used only in situations, such as buried pipe, in which the pipe is restrained from pullout. Elastomeric seal and compression fittings are available in special designs that incorporate end restraint.

cementing (of ABS, PVC, CPVC), and adhesive joining (of reinforced thermosetting resin pipe), all methods which produce monolithic joints, are preferred for applications requiring maximum strength and optimum chemical resistance (Fig. 3-4). Bell (sometimes referred to as socket) and spigot connections are utilized to join pipe by these three techniques. RTRP may also be joined by the butt-and-strap method whereby two pieces of pipe are butted together and overlays of a laminate are then applied over the butted section and allowed to cure (Fig. 3-5). Larger-diameter polyolefin pipe is joined by the heat fusion of the pipe butt ends.

Flanged connections are often used in industrial applications, particularly when making transitions to other materials, such as when connecting to a metal valve or to a tank outlet, and when it is advantageous to provide for easy removal of a pipe section or other component from the system for cleaning, maintenance, or other purpose.

Threading is also used with plastic pipe. However, molded threads are preferred for thermoplastics and are required for most thermosetting pipe. Molded threaded adapters are available and may be applied by solvent-cementing or with adhesive, whichever is applicable. Threads may not be cut on most thermosetting pipe for they may damage the structural integrity of the pipe wall. Thermoplastic pipe may be threaded, provided its wall thickness is not less than a prescribed minimum, normally at least that of Schedule 80 pipe.

For installations not excluding the use of elastomeric sealants such as neoprene or red rubber, there are mechanical-compression as well as bell and spigot connectors which incorporate such sealants into their design. Much thermoplastic and thermosetting piping specifically made for buried water and sewer lines is available with integral elastomeric-seal bell and spigot connectors. Such connectors greatly facilitate pipe construction, partly because of the ease of making the connection (a stab fit) and partly because the connection may be made under almost any weather or field condition.

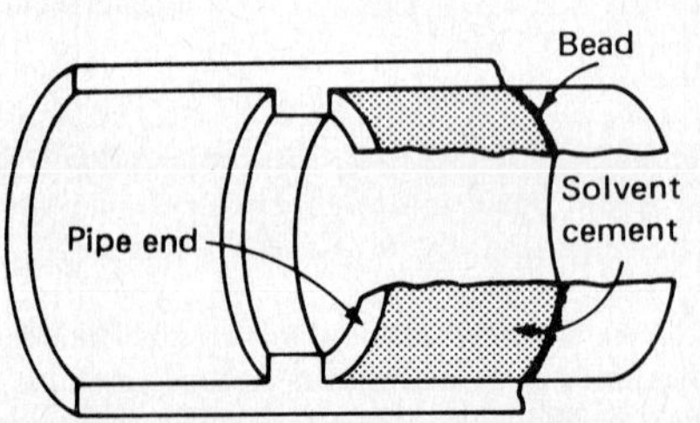

Figure 3-4 Threadless joint in PVC coupling (inside view).

Sometimes connectors utilizing grooved pipe are used with standard grooved-end

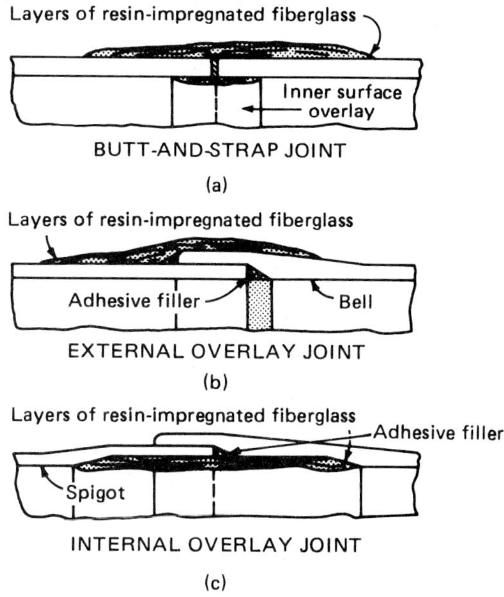

Figure 3-5 Butt and strap joints.

systems such as Victaulic or Gustin-Bacon. With thermoplastic pipe of sufficient wall thickness, the grooves may be cut or rolled in some cases. Cutting is not permitted with thermosetting pipe. Grooved adapters are available for both thermoplastic and thermosetting pipes.

COMMERCIALLY AVAILABLE PRODUCTS

Plastics piping is manufactured in an imposing array of materials, constructions, diameters, wall thicknesses, lengths, and fitting types. The more important products are listed in Table 3-8 which also reports for each product the available size range and its important end uses. Whenever the product is covered by a major standard, the applicable document is identified. Because of the dynamic rate at which new plastic piping standards are currently being written (and older ones revised), the reader is advised to check with the major standards issuing organizations for the most current listing.*

Fittings for larger-diameter pipes are not listed in Table 3-8 because they are often custom-fabricated rather than available from stock. Also not listed in Table 3-8 are piping components such as valves and flanges. In diameters ranging from ⅜ to 4 in, valves made from PVC, CPVC, PP, and PVDF are available in a great variety of different styles including check, ball, diaphragm, globe, gate, and needle. They are available with socket (for solvent-cementing or heat-fusion joining), butt, flanged, or threaded ends. Valves are also available in "Tru" union style and with multiports. A number of models may be obtained with pneumatic or electric actuators for automatic valve positioning. In the ⅜- to 4-in (0.95- to 10-cm) sizes a large array of other piping components, such as strainers, expansion joints, roof and floor drains, and line tapping fittings, is also available. Many

*The Plastics Pipe Institute, a Division of the Society of the Plastics Industry, 355 Lexington Avenue, New York, NY 10017, regularly updates its TR-5, "Standards for Plastics Piping," which lists standards issued by all major United States, Canadian, and international standards organizations (ISO) on both thermoplastic and thermosetting piping.

TABLE 3-8 Principal Commercially Available Plastic Piping Products

				End use																
				Water supply and distribution							Sewer and drain					Industrial			Duct	
Pipe material	Product standard*	Title (abbreviated) of standard or brief product description	Diameter range, in	Mains	Services	Drop pipe, wells	Well casing	Various: industrial, coml.,	Distributing: cold only	Distributing: hot & cold	Collecting system	Building connections	Drainage	Drain, waste, & vent	Natural gas distribution	Corrosives and abrasives	Compressed gases†	Liquid fuels	Conduit (above ground)	Duct (below ground)
ABS, PVC PE & PB	ASTM D 2513	Thermoplastic Gas Pressure Pipe & Fittings	¼–12												XX		X			
ABS, PVC & SR	ASTM F 480	Thermoplastic Water Well Casing	2–12				X													
ABS, PVC & PP	ASTM D 3311	DWV Plastic Fittings Patterns	1¼–6											X						
ABS & PVC	ASTM F 409	ABS & PVC Accessible & Replaceable Tube and Fittings	3–12											X						
ABS	ASTM D 1527	ABS Plastic Pipe, Sch. 40 & 80	⅛–12	X	X	X		X	X			X				X	X			
	D 2282	ABS Plastic Pipe, SDR-PR	⅛–12	X	X	X		X	X			X				X	X			
ABS	D 2465	ABS Plastic Pipe Fittings, threaded Sch. 80	¼–6	X	X	X		X	X			X				X	X			
	D 2468	ABS Plastic Pipe Fittings, socket, Sch. 40	¼–8	X	X	X		X	X			X				X	X			
	D 2469	ABS Plastic Pipe Fittings, socket, Sch. 80	⅛–8	X	X	X		X	X			X				X	X			
	D 2661	ABS DWV Pipe and Fittings	1¼–6										X	X						
	D 2680	ABS Composite Sewer Pipe & Fittings	8–15									X	X							
	D 2750	ABS Utility Conduit & Fittings	1–6																	X
	D 2751	ABS Sewer Pipe and Fittings	1–6									X	X							
	F 628	ABS Foam Core DWV	1¼–6										X	X						

PVC	ASTM																			
	D 1785	PVC Plastic Pipe, Sch. 40–80 & 120	⅛–12	X	X	X		X	X			X				X	X			
	D 2241	PVC Plastic Pipe, SDR-PR	⅛–24	X	X	X		X	X			X				X	X			
	D 2464	PVC Plastic Pipe Fittings, threaded sch. 80	⅛–6	X	X	X		X	X			X				X	X			
	D 2466	PVC Plastic Pipe Fittings, socket, sch. 40	⅛–8	X	X	X		X	X			X				X	X			
	D 2467	PVC Plastic Pipe Fittings, socket, Sch. 80	⅛–8	X	X	X		X	X			X				X	X			
	D 2665	PVC DWV Pipe & Fittings	1¼–6											X						
	D 2672	PVC Plastic Pipe, bell end	⅛–8	X	X	X		X	X			X				X	X			
	D 2729	PVC Drain Pipe & Fittings	2–6										X							
	D 2740	PVC Plastic Tubing	½–1¼		X	X		X												
	D 2949	3-in PVC Thin Wall DWV Piping	3											X						
	D 3033	PVC Sewer Pipe & Fittings, type PSP	4–15								X	X	X							
	D 3034	PVC Sewer Pipe & Fittings, type PSM	4–15								X	X	X							
	D 3036	PVC Line Couplings, socket type	1–8	X	X		X									X	X			
	ASTM F 512	PVC Conduit for Buried Installation	2–6																	
		PVC Sewer Pipe	8–27								X		X							
	AWWA C900	PVC Pressure Pipe for Water	4–12	X				X												
	UL 514	Electrical Outlet Boxes & Fittings	½–6																X	
	UL 651	Rigid Non-metallic Conduit	½–6																X	
	NEMA TC-2	Electrical Plastic Tubing & Conduit	½–6																X	
	TC-3	PVC Fittings for Conduit & Tubing	½–6																X	
PVC & CPVC	API 5LP	PVC and CPVC Line Pipe	½–12					X								X	X	X		
CPVC	ASTM																			
	D 2846	CPVC Hot Water Distribution Systems	⅜–2							X										
	F 437	CPVC Plastic Pipe Fitting, threaded, Sch. 80	¼–6					X		X						X	X	X		

TABLE 3-8 Principal Commercially Available Plastic Piping Products (*Continued*)

Pipe material	Product standard*	Title (abbreviated) of standard or brief product description	Diameter range, in	End use: Water supply and distribution							Sewer and drain					Industrial			Duct	
				Mains	Services	Drop pipe, wells	Well casing	Various: industrial, coml.,	Distributing: cold only	Distributing: hot & cold	Collecting system	Building connections	Drainage	Drain, waste, & vent	Natural gas distribution	Corrosives and abrasives	Compressed gases†	Liquid fuels	Conduit (above ground)	Duct (below ground)
	F 438	CPVC Plastic Pipe Fittings, socket, Sch. 40	¼–6					X		X						X	X	X		
	F 439	CPVC Plastic Pipe Fittings, socket, Sch. 80	¼–6					X		X						X	X	X		
	F 441	CPVC Plastic Pipe, Sch. 40 & 80	¼–12					X		X						X	X	X		
	F 442	CPVC Plastic Pipe, SDR-PR	¼–12					X		X						X	X	X		
	F 443	CPVC Bell End Pipe	⅛–8					X		X						X	X	X		
PE	ASTM																			
	D 2104	PE Plastic Pipe, Sch. 40	½–6	X	X	X		X	X		X	X				X	X	X		
	D 2239	PE Plastic Pipe, SDR-PR	½–6	X	X	X		X	X		X	X				X	X	X		
	D 2447	PE Plastic Pipe, OD-based, Sch. 40 & 80	½–12	X	X	X		X	X		X	X				X	X	X		
	D 2609	Plastic Insert Fittings for PE Pipe	½–4		X	X		X									X			
	D 2683	PE Fittings, socket-fusion type for OD-based pipe		X	X	X		X	X		X	X				X	X	X		
	D 2737	PE Plastic Tubing	½–2		X	X		X	X								X	X		
	D 3261	PE Fittings, butt-fusion type	½–10	X	X	X		X	X		X	X				X	X	X		
	F 405	PE Corrugated Tubing & Fittings	3–8										X							
	F 714	PE Plastic Pipe, SDR-PR, larger diam	3–63	X				X			X	X	X			X	X	X		

	AWWA C901	PE Pipe, tubing & fittings for water	½–3		X															
	‡	PE Pipe Spirally Wound	18–96								X	X	X							
	API 5LE	PE Line Pipe	½–12					X	X							X	X	X		
PB	ASTM-																			
	D 2662	PB Plastic Pipe, SDR-PR	½–6	X	X	X		X		X						X	X	X		
	D 2666	PB Plastic Tubing	½–2		X	X		X		X						X	X	X		
	D 3000	PB Plastic Pipe, SDR-PR, OD-controlled	½–6	X	X	X		X		X						X	X	X		
	D-3309	PB Plastic Hot-Water Distributing Systems	¼–2		X					X										
	‡	PB Plastic Pipe, SDR-PR, larger diam	3–42	X	X						X	X	X			X	X	X		
	AWWA C902	PB Pipe, tubing & fitting for water			X															
PP	‡	PP Plastic Pipe, Sch. 40 and 80	½–6									X	X	X		X	X	X		
	‡	PP Plastic Pipe Fittings, socket, Sch. 80	½–6									X	X	X		X	X	X		
	‡	PP Plastic Pipe Fittings, threaded, Sch. 80	½–6													X	X	X		
	‡	PP Chemical Drainage Pipe & Fittings	1½–4										X	X						
PVDF	‡	PVDF Plastic Pipe, Sch. 80	½–4													X	X	X		
	‡	PVDF Plastic Pipe, SDR-PR	½–4													X	X	X		
	‡	PVDF Plastic Pipe Fittings, socket, Sch. 80	½–4													X	X	X		
	‡	PVDF Plastic Pipe Fittings, threaded, Sch. 80	½–4													X	X	X		
RTRP	ASTM																			
	D 2517	RTR Pipe & Fittings for Gas	2–12												X		X			
	D 2996	Filament Wound RTR Pipe	1–12	X	X	X		X		X						X	X	X		
	‡	Filament-Wound Large Diameter RTR Pipe	12–160					X								X	X	X		
	D 2997	Centrifugally Cast RTR Pipe	1½–12	X	X	X		X	X							X	X	X		
	‡	RTR Pipe and Fittings	12–72					X								X	X	X		
	API 5AR	RTR Pipe Casing & Tubing	1½–10			X	X													
	API 5LR	RTR Line Pipe	1–12													X	X	X		
	‡	RTR Pipe Flanges	1–72	X				X									X			

TABLE 3-8 Principal Commercially Available Plastic Piping Products (*Continued*)

Pipe material	Product standard*	Title (abbreviated) of standard or brief product description	Diameter range, in	End use																
				Water supply and distribution							Sewer and drain					Industrial			Duct	
				Mains	Services	Drop pipe, wells	Well casing	Various: industrial, coml.,	Distributing: cold only	Distributing: hot & cold	Collecting system	Building connections	Drainage	Drain, waste, & vent	Natural gas distribution	Corrosives and abrasives	Compressed gases†	Liquid fuels	Conduit (above ground)	Duct (below ground)
TRP/RPMP	AWWA C950	RTR and RPM Pipe for Water	1–144	X	X			X		X						X	X	X		
RPMP	ASTM																			
	D 3262	RPM Sewer Pipe	8–108								X	X	X							
	D 3517	RPM Pressure Pipe	8–108					X			X					X	X	X		
	D 3754	RPM Sewer and Industrial Pipe	8–144					X			X	X	X			X	X	X		
	D 3840	RPM Non-Pressure Pipe									X	X	X							
		Fittings	8–144																	
	MIL-P-28584	Steam Condensate Lines, RTRP Filament Wound	2–6					X												
	MIL-P-22245	Pipe and Pipe Fittings, glass-fiber-reinforced plastic	2–16	X	X	X	X	X		X						X	X	X		

*Issuing agency identified in discussion on standards.

†Thermoplastic piping is not normally recommended for above-ground service because of safety considerations should the pipe fail. A special grade is available (see ABS piping).

‡Standard currently under development by ASTM.

of these items may be obtained up through about 8-in (20-cm) diam. Larger-sized components are sometimes specially fabricated, or metallic products are used which are connected to the plastic pipe by means of flanges or some other suitable connector.

Standard and special design manholes and access holes are available in both thermoplastic and thermosetting materials for sewerage and drainage applications. ASTM D 3753, "Specification for Glass Fiber Reinforced Manholes," is, to date, the only adopted standard for such products.

PRODUCT STANDARDS, CODES, AND APPROVALS

Standards

The primary source of standards on plastics piping is the American Society for Testing and Materials (ASTM), 1916 Race Street, Philadelphia, PA 19103. (See Table 3-8.) There are a number of other organizations that also develop such standards, generally on products related to their particular interest or activities. These include the following: American Water Works Association (AWWA), 6666 West Quincy Avenue, Denver, CO 80235, which issues standards for water distribution; the American Petroleum Institute (API), 300 Corrigan Tower Building, Dallas, TX 75201, on piping for oil and gas production; the National Electrical Manufacturers Association (NEMA), 2101 L Street, N.W., Washington, DC 20037, on conduit and ducting; and Underwriters Laboratories (UL), 333 Pfingston Road, Northbrook, IL 60062, on conduit and ducting. In addition, the federal government and some state agencies have also issued standards on plastic pipe, generally for projects in which they exercise regulatory or financial functions, or for purchases on their own behalf. Typical of these are the standards issued by the U.S. Department of Agriculture (USDA), the U.S. Department of Defense (DOD), the Federal Housing Administration (FHA), and the General Services Administration (GSA). Most of these documents parallel the basic requirements of ASTM and other listed standards.

The most frequently used dimensioning scheme for setting the outside diameter (OD) of plastic pipe is the traditional iron pipe size (IPS) system of commercial wrought steel pipe (ANSI B36.10). Most thermoplastic pipes are available with standard IPS outside diameters. Some PE and PB pipes which are designed to be joined with insert-type fittings are sized to the same inside diameters as Schedule 40 wrought steel pipe. In the case of thermoset pipes their outside diameters may be exactly, or approximately equal to the reference dimension depending on whether the pipe is sized from the outside in (as in centrifugally casting) or from the inside out (as when filament-winding over a mandrel). Other diameter systems that are utilized include:

- **Copper Tubing Size (CTS).** Based on standard outside diameters of copper tubes. CTS pipe is used for water and gas services and for hot- or cold-water plumbing.
- **Cast-Iron (CI) Pipe Size.** PVC and RTR water main piping is made for this outside-diameter basis and also conforms to the IPS system.
- **International Standards Organization (ISO) Sizes.** Some of the larger polyolefin pipes are made with these internationally set outside diameters.

Much pipe, especially in the larger sizes (for which compatibility with traditionally sized plastic piping components such as valves and fittings is not necessary), is made to fit into special diameter-sizing systems determined by the specific product standard or by the pipe manufacturer. Most of these systems have established diameter dimensions of such proportions that the resultant size of the inside bore is close to the pipe nominal diameter.

The wall thicknesses of most thermoplastic pipe of solid and homogeneous wall construction are defined in accordance with the standard dimension ratio (SDR) concept whereby the ratio of *average* outside diameter to *minimum* wall thickness is a constant value for each SDR pipe series over the entire range of pipe diameters. The standard diameter ratios that have been adopted by ASTM and other standard-writing organizations represent a series of preferred numbers that increase in steps of 25 percent, as follows: 11, 13.5, 17, 21, 26, 32.5, 41, etc. The advantage of establishing wall-thickness categories for thermoplastic pipe according to a constant ratio of diameter to wall thickness is evident from inspection of Eq. (1*b*): within each SDR category the pressure rating is the same for all pipe sizes. There is some pipe made to other than the established SDRs; for this diameter the actual wall thicknes ratio is referred to as diameter ratio (DR).

Thermoplastic pipes specifically intended for industrial uses are often made to the IPS Schedule 40, 80, and 120 system which sets not only the outside diameter but also the wall thickness for each nominal size in each schedule. Since in a given pipe schedule the ratio of diameter to wall thickness tends to decrease with increasing diameter, so too does the pipe pressure rating. Schedule 80 wall thickness, or greater, is required whenever pipe is threaded. The pressure rating of thermoplastic threaded pipe is reduced to half that for unthreaded pipe.

The wall thicknesses of thermoset pipe are not defined in the same way as for thermoplastic pipe because key properties such as strength and stiffness depend not only on material but also on exact wall construction, which can vary not only among manufacturers but even with pipe diameter. The resultant pipe wall thickness will generally be set by the performance requirements of the pipe. Some standards set minimum values for wall thickness. Thermosetting pipe may not be field-threaded. If the pipe is to be joined by threading, it is available with factory-applied molded threads. Threaded adapters are also available.

Codes

The use of piping for plumbing, fire protection, and for the transport of hazardous materials may be subject to the provisions of a code and/or to those of local, state, federal, or other regulations. All the major model plumbing codes which have become adopted, or referenced, by state and local jurisdictions permit and prescribe to a varying but fairly extensive degree the use of plastics piping for hot-cold water lines; water services; drain, waste, and vents (DWV); sewerage; and drainage. Plastics piping is also covered by other codes, such as the following which are of interest to industrial users:

American National Standards Institute Codes

ANSI B31.3-1980	Chemical Plant and Petroleum Refinery Piping
ANSI B31.8-1975	Gas Transmission and Distribution Piping Systems
ANSI Z223.1-1977	National Fuel Gas Code

Department of Transportation, Hazardous Materials Board, Office of Pipeline Safety Operations

Code of Federal Regulations (CFR), Title 49, Part 192, Transportation of Natural Gas and Other Gas by Pipeline: Minimum Federal Safety Standards

Code of Federal Regulations (CFR), Title 49, Part 195, Transportation of Liquids by Pipeline, Minimum Federal Safety Standards.

The National Fire Protection Association (Quincy, Mass.) Model Codes

NFPA 30	Flammable and Combustible Liquids Code
NFPA 54	National Fuel Gas Code
NFPA 70	***National Electrical Code®****
NFPA 70A	Electrical Code for One and Two Family Dwellings
NFPA 34	Outdoor Piping

****National Electrical Code®*** is a Registered Trademark of The National Fire Protection Association, Quincy, MA 02269.

Approvals

Some standards and various jurisdictions and authorities require that before a pipe may be used for certain applications it first must be approved for that use by a recognized, or specifically designated, organization. Organizations with listing and approval programs for plastic pipe include the following:

For Potable Water

The National Sanitation Foundation, NSF Bldg., P.O. Box 1468, Ann Arbor, MI 48105

Canadian Standards Association, 178 Rexdale Boulevard, Rexdale, Ontario, Canada, M9W 1R3

For Drain, Waste, and Vent

The National Sanitation Foundation and Canadian Standards Association (See above.)

For Meat- and Food-Processing Plants

U.S. Department of Agriculture, 14th and Independence S.W., Room 0717 South, Washington, DC 20250

For Underground Fire Protection Systems

Underwriters Laboratories, Inc., 333 Pfingston Road, Northbrook, IL 60062

Factory Mutual Research Corporation, 1151 Boston-Providence Turnpike, P.O. Box 688, Norwood, MA 02062

For Underground Gasoline and Petroleum Lines

Underwriters Laboratories Inc. (See above.)

DESIGN AND INSTALLATION

Standard piping products offered for specific uses such as cold water; hot or cold water; drain, waste, and vent; sewerage; and drainage are largely predesigned. For example, CPVC and PB hot- or cold-water tubing systems made in accordance with their respective standards ASTM D 2846 and D 3309 are pressure-rated at 100 lb/in^2 (690 kPa) for water at 180°F (83°C). Design and installation recommendations are included in an appendix to these documents. Since most standards for products dedicated to a specific application contain design and installation recommendations, such documents should be consulted by designers and installers. The installation of plastic plumbing piping products is often regulated by the applicable plumbing code. More detailed information on design and installation may be obtained from most manufacturers and plastic pipe trade associations (see "Additional Information").

When design is conducted to meet special requirements of a given application, particular attention must be given to the effects of solvents and corrosives on piping properties. Temperature and unusual loadings (such as cyclic pressure and vibration) must also be considered. Most manufacturers of industrial piping products can provide performance data and design and installation recommendations for special as well as ordinary conditions. Various references that provide design and installation information are listed under "Additional Information."

Fundamentally, the principles of the design and installation of plastic piping are the same as those applying to steel piping. However, because of differences in their properties, certain aspects of the design and installation of plastics may require different emphasis and solutions. The following paragraphs briefly discuss these more important aspects. Detailed design and installation recommendations for each specific product should be followed.

Pressure Rating

Continuous Pressure

The pipe pressure rating should be based on the design stress [see Eq. (1*b*)] that is established by taking into account the anticipated effect of time, temperature, and environment on the pipe strength properties. It should be recognized that because of material and fabrication differences, plastic fittings and joints may have a pressure rating lower than that for the pipes. The design stress should include adequate margin for safety considerations. With this in mind it should be recognized that most thermoplastics, excepting those specially formulated for this purpose, are not suitable for conveying compressed gases in above-ground service. In the event of accidental pipe failure, the large potential energy stored in compressed gases could precipitate a catastrophic-type failure mechanism that sometimes produces dangerous flying debris. Thermosets and certain specially formulated thermoplastics resist such failure and are suitable for this application.

Vacuum or External Pressure

The service capabilities of thinner-walled pipes made of less rigid materials may be determined in some cases not by internal pressure but by vacuum conditions created by transients (surges) or by the external pressure loading present inside a large tank. The buckling resistance of plastic pipes may be estimated using Timoshenkos's classic elastic buckling equation including the following adaptation which gives consideration to the effect of pipe ovality:

$$P_c = \frac{2E}{1-\mu^2}\left(\frac{t}{D_m}\right)^3 C \qquad (2)$$

where

P_c = collapsing pressure of unconstrained pipe, lb/in²
E = effective modulus of elasticity of pipe material, lb/in²
μ = Poisson ratio (approximately from 0.35 to 0.45 for thermoplastics for short-term loading)
t = pipe-wall thickness, in
D_m = pipe mean diameter, in
C = factor correcting for pipe ovality = $(r_o/r_i)^3$, where r_i is the major radius of curvature of the ovalized pipe, and r_o is the radius assuming no ovalization

For short-term loading conditions, the values of E and μ as obtained from short-term tensile tests yield reasonable correlations. For long-term loading, appropriate values as determined from long-term loading tests should be employed.

Cyclic Pressure

The shock load or high-pressure surge created by sudden closure of valves could exceed the pressure capabilities of a pipe and, if the pipe is not properly anchored, could result in fitting failure by overstraining of joints. Excessive surging should be eliminated by control of the rate of valve closure or the installation of accumulators. All plastic pipe can tolerate some surging in excess of working pressure, and because of its greater flexibility and viscoelastic properties the pressures generated by surging are of lower order than those for metal piping and are more quickly damped. However, under frequent and continuous surging the long-term strength of plastic pipe tends to be reduced by fatigue. Under these conditions an appropriately lowered value of hydrostatic design stress should be used.

Considerations for Above-Ground Uses

Supports, Anchors, and Guides

Most manufacturers supply information on support spacings. Typical recommendations are presented in Tables 3-9 and 3-10. Values are usually given for either single or con-

TABLE 3-9 Typical Recommended Maximum Support Spacing, in Feet, for Thermoplastic Pipe for Continuous Spans and for Uninsulated Lines Conveying Fluids of Specific Gravity up to 1.35

Pipe dimension		PVC			CPVC				PVDF				PP			
Nominal diam, in	Wall schedule	60°F	100°F	140°F	60°F	100°F	140°F	180°F	80°F	100°F	140°F	160°F	60°F	100°F	140°F	180°F
½	Schedule 40	4½	4	2½	5	4½	4	2½	3¾	3½	2	Continuous support recommended	1¾	1¾	1½	1¼
¾		5	4	2½	5½	5	4	2½	4	3¾	2½		2	2	1¾	1¾
1		5½	4½	2½	6	5½	4½	2½	4¼	4	2½		2	2	2	1¾
1¼		5½	5	3	6	5½	5	3	—	—	—		2½	2¼	2	2
1½		6	5	3	6½	6	5	3	4½	4¼	2½		2½	2½	2¼	2
2		6	5	3	6½	6	5	3	4½	4½	2¾		3	2¾	2½	2¼
3		7	6	3½	8	7	6	3½					3½	2¾	3	2¾
4		7½	6½	4	8½	7½	6½	4					4	3¾	3½	3
6		8½	7½	4½	9½	8½	7½	4½								
8		9	8	4½												
½	Schedule 80	5	4½	2½	5½	5	4½	2½	4½	4½	2½	Continuous support recommended	2	2	2	1½
¾		5½	4½	2½	6	5½	4½	2½	4½	4½	3		2½	2½	2¼	2
1		6	5	3	6½	6	5	3	5	4¾	3		2½	2½	2¼	2
1¼		—	—	—	—	—	—	—	—	—	—		3	2¾	2½	2½
1½		6½	5½	3½	7	6½	5½	3½	5½	5	3		3	3	2¾	2½
2		7	6	3½	7½	7	6	3½	5½	5¼	3		3½	3¼	3	2¾
3		8	7	4	9	8	7	4					4	4	3½	3½
4		9	7½	4½	10	9	7½	4½					4½	4½	4	3½
6		10	9	5	11	10	9	5								
8		11	9½	5½												

TABLE 3-10 Typical Recommended Maximum Support Spacing, in Feet, for Fiberglass-Reinforced Pipe for Temperatures up to 150°F (65°C) for Uninsulated Lines Conveying Fluids of up to 1.25 Specific Gravity

Nominal pipe size, in	Continuous span	Single span
1	7.5	6.3
1½	8.5	7.2
2	9.9	8.4
3	11.2	9.4
4	11.9	10.0
6	14.2	11.9
8	15.7	13.2
10	17.4	14.6
12	18.9	15.9

tinuous spans and for a given liquid specific gravity. A most important consideration is to ensure that the span distance between supports is based on the maximum system temperature. Depending on the piping material, pipe dimensions, and application, the support span may be dictated by the permissible deflection (generally about ½ in (1.3 cm) at midspan) or maximum allowable stress. The allowable stress should provide allowance for stresses generated by fluid pressure and by thermal and other loadings. Vertical pipe runs can be supported either in compression or tension. Long runs should be checked to ensure that the tensile or compressive load does not exceed the permissible design value. Standard strap, sling, clamp, clevis, and saddle supports providing at least 120° of contact are generally recommended (Fig. 3-6). Supports offering narrow or point contact should be avoided. Valves and other heavy piping components should be individually supported.

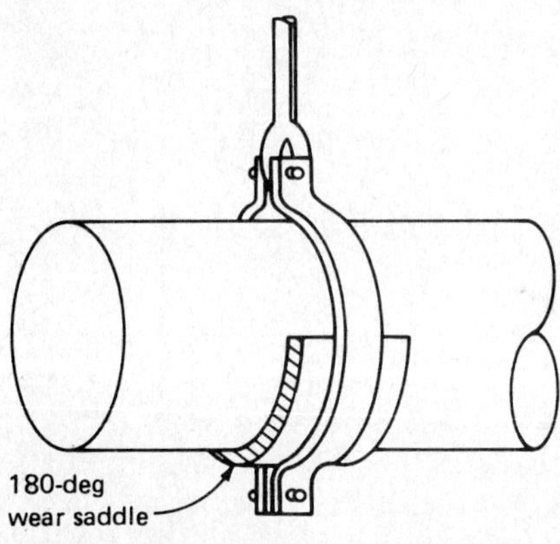

Figure 3-6 Wear-saddle prevents damage from standard hanger.

Anchors, which divide a pipe system into sections, must positively restrain the movement of pipe against all applied and developed forces, particularly dynamic loading. Anchors should generally be employed near changes of direction when transitioning to another piping material or when there is a change in line size. In long, straight runs, anchor spacing is generally recommended at about 200 to 300 ft (60 to 90 m). Anchor spacing and location will also be determined by the selection of anchoring for the control of expansion and contraction.

When the pipe is restrained against expansion and contraction it should be guided to prevent buckling. The guides should encircle the pipe but be loose enough to allow it to move freely in its axial direction.

Control of Expansion and Contraction

Because of the inherent flexibility of plastic piping, it is generally possible to design a pipe system so that no expansion joints are necessary. Their use should be avoided where possible, since they are expensive, and they remove the ability of the pipe to carry longitudinal loads. Offsetting this load with anchors can be an added problem with larger pipes. Preferred techniques for dealing with expansion and contraction are: (1) anchoring and guide spacing, (2) changing direction (offset legs), and (3) using expansion loops.

Although plastics expand more than steel, their relatively low modulus of elasticity results in significantly less end load for the same temperature change. The smaller thermal forces can generally be readily relieved by changes in direction. However, when using directional change to absorb thermal forces, neither the pipe nor any of its components should be subjected to a bending stress (or strain) in excess of that recommended by the manufacturer. The stress may be controlled by placement of anchors not closer than a calculated distance from the point of change of direction. The length of the legs of expansion loops should similarly be determined.

Oscillations and Vibrations

Because plastic pipe is so much more flexible than steel pipe, oscillations due to changes in velocity of fluid set up more easily and tend to be of greater amplitude. In long runs this is generally no problem but they could damage connected piping by subjecting it to excessive stresses or strains. The solution is to restrain the pipe by the use of anchors. High-amplitude vibrations from connected equipment, such as pumps, should be isolated from the piping by the use of flexible connectors.

Considerations for Below-Ground Uses

In terms of their underground performance all plastic pipes are classified as flexible, which signifies that when they are properly installed they are capable of developing sufficient diametrical deformation, without incurring material failure to fully activate soil support forces. Soil-assisted flexible pipes can easily support earth loads that would crush stronger rigid pipes for which total load bearing ability is almost entirely dependent on their own strength. To activate soil support, flexible pipes must be embedded in soils that are stable and that have been properly placed and densified around the pipe. ASTM documents D 2321, "Underground Installation of Flexible Thermoplastic Sewer Pipe," and D 3839, "Underground Installation of Flexible Reinforced Thermosetting Pipe and Reinforced Plastic Mortar Pipe," present detailed recommendations on the proper installation of flexible plastic pipe designed for nonpressure uses. Nonpressure pipe, which is generally of thinner wall construction and not rounded by internal pressure, requires somewhat more care in installation than heavier-wall pressure pipe. ASTM D 2774, "Underground Installation of Thermoplastics Pressure Piping," presents recommendations for solving this problem.

The basic principle of the installation of buried plastic piping is to embed it in a soil of such quality that the resultant ultimate pipe deflection is controlled to an acceptable value that is limited by either the pipe performance requirements or the pipe material capabilities. The former generally permits deflection up to about 10 percent (some engineers may set conservatively lower values) while the latter is determined by the maximum allowable stress, or strain, in the pipe wall for the given pipe material and construction. The pipe supplier can provide the limiting deflection values for a given pipe material and construction. These values may depend on the fluids being handled. With thermoplastic pipe of solid wall construction deflection will seldom be limited by material performance constraints.

The extent to which a flexible pipe will deflect when embedded in a given quality of soil may be estimated by a variety of methods. One of the better-known relationships, sometimes called the Iowa equation, was developed for flexible metal conduits at Iowa State University. A modification of this equation is

$$\frac{\Delta x}{D_i} = \frac{L_D K P}{EI/r^3 + 0.061E'}$$

where

Δx = horizontal deflection of the pipe, in (For relatively small deflections, the change Δy in vertical diameter of a circular section deforming elliptically is equal to 1.10 Δx. As an approximation, it is often assumed $\Delta y = \Delta x$.

D_i = pipe inside diameter prior to loading, in

L_D = deflection lag factor compensating for the time dependence of soil deformation, dimensionless

TABLE 3-11 Bureau of Reclamation Values of E' for Iowa Formula for Initial Average Deflection of Flexible Pipe

Soil type for pipe embedment material per ASTM D 2321	Soil type description (United Classification System, ASTM D 2487)	E', lb/in² for degree of compaction of embedment (proctor density, %)*			
		Dumped	Slight (>85%)	Moderate (85–95%)	High (>95%)
I	Manufactured angular, granular materials (crushed stone or rock, broken coral, cinders, etc.)	1000 (+4%)	3000 (+4%)	3000 (+3%)	3000 (+2%)
II	Coarse-grained soils with little or no fines	N.R.†	1000 (+4%)	2000 (+3%)	3000 (+2%)
III	Coarse-grained soils with fines	N.R.	N.R.	100 (+3%)	2000 (+2%)
IV	Fine-grained soils	N.R.	N.R.	N.R.	N.R.
V	Organic soils (peats, mulches, clays, etc.)	N.R.	N.R.	N.R.	N.R.

*Values in parentheses give the approximate limit of deflection beyond the average deflection that is computed by using the given E' values. These limits are for pipe of relatively low stiffness. As pipe stiffness increases, the limit is narrowed.

†N.R. indicates use not recommended by ASTM D 2321.

K = bedding constant which varies with the angle of bedding (i.e., bedding support), dimensionless (The bedding constant ranges from 0.110 for a point support on the bottom of a pipe to 0.083 for full support. For plastic pipe, the typical value is taken as 0.10.)

P = total vertical pressure acting on the pipe, lb/in²

r = pipe radius, in

E = modulus of elasticity of pipe material, lb/in²

I = moment of inertia of pipe wall per unit of length, in⁴/in (For round pipe $I = t_a^3/12$, in which t_a is the average wall thickness.)

E' = modulus of passive soil resistance, lb/in²

As a result of extensive field investigations of the load vs. deflection characteristics of various flexible pipes the U.S. Bureau of Reclamation* has developed a series of soil reaction E' values for use in the Iowa equation under the assumption that $K = 0.1$ and $D_L = 1.0$. These E' values may be used to estimate a pipe's initial average deflection. To assist in estimating the initial maximum (i.e., acceptance) deflection as a consequence of both soil loading and installation factors, the Bureau of Reclamation has also reported the observed upper limits of deflection values. Both these limits, which primarily apply to pipes of lower ring stiffness, and the values of E' are shown in Table 3-11 as a function of the embedment materials recommended by D 2321.

In actual practice, it is seldom necessary to go through the Iowa equation calculation, for if the recommended installation practices in D 2321 are followed, initial installed deflections can quite readily be held to approximately 5 percent and less. Burial of the thinner-walled, more flexible pipes may also require consideration of the adequacy of the pipe's wall compressive strength as well as its buckling stability (Fig. 3-7).

Recommendations for the design and installation of buried plastic pipe may be obtained from pipe manufacturers', trade associations, or from the listed information given under "Additional Information."

*Amster K. Howard, "Modulus of Soil Reaction Values for Buried Flexible Pipe," *Journal of the Geotechnical Division of the American Society of Civil Engineers*, vol. 103, No GTI, January 1977, pp. 33–43.

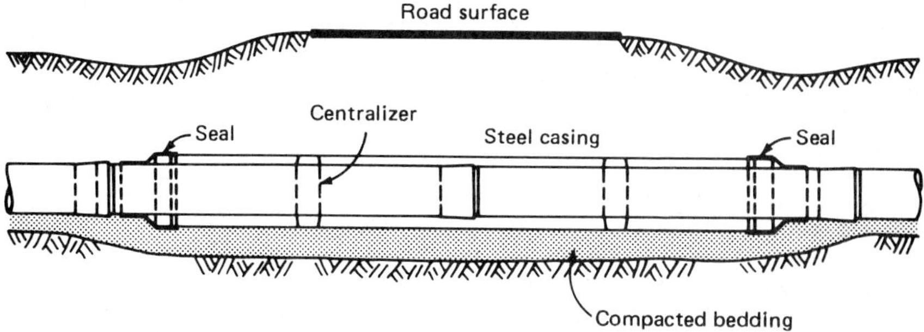

Figure 3-7 Steel casing protects pipe from concentrated loadings.

ADDITIONAL INFORMATION

The various plastic pipe trade associations issue reports, manuals, and lists of references on design and installation of their members' products. A list of current reports may be obtained by contacting each organization as follows:

Reinforced Thermosetting Piping

Reinforced Plastics/Composites Institute, a Division of the Society of the Plastics Industry, 355 Lexington Avenue, New York, NY 10017

The Materials Technology Institute of the Chemical Process Industries, Inc., 1380 Dublin Road, Columbus, OH 43215

Thermoplastic Pipe (Industrial, Gas Distribution, Sewerage, Water, and General Uses)

The Plastics Pipe Institute, a Division of the Society of the Plastics Industry, Inc., 355 Lexington Avenue, New York, NY 10017

Thermoplastics Pipe (Plumbing Applications)

Plastics Pipe & Fittings Association, 999 North Main Street, Glen Ellyn, IL 60137

PVC Piping (Water Distribution, Sewerage, and Irrigation)

Uni-Bell Plastics Pipe Association, 2655 Villa Creek Drive, Suite 164, Dallas, TX 75234

BIBLIOGRAPHY

The following references include useful information on plastics piping:

"PVC Pipe, Design and Installation," AWWA Manual No. M23, American Water Works Association, Denver, 1980.

"Standard for Reinforced Thermosetting Resin Pipe," AWWA C-950, American Water Works Association, Denver. (Appendix to standards includes much design and installation information.)

Britt, William F., Jr.: "Design Considerations for FRP Piping Systems," *Proceedings of the 1979 Conference on Managing Corrosion with Plastics,* National Association of Corrosion Engineers, Houston.

Cheremisinoff, Nicholas P., and N. Paul: *The Fiberglass-Reinforced Plastics Deskbook,* Ann Arbor Science, Ann Arbor, Michigan, 1978

Cooney, J. L.: "Guidelines for the Inspection and Maintenance of FRP Equipment and Piping," *Proceedings of the 1979 Conference on Managing Corrosion with Plastics,* National Association of Corrosion Engineers, Houston.

Escher, G. A.: "Transition to FRP, Basic Guidelines for Piping Designers and Users," *Proceedings of the 1975 Conference on Managing Corrosion with Plastics,* National Association of Corrosion Engineers, Houston.

Escher, G. A., and W. B. MacDonald: "Chemical and Mechanical Properties of Butt and Strap Joints," *Proceedings of the 1975 Conference on Managing Corrosion with Plastics,* National Association of Corrosion Engineers, Houston.

Greenwood: "Buried FRP Pipe—Performance Through Proper Installation," *Proceedings of the 1975 Conference on Managing Corrosion with Plastics,* National Association of Corrosion Engineers, Houston.

Kutschke, C. T.: "Use of Plastic Pipe for Industrial Applications," *Proceedings of the 1975 Conference on Managing Corrosion with Plastics,* National Association of Corrosion Engineers, Houston.

Launikitis, M. B.: "Chemically Resistant FRP Resins," *Proceedings of the 1977 Conference on Managing Corrosion with Plastics,* National Association of Corrosion Engineers, Houston.

Mallison, John H.: *Chemical Plant Design with Reinforced Plastics,* McGraw-Hill, New York, 1969.

Mruk, Stanley: "Thermoplastics Piping: A Review," *Proceedings of the 1979 Conference on Managing Corrosion with Plastics,* National Association of Corrosion Engineers, Houston.

Petroff, Larry J., and Luckenbill, Michael: "Flexibility of the Design of Fiberglass Pipe," *Proceedings of the 1981 International Conference on Plastic Pipe,* American Society of Civil Engineers, New York.

Plastics Piping Manual, The Plastics Pipe Institute, 355 Lexington Avenue, New York, 1976.

Rolston, Albert: "Fiberglass Composite and Fabrication," *Chemical Engineering,* January 28, 1980, pp. 96–110.

Rubens, A. C.: "Designing RTRP Systems Utilizing Published Engineering Data," *Proceedings of the 1979 Conference on Managing Corrosion with Plastics,* National Association of Corrosion Engineers, Houston.

Schrock, B. J.: "Thermosetting Resin Pipe," Preprint 3088, American Society of Civil Engineers, New York, 1977.

Proceedings of the International Conference on Underground Plastic Pipe, March 30–April 1, 1981, New Orleans, La., American Society of Civil Engineers, New York.

chapter 4-4

Asbestos-Cement Pipe and Fittings

by
Leo J. Horvath
Director, Technical Affairs
Association of Asbestos Cement Pipe Producers, Arlington, Virginia

DESCRIPTION AND APPLICATIONS

Asbestos-cement (A/C) pipe, composed of a mixture of portland cement and asbestos fiber, with or without silica, is completely inorganic and free from metallic substances. In the manufacturing process, these ingredients are combined, mixed with water, and formed as pipe on a rotating steel mandrel, creating a dense wall with a smooth interior surface. The pipe is cured in autoclave ovens for dimensional and chemical stability; during the curing time the silica reacts with the free lime, present in normally cured cement products, to form relatively insoluble calcium silicate compounds. These give A/C pipe added resistance to corrosive soils and fluids; as a nonconductor, it is immune to electrolysis.

The resulting A/C pipe is light in weight, resists corrosion, maintains good flow characteristics, and possesses high crush, flexural. and hydrostatic strength.

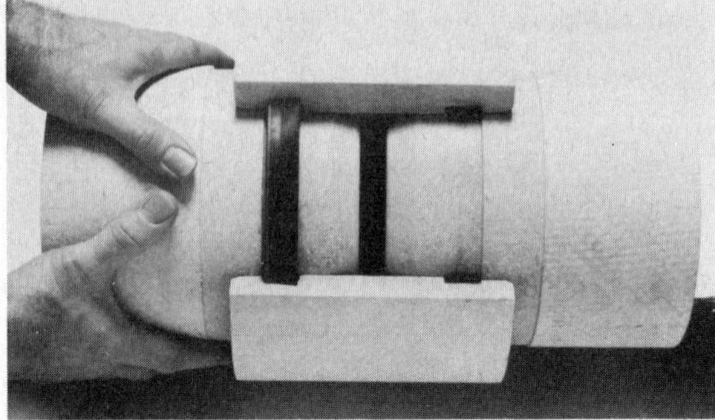

Figure 4-1 Asbestos-cement pipe coupling. *(CertainTeed Corporation.)*

The most common applications for A/C pipe are for water mains, sewage force mains, and gravity sewer systems. Other uses for A/C pipe are storm drains, perforated underdrains, electric and telephone conduits, irrigation systems, air and vent ducts, and building sewers. Because of its high corrosion resistance to many chemicals and freedom from rust and metallic oxides, A/C pipe is also used in many industrial services and, on occasion, for overhead process piping services. In addition, pre-insulated pressure pipe utilizes an outer asbestos-cement casing to contain an inner insulated core capable of handling underground installations of steam, chilled liquids, and high-temperature water.

FITTINGS AND COUPLINGS

Standard asbestos-cement or plastic fittings such as elbows, tees, wyes, adapters, and couplings are available for most sewer-pipe sizes and for building sewer lines. Metal fittings other than the couplings are normally used for pressure-pipe connections. A/C pipe products are designed for quick compatibility with cast-iron and ductile-iron fittings and easy interfacing with accessories made of these materials.

The joining of one A/C pipe to another is achieved easily with a push-together coupling, consisting of an asbestos-cement, plastic, or fiberglass sleeve with solid rubber rings as shown in Fig. 4-1. Compressing the rubber rings between the sleeve and the factory-machined pipe ends provides a tight seal that resists shock, vibration, and earth movement while compensating for the expansion and contraction of the pipe lengths. These joints, when used on pipe under pressure, withstand pressure equal to the rated pipe pressures with adequate safety factors. A coupling is assembled at the factory on one end of each standard section of A/C pipe.

AVAILABLE SIZES AND FORMS

Asbestos-cement pipe is manufactured in standard lengths of either 10 or 13 ft (3 or 4 m) and is also available in half or quarter lengths.

Sewer Pipe

Gravity Sewer Systems

A/C pipe is manufactured in five strength classifications which are designated as Classes 1500, 2400, 3300, 4000, and 5000. The class designation represents minimal crushing strength in pounds per linear foot of pipe, regardless of size. The pipe is manufactured with nominal inside diameters from 4 to 36 in (100 to 900 mm) and in some areas up to

TABLE 4-1 Asbestos-Cement Pressure Pipe, Minimum Crushing Load

Nominal pipe size, in	lb per linear ft Class 100	Class 150	Class 200
4	4,100	5,400	8,700
6	4,000	5,400	9,000
8	4,000	5,500	9,300
10	4,400	7,000	11,000
12	5,200	7,600	11,800
14	5,200	8,600	13,500
16	5,800	9,200	15,400

42 in (1050 mm). Larger diameters and higher strength classifications may also be obtained on special order.

Sewage Force Mains

A/C pipe is manufactured in three classes of operating pressures: 100, 150, and 200. The class designation represents the operating pressure in pounds per square inch. These three classes of pressure pipe, available in sizes from 4 to 16 in (100 to 400 mm), are capable of withstanding the crushing loads shown in Table 4-1.

Water Pipe

Distribution Systems

For distribution systems which have relatively unpredictable flows and many appurtenances, A/C pressure pipe is available in sizes from 4 to 16 in (100 to 400 mm) for operating pressures of 100, 150, and 200 lb/in^2 designated as Classes 100, 150 and 200, respectively. The crushing strengths of these three classes which are also used for pressure sewer systems are given in Table 4-1.

Transmission Systems

For transmission systems which have relatively steady flow and few appurtenances, A/C pipe is available in sizes from 18 to 42 in (450 to 1050 mm). The strength classifications of 30, 35, 40, 45, 50, 60, 70, 80, and 90 represent one-tenth of minimum allowable hydrostatic bursting strength in pounds per square inch for A/C transmission pipe. A/C transmission-piping systems are designed on the combined loading theory in which the design engineer takes into consideration and determines the operating and surge pressures, the earth and superimposed external loads, and a safety factor. The relationship of design internal pressures and external loads for A/C transmission pipe is shown in Table 4-2.

TABLE 4-2 Asbestos-Cement Transmission Pipe, Design Internal Pressure (P, lb/in^2) and Design External Loads (W, lb/linear ft)

Pipe size, in	Strength Classification 30 $P = 300$ W	35 $P = 350$ W	40 $P = 400$ W	45 $P = 450$ W	50 $P = 500$ W	60 $P = 600$ W	70 $P = 700$ W	80 $P = 800$ W	90 $P = 900$ W
18	2,500	3,000	4,000	5,000	6,500	8,500	11,000	14,000	18,000
20	2,500	3,500	4,500	5,500	7,100	9,500	12,000	15,000	20,000
21	2,500	3,500	4,500	5,800	7,300	9,700	12,500	16,000	21,000
24	2,800	3,800	5,000	6,200	8,100	11,000	15,000	19,000	24,000
27	3,500	4,200	5,500	7,000	8,800	12,500	16,500	20,500	27,000
30	3,500	4,500	6,000	7,500	9,700	13,500	18,000	22,500	30,000
33	3,500	5,000	6,500	8,000	10,500	14,500	19,500	24,500	33,000
36	4,000	5,000	7,000	9,000	11,200	16,000	21,000	26,000	36,000
39	4,200	5,300	7,500	9,700	12,000	17,200	22,500	28,000	39,000
42	4,300	5,700	8,000	10,500	13,000	18,500	24,000	30,000	42,000

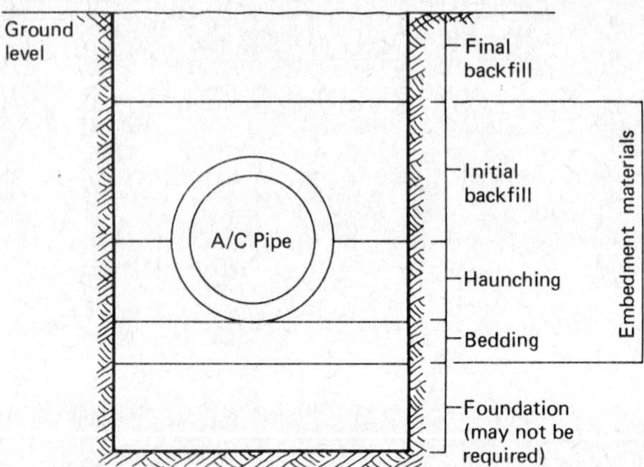

Figure 4-2 Trench cross section and terminology used in embedment and backfilling asbestos-cement pipe.

INSTALLATION AND MAINTENANCE

Asbestos fibers with a tensile strength four to five times greater than that of steel provide significant reinforcement in A/C pipe products to withstand internal hydrostatic pressure, external loads, or combinations of these forces. Even so, good construction procedures applicable for any underground piping systems are equally appropriate for A/C piping systems. Recommended practices for receiving, storage and handling, joint assembly, installation, and inspection and testing are covered in detail in A/C pipe manufacturers' literature, in American Society of Testing and Materials (ASTM) and American Water Works Association (AWWA) standards, and in the plans and specifications of the piping-system's design engineer. Terminology commonly used in a trench installation of A/C pipe is shown in Fig. 4-2. Following are highlights on installation of A/C pipe from these standards and specifications.

Receiving, Storage, and Handling

When receiving A/C pipe, each shipment should be inspected and inventoried. See Fig. 4-3. The pipe is inspected and carefully loaded at the factory using methods acceptable to the carrier whose responsibility it is to deliver the pipe in good condition. It is the responsibility of the receiver to ensure that there has been no loss or damage. Pipes should be stored, if possible, at the work site in the package units provided by the manufacturer. Rubber gaskets should be protected from oil and grease, direct sunlight, excessive exposure to heat, and electric motors which produce ozone. At all times, A/C pipe should be handled with care to avoid damage. Whether moved by hand, skidways, hoists, forklifts, or other handling equipment, A/C pipe should not be thrown, dragged, or bumped.

Trench Excavation

As a general rule, do not open the trench too far ahead of laying the pipe. In preparation for pipe installation, place (string) pipe as near the trench as possible but leave adequate space to perform necessary excavating and other related functions. If the trench is open, place the pipe on the opposite side from the excavated earth. When necessary to prevent caving, trench excavation should be sheeted and braced or sloped in accordance with

Figure 4-3 Asbestos-cement pipe shipment. *(CertainTeed Corporation.)*

applicable laws and ordinances. The trench width at the ground surface should be ample to permit the pipe to be laid and the backfill to be placed and throughly compacted. The trench should be kept free from water at all times. See Fig. 4-4.

Pipe should not be lowered into the trench until the pipe bed has been brought to correct grade. Any part of the trench bottom below grade should be backfilled with thoroughly compacted material. When an unstable subgrade condition is encountered, additional trench depth should be excavated and refilled with suitable, thoroughly compacted foundation material. All rocks, boulders, and large stones should be removed to provide a clearance of 6 to 9 in (150 to 225 mm) below and on all sides of pipe and fittings. When excavation is completed, a bed of sand, crushed stone, or earth that is free from rocks, frozen earth, or clods larger than 1 in (25.4 mm) should be placed and thoroughly compacted to properly regrade the trench bottom. A minimum clearance of 2 in (50 mm) below the coupling should be provided at each joint to permit proper joint assembly. The pipe should be provided continuous support between coupling holes.

Figure 4-4 Asbestos-cement pipe installation. *(CertainTeed Corporation.)*

Joint Assembly

Pipe and accessories should be lowered carefully into the trench by hand or with suitable equipment to avoid damaging the pipe and fittings or injuring the installers. Pipe ends, the coupling interior, and especially the exposed coupling groove and rubber gasket should be thoroughly cleaned prior to assembly. Joint assembly should be performed as recommended by the manufacturer. After alignment, insertion of the rubber gasket, and thorough lubrication of the pipe end as specified by the pipe manufacturer, the assembly of the joint is completed by a sliding action during which the lubricated pipe end slides under the rubber gasket located in the coupling groove and into the coupling to an automatic stop point. Each pipe joint is sealed with a coupling consisting of a sleeve and compressed rubber rings which keep the pipe ends separate, automatically providing for expansion, contraction, and joint flexibility. When pipelaying is not in progress, the open ends of installed pipe should be kept closed to prevent entrance of trench water, dirt, and other foreign matter into the pipeline.

Pipe Embedment and Backfilling

All pipe embedment material should be selected carefully, free from organic debris. The embedment material and its placement and compaction are important considerations in assuring that the pipe will satisfactorily resist trench loading conditions during construction and after the pipeline is completely installed.

Pipe-loading carrying capability is also greatly influenced by the degree of soil compaction around the lower half of the pipe section to provide satisfactory haunching. The initial backfill material should be placed to a minimum depth of 1 ft (30 cm) over the top of the pipe. After placement and compaction of pipe-embedment materials, the final backfill, which is usually placed by machine, need not be as carefully selected as the initial material but should contain no large stones or rocks, frozen soil, or debris.

Maintenance

Pipeline projects should be tested upon completion of installation to assure functional water and sewer systems construction. Before testing, all parts of the pipeline must be backfilled and braced to prevent movement under pressure. Since asbestos-cement will absorb some water, the line must be filled with water for a minimum of 24 h before being subjected to a hydrostatic pressure test. Approved methods of testing asbestos-cement pressure pipe (pressure-strength test and leakage testing) and nonpressure sewer pipe (infiltration, exfiltration, or low-pressure air-loss tests) are described in specifications listed in Table 4-3).

Implementation of proper installation, inspection, and testing procedures should assure maintenance-free, long-term performance of buried asbestos-cement piping systems.

In turn, good work practices complement proper construction procedures. Airborne asbestos fiber has been identified as a possible health hazard. The asbestos fiber in asbestos-cement pipe are not free, but are encapsulated, or locked into, the cement binder. Based on the best scientific data available to date, the proper use of A/C pipe does not pose a health hazard by reason of ingestion of asbestos fibers. Experience has shown,

TABLE 4-3 Specifications

AWWA	
C 400	Asbestos-Cement Distribution Pipe, 4 to 16 in
C 401	Selection of Asbestos-Cement Distribution Pipe, 4 to 16 in
C 402	Asbestos-Cement Transmission Pipe, 18 to 42 in
C 403	Selection of Asbestos-Cement Transmission Pipe, 18 to 42 in
C 603	Installation of Asbestos-Cement Pressure Pipe
ASTM	
C 296	Asbestos-Cement Pressure Pipe
C 668	Asbestos-Cement Transmission Pipe
C 428	Asbestos-Cement Nonpressure Sewer Pipe
C 644	Asbestos-Cement Nonpressure Small Diameter Sewer Pipe
C 663	Asbestos-Cement Storm Drain Pipe
C 508	Asbestos-Cement Underdrain Pipe
C 875	Asbestos-Cement Conduit and Fittings
D 1869	Rubber Rings for Asbestos-Cement Pipe

however, that minimizing exposure to airborne dust is the only effective method of preventing asbestos-related diseases. In 1978, AWWA published Manual No. 16, "Work Practices for Asbestos-Cement Pipe," which, based on results of field testing and study, recommends general guides to achieve safe and clean jobsite conditions when working with A/C pipe products.

STANDARDS AND SPECIFICATIONS

Asbestos-cement pipe is specified by pipe diameter, class or strength, and type of joint. Codes and specifications applicable to A/C pipe are issued by a number of organizations including various federal agencies, fire protection associations, and national and regional plumbing associations. A/C pipe should conform, as appropriate, to the standard specifications (Table 4-3) published by the American Water Works Association (AWWA) and the American Society for Testing and Materials (ASTM).

BIBLIOGRAPHY

American Water Works Association: "Work Practices for Asbestos-Cement Pipe," AWWA No. M16, 1978, Denver, Colorado.

Johns-Manville: "Pressure Pipe Installation Guide," 1979, Denver, Colorado.

ACKNOWLEDGMENT

For the preparation of this chapter, the author has drawn on standards published by the American Water Works Association and the American Society for Testing and Materials, as well as on source material supplied by asbestos-cement pipe manufacturers.

chapter 4-5

Pipe Insulation

by
Michael R. Harrison
Manager, Engineering and Technical Services
Johns-Manville Sales Corp., Denver, Colo.

HEAT-TRANSFER FUNDAMENTALS

Heat energy is transferred from one location to another by three different mechanisms: *conduction, convection,* and *radiation.* In insulation design theory, the objective is to minimize the contribution of each mode in the most efficient and economical manner. As temperatures vary, the relative importance of each transfer mechanism also varies, making different insulation designs appropriate for various applications.

Conduction

Energy transfer by conduction is a result of atomic or molecular motion. As molecules become heated, their vibration increases and energy is transferred to surrounding molecules. Conduction occurs in all three forms of matter: gas, liquid, and solid. Within most insulation, solid conduction is minimized by using an open-pore structure and a minimum amount of solid material. Gas conduction is more difficult to control, but to achieve much greater insulation efficiency, this mode must be limited. One method employs a vacuum, thus eliminating the gas from the system. This is very effective but costly, since the vacuum seal must be maintained in order to assure adequate performance. The second method of controlling gas conduction is to replace the air in the insulation by a heavier gas such as Freon®. Again, the seal must be maintained to avoid eventual air and moisture migration back into the cell structure.

Convection

Convective currents are established when a hot fluid (gas or liquid) rises from a heat source and is replaced by a cooler fluid which in turn is heated and rises, carrying the energy with it. In an insulation structure with many small cells, convection is minimized since the gas cannot freely pass through the structure. Most insulations are of sufficient density and formation to eliminate this mode within them, but convection plays a very important part in transferring energy from the insulation surface to the surrounding environment.

Radiation

As temperature increases, electromagnetic radiation gains in significance with regard to the total amount of energy transferred. Radiation occurs in a vacuum as well as in a gaseous environment, and its magnitude is dependent on the emittance of the radiating and receiving surfaces as well as the temperature difference between them. To control radiant flow, low-emittance surfaces are used in conjunction with absorbers and reflectors within the insulation itself. The mass density of the insulation is very important with a higher density reducing the level of radiation transfer.

There are obviously trade-offs that must be made in controlling the various heat-transfer mechanisms. Figure 5-1 illustrates the contribution to total conductivity of each mechanism at three different temperatures. The most efficient insulation design, both thermally and economically, will vary depending on the application conditions. References 1 and 2 are basic texts on heat transfer for further study.

Heat Flow

The level of heat flow to or from a system is directly proportional to the difference between the system and ambient temperatures and inversely proportional to the thermal resistance placed in the heat flow path:

$$\text{Heat flow} = \frac{\text{temperature difference}}{\text{resistance to heat flow}}$$

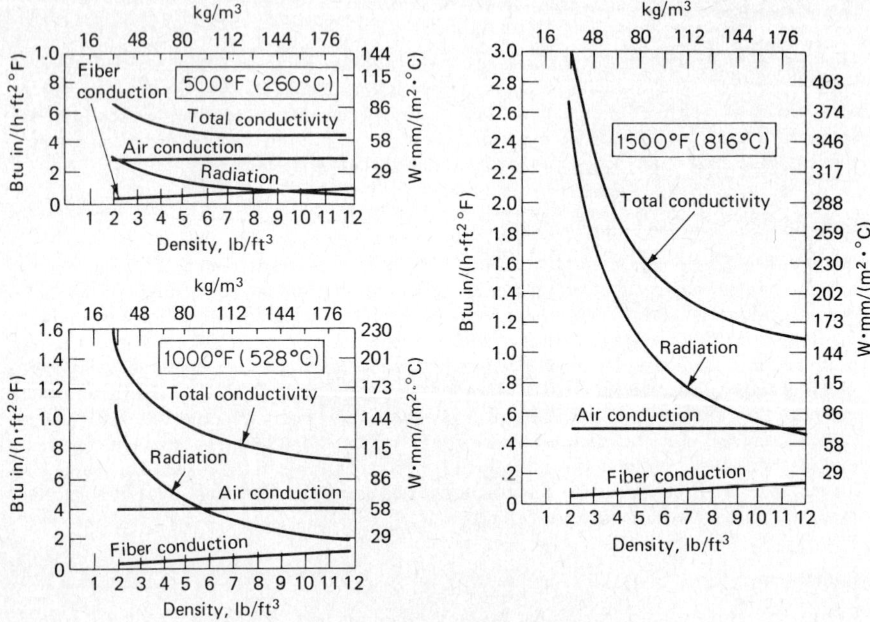

Figure 5-1 Contribution of each mode of heat transfer.[3]

In this light, the temperature difference is the forcing function, and as long as there is a differential, energy will flow. No amount of thermal resistance can completely stop the heat transfer; it can only slow the rate at which it occurs.

Total thermal resistance is generally composed of two distinct types of resistances: insulation and surface. Insulation resistance for a homogeneous material is determined by dividing the insulation thickness tk by its thermal conductivity

$$R_I = tk/k$$

The thermal conductivity, or k value, is an experimentally measured property of the insulation indicating the amount of heat transferred in 1 h through 1 ft^2 of 1-in-thick insulation with a temperature difference of 1°F. The units of k are (Btu)(in)/(h)(ft^2)(°F) [W·mm/m^2·°C]. The accurate measurement of thermal conductivity is very important since materials are often compared on this basis. The American Society for Testing and Materials (ASTM) has developed standardized test methods for measuring thermal conductivity; and for pipe insulation, the standard test is C 335, found in ASTM Part 18.[4] Since thermal conductivity increases with temperature, it is important to use the k value at the insulation mean or average temperature rather than the value at either the operating temperature (too high) or the ambient temperature (too low). For insulations of nonhomogeneous structure, a thermal conductivity based on 1-in (2.5-cm) thickness is not appropriate. Here, the C value on thermal conductance is used to represent the heat transfer through the actual thickness of insulation. Therefore,

$$R_I = 1/C$$

with C expressed in Btu/(h)(ft^2)(°F) [W/(m^2)(°C)].

The other type of thermal resistance is surface resistance

$$R_s = 1/f$$

where f represents the surface film coefficient. Surface emittance, air velocity across the surface, and temperature difference all influence the value of R_s. Table 5-1 lists various

TABLE 5-1 Values for Surface Resistance, R_s, (h)(ft²)(°F)/Btu [m²·°C/W]

A. Values for Still Air

$t_s - t_a$ °F	$t_s - t_a$ °C	Plain, fabric dull metal $\epsilon = 0.95$	Aluminum $\epsilon = 0.2$	Stainless steel $\epsilon = 0.4$
10	5	0.53 (0.093)	0.90 (0.158)	0.81 (0.142)
25	14	0.52 (0.091)	0.88 (0.155)	0.79 (0.139)
50	28	0.50 (0.088)	0.86 (0.151)	0.76 (0.133)
75	42	0.48 (0.084)	0.84 (0.147)	0.75 (0.132)
100	55	0.46 (0.081)	0.80 (0.140)	0.72 (0.126)

B. R_s Values with Wind Velocities

Wind velocity		Plain, fabric dull metal		Aluminum		Stainless steel	
mi/h	km/h	mi/h	km/h	mi/h	km/h	mi/h	km/h
5	8	0.35	0.06	0.41	0.07	0.40	0.07
10	16	0.30	0.05	0.35	0.06	0.34	0.06
20	32	0.24	0.04	0.28	0.05	0.27	0.05

*For heat-loss calculations, the effect of R_s is small compared with R_I, so the accuracy of R_s is not critical. For surface-temperature calculations, R_s is the controlling factor and is, therefore, quite critical. The values presented in Table 5-1 are commonly used values for piping and flat surfaces. More precise values based on surface emittance and wind velocity can be found in the referenced texts.

Source: Johns-Manville, Ref. 5

R_s values for three common surface types, temperature differentials, and wind velocities. These values will be used in subsequent heat-transfer calculations.

Since thermal resistances are additive, they are very convenient to work with in calculating heat transfer. From the basic definition,

$$Q = \frac{\Delta t}{R_I + R_s} = \frac{\Delta t}{tk/k + R_s} \tag{1}$$

becomes the basic equation for heat transfer through insulation. Use of this equation is illustrated later.

Insulation Effectiveness

Before leaving the fundamentals, the importance of insulation should be illustrated. Table 5-2 shows the amount of heat transfer from a bare surface at a given temperature differential. Essentially, these values are calculated from the fundamental heat-transfer equation [Eq. (1)] with the insulation thickness being zero. Listed below are three different sets of operating conditions for an 8-in (20-cm) pipe. To show the effectiveness of insulation, heat losses are shown for the bare pipe and for the pipe with 1 in of fiberglass insulation applied, even though a greater thickness would normally be used.

Operating temp., °F	*Ambient temp., °F*	*Bare heat loss Btu/(h)(ft)*	*Heat loss with 1 in fiberglass, Btu/(h)(ft)*	*Reduction in heat loss, %*
200	80	617	70	88.7
350	80	1882	188	90.1
500	80	3998	353	91.2

It is obvious that insulation should be used in all cases where the energy being lost is costly, useful energy that needs to be conserved.

TABLE 5-2 Heat Loss From Bare Surfaces*

Nominal pipe size, inches	Temperature difference, °F															
	50	100	150	200	250	300	350	400	450	500	550	600	700	800	900	1000
½	22	47	79	117	162	215	279	355	442	541	650	772	1047	1364	1723	2123
¾	27	59	99	147	203	269	349	444	552	677	812	965	1309	1705	2153	2654
1	34	75	124	183	254	336	437	555	691	846	1016	1207	1637	2133	2694	3320
1¼	42	94	157	232	321	425	552	702	873	1070	1285	1527	2071	2697	3406	4198
1½	49	107	179	265	367	487	632	804	1000	1225	1471	1748	2371	3088	3899	4806
2	61	134	224	332	459	608	790	1004	1249	1530	1837	2183	2961	3856	4870	6002
2½	74	162	271	401	556	736	956	1215	1512	1852	2224	2643	3584	4669	5896	7267
3	89	197	330	489	677	897	1164	1480	1841	2256	2708	3219	4365	5685	7180	8849
3½	102	225	377	558	773	1024	1329	1690	2102	2576	3092	3675	4984	6491	8198	10100
4	115	254	424	628	869	1152	1496	1901	2365	2898	3479	4135	5607	7304	9224	11370
4½	128	282	471	698	965	1280	1662	2113	2628	3220	3866	4595	6231	8116	10250	12630
5	142	313	524	776	1074	1424	1848	2350	2923	3582	4300	5111	6931	9027	11400	14050
6	169	373	624	924	1279	1696	2201	2799	3481	4266	5121	6086	8254	10750	13580	16730
7	195	430	719	1064	1473	1952	2534	3222	4007	4910	5894	7006	9501	12380	15630	19260
8	220	486	813	1203	1665	2207	2865	3643	4531	5552	6666	7922	10740	13990	17670	21780
9	246	542	907	1343	1859	2464	3198	4066	5057	6197	7440	8842	11990	15620	19720	24310
10	275	606	1014	1502	2078	2755	3576	4547	5655	6930	8320	9888	13410	17470	22060	27180
11	300	661	1106	1638	2267	3005	3901	4960	6169	7560	9076	10790	14630	19050	24060	29660
12	326	718	1202	1779	2463	3265	4238	5338	6701	8212	9859	11720	15890	20700	26140	32210
14	357	783	1319	1952	2703	3582	4650	5912	7354	9011	10820	12860	17440	22710	28680	35350
16	408	901	1508	2232	3090	4096	5317	6759	8407	10300	12370	14700	19940	25970	32790	40410
18	460	1015	1698	2514	3480	4612	5987	7612	9467	11600	13930	16550	22450	29240	36930	45510
20	510	1127	1885	2790	3862	5120	6646	8449	10510	12880	15460	18380	24920	32460	40990	50520
24	613	1353	2263	3350	4638	6148	7980	10150	12620	15460	18570	22060	29920	38970	49220	60660
30	766	1690	2827	4186	5795	7681	9971	12680	15770	19320	23200	27570	37390	48700	61500	75790
Flat	98	215	360	533	738	978	1270	1614	2008	2460	2954	3510	4760	6200	7830	9650

*Losses given in Btu per hour per linear foot of bare pipe at various temperature differences and Btu per hour per square foot for flat surfaces.
Source: Reference 3.

PIPE INSULATION MATERIALS

Table 5-3 lists the principal insulations being used along with their important properties. To assure accurate information, current data sheets from the manufacturer should always be consulted before specifying a particular product. Table 5-3 provides a brief description of each type of material together with the benefits and drawbacks of each.

Calcium Silicate

Lime, silica, and reinforcing fibers are used to form these rigid insulations. Since no organic binders are used, the products are noncombustible and maintain their physical integrity to very high temperatures. These materials are known for their exceptional durability and strength and are the high-quality standard in industrial plant environments where physical abuse is always a problem. The calcium silicate insulations are more costly than fibrous materials but are more thermally efficient at higher temperatures.

Cellular Glass

This product is also rigid and completely inorganic, being composed of millions of completely sealed glass cells. It is unique among the insulation materials in that it is a totally closed cell and will not absorb any liquids or vapors. Although its thermal conductivity is higher than most of the other insulations', the product is widely used in below-ambient-temperatures and buried applications where moisture is often a problem. Similarly, lines carrying volatile liquids often use cellular glass around valves and fittings to minimize the hazard of a saturated insulation. The product is load-bearing, but it is also brittle and subject to thermal shock at high temperatures.

Expanded Perlite

A naturally occurring material, perlite is expanded at high temperature to form a structure of tiny air cells within a vitrified product. Organic and inorganic binders are used in conjunction with reinforcing fibers to hold the perlite structure together. At temperatures below the organic oxidation point, the products have very low moisture absorption, but this increases at elevated temperatures. The products are rigid and load-bearing, but are more brittle and less thermally efficient than the calcium silicate materials.

Mineral Fiber and Rock Wool

Formed from molten rock or slag, these products are more refractory than fiberglass, but they utilize similar organic binders for structural integrity. The products are not load-bearing and contain varying amounts of unfiberized material. The greatest drawback of the mineral wool materials is the short fiber length which allows vibration and physical abuse to cause severe damage, particularly after the organic binder is oxidized. However, the products are less costly than most of the rigid insulations.

Glass Fiber

Fiberglass pipe insulations are formed from molten glass and bonded with organic resins. The principal products for commercial piping systems are the one-piece molded insulations which install rapidly by hinging open and then closing around the pipe. Flexible wraparound products are also available for large-diameter pipes and vessels. The fiberglass products are very thermally efficient and easy to work with. The only issue surrounding their use is for applications above the binder oxidation temperature of 400 to 500°F. Most manufacturers rate their products above the binder temperature and are confident of its performance due to the fiber matrix composed of long glass fibers. How-

TABLE 5-3 Properties of Various Industrial Pipe Insulation Materials

Insulation type	Temp. range, °F (°C)	Thermal Conductivity (Btu)(in)/(h)/(ft^2)/(°F) at T mean, °F(°C) (W·mm/m^2·°C)			Compresive strength, lb/in^2 (kPa) at % deformation	Classification or flame-spread smoke developed*	Cell structure (permeability and moisture absorption*
		75°F (42°C)	200°F (93°C)	500°F (260°C)			
Calcium silicate	to 1500 (815)	(41.6) 0.37 (52)	(93.3) 0.41 (59)	(260) 0.53 (76)	100 to 250 @ 5% (689 to 1722)	Noncombustible	Open cell
Cellular glass	−450 to 900 (268 to 482)	0.38 (54)	0.45 (65)	0.72 (104)	100 @ 5% (689)	Noncombustible	Closed cell
Expanded perlite	to 1500 (815)	—	0.46 (66)	0.63 (91)	90 @ 5% (620)	Noncombustible	Open cell
Mineral fiber	to 1900 (1038)	0.23 to 0.34 (33 to 49)	0.28 to 0.39 (40 to 56)	0.45 to 0.82 (65 to 118)	1 to 18 @ 10% (6.9 to 124)	Noncombustible to 25/50	Open cell
Glass fiber	to 850 (454)	0.23 (33)	0.30 (43)	0.62 (89)	.02 to 3.5 @ 10% (.13 to 24)	Noncombustible to 25/50	Open cell
Urethane foam	−100 to −450 to 225 (−73 to −268 to 107)	0.16 to 0.18 (23 to 26)	—	—	16 to 75 @ 10% (110 to 516)	25 to 75 140 to 400	95% closed cell
Isocyanurate foam	to 350 (177)	0.15 (22)	—	—	17 to 25 @ 10% (117 to 172)	25 55 to 100	93% closed cell
Phenolic foam	−40 to 250 (−40 to 121)	0.23 (33)	—	—	13 to 22 @ 10% (89 to 151)	25/50	Open cell
Elastomeric closed cell	−40 to 220 (−40 to 104)	0.25 to 0.27 (36 to 39)	—	—	40 @ 10% (275)	25 to 75 115 to 490	Closed cell

*FHC from ASTM E 84 and UL 723 which indexes flame and smoke development to that generated by Red Oak which has 100/100 FHC.

ever, there are certain applications which combine severe vibration with high temperature. In such cases, the fiberglass products, like the mineral wools, may tend to sag on the pipe and lose some of their efficiency.

Foams

Four products in the general foam category are now being used, primarily for cold service and plumbing applications. Polyurethane and isocyanurate foamed plastics offer the lowest thermal conductivity since the cells are filled with fluorocarbon blowing agents which are heavier than air. However, the products still need to be sealed to prevent the migration of air and water vapor back into the cells. There have been problems with urethanes regarding both dimensional stability and fire safety; the isocyanurates were developed, in part, to deal with these problems. However, a 25/50 FHC (fire hazard classification) is still being sought for these materials.

The phenolic foams have the required level of fire safety but do not offer thermal efficiencies much different from that of fiberglass. They are limited in temperature range, but their rigid structure allows them to be used without special pipe saddle-supports on small lines.

Elastomeric closed-cell materials are the fourth type and are used primarily in plumbing and refrigeration work. These products are both flexible and closed-cell, permitting rapid installation without an additional vapor barrier for moderate design conditions. The major problem is smoke generation in excess of the 25/50 FHC rating; this restricts their use in certain areas.

MATERIALS SELECTION AND APPLICATION

There are many factors involved in selecting the best insulation for a specific application. First, the service requirements and location must be analyzed to determine which products will, for example, meet the pipe temperature requirements and also be able to withstand the abuse anticipated. Also, special considerations such as fire protection, removability, and chemical resistance must all be reviewed in the selection process.

Finally, three insulation cost factors should be carefully analyzed. The *initial cost* of material and installation is straightforward; competitive products can be readily compared. *Maintenance costs* are not as clear-cut and vary greatly from one material to another, with the softer materials usually requiring more maintenance in industrial environments. The last cost element is that of the *heat lost* through the insulation system. If the same thickness is specified for several different materials, the product with the lowest thermal conductivity will provide the lowest lost-heat cost. Similarly, a material that requires less maintenance may well retain its original performance longer than a material which is easily damaged and degraded. All of these costs should be reviewed before choosing the lowest bid package alternative which only deals with initial costs.

The following paragraphs deal with insulation selection based on operating temperature, design considerations, and common industry usage.

Cryogenic

Cryogenic insulation systems [−455 to −150°F (−271 to −101°C)] are usually custom-engineered due to the critical nature of such service. The two problems encountered are moisture migration and insulation efficiency. Unequal vapor pressure serves to drive moisture to the cold pipe which results ultimately in ice formation and the subsequent deterioration of the insulation. Multiple vapor barriers are used to prevent this, and cellular glass adds additional security since it is totally closed-cell. The system must also be thermally efficient so as to keep the outside surface temperature above the ambient dew point. This may require outlandish thicknesses of conventional insulations, although the more efficient plastic foams are frequently used for this reason. The most severe applications employ vacuum products with multiple layers of reflective foil or vacuum cavities with powder fill.

Low-Temperature

Refrigeration, plumbing, and HVAC systems operate in the range of −150 to 212°F (−101 to 100°C). Cellular glass, fiberglass, and foam products are the most frequently used products. Vapor-barrier requirements are still important for all applications below ambient temperatures, but the level of protection needed becomes less as ambient conditions are approached. Applications above ambient require little special attention with the exception that the plastic foams begin to reach their temperature limits around 200°F (93°C).

Much of this temperature service is in residential and commercial construction. A variety of fire codes are in force for insulation—ranging from a requirement for noncombustibility to allowing smoke development up to 400. Many codes call for a flame spread of 25 or less, and products that carry a composite 25/50 FHC are suitable for virtually all applications. In general, foam products and cellular glass are used for most of the below-freezing applications while fiberglass is the primary product used for chilled water and above-ambient work. Also, premolded PVC fitting covers with fiberglass inserts are frequently used as a quick and efficient way of insulating the wide variety of fittings encountered.

Intermediate-Temperature

Almost all steam and hot-process piping systems operate in this temperature range: 212 to 1000°F (100 to 538°C). The products most frequently used are calcium silicate, fiberglass, mineral wool, and expanded perlite. Choices are usually based on thermal conductivity and resistance to physical abuse where applicable. The fiberglass products are the most thermally efficient in the lower temperatures, with calcium silicate being the best at higher temperatures. *It is important to use the insulation mean or average temperature when comparing thermal conductivities* in order to make the proper comparison.

The fiberglass pipe insulations reach their temperature limits within this range, usually at 500, 650, or 850°F (260, 343, or 454°C). However, they all have organic binders (as do the mineral wools) that begin to oxidize from 400 to 500°F (204 to 260°C). This should be recognized when applying the products to piping above this temperature, but should not be used as an arbitrary cutoff point for fibrous products unless the service conditions are severe enough to warrant it. Vibration and physical abuse conditions should be analyzed for each particular application.

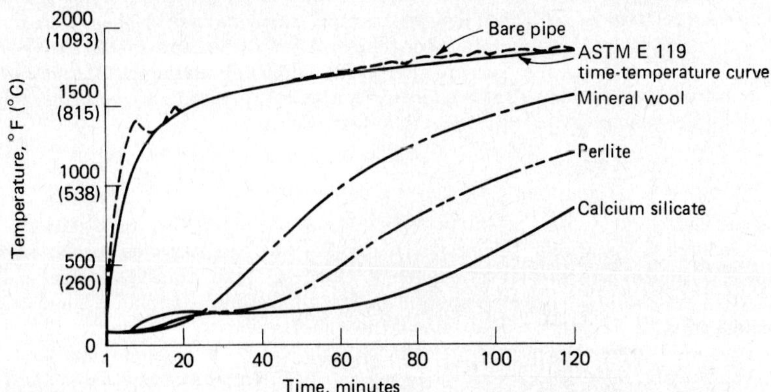

Figure 5-2 Fire-resistance test data for pipe insulation. *(Journal of Thermal Insulation.*[6]*)*

Fire safety is another concern, in this instance from the standpoint of fire protection rather than fire hazard. Figure 5-2 shows the results of fire tests run with three high-temperature materials. The water of hydration in the calcium silicate prolonged the time required for the steel pipe to reach an unacceptable temperature.

Valves and fittings are most often insulated with field-mitered sections of the standard pipe insulation or shop-fabricated fitting covers. There is limited usage of mesh-enclosed blankets which are laced around large valves. Also, cellular glass valve and flange covers are often used in conjunction with other insulations when the fluid being transported is volatile or has a low flash point.

Reviewing the entire range, fiberglass insulation is most often used in lower-temperature applications where physical abuse is not a problem, because of its thermal efficiency and rapid installation. Calcium silicate is the standard type for higher-temperature work and in severe service. Expanded perlite provides the benefit of moisture resistance at lower temperatures, and mineral wool offers initial cost economies for higher-temperature work.

High-Temperature

Superheated steam, exhaust ducting, and some process operations are the few piping applications in the temperature range from 1000 to 1600°F (538 to 871°C). Again, calcium silicate, expanded perlite, and mineral wool are the usual materials employed. These products reach their temperature limits within this range, and the higher temperature requirements are usually met by ceramic fiber blanket materials wrapped around the piping.

Installation Techniques and Specifications

The proper installation of pipe insulation is critical to the in-place performance of the product. References 7 and 8 provide detailed schematics and procedures for such work and should be consulted by plant maintenance workers involved in insulation application. Similarly, the proper selection of protective coatings and jackets will greatly influence the long-term product performance. These references, along with technical data from the coating and jacketing manufacturers, provide appropriate information for proper selection.

INSULATION THICKNESS AND HEAT-LOSS CALCULATIONS

After the material is selected, an appropriate thickness of the material as well as the heat loss or gain with that thickness must be determined. The proper amount of insulation to use depends on the *thermal design objective* of the system: What is the insulation supposed to accomplish? There are four broad categories which encompass most insulation objectives:

1. Personnel protection
2. Condensation control
3. Process control
4. Economics

Terminology

The following symbols, definitions, and units will be used throughout the calculations.

t_a = ambient temperature, °F (°C)

t_s = surface temperature of insulation next to ambient, °F (°C)

t_h = hot-surface temperature, normally the operating temperature (cold-surface temperature in cold applications), °F (°C)
k = thermal conductivity of insulation, always determined at mean temperature, (Btu)(in)/(h)(ft^2)(°F) [W·mm/m^2·°C)]
t_m = $(t_h + t_s)/2$, mean temperature of insulation, °F (°C)
tk = thickness of insulation, in (mm)
r_1 = actual outer radius of steel pipe or tubing, in (mm)
r_2 = $r_1 + tk$, radius to outside of insulation on piping, in (mm)
eq tk = $r_2 \ln (r_2/r_1)$, equivalent thickness of insulation on a pipe, in (mm)
f = surface air film coefficient, Btu/(h)(ft^2)(°F) [W/m^2·°C]
$1/f$ — surface resistance, (h)(ft^2)(°F)/Btu [m^2·°C/W]
R_I = tk/k, thermal resistance of insulation, (h)(ft^2)(°F)/Btu [m^2·°C/W]
Q_F = heat flux through a flat surface, Btu/(h)(ft^2) [W/m^2]
Q_p = $Q_F \times 2\pi r_2/12$, heat flux through a pipe, Btu/(h)(ft) (For SI, Q_p and $Q_F \times 2\pi r_2$, W/m)
Δ = difference by subtraction, unitless
RH = relative humidity, percent
DP = dew-point temperature, °F (°C)

SI Conversions from USCS Units

$$k_{\text{Eng}} \times 144.2279 = k_{\text{SI}}$$
$$R_{\text{Eng}} \times 0.17611 = R_{\text{SI}}$$
$$Q_{F,\ \text{Eng}} \times 3.155 = Q_{F,\ \text{SI}}$$
$$Q_{p,\ \text{Eng}} \times 0.962 = Q_{p,\ \text{SI}}$$

Specific SI units use kelvins rather than degrees Celsius, but in Δ calculations, there is no difference between the two. Degrees Celsius is used here for convenience.

Calculation Fundamentals

There are two concepts that must be understood before proceeding with the calculations. First, in a system in steady-state equilibrium, the heat transfer through each portion of the system is the same. Specifically, the heat transfer through the insulation is equal to the heat transfer from the insulation surface to the surrounding air.

Since the heat loss is proportional to both the temperature difference and thermal resistance of any portion of the system, the following equations result:

$$Q = \frac{t_h - t_a}{R_I + R_s} = \frac{t_h - t_s}{R_I} = \frac{t_s - t_a}{R_s} \tag{2}$$

The second principle relates to the equivalent-thickness concept of pipe insulation. Since the insulation surface area is greater than the bare pipe surface, the pipe surface "sees" a greater insulation thickness than is actually there. This equivalent thickness (*eq tk*) is used as the insulation thickness for the thermal calculation, but the actual thickness is used for calculating the physical number of square feet of surface area per linear foot. The equation for equivalent thickness is

$$eq\ tk = r_2 \ln \frac{r_2}{r_1} \tag{3}$$

where r_1 and r_2 represent the inner and outer radii of the insulation system, respectively. Figure 5-3 provides for easy conversion to equivalent thickness from any specific thickness and pipe size. Some materials are manufactured in true even thicknesses, whereas others follow the ASTM schedule of simplified thicknesses[9] to accommodate proper nesting of double-layer construction. In any event, the proper determination of insulation resistance for pipe insulation is

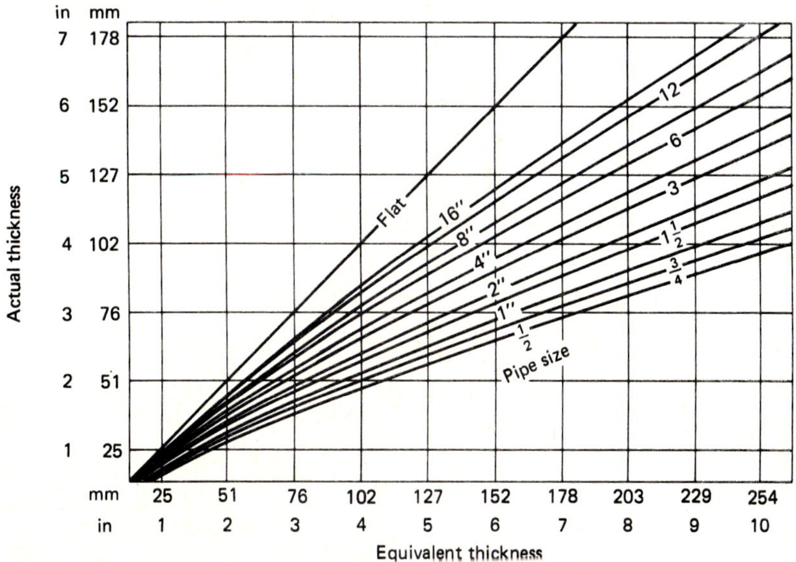

Figure 5-3 Equivalent thickness chart.[5]

$$R_I = \frac{eq\ tk}{k}$$

Figures 5-4 and 5-5 provide typical thermal conductivity values for sample calculations.

Personnel Protection

To protect workers from getting burned on hot piping, the insulation surface temperature should be within the safe touch range of 130 to 150°F; 140°F (60°C) is the temperature most often specified. Both ambient temperature and surface resistance R_s are important in this calculation. As noted in Table 5-1, the R_s values for aluminum are

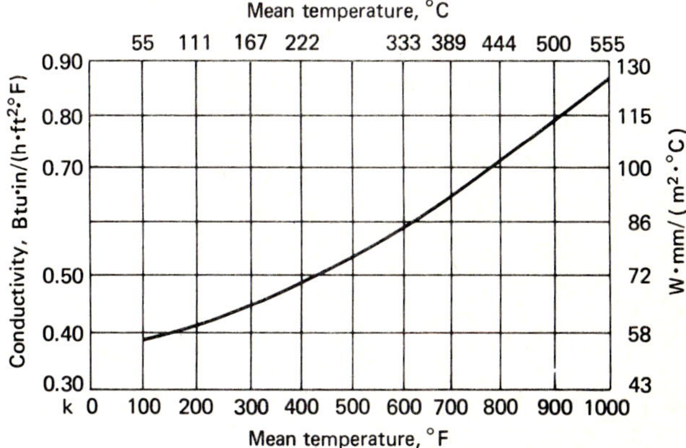

Figure 5-4 Thermal conductivity for typical calcium silicate pipe insulation.

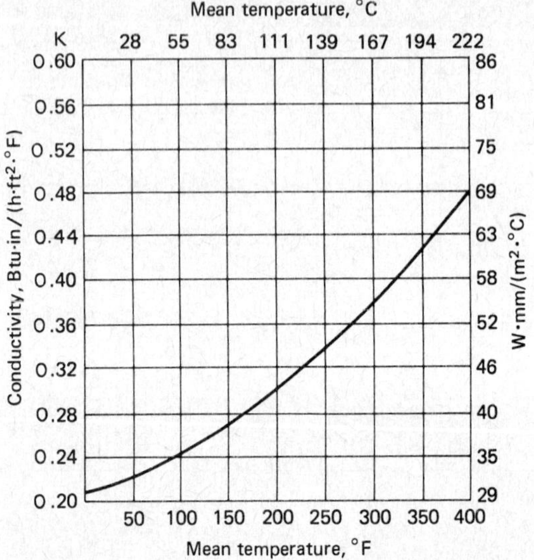

Figure 5-5 Thermal conductivity for typical fiberglass pipe insulation.

greater than for dull surfaces and will result in higher surface temperatures for the aluminum over the same thickness of insulation.

To calculate the thickness required to achieve a specific surface temperature, the basic heat loss equation is manipulated as follows:

$$Q = \frac{t_h - t_s}{R_I} = \frac{t_s - t_a}{R_s}$$

$$R_I = R_s \frac{t_h - t_s}{t_s - t_a} \qquad \text{and} \qquad R_I = \frac{eq\ tk}{k}$$

$$\therefore eq\ tk = kR_s \frac{t_h - t_s}{t_s - t_a}$$

EXAMPLE An 8-in pipe operating at 700°F (371°C) in an 85°F (29°C) ambient with aluminum jacketing. Determine the thickness of calcium silicate required to keep the surface at 140°F (60°C).

Step 1. Determine k at $t_m = (740 + 140)/2 = 420°\text{F}$:

$$k = 0.49\ (\text{Btu})(\text{in})/(\text{h})(\text{ft}^2)(°\text{F}) \qquad [71\ \text{W}\cdot\text{mm}/(\text{m}^2)(°\text{C})]$$

from Fig. 5-4 for calcium silicate.

Step 2. Determine R_s from Table 5-1 for aluminum:

$$t_s - t_a = 140 - 85 = 55°\text{F}$$

So, $R_s = 0.85\ (\text{h})(\text{ft}^2)(°\text{F})/\text{Btu}\ [0.1497\ (\text{m}^2)(°\text{C})/\text{W}]$

Step 3. Calculate

$$eq\ tk = (0.49)(0.85)\frac{700 - 140}{140 - 85}$$

$$= 4.24\ \text{in}$$

$$\text{In SI} = (71)(0.1497)\frac{371.11 - 60}{60 - 29.44}$$

$$= 10.8\ \text{mm}$$

Step 4. Determine from Fig. 5-3 that 4.24 in *eq tk* on an 8-in pipe can be accomplished with 3.25-in (83 mm) insulation. Always rounding off to the next half-inch increment, the recommendation would be 3.5-in (89 mm) calcium silicate.

Condensation Control

On cold systems, the insulation thickness must be sufficient to keep the insulation surface above the dew point of the ambient air. Table 5-4 gives the dew point for various temperature and relative-humidity combinations. The calculation is the same as for personnel protection with the surface temperature taking on the value of the dew-point temperature:

$$eq\ tk = kR_s \frac{(t_h - \mathrm{DP})}{(\mathrm{DP} - t_a)}$$

EXAMPLE A 4-in diam chilled water line is operating at 40°F (4.4°C) in an ambient temperature of 90°F (32.2°C) and 90 percent RH. Determine the amount of fiberglass pipe insulation with a kraft jacket required to prevent condensation.

Step 1. The dew point from Table 5-4 for 90°F, 90 percent RH is DP = 87°F (30.6°C)

Step 2. Determine k at $t_m = (40 + 87)/2 = 63.5$°F (17.5°C); $k = 0.23$ from Fig. 5-5 for fiberglass

Step 3. Determine R_s from Table 5-1 for fabric jacket and $t_a - \mathrm{DP} = 90 - 87 = 3$°F (−16.1°C). ∴ $R_s = 0.54$

Step 4. Calculate $eq\ tk = (0.23)(0.54)\ (40 - 87)/(87 - 90) = 1.95$ in (49.5 mm)

Step 5. Actual thickness required from Fig. 5-3 for a 4-in (101-mm) pipe is 1.5 in (38.1-mm). It would not be unreasonable to specify 2-in (50.8-mm) thickness for this application since 1.5 in is borderline.

Table 5-5 gives the thickness of fiberglass pipe insulation required to prevent condensation for various operating temperatures and three different ambient conditions.

Process Control

There are many complex flow calculations required to determine the amount of insulation required to maintain process temperature or allow a specific temperature drop. The simple calculation below is for determining heat loss only and is basic to all of the process control calculations. The heat loss from a pipe is

$$Q_p = Q_F \frac{2\pi r_2}{12} = \frac{t_h - t_a}{R_I + R_s} \frac{2\pi r_2}{12} \qquad \mathrm{Btu/(h)(ft)} \tag{4}$$

with

$$R_I = \frac{eq\ tk}{k}$$

In this calculation, a surface temperature t_s must first be estimated to arrive at a t_m, k, and R_s. Then, when the t_s estimate is checked, the calculation can be redone with the new estimate if necessary.

EXAMPLE A 16-in diam steam line operating at 850°F (454°C) in an 80°F (27°C) ambient temperature is insulated with 3.5 in (89mm) of calcium silicate with an aluminum jacket. Determine the heat loss per linear foot and the surface temperature.

Step 1. Assume $t_s = 140$°F, $t_m = \dfrac{850 + 140}{2} = 495$°F; k from Fig. 5-4 for calcium silicate is 0.53

Step 2. Determine R_s for aluminum from Table 5-1 for $t_s - t_a = 60$°F; $R_s = 0.85$.

Step 3. Determine *eq tk* from Fig. 5-3 for 3.5 in on a 16-in pipe, *eq tk* = 4.2 in:

$$R_I = \frac{\mathrm{eq}\ tk}{k} = \frac{4.2}{0.53} = 7.92$$

TABLE 5-4 Dew-Point Temperature

Dry-bulb temp, °F	Percent Relative Humidity								
	10	15	20	25	30	35	40	45	50
5	−35	−30	−25	−21	−17	−14	−12	−10	−8
10	−31	−25	−20	−16	−13	−10	−7	−5	−3
15	−28	−21	−16	−12	−8	−5	−3	−1	1
20	−24	−16	−11	−8	−4	−2	2	4	6
25	−20	−15	−8	−4	0	3	6	8	10
30	−15	−9	−3	2	5	8	11	13	15
35	−12	−5	1	5	9	12	15	18	20
40	−7	0	5	9	14	16	19	22	24
45	−4	3	9	13	17	20	23	25	28
50	−1	7	13	17	21	24	27	30	32
55	3	11	16	21	25	28	32	34	37
60	6	14	20	25	29	32	35	39	42
65	10	18	24	28	33	38	40	43	46
70	13	21	28	33	37	41	45	48	50
75	17	25	32	37	42	46	49	52	55
80	20	29	35	41	46	50	54	57	60
85	23	32	40	45	50	54	58	61	64
90	27	36	44	49	54	58	62	66	69
95	30	40	48	54	59	63	67	70	73
100	34	44	52	58	63	68	71	75	78
105	38	48	56	62	67	72	76	79	82
110	41	52	60	66	71	77	80	84	87
115	45	56	64	70	75	80	84	88	91
120	48	60	68	74	79	85	88	92	96
125	52	63	72	78	84	89	93	97	100
−15.0	−37.2	−34.4	−31.7	−29.4	−27.2	−25.6	−24.4	−23.3	−22.2
−12.2	−35.0	−31.7	−28.9	−26.7	−25.0	−23.3	−21.7	−20.6	−19.4
−9.4	−33.3	−29.4	−26.7	−24.4	−22.2	−20.6	−19.4	−18.3	−17.2
−6.7	−31.1	−26.7	−23.9	−22.2	−20.0	−18.9	−16.7	−15.6	−14.4
−3.9	−28.9	−26.1	−22.2	−20.0	−17.8	−16.1	−14.4	−13.3	−12.2
−1.1	−26.1	−22.8	−19.4	−16.7	−15.0	−13.3	−11.7	−10.6	−9.4
1.7	−24.4	−20.6	−17.2	3.0	−12.8	−11.1	−9.4	−7.8	−6.7
4.4	−21.7	−17.8	−15.0	−12.8	−10.0	−8.9	−7.2	−5.6	−4.4
7.2	−20.0	−16.1	−12.8	−10.6	−8.3	−6.7	−5.0	−3.9	−2.2
10.0	−18.3	1.0	−10.6	−8.3	−6.1	−4.4	−2.8	−1.1	0.0
12.8	−16.1	−11.7	−8.9	−6.1	−3.9	−2.2	0.0	1.1	2.8
15.6	−14.4	−10.0	−6.7	−3.9	−1.7	0.0	1.7	3.9	5.6
18.3	−12.2	−7.8	−4.4	−2.2	0.6	3.3	4.4	6.1	7.8
21.1	−10.6	−6.1	−2.2	0.6	2.8	5.0	7.2	8.9	10.0
23.9	−8.3	−3.9	0.0	2.8	5.6	7.8	9.4	11.1	12.8
26.7	−6.7	−1.7	1.7	5.0	7.8	10.0	12.2	13.9	15.6
29.4	−5.0	0.0	4.4	7.2	10.0	12.2	14.4	16.1	17.8
32.2	−2.8	2.2	6.7	9.4	12.2	14.4	16.7	18.9	20.6
35.0	−1.1	4.4	8.9	12.2	15.0	17.2	19.4	21.1	22.8
37.8	1.1	6.7	11.1	14.4	17.2	20.0	21.7	23.9	25.6
40.6	3.3	8.9	13.3	16.7	19.4	22.2	24.4	26.1	27.8
43.3	5.0	11.1	15.6	18.9	21.7	25.0	26.7	28.9	30.6
46.1	7.2	13.3	17.8	21.1	23.9	26.7	28.9	31.1	32.8
48.9	8.9	15.6	20.0	23.3	26.1	29.4	31.1	33.3	35.6
51.7	11.1	17.2	22.2	25.6	28.9	31.7	33.9	36.1	37.8

Percent relative humidity									
55	60	65	70	75	80	85	90	95	100
−6	−5	−4	−2	−1	1	2	3	4	5
−2	0	2	3	4	5	7	8	9	10
3	5	6	8	9	10	12	13	14	15
8	10	11	13	14	15	16	18	19	20
12	15	16	18	19	20	21	23	24	25
17	20	22	23	24	25	27	28	29	30
22	24	26	27	28	30	32	33	34	35
26	28	29	31	33	35	36	38	39	40
30	32	34	36	38	39	41	43	44	45
34	37	39	41	42	44	45	47	49	50
39	41	43	45	47	49	50	52	53	55
44	46	48	50	52	54	55	57	59	60
49	51	53	55	57	59	60	62	63	65
53	55	57	60	62	64	65	67	68	70
57	60	62	64	66	69	70	72	74	75
62	65	67	69	72	74	75	77	78	80
67	69	72	74	76	78	80	82	83	85
72	74	77	79	81	83	85	87	89	90
76	79	82	84	86	88	90	91	93	95
81	84	86	88	91	92	94	96	98	100
85	88	90	93	95	97	99	101	103	105
90	92	95	98	100	102	104	106	108	110
94	97	100	102	105	107	109	111	113	115
99	102	105	107	109	112	114	116	118	120
104	107	109	111	114	117	119	121	123	125
−21.1	−20.6	−20.0	−18.9	−18.3	−17.2	−16.7	−16.1	−15.6	−15.0
−18.9	−17.8	−16.7	−16.1	−15.6	−15.0	−13.9	−13.3	−12.8	−12.2
−16.1	−15.0	−14.4	−13.3	−12.8	−12.2	−11.1	−10.6	−10.0	−9.4
−13.3	−12.2	−11.7	−10.6	−10.0	−9.4	−8.9	−7.8	−7.2	−6.7
−11.1	−9.4	−8.9	−7.8	−7.2	−6.7	−6.1	−5.0	−4.4	−3.9
−8.3	−6.7	−5.6	−5.0	−4.4	−3.9	−2.8	−2.2	−1.7	−1.1
−5.6	−4.4	−3.3	−2.8	−2.2	−1.1	0.0	0.6	1.1	1.7
−3.3	−2.2	−1.7	−0.6	0.6	1.7	2.2	3.3	3.9	4.4
−1.1	0.0	1.1	2.2	3.3	3.9	5.0	6.1	6.7	7.2
1.1	2.8	3.9	5.0	5.6	6.7	7.2	8.3	9.4	10.0
3.9	5.0	6.1	7.2	8.3	9.4	10.0	11.1	11.7	12.8
6.7	7.8	8.9	10.0	11.1	12.2	12.8	13.9	15.0	15.6
9.4	10.6	11.7	12.8	13.9	15.0	15.6	16.7	17.2	18.3
11.7	12.8	13.9	15.6	16.7	17.8	18.3	19.4	20.0	21.1
13.9	15.6	16.7	17.8	18.9	20.6	21.1	22.2	23.3	23.9
16.7	18.3	19.4	20.6	22.2	23.3	23.9	25.0	25.6	26.7
19.4	20.6	22.2	23.3	24.4	25.6	26.7	27.8	28.3	29.4
22.2	23.3	25.0	26.1	27.2	28.3	29.4	30.6	31.7	32.2
24.4	26.1	27.8	28.9	30.0	31.1	32.2	32.8	33.9	35.0
27.2	28.9	30.0	31.1	32.8	33.3	34.4	35.6	36.7	37.8
29.4	31.1	32.2	33.9	35.0	36.1	37.2	38.3	39.4	40.6
32.2	33.3	35.0	36.7	37.8	38.9	40.0	41.1	42.2	43.3
34.4	36.1	37.8	38.9	40.6	41.7	42.8	43.9	45.0	46.1
37.2	38.9	40.6	41.7	42.8	44.4	45.6	46.7	47.8	48.9
40.0	41.7	42.8	43.9	45.6	47.2	48.3	49.4	50.6	51.7

TABLE 5-5 Minimum Insulation Thickness Of Fiberglass Pipe Insulation Needed To Prevent Condensation (Based on Still Air and AP Jacket)

Operating pipe temperature	80°F (26.6°C) & 90% RH			80°F (26.6°C) & 70% RH			80°F (26.6°C) & 50% RH		
	Pipe size, in	Thickness in	Thickness mm	Pipe size, in	Thickness in	Thickness mm	Pipe size, in	Thickness in	Thickness mm
0–34°F (−18–1°C)	Up to 1	2	51	Up to 8	1	25	Up to 8	½	13
	1¼ to 2	2½	63	9 to 30	1½	38	9 to 30	1	25
	2½ to 8	3	76						
	9 to 30	3½	89						
35–49°F (1–9°C)	Up to 1½	1½	38	Up to 4	½	13	Up to 30	½	13
	2 to 8	2	51	4½ to 30	1	25			
	9 to 30	3	76						
50–70°F (10–21°C)	Up to 3	1½	38	Up to 30	½	13	Up to 30	½	13
	3½ to 20	2	51						
	21 to 30	2½	63						

Source: Johns-Manville, Ref 5.

Step 4. Calculate heat loss per square foot:

$$Q_F = \frac{850 - 80}{7.92 + 0.85} = 87.8 \text{ Btu/(h)(ft}^2) \qquad [(227 \text{ W/m}^2)]$$

Step 5. Check surface temperature assumption by

$$t_s = t_a + (R_s \times Q_F) = 80 + (0.85 \times 87.8)$$
$$= 155°\text{F}$$

This leads to a new t_m = 502.5 which is close enough to assumed t_m = 495 that k will not change.

Step 6. Calculate heat loss per linear foot:

$$[\text{nt}Q_p = Q_F \frac{2\pi r_2}{12} = 87.8 \frac{2\pi(8 + 3.5)}{12}$$
$$= 529 \text{ Btu/(h)ft} \qquad (508 \text{ W/m})$$

Freeze Protection

Water lines must either be heat-traced or have a controlled flow to prevent freezing. These calculations can be performed for various combinations of water temperature, line length, and ambient conditions. Table 5-6 is presented as an estimating tool for freeze protection. It shows the number of hours until freezing occurs along with the gallons per minute flow required in a 100-ft (30.3 m) pipe run to prevent freezing. The calculations are based on fiberglass pipe insulation with k = 0.23, initial water temperature of 42°F (6°C), and an ambient air temperature of −10°F (−23°C). The flow rate for the 100-ft (30.3-m) length may be prorated for longer or shorter pipes.

Economics

With energy costs on the rise, insulating for the purpose of energy conservation and economics is becoming the standard for most insulation-thickness calculations. The first economic thickness equations were presented in a 1926 paper by L. B. McMillan.[8] Since then, many nomographs and computer programs have been developed to perform these complex calculations.[10] Discounted cash-flow techniques, fuel escalation, maintenance costs, installed insulation costs, and tax effects are all involved in the computations. However, the economic thickness of insulation (ETI) concept is still quite simple. Economic thickness is defined as that thickness of insulation at which the cost of the next

TABLE 5-6 Freeze Protection Data, Hours to Freeze and Flow Rate (gal/min) Required to Prevent Freezing*

Nominal pipe size IPS	Insulation thickness: 1 in, Hours to freeze	1 in, Flow rate, gal/min per 100 ft	2 in, Hours to freeze	2 in, Flow rate, gal/min per 100 ft	3 in, Hours to freeze	3 in, Flow rate, gal/min per 100 ft
½	0.30	0.087	0.42	0.282	0.50	0.053
¾	0.47	0.098	0.66	0.070	0.79	0.058
1	0.66	0.113	0.96	0.078	1.16	0.065
1½	0.90	0.144	1.35	0.096	1.67	0.078
2	1.72	0.169	2.64	0.110	3.31	0.088
2½	2.13	0.195	3.33	0.124	4.24	0.098
3	2.81	0.228	4.50	0.142	5.80	0.110
4	3.95	0.279	6.49	0.170	8.49	0.130
5	5.21	0.332	8.69	0.199	11.54	0.150
6	6.48	0.386	10.98	0.228	14.71	0.170
7	7.66	0.437	13.14	0.255	17.75	0.189
8	8.89	0.487	15.37	0.282	20.89	0.207

*Based upon 42°F (6°C) initial water temperature and −10°F (−23°C) ambient temperature.
Source: Johns-Manville, Ref. 5.

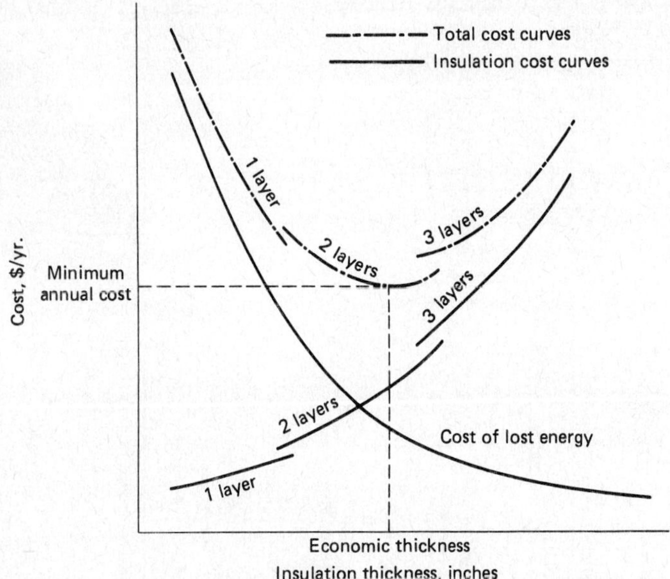

Figure 5-6 Economic thickness of insulation (ETI).

increment is just offset by the energy savings due to that increment over the life of the project.

Figure 5-6 graphically illustrates the ETI procedure which adds the installed insulation costs to the fuel costs to arrive at the point of minimum annual cost.

The latest version of the ETI program is sponsored by the Thermal Insulation Manufacturers Association (TIMA) and the National Insulation Contractors Association (NICA). The computer program is available from TIMA[12] and several insulation manufacturers offer to run the program for clients. Figure 5-7 shows the ETI program output. The first sections simply reproduce the input data used in the calculations. The output data begin with the seven columns of information. Column 1 is the insulation thickness which produces the other results. Column 2 is the annual cost associated with each thickness; the minimum value here corresponds with the economic thickness. Column 3 is the discounted payback period for the investment. Column 4 is the discounted present value of the energy saved per foot over the life of the project. Columns 5 and 6 are the heat loss and surface temperature, respectively, for each thickness. Column 7 is the cost of the installed insulation per linear foot.

The ETI program is quite sophisticated and allows for almost all the variables encountered in insulation design. Many corporate staff and consulting engineers have developed similar programs particularly suited to their specific needs. This is a very positive trend and establishes thermal insulation in its proper role, that of being a primary method of reducing energy consumption and improving overall plant efficiencies.

Economic Thickness of Insulation for Hot Surfaces (ETIHOT)
New and Retrofit Jobs

Project :	Project no:
System :	Location :
Date :	Engineer :
Contact:	

Fuel type: Steam
Present price: 6.00 $/MMBTU
Efficiency: 100.00%

Annual fuel inflation rate	=	10.0%
After-tax value of money	=	10.0%
Incremental heating equipment investment	=	2.00 $/1000 Btu/h
Effective income tax rate	=	46.0%
Heating-plant depreciation period	=	20.0 yr
Economic life of new insulation	=	20.0 yr
Emissivity of jacketing	=	0.90
Percent of new insulation cost for annual insulation maintenance	=	1.0%/yr
Percent of annual fuel bill for heating plant maintenance	=	5.0%/yr
Annual average wind speed	=	0.00 mi/h
Emissivity of present surface	=	0.80

Calculations are performed using linear cost and thickness data

Pipe description		=	8.00-in IPS-horizontal pipe
Operating temperature		=	250.0°F
Ambient temperature		=	80.0°F
Previous Insulation	Code 0	=	None
New insulation	Code 1	=	Fiberglass
Previous thickness of insulation		=	0.0 in
Hours of operation		=	5400.0 per yr
Reference thickness of insulation for payback calculation		=	0.00 in
Piping complexity factor		=	1.00
Conductivities supplied by user	New	=	None supplied
	Old	=	None supplied

Thick. of new insul., in	Annual cost, $/ft	Payback period, yr	Pres. value of heat saved, $/ft	Heat loss, Btu/(h)(ft)	Surf. temp., °F	Insul. cost, $/ft
0.00	35.99	——	0.00	830.40	249.6	0.00
1.50	4.22	0.61	28.03	72.60	96.6	10.04
2.00	3.81	0.74	28.57	58.00	92.7	12.09
2.50	3.64	0.87	28.90	49.09	90.1	14.15
3.00 S.L.	3.60	1.00	29.13	43.05	88.4	16.20
3.00 D.L.	3.83	1.14	29.13	42.92	88.4	18.34

ETI corresponds to the thickness with the lowest annual cost, 3.00 in.

Figure 5-7 TIMA ETI computer program output. Heat savings and present values are projected based on given input variables. Actual values may vary depending on the actual service conditions.

REFERENCES

1. McAdams, W. H.: *Heat Transmission*, McGraw-Hill, New York, 1954.
2. Sparrow, E. M., and R. D. Cess: *Radiation Heat Transfer*, McGraw-Hill, New York 1978.
3. Greebler, P.: "Thermal Properties and Applications of High Temperature Aircraft Insulation," American Rocket Society, 1954. Reprinted in *Jet Propulsion*, November-December 1954.
4. American Society for Testing and Materials: *Annual Book of ASTM Standards*, part 18, Thermal and Cryogenic Insulating Materials Building Seals and Sealants; Fire Tests; Building Constructions; Environmental Acoustics.
5. Johns-Manville Sales Corporation, Industrial Products Division, Technical Data Sheets, Denver, Colo.
6. Kanakia, M., W. Herrera, and F. Hutto, Jr.: "Fire Resistance Tests for Thermal Insulation," *Journal of Thermal Insulation*, vol. 1, no. 4, April 1978, Technomic Pub. Co., Westport, Conn.
7. *Commercial & Industrial Insulation Standards*, Midwest Insulation Contractors Assn., Inc., Omaha, 1979.
8. Malloy, J. F.: *Thermal Insulation*, Reinhold, New York, 1969.
9. ASTM part 18, op. cit., STD C 585.
10. McMillan, L. B.: "Heat Transfer Through Insulation in the Moderate and High Temperature Fields: A Statement of Existing Data," No. 2034, The American Society of Mechanical Engineers, New York, 1934.
11. *Economic Thickness of Industrial Insulation*, Conservation Paper No. 46, Federal Energy Administration, Washington, D.C., 1976. Available from Superintendent of Documents, U.S. Government Printing Office, Washington, DC 20402, Stock #041-018-00115-8.
12. TIMA ETI Computer Program, Thermal Insulation Manufacturers Association, 7 Kirby Plaza, Mt. Kisco, NY 10549.

chapter 4-6

Industrial Hose

by
Charles H. Artus
Product Application Engineer
Gates Rubber Company
Denver, Colorado

INTRODUCTION

There are many varieties of industrial hose and associated couplings. Each type is designed to give satisfactory and safe service life for a particular application where needed for flexible conveyance of liquids, gases, and certain solids in powder or granular form.

To obtain maximum service for any type of industrial hose, the user must: consider the application, select hose and couplings to match the application, take reasonable care of the hose assembly, and observe all applicable safety regulations.

Even regular care and maintenance would help very little to prolong the life of an industrial hose assembly *that was not originally selected to suit the application.* Hoses will soon fail regardless of the care given them in situations that are beyond the design of the hose.

For example: A hose with a natural rubber tube will fail in gasoline service. The tube stock swells, the hose loses its tensile strength and discolors the fluid. High pressures will burst a hose designed for low pressure; coupled assemblies will fail when low-pressure couplings are used on high-pressure hoses. And a hose with a thin tube not resistant to abrasion will soon wear through when handling abrasive products.

So that safest and most reliable service can be obtained, the plant engineer needs to know a few basic things about industrial hose—such as the kinds of hose available, limitations and applications, care and maintenance. We look at these factors one by one.

TYPES OF INDUSTRIAL HOSE

Following are brief descriptions of, and recommendations for, the basic kinds of industrial hose.

Air Hose

Air hose is used for efficient handling of air and compressed air in industrial applications. These hoses can be used for air tools, air drills, air and vapor ducts, agriculture, paint spraying, and welding.

Water Hose

Water hoses are designed for transport of water (but not recommended for use as vibration dampers or in closed water systems). Applications include water suction, water discharge, cleanup, and general-purpose water usage.

Steam Hose

This is a very specialized hose used to convey wet saturated steam, dry saturated steam, and superheated steam.

Materials-Handling Hose

These are specially designed to handle a wide variety of bulk commodities. Applications include: beverage and food, sand suction, dredge sleeves, cement, plaster, vacuums, sand blasting, fish handling, grains and flour.

Acid and Chemical Hose

These hoses are designed to withstand the corrosive effects of caustic, acidic, oxidizing, or chlorinated liquids as well as toxic substances such as anhydrous ammonia.

Petroleum-Transfer Hose

This is manufactured particularly to handle fuel oil, liquid petroleum gases, hot asphalt, gasoline, and diesel fuel.

HOSE CONSTRUCTION

Industrial hose generally consists of three parts—tube, reinforcement, and cover—plus all the necessary cements. See Fig. 6-1.

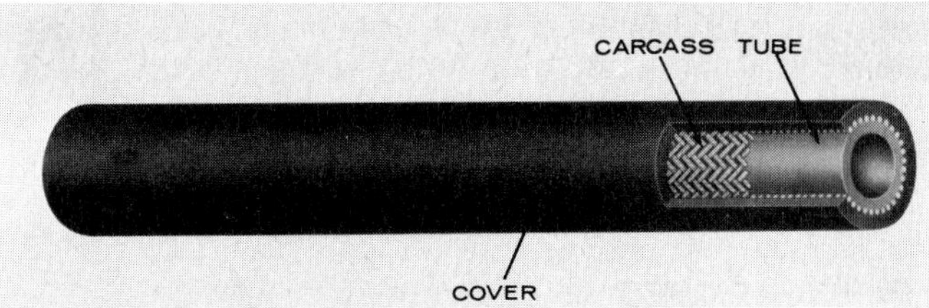

Figure 6-1 Reinforced-hose construction. (*Gates Rubber Company.*)

Tube

The *tube* is the inner layer and its function is to contain the substance being transported and evenly transmit forces, due to pressure, to the reinforcing element. Made of rubber or plastic, it is either extruded or made from sheet stock by spiraling or laminating.

Reinforcement

Outside the tube is the *reinforcement*. This is the material which provides the overall strength, pressure resistance, and sometimes the collapse resistance.

For some industrial hose applications, the service pressure is low enough that the strength of the tube stock is sufficient to meet the required service. This hose is called tubing, or nonreinforced, hose. All other hoses require some type of textile or wire reinforcement. The usual reinforcing materials are textile fibers or filaments or metal wires. Common textile reinforcements are made of cotton, rayon, nylon, and polyester; and they may take the form of a yarn- or fabric-like material. Wire reinforcements vary as to size and composition of the material used to make the wire. Identification of the hose usually relates to the method of reinforcement. For example, a hose with a braid reinforcement would be called a braided hose. Following are the principal forms of reinforcement.

- **Knit.** The knit reinforcement is applied around the tube by a knitting machine. A lock-stitch pattern provides greater resistance to diametrical hose growth. Knit hose is used in low-pressure ranges, seldom exceeding 100 to 120 lb/in^2.
- **Braid.** Braided hose has the reinforcement—either yarn or wire—applied by using horizontal or vertical braiders. The terms vertical and horizontal refer to the axis of the hose as it is being braided. Hose with a braid reinforcement finds applications in many areas, including air, water, hydraulic, and steam.
- **Spiral.** Spiral-reinforced hose has all the wire or textile strands in one layer laid parallel on the hose in one direction. At least two layers, or multiples of two layers, of reinforcement are normally required; with the layers spiraled in alternating directions to form a balanced construction. Additional layers of reinforcement may be applied to increase the service-pressure level, with a usual maximum of six layers. Typical applications of spiral hose range from low-pressure general-purpose to high-pressure hydraulic.
- **Wrapped.** The greatest variety of hoses are those in which woven reinforcement is wrapped onto the tube. These hoses range from ¼-in-ID air hose up to 36 to 48-in-diam suction and discharge hose. Wrapped hoses have the reinforcement applied by either spiraling or laminating the material onto the tube.

Cover

The *cover* is usually a rubber or plastic layer placed over the reinforcement. It protects the reinforcement and serves to keep elements such as weather, abrasion, and chemicals from weakening it.

The application of a cover over the reinforcement is completed in much the same manner as the tube is fabricated. The cover may be applied using an extruder to form a seamless, cylindrical tube around the hose or by spiraling or laminating calendered cover material onto a mandrel-supported hose.

SELECTION OF INDUSTRIAL HOSE

The industrial hose purchaser should determine the following:

1. Inside diameter of hose required
2. Length required
3. Material to be conveyed, including chemical makeup and temperature range
4. Suction and/or discharge rated working pressure
5. Hose ends or fittings required
6. External environment (temperature range, corrosive fluids, complete weather conditions)
7. External stresses (kinking, crushing, pulling, excessive bending)
8. Special requirements, i.e., government and safety regulations, special tests, etc.

When analyzing the cost differences among industrial hoses to be selected for in-plant piping, several factors must be considered.

First consider the method of construction. A vertical braided hose is the least expensive because it is built in long, continuous lengths and its reinforcement consists of yarn. Vertical braided hoses have small inside diameters, up to 1½ in. The horizontal braided hose construction provides larger hose diameters and greater working pressures than the vertical braided construction. Reinforcements consist of one or more layers of braided fibers or wire. Horizontal braided-wire reinforcement is similar to the standard horizontal braided hose with the addition of a reinforcing wire spiralled between the fiber braid reinforcement. Wrapped and wire-reinforced hoses are the most expensive, primarily because they are individually hand-built. Wire spiraled onto the hose between multiple plies of wrapped fabrics also adds to the cost of making these hoses.

Realizing that the mass-produced vertical braided hose is less expensive than wrapped hose and that yarn reinforcement usually costs less than wire, the next cost factor to be considered is the stock type used to construct the tube and cover of the hose. Table 6-1 shows the elastomers most commonly used in industrial hose in order of relative cost, from top to bottom.

The general properties listed in Table 6-1 provide a reliable guideline. However, the user should always follow the manufacturer's recommendations as to the use of any particular rubber composition. This is especially true with respect to the resistance of the rubber composition to the materials it is exposed to, from within and without.

Although the construction methods and materials used to make the hose provide a cost comparison, a true cost index must take into account certain variables. These include the dimensions of the hose, the quantity of materials used, and the changing price of many petroleum-based raw materials.

Table 6-2 is a section of hose types most commonly found in industrial plants. The hoses are grouped by general application, description, and cost index.

In applying any product, the more the plant engineer knows about the details of application, the better the product that can be selected to those needs. *This is particularly true with industrial hose because of the many variables involved.*

The process of selecting a hose usually involves one of two situations:

1. Making the optimum choice from several suitable selections for general application.
2. Making the one choice which satisfies all the requirements of a special application such as transferring anhydrous ammonia or LP gas.

TABLE 6-1 Characteristics of Hose Stock Types and Relative Cost Comparisons*

Common name	Usage in hose	Typical hose application	General properties
EPDM (Ethylene propylene diene)	Tube and cover	Steam (dry or wet), hot water, engine coolant, air (hot or cold), mild chemicals, and generally nonoily products	Excellent ozone, chemical, and aging characteristics; poor resistance to petroleum-based fluids
Natural rubber or synthetic (styrene-butadiene) rubber (SBR)	Tube and cover	Materials handling, water, air, chemicals, and generally nonoily products	Excellent abrasion and good low-temperature resistance; poor resistance to petroleum-based fluids
Buna-N	Tube	Fuel oils, diesel oils, aromatics, gasolines, and other petroleum products special applications	Excellent resistance to petroleum-based fluids; good physical properties; good resistance to heat and chemicals
Neoprene	Tube and cover	General-purpose air and water, saturated steam, materials handling, acid-chemical, and petroleum	Good to excellent abrasion, flame, petroleum, weathering, ozone, and heat resistance
Butyl	Tube and cover	Dry steam, engine coolant, hot air, chemicals, and generally nonoily products	Very good weathering resistance; low permeability to air; good physical properties
Hypalon	Tube and cover	Acid-chemical transfer	Excellent ozone, chemical, and aging resistance; good abrasion and heat resistance; fair resistance to petroleum-based fluids
Gatron	Tube	Acid-chemical transfer	Excellent general chemical, ozone, and weathering resistance; wide temperature flexibility range
FPM	Tube	Acid-chemical transfer	Excellent resistance to aromatic fluids and chlorinated hydrocarbons; outstanding heat resistance; excellent weathering resistance

*Chemical and physical properties are subject to a considerable amount of control through compounding. Characteristics given below for each stock type are therefore generalized to some degree. *These stocks are listed in order of relative costs,* with EPDM being the least expensive and FPM being the most expensive. Neoprene and Buna-N are in the same general cost range.

TABLE 6-2 Hose Types in Industrial Use

Application	Method of construction	Tube stock	Cover stock	Relative* cost index
General-purpose hose				
Low-pressure air or water	Spiral or vertical braided	EPDM	EPDM	1
High-pressure air or water (improved resistance to heat, oil, and abrasion)	Vertical braided	Buna-N	Neoprene	2
Petroleum-product transfer				
Aromatic fuels at full suction	Horizontal braided, wire reinforced	Buna-N	Neoprene	3
Butane-propane	Horizontal braided	Neoprene	Neoprene	4
Steam hose				
Saturated and superheated steam up to 250 lb/in^2 (1.72 MPa) and 450°F (232°C)	Wire braided (one or two layers)	EPDM	EPDM	5
Ssaturated steam up to 200 lb/in^2 (1.38 MPa) and 450°F (232°C)	Wire braided (one or two layers)	EPDM	EPDM	4
Materials-handling hose				
Bulk commodity, highly abrasion resistant	Wire reinforced	SBR	SBR	5
Food handling, primarily oils up to 150°F (65°C)	Wire reinforced	Food-grade neoprene	Neoprene	6
Dusty materials or abrasives in water suspension	Wire reinforced	Gum rubber (to absorb impact)	SBR	8
Acid-chemical transfer hose				
Strong, oxidizing solutions	Horizontal braided, wire reinforced	Hypalon	Neoprene	6
Highly aromatic fluids and chlorinated hydrocarbons	Horizontal braided, wire reinforced	FPM	Neoprene	12

*Using general-purpose hose as least expensive. Relative cost will vary, depending on size.

In the first case, final selection is usually made on the basis of cost. In the second, it is usually a matter of satisfactorily meeting the most important or critical conditions *first* and then the conditions of lesser importance.

SELECTION OF INDUSTRIAL COUPLINGS

In most hose catalogs, couplings are listed by name, and each is described as to construction, application, thread, and method of attaching it to the hose. Adapters and fittings are available for special applications. A complete up-to-date catalog should be obtained from the manufacturer before buying decisions are made.

Couplings can be grouped by a pressure rating of low, medium, or high. These terms have a general meaning as follows:

Low pressure	100 lb/in^2 (680. kPa) maximum
Medium pressure	315 lb/in^2 (2135 kPa) maximum
High pressure	800 lb/in^2 (5520 kPa) maximum

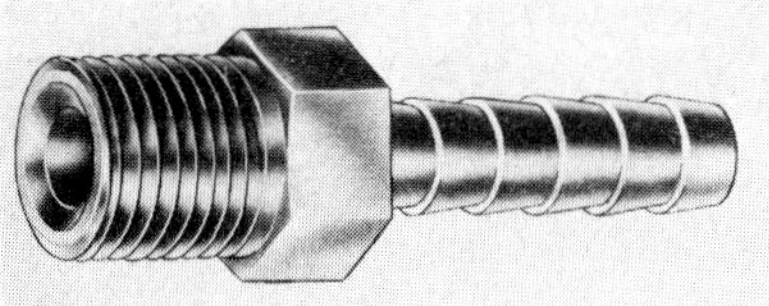

Figure 6-2 Air-hose coupling.

Figure 6-3 Combination nipple. (*Gates Rubber Company.*)

Coupling ratings apply only when the recommended clamps or bands are used. Care should be taken that the proper size of clamp is selected, since many clamps will fit only one size of hose or a narrow range of sizes.

GENERAL APPLICATIONS

Five different coupling types can be used to handle approximately 90 percent of all hose applications found in industrial plants. The standard air-hose coupling or insert (Fig. 6-2) is a machine brass, low- or medium-pressure coupling for use on air hose, general-purpose hose or food-handling hoses (except meat products).

The combination nipple (Fig. 6-3) is made of swaged steel pipe with scored or serrated shank and NPT (American Standard Taper Pipe Threads) threads on the male end. This nipple can be used with low- or medium-pressure water hose or materials handling hose.

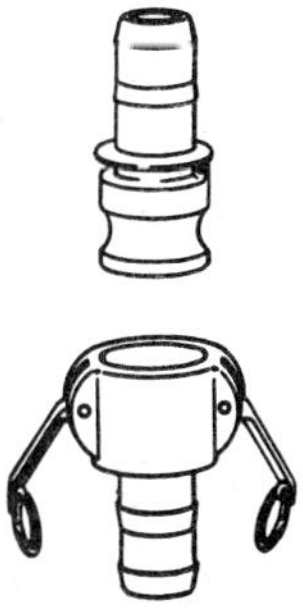

Quick-connecting couplings (Fig. 6-4) are used primarily in medium-pressure applications where the hose is frequently connected and disconnected such as at tank farms, bulk-liquid storage receptacles, and tank trucks. The two parts of this coupling fit snugly together and are held in place by two cams on the female shank coupler which rotate against a groove in the male adapter.

Figure 6-4 Quick-connecting coupling. (*Gates Rubber Company.*)

For maximum safety when handling high-pressure air, water, or steam, special high-pressure couplings must be employed. These are interlocking couplings, with a ground joint seal or with a washer joint seal (Fig. 6-5).

A coupling especially recommended for use on air hose is the universal quick-acting coupling. Several types of heads are available, but all have the same size attaching heads regardless of the hose size (Figs. 6-6 to 6-9).

Figure 6-5 Interlocking coupling. (*Gates Rubber Company.*)

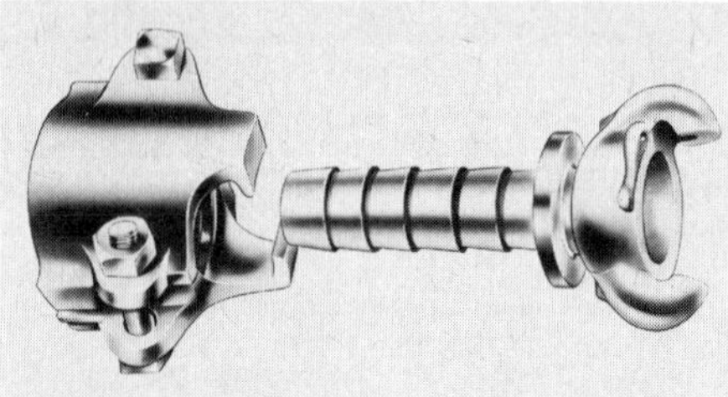

Figure 6-6 Quick-acting coupling, hose end (right), and clamp. (*Gates Rubber Company.*)

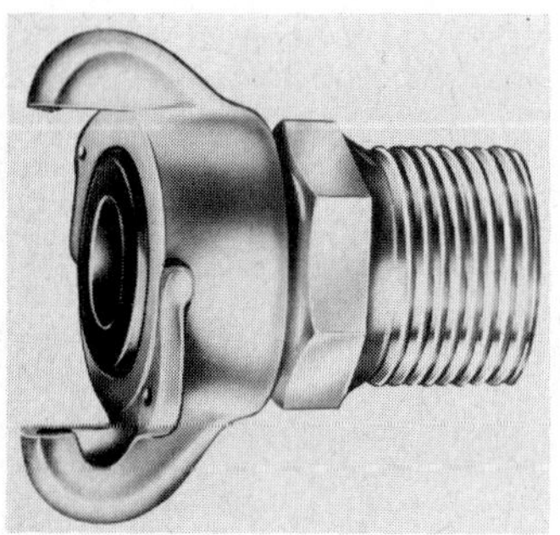

Figure 6-7 Quick-acting coupling, male end. (*Gates Rubber Company.*)

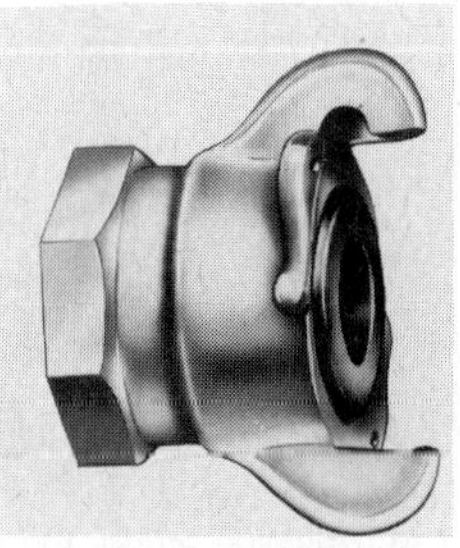

Figure 6-8 Quick-acting coupling, female end. (*Gates Rubber Company.*)

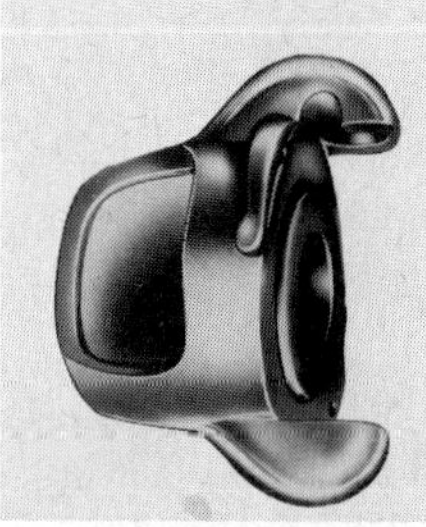

Figure 6-9 Quick-acting coupling, blank. (*Gates Rubber Company.*)

CRITICAL APPLICATIONS

For certain critical fluids, specific couplings are used, e.g., nonsparking for inflammable fluids. Below are guides to critical applications:

1. Use only the couplings recommended by the manufacturer for conveying:
   Steam
   LP gas
   Anhydrous ammonia
   Corrosive chemicals
   Petroleum products
2. For any high-temperature application, use only interlocking-type couplings.
3. For conveying flammable fluids, use couplings made of nonsparking materials, such as brass or aluminum.
4. For ground fueling of aircraft, use coupled assemblies only, as recommended by the supplier.

MAINTENANCE

Hose Care

Hose has definite service limitations and will certainly fail prematurely *if care is not taken to avoid exceeding these limitations.* General rules for caring for a hose are quite elementary but can be easily overlooked. They are especially important *because of the limited amount of maintenance or repair that can be made to extend the life of the hose assembly.*

Storage is important. If new hose is stored carefully, hose shelf life will be about 5 years before gradual deterioration begins. The hose should be stored in its original packing container or crate, out of direct sunlight. Avoid extremes of temperatures and exposure to ozone or direct heat. If the hose is shipped coiled, lay coils flat on the shelf; hose shipped straight should be stored straight.

Considerations for Hose Service

Environment

The general rule is to avoid those conditions which will accelerate hose aging. If conditions are unusually severe, a hose designed for that application must be used.

First, avoid extreme heat or cold unless the hose has been designed and built to withstand such extremes. The typical industrial hose will give satisfactory life in a temperature range of 0 to 150°F (−18 to 65°C). At −20°F (−7°C) the hose will lose some flexibility. At −30 to −40°F (−14 to −40°C) normal hose may crack if flexed sharply. Special hoses are available that will be serviceable down to −60°F (−50°C).

The limitations at about 150°F (65°C) vary with the type of elastomer and service, so general rules cannot be given. Always follow the manufacturer's recommendations for hose to be used in high ambient temperatures.

In addition, exposure to high concentrations of ozone will cause hose covers to crack. This is not a problem with routine applications, but the cracking can be severe in high smog areas, or near electric generating machinery.

Finally, consider weathering. Hoses in continuous service outdoors should have weather-resistant covers and, in all cases, should not be continuously exposed to oil or corrosive chemicals.

External Abuse

Do not over-bend the hose to the point of kinking. Always observe minimum-bend-radius recommendations. It is true that wire-reinforced hose may have greater rigidity, but it can be crushed or deformed by external weight or forces. Couplings and hose can also be damaged by too much end pull. A hard pull at any angle may kink the hose next to the coupling, especially in subzero temperatures.

Large-diameter hoses [4-in (10 cm) ID and greater] have some special considerations. One problem may be overstressing the hose carcass. Handle heavy hoses with slings every 6 to 10 ft (2 to 3 cm) and do not lift a long section from the middle with the ends hanging down.

Another concern—sometimes the hose cover is exposed to wear in one particular spot. In this case, the user should add a protective outside cover to avoid wearing through the cover and exposing the reinforcing.

Maintenance and Repair

Good maintenance programs including inspection, testing, and repair will ensure that maximum, safe service life is obtained from the hose and that damaged hose will be removed from service before it becomes a hazard.

All hoses should be inspected periodically. Hoses used in hazardous applications should be inspected at more frequent intervals. Basically, the user should be looking for evidence of stress or external abuse such as abrasion, cuts, and exposure to chemicals and oils. Obviously, the sources of these abuses should be corrected or eliminated.

The user should also look for evidence of failure that will necessitate taking the hose out of service or require recoupling or other repairs.

Inspection procedures for industrial hoses in nonhazardous applications are:

1. Lay hose out straight in a dry, light area.
2. Visually inspect hose for kinks, bulges, or soft spots in the cover, and excessive cover

wear that exposes reinforcing. Evidence of this kind usually means the hose should be removed from service.

3. Inspect couplings for signs of slippage. Examine hose adjacent to couplings for breakage. If necessary, hose can be cut off behind the couplings and recoupled. Retighten bolted clamps if necessary.
4. Excessive amounts of oil or chemicals on the cover should be wiped off. Periodic or specialized inspection and testing are necessary for hoses used in hazardous applications. These hoses should be inspected at definite intervals following established procedures. The *Hose Handbook* and individual bulletins published by the Rubber Manufacturers Association outline daily inspection procedures for hoses used for hard-to-handle anhydrous ammonia, liquid petroleum gas, oil suction and discharge, and aviation ground fueling, as well as for motor-vehicle hose.

Hose Repair

Hoses can be recoupled to replace damaged couplings or to rejoin the ends of a failed section of hose that is cut out of the original length. The extent of recoupling or coupling repair that can be done depends on the type of couplings and the size of the hose.

Hoses that have plain ends can be recoupled by cutting off the old coupling and applying a new one, provied the overall length is not too small. All braided hoses and some wrapped hoses up through 4-in (10 cm) ID have plain ends. Hoses having special ends or built-in couplings cannot normally be recoupled.

Large-diameter built-in nipples that have been worn thin by abrasives can be built up with plates. However, the rubber stock must be protected from any welding heat.

Beyond recoupling, there are very few ways in which a damaged or badly worn hose can be safely reclaimed, but the hose manufacturer should be consulted for recommendations.

STANDARDS

Table 6-3 lists applicable OSHA standards and their applications.

TABLE 6-3 OSHA Standards

Standards paragraph	Application
1910.106 Flammable and Combustible Liquids	General Transfer Curb pump, UL listed Flesible connectors, UL listed
1910.107 Spray Finishing Using Flammable and Combustible Materials	a. General, low, or medium pressure b. High pressure, airless, with static wire
1910.109 Explosives and Blasting Agents	As specified; special electrical properties
1910.110 Storage and Handling of LP Gas	As specified; UL listed
1910.111 Storage and Handling of Anhydrous Ammonia	Transfer service
1910.134 Respiratory Protection	Hoses for face masks
1910.158 Stand Pipe and Hose System for Fire Equipment	Various
1910.165a Sources of Standards	Fire extinguishers
1910.177 Indoor General Storage	Sprinkler systems
1910.243 Guarding of Portable Power Tools	Various air hoses
1910.1252 Welding, Cutting & Brazing	General welding and related

BIBLIOGRAPHY

OSHA standards are contained in *Federal Register,* October 18, 1972, vol. 37, no. 22, part II.

Hose Handbook, Rubber Manufacturer's Association, 1901 Pennsylvania Ave., N.W., Washington, DC 20006.

"Industrial Hose Finder," No. 39995. The Gates Rubber Company, P. O. Box 5887, Denver, CO 80217.

chapter 4-7

Valves

prepared by

Communications Committee
The Valve Manufacturers Association
McLean, Virginia

INTRODUCTION

A valve may be defined as a mechanical device by which the flow of liquid or gas may be started, stopped, or regulated by a movable part that opens, shuts, or partially obstructs one or more ports or passageways.

Plant engineers describe a valve as one of the most essential control instruments used in industry.

By the nature of their design and materials, valves can open and close, turn on and turn off, regulate, modulate, or isolate an extremely large array of liquids and gases, from the most basic to the most corrosive or toxic. They range in size from a fraction of an inch to 30 ft (9 m) in diameter. They can handle pressures ranging from vacuum to more than 20,000 lb/in^2 (140 MPa/m^2) and temperatures from the cryogenic region to 1500°F (815°C). Some applications require absolute sealing; in others leakage is not a factor.

CATEGORIES OF VALVES

Because of all these variables, there can be no universal valve; therefore, to meet the changing requirements of industry, innumerable designs and variations have evolved over the years as new materials have been developed. All these designs fall into nine major categories: gate valves, globe valves, ball valves, butterfly valves, pinch valves, diaphragm valves, plug valves, check valves, and relief valves.

These basic categories are described in the following paragraphs. It would be impossible to mention every feature of every valve manufactured, and we have not attempted to do this. Instead, a general overview of each type is presented in outline format, giving service recommendations, applications, advantages, disadvantages, and other information helpful to the reader. In many cases, a disadvantage inherent in a type of valve has been overcome or corrected by a particular manufacturer. Therefore, for specific applications, manufacturers' recommendations should be sought.

Gate Valves

A gate valve is a multiturn valve in which the port is closed by a flat-faced, vertical disk that slides at right angles over the seat (Fig. 7-1).

Recommended

- For fully opened or fully closed, nonthrottling service
- For infrequent operation
- For minimum resistance to flow
- For minimum amounts of fluid trapped in line

Applications

- General service, oil, gas, air, slurries, heavy liquids, steam, noncondensing gases and liquids, corrosive liquids

Advantages

- High capacity
- Tight shutoff
- Low cost
- Simple design and operation
- Little resistance to flow

Disadvantages

- Poor flow control
- High operating force
- Cavitates at low pressure drop
- Must be kept in full open or full closed position
- Throttling position will erode seat and disk

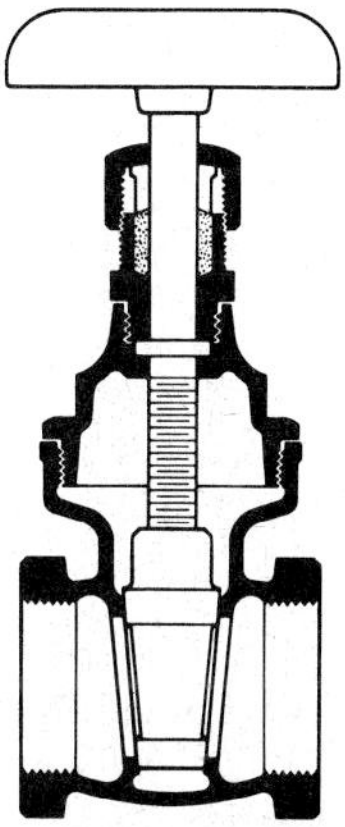

Variations

- Solid wedge, flexible wedge, split wedge, double disk

Figure 7-1 Gate valve.

Materials

- Body: bronze, cast iron, iron, forged steel, Monel, cast steel, stainless steel, PVC plastic
- Trim: various

Special Installation and Maintenance Instructions

- Lubricate on regular schedule
- Correct packing leaks immediately
- Always cool system when closing down a "hot" line and checking closed valves
- Never force valves closed with wrench or pry
- Open valves slowly to prevent hydraulic shock in line
- Close valves slowly to help flush trapped sediment and dirt

Ordering Specifications

- Type of end connections
- Type of wedge
- Type of seat
- Type of stem assembly
- Type of bonnet assembly
- Type of stem packing
- Pressure rating: operating and design
- Temperature rating: operating and design

Plug Valves

A plug valve is a quarter-turn valve that controls flow by means of a cylindrical or tapered plug with a hole through the center, which can be positioned from open to closed by a 90° turn (Fig. 7-2).

Recommended

- For full-open or full-closed service
- For frequent operation
- For low pressure drop across the valve
- For minimum resistance to flow
- For minimum amount of fluid trapped in line

Applications

- General service, slurries, liquids, vapors, gases, corrosives

Advantages

- High capacity
- Low cost
- Tight shutoff
- Quick operation

Disadvantages

- High torque for actuation
- Seat wear
- Cavitation at low pressure drop

Variations

- Lubricated, nonlubricated, multiport

Materials

- Iron, ductile iron, carbon steel, stainless steel, Alloy 20, Monel, nickel, Hastelloy, plastic-lined

Special Installation and Maintenance Instructions

- Allow space for operation of handle on wrench-operated valves
- For lubricated plug valves, lubricate before putting into service
- For lubricated plug valves, lubricate on regular schedule

Ordering Specifications

- Body material
- Plug material
- Temperature rating
- Pressure rating
- Port arrangement, if multiport valve
- Lubricant, if lubricated valve

Globe Valves

A globe valve is multiturn valve in which closure is achieved by means of a disk or plug that seals or stops the fluid on a seat generally parallel to the line flow (Fig. 7-3).

Recommended

- For throttling service or flow regulation
- For frequent operation
- For positive shutoff of gases or air
- Where some resistance to flow is acceptable

Applications

- General service, liquids, vapors, gases, corrosives, slurries

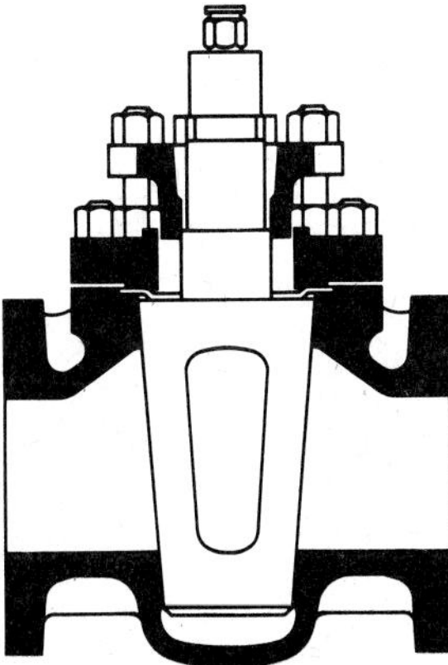

Figure 7-2 Plug valve.

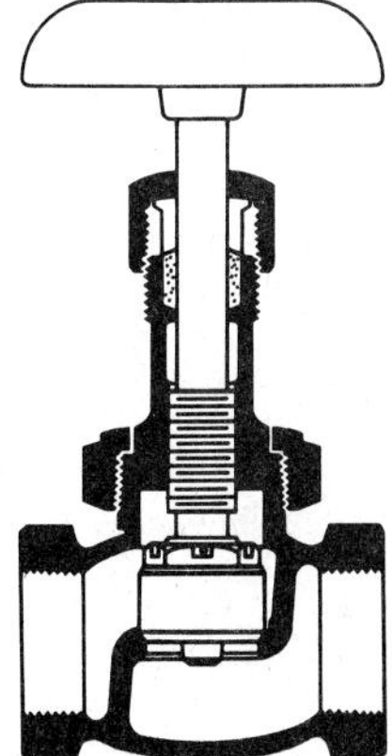

Figure 7-3 Globe valve.

Advantages

- Efficient throttling with minimum wire drawing or disk or seat erosion
- Short disk travel and fewer turns to operate, saving time and wear on stem and bonnet
- Accurate flow control
- Available in multiports

Disadvantages

- High pressure drop
- Relatively high cost

Variations

- Standard, Y pattern, angle, three-way

Materials

- Body: bronze, all iron, cast iron, forged steel, Monel, cast steel, stainless steel, plastics
- Trim: various

Special Installation and Maintenance Instructions

- Install so pressure is under disk, except in high-temperature steam service
- Lubricate on strict schedule
- Flush foreign matter off seat by opening valve slightly
- Correct packing leaks immediately by tightening the packing nut

Ordering Specifications

- Type of end connection
- Type of disk
- Type of seat
- Type of stem assembly
- Type of stem seal
- Type of bonnet assembly
- Pressure rating
- Temperature rating

Ball Valves

A ball valve is a quarter-turn valve in which a drilled ball rotates between resilient seats, allowing straight-through flow in the open position and shutting off flow when the ball is rotated 90° and blocks the flow passage (Fig. 7-4).

Recommended

- For on-off, nonthrottling service
- Where quick opening is required
- For moderate temperature requirements
- Where minimum resistance to flow is needed

Applications

- General service, high temperatures, slurries

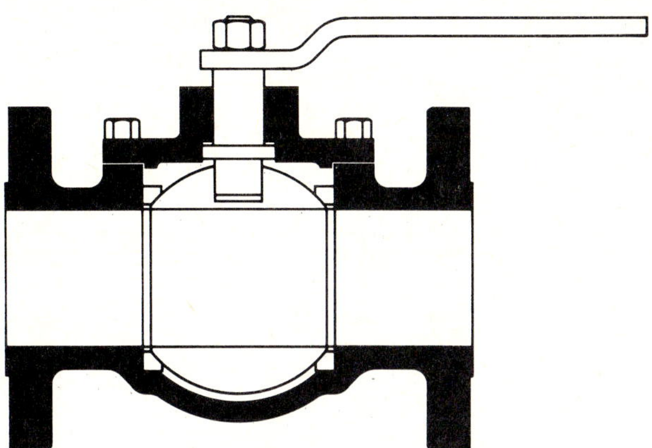

Figure 7-4 Ball valve.

Advantages

- Low cost
- High capacity
- Bidirectional shutoff
- Straight-through pattern
- Low leakage
- Self-cleaning
- Low maintenance
- No lubrication requirement
- Compact
- Tight sealing with low torque

Disadvantages

- Poor throttling characteristics
- High torque for actuation
- Susceptible to seal wear
- Prone to cavitation

Variations

- Top entry, split body or end entry, three-way, venturi, full-ported, reduced port

Materials

- Body: cast iron, ductile iron, bronze, brass, aluminum, carbon steels, stainless steels, titanium, tantalum, zirconium, and polypropylene and PVC plastics
- Seat: TFE, filled TFE, nylon, Buna-N, neoprene

Special Installation and Maintenance Instructions

- Allow sufficient space for operation of long handle

Ordering Specifications

- Operating temperature
- Type of port in ball
- Seat material
- Body material
- Operating pressure
- Full or reduced port
- Top entry or side entry

Butterfly Valves

A butterfly valve is a quarter-turn valve that controls flow by means of a circular disk with its port axis at right angles to the direction of flow (Fig. 7-5).

Recommended

- For full-open or full-closed service
- For throttling service
- For frequent operation
- Where positive shutoff is required for gases or liquids
- Where minimum amount of fluid trapped in line is allowed
- For low pressure drop across valve

Applications

- General service, liquids, gases, slurries, liquids with suspended solids

Advantages

- Compact, lightweight, low-cost
- Low maintenance
- Minimum number of moving parts
- No pockets
- High capacity
- Straight-through flow
- Self-cleaning

Disadvantages

- High torque for actuation
- Limited pressure-drop capability
- Prone to cavitation

Variations

- Wafer, lug wafer, flanged, screwed, fully lined, high-performance

Materials

- Body: iron, ductile iron, carbon steels, forged steel, stainless steels, Alloy 20, bronze, Monel
- Disk: all metals, elastomer coatings such as TFE, Kynar, Buna-N, neoprene, Hypalon
- Seat: Buna-N, Viton, neoprene, rubber, butyl, polyurethane, Hypalon, Hycar, TFE

Special Installation and Maintenance Instructions

- May be operated by lever, handwheel, or chainwheel
- Allow sufficient space for operation of handle if lever-operated
- Valves should remain in closed position during all handling and installation operations

Ordering Specifications

- Type of body
- Type of seat
- Body material
- Disk material
- Seat material
- Type of actuation
- Operating pressure
- Operating temperature

Diaphragm Valves

A diaphragm valve is a multiturn valve that effects closure by means of a flexible diaphragm attached to a compressor. When the compressor is lowered by the valve stem, the diaphragm seals and cuts off flow (Fig. 7-6).

Recommended

- For full-open or full-closed service
- For throttling service
- For service with low operating pressures

Applications

- Corrosive fluids, sticky and/or viscous materials, fibrous slurries, sludges, foods, pharmaceuticals

Advantages

- Low cost
- No packing glands

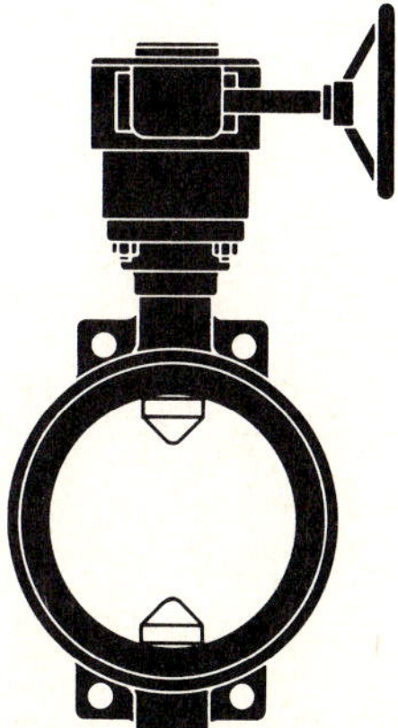

Figure 7-5 Butterfly valve.

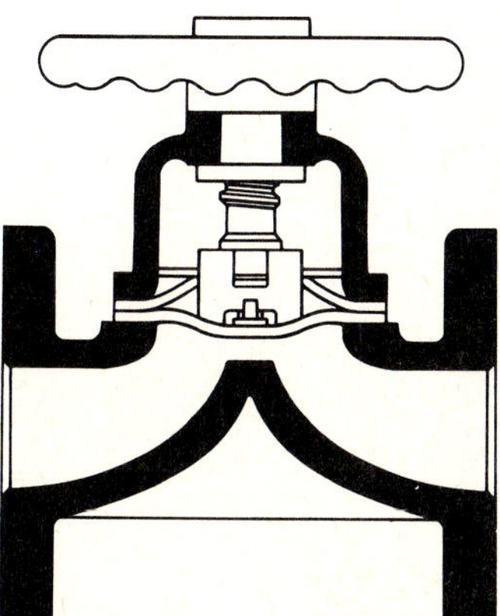

Figure 7-6 Diaphragm valve.

- No possibility of stem leakage
- Immune to problems of clogging, corroding, or gumming of media

Disadvantages

- Diaphragm subject to wear
- High torque under live-line closure

Variations

- Weir type and straight-through type

Materials

- Metallic, solid plastic, lined—wide variety of each

Special Installation and Maintenance Instructions

- Lubricate on a regular schedule
- Do not use bars, wrenches, or cheaters to close

Ordering Specifications

- Body material
- Diaphragm material
- End connections
- Type of stem assembly
- Type of bonnet assembly
- Type of operation
- Operating pressure
- Operating temperature

Pinch Valves

A pinch valve is a multiturn valve that effects closure by means of one or more flexible elements, such as diaphragms or rubber tubes, that can be pressed together to cut off flow (Fig. 7-7).

Recommended

- For on-off service
- For throttling service
- For moderate temperatures
- Where pressure drop through valve is low
- For services requiring low maintenance

Applications

- Slurries, mining slurries, liquids with large amounts of suspended solids, systems that convey solids pneumatically, food service

Advantages

- Low cost
- Low maintenance
- No internal obstruction or pockets to cause clogging

- Simple design
- Noncorrosive and abrasion-resistant

Disadvantages

- Limited vacuum application
- Difficult to size

Variations

- Exposed sleeve or body, encased metallic sleeve or body

Materials

- Rubber, white rubber, Hypalon, polyurethane, neoprene, white neoprene, Buna-N, Buna-S, Viton-A, butyl rubber, silicone, TFE

Special Installation and Maintenance Instructions

- Large sizes may require supports above or below the line if pipe supports are inadequate

Ordering Specifications

- Operating pressure
- Operating temperature
- Sleeve material
- Exposed or encased sleeve

Check Valves and Relief Valves

Two categories of valves are specific-purpose rather than general-service valves. These are check valves and relief valves. Unlike the other types described in this section, they are self-actuated valves and operate without outside control, depending for their operation on flow direction or pressures within the piping system. Since both types are nor-

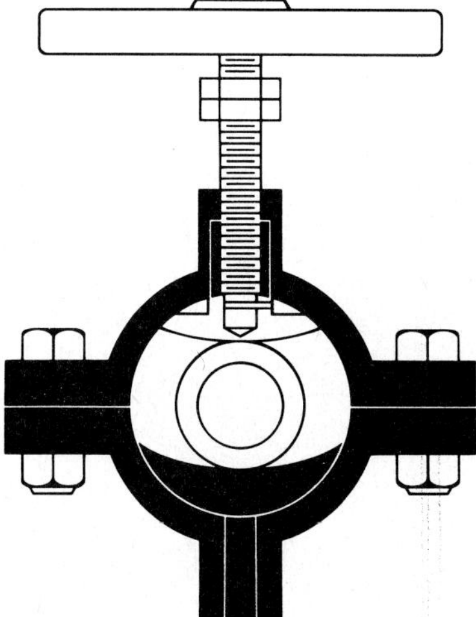

Figure 7-7 Pinch valve.

mally used in conjunction with flow-control valves, the choice of valve is often determined by the same conditions that determine the selection of the flow-control valve.

Check Valves

A check valve (Fig. 7-8) is designed to check reversal of flow. Fluid flow in the desired direction opens the valve; reversal of flow closes it. There are three basic styles of check valves: (1) swing check, (2) life check, and (3) butterfly check.

Swing Check Valve. A swing check valve has a hinged disk designed to open completely with line pressure and close when line pressure ceases and backflow begins. There are two designs: a Y pattern, which has an access opening in the body for easy regrinding of the disk without removing the valve from the line, and a straight-through pattern that has replaceable seat rings.

Recommended

- Where minimum resistance to flow is needed
- Where there is infrequent change of direction in the line
- For service in lines using gate valves
- For vertical lines having upward flow

Applications

- For low-velocity liquid service

Advantages

- Unobstructed view
- Turbulence and pressure within valve are very low
- Y-pattern disk can be reground without removing valve from line

Variations

- Tilting-disk check valve

Materials

- Body: bronze, all iron, cast iron, forged steel, Monel, cast steel, stainless steel, carbon steel
- Trim: various

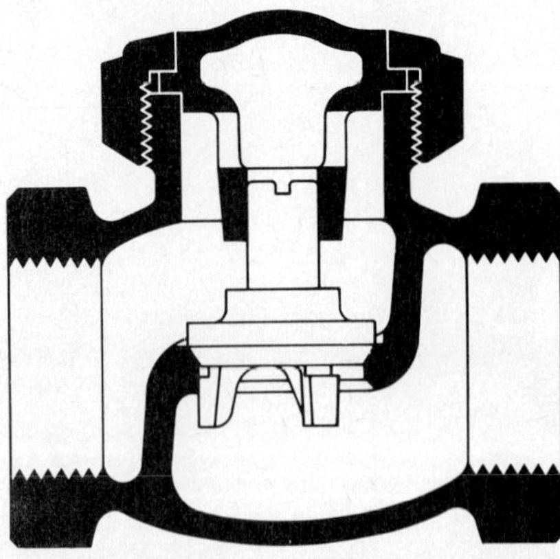

Figure 7-8 Check valve (lift type).

Special Installation and Maintenance Instructions

- In vertical lines, pressure should always be under seat
- If valve fails to seal, check seating surfaces
- If seat is damaged or scored, regrind or replace
- Before reassembling, clean internal portions thoroughly

Lift Check Valve. A lift check valve is similar in design to a globe valve except that the disk is lifted by the forward line pressure and closed by gravity and backflow.

Recommended

- Where there are frequent changes of direction in the line
- For use with globe and angle valves
- For use where pressure drop across valve is not a problem

Applications

- Steam, air, gas, water, and vapor lines with high flow velocities

Advantages

- Minimum travel of disk for full-open position
- Quick-acting

Variations

- Three body patterns: horizontal, angle, vertical
- Ball check, piston check, spring-loaded check, stop check

Materials

- Body: bronze, all iron, cast iron, forged steel, Monel, stainless steel, PVC, Penton, impervious graphite, TFE-lined
- Trim: various

Special Installation and Maintenance Instructions

- Line pressure should be under seat
- Horizontal pattern should be installed in horizontal lines
- Vertical pattern is used for vertical pipes with flow upward, from beneath seat
- If backflow leaks, check disk and seat

Butterfly Check Valve. A butterfly check valve has a split disk hinged on a shaft in the center of the disk so that a flexible sealing member attached to the disk is at a 45° angle to the valve body when the valve is closed. The disk then has to move only a short distance away from the body toward the center of the valve to open fully.

Recommended

- Where minimum resistance to flow in the line is needed
- Where there is frequent change of direction in the line
- For use with butterfly, plug, ball, diaphragm, or pinch valves

Applications

- For liquid or gas service

Advantages

- Body design lends itself to installation of various types of seat liners
- Less expensive for corrosion resistance
- Quiet operation
- Simplicity of design permits construction in large diameters
- May be installed in virtually any position

Variations

- Fully lined
- Soft-seated

Materials

- Body: steel, stainless steel, titanium, aluminum, PVC, CPVC, polyethylene, polypropylene, cast iron, Monel, bronze
- Flexible sealing members: Buna-N, Viton, butyl rubber, TFE, neoprene, Hypalon, urethane, Nordel, Tygon, silicone

Special Installation and Maintenance Instructions

- On lined valves, liner should be protected from damage during handling
- Make sure valve is installed so that forward flow opens valve

Relief Valves

A relief valve (Fig. 7-9) is a self-actuated valve designed to provide accurate automatic pressure regulation. The valve is used primarily for noncompressible fluid service and opens slowly as pressure increases, regulating the operating pressure.

Closely related to the relief valve is the safety valve, which opens quickly with a pop action to relieve excessive pressure caused by gases or compressible fluids.

Sizing is very important in relief valves and is determined by specific formulas.

Recommended

- In systems where a predetermined pressure range is required

Applications

- Hot water, steam, gases, vapor

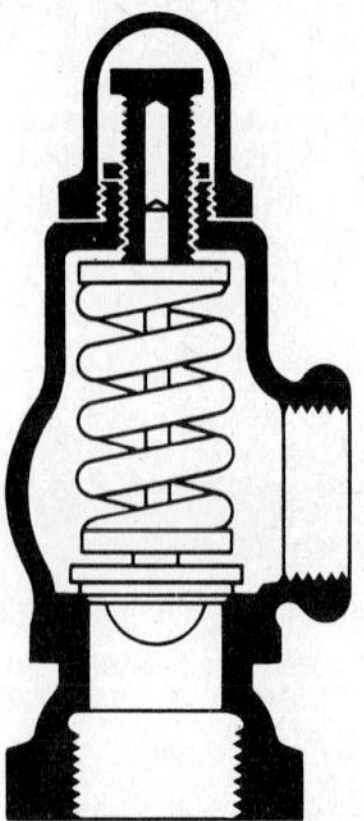

Figure 7-9 Relief valve.

Advantages

- Inexpensive
- No auxiliary power required for operation

Variations

- Safety, safety relief
- Diaphragm construction for valves used in corrosive service

Materials

- Body: cast iron, carbon steel, glass-TFE, bronze, brass, TFE-lined, stainless steel, Hastelloy, Monel
- Trim: various

Special Installation and Maintenance Instructions

- Should be installed in accordance with provisions of ASME Unfired Pressure Vessel Code
- Should be installed in readily accessible areas for inspection and maintenance

CODES AND STANDARDS

Various professional and industrial organizations have created codes and standards that are applicable to the design, selection, and use of industrial products. Certain of these standards are peculiar to specific industries, such as those set up by the American Water Works Association (AWWA), American Gas Association (AGA), American National Standards Institute (ANSI), American Society of Mechanical Engineers (ASME), and Manufacturers Standardization Society of the Valve and Fitting Industry (MSS). These are but a few of the existing organizations.

The valve specifier should be aware of those codes that apply to a specific industry and the application in question. For further information on the codes and standards available, these organizations may be contacted at the addresses listed below:

AWWA, 6666 W. Quincy Avenue, Denver, CO 80235

ANSI, 1430 Broadway, New York, NY 10018

AGA, 1515 Wilson Boulevard, Arlington, VA 22209

API, 2101 L Street N.W., Washington, DC 20037

ASME, 345 East 47th Street, New York, NY 10017

MSS, 1815 N. Fort Myer Drive, Arlington, VA 22209

BIBLIOGRAPHY

Beard, Chester S.: *Final Control Elements,* Rimback Publications Div., Chilton Co., Radnor, Pa., 1969.

Hutchison, J. W.: *ISA Handbook of Control Valves,* Instrument Society of America, Pittsburgh, 1971.

O'Keef, William, (Ed.): "Valves," *Power Magazine,* March 1971, pp. S-1 to S-16.

Schweitzer, Philip A.: *Handbook of Valves,* Industrial Press, New York, 1972.

section 5

Instrumentation and Automatic Controls

by

Ranjit S. Randhawa
The Foxboro Company,
Foxboro, Massachusetts

chapter 5-1

Instrumentation

TERMINOLOGY

Accuracy An ideal instrument would be one whose input-to-output relationship has a defined linear or nonlinear equation. Instruments, however, suffer from errors such as hysteresis, deadband, conformity, and repeatability. *Hysteresis* and *deadband* errors are illustrated in Fig. 1-1.

Hysteresis error Suppose, as input increases, the reading follows along curve *OBA*. If input returns to *O*, the reading may follow *AB'O*, a different path. The difference *BB'* at any input is the *hysteresis* error, a repeatable error due to energy absorption by the instrument element (typically a spring, diaphragm, or magnetic core).

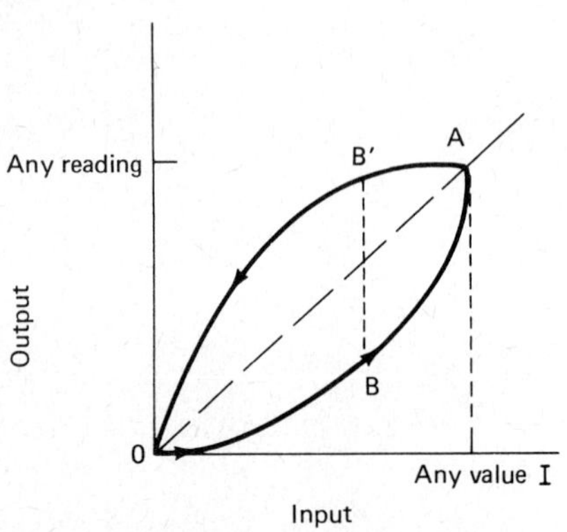

Figure 1-1 Hysteresis and deadband errors.

Deadband error Suppose input increases from *O* to any value *I*, and the output reading does not change (due to friction, for example) until the input is at *I*, whereupon the output reading jumps to *A*. If the input reverses, returning to *O*, the output may remain at *A* until the input returns to *O*; then the output jumps to *O*. The value of *I* is the *dead band*, the amount by which the input changes for a change in output.

Conformity The maximum deviation of an instrument's actual calibration curve as compared to its specified characteristic curve is called its *conformity*. There are two methods by which a numerical value for conformity is derived. First is the terminal-based method whereby the calibration curve is forced to coincide at the end points with the specified characteristic curve. With the second method, the calibration curve is forced to concide only with the lower-range value of the characteristic curve of the instrument.

Repeatability The closeness of agreement among a number of consecutive measurements of the output for the same value of the input under the same operating conditions, i.e., approaching from the same direction for full-range traverses, is called repeatability. It is usually measured as nonrepeatability and expressed as repeatability in percent of span. It does *not* include hysteresis since input measurements are varied in only one direction. See Fig. 1-2.

Accuracy rating This is a number or quantity that defines a limit which errors will not exceed when a device is used under specified operating conditions. When operating conditions are not specified, reference conditions are assumed. For example, a primary flow device may lose accuracy at higher fluid temperatures. Unless specified otherwise, however, a reference or standard test temperature is assumed.

As a performance specification, accuracy (or reference accuracy) is assumed to mean the accuracy rating of the device when used at reference operating conditions. The units being used must be stated explicitly. Preferably, a ± sign should precede the number or quantity, but absence of a sign is taken to mean ±.

Accuracy rating can be expressed in a number of ways. The following examples are typical:

1. Accuracy rating expressed in terms of the *measured variable itself*. Typical expression: The accuracy rating for a certain temperature recorder-indicator is ±2°F or ±1°C.
2. Accuracy rating expressed in terms of *span*. Typical expression: The accuracy rating is ±0.5 percent of span. (This percentage is calculated using scale units such as degrees Fahrenheit, pounds per square inch gauge, etc.). With that accuracy rating, a pressure transmitter with a span of 200 psig (1400 kPa) would be inaccurate by ± 1 psig (6.89 kPa).

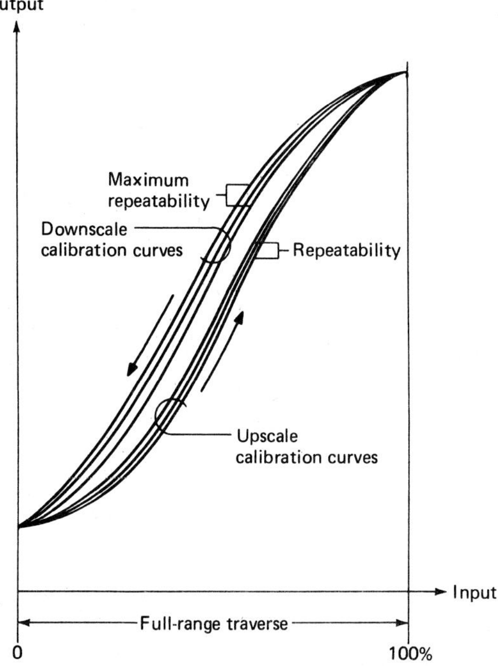

Figure 1-2 Repeatability.

3. Accuracy rating expressed in percent of the *upper-range value.* Typical expression: The accuracy rating is ±0.5 percent of the upper-range value. (This percentage is calculated using scale units such as kilopascals, degrees Fahrenheit, etc.) Thus, a temperature recorder with an upper range of 50-mV thermocouple input would be inaccurate by ±2.5 mV.
4. Accuracy rating expressed in percent of *scale length.* Typical expression: The accuracy rating is ±0.5 percent of scale length. A manometer with a scale length of 20 in (0.5 m) would have an error of 0.1 in (±0.25 cm).
5. Accuracy rating expressed in percent of *actual output reading.* Typical expression: The accuracy rating is ±1 percent of actual output reading. Thus, a manometer reading of 0.2 m with a ±1 percent reading error would have an error of ±0.2 cm *at that reading.*

Span The algebraic difference between the upper and lower range values. For example:

1. Range 0 to 150°F—span 150°F
2. Range −20 to 200°F—span 220°F
3. Range 20 to 150°C—span 130°C

For multirange devices, this definition applies to the particular range that the device is set to measure.

Span adjustment Means provided in an instrument to change the slope of the input-output curve. See Fig. 1-3.

Zero adjustment Means provided in an instrument to produce a parallel shift of an input-ouput curve. See Fig. 1-3.

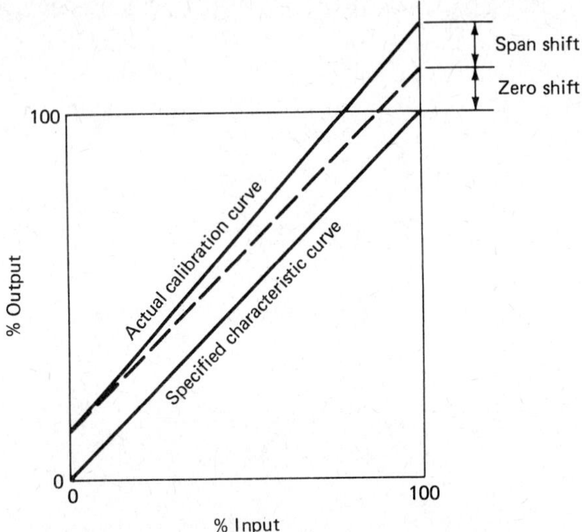

Figure 1-3 Accuracy rating span shift and zero shift.

Elevated zero range A range in which the zero value of the measured variable, measured signal, etc., is greater than the lower range value. The zero may be between the lower- and upper-range values, at the upper-range value, or above the upper-range value. Examples: range −20 to 200°F, −100 to −10°C. The terms *suppression, suppressed range,* or *suppressed span* are also frequently used to denote elevated zero range.

Suppressed zero range A range in which the zero value of the measured variable is less than the lower range value. For example: 20 to 100 scale range. The terms *elevation, elevated range,* and *elevated span* are also used to denote suppressed zero range.

Reproducibility The closeness of agreement among repeated measurements of the output for the same value of input made under the same operating conditions over a period of time, approaching from both directions. It is usually measured as a nonreproducibility and expressed as reproducibility in percent of span for a specified time interval.

INTRODUCTION

Measurement of physical phenomena is the basis of all technical activity. Obtaining, displaying, and conveying the quantitative facts of physical processes are the prime functions of instrumentation. This chapter presents the practical fundamentals of this technology.

LEVEL-MEASUREMENT METHODS

Level, as a process variable, is a common measurement both for control and indication. Various methods are used, and the selection of any one is based on many factors. Refer to Table 1-1.

Float Method

This is the simplest of level-measuring methods and makes use of a float which essentially follows the level in a closed or open vessel. The position of the float can be used to

TABLE 1-1 Level-Measurement Methods

Method	Type of liquid: Clean	Hard to handle	Solid	Range	Relative cost	Output type
Float	Good	Fair	——	75 mm–15 m (3 in–50 ft)	Low	Contact or tape
Displacement	Fair	Poor	——	150 mm–4 m (6 in–12 ft)	Med	Contact and/or signal
Head (pressure)	Good	Excellent	——	50 mm on (2 in on)	Med	Signal
Differential pressure	Good	Excellent	——	130 mm on (5 in on)	Med	Signal
Air (gas) bubbler	Fair	Fair	——	250 mm–75 m (10 in–250 ft)	Low	Signal
Capacitance	Good	Fair	Good	Wide	Med	Signal
Conductivity	Fair	——	Poor	Point	Med	Contact
Thermal	Good	——	——	Point	Med	Contact
Radiation	Good	Excellent	Good	Wide	High	Signal
Weighing	Good	Excellent	Good	Wide	High	Signal
Ultrasonic	Good	Excellent	Good	Wide	High	Signal

sense the level at a predetermined point by magnetically coupling the float to a mercury switch or miniature-type switch. This method is normally used for sensing high and/or low levels in a vessel. In large storage tanks, where a local indication or recording of the level is required, the float is coupled with a tape or cable which then, through a pulley mechanism, positions a local pointer for indication or applies a torque in a recorder. In both applications turbulence is kept to a minimum.

Displacement Method

The method employs a displacer which is located so that it is totally immersed when the level is at its predetermined maximum point (Fig. 1-4). The amount of force acting on the displacer is equal to the weight of the liquid displaced. The displacer weighs more than the maximum amount of liquid it can displace and is normally cylindrical in shape so that the relationship of buoyancy to submersion is linear. The displacer is linked to a torque tube which twists linearly with the buoyance of the displacer.

The buoyant force can be determined by

$$F = V\frac{L_w}{L}\mathrm{D} \tag{1}$$

where

V = total displacer volume
L_w = working length of displacer
L = total length of displacer
D = density of fluid

The level measurement is independent of the pressure in the vessel but may be subject to problems where extreme turbulence is encountered. Special precautions have to be taken under these circumstances.

The same principle of measurement can be used for measuring density. Here the displacer is kept fully immersed; the buoyant force is then a function of the span of density to be measured. A further application of this type of measurement is the interface level between two liquids. The interface level is allowed to vary over the length of the displacer while it is fully immersed. The force now depends upon the difference in densities of the two liquids. Standard displacers are designed for a range of buoyancies from 1.47 lb (0.67 kg) to about 12 lb (5.45 kg), including the weight of the hanger assembly.

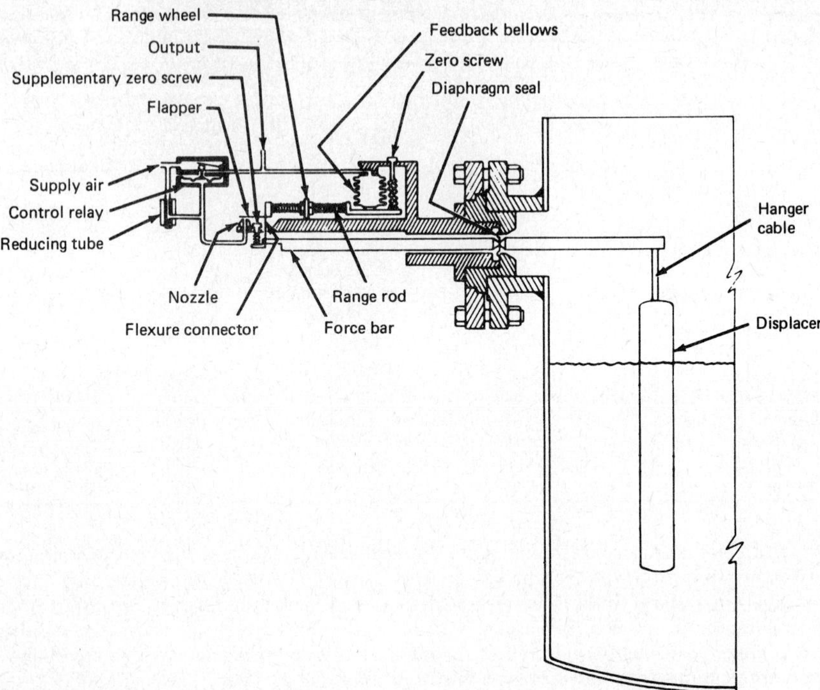

Figure 1-4 Schematic diagram of buoyancy transmitter with displacer.

Head-Pressure Method

This is a useful means of measuring level when a pressure transmitter is used to convert the head pressure in an open tank to an equivalent level. The transmitter output can be used for remote indication or control. The difficulties in this type of measurement usually arise because most tanks are closed vessels that are also pressurized, thus causing variations in pressure to affect level measurements. Furthermore, temperature variations

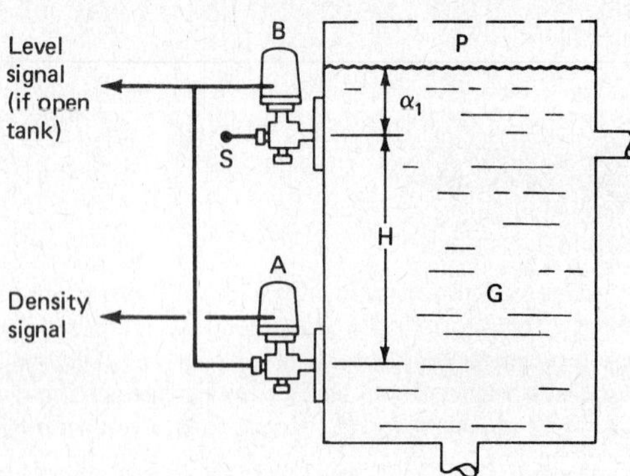

Figure 1-5 Differential pressure method of density measurement. *Key: A,* differential pressure transmitter; *B,* 1:1 repeater; *G,* specific gravity of liquid; *P,* pressure.

may cause density, too, to vary, which will then affect the level measurement. These problems limit the usefulness of this method for level measurements, but it is an excellent method for determining the density of liquids by keeping the level of the liquid constant in either a vessel with constant pressure or an open vessel with no pressure. An installation where pressure variations in the vessel can be taken into account will use a differential pressure measurement where the tank pressure is subtracted from the measured head. The differential pressure transmitter should be calibrated for a range of $H(G_2 - G_1)$, and its elevation is HG_1. Refer to Fig. 1-5.

Differential-Pressure Method

This is the most common method of measurement for both control and indication. Refer to Fig. 1-6.

A typical force-balance type of differential pressure (*d/p*) pneumatic transmitter (the d/p Cell*) is shown in Fig. 1-7. An electric transmitter would work essentially on the

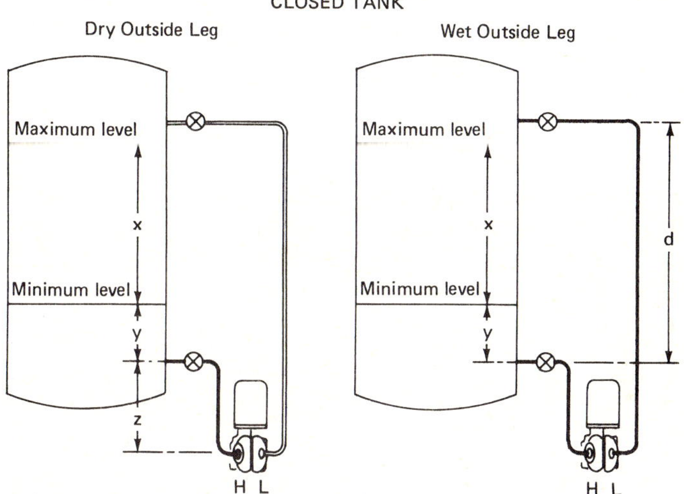

Figure 1-6 Differential-pressure level measurement. For dry outside leg, Span = xG_L and Elevation = $yG_L + zG_S$; for wet outside leg, Span = xG_L and Suppression = $dG_S - yG_L$. In both cases, G_L is the specific gravity of the liquid in the tank and G_S is the specific gravity of the liquid in the outside filled line. If the transmitter is at the level of the lower tank tap or if an air purge is used, $z = 0$.

same principle (force-balance) except that the detection, output, and feedback-balancing force would be accomplished with electric components, e.g., a differential transformer and a feedback motor. The output is normally a current signal, the standard being 4 to 20 mA.

Air Bubbler Method

This is a common measuring method for large, open-storage vessels. The level is measured by determining the pressure required to force air or gas into the liquid at a point beneath the surface. Figures 1-8 to 1-10 indicate various types of installations. The bubble pipe can be of any material and is notched at the end, as shown in Fig. 1-11. This prevents large bubbles from forming.

*d/p Cell is a trademark of The Foxboro Company, Foxboro, Massachusetts.

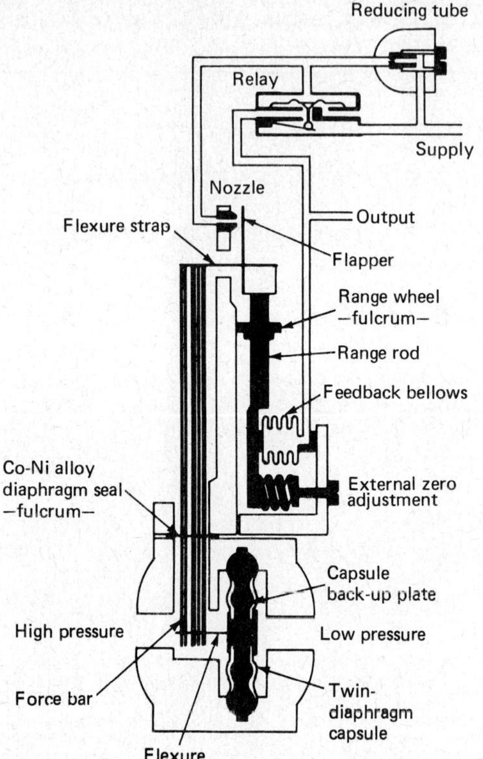

Figure 1-7 Differential-pressure transmitter. (*The Foxboro Company.*)

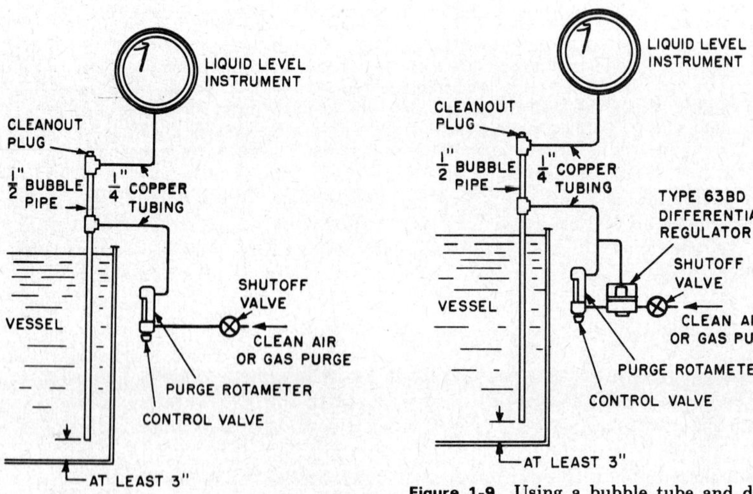

Figure 1-8 Using a bubble tube.

Figure 1-9 Using a bubble tube and differential-pressure regulator.

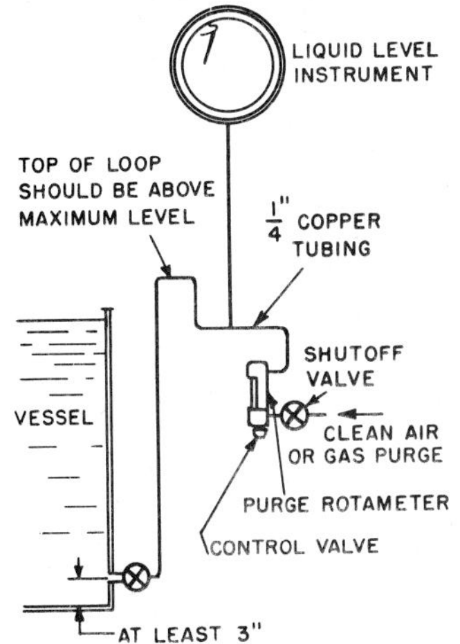

Figure 1-10 Purging directly into the side of a vessel.

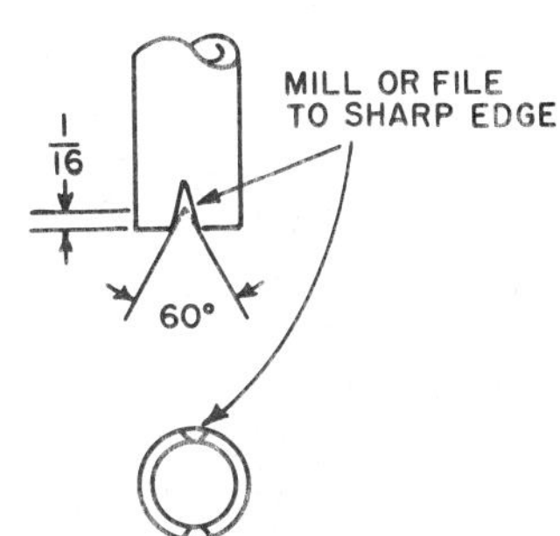

Figure 1-11 Detail of the notch in a bubble pipe.

Capacitance Method

The capacitance between two concentric cylinders is a direct function of the dielectric material between the two cylinders. Level measurements are accomplished by using a probe (one plate of the capacitor) and the tank, which acts as the second plate. As the level varies, the capacitance varies linearly, and this change in capacitance can be detected by using a bridge excited by a high-frequency oscillator. For nonconductive materials, an uninsulated probe can be used. Conductive materials require that the probe be coated with an insulator.

A number of possible errors can occur. For example, changes in the composition and in the temperature of a material will cause changes in its dielectric constant. The amount of water in a material has a profound effect on its dielectric constant. These changes will directly affect the level measurement. If the liquid tends to wet or adhere to the probe, then the level is measured only as far as the the probe is wetted, especially if the material is conductive. This is one of the serious drawbacks of this measuring method. The accuracy depends upon the differential capacitance, which is usually designed to be greater than 10 pF. The narrower this capacitive span, the more difficult the measurement.

Conductivity Method

This method is applicable for single-point measurement with a conductive process material. Two probes are used and placed such that an electric path is provided through the conductive material when both probes are covered. This is used to close the circuit of a relay, thereby providing a contact output for a certain level. Multipoint measurements can be made to provide point measurements for different levels. The method is inexpensive and the equipment fairly rugged, but it is not intrinsically safe unless special precautions are taken.

Thermal Element Method

The thermal element method is again used for point measurements and depends upon the fact the thermal conductivity of most process liquids is much higher than its associated vapor. When the thermal element comes in contact with liquid, the rate of heat transfer increases, causing its resistance to increase. This increased resistance is detected by an increased voltage drop which can then be used to activate a relay. The equipment used in this method is rugged and not affected by process changes, but the method is not intrinsically safe, and the heat input may cause changes in the quality of the measured product. This method, in general, has limited use in practice.

Radiation-Difference Method

This method is very useful in those applications in which the dangerous or extremely corrosive nature of the process fluid requires a noncontact type of measurement. The method depends upon the absorption of radiation by the process material. As the level of this material increases or decreases, the amount of radiation absorbed increases or decreases. This can then be determined by using a low-level detector, such as the Geiger-Müller tube. This method is applicable both for point measurement or a range of measurements. Safety considerations have to be kept in mind when considering installation, especially in terms of radiation exposure. The normal sources of radiation used are radium, cesium 13, and cobalt.

Weighing Method

This method, like radiation, is well suited for measuring process materials which do not allow contact-type measurement methods. Basically, the vessel weight is continuously measured by using either strain-gauge-type weigh cells or hydraulic pressure. Knowing the weight of the empty vessel, the weight of the process material in the vessel can be related to the depth of the material in the tank. Special precautions have to be taken to balance the vessel so that associated piping connected to the vessel does not affect the actual weight. The method has limited use in actual process control compared with some of the other methods, e.g., differential-pressure measurement using isolating seals for corrosive process materials.

Ultrasonic Method

This method uses an ultrasonic oscillator (20,000 Hz and higher) to excite a sensor. For point measurements, the signal is picked up by another sensor as long as a transmission path is available. When the level rises, the path is interrupted, thereby indicating the level. Another method is to allow the process material to damp the vibration of the sensor. This damping can be detected to provide a point measurement. Continuous measurements can be accomplished by using intermittent transmission and measuring electronically the time taken for the reflected signal to return to the sensor. This method is applicable to both liquids and solids; measurement of the level of solids in silos or storage bins has been the most popular application.

PRESSURE-MEASUREMENT INSTRUMENTS

Overview

Pressure is a fundamental process variable and its measurement can be used either directly for control or to infer other measurements, for example, level, flow, and temperature. There are many types of transducers that can be used. The most common are listed in Table 1-2.

The transducers can be linked to either a pneumatic or an electronic transmitter to develop a signal 3 to 15 psig (0.02 to 0.1 MPa) or 4 to 20 mA. The heart of the pneumatic transmitter is the flapper-nozzle assembly, including the pneumatic relay. The output from the relay is controlled by the back pressure developed in the nozzle which, in turn, is developed by the relationship of the flapper to the nozzle opening. Figures 1-12 and 1-13 indicate this relationship and a typical relay.

TABLE 1-2 Pressure-Measurement Elements*

Element	Local	Remote	Range	Contact with process	Relative cost	Accuracy, % of span
Bourdon tube	✓	✓	0–12,000 psig (0–83 MPa)	Yes*	Low	±0.25–5
Bellows		✓	0–3000 psig (0–20 MPa)	Yes*	Med	±0.5–1
Manometer	✓		0.05 in Hg–1 atm (3–100 kPa)	Yes	Low	0.1–1
Diaphragm	✓	✓	0.05 in Hg–1 atm (3–100 kPa)	Yes	Med	±0.5
Strain gauge		✓	1 atm on (0.1 MPa on)	No	High	±1
Dead weight	✓		1 atm on (0.1 MPa on)	No	High	High

*A chemical seal can be used for isolation.

There are two arrangements for making use of the flapper-nozzle relationship. These arrangements lead to either *motion-balance* or *force-balance* instruments. Figures 1-14 and 1-15 indicate schematics of the two types. Force-balance instruments tend to be more widely used because of their ruggedness, high degree of insensitivity to vibration, and method of mounting.

Electronic transmitters generally tend to be of the force-balance type. Figure 1-16 indicates a typical layout of a differential transformer type.

The pressure applied to the bellows capsule is applied to the force bar through a flexure and is transmitted to the detector. The lever system moves the ferrite disk (part of the detector), causing a change in the output of the differential transformer. This change in output is sensed by the amplifier oscillator circuit, and its output is rectified to provide the transmission signal. The feedback coil, in series with the output signal, supplies the balancing feedback force.

There are other types of detectors that have also been used: the inductance type, where the movement of the ferrite piece changes the air gap, which then changes the inductance of an oscillator circuit and thereby changes its output; and the variable reluctance type, where the movement of an armature causes the inductance ratio between two coils to change. The two coils are part of a bridge circuit which is rebalanced by a feedback amplifier, thereby changing the capacitance in the bridge. These devices can be considered to be motion-balance instruments.

Bourdon Tube

Simplest of the pressure-sensing elements, the Bourdon tube consists of a curved flattened tube with one end sealed and either connected to an indicator through linkage and gears or applied through a flexure mechanism to the force bar of a transmitter. The other end is connected to the process. Figure 1-17 shows a typical layout of an indicator. As pressure is applied, the tube tends to straighten, causing a small movement of the tip.

There are a number of other types of tube designs that have been developed. Figure 1-18 illustrates some of them.

The material of the tube can vary from bronze for some low-pressure ranges [up to 400 lb/in^2 (2.76 MPa)] to Ni-Span C* alloy which allows a C Bourdon tube to range up to 12,000 lb/in^2 (83 MPa). Absolute pressure measurements are limited to a maximum of 100 lb/in^2 (0.7 MPa).

In applications where the process fluid is extremely corrosive, it is normal to use pressure seals, connected directly or through capillary tubing. The system is solidly filled

*Ni-Span C is a trademark of Huntington Alloys, Inc., Huntington, West Virginia.

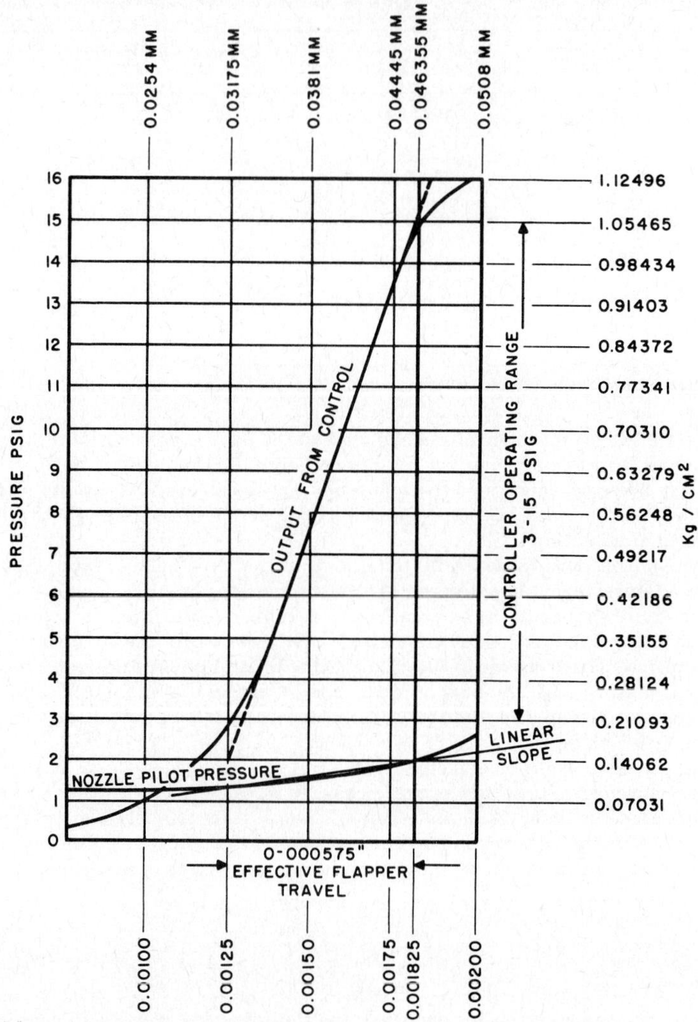

Figure 1-12 Curves for flapper-nozzle relation vs. operating pressure.

with a suitable liquid transmission medium. Process pressure is applied to the flexible member of the seal cavity in the measuring element, causing element movement in proportion to applied pressure. The spring rate of the flexible member must be less than that of the measuring element to allow full-range operation. A number of filling fluids have been used, the standard being DC-704, a silicone-base (Dow Corning Co.) product. This fluid has a low thermal coefficient of expansion with temperature limitation of 0 to 700°F (−18 to 370°C). The fluid must completely fill the system, as pockets of gas or air can cause large errors. There are many different types of seals available. Figure 1-19 indicates some of them.

Bellows

Refer to Fig. 1-20. Widely used both for measurement and control, bellows can be arranged as motion-balance or force-balance instruments. Figure 1-20 indicates the use

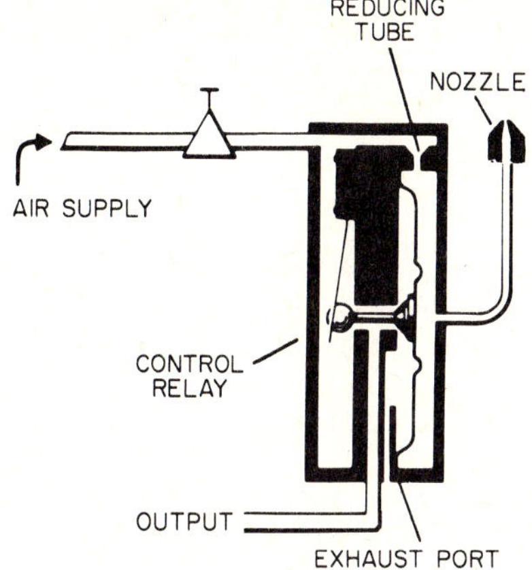

Figure 1-13 Pneumatic relay G.

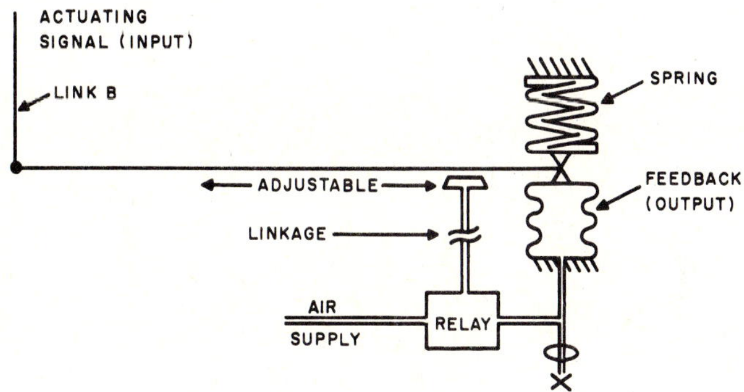

Figure 1-14 Motion-balance schematic.

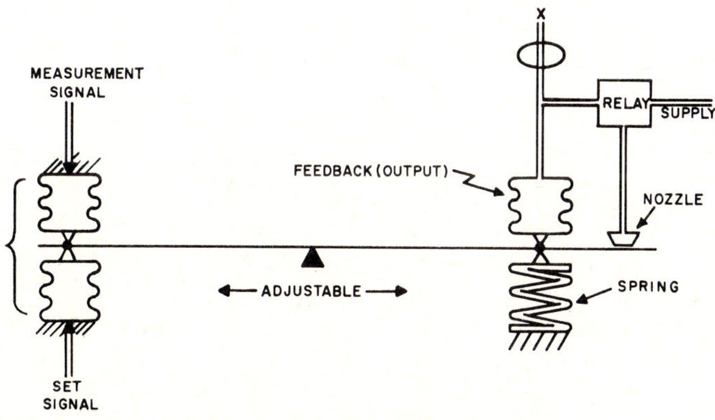

Figure 1-15 Force-balance schematic.

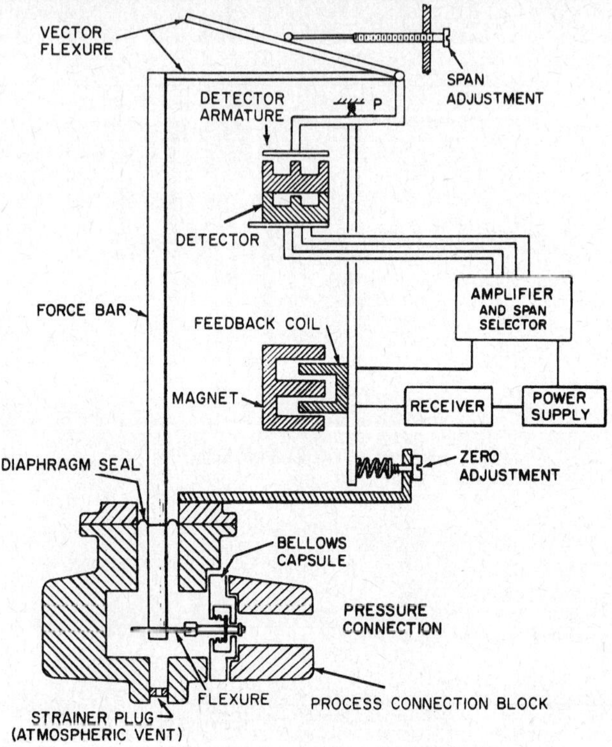

Figure 1-16 Differential transformer-type pressure transmitter.

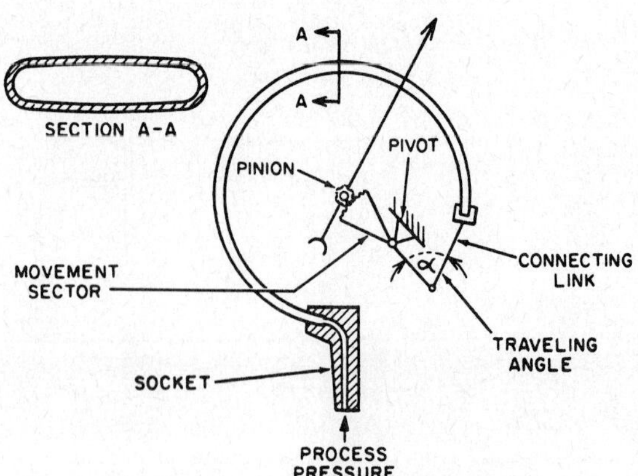

Figure 1-17 C Bourdon pressure element.

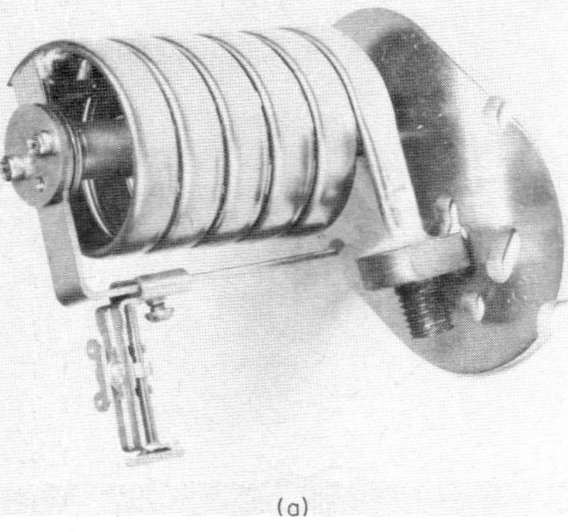

(a)

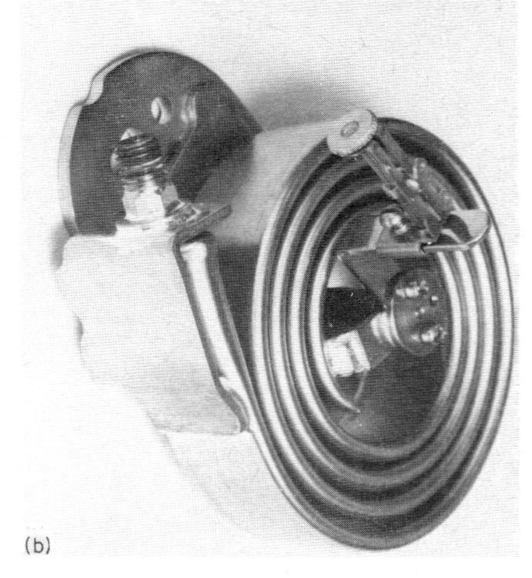

(b)

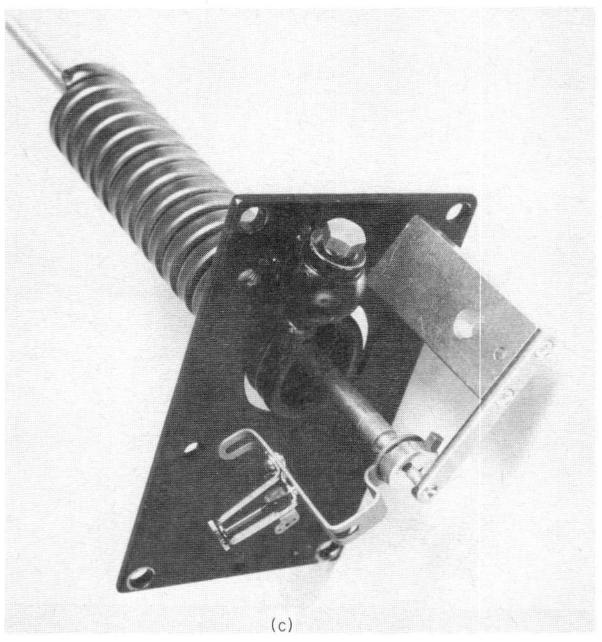

(c)

(d)

Figure 1-18 Bourdon-tube pressure elements. (*The Foxboro Company.*)

(a)

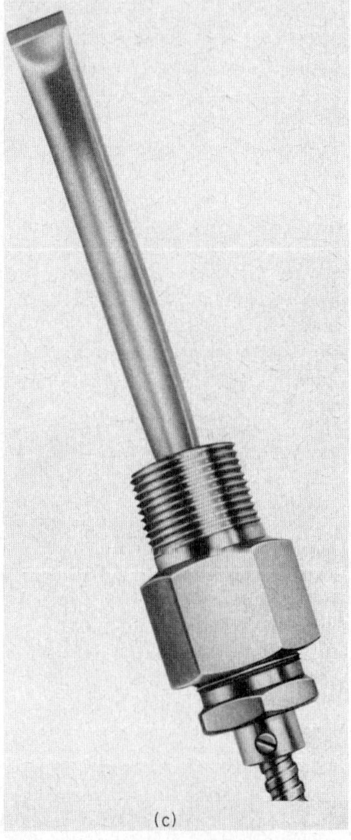
(c)

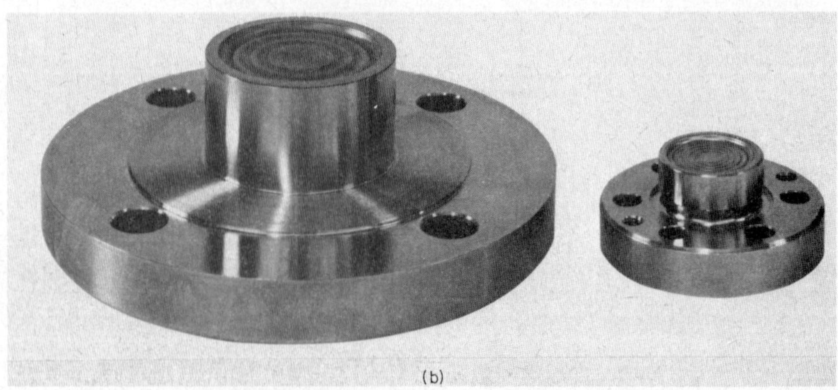
(b)

Figure 1-19 Pressure seals. (*The Foxboro Company.*)

(d)

(e)

Figure 1-19 (*continued*) Pressure seals. (*The Foxboro Company.*)

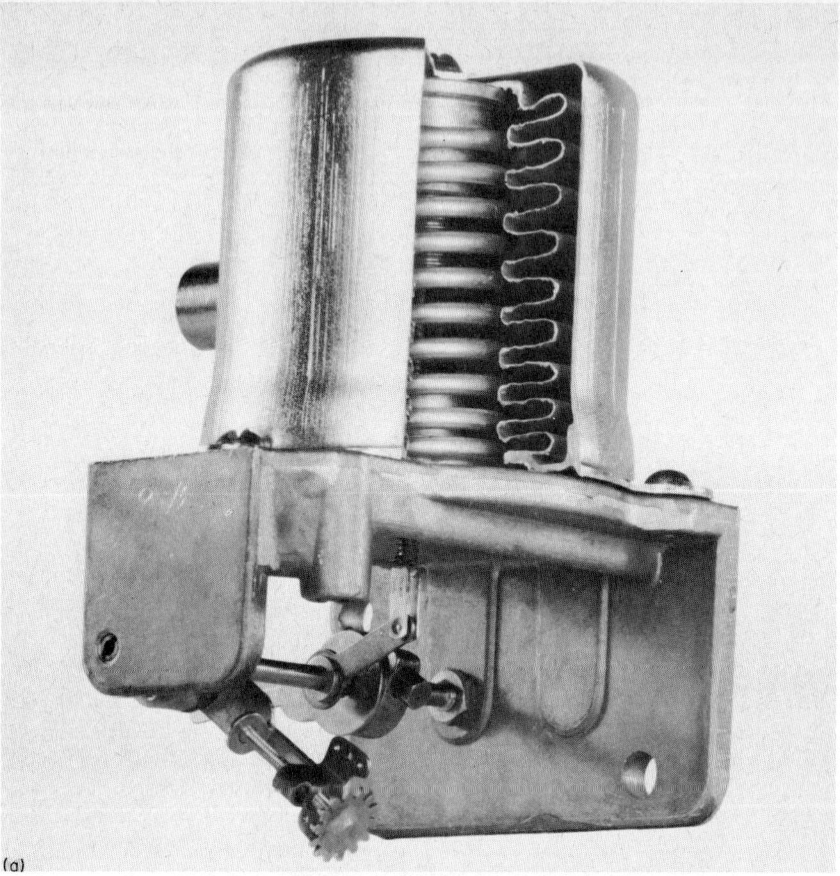

(a)

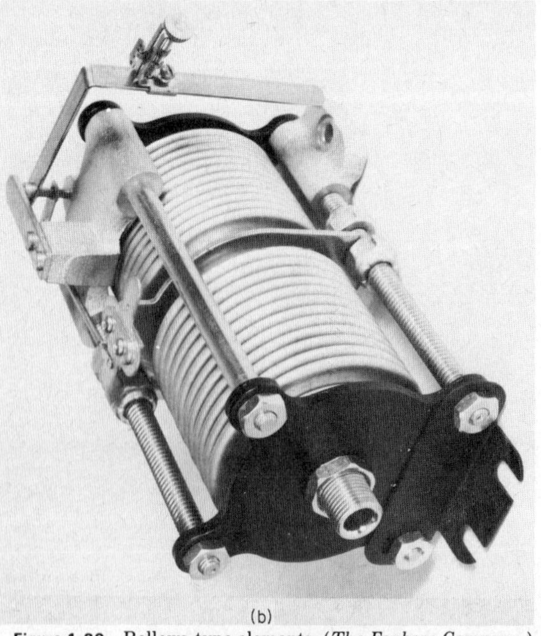

(b)

Figure 1-20 Bellows-type elements. (*The Foxboro Company.*)

of a bellows capsule for pressure measurement. Two-bellows systems are used when compensation is required, e.g., in absolute pressure measurement. One of the bellows is completely evacuated and acts in opposition to the measurement bellows. As atmospheric pressure varies, the evacuated bellows expands or contracts, thereby adding or subtracting from the measured absolute pressure. The size of the bellows can be varied for different pressure ranges from 0 to 7 lb/in² (0 to 50 kPa) to 0 to 2000 lb/in² (0 to 14 MPa).

Diaphragm

Refer to Fig. 1-21. The pressure-measuring element consists of a series of diaphragms which are connected together. Application of pressure causes the elements to expand,

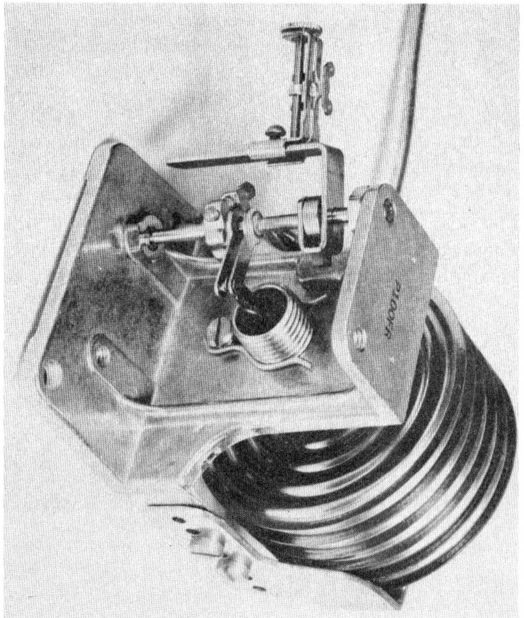

Figure 1-21 Diaphragm element. (*The Foxboro Company.*)

and the movement is then used for indication. These elements are typically used for low pressure applications of less than 5 lb/in² (1 kPa).

Figure 1-22 indicates a differential-pressure transmitter using a diaphragm capsule. The size of the capsule determines the range. Differential-pressure ranges from as low as 0 to 5 in H_2O (0 to 20 mm H_2O) to 0 to 850 in H_2O (0 to 330 cm H_2O) under static pressures up to 6000 lb/in² (41 MPa) are available.

Special Instruments

There are other types of pressure-measuring instruments that are available but have limited application in process industries. A manometer and a deadweight type are still quite popular, primarily for calibration of other pressure elements. The deadweight tester converts a known weight into a given liquid pressure which is then used in the calibration of an instrument.

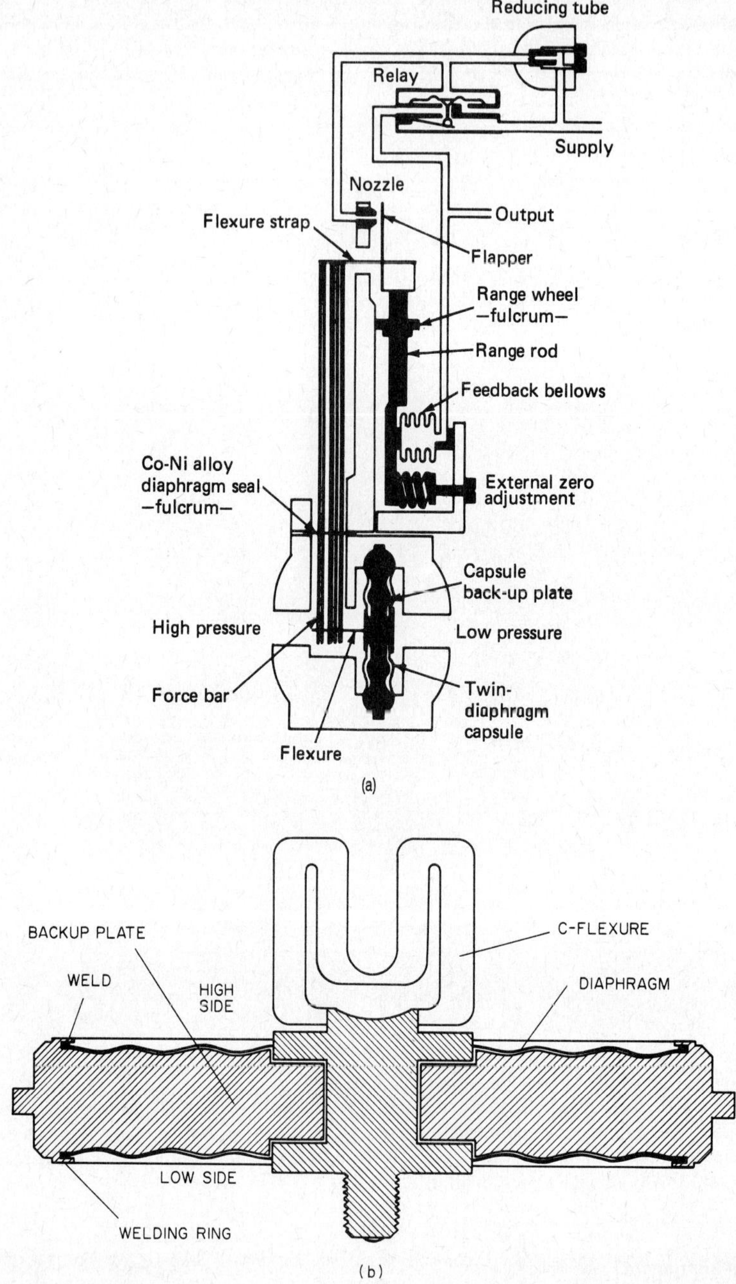

Figure 1-22 Diaphragm capsule for differential pressure measurement: (*a*) in situ in transmitter; (*b*) detailed structure.

TEMPERATURE-MEASURING METHODS

Refer to Table 1-3 for the more common methods used for measuring temperature in an industrial process environment. Each of these is discussed in detail in the following text.

TABLE 1-3 Temperature-Measuring Methods

	Local	Remote	Range, °F (°C)	Accuracy ±% span	Cost	Ease in replacement of components
Filled thermal system	✓	✓	−450 to 1400 (−260 to 760)	0.5 to 2	Low–med	Intermediate
Thermocouples		✓	−420 to 2000 (−250 to 1100)	0.3 to 1	Med	Easy
Resistance measurement		✓	−450 to 1625 (−260 to 900)	0.1 to 0.5	High	Difficult
Thermistors		✓	−200 to 500 (−130 to 260)	0.1	Med	Difficult
Bimetallic	✓		−80 to 800 (−60 to 420)	1 to 2	Low	Easy
Pyrometers	✓		High temperature	1 to 2	High	—
Paints	✓		100 to 2000 (40 to 1100)	—	Low	—

Filled Thermal Systems

The basis of measurement consists of a bulb connected by a capillary to a helical or C Bourdon tube element. The system is filled under pressure so that an increase in pressure causes a movement of the helical or Bourdon tube element. This movement can then be linked for local indication or through a transmitter for an electric or pneumatic signal.

Filled systems are classifed by SAMA (Scientific Apparatus Makers Association) into four classes with subclassifications in each class.

Classes IA and IB

These are liquid-filled systems. Class IA refers to fully compensated systems while Class IB refers to case-compensated systems. Fully compensated refers to compensation of ambient temperature variation along the length of the connecting capillary tubing. It consists of (in addition to the normal measuring bulb) a connecting capillary and pressure element, with a second capillary tube and pressure element. The two capillary tubes are run together while the pressure elements are connected such that equal amounts of motion nullify each other. Thus ambient-temperature variation along the length of the capillary tubes and at the case produce equal movements of the pressure elements and hence cancel the effect on the output of the instrument. Class IB, *case compensation,* nullifies ambient temperature variations at the case in a similar fashion but may only use a bimetallic strip instead of a pressure element. These instruments are generally mounted very close to the bulb.

The initial pressure in this class of instruments is very high, making it a volumetric instrument, free of static errors from bulb position with respect to the instrument case. Furthermore, being volumetric devices, the amount of available volume for expansion is limited, also limiting the amount of available overrange.

These instruments can also be used for measuring temperature differences by using two bulbs with associated capillary tubing and pressure elements. The pressure elements are arranged in a manner similar to Class IA fully compensated systems. Ambient-temperature variations are minimized by running the capillary tubes as close to each other as possible. Temperature-difference spans greater than 36°F (20°C) are normally required.

Classes IIA, IIB, IIC, and IID

In these systems, the bulb, capillary tube, and pressure-measuring element are partially filled with a volatile fluid such that the vapor-liquid interface level occurs in the bulb. As the temperature rises, the vapor pressure increases and this increase can then be sensed via the pressure element. The system is stable and unaffected by ambient-temperature variations. They are the simplest to manufacture and therefore least expensive, both initially and in use.

The difference characterizing the four subclassifications is as follows: In a Class IIA system the capillary tube and the pressure element are filled with liquid while the bulb has both vapor and liquid. These systems are used when the bulb temperature is higher than the ambient temperature. Class IIB systems are the reverse of Class IIA, with the bulb used for temperatures less than the ambient temperature. In Class IIA systems, the effect of the liquid head in the capillary has to be considered if the bulb is installed above or below the instrument case. Class IIC systems are systems where the bulb temperature crosses the ambient temperature over its operating range. This causes the capillary to be filled with liquid or vapor, resulting in static head errors unless the bulb is mounted at the same level as the case. Class IIC instruments are normally used as local indicators. Class IID systems are like Class IIC systems, except that the capillary tube and the pressure element are filled with a nonvolatile fluid, while the bulb contains liquid and vapor.

In all these classes the output is nonuniform and depends upon the vapor-pressure-vs.-temperature curve of the volatile fluid. Applications are geared such that normal operating temperatures are at the higher end of the scale where the scales are much wider, allowing greater readability.

Class III

These are gas-filled systems which are, like Class II systems, simple to construct, with wide applicable range limits. They are normally used with large bulbs which override the pressure variations in the capillary tube due to ambient-temperature changes. Gas pressure allows less power to be developed in the pressure-sensing element, making this class suited for wide temperature spans. The large bulbs are also well suited for average temperature measurements. Case compensation using a bimetallic strip is used in Class IIIB systems. Class IIIA systems use the normal parallel thermal system without a bulb.

Thermocouples

These sensors have in recent years achieved greater popularity with increasing use of electronic instruments. Basically, a thermocouple consists of two dissimilar metal wires, such as iron and constantan, joined to produce a thermal electromotive force (emf) when the junctions are at different temperatures. The measuring or hot junction is the end inserted in the medium where the temperature is to be measured. The reference or cold junction is the open end that is normally connected to the measuring-instrument terminals. The emf generated is a function of the difference in junction temperatures. Refer to Fig. 1-23.

The thermal emf developed by a thermocouple of two homogeneous metals is independent of the temperature gradient and distribution along the wires. Inhomogeneity in a thermocouple passing through a temperature gradient will produce undesirable emf's with resultant errors in temperature readings.

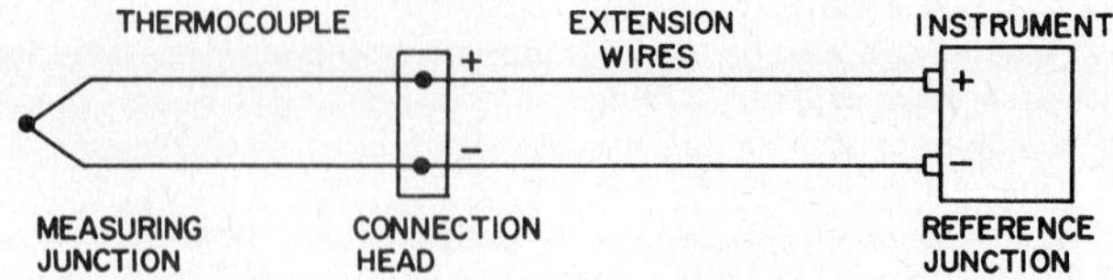

Figure 1-23 Thermocouple connection with extension wires.

TABLE 1-4 Thermocouple Comparison

ISA type	Positive wire	Negative wire	Recommended temp °F (°C)	Atmospheric conditions	Standard limits of error
T	Copper	Constantan	−450 to +750 (−270 to 400)	Oxidizing reducing	±2% of reading at low temperatures to ±¾% of reading at high temperatures
J	Iron	Constantan	0 to +1650 (−18 to +900)	Reducing	±¾% of reading at high temperatures
K	Chromel	Alumel	0 to 2300 (−18 to 1260)	Oxidizing or neutral	±¾% of reading at high temperatures
E	Chromel	Constantan	−300 to +1600 (−180 to 870)	Oxidizing	±¾% of reading at high temperatures
R, S	Platinum-rhodium	Platinum	0 to 2800 (−18 to 1530)	Oxidizing	0.5% of reading
B	Pt70–Rh30	Pt94–Rh6	0 to 3200 (−18 to 1760°C)	Inert or slow oxidizing	0.5% of reading

Introduction of intermediate metals into a thermocouple circuit will not affect the emf of the circuit, provided that the new junctions remain at the same temperature as the original junction. This fact is used in applying extension wires to cut costs, since some of the thermocouples are made of expensive metals. These extension wires must be made of materials having the same thermal emf characteristics as the thermocouple. If copper wires are used, temperature fluctuations at the reference junction (in this case the connection head) will introduce errors in proportion to the magnitude of the temperature changes. Use of proper extension wires moves the reference junction to the measuring instrument where temperature compensation can be applied. Figure 1-24 indicates the relationship of five standard thermocouples.

Table 1-4 highlights the different features of some standard thermocouples.

Figure 1-25 illustrates a simplified circuit of a Foxboro Series 33B transmitter. TC is

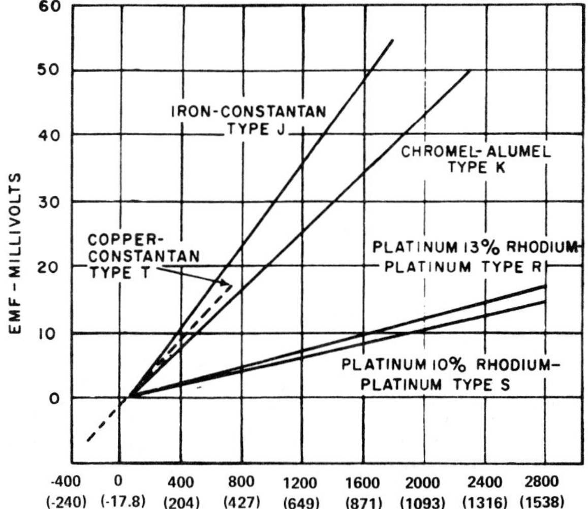

Figure 1-24 Thermocouple relationships.

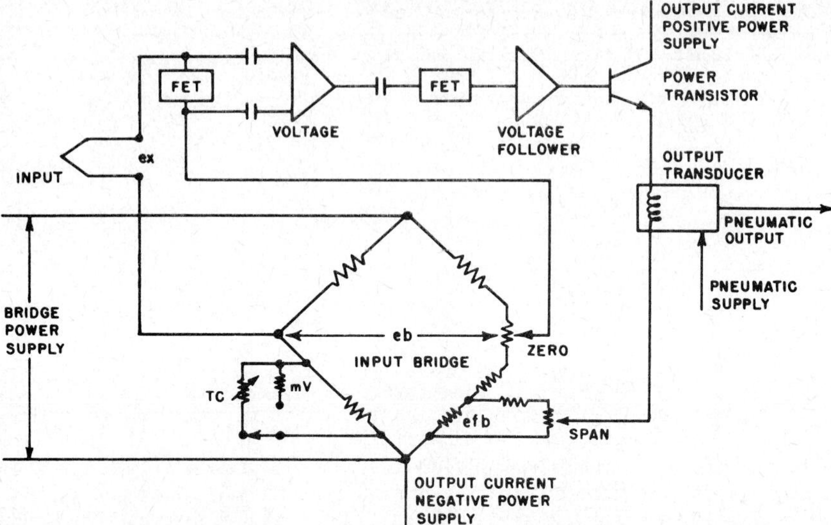

Figure 1-25 Simplified circuit diagram of Foxboro 33B series transmitter. (*The Foxboro Company.*)

a temperature-sensitive resistor which varies the bridge output as the actual reference-junction temperature varies from the design reference-junction temperature.

It should be noted that thermocouples are point-measuring devices. To get average temperature one should connect a number of thermocouples in parallel as shown in Fig. 1-26. Figure 1-27 shows the connection of two thermocouples for differential pressure measurement. The following points should be considered:

1. Swamping resistors are required to compensate for varying resistance of the thermocouple and extension wires.
2. In order to prevent ground loop currents the thermocouples must *not* be grounded.
3. The thermocouples and extension wires *must* be of the same type.
4. In differential measurement no reference-temperature compensation is required. Copper leads can be used between the connection box and the instrument.

Resistance Temperature Detectors

Based on the change of electric conductivity with temperature, the resistance temperature detector (RTD) consists of a coil of wire made of materials such as nickel and platinum. Spans as low as 5°F (3°C) can be achieved with a nickel RTD with a certified accuracy of 0.1°F (0.06°C). The sensor can be constructed with two, three, or four leads. For most industrial applications, two- or three-lead sensors are used in a Wheatstone bridge as shown in Fig. 1-28.

Differential temperature differences can also be measured by using two equal RTDs in adjacent sides of a Wheatstone bridge. The linearity of measurement is to an extent a function of the span of measurement. For wide-span instruments and extreme accuracy, nonlinear scales and charts are recommended. A certain amount of linearity can be achieved by connecting large-value equal resistors in series with the leads of the sensor.

Thermistors

These are metal oxide resistors having high temperature coefficients (usually negative), with resistance being a function of absolute temperature. They are used much like the

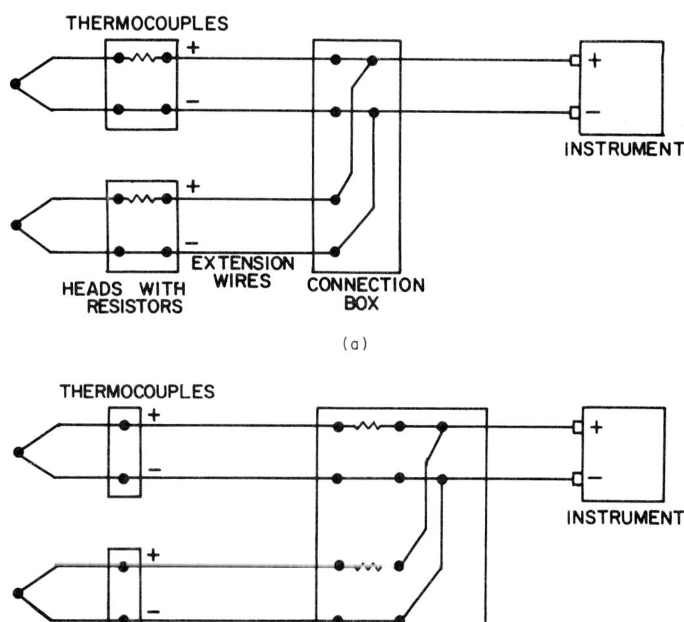

Figure 1-26 Average temperature circuits.

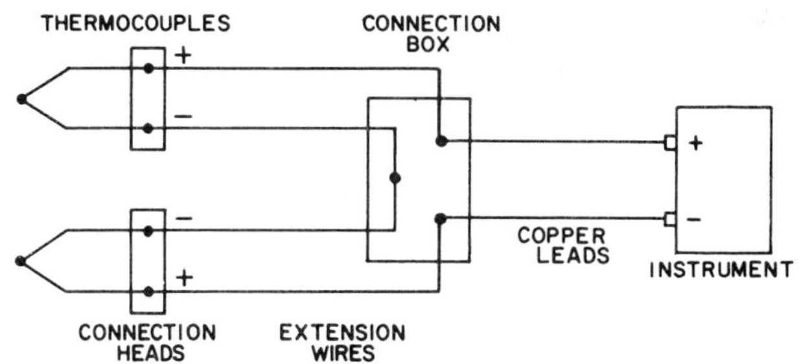

Figure 1-27 Differential temperature circuit. (*The Foxboro Company.*)

RTDs. The large temperature coefficient makes them useful for very narrow-span temperature measurement. Self-heating is a problem to be considered when current flow through the sensor causes sensor temperature to be higher than ambient temperature.

Though not widely used in process environment, the application of thermistors in various types of electronic circuitry for temperature compensation has been widespread. The temperature-resistance characteristic curve tends to be fairly nonlinear, making interchangeability of sensors a problem.

Radiation Pyrometers

These instruments are a special class used in those applications where very high temperatures are to be measured and normal contact-type measurement is not possible. The

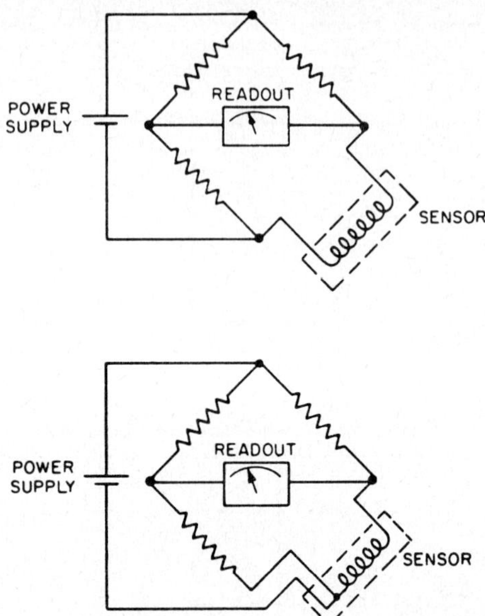

Figure 1-28 Two- and three-wire Wheatstone bridge.

basic principle is the use of an optical system to focus energy radiated to a detector. The system can be manual where this focused energy is compared with a calibrated optical filament, or automatic by using thermopiles or photomultiplier tubes. The accuracy depends on many factors, such as the emissivity of the source, reflections from other sources, and the line of sight. The temperature scale is nonlinear and the system tends to be expensive.

FLOW MEASUREMENT

The primary purpose of industrial control systems is to balance the material and energy flows in a process. Flow is the most common of the process variables. Accurage measurement and control are the two most important instrumentation functions. Table 1-5 lists some of the more common measurement methods and their characteristics.

Head-Type Devices

These are the most common types of measurement devices. Basically they depend upon a constriction in the fluid flow, thereby causing a pressure drop which can then be measured by a differential-pressure type of instrument.

Orifice Plates

Figure 1-29 indicates the pressure profile for an orifice plate.

The *vena contracta* is the location where the downstream flow has the maximum velocity and the minimum cross-sectional area. Figure 1-30 indicates this location with respect to the diameter ratio of orifice opening to pipe inside diameter. Figure 1-31 illustrates the various types of orifice plates.

The location of the orifice plate depends upon the type of taps used for the measurement of differential pressure. A certain section of straight pipe is required both upstream and downstream. Figure 1-32 is an example of the dimensions required using straight-

TABLE 1-5 Flow Measurements*

Head type	Liquids	Viscous liquid	Slurry	Gas	Solids	Linear	Rangeability	Cost	% full-scale accuracy	Indirect totalizer	Pressure loss
1. Orifice plates	✓	L		✓		SR	4:1	Low	¼–2	✓	High
2. Rotameters	✓	L	L	✓		✓	10:1	Med	½–2	—	F
3. Venturi tubes, nozzles	✓	L	✓	✓		SR	4:1	High	¼–3	✓	Med
4. Pitot tubes	✓			✓		SR	3:1	Low	2–5	—	L
5. Elbow	✓	L	L	✓		SR	3:1	Low	5–10	—	No
6. Target meters	✓	L	L			SR	4:1	Med	½–2	✓	High
7. Weirs, flumes	✓	L	L			NL	100:1	Low	2–5	—	Med
Velocity type											
1. Magnetic	✓	✓	✓			✓	20:1	High	½–1	✓	No
2. Vortex	✓	L				✓	10:1	Med	½–2	✓	Med
Displacement											
1. Positive displacement	✓	L				✓	20:1	Med	¼–1	✓	Med
2. Turbine	✓	L	L	✓		✓	20:1	Med	¼–1	✓	Med
Mass flow											
1. Weight types					✓	✓	20:1	Med	½–3	—	—
2. Solids flowmeters					✓	✓	20:1	Med	½–3	—	—

*L, limited; NL, nonlinear; SR, square root; F, fixed.

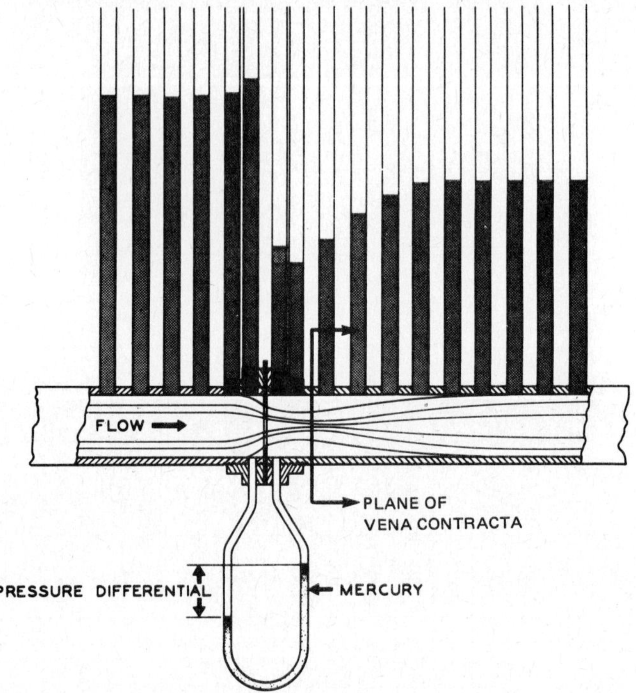

Figure 1-29 Pressure profile using an orifice plate.

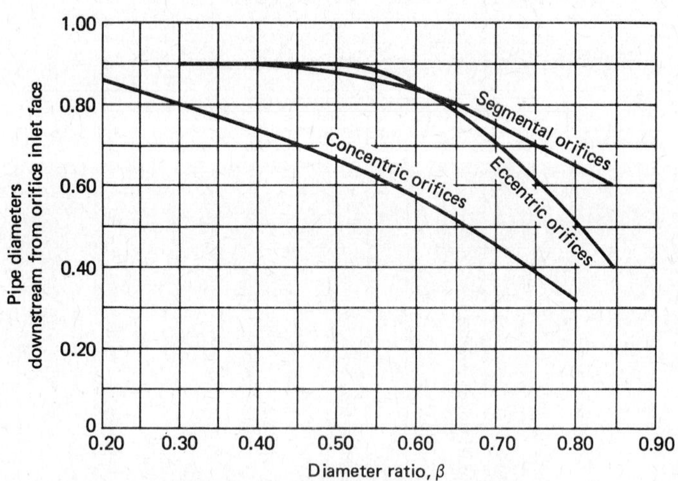

Figure 1-30 Location of the vena contracta.

ening vanes, which are devices that eliminate swirls, crosscurrents, and eddies set up by pipe fillings and valves in the upstream run.

Other types of primary elements that are used are shown in Fig. 1-33.

Figure 1-29 indicates that the pressure downstream of the primary element does not recover to its full value. Figure 1-34 indicates the permanent head loss as a percent of measured differential.

CONCENTRIC ORIFICE PLATE ECCENTRIC ORIFICE PLATE SEGMENTAL ORIFICE PLATE

Figure 1-31 Types of orifice plates.

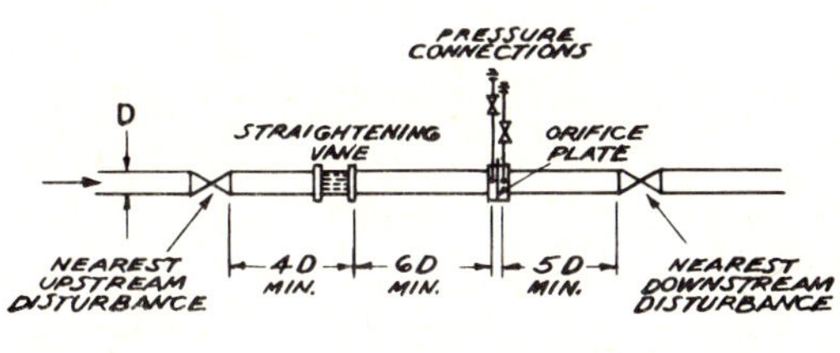

FLANGE TAPS

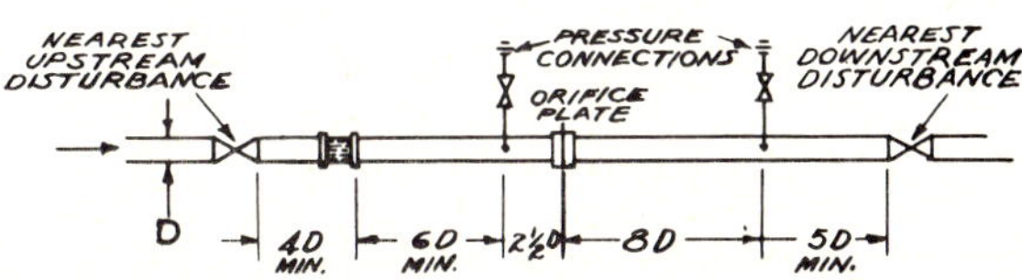

2 1/2 D AND 8 D TAPS

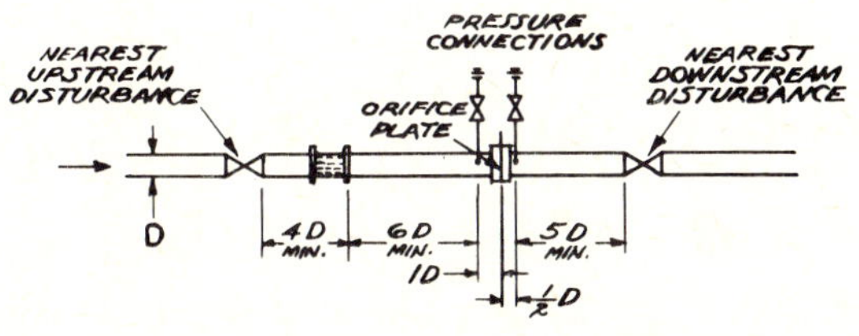

1 D AND 1/2 D TAPS

Figure 1-32 Liquid flow installation.

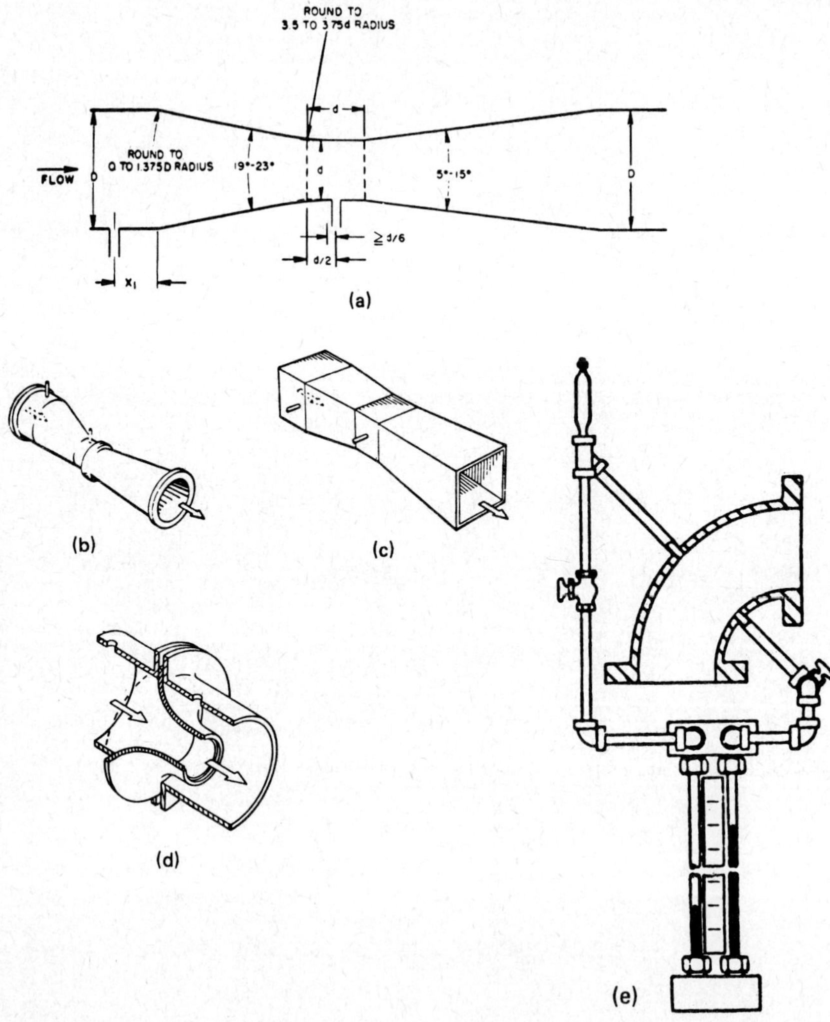

Figure 1-33 Flow nozzles, venturi tubes, and elbows. (*a*) The critical dimensions of the classical venturi tube. (*b*) An eccentric venturi tube. (*c*) A rectangular venturi tube. (*d*) Flow nozzle. (*e*) Elbow (used as primary device).

The following equations can be used for calculating flow rates for different services. Representative values for the S factor are given in Fig. 1-35.

Liquids:

$$\text{gal/min} = \frac{5.67 S D^2 \sqrt{h_w}}{G} \qquad \text{(to get L/min use 0.0066 instead of 5.67)}$$

Steam or other vapors:

$$\text{lb/h} = 359 S D^2 \sqrt{w h_w} \qquad \text{(to get kg/h use 0.01251 instead of 359)}$$

Gas:

$$\text{scfh} = 338.17 S D^2 F_g F_{tf} \sqrt{p h_w} \qquad \text{(to get Nm}^3\text{/h use 1 instead of 338.17)}$$

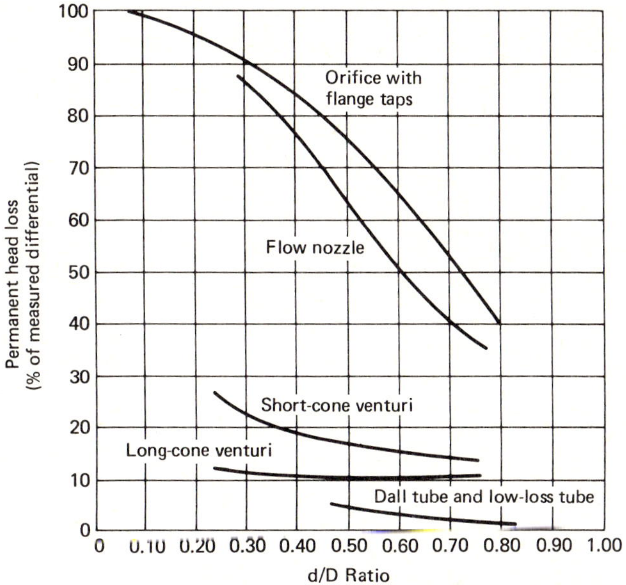

Figure 1-34 Permanent head loss for various devices.

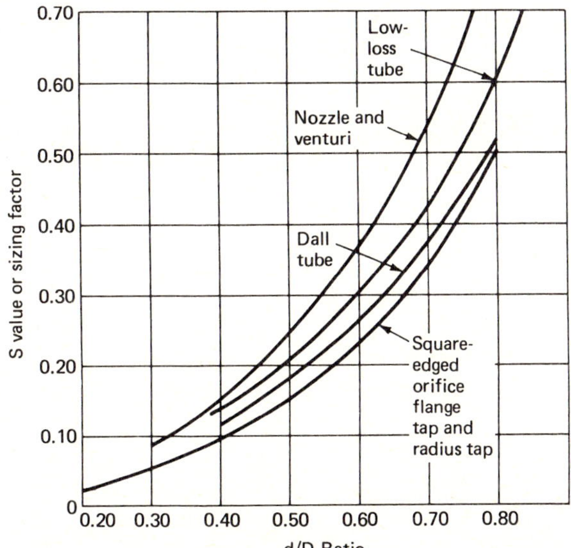

Figure 1-35 *S* values for various differential devices against beta ratios (d/D).

The standard condition is defined to be at 60°F (16°C) and 14.73 psia (760 mmHg), and the symbols in the equations are defined as follows:

S = flow coefficient (see Fig. 1-35)
G = specific gravity of the process fluid
h_w = differential pressure, mm H_20
w = steam or vapor density, lb/ft^3 (kg/m^3)

$F_g = \sqrt{1/G}$
F_{tf} = factor for flowing gas temperature, $\sqrt{520/(460 + °F)}$ or $\sqrt{228.7/(273.1 + °C)}$
P = static pressure psia (kg/cm² abs) [either p_1 (upstream) or p_2 (downstream) since the differential h_w is ordinarily chosen to be less than 4 percent of p]

Target Flowmeter

Another constriction-type device that has become popular recently is the target flowmeter. Figure 1-36 indicates a typical assembly with a pneumatic transmitter.

Steam- and water-flow measurements in outdoor installations are common applications. The complications of condensate pots, heat tracing, seal fluids, and antifreeze compounds to prevent freezing in cold weather are avoided.

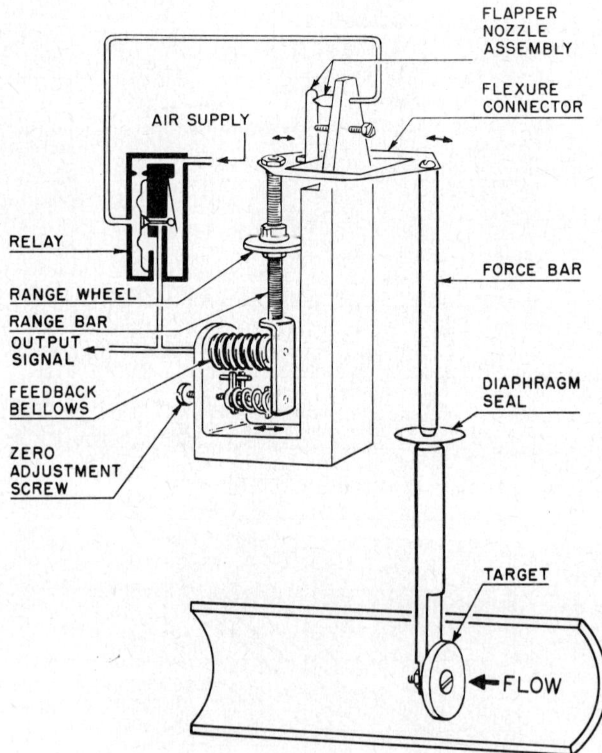

Figure 1-36 Target flowmeter. (*The Foxboro Company.*)

The flow in volume or mass units is proportional to the square root of the force on the target.

Measurement accuracy of the head-type meters discussed so far depends on the Reynolds number R_D, which can be defined by

$$R_D = \rho \frac{VD}{\mu}$$

where

ρ = density

V = velocity
D = pipe inside diameter (I.D.)
μ = viscosity

The higher the Reynolds number, the flatter the velocity profile of the flow across the pipe inside diameter. If we define a flow coefficient K_a as a factor that when multiplied

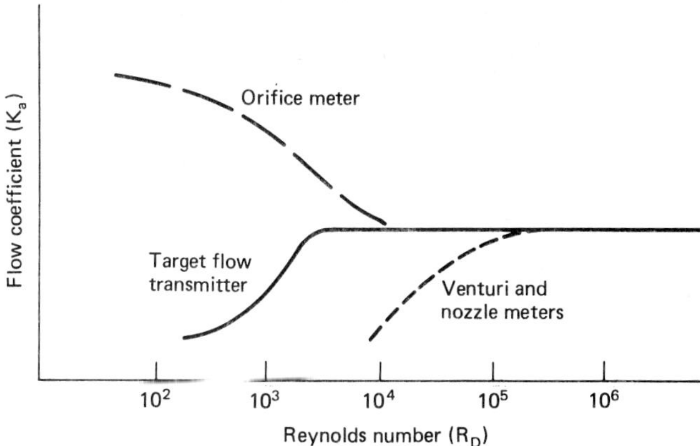

Figure 1-37 Typical flow-coefficient curves.

by the measured phenomena gives the required flow, then the plot of K_a vs. R_D for head-type meters is given in Fig. 1-37.

Weirs and Flumes

Open-channel measurements are important, especially in the waste- and water-treatment fields. Head H_a is developed by placing a weir in the flowpath as shown in Fig. 1-38. Aeration under the nappe is required for accurate flow measurement.

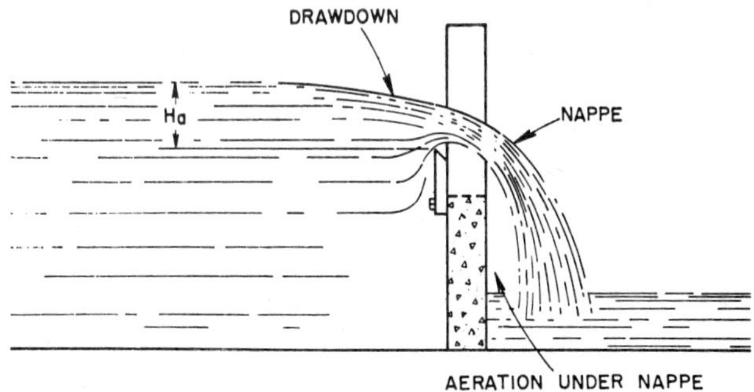

Figure 1-38 Aeration under nappe of weir.

Formulas for different types of weirs are given below. For a V-notch weir (Fig. 1-39*a*),

$$\text{Flow} = Q_{cfs} = 2.48 \tan(\theta/2)\, H^{5/2} \qquad \text{ft}^3/\text{s}$$

For a rectangular weir,

$$Q_{cfs} = 3.33\,(L - 0.2H)\, H^{3/2}$$

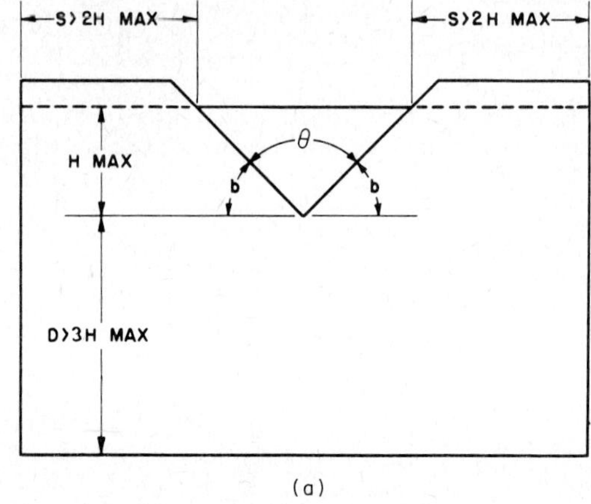

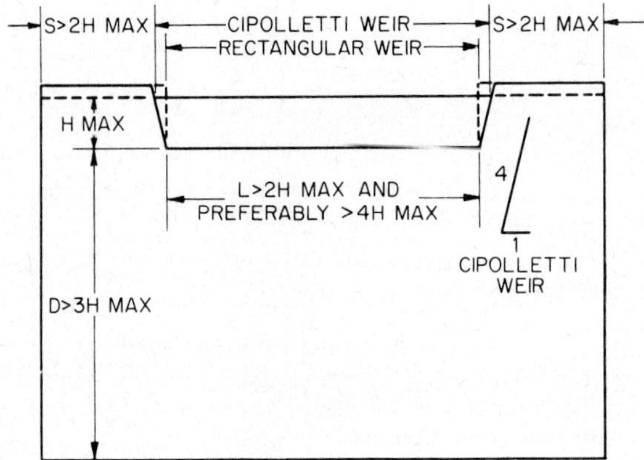

Figure 1-39 Weirs: (*a*) V-notch weir; (*b*) rectangular and Cipolletti weirs.

For a Cipolletti weir (Fig. 1.39*b*),

$$Q_{cfs} = 3.367LH^{3/2}$$

Figure 1-39 indicates *L, H,* and *D,* the typical dimensions for referenced weirs.

Flumes are low-head-loss measuring devices where a formed channel restriction changes static head to velocity head. Figure 1-40 illustrates the Parshall flume.

The three basic methods for measuring the head in a weir or flume are the float-and-cable, the in-flume float device, and the bubble tube.

Rotameters

Our discussion so far has emphasized meters having a variable differential head and a constant restriction area. Rotameters, however, are devices using a constant differential and a variable restriction area. These instruments are typically used in measurement of small liquid or gas flows with local indication. Transmitter-type instruments for measuring large amounts of liquid flow (e.g., oil) are commercially available.

This device consists of a vertical tapered tube through which the flow passes in an upward direction. A float moves up until the upward force acting on the float is balanced by the downward gravitational force. If the tube is made out of glass, the position of the

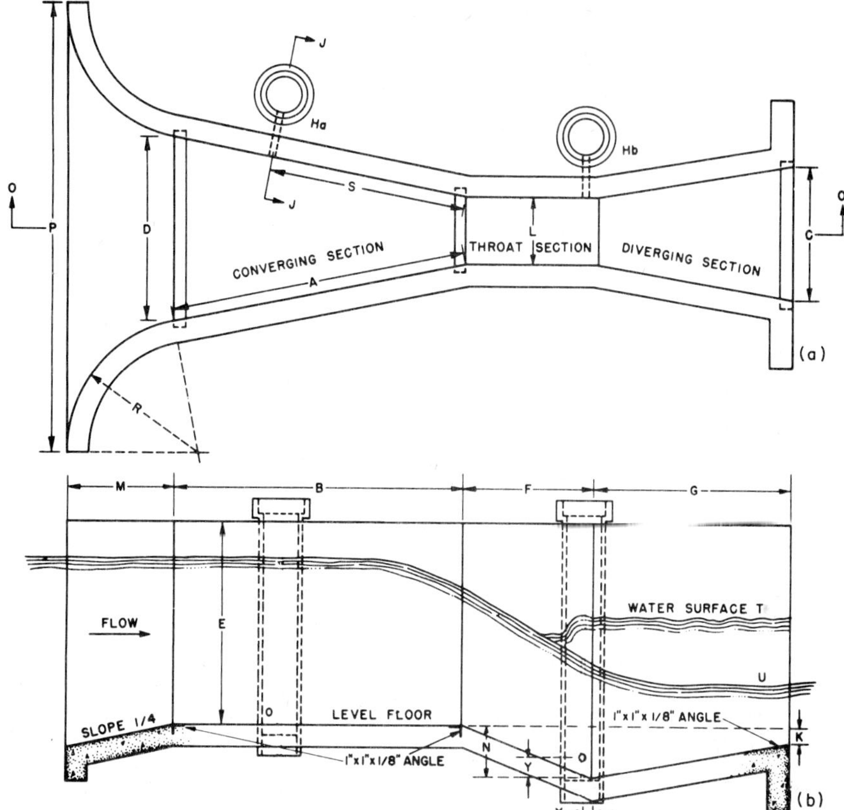

Figure 1-40 (*a*) Diagram and (*b*) dimensions of Parshall flume.

float is a direct and linear measurement of the flow. Transmitter-type instruments make use of a metallic tube in which the float is mechanically linked to the transmitter mechanism. Its accuracy is comparable to that of other head-type meters, with lower accuracy for indicating-type glass-tube meters. An advantage in transmitter-type instruments is the area available around the float for entrained fluid particles. The disadvantage is the mechanical linkage and its associated maintenance problems.

Velocity-Type Devices

The most common velocity device is the magnetic flowmeter. Its advantages are that no head loss occurs, it handles solids in suspension, no liquid connections are required, and an electronic output suitable for in-plant transmission is produced. Figure 1-41 shows a schematic drawing (*a*) and a cross section (*b*) of a typical instrument. Symbols used in Fig. 1-41 are listed:

E = generated voltage
C = meter constant
H = magnetic field
D = distance between conductors (pipe I.D.)
V = velocity of flow

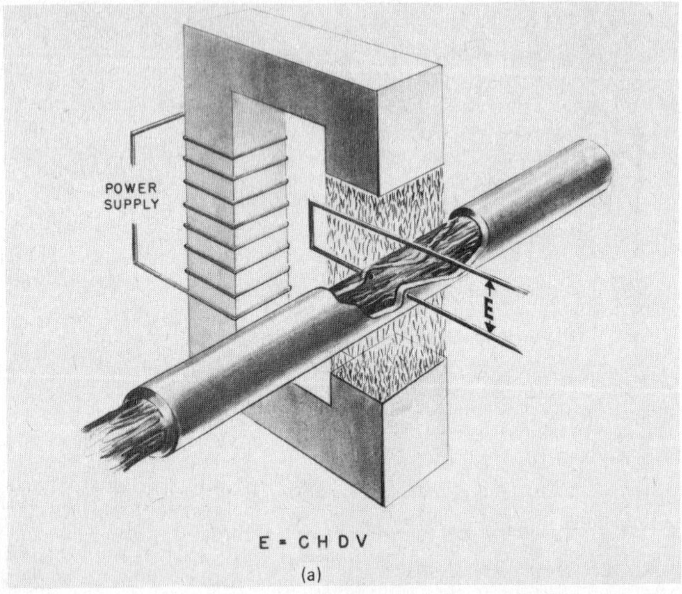

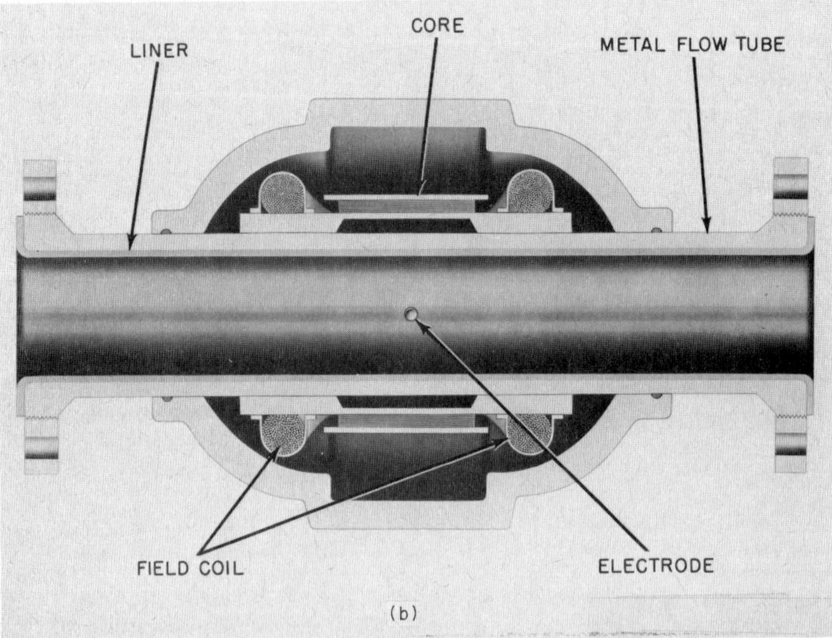

Figure 1-41 Magnetic flowmeter. (*a*) Schematic diagram. (*b*) Cross-sectional diagram. (*The Foxboro Company.*)

The output is linear with velocity for a constant magnetic field. The instrument is fairly rugged and can measure wide ranges of flow. To minimize current flow in the measuring circuit, a high-input impedance amplifier is used, with special precautions for shielding the input circuit.

Another velocity meter of recent development is the vortex-shedding meter. Liquid flowing through the meter housing passes a specially shaped vortex element which causes vortices to form and shed (separate) from alternate sides of the element at a rate proportional to the flow rate of the liquid. These vortices create an alternating differential pressure which is sensed by a detector located at the "tail" of the vortex generator. An ac voltage signal is produced in the flowmeter with a frequency synchronous with the vortex-shedding frequency.

The meter offers accuracies on a par with turbine and positive displacement meters but has the advantage of requiring no moving parts in the liquid stream. Therefore, the problems arising from overspeeding the turbine with two-phase flows, or from damage when slugs of liquid impinge upon it, do not exist. The meter equation is

$$f = kQ$$

where

f = frequency, pulses per minute
k = meter constant, pulses per unit volume
Q = flow rate

Figure 1-42 indicates the variation of k with respect to Reynolds number (signature curve) for a vortex flowmeter.

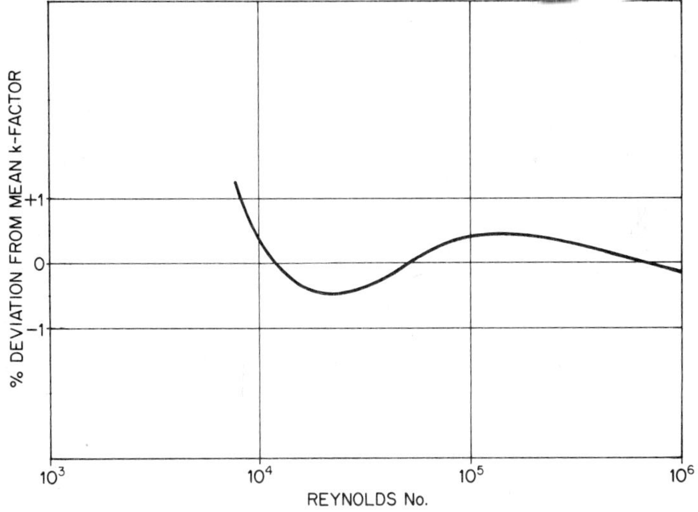

Figure 1-42 Typical signature curve of E83 vortex flowmeter.

Displacement Meters

There are many configurations used for these types of meters. A rotor is placed in the flowpath and turns as a function of the force imparted to it by the flowing fluid. This motion can either be mechanically linked to a totalizer indicator or magnetically coupled so that each rotation produces a pulse. The meter output is linear with flow and is capable of being used over a wide range. The advantages of this meter lie in its accuracy and ruggedness in clean fluid application. The two disadvantages are the susceptibility of the rotor bearing to dirt and its limitation under high-velocity gas flows. These may occur due to flashing of the liquid under certain conditions of process operation. *Turbine meters* have to be designed such that pressure drop across the meter does not cause the flowing fluid to flash (Fig. 1-43*a*). The meter accuracy is a function of viscosity, with the smaller-sized meters being affected more. Variation of k (equation similar to that for a vortex-shedding meter) is approximately ± 3 percent for a 1-in (2.5 cm) meter while it

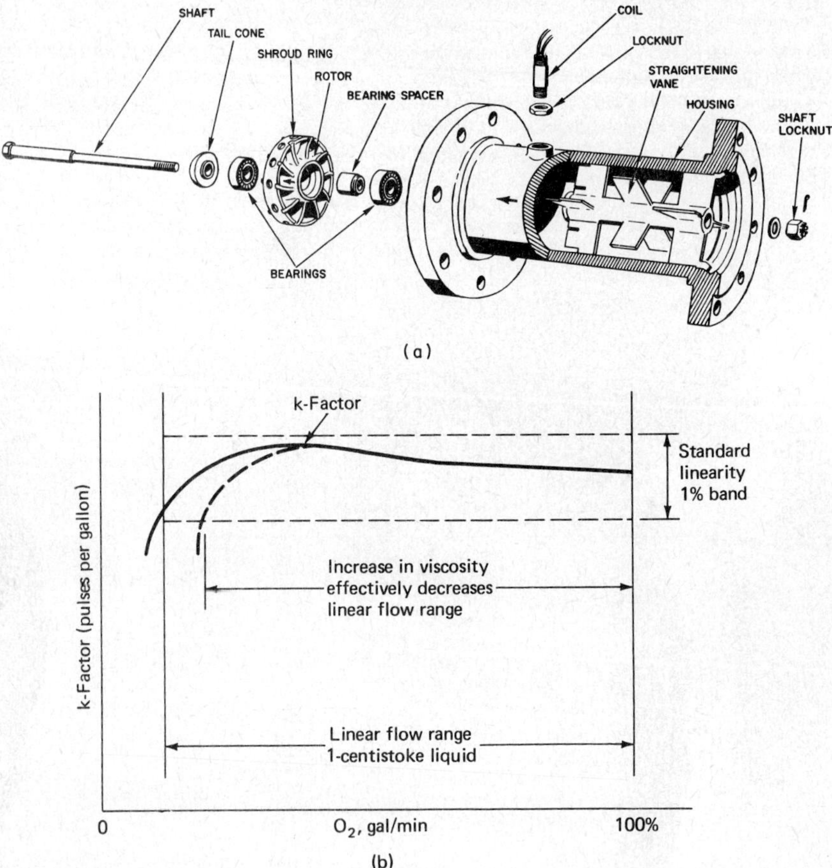

Figure 1-43 (*a*) Schematic of a turbine flowmeter. (*b*) A typical signature curve.

is approximately ±0.75 percent for a 6-in (10.5 cm) meter. Figure 1-43*b* is a typical signature curve of a turbine meter showing the effect of viscosity.

ANALYTICAL MEASUREMENTS

The successful operation of some complicated chemical processes is partially dependent upon analytical measurements and their use in process control. This discussion is limited to those measurements that can be made directly using a sensing electrode or other detectors. The accuracy and repeatability, to a certain extent, is a function of the "known" sample used for calibration. The application of the sensor is based to a great extent on the background chemistry of the process and, therefore, careful study of the alternatives is required before a particular choice can be made.

Chromatographs

These are general-purpose instruments used both for on-line process control and laboratory composition analysis. It would be difficult to run a modern chemical process without these instruments.

Components

The three basic components of a chromatograph are (1) the sample injection mechanism, (2) the separation column, and (3) the detector. Figure 1-44 shows the basic operation. As long as conditions within the analyzer remain the same, the three components indicated will appear at the same instant of time as measured from the start of the analysis. The height of the peak identifies the percent of that component present in the stream. The chart record at the end of the cycle is called a chromatogram.

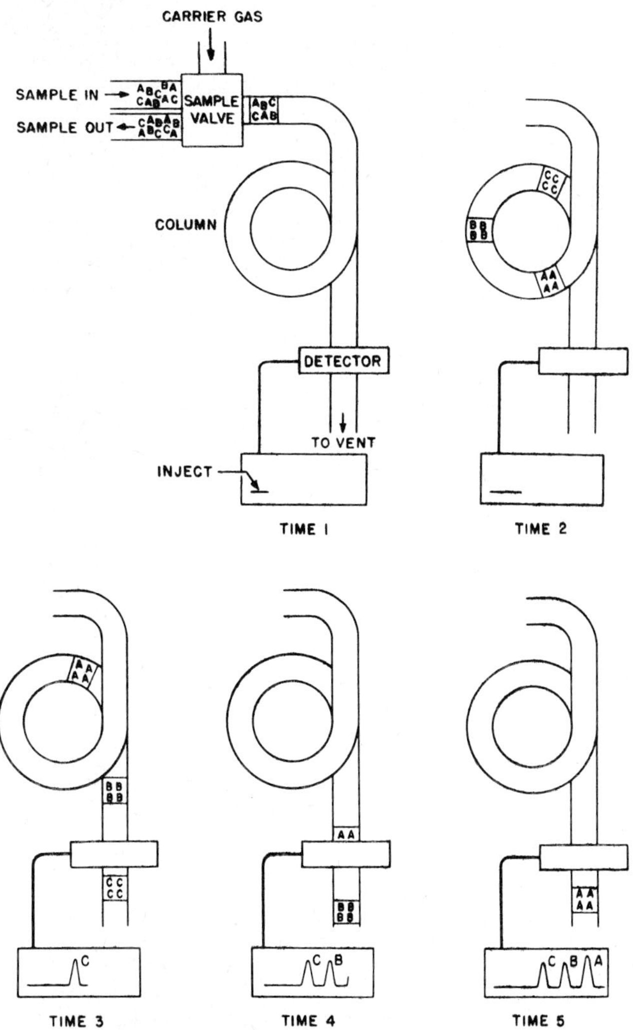

Figure 1-44 Chromatographic separation.

Analytical Conditions

The four major conditions that the analysis depends upon are (1) sample size, (2) carrier-gas flow rate, (3) analyzer temperature, and (4) carrier-gas pressure. Sample size must remain constant because detectors "see" only the actual amount of a given component, not a percent of the whole sample. If a sample twice as large as an earlier sample were used, the detector would show twice as much of a given component, assuming the sample was taken from the same source. Therefore, to get a true percent analysis the sample size must remain constant.

Carrier-gas flow rate and carrier-gas pressure are interrelated. A more rapid carrier-gas flow rate will "carry" the sample and components through the column faster, causing the time of elution to be shorter. Analyzer temperature must be held constant, since a higher temperature will cause the column elution rate to increase.

Columns

The column is used for separation on a time basis. It consists of a small-diameter tube made of type 316 stainless steel, varying in length from several inches (centimeters) to a few yards (meters). Three basic types of columns are used.

Partition Columns. The partition type uses a solid support, such as crushed firebrick coated with a high-boiling-point liquid (the liquid phase). Separation is achieved by the relative solubility of each component of the sample in the film of liquid (usually on oil) coating the support. A component of low solubility passes through the column much more quickly than one with high solubility. If the liquid phase had a high vapor pressure at the operating temperature, the flow of carrier gas would strip this liquid, leaving the column bare and thereby affecting the separating ability of the column. This phenomenon is known as *column bleed.*

Adsorption Columns. A second type of column is the adsorption column, where separation is based upon relative difference in adherence to an adsorbent material used to pack the column. These packing materials are surface-active solids such as charcoal, silica gel, and activated alumina. Components which adhere the least are eluted first.

Molecular Sieve. The third type of column includes Molecular Sieves, where the variation in molecular size of the components is used for the separation. The larger molecules

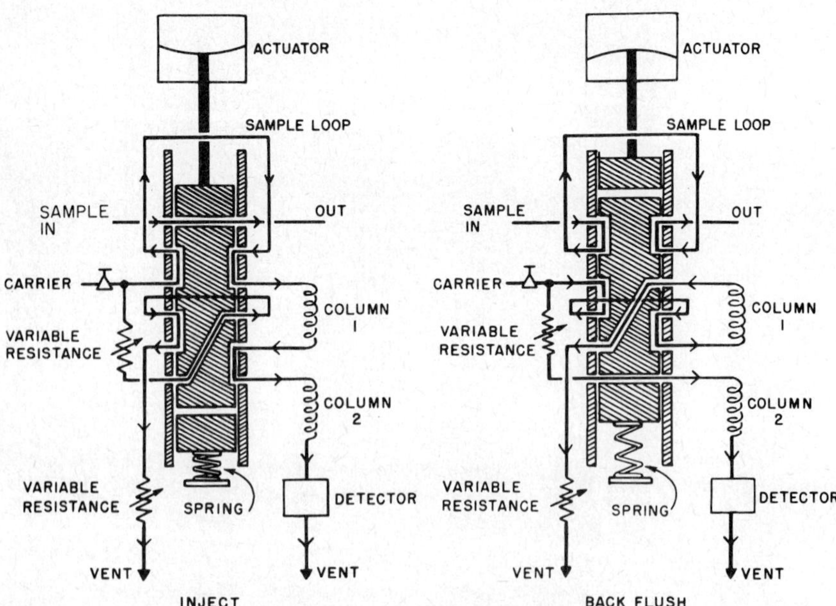

Figure 1-45 Schematic diagram of sample and back-flush valves.

are slowed down more than the smaller ones. These columns are most often used for hydrogen, oxygen, argon, nitrogen, carbon monoxide, and methane.

Backflushing. Figure 1-45 shows a schematic diagram of a sample and a backflush valve. Backflushing is used to preserve the life of a column and/or decrease the total cycle time for the particular application.

Detectors

There are many types of detectors that have been used. The thermal-conductivity type is based upon the difference in conductivity of the binary mixture flowing across the element. The flame-ionization type is based upon the ion current generated when the binary mixture of carrier gas and component is burned upon mixing with fuel, e.g., hydrogen. It does not work for inorganic compounds.

The gas-density type senses the change in density between the carrier gas and the binary mixture by using the change in differential density across an orifice or a "pneumatic" Wheatstone bridge.

The output of the detector is the chromatogram. Though this type of output is useful for laboratory-type analysis, a continuous-trend output is more useful for control. This is accomplished by incorporating extra circuitry to detect the peak for a given component and a memory circuit to hold this peak value between column cycles. The output can be a continuous signal which is used in the control loop.

Ion-Based Measurements

The most common of these is the measurement of pH. There are many applications for this measurement, with the field getting wider in scope. Some examples are:

1. Those chemical processes where the pH determines the reaction rate
2. Uniformity in product quality based upon maintaining correct pH valve
3. Neutralization of waste effluent for discharge into sewers and rivers (controlled by law)
4. Corrosion control in high-pressure boilers

The principle of measurement is based upon dissociation of chemical compounds into positively and negatively charged particles (known as cations and anions) in an aqueous solution. For a hypothetical chemical compound MA the chemical reaction may be written as

$$MA \rightleftharpoons M^+ + A^-$$

The amount of dissociation depends upon the compound and the temperature of the solution. At any fixed temperature, a fixed relationship exists between the ions and the undissociated compound. This relationship is based on the *activity* of the ions, a term which indicates the ability of an ion to take part in a reaction and is related to the *concentration* of the ions by

$$a = vc$$

where

a = activity of the ion
v = activity coefficient
c = concentration of the ion

The dissociation is related to the activity of the ions by the dissociation constant k as follows:

$$k = \frac{[a_{M^+}][a_{A^-}]}{a_{MA}}$$

where a_{M^+}, a_{A^-}, and a_{MA} are the activity of the positive and negative ions and the undissociated compound, respectively.

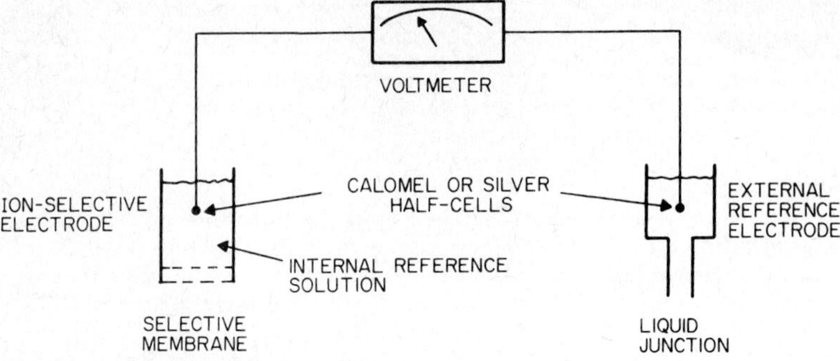

Figure 1-46 Basic arrangement for ion-selective measurement.

Therefore, if one could measure the activity of a particular ion, by knowing its activity coefficient the concentration of the particular ion in solution can be determined. If the ion of interest happens to be hydrogen [H^+], the measurement is known as pH, which is a short form for $1/a_{H^+}$.

Recent research has been fruitful in developing sensors which can measure other types of ions, such as fluoride, silver, sulfide, chloride, and cyanide. Figure 1-46 shows a basic setup for measurement using two half-cells, one formed by the measuring electrode and the other by the reference electrode.

The potential generated by a measuring electrode is related to the ionic activity, and is expressed by the Nernst equation,

$$E = E^1 + \frac{2.3RT}{nF} \log \frac{a_1}{a_2}$$

where

E = electrode potential
E^1 = constant for a given electrode at a fixed temperature
R = gas law constant, 1.986 Btu/(lb)(mol)(°R) [1.986 cal/(g)(mol)(K)]
T = absolute temperature, K or °R
F = Faraday's constant, 96,490 C/g-ion
n = charge of an ion, including sign
a_1 = activity of measured ion in process solution
a_2 = activity of measured ion in the internal solution

If the internal solution is made up such that it has constant activity, the equation can be simplified to

$$E = E^\circ + \frac{2.3RT}{nF} \log a_1$$

where E° is the standard emf of the hydrogen cell.

For measurement of hydrogen activity the equation can be simplified to

$$E = 414.12 - [59.16 \text{ pH at } 77°\text{F } (25°\text{C})] \qquad \text{in mV}$$

Since the simplified Nernst equation has a temperature term, errors can exist for pH measurements at any other point than pH = 7.0, which is the isopotential point for this ion measurement. Figure 1-47 indicates the errors that can be expected.

Figure 1-48 shows the construction details of a glass pH measuring electrode with the internal solution buffered to have constant ph of 7.0. Figure 1-49 shows a flowing reference electrode with the ceramic tip acting as the liquid junction. The silver electrode with the silver chloride coating forms a silver–silver chloride half-cell in the reference electrode. To eliminate this potential in the overall measuring circuit (Figure 1-46), a

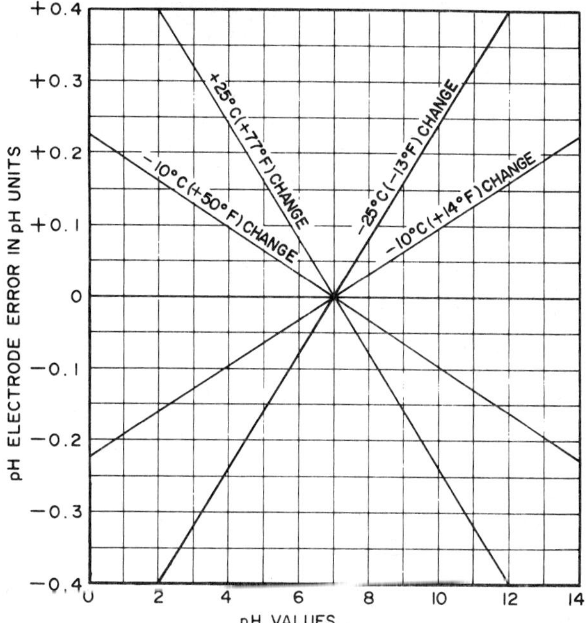

Figure 1-47 Graph of pH values at various solution temperatures vs. pH measurement errors due to temperature differences across the measurement electrode tip. No temperature error and no compensation required at pH 7.

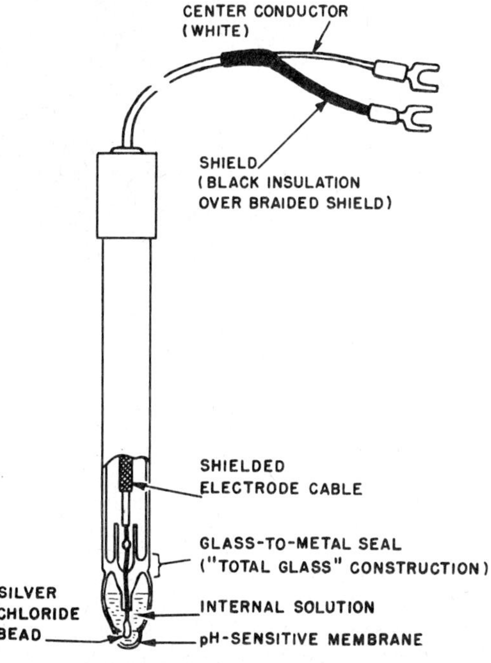

Figure 1-48 Construction of details of measuring electrode.

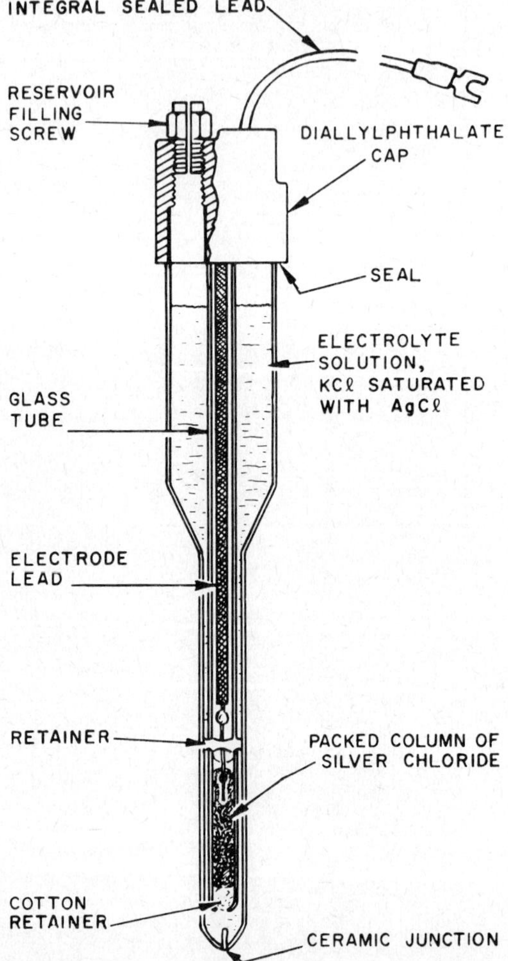

Figure 1-49 Construction details of flowing reference electrode.

similar half-cell is created in the measuring electrode, thereby having the meter read only the potential across the measuring electrode's ion-sensitive membrane. These electrodes can also come in the solid-state form, where the flow at the liquid junction is decreased to very small amounts.

It should be noted that the reference electrode does not create a measurement potential at the liquid junction since the flow of liquid prevents that. However, a small potential is still created by the ionic flow, and this may be compensated for in the measuring instrument. The measuring instrument must have a high input impedance to prevent current flows which can cause polarization problems in the electrodes. This high impedance requirement and millivolt measurement require special care in shielding of the measurement leads and grounding of the system.

While the Nernst equation was applied to the activity of specific ions, a more general form would be to consider it as follows:

$$E = E_0 + \frac{2.3RT}{nF} \log \frac{[\mathrm{a_{ox}}]}{[\mathrm{a_{red}}]}$$

where we consider the ratio of the activity of all the oxidized ions in solution to that of all the reduced ions. Oxidation here means the loss of an electron; reduction means a gain. The measuring electrode is a metallic plate (could be platinum), and the reference electrode is similar as for ion-selective measurement. E_0 is a constant which varies for differing types of reactions. These types of measurements are called oxidation-reduction-potential (or redox) measurements and are used when an oxidation or reduction type of chemical reaction is taking place, e.g., oxidation of cyanide or reduction of chromate in effluent treatment. The installations tend to be unique and depend greatly upon the background chemistry.

Conductivity

Aqueous solutions are electrically conductive, the amount being a function of temperature and concentration of ions in solution. It is expressed as specific conductance or conductivity in siemens. The conductance between two electrodes can be given by

$$C = \frac{k}{(1/A)} \quad \text{or} \quad \frac{K}{F}$$

where

F = cell factor in cm^{-1}, the ratio of the distance in cm between the two electrodes
A = area of the electrodes, cm^2
C = conductivity, normally given in microsiemens

The cell factor allows the spanning for the measuring instrument, which is basically a Wheatstone bridge. It should be noted that the measurement is non-ion-specific but is an indication of all ions present in solution. For calibration, known samples are used, thereby allowing the cell to be calibrated to read directly in concentration units.

The conductivity of a solution is a function of its temperature. Compensation can be provided if the solution's conductivity temperature curve is known. This limits the cell's application only to the solution for which the calibration was provided. When an electric current is passed through a solution, polarization can occur. One of the effects is electrolysis, whereby a gaseous layer can form on the electrode surface, thus increasing the relative resistance of the cell. For this reason, alternating currents are normally used.

Figure 1-50 illustrates the cell factor for various configurations of electrodes.

There are many applications where conductivity cells can provide a relatively inexpensive method for controlling processes like detecting impurities in boiler feedwater, concentration of black liquor, and other applications where the concentration of a known compound in solution has to be determined.

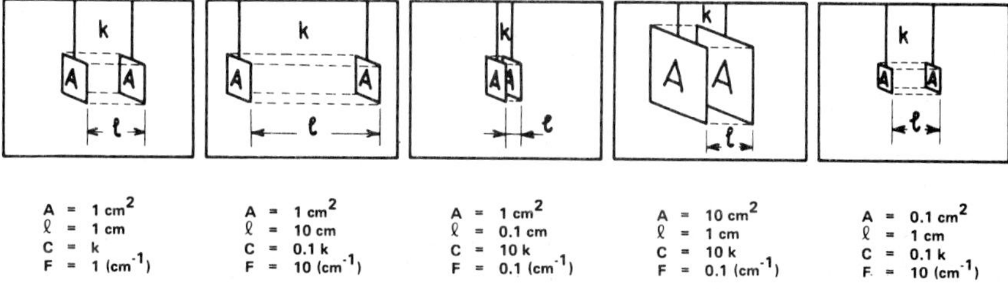

Figure 1-50 Electrode cell factors.

Oxygen and Dissolved Oxygen

These measurements have become increasingly important in the combustion and water-treatment fields. The most popular method for oxygen measurement is the electrochemical detector, which works in principle by using the Nernst equation. Figure 1-51 indicates a schematic drawing of the basic cell, with P_1 the partial pressure of oxygen in the reference gas and P_2 the partial pressure of oxygen in the measured gas. The equation is

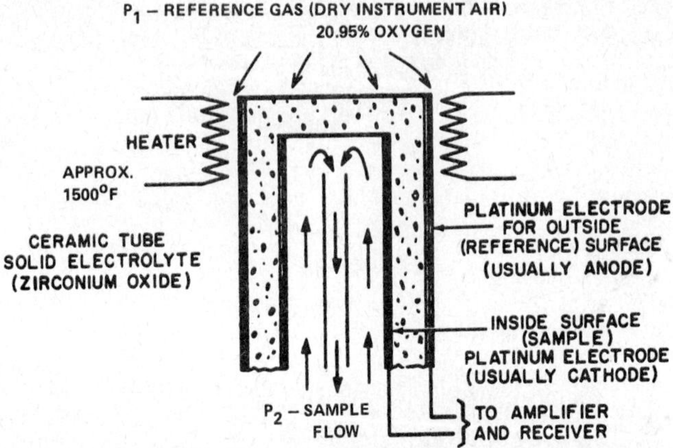

Figure 1-51 High-temperature electromechanical oxygen detector.

$$E = E^\circ + \frac{2.3RT}{nF} \log \frac{P_1}{P_2}$$

and the resulting logarithmic output voltage is approximately 53 mV per decade change in oxygen content. The output decreases as the concentration (actually activity) of measured oxygen increases. The measurement is a percent measurement as related to all components in the sample gas, including water vapor. The probe can be installed on-line with no sample preparation system required.

There are many devices available for measuring dissolved oxygen (DO). In general, the construction requires two electrodes in an electrolyte enclosed by a permeable membrane. Depending upon the electrodes, a galvanic reaction or polarization takes place when a small voltage is applied to the electrodes. The dissolved oxygen diffuses through the membrane and is dissolved in the electrolyte. An oxidation-reduction reaction causes

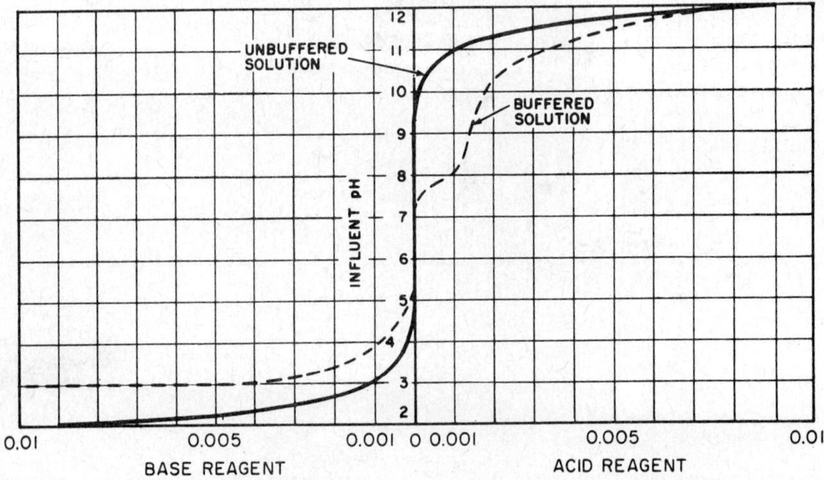

Figure 1-52 Typical neutralization curves for unbuffered solutions (strong acid or strong base) and buffered solutions.

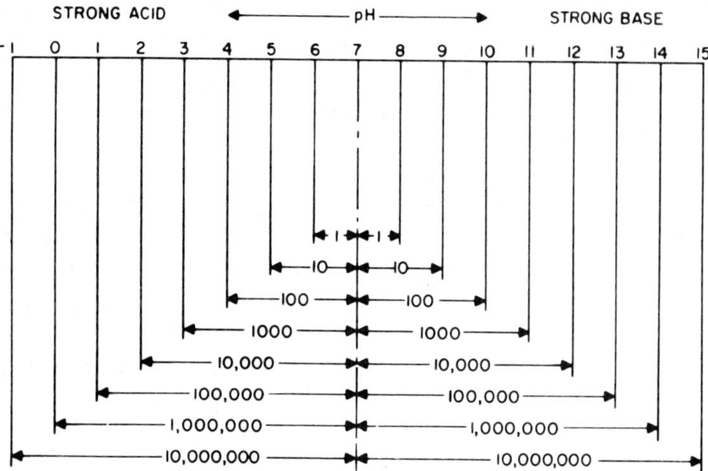

Figure 1-53 Graph of reagent demand. Reagent addition units are 10^{-6} mol/Li.

a current flow which can then be measured as a percentage of the maximum amount that could be present at the existing temperature. Since temperature determines the saturation capability of the solution to dissolve oxygen, automatic compensation is usually required.

General

From a control-application viewpoint, all instruments based upon the Nernst equation have to take into account the nonlinear nature of the measurement. The logarithmic nature of the measurement requires special consideration of the rangeability requirements of the system. Figure 1-52 illustrates the nonlinear nature of the neutralization curve for strong acid or strong base solutions and the effect of buffering in the solution.

The rangeability requirements can be noted in Fig. 1-53. It can also be noted that while neutralization of an acid from pH 6.0 to 7.0 requires one unit of base, to go from pH 2.0 to 7.0 requires 10,000 units. Therefore, if the influent stream varies from pH 2.10 to 7.10, the control system must have a rangeability of 10,000:1.

chapter 5-2

Automatic Controls

INTRODUCTION

Industrial processes are generally characterized by mass and energy flows. Application of automatic controls is concerned with the behavior of the process under dynamic or unsteady-state conditions where the accumulation of mass or energy cannot be tolerated. Hence the control problem is one in which the designer uses various engineering tools to match the supply against the demand over a period of time. The three terms, supply, demand, and time, can vary and depend to a great extent upon the type of processes considered. As processes get more and more complex and the availability of raw materials, including energy, becomes limited, the application of automatic control theory takes on greater significance.

Types of Controls

There are two basic types of control: open-loop and closed-loop.

Open-Loop Control

This is the simplest type of control that can be applied. It involves making an estimate of the amount of control action required based upon achieving a desired objective without regard to the actual conditions of the process. For example, a washing machine operates without regard to the actual condition of the clothes. The amount of detergent used and the settings on the machine are an estimate of the control action required in achieving the objective (clean clothes), they are not based on the actual state of our objective. This type of control is generally inadequate and is seldom encountered in industrial process work; however, this type of control cannot be completely ignored. With the advent of computers and their memory capability, control sequences based upon historical data may be considered in the future, especially for the kinds of processes where a measurement of the final objective may be difficult, if not impossible.

Closed-Loop Feedback

This is the most common type of control mechanism used. Any process in which the process variable under control is measurable allows the use of such control strategy. The importance of such a loop can be judged by the fact that most bio-socioeconomic processes incorporate some kind of closed-loop feedback mechanism, also known as a *servomechanism* (Fig. 2-1).

The controlled variable is the process variable which we are trying to maintain at some desired value (called the set point). Industrial processes are characterized by the

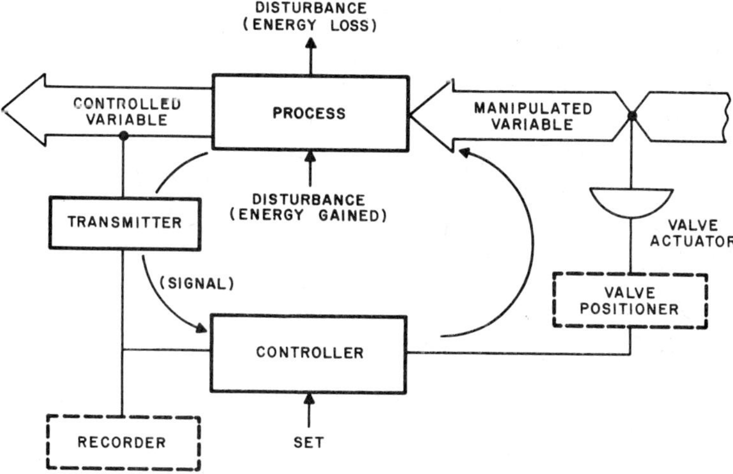

Figure 2-1 Closed-loop control. (*The Foxboro Company.*)

many types of control variables that are encountered, e.g., temperatures, flows, and levels. The function of the transmitter is to quantify this variable in terms of signals, which could be pneumatic, electric, hydraulic, or just a mechanical output like the position of a lever. It should be remembered that not all these measurements are linear, i.e., the output signal is not necessarily linear with respect to the controlled variable or, even if it is linear, is not a direct reading of the controlled variable, e.g., the controlling flow based directly on the pressure drop ΔP across an orifice meter. In all these cases, linearization can be achieved, thereby making the transmitter output a direct indication of the controlled variable.

The manipulated variable is that which the controller varies in its efforts to maintain the controlled variable at set point. The controller output is a signal to, say, a valve actuator, causing the valve to move to a position which would depend upon the value of the signal, type of valve, and the process conditions under which it is operating. The valve positioner causes a fixed relationship between the controller signal and valve position. Even though this relationship could be linear, the relationship between valve position and flow-through is generally complex and nonlinear. The nonlinearities in measurement and in the control of the manipulated variable have a significant influence upon the performance of the control loop.

The feedback control loop operates in an environment where constant disturbances are taking place. These disturbances affect the controlled variable and could be due to changes in the manipulated variable other than those instituted by the controller; e.g., changes in mass or energy flows to the input of the process, or changes of the same variable on the output side of the process. Changes can also be initiated by the operator when changing the set point of the loop. The controller loop must therefore perform both as a regulator and as a servomechanism. The strategy used is straightforward and makes use of the error (the difference between set point and the measured controlled variable) to develop a control signal which will drive the error to zero and thereby achieve the objective. The performance of the loop can be judged as some integral function of error and time, the controller and its action then being set to minimize this function.

The existence of this error means a certain economic loss in the production process in terms of either the production of off-specification product, the use of excessive amounts of energy, or both. As these considerations become more and more important in the future, control loops will tend to get complex, and input disturbances will be measured and used in strategies called *feed-forward.* Here the measured inputs are used to compute the required positioning of the manipulated variable based upon some mathematical model of the process. The amount of correction required is then based on the variation of the inputs to the process. As the models tend to be simplified representations of the process, feed-forward strategies are not accurate over the entire operating range of the process. Feed-forward control strategies are normally used in conjunction with feedback, as shown in Fig. 2-2.

It should now be noted that the feedback controller does not have any direct control of the process but acts to take minor trim action on the feed-forward model. The major control function is accomplished by the feed-forward scheme. The limited action by the feedback controller allows a much more stable operation of the process, allowing set

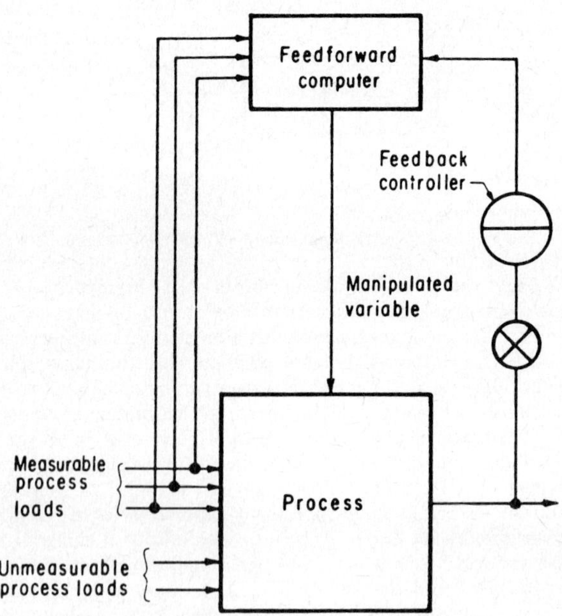

Figure 2-2 Feed-forward with feedback.

points to be set closer to the required specification values and also allowing increased throughput.

The time response of the entire control loop is made up of the sum of the responses of the primary element, the transmitter, all receivers in series with the controller, the controller, the final operator, and the process itself. There are two types of time elements which occur in industrial processes: first, dead time (or transportation lag) and, second, the resistance-capacitance (*RC*) time constant. To determine how much of each exists, a simple open-loop test can be performed as follows. With the loop in steady-state operation, the controller output is manually changed by a step amount. The response of the controlled variable is plotted as a function of time, as shown in Fig. 2-3.

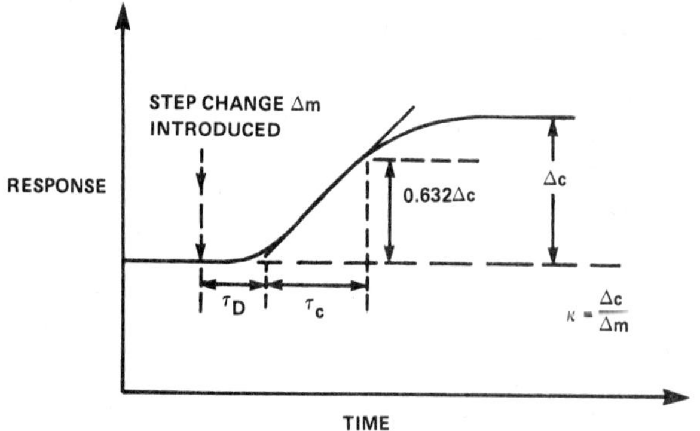

Figure 2-3 Open-loop response.

The graphical construction is designed to approximate a process consisting of a dead time τ_D and a capacitance time τ_c. Most processes will have some combination of these two elements, with the control problem becoming easier as the ratio of τ_D/τ_c becomes smaller. While this example is a simple model of the process, more complex models are available which approxmiate the response curve more closely. These, however, become cumbersome for practical use. Time elements introduce a phase lag; i.e., the output is delayed with respect to the input. The phase lag varies as the frequency of the input signal and becomes larger as the frequency increases.

If a loop is to oscillate with a sustained cycle, the phase shift of an upset, after going through all the elements in the loop, must be exactly 360 degrees.

All feedback controllers contain an element called *negative feedback,* which introduces a phase shift of 180 degrees. The remainder of the control loop must create an additional 180 degrees for sustained oscillations to exist. The point at which this occurs is the natural period of the loop as long as the controller contributes no phase shift.

In addition to the phase shift in a loop, one must consider the gains of the various elements in a control loop. There are two types of gains, static and dynamic. Figure 2-3 introduces the static gain κ as the change in output divided by the change in input. Dynamic gain would be the same ratio but would be a function of the frequency of the input signal. The gain at any given frequency would be the product of the static and dynamic gains. A loop gain could then be defined as the product of the gains of the individual elements of a loop. It is a dimensionless number which is an indication of the stability of the loop. As the loop gain becomes greater than 1, the loop tends to oscillate with larger and larger amplitude, while a value less than 1 would cause a certain amount of damping of the amplitude of oscillation. Normal industry practice is to achieve a loop gain of 0.5, which is also called ¼ amplitude damping, where every amplitude is half the previous one. These criteria do not necessarily prevail in all industrial control loops, but they are typical. Particular process requirements may mandate other dynamic responses.

CONTROLLERS

The introduction to this chapter described a feedback loop. Figure 2-1 indicated the four basic elements, one of which is the controller. The actual hardware used could be pneumatic, electronic (analog or digital), and hydraulic. Any controller can be considered to be composed of two parts, as shown in Fig. 2-4.

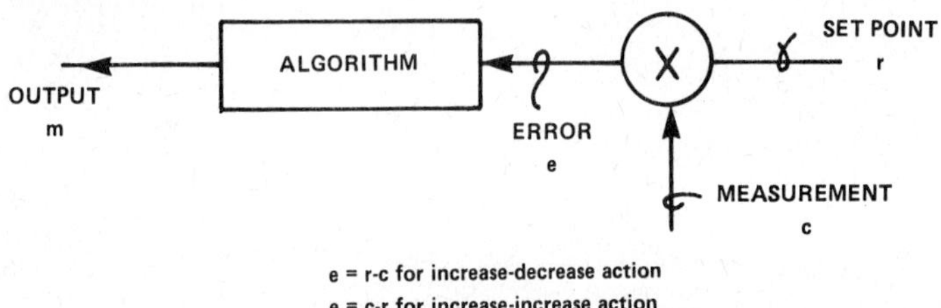

e = r-c for increase-decrease action
e = c-r for increase-increase action

Figure 2-4 Basic controller elements.

The error detector determines the magnitude and, more important, the direction of the error; the algorithm defines the specific action of any one particular type of controller. The type of error signal implemented defines the direction of the output as a function of the measurement. For example, if the measurement c exceeds the set point r, the error would be negative, which then causes the algorithm to decrease the output m. If an "air-to-open" type of valve had been used, the decrease in output would tend to close the valve, which would decrease the flow to the process and thereby decrease the measurement back toward the set point. There are occasions in which an "air-to-close" valve may have been used, in which case the action of the controller would have to be changed. This option to change the action of the controller is a normal feature of the controller.

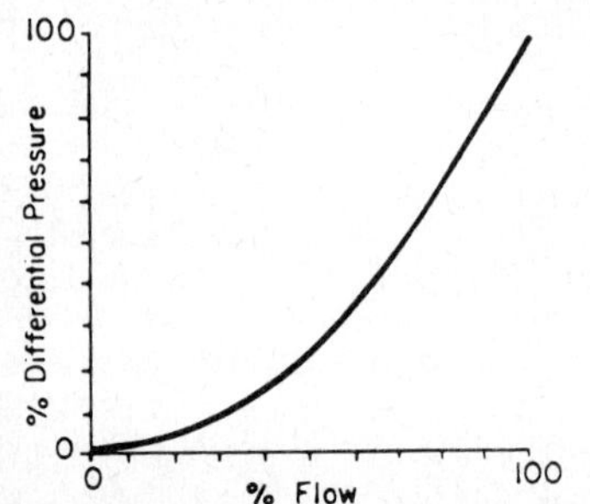

Figure 2-5 Flow measurement by differential pressure means.

While most control theory is based upon the linearity of the gain of various elements which make up a loop, practical design and operation must take into account the nonlinearities that occur due to variations in set point and/or variations in process load. For example, we have noted in Chapter 5-1, under "Flow Measurement," that the flow has a square-root relationship with the differential pressure measured. Figure 2-5 indicates this relationship.

The effect of this type of nonlinearity (it could occur in any of the four elements of a control loop) is normally to give good control at one set of operating conditions only. This could cause damaging results in those cases in which the controller had been set during low system gain because, at high system gain, the loop would tend toward oscillation. Elimination of these nonlinearities, by using special hardware or by selecting loop elements having equal and opposite characteristics, goes a long way in the overall stability of the loop, e.g., selecting a quick-opening valve whose characteristics would be opposite to a head-type flow measurement.

Selection of a controller is based to a great extent on the process characteristics and the precision needed to control it at an exact set point. The four basic control actions generally used are discussed in the text that follows.

Two-Position (ON-OFF)

This is the simplest of all control actions available. The output of the controller is at either 100 or 0 percent and could be either an analog signal or a contact actuation. The result in either case is that the final actuator is completely open or completely closed. The effect on the measured variable depends upon the type of process, specifically the capacity time constant. For a large-capacity process, the measured variable would change slowly to allow reasonable final actuator action. The measurement oscillations would be correspondingly small. Compare this to a process having very little capacity; the measured variable would tend to change very quickly, which would cause rapid actuator action, and this in turn would cause large and rapid oscillations in the measurement. Under these circumstances and where precise control is required, a throttling type of control action would be preferred.

There are three basic methods of achieving two-position control. Figure 2-6 illustrates the control action.

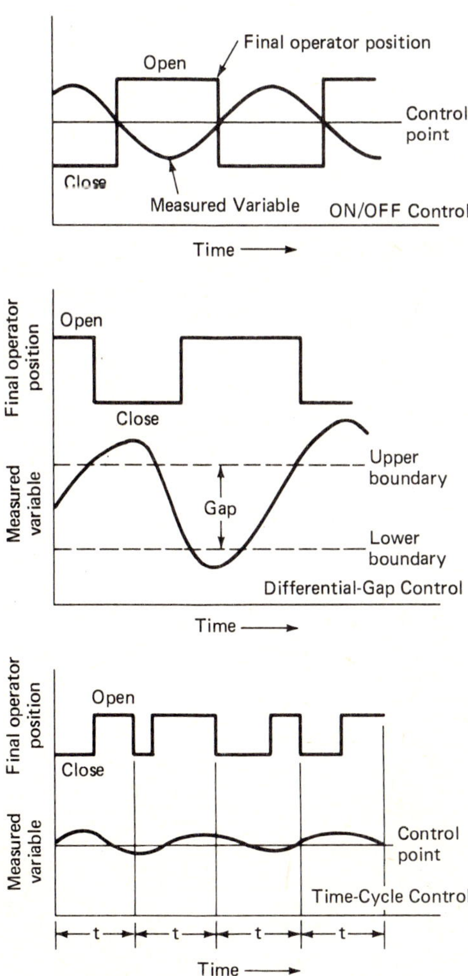

Figure 2-6 Two-position control.

A thermostat for an electric heater is an example of an ON-OFF control system. Here, the controller operates continuously to control the temperature of the process. Another application is for safety shutdown of a process when safe operating conditions have been exceeded.

To avoid frequent cycling and maybe damaging results to the final actuator, a differential-gap controller is used. A band or gap exists around the control point. When the measured variable exceeds the upper boundary of the gap, the final actuator is closed and remains closed until measurement drops below the lower boundary of the gap, at which point the final actuator opens. The control is not precise, but it does prevent excessive wear on the final actuator. In industry, this type of control is often found in noncritical-level control applications where the level could be anywhere between two limits.

In time-cycle control, a time base t is established. During this time period, the final operator is closed for a certain percentage of the time and open for the remainder. The ratio of closed time to open time is determined by the relationship between the measured variable and the control point.

A time-cycle controller is normally set up so that, when the measured variable equals the desired control point, the final operator will be open for half the time cycle and closed for the other half. As the measured variable drops below the control point, the final operator will remain open longer than it is closed. This type of control is often found on electrically operated heaters and on dry-solid control gates where a throttling gate position would cause buildup and clogging.

Proportional

To better understand this and the control actions described later, an open-loop response will be considered first. The open loop means that the controller is considered by itself with an artificially generated measurement signal. A change in input can be generated by changing either the set point or the measurement signal.

Figure 2-7 is a schematic diagram of a force-balance-type pneumatic proportional control; Fig. 2-8 is a schematic of an equivalent electronic controller using an operational amplifier.

The basic equation implemented is

$$m = (100/\%\text{PB})(r - c) + \text{bias}$$

where

r = set point
c = measurement
$\%$PB = proportional band, in percent

The factor 100/%PB is also termed the *gain* of the controller. The *bias* term allows the output to be approximately 50 percent of the signal range of the controller when measurement equals set point. This condition can occur only under one set of operating conditions, i.e., when the output from the controller positions the final actuator such that the manipulated process variable exactly matches the process requirements for the given load. For different operating conditions, the output from the controller is a function of the error (the difference between the set point and measurement) and the set proportional band (PB). Figure 2-9 shows the effect of PB on valve travel for various settings.

Line A indicates that for a PB of 100 percent, the measurement has to change from 0 to 100 percent of its span for a 0 to 100 percent throttling action. As the PB is decreased, lines B, C, and D indicate that smaller changes in measurement allow full throttling action. When the PB is greater than 100 percent, the throttling action is reduced to less than 0 to 100 percent.

It should be noted that a limited type of proportional control—with correspondingly limited possibilities of application—can be handled by a transmitter. From the preceding discussion, it can be deduced that a penumatic transmitter can perform as a 100

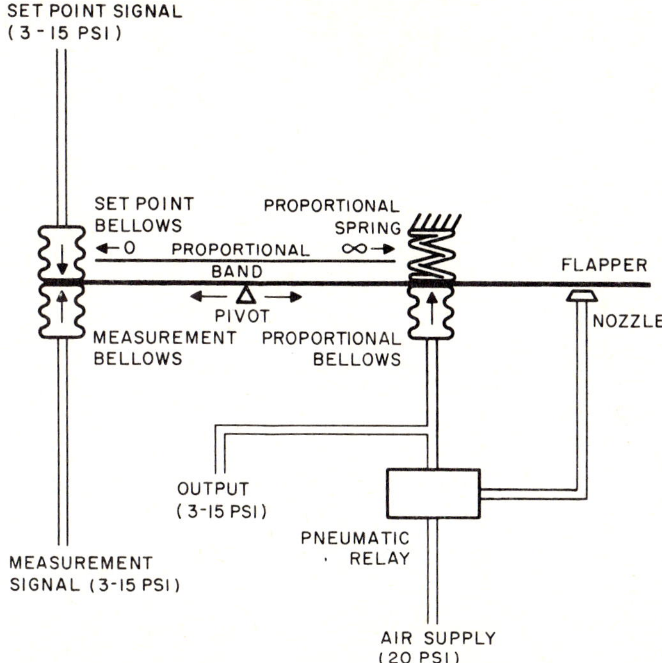

Figure 2-7 Pneumatic force-balance proportional controller.

percent PB *controller* when connected directly to a pneumatic final actuator. Some tank-level controls are implemented this way, especially when strict level control is not required.

The controller algorithm indicates that, for an output other than 50 percent, an error must exist; the size of this error is dependent upon the PB (gain setting in the controller). There is nothing in the algorithm which would cause this error to be eliminated. This error is called the *offset*, and its size dependent upon the particular PB (gain) for a given set of conditions. Figure 2-10 illustrates the types of responses that can be achieved.

As the proportional band is decreased, the amount of offset decreases, but the response tends to become more oscillatory. This should be obvious when the loop gain is considered. As the PB is decreased, the controller gain increases, which causes the loop

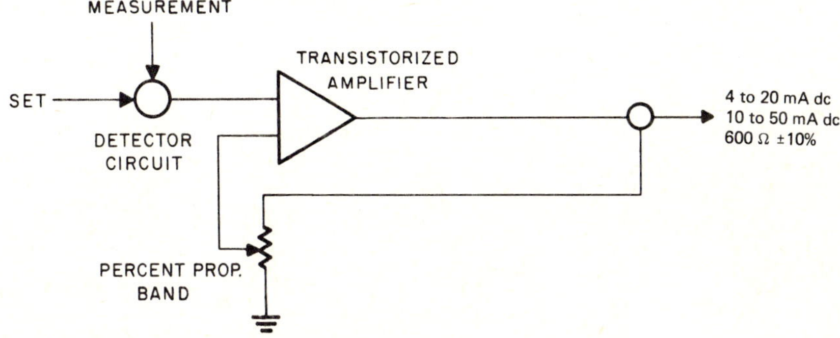

Figure 2-8 Electronic proportional controller.

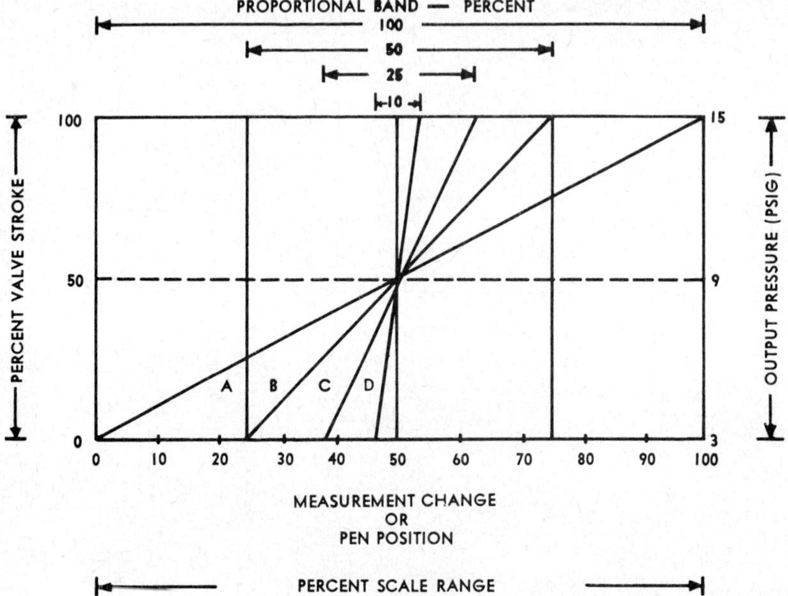

Figure 2-9 Proportional response to changes in proportional band (gain).

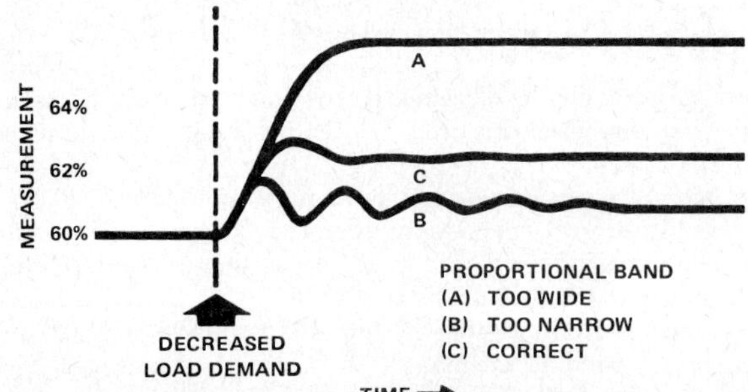

Figure 2-10 Proportional control response curves.

gain to increase thereby causing larger oscillations in the response. Large-capacity processes in general require narrow proportional bands, while processes having fast reaction times can accommodate wide proportional band settings only.

Proportional-Plus-Integral

These types of controllers are also known as *two-mode* controllers. In the discussion of proportional control, it was shown that the wide proportional bands required for some processes lead, in turn, to large offset errors. This is an intolerable condition in the majority of process applications. The application of the integral (also known as reset action) allows the final actuator-to-measurement relationship to change. This relationship changes when the measurement is not precisely at its set point. The effect is to cause the final actuator position to change in the direction, and at a predetermined rate, that will allow the measurement to equal the set point. Figures 2-11 and 2-12 are schematics of pneumatic and electronic proportional-plus-integral controllers.

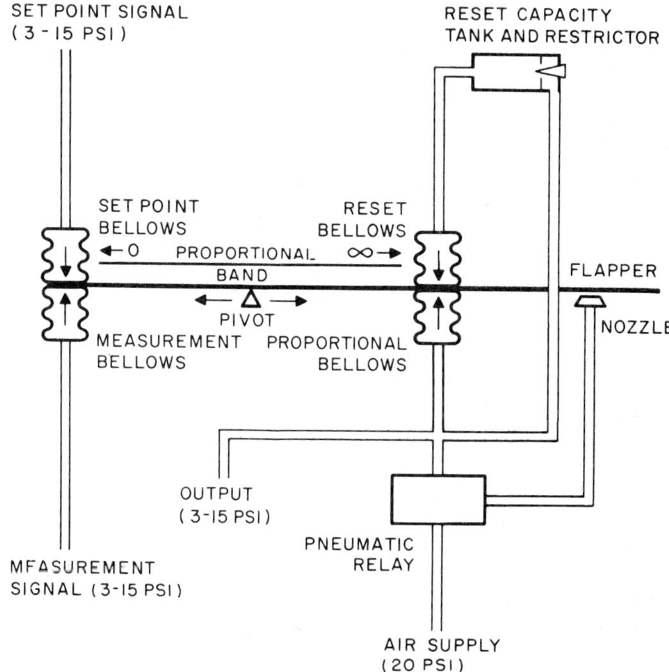

Figure 2-11 Pneumatic force-balance proportional-plus-integral controller.

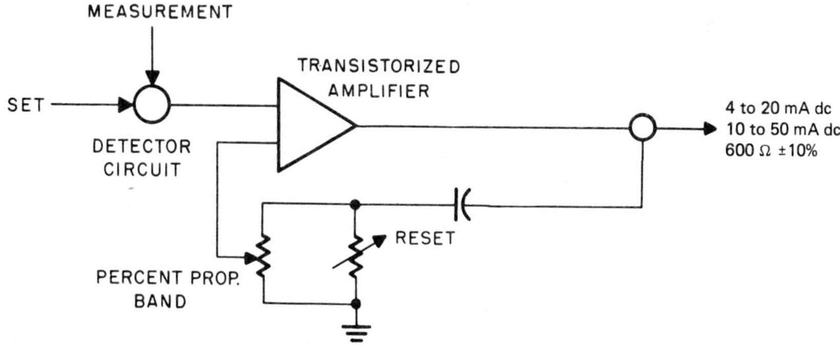

Figure 2-12 Electronic proportional-plus-integral controller.

The basic equation implemented in these controllers is

$$m = (100/\%\text{PB})\left(e + \frac{1}{R_T}\int e\,dt\right)$$

where

e = error $(r - c)$

R_T = the integral time constant

We see from the above equation that the output of the controller is the sum of both the proportional action and the integral (reset) action. Figure 2-13 illustrates open-loop response of a proportional-plus-integral controller.

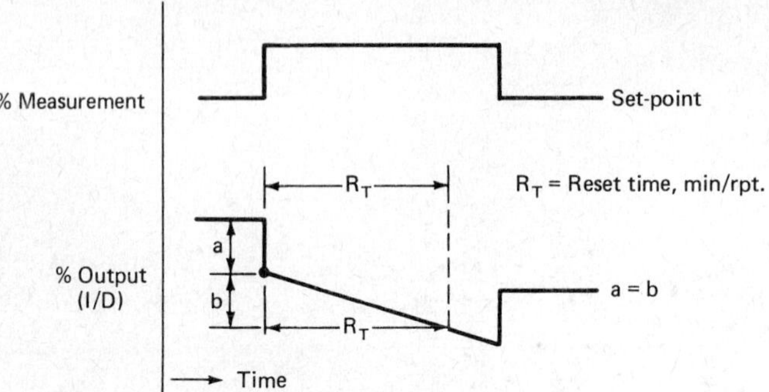

Figure 2-13 Open-loop response of proportional-plus-integral controller.

It should be noted that the integral action R_T can be defined as the time (in minutes) taken by the controller to change its output by an amount equal to the proportional action. This definition allows R_T to be calibrated in terms of minutes per repeat. An alternative way to set this calibration is the reciprocal of minutes per repeat, i.e., repeats per minute. Both these terms are equally popular among the various manufacturers. Though the introduction of integral action eliminates the offset error, it introduces a phase lag in the controller. The amount of gain is no longer the simple (100/%PB) term of the proportional controller, but the vector sum of the proportional part and integral part which gives a resultant larger than either. The phase lag introduced is the amount by which this resultant lags the proportional action. Two results can be deduced from this discussion. First, the addition of integral action for a given proportional action will cause the control-loop oscillations to increase; second, this oscillation will have a longer period than the original oscillations. Figure 2-14 shows response curves for a temperature loop where the integral is measured in minutes per repeat.

If the controller is unable to return the measurement to the set point, the integral action continues to increase or decrease the output until the saturation limit of the controller is reached. If at this time the measurement starts to change, the controller output does not change from its saturation limit until the measurement crosses the set point. If the system has any capacity, the measurement will overshoot the set point by an amount which depends upon the capacity and integral setting. This limitation is called *integral windup* and has to be seriously considered on discontinuous or batch-type processes.

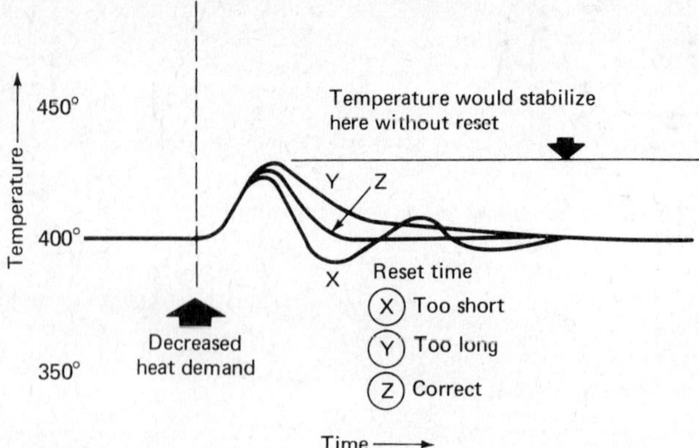

Figure 2-14 Proportional-plus-integral control response curves.

There are special batch-type controllers available which prevent the output of the controller from reaching saturation limits, thereby causing less or no overshoot in the measurement for the above conditions.

Proportional-Plus-Derivative

The addition of derivative action to the proportional controller can improve the response of the system, especially in those batch-type processes where integral windup can cause problems. The amount of derivative action is proportional to the rate-of-change of the measurement. Figure 2-15 illustrates the open-loop response and defines the derivative time constant.

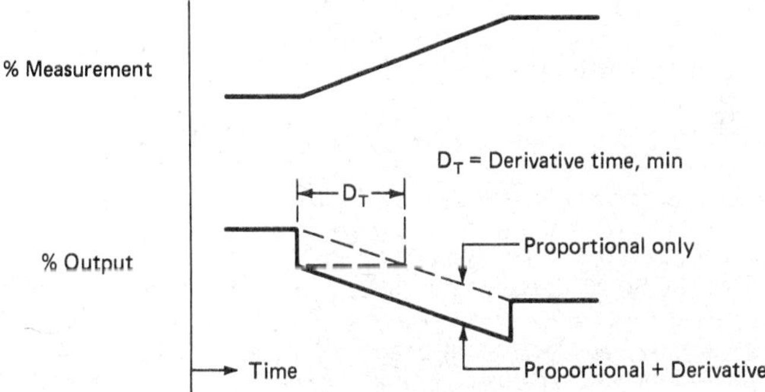

Figure 2-15 Open-loop response of proportional-plus-derivative controller.

Since the derivative control is a function of the rate of change of measurement, introduction of this mode in applications where the measurement can change very rapidly, or if the measurement is noisy, can degrade the controllability of the loop. The most useful applications are those where the system has large capacities with little dead time.

Proportional-Plus-Integral-Plus-Derivative

This is also called the three-mode controller. The open-loop response indicates the sum of all three modes in its output. Figures 2-16 and 2-17 are schematics of a pneumatic and an electronic controller, respectively.

In our previous discussion, integral action was considered to cause an additional lag in the control loop. The derivative has an opposite effect, i.e., it causes a phase *lead* and is, therefore, sometimes also called *preact.* The total gain of the controller is the vector sum of the individual mode gains and the phase lag or lead introduced in the controller is the angle the resultant-gain vector makes with respect to the proportional-gain vector. If the resultant-gain vector has a phase lead, the resultant response of the system will have a period of oscillation smaller than the natural period. In general, if the integral time in minutes per repeat is made equal to the derivative time in minutes, the response is quite satisfactory. For large-capacity processes, the amount of derivative action may be increased.

Table 2-1 gives a brief summary of some standard process loops that are common in industrial process plants. Indicated are some major characteristics of the processes and the control systems.

Multivariable Control

The four common types of multivariable control loops are *ratio, cascade, feed-forward,* and *override.* The importance of these types of loops can be appreciated by considering

TABLE 2-1 Summary of Common Loops

Variable	Process	Control System
Flow	Very fast Most lags are in the control system Nonlinear (square) measurement common Noisy	Proportional-plus-reset controllers Low gain, fast reset Derivative hurts Linear valves for differential pressure measurement Equal percentage valves for linear measurement Valve is the major dynamic element
Pressure, liquid	Fast Most lags are in the control system Nonlinear (square) Noisy	Proportional-plus-reset controllers Gain near 1, fast reset rate Derivative of no value Linear valve
Pressure, gas	Single capacity No dead time Linear, no noise Simple process	Self-acting or high gain proportional controllers Reset seldom necessary Derivative unnecessary Valve characteristic relatively unimportant
Pressure vapor	Dynamics vary Dead time possible Slow compared to other pressure processes Linear, no noise	Three-response controllers Settings vary Equal percentage valves
Level	Single capacity (integrating) No dead time Linear Infrequent noise	Precise control: High gain or proportional-plus-reset controllers Averaging control: Low gain proportional plus reset or specialized controllers Valve characteristic unimportant
Temperature	Multiple-capacity system Dead time possible (especially in heat exchangers) Nonlinear No noise	Three-response controllers Settings vary, but gain usually above 1 Derivative of limited value if dead time is large Equal percentage valves Measurement dynamics are important
Composition	Dynamics vary Dead time usually present Usually linear Sometimes noisy due to poor mixing	Proportional-plus-reset controller Low gain, variable reset rate Derivative sometimes useful On-line analyzers fast, often noisy, pH nonlinear Sampling systems complicate both measurement and control, add dead time Linear valves

that the response of any controlled variable in the industrial process environment will, in general, be a function of a number of other variables that could be either on the load side or on the input side of the unit. To account for these variables, multivariable control loops have to be considered.

Ratio

Here the controlled variable is the ratio of two measured variables, where one of these is the controlled variable and the other is the "wild" variable. The wild variable is not necessarily uncontrolled, it could be a controlled variable of another loop. These systems need not be limited to two components but can have the wild variable set the ratio, through separate relays, to several controlled variables. An alternative to this method

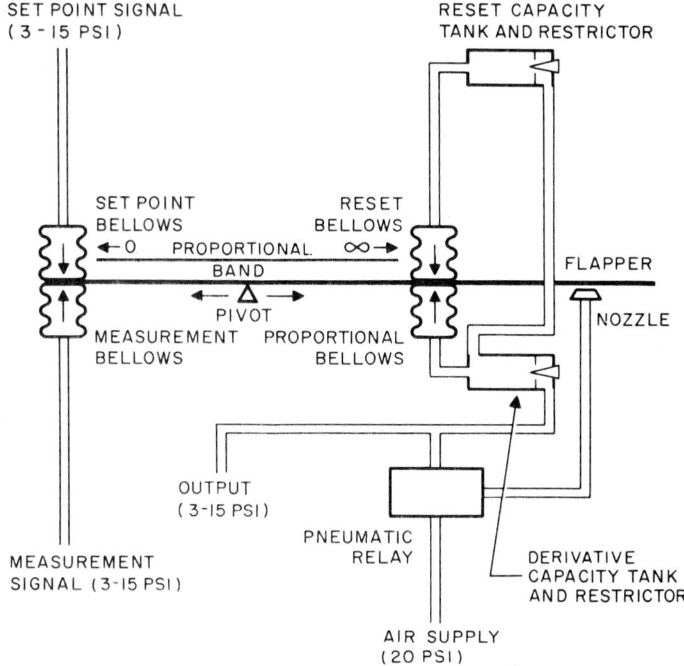

Figure 2-16 Pneumatic force-balance proportional-plus-integral-plus-derivative controller.

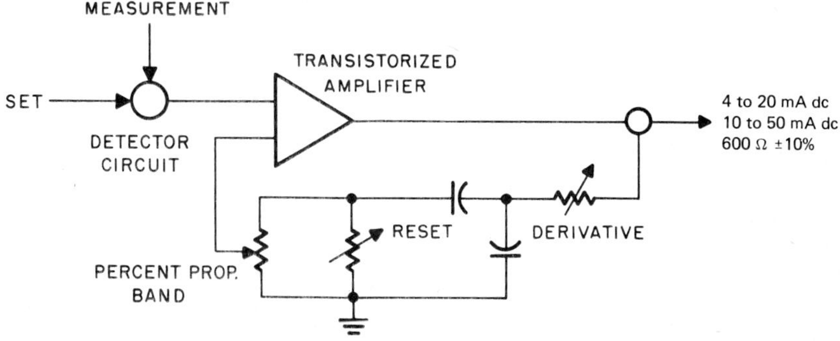

Figure 2-17 Electronic proportional-plus-integral-plus-derivative controller.

would be to set the ratio as a percentage of the required total flow, which itself is set by a master demand signal. The two types of systems are called *series* and *parallel*, respectively. The series approach has one basic advantage in having an inherent interlock between the wild variable and controlled variable.

There are a number of practical applications where these control schemes have been found useful. Examples are blending, control of air-to-fuel ratio in boilers, and ratioing of reactant flow in chemical processes. In steady-state operations, the system operates like an ordinary flow loop, i.e., dynamically quite fast. However, under changing loads, nonrecoverable errors can exist that can be removed in memory-type blending systems. Figure 2-18 indicates an example of the required setup.

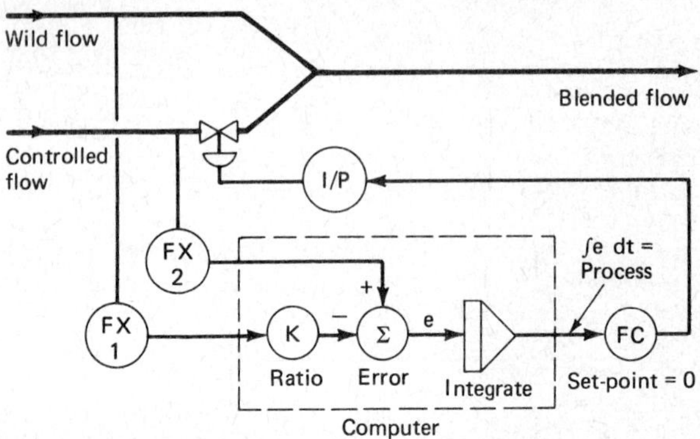

Figure 2-18 Memory-blending system controls accumulated flow ratio.

These systems can be either analog or digital, and frequently employ pulse-type measuring instruments for precise measurement of total flow. It should be noted that a certain amount of downstream capacity is required so as to be able to eliminate the accumulated error.

Cascade

In a single feedback loop, the output of the controller is directly used to manipulate the final actuator. Under these conditions, fluctuations on the input side of the manipulated variable will be noticed by the controller when the measurement changes. An example would be the header pressure change on a steam supply being used to control the outlet temperature from a heat exchanger. This after-the-fact control can be avoided by setting a secondary control loop which gets its set point from the primary controller. Figure 2-19 is a typical setup.

To properly apply these systems, the response of the secondary loop must be considerably faster than that of the primary loop. If this condition is not met, the resulting instability is the result of nearly simultaneous set-point and measurement changes in the secondary loop. It is for this reason that it is not recommended that a positioner be used on a final actuator in flow loops. The valve actuator contributes most of the capacity lag in a flow loop. With a positioner, this capacity lag exists in the secondary loop, with resultant instability of the entire flow loop.

A note of caution in actual operation of cascade loops is the integral windup that can occur in the primary controller, if the secondary controller were to be manual or if it were to fail. Special precautions have to be taken, and these could be either to interlock the stations or to use external integral feedback of the primary controller. The latter is the more useful method and is implemented by using the manipulated variable (steam flow in Fig. 2-19) as the external integral feedback signal.

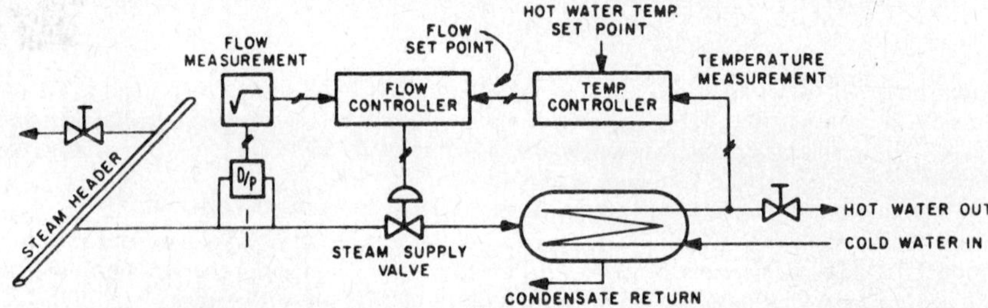

Figure 2-19 Cascade control of a heat exchanger.

Feed-Forward

The introduction to this chapter included the basic strategy and philosophy of feed-forward control systems. With the availability of both analog and digital computation elements such as multipliers, adders, etc., it becomes relatively easy to physically put together the hardware to perform the feed-forward calculation. The difficulty arises in setting up the required process model based on energy- or mass-balance equations. These equations must also take into consideration the dynamics involved for each disturbance variable considered in the model. For example, in the previous discussion on cascade control of a heat exchanger, it is observed that the load disturbances due to changes in both the temperature of the incoming cold water and the amount of flow would affect the controlled variable (the outlet temperature).

A simple model based upon *energy in* equalling *energy out* could be set up with dynamic compensation for the rate of change of product coming in. This model could then be tuned for a given set of conditions by adjusting the value K as shown in Fig. 2-20.

If our process model had been perfect, a perfect control would have been possible. But this perfection is difficult to achieve, because of first the large number of variables that can affect the control variable and second the difficulty of achieving exact mathematical representation of the process. To take into account these inaccuracies, a feedback loop is normally used and is called the *feedback trim*. As the name suggests, the action of this loop is not to take over complete control of the manipulated variable but

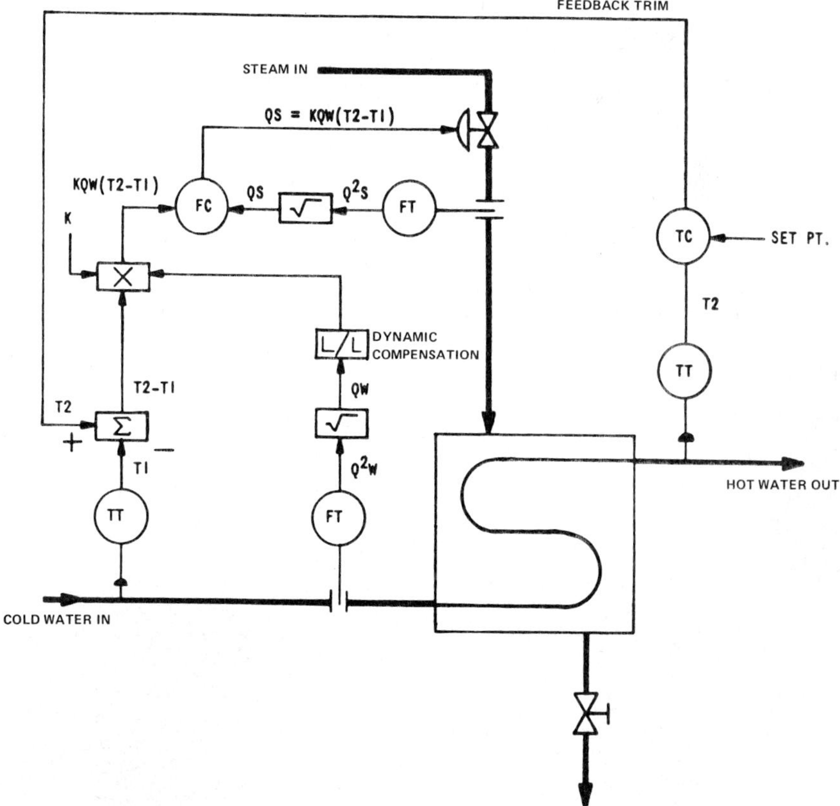

Figure 2-20 Feed-forward control with dynamic and feedback trim compensation.

to adjust the feed-forward model for varying operating conditions. Since this action can be slow, the "tuning" constants of the feedback trim controller are normally set much wider than if it had been the main control mechanism.

There are numerous examples of the use of feed-forward control in varying industrial processes. The oldest practical application is the setting of feedwater flow via steam flow for the control of drum level in a boiler. There is a certain amount of inherent safety in most feed-forward control systems. For example, in drum-level control the loss of steam flow automatically shuts off the feedwater flow. This action is the opposite of what a feedback controller would do—open the feedwater actuator if drum level were below set point. This inherent safety feature was also described in our discussion of series-type valve-control systems, which can be considered to be feed-forward control systems. Distillation towers, evaporators, and waste-steam neutralization are some examples of processes where feed-forward systems have been successfully applied.

Override

These systems are based on relays which allow the selection of the lowest or the highest input signals. The systems are used in the protection of equipment, in auctioneering, and in areas where redundant instrumentation has been used.

In the protection of equipment, it may be necessary to limit one variable to maintain safe operation. For example, a pump may have to be protected from both high discharge pressure and motor overload. If discharge pressure is the primary control, it will maintain control until an overload condition is detected. At that instant, the overload controller is selected for control by a low-select relay. Another example is the interlocking of fuel and air in boiler combustion control. Here, fuel is interlocked by air through a low-select relay, thereby preventing fuel flow from exceeding airflow; airflow is interlocked by fuel flow through a high-select relay, thereby preventing airflow from decreasing below fuel flow.

Auctioneering simply describes the selection of the highest or lowest of a set of signals. For example, if it is necessary to know the hottest spot in a reactor bed, a selection through a high-select relay for a number of inputs measuring temperature across the reactor bed could be used. In redundant instrumentation, the controls are set up to select a signal from a number of transmitters measuring the same variable. These are used in situations where loss of signal could produce an awkward control-system response.

Multiple-Loop Systems

Most industrial processes are composed of many individual control loops where one variable controls one manipulated variable or many input variables but still only one manipulated variable.

The first general problem is selection of the input variable that should control a selected manipulated variable. This is not obvious in some complex situations. However, a general rule is to select a manipulated variable having the greatest influence on the controlled variable. Other variables of interest would have to be taken into account by setting up cascaded systems or by operating them at fixed set points.

The second problem, the interaction between control loops, is a little more complex. An example of severe interaction is controlling both flow and pressure of gas in a pipe. For the smallest of upstream or downstream load upsets, the two control loops would interact because of the approximately equal natural period of oscillation of the two loops. The solutions are complex and use various decoupling circuits or detuning of controllers to prevent interaction. A physical solution is to structure the plant such that the dynamics of one variable could be changed by introducing capacity in one loop. In most cases the solutions are expensive and require a thorough knowledge of the process under consideration. The type of control hardware used also affects the problem. Although analog, pneumatic, and electronic components are available for most computing functions, a digital computer system provides the greatest ease and flexibility.

ELECTRONIC DIGITAL COMPUTERS

The present revolution in very large scale integrated (VLSI) circuit design is leading to a host of applications in the process-control field. These applications fall into three broad categories—smart measuring instruments, communications links, and controllers. These can be considered either in terms of dedicated application or shared application, where one controller can perform many functions on a time-shared basis.

Hardware

The foundation for these applications is the microprocessor, a basic building block that allows the application designer flexibility in configuring electronic circuitry to particular requirements. This flexibility is achieved because the microprocessor works somewhat similarly to a computer; it has a memory, an arithmetic and logic unit, a controller, and input-output circuitry. Refer to Fig. 2-21.

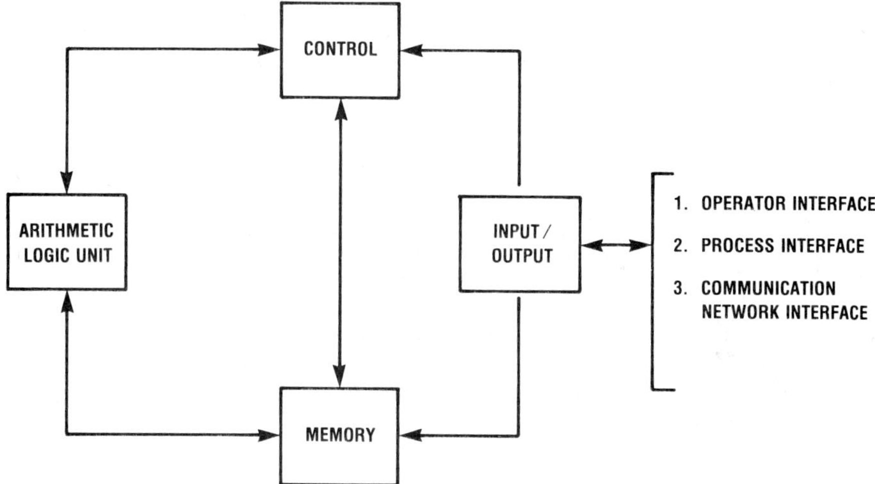

Figure 2-21 Basic processor.

The memory could be dedicated and part of the microprocessor or it could be external but under the processor's control. In either case, this fixed piece of memory (usually an ROM—read-only memory) contains words where the bit pattern for a particular set of words contains the requisite logic for the controller in the microprocessor to do a particular function. For example, this could be to read a word from the input and place it in the arithmetic logic unit (ALU). Another instruction could get another word from the input circuitry and also take it to the ALU. A third set of instructions could then add the two words together to give the required answer. Finally, one more instruction could take this answer and place it in the output circuitry. Given this basic instruction set, one could then develop the necessary logic for the microprocessor to accomplish the required functions.

This logic can now be implemented in two ways: The first is to design the microcode in such a manner that the microprocessor cycles through this code continuously, thereby dedicating the circuitry to performance of a predetermined function. The second is to define the microcode in such a manner that various sections of it could be used independently, and a particular function could then be externally selected. This external selection is accomplished by having an external memory (a RAM—random access memory) which contains words that, when decoded by the microprocessor, will point to the internal microcode instruction set that is required to be run. This allows the user to define requirements in a higher-level language for a given processor, rather than to work in terms of the machine-level microcode. Further, it now allows the processor to possibly do multiple tasks, where each task could be defined as a unique set of a higher-language code, basically a program. Thus, many programs could exist in the external memory and the processor would then run through each program, thereby performing many functions. The processor-memory combination looks basically like a regular computer, at least from the hardware description presented so far. It should be noted that there are many identifying characteristics of the various microprocessors now available. The first is word size, i.e., how many bits define a word? This could range from 4 bits to 16 bits. The 4-bit processors can be combined together to form larger word sizes. This is known as *bit slicing*. The size of the instruction set determines the power and flexibility in programming. Associated with the increased flexibility of the instruction set is the number of independently addressable (and therefore accessible) registers in the processors. This allows more complex programming. Another feature is the speed at which the microprocessor can perform the required function defined for each instruction in its higher-language repertoire.

In using a microprocessor as a computer, while considerations of the characteristics of the hardware are necessary, the major component that defines the uniqueness of each microprocessor is the operating system. In process control, the operating system complexity increases manyfold due to the real-time operation of the process. The operating system, also called the real-time executive, is a block of program logic that has five major functions to solve:

1. Scheduling the time of the processor among the many control programs
2. Managing the external main memory
3. Handling input and output for the various programs
4. Maintaining a data base for use by the tasks (programs) and also for reporting to management control
5. Allowing some mechanism whereby tasks can communicate with each other

As hardware costs of the microcomputer have steadily fallen, the software (executive programs) costs have taken on the major share in the overall cost of the computer installation. These needs lead to a large section of the main memory being taken by the real-time executive and up to 30 percent of the computer time. The rest of the memory space and processor time is available for actual application software.

These constraints are not necessarily a drawback. With the price of semiconductor memories falling rapidly, the expense of a large main memory is of minor consideration. Moreover, a limited memory space can be shared among many tasks or control loops, thus making even small microprocessor-based computers cost-effective compared with analog systems. With this price effectiveness is the increased reliability of the semiconductor components, allowing for an increased faith in the computer and thereby decreasing the requirements for backup. This backup could be either a duplicate computer or an analog system. In either case, the backup takes over control of the process on failure of the main system.

These computer-based systems are available in many sizes and varying capabilities. They can vary from being rather large and capable of controlling many hundreds of loops plus data-base management for report generation to being small dedicated units controlling a few loops. These smaller dedicated units can be linked together through a communication network, thereby leading into the beginning of application, i.e., distributed control. Each dedicated unit will perform its specialized function, which could be either measurement or control, or both. Normally, one large computer in the network would exist to act as the supervisor or master station.

In the application of computers for process control, some additional requirements

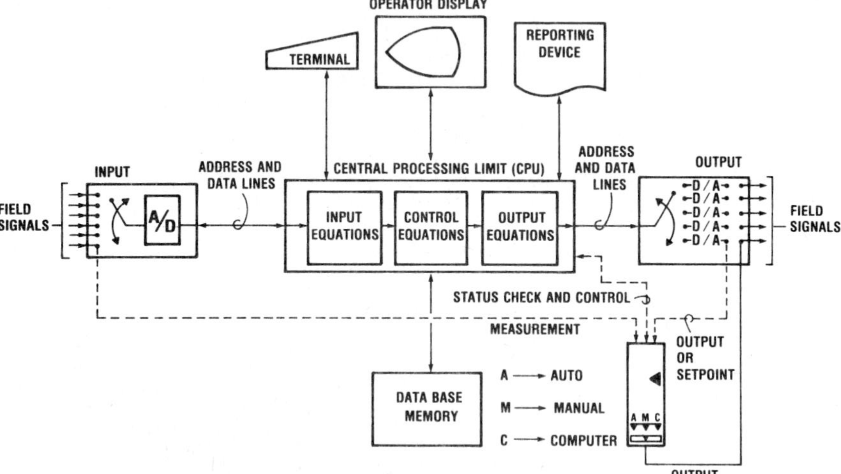

Figure 2-22 Basic elements of a computer for process control.

have to be considered. Primary is the additional hardware required to multiplex a large number of inputs and outputs, which are analog in nature. A conversion process must be performed on these analog signals to present them in a digital word recognizable by the computer and then to convert the digital value back to an analog form useful to the final actuator. Refer to Fig. 2-22.

Software

The software, the second requirement, should be capable of operating in a real-time mode. For example, if one were to schedule a given control program to run every 5 s, the operating system would have to be smart enough to recognize the priority of this program with respect to others and ensure that this schedule is met. This priority positioning of the real-time programs allows many control loops to be set up, each working at a certain given scan cycle. This flexibility allows certain loops to be scheduled as a function of the process characteristics. Thus, a process which is inherently fast (as given by its natural period) should also have its control program scheduled fast. A general rule is to have this schedule 4 to 10 times faster than the natural period of the process. An additional feature that the application software should have is the easy configuration of the various control loops required by a particular plant. This feature distinguishes the use of a computer from an analog system. No physical wiring or tubing changes are necessary. Hence, it is possible to change the control configuration as plant conditions change or as more sophisticated control strategies are implemented. The number of control loops and the ease of configuration or modification are functions of a particular manufacturer's software package and should be considered in any evaluation.

Operator Interface

The third important point for process control computers is the operator interface with the process. It is necessary for the operator of the process to have a clear, up-to-date picture of the conditions in the plant. Conditions requiring immediate attention—various alarms, for example—should be clearly and immediately brought to the operator's attention. Next, the software should be able to recognize any action taken and update outputs to the plant. Graphic displays on CRTs (cathode-ray tubes) are now the most common method by which the operator interfaces with the plant. Three major factors have to be considered. First, the ease with which these displays can be created and if there are any limitations to the number of different displays a system allows. Second, the speed with which (1) various displays can be accessed by the operator and (2) the programs that update and read actions taken by the operator from a given display can be obtained. Third, the ease with which various levels of information can be accessed or changed. For example, one may want to go from a plant unit display to accessing a particular loop of the unit, to a particular section of the loop, to ultimately change some parameter value which could then reflect the alarm value. This could possibly affect the alarm limits, tuning constants, or some calibration or characterization curves. Again, levels of sophistication vary among manufacturers.

Loop Control

Control of loops on a sample basis affects the controllability of the loop. As the sample time is increased, the loop control degrades. This is related directly to the dollar cost of the operation. However, a decrease in sample rate increases the load on the computer which decreases its ability to do other computations. This can be related to the dollar cost of the computing effort. Figure 2-23 shows the relationship between the two costs.

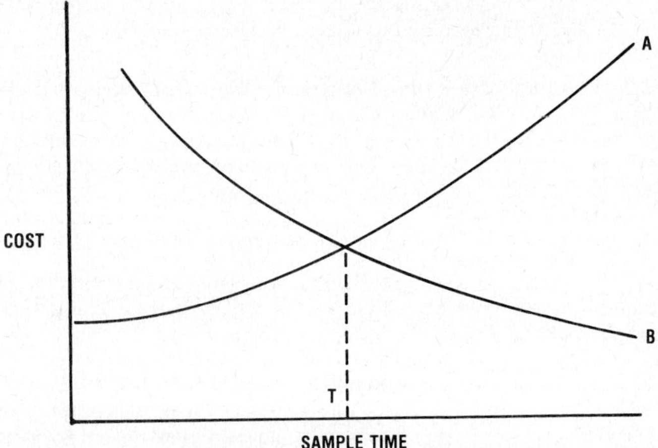

Figure 2-23 Sample rate costs.

Loop control is accomplished by basically solving equations for each loop. Normally there are three sections to these equations:

1. Input equation which reads the digital value from the input hardware (A/D), checks for validity, characterizes for some preset or calculated calibrations, and converts these into numbers meaningful to both the computer and the operator. They can be used to check for alarm limits and also can do some filtering of noisy measurement signals.
2. Control equation which basically gets its measurement from the input section, a set-point value either from the operator or from some other program which may be calculating the required set point, and using the two basic values, calculating an output value for the final actuator.
3. Output value calculated could be checked for limit or alarm conditions, characterized for the final actuator and output hardware, and finally sent to the D/A converter and a sample-and-hold device which is required to retain the output value between sample times.

It should be appreciated that the inherent computational ability of the computer allows the use of complex control algorithms. This enhances the controllability of the plant. The control software should offer some standard algorithms (e.g., PID) and the capability for the user to create and use his or her own.

Backup

An important consideration in any computer application for continuous process control is safety. This involves computer backup. There are basically two options. One is to back up the main system with an identical system. The second is to have an analog backup for each loop. The analog backup would have the capability of either taking over full automatic control or being simply a manual backup in case of computer failure. This requires that the interface hardware recognize the computer failure status and transfer the analog loop to a predefined mode, auto or manual. On the other hand, the computer should be able to recognize the status of the backup station and any action the operator takes with it. Thus, when the operator does transfer to computer control, a smooth bumpless transfer should take place, again a function of the control software. The backup station need not be analog, but could be another computer (mostly dedicated), leading to the concept of distributed control and network hierarchies. Each of these dedicated computers is a particular plant unit control and a number of them are capable of talking to the main computer.

Whatever backup method is implemented, there are two basic methods of operating. First, the main computer does the full control with the backup simply tracking. This is called the DDC (direct digital control) mode. In the second, the main computer does not actually control the final actuator, but provides only a set point to the backup, which becomes the main controller. This is called SPC (supervisory control or set-point control). There are certain advantages in this mode. Failure of the supervisory computer does not affect the plant operation. The computation load decreases quite a bit since the dedicated controller provides the main control action. This decrease could be effectively utilized to run management reporting programs, build an historical data base, and institute both complex feed-forward control strategies and adaptive tuning techniques. In this mode of operation, the computer's advantage of computational power and its memory capability will be fully utilized.

Historical Data Base

The historical data base is the ability to store past plant operating data on a real-time basis, i.e., the data must relate to the actual time of plant operations to serve any useful purpose in either reporting for management control, for cost and efficiency calculations, or in using the data to plot or display trends of the various plant variables. Flexibility and memory size play an important part in the software that performs these functions. Flexibility in data-base development is the ability to add or delete variables to the data base. Furthermore, the data base should allow easy access for the reporting or trending software. When trying to determine the number of variables that can be used, one must take into account size considerations, how often the data collection is performed, and for how long a period the data are to be stored on-line for a given system. These factors have to be considered to determine the amount of on-line memory that is required. The reporting function may have three modes of operation: first, reports scheduled on a regular basis; second, reports called for on demand by operating personnel; and third, exception reports that are printed under given plant conditions.

Tuning Techniques

Feed-forward and adaptive tuning techniques lend themselves to easier treatment since complex equations and algorithms can easily be solved. The availability of memory allows past loop responses to be used to update terms of a given process model. These updated terms can be used to calculate tuning constants of a given control algorithm (for example, proportional gain, integral, and derivative of a PID control equation).

Whereas our discussion has been concentrated on the control of continuous loops, the use of a computer for logical operations can also be easily accomplished. For example, it can be used to control a batch process where very little continuous control may be needed but a large number of logical operations have to be performed. The computer inputs and outputs will be primarily digital (i.e., contacts) with some analog signals. The computer algorithms have to be set up to sequence through the required batch cycle while continuously monitoring the plant conditions and ensuring the completion of each batch cycle. Again, flexibility should be available to stop the sequence at any step, to continue the sequence, and to allow the operator manual sequencing of the process. Further, the operator should be allowed to change operations in a given sequence, for example, change the temperature ramp profile of a reactor or change the given product recipe. The software should also be able to perform relay-type logic, thus eliminating the need for hardware relay logic. Specialized computers specifically designed for these functions are called programmable logic controllers (PLC). These devices have a limited operating system which allows the running of one preprogrammed application program. Operator interface, if provided, is limited to some predefined function keys. The present trend is to make these devices more flexible, bringing them into the domain of minicomputers.

Because the technology is changing so rapidly now, the plant engineer should make a careful comparative study of available program controller units before making any key purchase decisions.

section 6

Noise and Vibration Control

chapter 6-1

Noise Control

by

Paul Jensen

Manager, Industrial Services
Bolt Beranek and Newman Inc.
Cambridge, Massachusetts

INTRODUCTION

Industrial plants are inherently noisy. Not until the 1960s and 1970s, however, did concern for public health and welfare lead to the intensive study and analysis of industrial noise and promulgation of regulations to control it. Most important of the regulations was the Occupational Safety and Health Act of 1970, whose Standard 1910.95 established limits for occupational noise exposure and thus gave impetus to the development of industrial noise control. Permissible noise exposures for employees, according to the act, cannot exceed those given in Table 1-1.

The act also covers combined exposures to noise and means of controlling noise:

> When the daily noise exposure is of two or more periods of noise exposure at different levels, their combined effect should be considered, rather than the individual effect of each. If the sum of the following fractions

TABLE 1-1 Permissible Noise Exposure Levels

Duration per day, h	Sound level, dBA,* slow response
8	90
6	92
4	95
3	97
2	100
1½	102
1	105
½	110
¼ or less	115

*See definition of dBA later in this chapter.

$$E = \frac{C_1}{T_1} + \frac{C_2}{T_2} + \cdots + \frac{C_n}{T_n}$$

exceeds unity, then, the mixed exposure should be considered to exceed the limit value. C_n indicates the total time of exposure at a specified sound level, and T_n indicates the exposure limit at that level.

Exposure to impulsive or impact sound should not exceed 140 dB peak sound pressure level.

When employees are subjected to sound levels exceeding those listed in the table, feasible administrative or engineering controls shall be utilized. If such controls fail to reduce sound levels within the levels of the table, personal protective equipment shall be provided and used to reduce sound levels within the levels of the table.

If the variations in sound level involve maxima at intervals of 1 second or less, it is to be considered continuous.

In all cases where the sound level exceeds the values shown herein, a continuing, effective hearing conservation program shall be administered.

To enforce the standard, OSHA inspectors are empowered to enter and inspect any workplace in the United States, after presenting their credentials to the person in charge. Representatives of the employer and of the employees may accompany the inspector on the tour of the work place.

If a violation is found during an inspection, an attempt will be made to settle the issues without going to court. However, employers who fail to comply with the standard may be taken to court by OSHA. Penalties range from civil fines up to $10,000 through criminal penalties up to $20,000 plus a year of imprisonment.

The OSHA standard has been adopted by other federal agencies, notably the Bureau of Mines. In 1978, the Department of Defense issued its own noise standard, Instruction No. 6055.3, which limits exposure of members of the armed forces to hazardous noise.

The present OSHA standard described above is being considered for revision now. In 1974, the Department of Labor published proposed rules that, if accepted, will keep the permissible sound level at 90 dBA for an 8-h exposure but will incorporate in the dosage calculation the allowance of 85 dBA for 16 h.

In addition, the proposed rule spells out very specifically the adoption of a stringent hearing conservation program for all workers who are exposed to an equivalent sound level of 85 dBA for 8 h (50 percent of allowable noise exposure).

OSHA proposes to limit exposure to impulses at 140 dB to 100 per day and to permit a tenfold increase in the number of impulses for each 10-dB decrease in the peak pressure of the impulse. For example, the number of impulses allowed at 130 dB would be 1000 per day, and the number of impulses allowed at 120 dB would be 10,000 per day.

As with the existing standard, OSHA states in the proposed regulation the methods that must be used to ensure compliance. Engineering and administrative controls, to the extent that they are technically and economically feasible, are to be used to reduce employee noise exposure to permissible limits. When these controls are unable to reduce noise to permissible limits, they nonetheless are to be used to reduce noise to the lowest level feasible and are to be supplemented by personal protective equipment. One exception is noted: The proposed rule would allow hearing protectors to be the sole means of noise control if an employee is exposed to noise beyond permissible limits for only one day each week.

CONCEPTS AND VOCABULARY OF NOISE CONTROL

Sound Noise is simply unwanted sound; it is a series of vibrations in the air. Human ears are sensitive to these vibrations; they sense them and pass them on to the brain. Acoustics—the branch of physics that deals with the production, transmission, and control of sound—has its own concepts and vocabulary. The following items are the key terms and concepts needed by an engineer working in noise control.

Frequency The frequency f of a sound is the number of its vibrations. Units of frequency may be expressed in cycles per second (cps) or in hertz (Hz). Since the frequency is a description of a periodic phenomenon, it is reciprocal of the time period, or the time necessary for the phenomenon to repeat.

The audible frequency range (16 to 20,000 Hz) has been divided into a series of octave bands and one-third-octave bands. Just as with an octave on a piano keyboard, an octave in sound analysis represents the frequency interval between a given frequency and twice that frequency. The interval is identified by the center frequency, representing the geometric mean of the bounds of that interval. The internationally agreed upon 1000-Hz center frequency determines the center frequencies of the remaining bands. The center frequencies and approximate cutoff frequencies are listed in Table 1-2.

Velocity of sound A sound wave travels at a velocity that depends primarily on the elasticity and the density of the medium. In air at normal temperature, the velocity of sound c is approximately 340 m/s (1100 ft/s).

Wavelength Under normal conditions, sound waves always travel through air at essentially the same velocity. Therefore, at any frequency, the wavelength λ of the sound (i.e., the distance that the sound travels in one cycle) can be calculated by

$$\lambda = \frac{c}{f}\, m \qquad \text{(ft)}$$

Sound power Sound power in watts (W) describes the energy of the sound source. This sound power may be

1. The total power radiated by the sound source over its entire frequency range
2. The power radiated in each of a series of frequency bands
3. The power radiated in a limited frequency range

Sound intensity Sound intensity I is equal to the sound power radiated in a specified direction through a unit area normal to this direction. If a steady-state emission of sound energy (power) is assumed from the sound source, in free space the energy will spread itself thinner as it travels further from the sound source.

Intensity, though useful when discussing wave motion and energy radiation and spreading, has the practical disadvantage of employing very small numbers and failing to relate directly to human hearing experience. A more convenient form of expressing people's hearing sensation is *sound pressure.*

Sound pressure Sound waves produce changes in the density of air as they travel through it. These changes in the air density cause pressure fluctuations around the ambient static pressure, or sound pressure. In most cases, these disturbances created in the air cannot be easily expressed mathematically in time and space. The reason is that,

TABLE 1-2 Center and Approximate Frequency Limits for the Standard Set of Contiguous Octave and One-Third Octave Bands Covering the Audio Frequency Range

	Frequency, Hz					
	Octave			One-third octave		
Band	Lower band limit	Center	Upper band limit	Lower band limit	Center	Upper band limit
12	11	16	22	14.1	16	17.8
13				17.8	20	22.4
14				22.4	25	28.2
15	22	31.5	44	28.2	31.5	35.5
16				35.5	40	44.7
17				44.7	50	56.2
18	44	63	88	56.2	63	70.8
19				70.8	80	89.1
20				89.1	100	112
21	88	125	177	112	125	141
22				141	160	178
23				178	200	224
24	177	250	355	224	250	282
25				282	315	355
26				355	400	447
27	355	500	710	447	500	562
28				562	630	708
29				708	800	891
30	710	1,000	1,420	891	1,000	1,122
31				1,122	1,250	1,413
32				1,413	1,600	1,778
33	1,420	2,000	2,840	1,778	2,000	2,239
34				2,239	2,500	2,818
35				2,818	3,150	3,548
36	2,840	4,000	5,680	3,548	4,000	4,467
37				4,467	5,000	5,623
38				5,623	6,300	7,079
39	5,680	8,000	11,360	7,079	8,000	8,913
40				8,913	10,000	11,220
41				11,220	12,500	14,130
42	11,360	16,000	22,720	14,130	16,000	17,780
43				17,780	20,000	22,390

over an extended portion of the vibrating surface that is creating the airborne disturbance, some portions are compressing the surrounding air, while other portions are causing rarefactions.

The unit approved by the International Standards Organization (ISO) for measuring sound pressure is the pascal (Pa), though the terms microbars (μbar), dynes per square centimeter (dyn/cm^2), and newtons per square meter (N/m^2) have all been used.

Conversion factors are

$$1 \text{ bar} = 10^5 \text{ Pa}$$
$$1\ \mu\text{bar} = 1 \text{ dyn/cm}^2$$
$$1 \text{ dyn/cm}^2 = 10^{-1} \text{ Pa}$$
$$1 \text{ N/m}^2 = 1 \text{ Pa}$$

Sound power level The sound power level (PWL) is the designation in decibels of the ratio of two sound powers. The PWL is independent of the environment and the distance from the sound source.

When the PWL of a sound source is being given, it is customary to use a reference level of 10^{-12} W:*

$$\text{PWL} = 10 \log \frac{W}{10^{-12}} \quad \text{dB}$$

Sound intensity level The sound intensity level (IL) is the designation in decibels of the ratio of two intensities. The IL is influenced by the environment and the distance from the sound source.

When indicating IL, it is customary to use a reference level of 10^{-12} W/m^2:

$$\text{IL} = 10 \log \frac{I}{10^{-12}} \quad \text{dB}$$

Sound pressure level The sound pressure level (SPL) is the designation in decibels of the ratio of two pressures squared. The reason for using pressures squared is that all sounds consist of positive pressure disturbances (compressions) and negative pressure disturbances (rarefactions) measured from the static pressure. The mean value of the sound pressure disturbances would not be a meaningful value (it would be zero, since there are as many compressions as rarefactions). When indicating SPL, it is customary to use a reference level p_0 of 2×10^{-5} Pa. This reference value is roughly equivalent to the threshold of hearing at a frequency of 1000 Hz:

$$\text{SPL} = 10 \log \left(\frac{p}{2 \times 10^{-5}} \right)^2 = 20 \log \frac{p}{2 \times 10^{-5}} \quad \text{dB}$$

Some relationships between sound pressure and sound pressure level ($p_0 = 2 \times 10^{-5}$ Pa) are shown in Table 1-3.

TABLE 1-3 Sound Pressure and Sound Pressure Level ($p_0 = 2 \times 10^{-5}$ Pa)

Sound pressure, p	Sound pressure level, L, dB
p_0	0
$1.12p_0$	1
$1.26p_0$	2
$2p_0$	6
$3.16p_0$	10
$10p_0$	20
10^2p_0	40
10^4p_0	80
10^6p_0	120

Sound level There are many single-number schemes for evaluating sounds according to people's response to noise. The simplest of these single-number schemes is the sound level in dBA. The "A" means that the frequency spectrum has been weighted by an electrical network in the sound-measuring equipment before the single number was derived. It is simple to measure the A-weighted sound level and, fortunately, this single number correlates as well as or better than other commonly known noise ratings. The frequency response of the A-weighting is shown in Table 1-4.

Table 1-5 shows some typical sound pressure levels in industrial environments.

Decibel The decibel (dB) is a dimensionless unit for expressing the ratio of two numerical values on a logarithmic scale. It is convenient to use decibels in dealing with sound power, sound intensity, or sound pressure because of the tremendous range of values of these quantities that can be perceived by the ear. Audible intensities range from 10^{-12} to 10 W/m^3.

*Originally, 10^{-13} was used as a reference power. When this is the case, the power level must be reduced by 10 dB for use in the formulas of this chapter.

TABLE 1-4 A Scale Weighting

Octave band center frequency, Hz	One-third octave band center frequency, Hz	A-weighting relative response, dB
	20	−50.5
	25	−44.7
31.5	31.5	−39.4
	40	−34.6
	50	−30.2
63	63	−26.2
	80	−27.5
	100	−19.1
125	125	−16.1
	160	−13.4
	200	−10.9
250	250	− 8.6
	315	− 6.6
	400	− 4.8
500	500	− 3.2
	630	− 1.9
	800	− 0.8
1000	1000	0
	1250	+ 0.6
	1600	+ 1.0
2000	2000	+ 1.2
	2500	+ 1.3
	3150	+ 1.2
4000	4000	+ 1.0
	5000	+ 0.5
	6300	− 0.1
8000	8000	− 1.5
	10000	− 2.5
	12500	− 4.3
16000	16000	− 6.6
	20000	− 9.3

TABLE 1-5 Typical Sound Pressure Levels in Industrial Plants

	Octave band center frequency, Hz							
	A	125	250	500	1000	2000	4000	8000
Synthetic spinning machine	95	76	79	84	85	88	90	89
Rock crusher	101	102	100	99	96	93	85	77
Letterpress	96	93	94	94	93	89	84	80
Hammer mill	96	94	92	92	89	87	85	79
Hand-held sand blaster	95	85	82	82	87	88	90	96
Plastic extruder	97	103	99	94	89	88	85	82
Candy wrapper	90	78	82	83	84	84	80	75
Fly shuttle loom	102	86	87	91	95	98	95	89
Can filling/seamer	98	86	87	92	92	93	89	87
Ring twister	94	88	86	88	90	89	82	78
Chipper	105	92	98	100	100	99	99	106
Wood chipper	109	104	110	110	106	96	88	80
Billet heater	102	96	107	101	91	84	78	75
Buffing machine	101	88	85	89	98	94	89	83
Punch press	102	94	96	96	96	97	95	87
Wire-drawing machine	96	91	93	92	92	90	82	76
Molding machine	98	80	82	98	90	91	89	81
Concrete-block machine	106	100	102	101	101	101	96	90

Addition of decibels Since decibels are logarithmic units, they are not added arithmetically. Decibels are added by converting them to power, intensity, or pressure, adding these quantities, and then converting them back to decibels. In other words, 80 dB and 80 dB do not add up to 160 dB, but to 83 dB. Figure 1-1 gives the most convenient way to combine decibels.

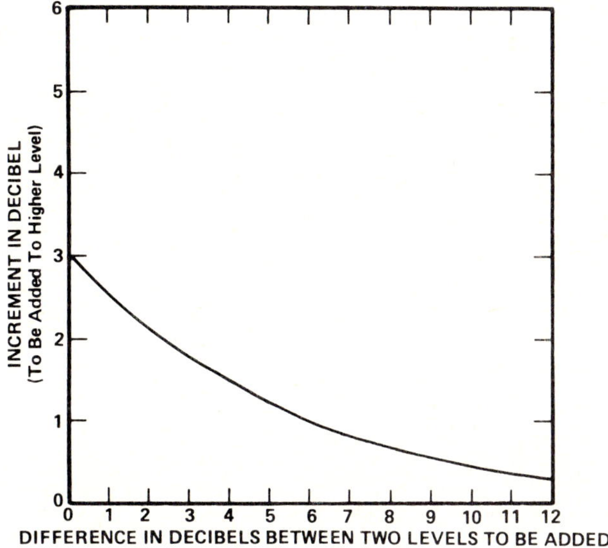

Figure 1-1 Chart for combining sound levels.

The inverse square law Under free-field conditions of sound radiation, the sound intensity is proportional to the square of the sound pressure:

$$I = \frac{p^2}{c\rho_0} \qquad \text{W/m}^2$$

where

p = sound pressure, Pa
c = velocity of sound, m/s
ρ_0 = density of air, kg/m³

When the sound intensity is related to the sound power of a source, the result is

$$I = \frac{p^2}{c\rho_0} = \frac{W}{4\pi r^2} \qquad \text{W/m}^2$$

$$p^2 = c\rho_0 \frac{W}{4\pi r^2} \qquad \text{Pa}^2$$

The relationship between the sound pressures squared at two distances from the source becomes

$$\frac{p_1^2}{p_2^2} = \frac{r_2^2}{r_1^2}$$

This classic relationship is called the *inverse square law.*

Reverberation and reverberation time Reverberation is the persistance of sound after the source has stopped.

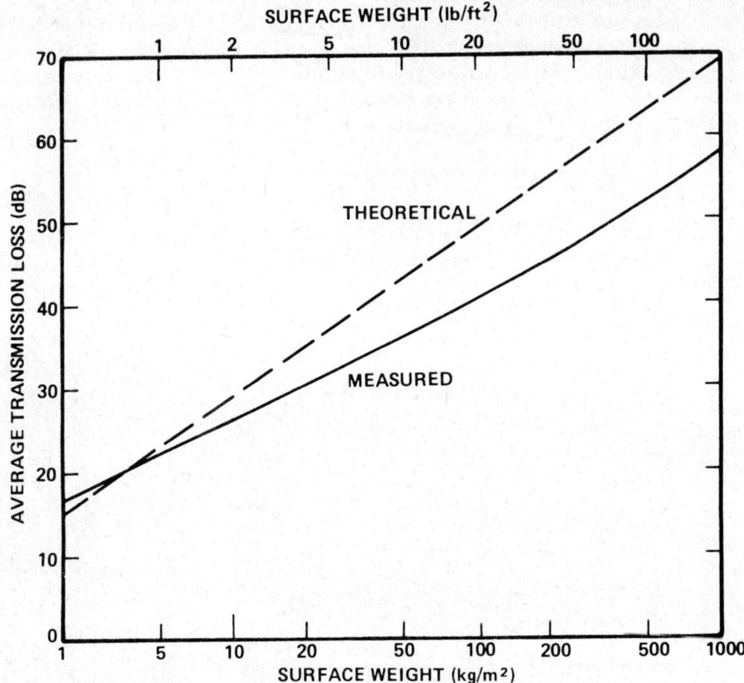

Figure 1-2 Average transmission loss of single homogeneous walls in the 100- to 3150-Hz frequency range.

The reverberation time t in a room has been defined as the time required for the sound level to decrease 60 dB after the source has stopped.

Transmission coefficient The transmission coefficient τ is the fraction of incident sound energy that is transmitted through a barrier. Thus

$$\tau = \frac{W_2}{W_1}$$

where W_1 is the incident sound energy in watts and W_2 is the sound energy transmitted in watts.

Transmission loss The transmission loss TL of a barrier is defined as the ratio of transmitted sound energy to the incident sound energy.

In logarithmic form, this ratio is expressed as

$$\text{TL} = 10 \log \frac{W_1}{W_2} \quad \text{dB}$$

This can be written as

$$\text{TL} = 10 \log \frac{1}{\tau} \quad \text{dB}$$

To calculate in detail the transmission loss for even simple constructions is difficult. As a result, engineers rely in most cases on laboratory data obtained in tests between specially constructed rooms.

Mass law The mass law says that for each doubling of the weight of a single-thickness wall, the average transmission loss increases by 6 dB. Figure 1-2 shows a graphic presentation of the transmission loss of a single wall. The empirically determined curve (solid) is lower than the theoretical curve (dotted). Walls with surface weight less than 200 kg/m^2 (40 lb/ft^2) gain only 5 dB per doubling, while for walls whose surface weight is above 200 kg/m^2, the mass law is valid again.

Composite transmission loss The composite transmission loss of a barrier is determined when the transmission coefficients for each part are known. The sound power transmitted by the two elements with a common incident sound power is

$$W_{trans} = \tau_1 S_1 + \tau_2 S_2 \qquad \text{W}$$

where τ_1 and τ_2 are the transmission coefficients for the individual parts, S_1 and S_2 are the areas of the individual parts in square meters (square feet), m^2 (ft^2), and W is watts.

The composite transmission loss becomes

$$\text{TL} = 10 \log \frac{S_1 + S_2}{\tau_1 S_1 + \tau_2 S_2} \qquad \text{dB}$$

Noise reduction The difference in sound pressure levels between two rooms is called the noise reduction (NR). The NR accounts for all sound paths and is thus more inclusive than the transmission loss.

The noise reduction can be expressed as

$$\text{NR} = \text{TL} - 10 \log \frac{S}{A_2} - C_1 + C_2 \qquad \text{dB}$$

where

S = area of the common wall, m^2 (ft^2)
A_2 = total absorption in the receiving room, m^2-sabin (ft^2-sabin)
C_1 = correction which depends on air leaks*
C_2 = correction which depends on flanking transmission.*

THEORY AND PHYSIOLOGY OF NOISE

Human Response to Noise

The psychophysical characteristics of a noise—those that determine how people react to it—cannot be measured with any currently available instrumentation. The only method that can be used is to expose a sufficiently large group of people *(test population)* to the noise in question in physical, psychological, and social situations similar to that of the people influenced by the noise—obviously a cumbersome and time-consuming method.

Because of such difficulties in developing data, acousticians, sociologists, and others have tried to develop objective measurement procedures for noise. But people obviously do not respond the same way to the same noise. Their response depends on their previous noise exposure, psychological attitude, socioeconomic status, the activity they are engaged in when they hear the noise, and other factors.

However, rating scales derived from physical measurements of the noise have been developed. These correlate reasonably accurately with the *average* human response to noise, particularly when the tests are confined to the laboratory environment. These scales rate human response in terms of loudness, perceived noise level (related to annoyance), articulation index (related to the ability to converse with ease), and other factors.

Numerous single-number rating schemes have been proposed for evaluating different noises according to one aspect or another of people's subjective response to the noise. Some of the ratings are quite simple; others are very complicated. Among the better known are

- Overall sound pressure level
- A-weighted sound level (dBA)

*C_1 and C_2 are zero for no air leaks and no flanking transmission.

- Equal energy sound level (L_{eq})
- Loudness level (LL)
- Articulation index (AI in percent)
- Speech interference level (SIL)
- Noise criterion curves (NC)
- Perceived noise level (PNdB)
- Noise and number index (NNI)
- Traffic noise index (TNI)
- Noise pollution level (NPL).

Acousticians and engineers have responded to the demand for noise control by developing the *systems approach*. Each problem of noise control is viewed as having three components:

- A noise *source*
- A noise *path*
- A noise *receiver*

Each component is considered in the solution of the problem. Engineers must

- Determine the characteristics of the noise sources
- Determine the tolerance of the receiver for intruding noise
- Determine how much and what kind of control must be used at the source and/or along the path to achieve the desired level of sound

Acoustical Treatments

Sound that originates in an enclosed space, such as a room or a factory, will spread until it reaches the surfaces, where it will either be absorbed or reflected. If room surfaces are hard, sound will reverberate, intermittent sounds will be mixed together, and steady sounds will add up. The result will be a relatively noisy space. If room surfaces are soft, the space will be relatively quiet.

It is important, then, that the amount of sound either absorbed or reflected can be quantified. The sound absorption efficiency of a material is determined as the fraction of incident sound energy that is absorbed by the surface. This is called the sound absorption coefficient α:

$$\alpha = \frac{I_a}{I_i}$$

where I_i is the sound intensity impinging on the material in watts per square meter and I_a is the sound intensity absorbed by the material, in watts per square meter. If $\alpha = 1.0$, all impinging sound energy is absorbed. If $\alpha = 0.0$, all impinging sound energy is reflected.

If 1 m^2 (ft^2) of material absorbs 20 percent of the impinging sound energy, 5 m^2 (ft^2) will absorb as much as 1 m^2 (ft^2) having complete efficiency. The sound absorption A provided by a material can be determined by

$$A = \alpha S \qquad \text{m}^2\text{-sabin}$$

where S is the surface area in square meters.

But rooms are constructed of several materials, each having different absorption coefficients. The total sound absorption then becomes

$$A = \alpha_1 S_1 + \alpha_2 S_2 + \cdots + \alpha_n S_n = \Sigma\, a_n S_n$$

Table 1-6 lists absorption coefficients of many typical building materials, and Table 1-7, the absorption coefficients of commonly used acoustic—or sound-absorbing—materials.

TABLE 1-6 Sound Absorption Coefficients of General Building Materials and Furnishings*

Materials	Coefficients, Hz					
	125	250	500	1000	2000	4000
Brick, unglazed	0.03	0.03	0.03	0.04	0.05	0.07
Brick, unglazed, painted	0.01	0.01	0.02	0.02	0.02	0.03
Carpet, heavy, on concrete	0.02	0.06	0.14	0.37	0.60	0.65
Same, on 40-oz hair felt or foam rubber	0.08	0.24	0.57	0.69	0.71	0.73
Same, with impermeable latex backing on 40-oz hair felt or foam rubber	0.08	0.27	0.39	0.34	0.48	0.63
Concrete block, coarse	0.36	0.44	0.31	0.29	0.39	0.25
Concrete block, painted	0.10	0.05	0.06	0.07	0.09	0.08
Fabrics						
Light velour, 10 oz/yd², hung straight, in contact with wall	0.03	0.04	0.11	0.17	0.24	0.35
Medium velour, 14 oz/yd², draped to half area	0.07	0.31	0.49	0.75	0.70	0.60
Heavy velour, 18 oz/yd², draped to half area	0.14	0.35	0.55	0.72	0.70	0.65
Floors						
Concrete or terrazzo	0.01	0.01	0.015	0.02	0.02	0.02
Linoleum, asphalt, rubber, or cork time on concrete	0.02	0.03	0.03	0.03	0.03	0.02
Wood	0.15	0.11	0.10	0.07	0.06	0.07
Wood parquet in asphalt on concrete	0.04	0.04	0.07	0.06	0.06	0.07
Glass						
Large panes of heavy plate glass	0.18	0.06	0.04	0.03	0.02	0.02
Ordinary window glass	0.35	0.25	0.18	0.12	0.07	0.04
Gypsum board, ½ in nailed to 2 × 4's 16 in. o.c.	0.29	0.10	0.05	0.04	0.07	0.09
Marble or glazed tile	0.01	0.01	0.01	0.01	0.02	0.02
Openings						
Stage, depending on furnishings			0.25–0.75			
Deep balcony, upholstered seats			0.50–1.00			
Grills, ventilating			0.15–0.50			
Plaster, gypsum or lime, smooth finish on tile or brick	0.013	0.015	0.02	0.03	0.04	0.05
Plaster, gypsum or lime, rough finish on tile or brick	0.14	0.10	0.06	0.05	0.04	0.03
Same, with smooth finish	0.14	0.10	0.06	0.04	0.04	0.03
Plywood paneling, 3/8 in thick	0.28	0.22	0.17	0.09	0.10	0.11
Water surface, as in a swimming pool	0.008	0.008	0.013	0.015	0.020	0.025
Air, sabins per 1000 ft³ at 50% RH				.9	2.3	7.2

*Complete tables of coefficients of the various materials that normally constitute the interior finish of rooms may be found in the various books on architectural acoustics. The following short list will be useful in making simple calculations of the reverberation in rooms.

Materials Selection

The most commonly used materials for control of noise in industry are absorbers and transmission-loss materials for airborne sound and vibration isolators and dampers for solid-borne sound. Selection of materials is governed by factors other than acoustical. These factors may be broadly classified as environmental and regulatory. Environmental factors include:

- Moisture, water spray, water immersion
- Oil, grease, dirt
- Vibration

TABLE 1-7 Sound Absorption Coefficients of Common Acoustic Materials

Materials*	Frequency, Hz 125	250	500	1000	2000	4000
Fibrous glass (typically 4 lb/ft³) hard backing						
1 in thick	0.07	0.23	0.48	0.83	0.88	0.80
2 in thick	0.20	0.55	0.89	0.97	0.83	0.79
4 in thick	0.39	0.91	0.99	0.97	0.94	0.89
Polyurethane foam (open cell)						
¼ in thick	0.05	0.07	0.10	0.20	0.45	0.81
½ in thick	0.05	0.12	0.25	0.57	0.89	0.98
1 in thick	0.14	0.30	0.63	0.91	0.98	0.91
2 in thick	0.35	0.51	0.82	0.98	0.97	0.95
Hair felt						
½ in thick	0.05	0.07	0.29	0.63	0.83	0.87
1 in thick	0.06	0.31	0.80	0.88	0.87	0.87

*For specific grades, see manufacturer's data.

- Temperature
- Erosion by fluid flow

Regulatory factors include:

- Lead-bearing material forbidden near food processing lines
- Restrictions on materials that may be in contact with foods being processed—glass, Monel, or stainless steel permitted
- Requirements for material not to be damaged by disinfecting
- Firebreak requirements on ducts, pipe runs, shafts
- Flame-spread rate limits on acoustically absorbing materials
- Fire-endurance limits on acoustically absorbing materials
- Restrictions on shedding of fibers in air by acoustically absorbing materials
- Elimination of uninspectable spaces in which vermin may hide
- Requirements for secure anchoring of heavy equipment
- Restrictions on hole sizes in machine guards (holes can reduce radiated noise of vibrating sheets)

NOISE-CONTROL TECHNIQUES

As previously noted, noise can be controlled at its *source,* along the *paths* it travels through air or through structures, and at the ears of the *receiver.* Industry today uses techniques that include both noise-source treatments and transmission path treatments. Personal protective equipment, such as earmuffs and earplugs, is often useful and efficient in reducing a worker's daily noise dose, although it is no substitute for engineered noise control. Often overlooked in general discussions of controlling industrial noise are methods that require no machine modifications or additions—good maintenance and operating procedures and planned replacement of machines or machine parts. These methods are described at the end of this section.

The primary techniques used in controlling noise in industry are the following. Noise-source treatments:

- Elimination or modification of specific equipment elements

- Substitution or redesign of entire machines or processes
- Surface damping

Transmission-path treatments:

- Mufflers
- Addition of room absorption
- Construction of noise shelters or other forms of personnel enclosures
- Vibration isolation
- Barriers
- Lagging

Receiver treatments:

- Earplugs
- Earmufflers

Noise-Source Treatments

Modifications or Substitutions

Quiet components are available to replace the noise-generating parts of certain machines. The equipment or components may need some engineering design and analysis to ensure that the machines are compatible with the manufacturing process. However, in many instances, off-the-shelf replacement items can be purchased for direct application.

The use of alternative materials for selected components can also reduce the noise generated by a machine or part of a machine. For example, the noise produced by the impact and subsequent vibration of metal parts can be significantly reduced by the addition of soft buffers or by the replacement of some non-load-bearing parts by plastic or other heavily damped materials. (*Damped* materials do not respond greatly or "ring" when struck.)

Surface Damping

Machinery housings often consist of large areas of flexible metal plate; if they are set into vibration, such plate areas may become significant acoustical radiators of airborne sound. Vibration is caused by some forcing mechanism such as the internal oscillating, rotating, or reciprocating components of the machine. The action of these components will tend to excite the surfaces at their resonant (or natural ring) frequency. The adding of damping (vibration-energy-absorbing) material to the surface will cause the amplitude of vibration at the resonant frequency to be reduced. Adding more material can stiffen and change the natural frequency so that it is no longer so easily excited.

Damping (viscoelastic) materials can be applied either as a free layer or as a constrained layer. (See Fig. 1-3*a* and *b*.) The material can be either troweled on, glued and baked on, or (for intermittent use) attached by a magnetic layer. Practical concerns regarding the specific material are ease of cleaning, toxicity, durability, and temperature-independent efficiency in the required frequency ranges. As a rule of thumb, for a damping material to have any effectiveness, it must be at least as thick as the material to be damped.

Transmission-Path Treatments

Mufflers

Mufflers are used to reduce noise associated with air- or gas flow. Examples are noise at the intake or exhaust of engines, at the inlet and outlet of fans, or generated by a high-velocity air or stream jet. Mufflers are also used in passageways to control the escape of

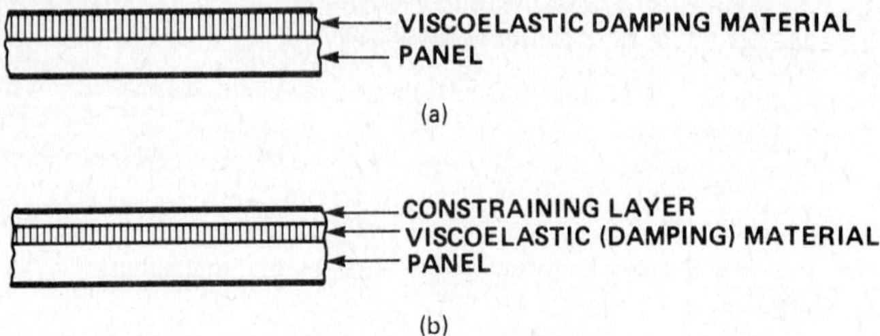

Figure 1-3 (*a*) Panel with free layer of viscoelastic (damping) material. (*b*) Panel with constrained layer of viscoelastic (damping) material.

noise in paths that must carry materials or personnel. A piece of acoustically lined duct, through which punch press punchings may pass, will also serve as a simple muffler to reduce noise escaping from the impacting region of the press.

There are two basic types of muffler: one is a dissipator of the acoustic energy; the other relies on reflection to confine the acoustic energy.

Figure 1-4 shows the acoustical performance of standard dissipative mufflers. The pressure drop through a muffler (the added drag the muffler puts on the flow) can be minimized by maintaining large openings and low velocities of flow through the muffler. (Indeed, if the velocity is not sufficiently reduced, the muffler can become, in turn, an

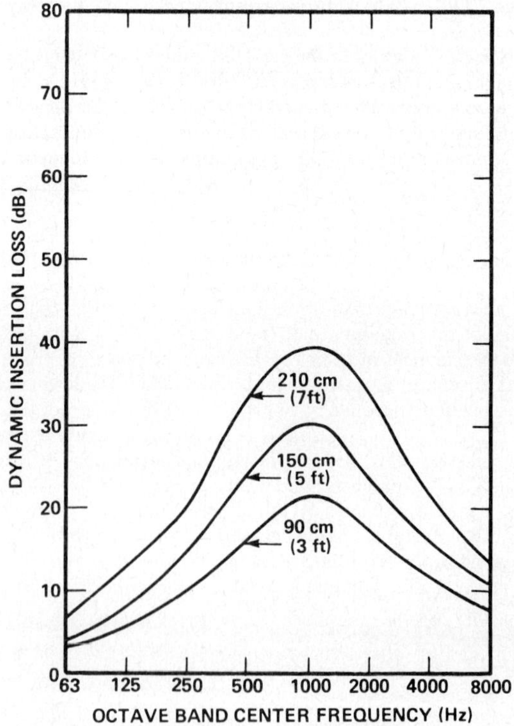

Figure 1-4 Acoustical performance of mufflers of different lengths.

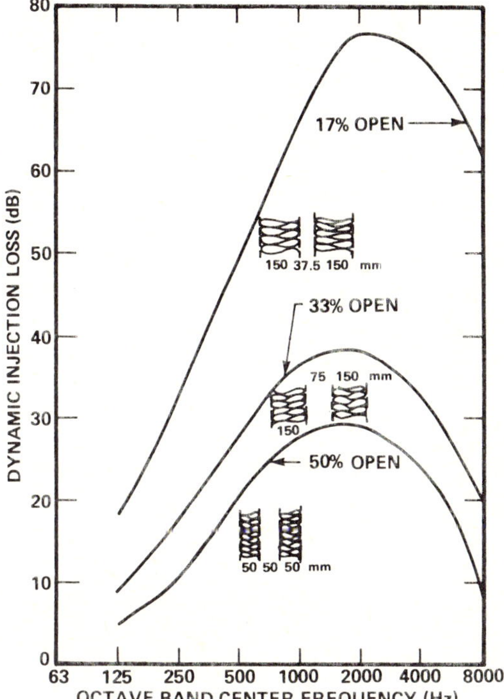

Figure 1-5 Acoustical performance of 90-cm (3-ft) long mufflers with different spacing and thickness of baffles.

additional noise source.) However, the distance between absorptive surfaces of a dissipative muffler is critical to the band of frequencies of sound absorbed. Sets of mufflers in series may sometimes be necessary to attenuate sound in different frequency bands.

Figure 1-5 shows the acoustical performance of dissipative mufflers with different spacing and thickness of baffles. The problem of contamination of the porous surfaces of the sound-absorbing material can be solved by covering the surface with a very thin, nonporous, limp, but durable skin, which is designed to allow sound to pass through it, yet still contain the flow.

The practical effectiveness of any design is likely to be limited by the available space, since it is theoretically possible to lengthen a muffler to produce any required attenuation (fading of sound). Practical design may range from an array of 15-ft- (5-m-) long absorptive splitters for the large duct of an induced-draft fan on a boiler to the 3-ft- (1-m-) long device (reactive muffler) presently fitted to an automobile engine.

Addition of Room Absorption

One commonly used method for noise reduction in a room is to clad the interior surfaces (ceiling and walls) with highly absorptive materials. The acoustical materials will have a minor noise reduction effect close-in to the noise sources, while at distances several meters from the noise sources, significant noise reduction (5 to 10 dB) can be obtained. Figure 1-6 shows the sound pressure level drop-off as a function of distance from a sound source for rooms with various amounts of acoustical materials. As Fig. 1-6 shows, a 3-dB drop in sound occurs for each doubling of the total amount of absorption at distances away from the noise sources (in the reverberant field).

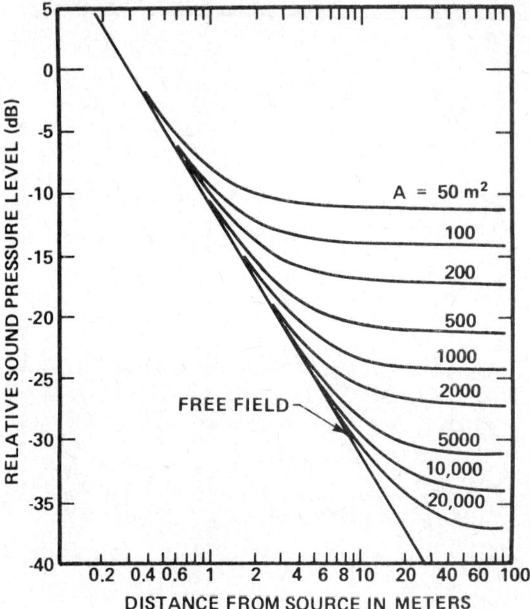

Figure 1-6 Drop-off of sound pressure level as a function of distance from sound source.

Enclosures

Enclosures, another useful means of controlling noise, are structures that completely surround the noise source and thus contain the sound it generates. However, enclosures can cause a buildup of high-level acoustic energy within it. Enclosures, therefore, usually consist of a wall with mass chosen to provide the required attenuation with a inner lining of porous material to dissipate the buildup of acoustic energy produced. In some cases where complete enclosures are built, the machines inside may require placement on vibration isolators—devices that prevent the transmission of structure-borne noise to the outer surfaces of the enclosure.

Employees exposed to high-level noise from a number of sources can also be protected by acoustic booths. These can range from small open-fronted telephone-booth-sized cabinets (into which the operator steps while observing the operation of a semiautomatic machine) to completely enclosed control consoles. In many cases, acoustic booths for personnel can provide an island of protection; an employee exposed to high-level noise during part of the workday can be protected well enough to reduce total exposure during normal working hours to less than the permissible exposure.

The design of worker enclosures requires suitable walls to produce the necessary sound reduction and also some internal sound absorption to prevent any reverberant buildup of transmitted sound. The use of fixed windows allows the necessary orientation of the booth, which can be important if an open entrance is to be used. The design and location of such booths can require a careful review and measurement program of the acoustical situation and may entail some local room acoustical treatment.

If there are gaps in the enclosure, sound can escape. The greater the percentage of open area of an enclosure, the smaller the reduction in radiated sound. Table 1-8 gives an indication of the effects of openings in otherwise well-designed enclosures.

In a practical sense, enclosure problems are not difficult to overcome but do require some accommodation. For ease of maintanance, enclosures can be constructed to lift vertically upward and off the machinery by use of overhead cranes. Access openings can be

TABLE 1-8 Noise Level Effects of Openings

Percentage of open area in enclosure	Maximum average noise reduction, dB
50	3
25	6
10	10
1	20

provided by use of tunnels lined with acoustically absorbent material (in effect, mufflers). Access for controls can be designed with hinged covers that lift easily, or the controls can be relocated. Some enclosure panels can be lifted automatically at the correct point in the machine cycle to provide access. Ducts can be provided with small ventilating fans to produce a controlled cooling airstream and, if combined with filters, can allow a controlled environment for some operations.

Vibration Isolation

Large vibrating areas (machinery housings or large sections of framing) may radiate large amounts of sound, even when the vibration is almost imperceptible. For noise control, the surfaces can be acoustically "isolated" from the vibrating drive mechanism by use of vibration isolation mounts, breaks, or pads installed between the vibrating source and the radiating surface. (See also Chap. 6 2, "Vibration Control.")

Barriers

A barrier is a solid wall used to shield a receiver from the direct radiated sound waves of a machine. To be effective, it must be sufficiently massive to provide the required reduction in the sound transmitted through the barrier, and it must be sufficiently wide and high to prevent the sound refracted around the edges from becoming significant. A barrier positioned halfway between the source and the receiver will require the greatest dimensions to produce the same acoustic shielding. Barriers, therefore, should be placed close to either the source or the receiver.

Barriers can also be used to shield a group of workers not working on noisy machines from an area containing noisy machines. Of course, the effectiveness of such barriers can be significantly reduced by reflected sound, from a ceiling or nearby sidewalls, for example. Therefore, barriers should be combined with ceiling and possibly wall treatment to reduce this alternative sound path. Figure 1-7 shows the acoustical performance of a barrier.

Barriers can be provided with acoustically absorptive material on the side facing the source to avoid a reflected buildup of sound near the machine.

Simple small barriers can be located to provide individual operator protection on some machinery. In this case, a transparent material may allow visual monitoring. On small machines the controls may still be reached under or around the barrier. Such arrangements can also act as safety shields.

Although a barrier is designed to stop the sound from reaching a given receiver, the barrier does not necessarily have to impede the passage of materials and products. The use of overlapping entrances to produce a visual and acoustic blockage can allow easy access for forklift trucks and the location of conveyors.

Lagging

Lagging of pipes, ducts, and other radiating surfaces requires the application of a double-layer construction, consisting of a resilient inner layer, for example, 1 to 4 in (2.5 to 10 cm) of glass fiber material and an airtight heavy outer layer. In many cases, lagging can be combined with or can replace the thermal insulation on a pipe and, as such, presents no installation or maintenance problem. The outer coating of sheet aluminum or mild steel can often be retained, as long as this outer layer is isolated everywhere from the inner vibrating structure. This separation is essential since the wrapping effectively pro-

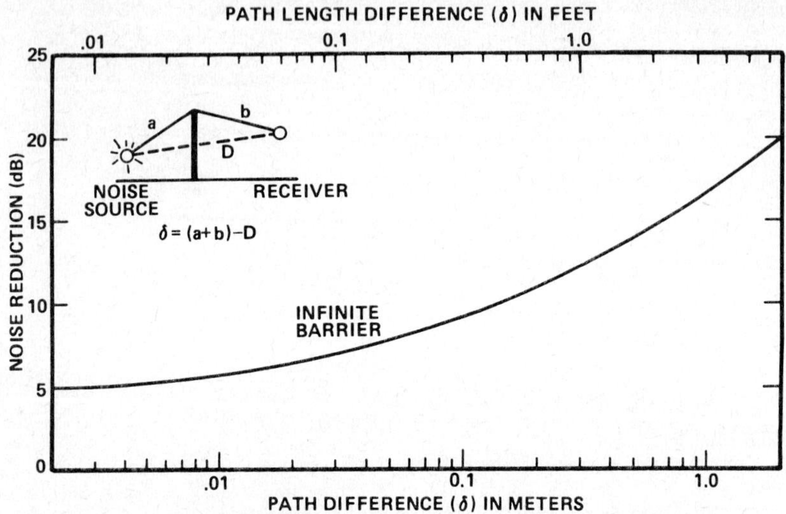

Figure 1-7 Average noise reduction of acoustical barrier of infinite length.

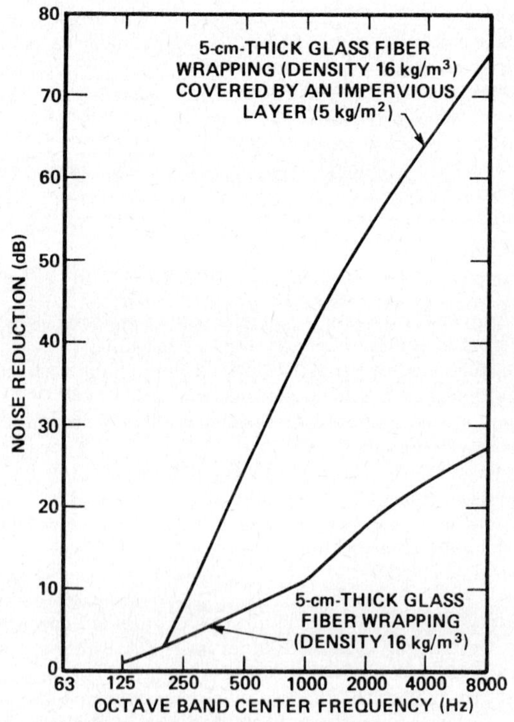

Figure 1-8 Acoustical performance of pipe wrapping.

vides an outer nonvibrating surface, separated by an absorbent material to dissipate sound energy produced by the pipe. An acoustic "short-circuit," such as at a valve where the outer jacket touches the pipe work, can allow the vibration to be transmitted directly and, hence, sound to be radiated.

Acoustic lagging requirements can be specified as part of the normal design supply and procurement process.

Figure 1-8 shows the acoustical performance of an acoustic lagging of a pipe.

Receiver Treatments

Earplugs

Earplugs are the simplest and least-expensive form of ear protection against excessive sound levels. All plugs share the feature of being inserted into the ear canal; they fall into three general categories:

1. Prefabricated earplugs of either rubber or plastic attempt to provide complete occlusion of sound from the inner ear by making an airtight seal over the ear canal.
2. Individually molded earplugs, made of silicone or synthetic rubber, are inserted in liquid form ½ in into the user's ear. The liquid can cure into a soft, customized earplug in 10 min. The greater weight of these plugs (11 to 13 g) is distributed over a larger area, so they are still comfortable.
3. Temporary or disposable ear plugs—made from materials such as wax-impregnated cotton, glass fiber, down, or foam (which are better than just cotton balls)—constitute the simplest form of plug.

Earplugs are not a permanent investment. They are easily lost, and the shape of the plastics is often deformed over time.

Earmuffs

Earmuffs are designed to isolate excessive sound levels from the ear by placing an airtight shell over the outer ear canal and attenuating the sound by the cover and cavity. Most earmuffs share a similar form of construction, differing mainly in materials and details.

An outer dome, made of hard shatterproof thermoplastic or the like, provides an initial shield. The heavier this dome (more mass) and the deeper the cavity around the ear (more air space separation), the more low-frequency sounds are absorbed.

Within this dome, an open-cell polyurethane filler attenuates high-frequency sounds (absorption occurs in the pores) and also eliminates the sea-shell effect of a hollow cover (high percent of absorption in a reflective space).

The seal with the ear must be airtight and yet flexible enough to prevent direct transmission of vibration from the rigid dome. For this, earmuff manufacturers use a vinyl-covered fluid-filled sack or a flexible semisoft plastic foam. The liquid seal is more effective for low-frequency attenuation.

The force of the headband must be enough to secure the seal yet be loose enough not to cause discomfort. This force ranges from 1 to 2.4 lb; an average is 2 lb (1 kg). The force also varies with head size. Values above 2.4 lb have been found to be uncomfortable and unacceptable.

The weight of earmuffs ranges from 6.75 to 16.0 oz (200 to 500 g); the average is 10 to 11 oz (300 to 330 g). The heavier muffs are generally more efficient. The weight should be broadly distributed on the skull by a wide headband. Earmuffs can also be held by a strap across the nape of the neck or by a safety helmet, which may offer some shielding of direct sound from the skull in certain frequencies. This reduction of bone-conducted sound can be a factor in extremely loud situations.

All surfaces in contact with the skin must be nontoxic and immune to degradation or irritation caused by perspiration, ear wax, humidity, skin oils, or dirt.

With any protector placed in or over the ear, acceptance by the user is an important factor. At sound levels that are painful to the ear, workmen seldom need prodding to wear protection. However, earmuff comfort becomes a determinant as sound levels decrease because workers can readily adapt to noise that—while sufficient to cause hearing deterioration over time—does not itself cause physical pain.

Since individual components of a set of earmuffs may become damaged or worn out, manufacturers usually provide individual replacement parts. The initial cost investment, durability, maintenance, and availability of replacement parts are factors that, along with noise-reduction characteristics and user comfort, will aid in the choice of a hearing protector.

At moderate and high noise levels, communication is enhanced when either earplugs or earmuffs are worn. This is a valuable safety aid so workmen can hear warnings.

Techniques That Require No Modifications or Additions

No discussion on noise-control techniques should omit a brief discussion of ways to reduce noise that do not involve equipment modification or addition. Examples are: proper maintenance, good operating procedures, and equipment replacement.

Proper Maintenance

Malfunctioning or poorly maintained equipment makes more noise then properly maintained equipment. Steam leaks, for example, generate high sound levels (and also waste money). Bad bearings, worn gears, slapping belts, improperly balanced rotating parts, or insufficiently lubricated parts can also cause unnecessary noise. Similarly, improperly adjusted linkages or cams or improperly positioned machine guards often make unnecessary contact with other parts and result in noise. Missing machine guards can also allow noise to escape. These types of noise sources share one characteristic. Their noise emissions can be readily controlled, though there is no simple way to predict how much noise reduction can be achieved through proper maintenance.

Operating Procedures

The way an operation is performed can cause workers to be overexposed to noise. Some operations are monitored by workers stationed near a noise source. At times, the distance is more critical in terms of noise exposure than the operation necessitates. In other words, the operator can be stationed at some other, quieter location without degrading work performance. Some operations can be monitored or performed from inside an operator "refuge," a booth or a room. Relocation of machine control systems can often augment this type of noise control.

Noise reduction obtained by relocating operators can be estimated by measuring sound levels at the existing station and the planned new station. If an operator booth is employed, noise reductions can be expected to range from 10 to 30 dB, with the higher value for booths with good windows and doors and the lower value for booths that are open to the environment on one or two sides.

Equipment Replacement

In some cases, the modification most readily available is quieter equipment that can be used to perform the same task. For example, several major manufacturers now sell quieted electric motors or quieted compressors. Quieted versions of equipment typically are more expensive than unquieted ones. Certainly, situations will arise when the purchase of different or newer equipment may be appropriate for production purposes, and these situations may be effectively combined with noise considerations. *Be aware that new equipment may not necessarily be quieter just because it is new.* Noise specifications can play a significant role in quieting an environment when an upgrading or expansion program is undertaken, and these specifications will be more important as pressure increases on equipment manufacturers to produce quieter equipment.

INSTRUMENTATION

Before a noise measurement program is started, its objective should be clearly understood. The objective could be to determine the approximate value of the noise during a short period of time or to find out whether the noise environment is hazardous to the health and welfare of the worker. The objective might also be to see if the noise exceeds some locally adopted sound level limits. Depending on the purpose and the desired accuracy, a wide range of descriptors and instrumentation is available.

A basic research program, that is, one that would assess noise throughout an entire plant, requires a large amount of detailed data. A noise-control investigation of a particular area or machine, on the other hand, normally requires much less detailed information, and many monitoring systems are set up to detect only relatively simple changes. For example, a system might be used to set off a flashing light whenever a sound level exceeds 90 dBA for periods longer than ½ s.

The simplest instrumentation available to determine sound pressure levels and sound levels is a sound-level meter. There are four different classifications of portable sound-level meters:

1. Precision
2. General purpose
3. Survey
4. Special purpose

The precision and tolerances of the indicating meter and the weighting networks vary significantly for the various types.

A sound-level meter must be kept calibrated if it is to provide meaningful data. Most equipment is battery-operated; the batteries must be fresh and capable of supplying the instrument with sufficient power. The manufacturer's instructions on how to check batteries appear on every instrument. A battery check is followed by a calibrator-pistonphone check. The calibrator-pistonphone is placed over the microphone and turned on. It will supply a pure tone at a known sound pressure level, which will allow appropriate adjustments to the meter reading, in accordance with the manufacturer's procedure.

Sound-level meters measure noise only at a given point at the time of observation. If the noise being measured is constant in both space and time, meters will give an accurate representation of the situation. However, if the sound level changes with time and location (for instance, as an operator moves around), it will be necessary either to record the sound level manually with short time intervals (5 to 10 s) or tape record the noise data, and later analyze the time history of the noise. The second approach is preferable when a worker's noise exposure is related to duty cycles or product flow. In this case, extrapolations can be made, on the basis of total day production, to determine the noise exposure of an employee over a full day.

In industry, there are situations in which time pressure is great, more than one person must be monitored, and duty cycles are not easily definable. In such cases, audiodosimeters can be used to measure an employee's noise exposure. The *audiodosimeters,* devices about the size of a cigarette pack, are worn by employees to record the noise exposure of the wearers wherever they go. The microphone can be fixed to the unit or detached from the unit and placed in the hearing zone of the wearer. The audiodosimeters can be obtained with an internal circuit that integrates the sound level and time in accordance with the OSHA regulation (halving of the exposure time for each 5-dB rise in sound level), the Department of Defense instruction (halving of the exposure time for each 4-dB rise in sound level), or the International Standard Organization standard (halving of the exposure time for each 3-dB increase in sound level).

STANDARDS

Two internationally known organizations have produced standards that acousticians use almost daily. Table 1-9 lists the most important of these standards and indicates the agencies that created them—the American National Standards Institute (ANSI) and the American Society for Testing and Materials (ASTM).

TABLE 1-9 Sound-Level Standards

American National Standards Institute (ANSI)	
S1.1-1960 (R1976)	Acoustical Terminology (including mechanical shock and vibration)
S1.2-1962 (R1976)	Method for the Physical Measurement of Sound (partially revised by S1.13-1971 and S1.21-1972)
S1.4-1971 (R1976)	Specification for Sound Level Meters (IEC 123)
S1.6-1976 (R1976)	Preferred Frequencies and Band Numbers for Acoustical Measurements (ISO 454; agrees with ISO 266)
S1.7-1970 (R1975)	Standard Test Method for Sound Absorption Coefficients by the Reverberation Method (See ASTM C 423-77)
S1.8-1969 (R1974)	Preferred Reference Quantities for Acoustical Levels
S1.10-1966 (R1976)	Method for the Calibration of Microphones
S1.11-1966 (R1976)	Specifications for Octave, Half-Octave, and Third-Octave Band Filter Sets
S1.13-1971 (R1976)	Methods for the Measurement of Sound Pressure Levels
S1.21-1972	Methods for the Determination of Sound Power Levels of Small Sources in Reverberation Rooms
S1.23	Method for the Designation of Sound Power Emitted by Machinery and Equipment (See ASA Standard Catalog No. 5-1976)
S2.9-1976	Nomenclature for Specifying Damping Properties of Materials
S1.26-1978	Method for the Calculation of the Absorption of Sound by the Atmosphere
S1.36-1979	Survey Methods for the Determination of Sound Power Levels of Noise Sources
S3.17-1975	Method for Rating the Sound Power Spectra of Small Stationary Noise Sources (See ASA CAT. NO. 4-1975)
S3.19-1974	Method for the Measurement of Real-Ear Protection of Hearing Protectors and Physical Attenuation of Earmuffs (See ASA CAT. NO. 1-1975)
American Society for Testing and Materials (ASTM)	
C 423-77	Standard Test Method for Sound Absorption Coefficients by the Reverberation Room Method
C 634-77	Standard Definitions of Terms Relating to Environmental Acoustics
E 90-75	Standard Method for Laboratory Measurement of Airborne Sound Transmission Loss of Building Partitions
E 336-77	Standard Test Method for Measurement of Airborne Sound Insulation in Buildings
E 413-73	Standard Classification for Determination of Sound Transmission Class
E 477-73	Standard Method of Testing Duct Liner Materials and Prefabricated Silencers for Acoustical and Airflow Performance
E 596-78	Standard Method for Laboratory Measurement of the Noise Reduction of Sound-Isolating Enclosures

BIBLIOGRAPHY

American Industrial Hygiene Association: *Industrial Noise Manual,* AIHA, Akron, Ohio, 1975.

American Institute of Physics: *Glossary of Terms Frequently Used in Acoustics*, AIP, New York, 1960.

Bell, L. H.: *Fundamentals of Industrial Noise Control,* Harmony, Trumbull, Conn., 1973.

Beranek, L. L.: *Noise and Vibration Control,* McGraw-Hill, New York, 1971.

Cheremisinoff, P. N., and P. P. Cheremisinoff: *Industrial Noise Control Handbook,* Ann Arbor Science, Ann Arbor, Mich., 1977.

Diehl, G. M.: *Machinery Acoustics,* Wiley, New York, 1973.

Jensen, P., C. R. Jokel, and L. N. Miller: *Industrial Noise Control Manual,* U.S. Department of Health, Education, and Welfare, NIOSH, Cincinnati, Ohio, 1978.

Lipscomb, D. M., and A. C. Taylor: *Noise Control Handbook of Principles and Practices,* Van Nostrand Reinhold, New York, 1978.

Morse, P. M., and K. U. Ingard: *Theoretical Acoustics,* McGraw-Hill, New York, 1968.

chapter 6-2

Vibration Control

by

Eric E. Ungar
Principal Engineer
Bolt Beranek and Newman Inc.
Cambridge, Mass.

CHARACTERIZATION OF VIBRATIONS

Vibration refers to oscillatory (back and forth) motions of structures, mechanical systems, or components of these. A vibration generally is characterized by the displacement, velocity, or acceleration measured at one or more points on the item of interest in specific directions of interest (e.g., perpendicular to a floor or wall).

An observer of the time variation of a vibration—for example, one who watches the displayed signal obtained from a sensor on an oscilloscope or chart recorder—often obtains a record that approximates the one shown in Fig. 2-1*a*. Such a regular curve, which corresponds mathematically to a sine or cosine, is called *sinusoidal* or *simple harmonic*. Note that it deviates from zero (the middle position) equally in both directions; the maximum excursion from zero in one direction is called the *amplitude A*, the total excursion in both directions is the *double amplitude 2A* (or sometimes the peak-to-peak value). Amplitudes may be given in units of displacement, velocity or acceleration, depending on how the vibration is measured.

The time interval T between successive peaks is called the *period* and usually is measured in seconds. The number of vibration cycles (i.e., the number of periods) that occur per second is called the *frequency* f and is generally measured in hertz (Hz), which is the internationally standardized name that has replaced cycles per second (cps).

In practice one rarely obtains a simple record like that of Fig. 2-1*a*. One is more likely to obtain a record that looks like Fig. 2-1*b*. This record may consist of a basic sinusoid like that of Fig. 2-1*a*, to which there are added one or more sinusoids with higher frequencies (shorter periods) and generally smaller amplitudes. Fig. 2-1*b* is said to represent a *multifrequency* or *complex* vibration. The component with the lowest frequency (greatest period) is called the *fundamental* component. Components that occur at frequencies that are integer multiples of that of the fundamental one are called *harmonics*.

Vibrations that have no well-defined period or amplitude—i.e., in essence, when the record never repeats itself—are called *nonperiodic* or *irregular*. A sample of a nonperiodic vibration record is given in Fig. 2-1*c*.

A vibration that has essentially the same amplitude over an extended time period is called *steady*, whereas a vibration with time-varying amplitude is called *transient*. Figures 2-1*a*, *b*, and *c* illustrate steady vibrations, whereas Fig. 2-1*d* illustrates a typical decaying transient containing a single-frequency component; similar transients with multiple-frequency components or with irregular behavior may readily be visualized.

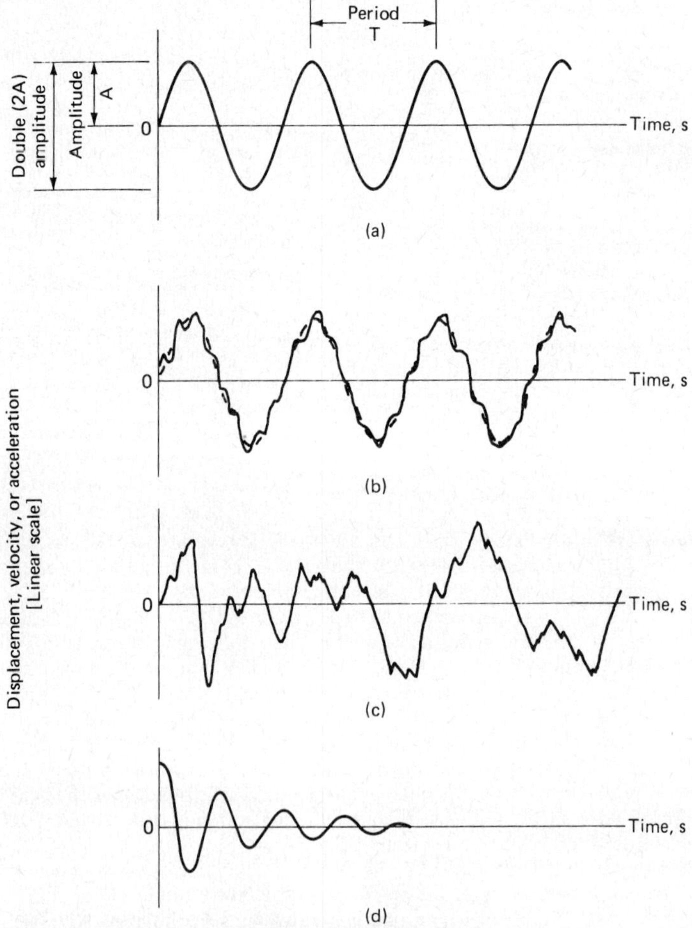

Figure 2-1 Typical vibration records: (*a*) steady sinusoidal or simple harmonic vibration; (*b*) steady multifrequency vibration; (*c*) irregular (nonperiodic) vibration; (*d*) decaying transient single-frequency vibration.

It is often useful to characterize a vibration in terms of a plot of amplitude vs. frequency. Such a plot is called a *spectrum.* One may obtain such a plot by passing the (electric) vibration signal through a variable filter to a readout device. The filter permits only selected frequency components to reach the readout, which then indicates the amplitudes of these selected components. There are also available instruments called *spectrum analyzers* that automatically provide amplitude-frequency displays for vibration signals fed into them.

Knowing the amplitude at a given frequency in terms of any one of the three motion quantities (displacement d, velocity v, acceleration a), one can calculate the amplitude in terms of the other two from:

$$
\begin{aligned}
a &= 2\pi f v = (2\pi)^2 f^2 d \\
v &= \frac{a}{2\pi f} = 2\pi f d \\
d &= \frac{a}{(2\pi)^2 f^2} = \frac{v}{2\pi f}
\end{aligned}
\tag{1}
$$

where a, v, and d always contain the same length units and seconds and f is in hertz. For example, to d given in mils there corresponds v in mils/s and a in mils/s^2. For $v = 10$ ft/s and 20 Hz, $a = 2\pi(20)(10) = 1256$ ft/s^2 and also $d = 10/2\pi(20) = 0.080$ ft. This can readily be converted to other units, e.g., to g's for acceleration (where $1g = 32.2$ ft/s^2 = 386 in/s^2) or to metric units.

Figure 2-2, which is based on Eq. (1), is a convenient chart for the approximate conversion between motion quantities.

CAUSES OF VIBRATIONS

Vibrations are always caused by unsteady forces, that is, by forces that may be oscillatory in magnitude or direction or by forces that are suddenly applied or released. These forces need not be due to mechanical causes; electromagnetic, aerodynamic or fluid-related forces also are often encountered in practice.

Unbalances in rotating machines produce net centrifugal forces that change direction in space as the machine rotates. For a machine with a horizontal shaft, such a force acts upward at one instant and downward a half-rotation later, thus producing a force that acts on the floor sinusoidally at a frequency that corresponds to the *shaft rotation frequency* f_r (hertz) $= N/60$, where N denotes the shaft rotation speed in revolutions per minute. As such a machine comes up to speed, it produces transient vibrations that increase in frequency and amplitude until steady operating conditions are reached; when such a machine is turned off and coasts toward a stop, it produces decaying transient vibrations with ever-decreasing amplitude and frequency.

Reciprocating machines also produce unbalanced inertia forces which are transmitted to the machine housing and supports. The primary components of these forces occur at the crankshaft's rotational frequency and at the first few integer multiples of that frequency. These forces generate vibrations along the direction of piston travel, as well as perpendicular to that direction, in the plane of the crank, and also produce vibratory moments or couples; the relative magnitudes depend on the cylinder arrangement and the degree of dynamic balance.

Fans, blowers, and pumps tend to generate steady vibrations, due to both unbalances and repetitive fluid pulses. The latter occur primarily at the *blade passage frequency,* which is the frequency with which blades pass a fixed point. For a rotor with n blades, rotating at N r/min, the blade-passage frequency is given by f_b (hertz) $= nN/60$, where N is in revolutions per minute.

Turbulent flows of water or air in a duct, or flows from a blower or airjet impinging on a surface, typically produce irregular forces on the structural surfaces. Similarly, irregularly repeated impacts, e.g., due to footfalls produced by many people walking on a floor, tend to produce irregular vibrations.

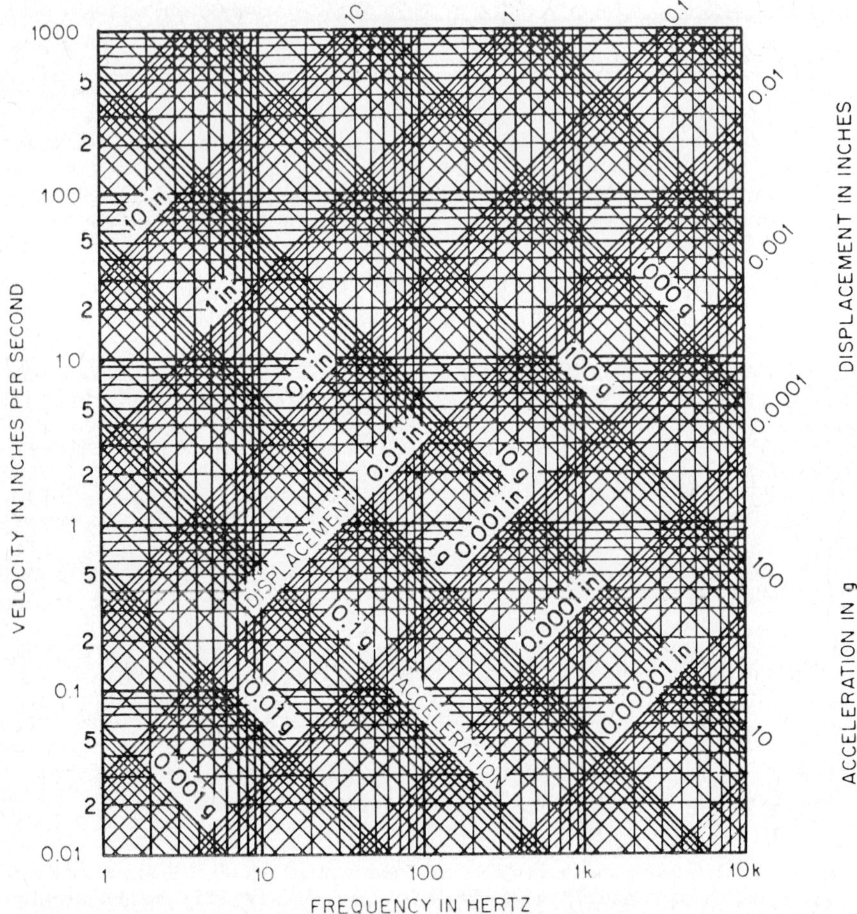

Figure 2-2 Chart for conversion between displacement, velocity, and acceleration. Reproduced with permission from *Handbook of Noise and Vibration Control*, C. M. Harris (ed.), 2d ed., McGraw-Hill, New York, 1979.[1]

Single impacts, such as are produced by a single operation of a punch press, generate force pulses that typically result in decaying transient vibrations. Repetitive impacts result in repetitive transients, but if these impacts are repeated so rapidly that the vibrations due to one impact do not decay much before the next impact occurs, then the vibrations tend to have more of a steady irregular character. Thus, impacts of materials in a gravity chute against the chute bottoms and sides tend to produce essentially nonperiodic (irregular) vibrations.

Rattling, slippage, and nonlinearities (deviations from proportionality between forces and displacements) associated with large excursions generally introduce higher-frequency components than those associated with the basic forces and motions.

EFFECTS OF VIBRATIONS; THE NEED FOR VIBRATION CONTROL

Excessive vibrations can have adverse effects on personnel, equipment, and structures. Vibrations can annoy people, can interfere with their ability to perform or concentrate on mental tasks, can make it difficult for people to carry out precise movements or to make precise readings of instruments, and in extreme cases can lead to physical disabilities and injuries. Vibrating surfaces act somewhat like loudspeaker membranes and radiate sound, which again may range from annoying to painful to injurious, depending on its intensity. Vibrations may also produce such secondary effects as rattling of windows or motion of lights, which also tend to be annoying or distracting.

Vibration of a machine may reduce the life of its components, particularly those most highly loaded. Oscillatory stresses induced in machine parts, supports, building structures, and also in connections (hold-down bolts, pipes, cables) tend to produce failures of these items due to structural fatigue. Machine tools subjected to excessive vibrations may produce poor finishes; some precision equipment (optical systems, microscopes, gauges, microassembly equipment) cannot be effectively used at all in the presence of vibrations.

The need for vibration control occurs wherever there exist adverse effects due to vibrations. The amount of reduction that is required depends on the existing vibration and on what level of vibration is acceptable; zero vibration is as much an impossibility as an immovable object or an irresistible force. For many items of sensitive equipment, the manufacturers indicate what vibrations should not be exceeded. The vibration limits acceptable to people are available in handbooks.[2] In many other cases, unfortunately, no solid vibration criteria are available, so that one is forced to proceed by trial and error.

Control of vibration is a highly developed specialized branch of engineering, which has been the subject of numerous papers, many texts, and several handbooks. Many of these publications deal primarily with analyses or specialized problems and thus require study and interpretation before they can be put to practical use. The list of references included at the end of this chapter is limited to items believed to be most directly applicable to plant engineering.

DIAGNOSING A VIBRATION PROBLEM

It is usually convenient to consider any vibration problem in terms of: (1) the *source* of the undesirable vibrations, (2) the *receiver,* i.e., whatever is adversely affected by vibrations and requires protection, and (3) the *path* along which vibrations from the source reach the receiver. In practical situations, many sources often contribute to the vibrations experienced by a single receiver, and vibrations from a given source often reach a single receiver via several paths.

The most convincing approach to identify the vibration source responsible for a given problem consists of turning off all possible sources, then turning them on one at a time while observing the resulting effects at the receiver of interest. Similarly, one can best identify the predominant paths by interrupting one at a time or by disabling all and then reestablishing one at a time. These procedures can only rarely be carried out in practice to their full extent, but they can often be carried out sufficiently to provide valuable partial, if not full, insight into the problem.

It is often useful to supplement the on-off procedures discussed in the foregoing paragraph by comparison of the vibration spectra measured on or near various sources with the spectrum measured at the receiver. If the receiver, for example, experiences adverse vibrations only at 50 and 80 Hz, then a source that generates vibrations at only 30 Hz cannot be responsible for these receiver vibrations; one would look for sources that produce vibrations near 50 and 80 Hz. Similarly, measurements made along the important contributing paths would reveal 50- and 80-Hz components, whereas these components would be present to a lesser extent along the less significant paths.

There are also available *correlation methods* for source and path identification. These methods, however, require specialized equipment and expertise.

In dealing with any vibration problem, one must keep in mind the phenomenon called *resonance.* Any mechanical system or structure has a number of frequencies at which it can be set into vibration very easily; these are called *natural frequencies.* (The lowest of these, called the *fundamental natural frequency,* is often most easily excited and of greatest importance.) Resonance occurs if a system is subjected to a vibratory force or motion at one of its natural frequencies; large vibrations can then occur *even with small inputs.* The natural frequencies of a system can be determined readily; if a system is deflected and released, or if it is struck, it will vibrate at one or more of its natural frequencies. In order to be able to identify these, however, it is usually necessary to have the items of concern turned off, so that the natural frequencies will not be masked by the excitation frequencies.

A *stroboscope* is often useful for diagnostic purposes. This consists of a light that is made to flash at precisely timed intervals. The light is aimed at a vibrating part and the flashing frequency is adjusted manually until the vibrating part appears to stand still; the frequency of the vibration can then be read from the instrument. Then, by changing the flashing frequency slightly, one can observe the vibration in apparent slow motion to see where the largest excursions occur.

VIBRATION CONTROL STRATEGY[3]

It is generally best to control a vibration at its source, because this approach avoids problems at all potential receivers. However, in cases where only a limited number of receivers are of concern and where control at the source(s) is not feasible, control at the receiver(s) may be preferable. Reduction or elimination of vibrations at the source typically involves improving the dynamic balance of rotating or reciprocating equipment, substituting items with lesser vibrations for those with more (e.g., centrifugal pumps for reciprocating pumps), or changing operating speeds to eliminate resonance conditions. Reduction of the adverse effects of vibrations at the receiver generally involves substitution of less-vibration-sensitive items or processes, or adding stiffening or mass judiciously in order to eliminate resonances, if any are present.

Vibration *isolation* generally turns out to be the most cost-effective means for vibration control. Isolation involves insertion of soft flexible elements in the propagation path so as to reduce the transmitted forces and motions. Because of the multitude of paths that can begin at any source or terminate at any receiver, isolation is best accomplished near a source or receiver.

Some other vibration control methods that are useful only under certain specific circumstances are described at the end of this chapter.

Vibration Isolation at the Source

Basic Principles

The basic concepts of vibration isolation can be understood with the aid of the schematic sketch of Fig. 2-3, which shows a machine that generates a vertical oscillatory force of amplitude F (e.g., due to an imbalance) rigidly attached to a base, which in reality may be a machine, a foundation, an inertia block, or the floated floor of a building. This base, in turn, is mounted atop a supporting structure via a series of springs.

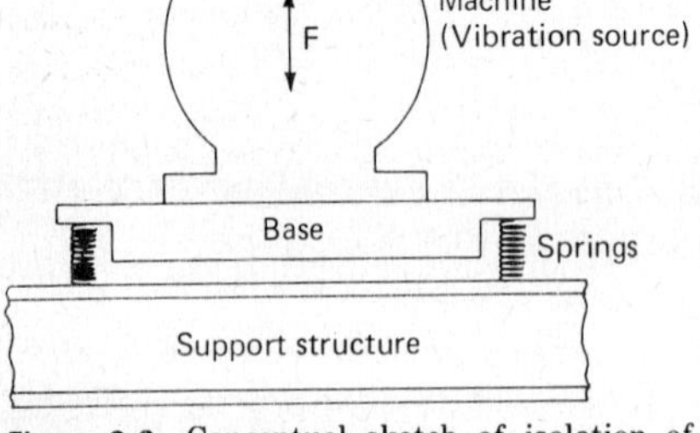

Figure 2-3 Conceptual sketch of isolation of vibration source.

The oscillatory force produces motion by accelerating the combined mass m of the machine and the base. This motion of the base produces oscillatory compression (and extension) of the springs (superposed on the static compression due to the weight they support), which in turn gives rise to oscillatory forces on the support.

If the force varies slowly, i.e., at low frequencies, then the inertia of the mass offers little opposition to the motion, and the force essentially acts directly to compress the springs. The machine-base mass here moves just enough for the total spring force to match the externally applied force, as it would if the force were applied statically, and thus the entire applied force is transmitted to the support structure. On the other hand, if the force F varies rapidly, i.e., at high frequencies, then the inertia of the mass opposes the motion to such an extent that the inertia's effect is much greater than that of the springs. The springs then compress and extend very little, and only the spring forces resulting from these small spring deflections are transmitted to the support structure.

The ratio of the amplitude F_S of the total force that the springs exert on the support (assumed rigid) to the amplitude F of the exciting force is called the *transmissibility* T. The transmissibility also is equal to F_S/F_R,* where F_S denotes the amplitude of the oscillatory force that acts on the support if the base is resiliently supported and F_R represents the corresponding oscillatory force that acts on the support structure if the base is rigidly fastened to it. Thus, the transmissibility also indicates the relative reduction of the force transmitted to the support structure (and of its motion) that results from the insertion of the isolation springs. The *isolation efficiency* E, which is defined by $E = 1 - T$ indicates what fraction of the exciting force is prevented from acting on the support, or, equivalently, by what factor the motion of the support is reduced due to use of the springs or springlike elements. For example, if a given isolation system results in a transmissibility of 0.05, then its isolation efficiency is 0.95, indicating that the use of the isolators reduces the vibrations of the support structure by 95 percent.

In order to have a reasonable beneficial effect, the springs (or other resilient elements) must be soft enough so that their total stiffness k is less than

$$k_{max}\ (\text{lb/in}) = \frac{m\ (\text{lb}) \times f^2(\text{Hz})}{40} \tag{2}$$

where m represents the total mass of the machine and base (in pounds) and f denotes the disturbing frequency (in hertz) of concern. For springs that have a straight-line force-deflection characteristic (the slope of which corresponds to the stiffness k)—for example, for steel coil compression springs that do not become solid at the greatest expected deflection — another way of specifying a soft enough system is to require that the *static deflection* s of the springs (that is, the deflection under the static load composed of the machine and base)

$$s\ (\text{in}) = \frac{m\ (\text{lb})}{k\ (\text{lb/in})}$$

be greater than

$$s_{min}\ (\text{in}) = \frac{40}{f^2\ (\text{Hz})} \tag{3}$$

If the resilient elements satisfy the maximum-stiffness (or minimum-static-deflection) requirement of the foregoing paragraph, then the transmissibility T and isolation efficiency E may be found from

$$T = 1 - E \approx \frac{10k\ (\text{lb/in})}{m\ (\text{lb}) \times f^2\ (\text{Hz})} = \frac{k}{4k_{max}} = \frac{s_{min}}{4s} \tag{4}$$

Practical Considerations

As evident from the foregoing expression, the lowest disturbing frequency f corresponds to the greatest vibration transmission. Therefore, an isolation system must be designed for the lowest disturbing frequency of concern.

In general, the vibration transmission can be reduced in two ways: (1) by using softer resilient elements (reducing the total stiffness k) or (2) by increasing the supported mass

*This equality holds only if the force produced by the source is unaffected by the degree to which the motion is restrained.

m. It should be noted that the use of softer springs (beyond the $k < k_{max}$ requirement) leads to greater static deflection, but produces practically no change in the vibratory motion of the base. Because the static deflection is proportional to the supported mass and the motion of that mass is inversely proportional to the mass, an increase in that mass results in increase of static deflection and in a reduction of the vibratory motion of the machine and base.

It is important to note that useful isolation can be achieved only if the resilient elements (springs) are considerably softer than the supporting structure, i.e., only if the resilient elements deflect considerably more under a given load than does the support structure. Otherwise, the support structure provides the predominant resilience and the springs merely serve to transmit the forces to the support essentially without attenuating them.

Selection of an isolation system must account for the fact that an excitation frequency f that is produced by a machine usually varies with the machine's rotational speed. As a machine is brought up to speed or slowed to a stop, a speed at which the exciting frequency matches the natural frequency of the machine on its resilient supports may be encountered. At this speed, which may be estimated from

$$N_r \approx 188 \sqrt{\frac{k\ (\text{lb/in})}{m\ (\text{lb})}} \approx \frac{188}{\sqrt{s\ (\text{in})}} \tag{5}$$

there may occur intense vibrations. Their magnitude depends on how fast the machine passes through this resonance speed and on the damping characteristics of the isolation system. If the machine accelerates or decelerates rapidly, vibrations do not have time to buildup. For machines that accelerate or decelerate slowly, the magnitude of the vibration produced at resonance is inversely proportional to the damping in the system.

Figure 2-4 is a convenient chart for estimating the major isolation system parameters one needs in order to achieve a desired vibration reduction or transmissibility. The dashed line in the figure illustrates the case where there exists a disturbing frequency of 3000 r/min (corresponding to 50 Hz), at which it is desired to reduce the vibration by 99.9 percent, i.e., to 0.01 times its original value. The chart shows that here a value for k/m of about 2.5 (lb/in)/lb is needed; thus, for a 2000-lb mass, a total stiffness of $2.5 \times 2000 = 5000$ lb/in is required. (Of course, a lesser stiffness would provide better isolation than that prescribed, whereas a greater stiffness would result in poorer isolation.) One may also read from the chart that to $k/m = 2.5$ (lb/in)/lb there corresponds a static deflection of about 0.4 in and a resonance frequency of about 300 r/min or 5 Hz.

Damping refers to a spring's or a structure's capability for dissipating oscillatory energy; a bell or a steel spring rings (i.e., vibrates) for a long time after it is struck; that is, it takes a relatively long time to dissipate the vibratory energy imparted to it. On the other hand, a rubber or cork rod vibrates only briefly after an impact; it dissipates vibratory energy rapidly and thus is highly damped. For machines that accelerate or coast down slowly, isolation elements of highly damped materials (e.g., rubber or cork) should be used, or energy-dissipation devices (e.g., friction pads or dashpots) should be added in parallel with the resilient members. In some cases, the incorporation of snubbers in the isolation systems may suffice to limit the excursions of the base as the machine passes through resonance. Such snubbers may, for example, be in the form of rubber cones that are mounted so that the base bumps against them when its excursion exceeds a given amount.

If slow acceleration or coast-down is no problem, then the type of isolator material and the isolator configuration essentially make no difference; the only parameter that counts is the total stiffness k under the expected load conditions and operating frequencies. For many resilient elements, particularly metal springs, the stiffness is practically independent of frequency and also of load within the design load range. Whenever vibrations in more than one direction (e.g., in the horizontal, as well as the vertical direction) are of concern, however, the proper resilience for isolation in all directions must be pro-

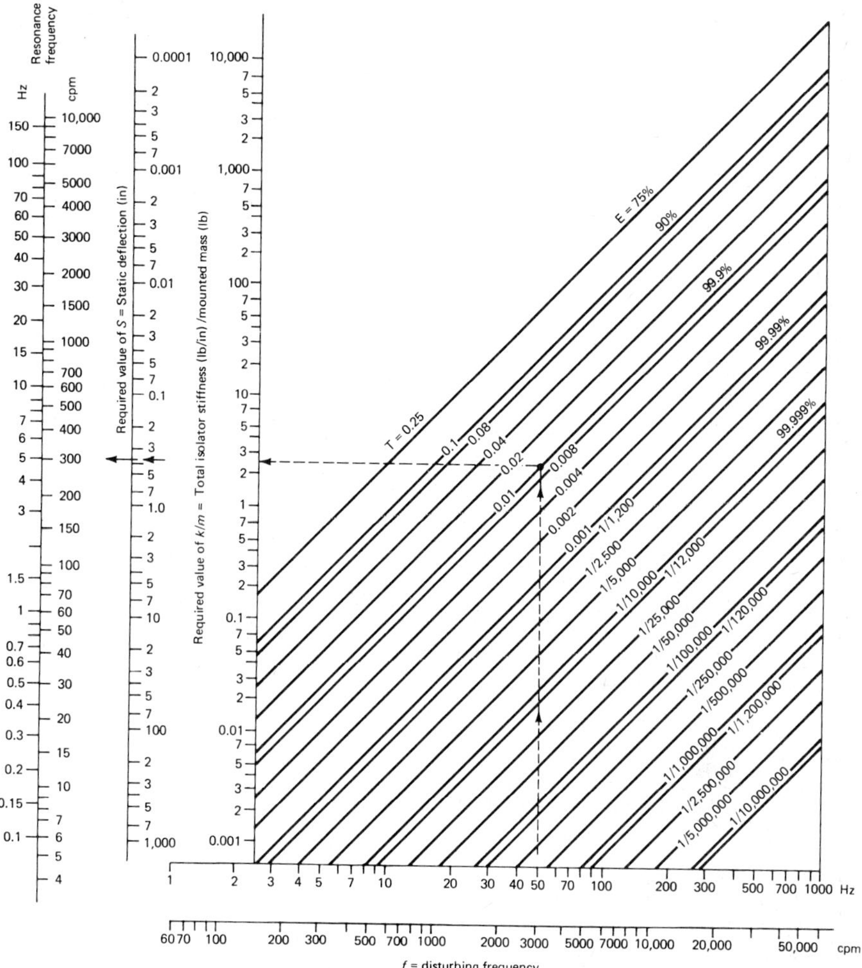

Figure 2-4 Chart for estimation of isolation-system requirements.

vided. In this case, appropriately selected and aligned springs, specially configured rubber pads, or suitably chosen commercial isolators should be employed.

Because of the beneficial effect of increased mass of the machine base, it usually is desirable to mount several machines on the same base. In this way, the vibration transmitted from each machine is attenuated by the mass of the base and also by the masses of the other machines. If such a common base is to be effective, it must not have any resonances of its own at or near the operating speeds of all machines mounted on it; ideally, its fundamental resonance should occur above all operating frequencies. This implies that such bases should be as stiff as possible.

Attention must also be given to avoidance of excessive rocking vibrations of isolated bases. For this purpose it is usually useful to employ wide bases, so that the springs act with large moment arms. It is also beneficial, particularly for bases supporting several items, to have the springs distributed so that they "pick up" the loads locally, e.g., so as to have stiffer or more closely spaced springs near the heavier items and less stiff or more widely spaced springs near the lighter items.

In practice one usually needs to make provisions to avoid the transmission of vibrations via paths that bypass the isolated base. For this purpose it usually is desirable to provide flexible sections of piping, electrical conduit, cable, and ducts between the isolated machines and the unisolated surroundings. (Appropriate devices especially designed for vibration isolation are commercially available; devices such as loops and bellows that are primarily designed to accomodate thermal expansion rarely are adequate for vibration isolation.) Similarly, hold-down bolts through isolators, and similar arrangements that in effect serve as vibration "short-circuit" paths around the isolators, must be removed or themselves isolated, e.g., with soft rubber sleeves and washers.

Isolation of Vibration-Sensitive Items

Basic Principles

The fundamental concepts pertaining to isolation of items to be protected from vibrations may be visualized with the aid of the schematic sketch of Fig. 2-5. This shows a

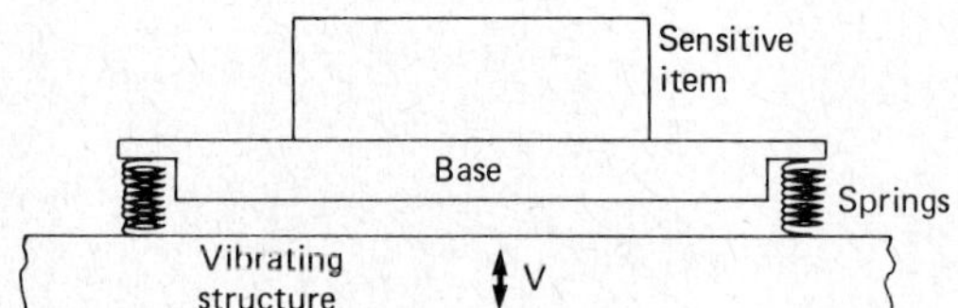

FGigure 2-5 Conceptual sketch of isolation of a sensitive item.

vibration-sensitive item attached to a base, which is fastened to a vibrating structure (e.g., a floor of a building housing vibration sources) via resilient elements, represented in the sketch by springs. In reality, the base may consist of a machine frame or foundation, an inertia block, or a floated (isolated) slab or floor.

If the sensitive item is attached to the vibrating structure directly, without springs, then it vibrates with the same amplitude as the vibrating structure. The same is true if the attachment springs are stiff. However, if the springs are soft, then the oscillatory deflection of the vibrating structure leads to only relatively small forces acting on the base; the inertia of the mass of the base resists acceleration and thus keeps the resulting motion small. This inertia effect is more pronounced for larger accelerations, i.e., higher frequencies, and therefore better isolation is obtained at higher frequencies, all other things being equal.

The ratio of the amplitude of the motion V_I (displacement, velocity, or acceleration) of the sensitive item to that of the vibrating structure, V (measured in terms of the same quantity as is V_I), is called the *motion transmissibility* T_v, or just *transmissibility*.* This transmissibility also turns out to be equal to V_R/V_I, where V_R represents the amplitude of the motion that the sensitive item experiences if it is rigidly fastened to the vibrating structure and V_I represents the corresponding amplitude that the sensitive item experiences if the isolators (springs) are used.† Thus, the motion transmissibility indicates the relative reduction in the motion of the sensitive item that results from use of the isolation elements (e.g., springs).

*It is somewhat unfortunate that the same word, "transmissibility," is used to refer to the force transmissibility T associated with source isolation as is used to refer to the motion transmissibility T_v associated with receiver (sensitive-item) isolation. However, because T and T_v obey the same relation, this usage tends to cause little confusion.

†This equality holds only if the motion of the vibrating structure is unaffected by the magnitude of the spring forces that act on that structure.

The *isolation efficiency** E_v is defined by $E_v = 1 - T_v$ and thus indicates what fraction of the motion of the vibrating structure is kept from the sensitive item. For example, if $T_v = 0.01$, then the isolation efficiency is 99 percent; the sensitive item vibrates only 1 percent as much as it would if it were rigidly attached to the vibrating structure.

In order for the isolation system to be significantly effective, the resilient elements must be soft enough so that their total stiffness k is less than k_{max} as given by Eq. (2), with m here representing the total mass of the sensitive item and its base and f denoting the lowest frequency of concern. For resilient elements, such as steel coil springs, whose force-deflection characteristic plots as a straight line, the "soft enough spring" requirement corresponds to the requirement that the static deflection s of the springs under the total static weight of the sensitive item and its base be greater than s_{min} as given by Eq. (3). If the resilient elements satisfy these requirements, then the motion transmissibility T_v is given by the same expression as is the force transmissibility T, namely, Eq. (4).

Practical Considerations

Because the lowest disturbing frequency f corresponds to the greatest transmissibility T, an isolation system must be designed with respect to the lowest disturbing frequency of interest.

Figure 2-4 is a convenient chart for estimating the major isolation system parameters needed in order to achieve a desired vibration reduction or transmissibility. The dashed line in the figure illustrates the approach for a case where there exists a disturbing frequency of 3000 r/min (corresponding to 50 Hz). At this frequency it is desired to reduce the vibration by 99.9 percent, i.e., to 0.01 of its original value. The chart shows that here a value of k/m of about 2.5 (lb/in)/lb is needed; thus, for a 2000-lb mass, a total stiffness of $2.5 \times 2000 = 5000$ lb/in is required. (Of course, a lesser stiffness would provide better isolation than that prescribed, whereas a greater stiffness would result in poorer isolation.) One may also read from the chart that to $k/m = 2.5$ (lb/in)/lb there corresponds a static deflection of about 0.4 in and a resonance frequency of about 300 r/min or 5 Hz.

Once the resilient elements are soft enough to satisfy the requirements indicated in the foregoing paragraph, inprovement in the isolation (reduction in the transmissibility) can be obtained by (1) using softer springs and/or (2) increasing the mass of the base. Reducing the spring stiffness by a given factor has the same effect on reducing the motion of the isolated item as increasing the combined mass of the item and its base by the same factor. Use of softer springs is usually less difficult and less costly than obtaining significantly increased base mass; thus, use of softer springs is generally preferable. However, softer springs are subject to greater static deflection and also tend to permit the isolated item to vibrate or to rock more due to motions of its own parts (e.g., of pistons, slides, spindles). In order to minimize this rocking and secondary vibration, it is often useful to balance the isolated items or to distribute the springs so as to oppose the dynamic loads directly (by placing more or stiffer springs where the greater loads act and having the spring force axes aligned with the direction of the loads).

In all cases, in order to be able to provide useful isolation, the resilient elements (springs) must be more flexible than the vibrating structure to which they are attached and the base supporting the sensitive item. That is, for a given force applied at the center of gravity of the isolated item, the resilient elements must deflect more than either one of the structures to which they are attached. The base of the sensitive item, in addition to being as heavy as possible, should be as stiff as possible.

It often is advantageous to have several sensitive items mounted on the same base; then the total mass of all items and the base act together to assist in the isolation of each item. In this case it is important that the base be as rigid as possible, so that all masses are well interconnected dynamically. In selecting items to be mounted on a common base, it must be kept in mind that items rigidly fastened to the same base can transmit vibra-

*Again, "isolation efficiency" is common usage, although *motion isolation efficiency* is more appropriate.

tions and rocking motions to each other; thus it is generally inadvisable to mount extremely sensitive items together with items that may cause dynamic disturbances.

Any connections between the isolated item or its base and the vibrating structure must be less stiff than the resilient (spring) elements if these connections are not to transmit more vibration than the springs. Thus, care must be taken to use flexible bellows, tubing, hoses, conduits, cables, etc., with coils and loops where appropriate, to reduce the vibration transmission via these connections.

Some Specialized Control Techniques

Two-Stage Isolation

In the cases illustrated by Figs. 2-2 and 2-3, a rigid connection is implied between the base and the equipment it supports. If resilient connections (e.g., rubber pads or springs) are used instead, then there are two stages of isolation: one below the base and one above it. Two-stage isolation systems can provide increased isolation at high frequencies, but they introduce additional system resonances at which there may occur large motions and dynamic forces. Careful isolation-system design is important to avoid these potential adverse effects.

Structural Damping

The large vibrations that occur if any system or structure vibrates at a resonance may best be reduced by *detuning*—changing the excitation or the system so as to avoid operation at resonance. If such changes cannot be made—either for operating reasons or because a system is subject to multifrequency excitation or because it may have a multitude of resonances at closely spaced frequencies—then increasing the damping constitutes essentially the only means for obtaining useful vibration reductions.

For thin metal panels, increased damping can usually be obtained most conveniently by means of layers of viscoelastic (tacky, rubbery) damping material that may be sprayed, troweled, or glued onto the metal panels. A variety of such materials is commercially available, and suppliers of these materials have experience and data pertaining to desirable material thicknesses.

For thick plates of metal or concrete, or for beams, it is often useful to employ a viscoelastic damping material as the middle layer of a sandwich configuration, with the outer layers consisting of the structural material. Careful design of such damped sandwich configurations, based on well-known but somewhat intricate specialized procedures, is required if these are to work properly.

In cases where large structures vibrate severely, it is sometimes useful to fasten boxes filled with sand or some other granular material at the locations where the greatest motions occur. Agitation of grains—and their "rattling" against each other and against the sides of the box—extracts energy from the vibration, thus reducing it.

Vibration Absorbers

Vibration absorbers, often also called *dynamic absorbers* or *dampers,* are useful primarily where a single fixed-frequency component is of concern. Such an absorber consists of a mass that is attached to a vibrating system via a spring, with the mass and spring values chosen so that the resonance frequency of this absorber system coincides with the frequency of concern. When this mass-spring system is attached to the base or housing of the vibrating machine or to the vibrating structure, it acts to reduce the vibratory motion of the item to which it is attached. The motion of the added mass is opposite to that of the structure to which it is attached; thus, the mass makes the spring push down on the vibrating structure when the latter moves upward, and vice versa.

For a vibrating component or structure with an attached absorber, the product of the absorber mass and its excursion amplitude needs to be equal to the product of the mass of the component and its excursion. Thus, absorbers with small masses need to operate at large excursions; trading off between absorber masses and excursions is part of the design process.

REFERENCES AND BIBLIOGRAPHY

1. Harris, C. M. (ed.) *Handbook of Noise and Vibration Control,* 2d ed., McGraw-Hill, New York, 1979.
2. Harris, C. M., and C. E. Crede (eds.): *Shock and Vibration Handbook,* 2d ed., McGraw-Hill, New York, 1976.
3. Harris, C. M.: "Vibration Control Principles," Chap. 19 of Ref. 3.

Den Hartog, J. P.: *Mechanical Vibrations,* 4th ed., McGraw-Hill, New York, 1956.

Peterson, A. P. G., and E. E. Gross, Jr.: *Handbook of Noise Measurement,* 7th ed., General Radio Co., Concord, Mass., 1972. Explains noise and vibration, their effects and their measurement and outlines vibration control approaches.

Tonndorf, J., H. E. von Gierke, and W. D. Ward: "Criteria for Noise and Vibration Exposure," Chap. 18 of Ref. 3.

Ungar, E. E., and R. Cohen: "Vibration Control Techniques," Chap. 20 of Ref. 3.

section 7

Pollution Control and Waste Disposal

chapter 7-1

Air-Pollution Control

by

Karl J. Danz, P.E.
Principal Engineer

Richard B. Ruch, Jr.
Project Manager

Benjamin A. Schranze, P.E.
Roy F. Weston, Inc.
West Chester, Pennsylvania

REGULATORY REQUIREMENTS

Although considerable federal and state legislation concerning air quality was developed during the 1960s, the primary statutory framework now in place was established by the Clean Air Act of 1970. This legislation was amended in 1974 to incorporate a program to ensure the prevention of significant air quality deterioration (PSD) in existing "clean" air regions; it was amended again in 1977 to include provisions which formalize the permitting of new (or to be modified) facilities. The Act was up for reauthorization in 1981, but will probably not be amended in final form until 1982.

TABLE 1-1 National Ambient-Air Quality Standards: Primary Standards, $\mu g/m^3$

Pollutant	Annual standard	Short-term standard	Short-term time, h
Sulfur dioxide (SO_x)	80	365	24
Particulates	75	260	24
Carbon monoxide (CO)	—	10 000	8
		40 000	1
Ozone (O_3)	—	240	1
Hydrocarbons (HC)	—	160	3 (6–9 A.M.)
Nitrogen dioxide (NO_x)	100	—	
Lead (Pb)	1.5 (3 mo)	—	

Regulations and Standards

National Ambient Air Quality Standards

The National Ambient Air Quality Standards (NAAQS) define the quality of air which must be achieved to prevent adverse effects. The primary air quality standards specify levels of pollution which cannot be exceeded without threatening adverse effects on human health. The secondary air quality standards set limits not to be exceeded without adverse effects on public welfare, including property, vegetation, and other influences. For each criterion pollutant there is a set of these standards, each with its own regulatory program to limit atmospheric concentration levels at or below the levels set by its standards. Refer to Table 1-1.

The secondary standards are identical to these primary standards, except for sulfur dioxide (for which there is a 3-h, short-term secondary standard of 1300 $\mu g/m^3$) and particulates (for which there is an annual secondary standard of 60 $\mu g/m^3$ as a guide to assessing achievement of the 24-h, short-term secondary standard of 150 $\mu g/m^3$).

State Implementation Plan

The State Implementation Plan (SIP) provides a scheme under which the National Air Quality Standards are expected to be achieved and maintained. These plans, which must be approved by federal EPA, provide comprehensive programs to reduce pollution through a variety of specific abatement measures, some of which are delineated in company-specific compliance programs. Many deficiencies in the SIPs are apparent, and modifications can be made, using specific legal procedures set up by EPA.

Air Quality Control Region

The Air Quality Control Region (AQCR) is a basic geographic delineation of areas having specific pollution control programs. There are 247 AQCRs in the United States in which air quality is defined by ambient-air monitoring data, estimates made from mathematical dispersion analyses, or both. Each AQCR has its own program, specified in the SIP, which has been developed to maintain acceptable air quality.

New Source Performance Standards

The New Source Performance Standards (NSPS) are an indirect way of limiting emissions from new sources and improving air quality.These standards now cover a number of basic industrial and process categories. A new plant is subject to the NSPS only if these standards are proposed by the authorities before the plant is under construction. Thus, before proceeding with a new plant, a company should determine which categories are covered by the NSPS. Industries affected by the NSPS are shown in Table 1-2. Industrial categories scheduled for NSPS development are listed in Table 1-3.

Prevention of Significant (Air Quality) Deterioration

Prevention of significant (air quality) deterioration (PSD) is a regulatory program requiring preconstruction approval of new plants which may have significant emissions levels. In some states that have not adopted their own PSD-permitting program, the federal EPA Regional Office has the reviewing and approval authority. The major aspects of this program include:

- **Area Classification System.** All areas in the United States undergo an area classification scheme. This permits moderate amounts of industrial development routinely, but it does not allow air quality degradation to the point where air quality standards are approached or exceeded. The *Class I* category involves pristine areas subject to the tightest control. *Class II* covers areas of moderate growth. *Class III* covers areas of major industrialization.
- **Increments of Air Quality.** Numerical limitations on the additional pollution which may be allowed above existing baseline concentrations of sulfur dioxide and total suspended particulates.
- **Use of Best Available Control Technology (BACT).** Each major new plant must install a type of pollution control device which has been deemed acceptable for the particular discharge involved. This is determined on a case-by-case basis, but EPA provides guidelines concerning acceptable control methods.
- **Preconstruction Review.** A company is prohibited from commencing construction on a new source until a review by the authorities has been completed and an opportunity has been provided for public response on any disputed item.

It is likely that the PSD regulations contained in the Clean Air Act will be significantly amended in 1982.

Major Emitting Facilities

Plants subject to PSD review include all major emitting facilities. A major emitting facility is defined as any one of 28 source categories (Table 1-4) that have the potential to emit 100 tons/year or more of any pollutant regulated under the Clean Air Act. Any other source not on the list of 28 source categories, but having the potential to emit 250 tons/year or more of any regulated pollutant, is included in the definition of major emitting facilities. These emission cutoffs apply to both new sources and modifications to existing sources. Modified major stationary sources that increase the potential to emit greater than the significant limits shown in Table 1-6 are subject to PSD review.

Sources subject to PSD require:

- Monitoring of ambient air quality or the use of available representative data
- Demonstration that the plant will not violate the applicable PSD increment or any other air quality standard, based on ambient air quality modeling analysis
- Installation of BACT
- Commitment to conduct postconstruction monitoring

TABLE 1-2 Industries Affected By New Source Performance Standards Issued Under the Clean Air Act

Steam generator	Primary zinc smelter
Municipal incinerator	Primary lead smelter
Portland cement plant	Primary aluminum reduction plant
Nitric acid plant	Coal cleaning plant
Asphalt concrete plant	Lime plants
Petroleum refinery	Grain elevators
Petroleum storage	Kraft pulp mills
Secondary lead smelter	Lignite-fired steam generators
Secondary brass and bronze smelter	Sulfur-recovery plants and refineries
Iron and steel mill	Electric steam generators
Sewage treatment plant	Primary aluminum reduction plants*
Ferroalloy production	Stationary gas turbines*
Phosphate fertilizer	Internal combustion engines*
Primary copper smelter	Glass manufacturing

*Proposed

TABLE 1-3 Industrial Categories for Which New Source Performance Standards Are To Be Developed*

Stationary fuel combustion
- 14. Stationary internal combustion engines

Metallurgical processes
- 10. By-product coke ovens
- 23. Foundries: gray iron
- 41. Foundries: steel
- 42. Secondary aluminum
- 20. Secondary copper
- 66. Secondary zinc
- 67. Uranium refining

Mineral products
- 57. Asphalt roofing
- 49. Brick and related clay products
- 60. Castable refractories
- 58. Ceramic clay
- 48. Fiberglass
- 38. Glass
- 45. Gypsum
- 19. Metallic mineral processing
- 13. Mineral wool
- 18. Nonmetallic mineral processing
- 64. Perlite
- 21. Phosphate rock preparation
- 43. Sintering: clay and fly ash

Polymers and resins
- 54. ABS-SAN resins
- 12. Acrylic resins
- 50. Phenolic resins
- 62. Polyester resins
- 30. Polyethylene
- 55. Polypropylene
- 53. Polystyrene
- 51. Urea-melamine resins

Food and agricultural
- 68. Alfalfa dehydrating
- 44. Ammonium sulfate
- 59. Ammonium nitrate fertilizer
- 69. Animal feed defluorination
- 63. Starch
- 70. Urea (for fertilizer and polymers)
- 27. Vegetable oil

Waste incineration
- 11. Incineration: industrial-commercial

Basic chemical manufacture
- 1. Synthetic organic chemical manufacturing
- 61. Borax and boric acid
- 47. Hydrofluoric acid
- 65. Phosphoric acid: thermal process
- 40. Potash
- 46. Sodium carbonate

Chemical products manufacture
- 52. Ammonia
- 2. Carbon black
- 31. Charcoal
- 71. Detergent
- 17. Explosives
- 7. Fuel conversion
- 34. Printing ink
- 35. Synthetic fibers
- 28. Synthetic rubber
- 29. Varnish

Evaporative Loss Sources
- 6. Dry cleaning
- 9. Graphic arts
- 15. Industrial surface coating: autos
- 3. Industrial surface coating: cans
- 8. Industrial surface coating: fabric
- 37. Industrial surface coating: large appliances
- 32. Industrial surface coating: metal coils
- 5. Industrial surface coating: paper

Petroleum industry
- 25. Crude oil and natural gas production
- 72. Gasoline additives
- 4. Petroleum refinery: fugitive sources
- 33. Transportation and marketing

Wood processing
- 24. Chemical wood pulping: acid sulfite
- 22. Chemical wood pulping: neutral sulfite (NSSC)
- 36. Plywood manufacture

Consumer products
- 56. Textile processing

Minor source categories
- Lead acid battery manufacture
- Solvent metal cleaning (degreasing)
- Industrial surface coating: metal furniture

*This list was issued under Section 311 of the Clean Air Act by EPA on 31 August 1978 (43 *Federal Register* 38872-77). The numbers were assigned by EPA to reflect priorities, the lowest numbers indicating highest priority.

- Public hearing
- Issuance of construction permit

Source applicability under PSD review must be determined for each new or modified facility with a potential increase of emissions of any regulated pollutants. Because of the complexity of these regulations, it is suggested that a qualified engineer familiar with the regulations be consulted.

TABLE 1-4 Major Emitting Facilities

1. Fossil-fuel-fired steam electric plants of more than 250,000,000 Btu per hour of heat input	15. Phosphate-rock processing plants
2. Coal-cleaning plants (with thermal dryers)	16. Coke-oven batteries
3. Kraft pulp mills	17. Sulfur-recovery plants
4. Portland cement plants	18. Carbon black plants (furnace process)
5. Primary zinc smelters	19. Primary lead smelters
6. Iron and steel mills	20. Fuel-conversion plants
7. Primary aluminum ore reduction plants	21. Sintering plants
8. Primary copper smelters	22. Secondary metal production facilities
9. Municipal incinerators capable of charging more than 250 tons of refuse per day	23. Chemical process plants
10. Hydrofluoric acid plants	24. Fossil-fuel boilers of more than 250,000,000 Btu per hour of heat input
11. Sulfuric acid plants	25. Petroleum storage and transfer facilities with a capacity exceeding 300,000 barrels
12. Nitric acid plants	26. Taconite ore processing facilities
13. Petroleum refineries	27. Glass fiber processing plants
14. Lime plants	28. Charcoal production facilities

Nonattainment Area

A nonattainment area is an area where any ambient-air quality standard is being violated. The 1977 Amendments to the Clean Air Act require each state to tighten abatement requirements to assure compliance of all primary standards by 1982. Extensions to 1987 are allowed for photochemical oxidants (ozone) and carbon monoxide. Unless SIP revisions have been approved by EPA which reflect the method to eventually attain air quality standards, no new sources may be constructed in such areas. In areas covered by SIP revisions, any new plant (or modification to an existing plant) which can emit more than a specified emission level will require, for permitting purposes:

- **Air Quality Impact.** Sources located in attainment area cannot significantly impact a nonattainment area above established threshold concentrations (Table 1-5).
- **Emission Offset.** Reductions from existing sources which exceed the anticipated emissions from the new source are required.
- **Use of Lowest Achievable Emission Rate (LAER) Control Technology.** The most stringent emissions limitation scheme will be required for the new facility. The LAER requirement is determined individually, but unlike BACT does not need to take other factors into account (such as capital and outside maintenance costs).
- **Other Sources in Compliance.** All other plants in the state owned by the same company must be on approved emissions compliance schedules.
- **Assurance.** State assurance that the applicable SIP is being carried out.

Offsets and LAER are required for those pollutants whose potential emissions will equal or exceed 50 tons/year for all pollutants except carbon monoxide, which is 100 tons per year.

TABLE 1-5 Threshold Increases in Ambient-Air Concentrations for Nonattainment Areas

	Averaging time				
Pollutant	Annual	24 hours	8 hours	3 hours	1 hour
SO_2	1.0 $\mu g/m^3$	5 $\mu g/m^3$		25 $\mu g/m^3$	
TSP	1.0 $\mu g/m^3$	5 $\mu g/m^3$			
NO_2	1.0 $\mu g/m^3$				
CO			0.5 mg/m^3		2 mg/m^3

Air Quality Models

Air quality dispersion models are analytical tools used by industry and regulatory authorities to simulate the air quality impact of individual or multiple emission sources. It is recognized that, in the absolute sense, the performance of these air quality simulation models is generally poor. In a relative sense, the models provide a creditable indication of air quality impact over various averaging periods of concern. However, these models, which must be approved by the appropriate regulatory agencies prior to use, constitute one of the few available means of evaluating a source's compliance with PSD increments or NAAQS in the absence of measured air quality data in the vicinity of an emission source.

The U.S. EPA has specified certain basic modeling procedures in order to develop consistency in the application and use of the various EPA-approved models. The actual model selected and used will depend on a variety of factors including the source type and emission characteristics. These procedures are described and a list of the EPA-approved models is given in the EPA publication, *Guideline on Air Quality Models*, issued in April 1978 and revised in December 1980. Because each case may have some special or unique considerations such as proximity to complex terrain, stack downwash, etc., it may be in your best interest to consult an individual with more intimate familiarity with details of such modeling procedures.

Preliminary Screening Analysis

In order to reduce the number of cases in which more sophisticated and expensive modeling analysis is required, EPA allows the use of preliminary screening tests to isolate those cases where there appears to be little prospect of exceeding applicable the terms of NAAQS. The screening tests use conservative estimates of emission(s) characteristics and diffusion phenomenology in simple algorithms. If the screening computation shows that emissions from the proposed new source would not exceed the significant impact levels or one-half the PSD increments, the ambient-air quality analysis is considered satisfied and no further modeling is required.

Procedure in Modeling

If more sophisticated modeling is required, the following steps are required.

- Define *baseline* ambient air quality which corresponds to conditions existing on August 7, 1977. In some cases, a hypothetical baseline air quality must be projected for the specific locality of a proposed new plant, using representative data which may be available from ambient-air monitoring stations in the same general area. Data from the most recent 3-year period should be used.
- Determine to what extent the *increment* above the baseline available for new growth may have been consumed by prior PSD applicants. Major source construction commencing after August 7, 1977 cannot be included in the baseline, but must be counted against the increment. Additional emissions caused by fuel switches (by coal conversion or gas curtailment) can also be counted against the PSD increments.
- Determine the incremental impact of all significantly emitted pollutants. In this case, consideration will be limited not only to the computation of the worst short-term and annual concentrations but also to estimating the effect on nearby nonattainment or Class I areas. If concentrations exceed the significant impact levels (Table 1-6), then an emission offset/LAER analysis must be conducted. If it is predicted that concentrations in the attainment area will be in excess of "de minimis" levels, ambient-air monitoring may be required. (See Table 1-7.)

Recommended Models

EPA has recommended various models which it feels have had acceptable validation and which will provide replicate results. For convenience, EPA has made numerous air quality simulation model codes available to anyone wanting to make modeling estimates; these codes are collectively referred to as the *Users' Network for Applied Modeling of Air Pollution (UNAMAP)*. They are available on nine-track magnetic tape from the

TABLE 1-6 Significant Net Emissions Increase

Pollutant	Increase, ton/yr
CO	100
NO_x (as NO_2)	40
SO_2	40
Particulate matter	25
Ozone	40 of volatile organic compounds
Lead	0.6
Asbestos	0.007
Beryllium	0.0004
Mercury	0.1
Vinyl chloride	1
Fluorides	3
Sulfuric acid mist	7
Hydrogen sulfide	10
Total reduced sulfur	10
Reduced sulfur compounds	10
Other CAA pollutants	> 0

TABLE 1-7 Concentration Impact Values Above Which Ambient-Air Monitoring May Be Required

Pollutant	Concentration, $\mu g/m^3$	Avg. time (max), h
CO	575	8
NO_2	14	24
TSP	10	24
SO_2	13	24
Lead	0.1	24
Mercury	0.25	24
Beryllium	0.0005	24
Fluorides	0.25	24
Vinyl chloride	15	24
Total reduced sulfur	10	1
H_2S	0.04	1
Reduced sulfur compounds	10	1

National Technical Information Service (NTIS), Springfield, Va. Some of the available models include:

- **CDM (climatological dispersion model).** This determines long-term (seasonal or annual) quasistable pollutant concentrations at any ground-level receptor using average emission rates from point and area sources and a joint frequency distribution of wind direction, wind speed, and stability for the same period.
- **PTPLU (point-plume).** An interactive program which performs an analysis of the maximum short-term concentrations from a single point source as a function of stability and wind speed. The final plume height is used for each computation.
- **PTDIS (point-distance).** An interactive program which estimates short-term concentrations directly downwind of a point source at distances specified by the user. The effect of limiting vertical dispersion by a mixing height can be included, and gradual plume rise to the point of final rise is also considered. An option allows the calculation of isopleth half-widths for specific concentrations at each downwind distance.
- **PTMTP (point-multipoint).** An interactive program involving multiple point sources for a number of arbitrarily located receptor points at or above ground level. Plume rise is determined for each source. Downwind and crosswind distances are determined for each source-receptor pair. Concentrations at a receptor from various sources are assumed additive. Hourly meteorological data are used; both hourly concentrations and averages over any averaging time from 1 to 24 hours can be obtained.

Other Useful Models

Other models which have been found useful and are acceptable to the regulatory authorities include:

- **Valley Model.** A point-source model which is applicable to some complex terrain situations
- **CRSTER.** Provides a refined analysis of a point source where there are no significant meteorological or terrain complexities
- **Multiple-point Terrain (MPTER).** Similar to CRSTER, except that it is capable of simulating the air quality impact of multiple point sources at more than one location.
- **Texas Climatological Model (TCM).** A multisource model used to evaluate the long-term impacts of complex source configurations
- **RAM.** A multisource model used to evaluate the short-term impacts of complex sources on air quality
- **Texas Episodal Model (TEM).** Similar in capability to RAM
- **Industrial Source Complex (ISC).** A multisource model which is capable of dealing with the aerodynamic downwash of effluent from a stack that does not meet "good engineering practice" (GEP) requirements as well as the impact from a variety of point, area, and line sources.

Good Engineering Practice (GEP) Stack Height Requirements

In October 1981, EPA proposed revisions to stack height regulations which were originally proposed in January 1979 (Federal Register, October 7, 1981). The newly proposed revisions redefined a variety of terms including "stack" and "establishing a de minimis stack height."

Under the revised regulations, a stack is any point in a source designed to emit solids, liquids, or gases into the air, including a vent or duct. Flares have been omitted from this definition.

The previously proposed regulations defined a minimum stack height (i.e., the distance from the ground-level elevation of a plant to the elevation of the stack outlet) as 100 ft (30 m). Otherwise, for a source to get credit for that protion of its stack height built higher than this level in order to avoid excessive concentrations caused by nearby structures, a GEP stack formula was specified. The formula equals the stack height (H_g) to the height of the structure (H) plus one and one half times the height or width of the structure (L), whichever is less ($H_g = H + 1.5L$). The formula was derived from the traditional engineering practice of building a stack to approximately two and one-half times the height of a nearby structure.

Like the previously proposed regulations, the newly revised regulations do not limit the physical stack height of any source. See also Chap. 4-7.* Instead, they set limits on the stack height to be used in ambient-air quality modeling for the purpose of setting an emission limitation.

Under the newly proposed regulations, EPA has defined a GEP stack height as the greater of (1) 65 m; or (2) $H_g = H + 1.5L$, where H_g = GEP stack height, H = height of nearby building structure(s), and L = lesser dimension (height or projected width) of nearby building structure(s).

A stack height less than GEP should be evaluated to ensure that excessive concentrations will not routinely occur offsite and cause the potential of exceeding any NAAQS or PSD increment.

Air Quality Monitoring

Before beginning any monitoring program, the applicant must obtain approval from federal EPA or the state. This prior approval is necessary to prevent possible delays in the application review caused by collection of unacceptable data.

Consideration should be given to the effects of existing sources, terrain, meteorological conditions, existence of fugitive or re-entrained dusts, etc. The number of sites is

*See Rosaler and Rice: *Standard Handbook of Plant Engineering,* 1983.

directly related to the expected spatial variability of the pollutant in the area(s) of study. The suggested approach is first to use appropriate dispersion modeling techniques to estimate the air quality impact of the proposed source and any existing sources within the impact range of the proposed source for each pollutant. The modeled pollutant contribution of the proposed source should be analyzed in conjunction with the model results for existing sources to determine the general location of maximum concentration. Monitoring should then be conducted in or as close to these areas as possible. Generally, one to four sites would be sufficient in most situations with multisource settings. Preconstruction monitoring may be required for periods ranging from 4 to 12 months.

Air Quality Monitoring Equipment

As specified by EPA, all ambient-air quality monitoring must be done with continuous reference or equivalent methods, with the exception of total suspended particulate (TSP) matter for which continuous reference or equivalent methods do not exist. For TSP, samples must be taken in accordance with the reference method which is normal technique.

The TSP reference method is described in 40 CFR 50. The identification of continuous reference or equivalent methods for SO_2, CO, O_3, and NO_2 which have been designated in accordance with 40 CFR 53, can be obtained by writing the Environmental Monitoring and Support Laboratory, Department E (MD-77), U.S. Environmental Protection Agency, Research Triangle Park, NC 27711.

For SO_2, CO, NO_2, O_3, and meteorological parameters, continuous analyzers must be used. Thus, continuous sampling (over the time period determined necessary) is required. For TSP, daily sampling (i.e., one sample every 24 h) is required except in areas where the applicant can demonstrate that significant pollutant variability is not expected. In these situations, a sampling schedule less frequent than every day would be permitted. A minimum of one sample every 6 days will be required for these areas, however.

At least one year of meteorological data should be available for input to dispersion models used in analyzing the impact of the proposed new source on ambient-air quality and for the analysis of effects on soils, vegetation, and visibility in the vicinity of the proposed source. In some cases, representative data are available from sources such as the National Weather Service. In many situations, however, on-site data collection will be required.

A rigorous quality assurance program is required as an integral part of any ambient-air monitoring plan. The specific quality assurance procedures to be followed, including the siting of ambient monitoring stations, are described in an EPA document titled, "Ambient Air Monitoring Guidlines for Prevention of Significant Deterioration (PSD)," November 1980.

Reporting

Air quality and meteorological data must be reported to the permit-granting authority at the time of the permit application, but this could be done every 3 months. The actual reporting frequency and data format, however, should be based on an agreement between the applicant and the permit-granting authority.

The quality assurance data, including precision checks and accuracy audits, should be submitted for each site along with the summarized air quality data.

BACT/LAER Determination

In the preparation of a federal PSD permit application as well in as filing most applications for state permits for attainment pollutant emissions from any new or modified source, a control technology review in terms of BACT determination is required. The rationale for the selection of a particular control technology for a certain emission source has to be based on economic, environmental, and engineering considerations in the BACT review process. Generally, the control technology specified for a particular pollutant which an NSPS emits is considered BACT.

For example, BACT for the control of particulates from a municipal-size incinerator is a wet venturi scrubber system. A venturi scrubber system to limit particulate emissions from an industrial boiler capable of being cofired on fuels including low-sulfur coal, wood wastes, and/or sludge is also considered BACT. However, if this incinerator or industrial boiler were located within a particulate nonattainment area or BACT-controlled emissions would have the potential result of a significant impact on a nearby particulate nonattainment area, an LAER technology evaluation would be required. For the two sources cited above, LAER technology is considered the use of a fabric filter (baghouse) system. A summary of previously determined BACT/LAER control technology for various source categories is contained in an EPA document titled, "A Compilation of BACT/LAER Determinations Revised", May 1980.

EMISSIONS SURVEY

The initial step in defining the emissions levels and determining the BACT or LAER technology for a projected facility, or to control atmospheric emissions from an existing plant, is an emissions survey. All applicable pollutant sources and quantities of emissions must be identified. The results of this survey will provide management with a representative and comprehensive overview and with sufficient detail from which to formulate abatement plans and design programs and to file permits. The basic steps in such a survey include:

- Cataloging emission sources
- Quantifying emissions
- Preparing a source identification file

Cataloging Emission Sources

General Plant Information

The first step in such cataloging activities is to provide general plant information. This information is required to provide a general background and overview of the proposed or existing plant and/or its modifications. The following data should be provided:

- A schematic block flow diagram of the process(es) for the plant showing the flow of raw materials into and out of the process(es), and the design of the air pollution control equipment. This diagram should allow for release of all emissions, process and fugitive, to the atmosphere.
- A materials balance across the process(es) and across the air pollution control equipment.
- Airflows, either shown on a schematic block flow diagram for the process(es) and the control equipment, or provided separately.
- A floor plan of all of the plant building and exit stacks, showing width and height of buildings and stacks, *present* and *proposed.*
- Details of proposed stack(s): height, diameter, exit gas flow, and temperature.
- Details of the design characteristics of the air pollution control equipment.
- Details of the dust generated by the raw and finished materials for assessment of fugitive emissions from the new operation.

Process Flowsheets

The second step is to prepare process flowsheets in sufficient detail to indicate the flow of all raw materials, additives, by-products, and waste streams. The flowsheet should identify all points of feed input, and all points at which atmospheric, liquid, and solid wastes are discharged. The engineer or supervisor responsible for each process should

POWER PLANT SURVEY FORM

Type of Heat Exchanger

	Primary	Standby
Coal-fired	☐	☐
Oil-fired	☐	☐
Gas-fired	☐	☐

If multiple-fired, check appropriate boxes

Rated Input Capacity________________Btu/hr
Maximum Operating Rate________________Btu/hr
Rated Steam Output________________lb/hr____(a)______Btu/lb steam
Maximum Steam Output________________lb/hr____(a)______Btu/lb steam
Furnace Volume width____ft x depth____ft x height____ft = ______cu ft
Operating Schedule________________hr/day____day/wk____wk/yr

Coal Firing

Type of Firing ☐ Grate ☐ Spreader stoker ☐ Pulverized coal ☐ Cyclone Type ________________ ☐ Dry bottom ☐ Wet bottom

Fly Ash Reinjection ☐ Yes ☐ No

Soot Blowing

☐ Continuous
☐ Intermittent

Time Interval Between Blowing______minutes

Duration______minutes

Outside Coal Storage ☐ Yes ☐ No

Maximum Amount Stored Outside______tons

Figure 1-1 Sample presurvey data sheet for fossil-fuel-fired steam generators.

verify that the flowsheets identify all sources. Many plants have numerous sources that must be identified correctly for identification purposes.

Survey Data Sheets. Analysis of process flowsheets can be further verified by review of prior permit applications, process blueprints, photographs, and inspection manuals. With the aid of these and any other resources available, the emissions surveyor can then develop checklists in the form of survey data sheets that will be used in the plant tour to ensure complete and efficient gathering of pertinent data. These survey data sheets will pertain chiefly to process and feed data, and control equipment and emissions data. Figure 1-1 shows an example of a process survey data sheet, representing presurvey eval-

TABLE 1-8 General Control Equipment Data

A. At maximum continuous production rate (MCPR)
   1. Inlet and outlet absolute cubic feet per minute (ACFM).*
   2. Inlet and outlet gas temperatures.
   3. a. Inlet and outlet percent of H_2O.
      b. Dew point.
   4. Inlet pollutant levels for
      a. TSP, lb/h, gr/scf†, etc., also provide particulate size distribution.
      b. SO_2, lb/h, gr/scf, etc
      c. NO_x, lb/h, gr/scf, etc.
      d. HC, lb/h, gr/scf, etc.
      e. CO, lb/h, gr/scf, etc.
   5. Expected and guaranteed efficiencies for each of the above pollutants.
   6. Expected and guaranteed pollutant levels at outlet for each of the above pollutants.
   7. If available, make, model, and type of control system(s).
   8. Explain basis for ACFM and °F selection for design. **CAUTION**—*These may be different from data used in dispersion modeling.*
   9. Material balance across control equipment.
   10. Block flow diagram of control equipment.
   11. Inlet particulate size distribution.

B. For system(s) which may have a lower efficiency at normal operating rates as compared to MCPR, provide all data outlined above at normal operating rate.

*Actual cubic feet per minute.
†Standard cubic feet.
Note: If available, provide specification sheets. Provide drawings of internals.

uation of a fuel-fired combustion source. Such an example can be modified to apply to most types of control equipment now in operation.

The process survey data sheet should include the following:

- Detailed information on operating conditions for the process as designed
- Identification of normal operation as continuous, batch, or intermittent, with frequency of emission discharges for each operation
- Description of raw materials, products, and wastes
- Values for normal operating temperature, equipment performance ratings, flows, pressures, and similar data that are routinely monitored and/or recorded

Control Equipment Survey Information. The information given in a control equipment survey sheet is outlined in Tables 1-8 to 1-12. Table 1-8 shows information for a general survey evaluation sheet, while Tables 1-9 to 1-12 outline data requirements for control methods.

Identical information and data must be developed and evaluated for any new or modified facility. This information and data are needed to select the control device or system that represents BACT or LAER control technology.

TABLE 1-9 Multiclone and Cyclone Data

1. Number of tubes for multiclones (multiple cyclones)
2. Length of tubes
3. Fractional size efficiency vs. P curves
4. Diameter of tubes for multiclone or diameter of cyclone
5. Pressure drop, in H_2O column
6. Particulate size distribution into cyclone and/or multiclone
7. Grain loading (gr/dscf)* at inlet and/or outlet
8. ACFM at °F
9. Design efficiency
10. Disposal and handling of dry, collected dust
11. Preventive maintenance program

*Dry standard cubic feet.

TABLE 1-10 Scrubber for SO_2 and/or Other Pollutants

1. Design inlet and outlet volume flows, temperatures, percent moisture, and SO_2 loadings.
2. Type of scrubber. Wet or dry. If wet, operational mode.
3. Is reheat necessary? Describe.
4. Percentage of exhaust gases? Describe.
5. Any prequench section? Describe.
6. Type of demister.
7. Type of internal construction.
8. Minimum number of isolatable modules.
9. Minimum liquid-to-gas ratio.
10. Minimum-maximum pH of scrubbing liquor.
11. Maximum gas face velocity.
12. Minimum ratio of reagent to SO_2 and chemical composition solution.
13. Pressure drop, in H_2O column.
14. Is any bypass to be provided?
15. Outline any measures to be taken to produce more uniform SO_2 loading to the scrubber (e.g., precleaning, coal blending, etc.).
16. Design efficiency for PM and/or SO_2 and/or other pollutants.
17. Redundancy of control equipment.
18. Parameters for packed-bed scrubbers.
    a. Type of packing
    b. Packing size and/or shape
    c. Packing height
19. Parameters for spray scrubbers:
    a. Number of nozzles
    b. Nozzle droplet size
    c. Nozzle design and/or shape
    d. Nozzle pressure psig
20. Materials of construction.
21. Disposal of sludge or wet dust.
22. Effluent to streams from sludge pond.
23. Percolation from sludge ponds to aquifers.
24. Preventive maintenance program.

Tour of Plant Facilities

The third step is a tour of the plant facilities, if an existing plant is involved. The tour should include discussions with the process engineer or supervisor in order to ensure identification of all sources, verify the process and control-equipment flowsheets, and account for any equipment modifications already made. Sufficient data should be gathered to allow computation of material balances in order to have a basis for quantifying each emission source.

The plant tour starts at the files. There the survey will gather design specifications

TABLE 1-11 Fabric Filter or Baghouse

1. Type of filter media.
    a. Chemical composition
    b. Woven cloth or felt
    c. Porosity of new cloth
    d. Weight of cloth in ounces per square yard
    e. Napped or nonnapped fibers in woven bags
    f. Life of bags
2. Air-to-cloth ratio.
3. Cleaning mechanism.
4. Number of compartments. Can one compartment be shut down for maintenance while rest of the baghouse continues operation? How is overall collection efficiency affected?
5. How is damage from gas dew point circumvented? What is dew point?
6. Overall collection efficiency.
7. Materials of construction.
8. Disposal of collected materials.
9. Preventive maintenance program.

TABLE 1-12 Electrostatic Precipitator (ESP)

1. Collection area, square feet per 1000 ACFM.
2. Number of fields.
3. Number of compartments. Can one compartment be shut down for maintenance while rest of ESP continues operation? How is the overall collection efficiency affected during maintenance?
4. Gas velocity through ESP, feet per second.
5. Description of gas conditioning techniques.
6. Design of charging area: weighed wires, rigid frames, etc.
7. Temperature of operation.
8. Resistivity of fly ash or particulate matter at temperature of operation.
9. Inlet gas details: How is turbulence minimized? Any aerodynamic gas flow studies?
10. Rapping or cleaning details: provisions to minimize reentrainment when cleaning.
11. Design efficiency for preventive maintenance.
12. Design of collecting rappers to prevent bridging, plugging, or improper operation.
13. Materials of construction.
14. Disposal of collected materials.
15. Preventive maintenance program.

for each process and control device. Correspondence may also yield pertinent information relating to operation and maintenance of process and control equipment, current status of compliance, comments of control agencies, public complaints, and the like. This kind of background information can enhance the understanding needed for a meaningful on-site inspection of each process and control device. If the particular (existing) emission source requires an air permit, most of the pertinent information on the operating and emission characteristics are contained in the permit. In most states, permits are renewed every 2 or 5 years.

Each air contaminant source has a duct that vents from the process to an outside chimney or stack. The exhaust gas is moved by a fan or, in some instances where heat is applied, by natural draft. For each operation, the ducting should be followed from the process to the point of entry to the atmosphere. In some instances, the exhaust gas stream is difficult to follow. Introduction of makeup air, splitting of gas streams, and ducting of several operations to a common stack complicate the overall exhaust system and require careful tracing to ensure that exhaust-gas paths are properly defined. Placement of fans and control devices must be noted. Air-conditioning, heating, and makeup vents must not be mistaken for process stacks. After defining all process systems, the surveyor should check the roof to identify any emission points that are left over. The check ensures that all egress points are accounted for.

Quantifying Emissions

All of the data obtained thus far from the process flowsheets, process survey forms, control equipment survey forms, stack survey forms, photographs, correspondence, discussions, and the plant tour can now be organized for development of an emission survey plan. This plan will indicate the quantity of emissions to be expected from each source, with possible variations due to season, time of day, feed materials, and similar variables. The emissions characterization should identify all important parameters affecting control of the pollutants and possible sampling techniques. These data will be used to establish a compliance program for each source. These programs will describe the plans that will be implemented by the company to achieve compliance and should contain the following increments of progress or milestones:

- Date of submittal of the final control plan or PSD permit to the appropriate air pollution control agency.
- Date of award of accepted control plan or PSD permit.
- Date by which contracts for emission control systems or process modifications will be awarded, or date by which orders will be issued for purchase of component parts to accomplish emission control or process modification.

- Date of initiation of on-site construction or installation of emission control equipment or process change.
- Date by which on-site construction or installation of emission control equipment or process modification is to be completed.
- Date by which final compliance or full operation is to be achieved.

Figure 1-2 presents an example of the activities that must be completed before compliance can be achieved. Depending on the nature of the emission source and the complexity and size of the modifications required, the time requirements for compliance can range from a few months to several years.

Quantification techniques can then be applied in order to develop and utilize the emission survey information. Mass balances usually can be established around each process, particularly when the throughputs and the composition of raw materials are known. The materials balance will indicate the extent of solid and liquid, as well as gaseous wastes.

A search of applicable air pollution control regulations will provide the basis for the mass balance. The control regulations state what pollutants are regulated and define each pollutant. The definition of each pollutant determines the conditions under which the pollutant is sampled and its chemical or physical makeup. For instance, because water vapor is not considered an air pollutant, it is not necessary to accurately account for it in the mass balance. Emissions of SO_2 or organic substances are usually regulated and must be estimated in the materials balance.

A materials balance for gaseous pollutants can be determined by analysis of raw materials, fuels, and products to give the gaseous pollutant potential of many of the compounds liberated during a combustion or chemical process. Knowledge of fuel composition is especially useful in estimating emissions, since many gaseous compounds in the fuel become airborne after combustion (e.g., sulfur in fuel-oil exhausts as sulfur dioxide). Other constituents, such as ash and volatile matter, directly affect the quantity of particulate emissions. The results of stack sampling tests can also be used to quantify pollutant emissions.

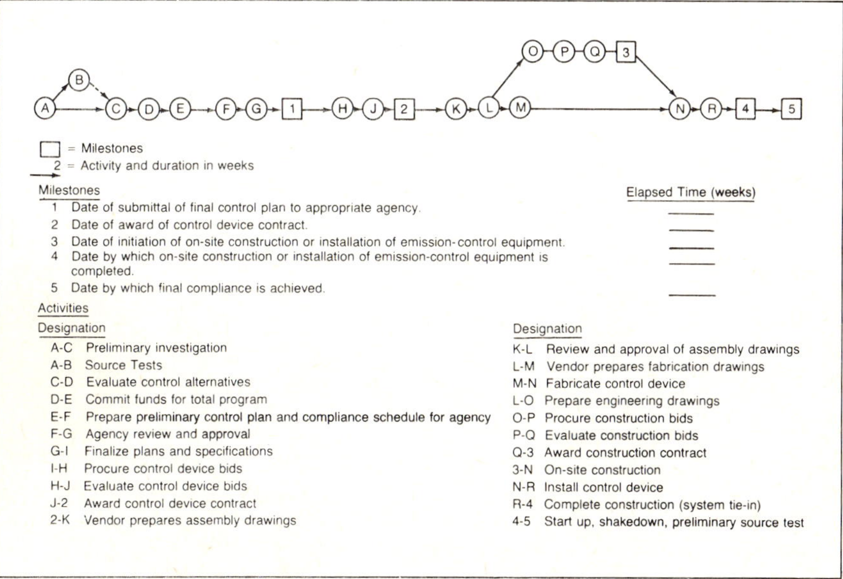

Figure 1-2 Sample compliance schedule chart.

In reviewing any permit application submitted, the regulatory authorities typically use an EPA publication titled, "A Compilation of Emission Factors," (AP-42) in order to estimate the potential uncontrolled and controlled emissions of a particular pollutant. This document and all its supplements should be obtained and maintained as reference material by the plant engineer.

Preparing a Source Identification File

A source identification file provides a means of standardizing data for the emission survey. For each pollutant source, a standard identification form gives a description of the process, a summary of emission data, the current compliance status, and proposed actions, if any are intended. A basic source identification form is shown in Fig. 1-3. Some sources involve several emission points with more than one pollutant at each point.

The source identification file should be indexed to provide easy access by any concerned party. The index should list all sources and identify each emission point for each source. Assignment of a number for each emission point will facilitate an alphanumeric search for emission points in the source identification file.

EMISSION CONTROL METHODS

The control equipment alternatives might be broadly grouped under the category of process or raw material changes to reduce the quantity of pollutant, eliminate production of a particularly undesirable pollutant, or collection and removal of the pollutant from the gas stream. Current designs employ some combination of these alternatives, with interaction between the alternatives. For example, using a scrubber to selectively remove a pollutant may lower an exhaust-gas temperature which could significantly reduce the thermal updraft effect, thereby reducing the dispersion of a conventional stack and necessitating the installation of a much taller stack. The problem of disposal of an undesirable liquid or solid waste removed from a gas stream may be greater than the problem of handling the gaseous waste itself. Clearly, in these situations, there must be a compromise. The control system must be considered to be a whole rather than individual parts.

The most important process parameters that must be collected for selection of control equipment are the following:

Flue gas characteristics

- Total flue gas flow rate
- Flue gas temperature
- Control efficiency required
- Particle size distribution
- Particle resistivity
- Composition of emissions
- Corrosiveness of flue gas over operating range
- Moisture content
- Stack pressure

Process or site characteristics (field survey)

- Reuse or recycling of collected emissions
- Availability of space
- Availability of additional electrical power
- Availability of water
- Availability of wastewater treatment facilities
- Frequency of start-up and shutdown

Emission Point No. ____________________

Emission Point Name ____________________

Date of Record ____________________

Source Name ____________________

Description of Source ____________________

Type of Permit ____________________

Date of Permit ____________________

Applicable Regulation(s) ____________________

Particulate Emissions ____________________ units ____________________

Allowable Emissions ____________________ units ____________________

Method of Determination ____________________

Gaseous Emissions ____________________

type ____________________ units ____________________

Allowable Emissions ____________________ units ____________________

Method of Determination ____________________

Compliance Status ____________________

Date Contract Awarded ____________________

Date Construction Began ____________________

Monitoring ____________________

ambient ____________________

stack ____________________

Figure 1-3 Source identification form.

Mechanical Collectors

The most familiar and widely used mechanical collector is the cyclone separator.(Fig. 1-4). Gas enters tangentially at the top of a cylindrical shell and is forced downward in a spiral of decreasing diameter in a conical section. Particles are centrifugally thrust outward and forced to spiral downward to the bottom, which is closed by an air lock. Since gas cannot escape at the bottom, it is forced to turn and travel, still whirling, back up the center of the vortex and out the top. Particles are discharged from the bottom through the air lock.

The tighter the spiral in which gas must flow, the greater the centrifugal force acting on a particle of given mass, and thus the more efficient a cyclone can be. Top diameters of cyclones range from more than 120 in to as small as 24 in, and capabilities reach 85 percent efficiency with particles as small as 10 μm at pressure drops from 0.5- to 3-in water gauge.

Cyclones alone are seldom adequate for pollution control, except where the load consists almost entirely of coarse particles, as in woodworking shops, or where particle density is unusually high. Cyclones are often used when there is a special reason to separately collect reusable coarse particles from useless fines (in fluid-bed catalyst regenerators, for example, where the fines pass through for subsequent separation by more efficient means). Cyclones are sometimes applied where gas cooling is necessary, where the cyclone is followed by a fabric filter; the cyclone scalps the coarse fractions of the dust load and at the same time provides cooling.

Cyclones are either applied singly, or manifolded to divide large gas flows between several individual units. They are also applied as two or more units of decreasing diameter in series for large loadings with a substantial percentage of small particles. The simple cyclone design, however, is not practical in very small diameters for capture of very small particles. Instead, the cyclone principle is employed in a different way in what is called a *multitube mechanical collector*.

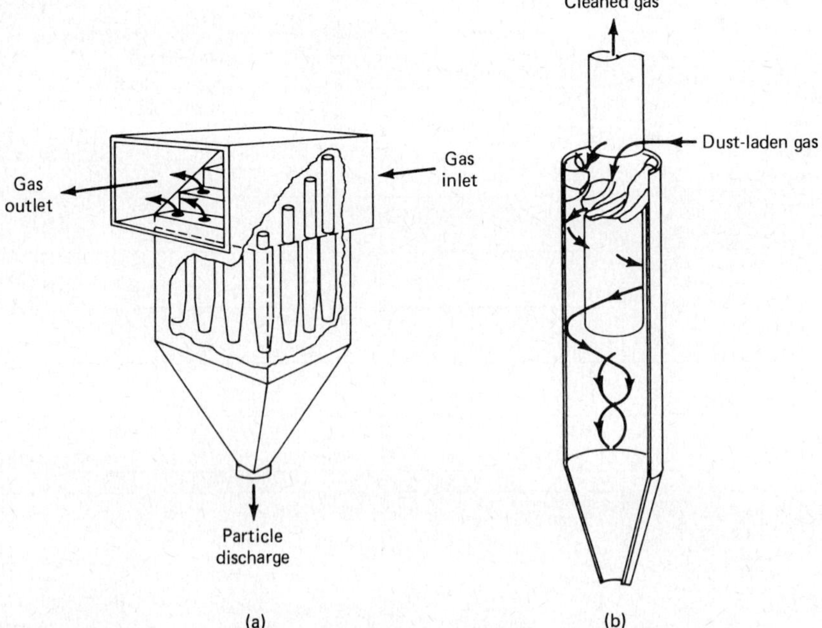

Figure 1-4 Mechanical collector-cyclone.

As the name implies, the multitube collector shown in Fig. 1-4 involves a large number of individual cyclone types in parallel within a single collector housing or shell. Dividing the gas stream among many tubes permits small tube diameters without requiring excessive gas velocity or pressure drop through the individual cyclone tubes. Tube diameters range from 4 to 10 or 12 in, with the larger sizes more common. Particles as small as 5 μm can be collected; efficiencies range up to 95 percent.

In a tube of a typical multitube collector, gas flows from the inlet section of the housing directly downward through an annular space between the outer tube wall and a smaller, concentric, exit tube. Fixed vanes in the annulus impart helical motion to the gas stream, thus applying centrifugal force for particle separation. As the gas stream slows down, reverses, and flows up to the exit tube, exit vanes may straighten out the cleaned-gas flow (which accomplishes a measure of gas pressure recovery to minimize overall pressure drop across the collector).

The smaller the tube diameter, the greater the tendency for large particles to bounce back off walls into the rising exit-gas stream. Tube diameter, inlet velocity to the collector, and the number of tubes in parallel (which determines the velocity through the individual tube) must be carefully selected for each application. It is vital that careful design ensure equal distribution of the inlet gas among all tubes to avoid excessive velocity in some and inadequate velocity in others, a special concern where angled or vertical gas flues are used to or from a multitube collector. In applications where particles have a tendency to be sticky, the smallest tube diameters cannot be used, and it may be necessary to eliminate exit recovery vanes.

Multitube mechanical collectors can be used alone for pollution control where the load fraction represented by particles smaller than 5 μm is small and the strictest emission codes do not apply. Increasingly, however, multitube collectors are used to scalp coarse particles from a gas stream ahead of a more efficient type of collector. Application to boilers that have been converted from oil to gas to pulverized coal is an example. Here, multitube collectors separate, for return to the boiler, coarse fuel particles with their content of unburned carbon, while ash fines pass through to a more efficient particulate collector.

Scrubbers

Scrubbers are compact inertial collectors capable of separating solid or liquid particles from a gas stream. They are also used to separate a chemically reactive or soluble gas constituent from other gas constituents in a flue gas stream. The most common use for gas separation involves its use in *flue gas desulfurization* (FGD).

In the simplest application of scrubbing, liquid is sprayed in at the top of a column and collision for particulate wetting or capture occurs as drops fall through a rising gas stream. Pressure drop is low, but application is limited to situations where 50 percent efficiency or less is acceptable and the percentage of particles smaller than 10 μm is low, or to scalping coarse particles ahead of a precipitator or a more efficient scrubber. A spray column is often used primarily for quenching hot gas, with coarse particulate removal a useful but incidental effect.

Centrifugal Scrubbers

In a centrifugal scrubber, column design and directed sprays cause drops and gas to mix in a rising vortex so that centrifugal force increases the momentum of collisions between particles and drops. Thus, smaller particles can be captured, and efficiencies as high as 90 percent can be achieved with particles as small as 5 μm, at pressure drops from 2- to 6-in water gauge.

Packed Scrubbers

A column may be fitted with impingement plates, wetted mesh, or fibrous packing, or packed with saddles, rings, or other solid shapes. In such scrubbers, at typically 1- to 10-in water-gauge pressure drop, efficiencies can range up to 95 percent with particles as small as 5 μm. Packed beds designed for gas absorption, however, are subject to fouling if the gas stream contains a significant fraction of solid particulates. In designs that use sprays to wash the packing, or that are packed with small spheres agitated by the gas flow, the fouling problem is reduced. A recent development employs a moist chemical-foam packing, which drains slowly from the scrubber with captured particulates and is replaced with fresh material.

Venturi Scrubbers

A venturi tube operating on the eductor or ejector principle, with scrubbing liquid as the motive fluid, can collect particles down to submicrometer size with efficiencies as high as 90 percent, if grain loading is low. Gas-pressure drop is not depended on for power input, and there can even be a gain in gas pressure across such a scrubber. The disadvantages are the requirement for substantial scrubbing-liquid flow at high pressure and the scrubber's inability to remove large particles because the induced gas-stream velocity is low.

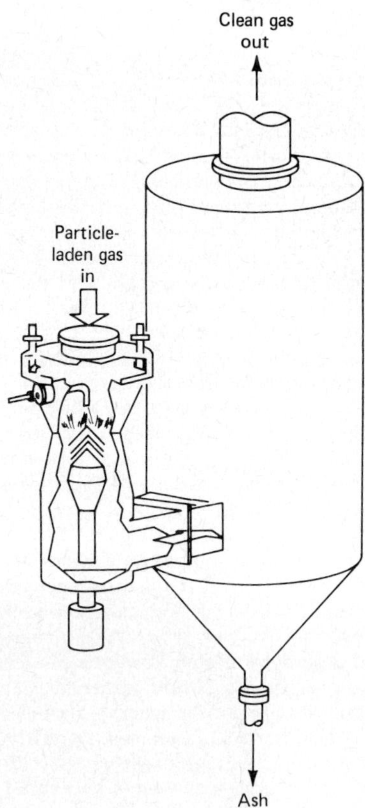

Figure 1-5 Scrubber.

A scrubber sometimes referred to as a *submerged-jet* type actually discharges gas through an orifice at, rather than under, the surface of a pool of scrubbing liquid. As it skims the surface, the gas atomizes some of the liquid into droplets for particle capture. A baffled flowpath causes the heavier particle-droplet agglomerations to drop back into the pool. Efficiencies can reach 90 percent, with particles down to 2 μm in size, at pressure drops from 2- to 6-in or more, water gauge. The special characteristic of this type of scrubber is its almost negligible power requirement for liquid pumping.

In what is usually called the *venturi scrubber,* Fig. 1-5, the drop in gas pressure across a venturi tube or orifice accelerates the gas stream. Scrubbing liquid is sprayed in at the throat or orifice (or is flowed in to be atomized by the passing gas stream) and mixes turbulently with the gas. Turbulence is ordinarily undesirable in fluid flow because it represents a useless consumption of power. But, in these scrubbers, turbulence is the cause of collisions between liquid droplets and particulates at high relative velocity, permitting capture of very small particles.

If the venturi throat or orifice area of a venturi scrubber is fixed, pressure drop (and, therefore particulate capture capability) will vary with gas-flow rate. If gas flow drops significantly below the design value, pressure drop will be inadequate to achieve design efficiency. Therefore, venturi-type scrubbers have been designed to permit variation of the throat or orifice area, manually or under control of pressure-differential sensors, to maintain optimum pressure drop for stable operation and constant collection efficiency.

When a venturi scrubber is employed, mechanical means can be provided for adjustment of throat area while maintaining the proper throat configuration. In the flooded-disk scrubber, the orifice through which gas is accelerated is the annular space between a horizontal disk positioned in a tapered duct and the duct wall. Pressure drop is maintained, despite flow variations, by raising or lowering the disk in the tapered duct to increase or decrease the orifice area.

Scrubbing liquid is supplied at the center of the upper face of the disk, where gas pressure forces it outward. At the disk periphery, liquid is sheared off and atomized by the passing high-velocity gas stream. Since this type of unit involves no spray nozzles or other constricted liquid passages, it is particularly suitable for recirculating liquid with a high solids content; high liquid pressure is not required. Because the orifice is a narrow annulus, its area can be large, where a high rate of gas flow is involved, without requiring distribution of scrubbing liquid over an excessive area.

Fabric Filters

Fabric filters collect solid particulates by passing gas through cloth bags that most particles cannot penetrate. As the layer of collected material builds, the pressure differential required for continued gas flow increases; consequently, the accumulated dust must be removed at frequent intervals. Cotton, wool, glass, and various synthetic fibers are used in fabric filters, which are shaped in the form of cylindrical bags or envelopes of roughly elliptical cross section.

Fabric filters are capable of 99+ percent collection efficiency with particles down to submicrometer size. High efficiency is attained with moderate pressure drops, typically in the range from 2- to 4-in water gauge. Power input is thus comparable to that of multitube mechanical collectors, while capability in collection of fine particle sizes is much greater. Operating cost, including maintenance, is somewhat higher, however, because some moving parts are involved and bags must be periodically replaced. Unlike wet scrubbers, the performance of fabric-filters is relatively unaffected by variations in gas-flow rate. Fabric filters are sometimes preceded by mechanical dust collectors, or settling chambers in the baghouse, when excessive grain loading or abrasive, coarse particles, or both, are involved.

The simplest type of fabric filter is the mechanical, or more properly, mechanically cleaned type. Near the bottom of the filter housing is a horizontal tube sheet separating the inlet dusty-air plenum from the clean-air plenum into which the bags discharge. Bags, closed at the top, are suspended with their open bottom ends tightly connected to openings in the tube sheet. Gas flow is upward, into the bags, through the fabric, into the clean-air plenum, and out. Particles accumulate inside the bags and, when dislodged, fall down into hoppers for disposal.

The traditional method of surface filter cleaning is mechanical shaking of the structure from which the bags are suspended. Design and operation are simple, but maintenance requires access to the shaking mechanism within the housing. The same gas-flow and bag arrangement is employed in the *shakerless*, or *reverse-air*, fabric filter. A fan, ducting, and dampers are provided to supply a reverse flow of air from the clean side of bags, through to their interiors, to dislodge accumulated particles. Bags are usually stiffened by wire rings to prevent their complete collapse when the airflow reverses.

The mechanical-shake action is violent enough to accelerate bag wear if the fabric is not resistant to abrasion or if unusually abrasive particles are being collected. In fact, the action can be positive enough for cleaning felted fabrics in some applications, although most mechanical fabric filters are surface filters. The gentler reverse-airflow cleaning action is preferred where abrasion will be a problem and is usually required when high gas temperature dictates use of glass-fiber bags; glass fabrics cannot be expected to withstand the mechanical shaking action employed in most applications, even when the fibers have been lubricated with graphite or silicone.

If air jets (Fig. 1-6) are used for cleaning instead of low-velocity reverse airflow, felted fabrics can be used and bags can be cleaned while contaminated gas continues to flow to them. A reverse-jet-type fabric filter employs a ring of nozzles around each bag and a mechanism to traverse the rings up and down the bag length. It does not require the elaborate inlet duct-and-damper arrangement necessary for continuous duty with mechanical fabric filters. But this advantage is substantially offset by the cost of providing and maintaining the complex cleaning mechanism, all of which is inside the filter housing, and by the possibility of misalignment that will allow abrasion of bags by the moving rings.

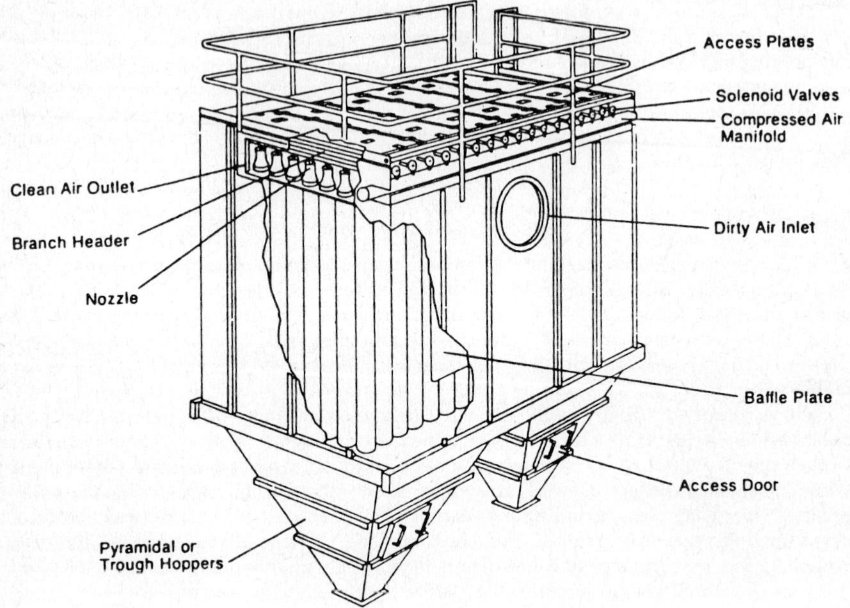

Figure 1-6 Fabric filter.

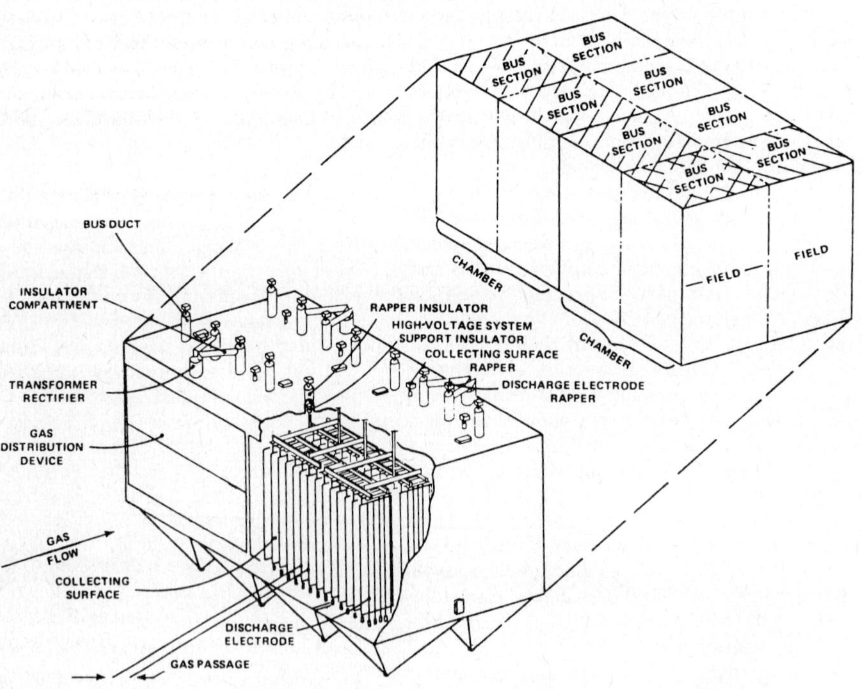

Figure 1-7 Electrostatic precipitator.

Electrostatic Precipitators

The electrostatic precipitator, Fig. 1-7, an extremely efficient air pollution control device, can remove more than 99 percent of the undesirable particulates in a gas stream. This high efficiency is possible because, unlike other pollution control devices, the precipitator applies the collecting force only to the particles to be collected, not to the entire gas stream. Thus, an extremely low, energy-saving power input, about 200 W/1000 ft^3/min (0.5 m^3/s), is required.

In operation, a voltage source creates a negatively charged area, usually by means of wires suspended in the gas-flow path. On either side of this charged area are grounded collecting plates. The high potential difference between these plates and the discharge wires creates a powerful electric field. As the polluted gas passes through this field, particles suspended in the gas become electrically charged and are drawn out of the gas flow by the collecting plates. They adhere to these plates until removed for storage or disposal. Removal is accomplished mechanically by periodic vibration, rapping, or rinsing. The gas stream, now free of particulate pollution, continues on for release to the atmosphere.

Precipitators may be designated as either plate or pipe varieties. The plate type, just described, is widely used for dry dust. The pipe type, for removal of liquid or sludge particles and particulate fumes, uses the same principle, but with a different mechanical setup; the discharge electrodes are suspended within a series of pipes (collecting plates) contained in a cylindrical shell, under a header plate.

As noted earlier, the precipitator applies the separating force directly to the particles, regardless of gas-stream velocity, thereby requiring less power than other control devices. And, because they can operate completely dry, they are ideal for use in the following situations:

- Where water availability and disposal are problems.
- Where very high efficiency on fine materials is required; for example, electrostatic precipitators are removing particulate matter as small as 0.05 μm from zinc oxide fumes with 97 to 98 percent efficiency.
- Where large volumes of gas must be treated. In nonferrous metals production, gas flows of 600 000 ft^3/min (250 m^3/s) at temperatures up to 800°F (425°C) are treated with up to 99.6 percent efficiency.
- Where valuable dry materials must be recovered, as in rotary-kiln or spray-drying operations, and in calcining.

Process Modifications

Pollutant generation can frequently be limited by using different processes of raw materials to produce the same end product. This is by far the best method of control if it can be used. In many cases, it is the most cost-effective way to reduce the emission of pollutants.

As an example, modifications to combustion equipment are used to regulate the combustion process to reduce emissions of nitrogen oxides. The amount of nitrogen oxides emitted from combustion is largely a function of the control of temperature and residence time in the primary flame zone. Accordingly, burners are being designed which produce significantly lower quantities of nitrogen oxides or are arranged in the furnace to minimize nitrogen oxides formation.

Low excess-air firing can reduce nitrogen oxide emissions by 10 to 30 percent. Essentially, oxygen input to the burners is reduced, thus reducing the chance of nitrogen oxides formation.

Nitrogen oxides may also be reduced 30 to 70 percent by using a two-stage combustion process. Substoichiometric quantities of primary air are supplied at the burner. Complete burnout is accomplished by using secondary air supplied at a lower temperature.

Flue gas recirculation is another technique used to reduce nitrogen oxides by 20 to 60 percent. The maximum flame temperature is reduced by dilution of the primary flame zone with the recirculated combustion flue gases. The oxygen concentration is also reduced, making nitrogen oxides formation less likely.

Not all of these processes can be used simultaneously, but by designing a combustion system using a judicious combination of the alternatives, nitrogen oxide emissions may be reduced by as much as 70 to 90 percent.

BIBLIOGRAPHY

1. EPA: "Compilation of BACT/LAER Determinations—Revised," EPA 450/2-80-70, May 1980.
2. EPA: "Compilation of Emission Factors," AP-42 and Supplements, Research Triangle Park, North Carolina, April 1981.
3. EPA, OAQPS: "Ambient Monitoring Guidelines for Prevention of Significant Deterioration (PDS)," EPA 450/4-80-012, November 1980.
4. EPA, OAQPS: *Guidelines for Air Quality Maintenance Planning and Analysis,* vol. 10, "Procedures for Evaluating Air Quality Impact of Near Stationary Sources," EPA-450/4-77-001, Research Triangle Park, North Carolina, October 1977.
5. EPA, OAQPS: "Guideline on Air Quality Models," EPA-450/2-78-027, Research Triangle Park, N.C., April 1978.
6. EPA, OAQPS, Region III: "Guidelines for Determining BACT," Philadelphia, December 1978.
7. EPA, Region III: "Permit Application Kit, Prevention of Significant Air Quality Deterioration," Air Programs Branch, Philadelphia, November 1978.
8. EPA, Technology Transfer: "Industrial Guide for Air Pollution Control," Contract 68-01-4147, by PED Co. Environmental, Inc., June 1978.
9. *Federal Register,* 1977 Clean Air Act, PSD, SIP Requirements, Part 52, June 19, 1978.
10. *Federal Register,* 40 CFR Parts 51 and 52, September 5, 1979.
11. *Federal Register,* 40 CFR Parts 51 and 52, February 5, 1980.
12. Quarles, J., Jr.: "Federal Regulation of New Industrial Plants," P.O. Box 998, Ben Franklin Station, Washington, DC 20044, January 1979.
13. Rymarz, T. M., and D. H. Klipstein: "Removing Particulates from Gases," *Chemical Engineering Deskbook,* vol. 82, no. 21, October 6, 1975, pp. 113–120.

chapter 7-2

Liquid-Waste Disposal

by
William M. Throop, P.E.
Manager, Market Development
Envirex Inc.
Waukesha, Wisconsin

GLOSSARY

Alkalinity Ability to neutralize acids—determined by the water's content of carbonates, bicarbonates, hydroxides, and borates, silicates, and phosphates, if present. Expressed in milligrams per liter of calcium carbonate ($CaCo_3$).

BOD_5 (biochemical oxygen demand) A measure of oxygen metabolized, in milligrams per liter, in five days by microorganisms that consume biodegradable organics in wastewater under aerobic (with air) conditions.

COD (chemical oxygen demand) The amount of oxygen, in milligrams per liter, needed to oxidize both organic and oxidizable inorganic compounds.

Effluent The liquid end product discharging from a process.

Floating matter Matter that passes through a 2000-μm sieve and separates by flotation in an hour.

Settleable solids Solids larger than 0.01 mm in diameter settling in two hours under quiescence.

Suspended solids Small filterable particles of solid pollutants in wastewater. The examination of suspended solids and the BOD_5 test constitute the two main determinations for water quality.

Total solids All dissolved, suspended, and settleable solids contained in a liquid.

Turbidity The amount of suspended matter in wastewater; quantity obtained by measuring its light-scattering ability.

INTRODUCTION

Generally, pollution is the addition of harmful or objectionable material to water in concentrations of sufficient quantity to result in measurable degradation of water quality.

Industrial wastewater pollution is not only a liability; it is a total systems problem. Various interactions between air, water, and solid wastes must be evaluated on an overall plant basis. The treatment of a water pollutant, for example, may create an air pollutant. Pollution is pollution de facto; as long as the deleterious material does not meet the standards of the regulatory agency, it is pollution. Harm or injury no longer has to be proved.

DESCRIPTION OF THE PROBLEM

The passage of the Federal Water Pollution Control Act as amended in 1977 under Public Law 92-217 forced the plant engineer to become familiar with its many ramifications. Among the provisions of this act that are of direct interest to industry is the establishment of water quality standards typified by Table 2-1, column A, for maintaining aquatic life.[1] These restrictions are enforced by the requirement that a permit be issued before discharges are permitted under the National Pollutant Discharge Elimination System (NPDES). Every holder of a NPDES permit is required to comply with monitoring sampling, recording, and reporting requirements.

Many local ordinances have pretreatment requirements limiting high effluent concentrations of wastes and toxic materials which might adversely affect treatment processes of publicly owned treatment works (POTW). A typical list of effluent limitations is shown in Table 2-1, column B.[2] When discharging to POTWs, industry is expected to pay its proportionate share of capital cost of the POTWs collection and treatment equipment. These user fees are usually based on a multiplier of BOD_5, suspended solids, and liquid volume, and vary with each municipality.

For specific pollutants, effluent guidelines for specific industrial categories are published in the *Code of Federal Regulations* (40 CFR 401).

Different effluent levels are allowable depending on the following:

1. Industrial subcategory
2. Control technology required (e.g., best available technology economically achievable)
3. Existing or new source (new sources are more severely regulated than existing sources)

TABLE 2-1 Maximum Discharge Limits

Constituent	A Direct discharge or recycle, mg/L	B* To POTW
Ammonia nitrogen (as N)	1.5	——
Arsenic (total)	1.0	——
Barium (total)	5.0	——
Boron (total)	1.0	1.0
Cadmium (total)	0.05	2.0
Chromium (total)	——	25.0
Chromium (total hexavalent)	0.05	10.0
Chromium (total trivalent)	1.0	——
Copper (total)	0.02	3.0
Cyanide (total)	0.025	@150°F & pH 4.5 = 2.0
Fluoride (total)	1.4	——
Iron (total)	1.0	50.0
Iron (dissolved)	0.5	——
Lead (total)	0.1	0.5
Manganese (total)	1.0	——
Mercury (total)	0.0005	0.0005
Nickel (total)	1.0	10.0
Fats, oils, and greases (FOG)	15.0	100.0 total
pH	5.0–10.0	4.5–10.0
Phenols	0.1	——
Phosphorus (as P)	1.0	——
Selenium (total)	1.0	——
Silver	0.005	——
Sulfate	500.0	——
Temperature	——	150°F (65°C)
Zinc (total)	1.0	——
Total dissolved solids	1000.0	No limit

*Units are milligrams per liter unless otherwise specified.

4. Where the effluents are discharged (effluent levels discharged into POTWs are different from direct discharges into navigable water)

For details pertaining to emissions by specific industry, the *Code of Federal Regulations* should be consulted. Because the promulgation of effluent guidelines is an ongoing process, the EPA should be contacted for the latest information.

All pollutants are classified as either conventional, toxic, or nonconventional. Conventional pollutants include BOD_5, TSS (total suspended solids), and pH. There are 129 priority pollutants that appear on the toxics list in the *Federal Register* 43(164)4108 (February 1978). Nonconventional pollutants are those that are neither toxic nor conventional, such as nitrogen, oil, and grease. Best conventional pollutant control technology will be required for conventional pollutants by July 1, 1984. Best available technology economically achievable will be required for toxic and nonconventional pollutants by the same date.

PLANT SURVEY

The accurate measurement of flow volume and pollutants in the waste flow are essential in assessing any wastewater problem, and in designing a wastewater treatment system. Limitations on effluent (see Table 2-1) make it imperative to analyze flows and impurities quickly, accurately, and at reasonable cost.

The best approach is to make a comprehensive wastewater survey that will (1) determine the quantity of wastewater discharge, (2) locate the major sources of waste within the plant, (3) determine wastewater composition, (4) explore in-plant or process changes

TABLE 2-2 Typical Process Discharge Volumes, BOD_5, and Suspended Solids for Industrial Wastewater Before Treatment

	Unit processed	Discharge per unit		BOD_5, mg/L	Suspended solids, mg/L
		Gallons	Liters		
Aluminum & copper	lb (kg)	12–13	45–50	N/A	300–500
Automotive	Car	10800	40900	190	215
Beverage, malt	bbl (L)	330	1250	390–1800	70–100
Canning					
Fruit	Case	20–40	75–150	300–1600	200–500
Vegetable	Case	50–100	190–380	700–2000	300–2000
Coal washing	ton (tonne)	125	138	N/A	2000–3000
Cooking	ton (tonne)	1500–2800	1650–3090	50–200	90
Dairy, milk	gal (L)	4–12	15–45	1800	560–4000
Electrical	kWh (kJ)	80	110	N/A	50–2000
Laundry					
Commerical wash	ton (tonne)	8600	36000	600–1860	400–2200
Industrial	ton (tonne)	5000	20000	650–1300	4900–8600
Manufacturing, gen. fabr.	ton (tonne)	700	3000	50–1500	200–15 000
Meat packing					
Cattle	Animal	400–2000	1515–7575	400–900	400–800
Chicken	Bird	8–9	30–34	150–2400	100–1500
Hogs	Animal	300–600	1136–2273	1000	650
Office building	Person	30–45	114–170	117	176
Paint, latex	gal (L)	3	11	2000–3000	15 000–60 000
Paper making	lb. (kg)	65	108	200–800	500–1200
Pharmaceutical	——	——	——	600–2500	500 1000
Phenolic resins	ton (tonne)	75 000	313 000	11500	40
Railway maintenance	Locomotive	3000	11 360	500–800	200–600
Refining	bbl crude	770	2900	100–500	300–700
Rubber, synthetic	Car tire	500	1890	25–1600	60–2200
Steel					
Cold-rolled	lb (kg)	9	15	150	100–300
Hot-rolled	lb (kg)	18	30	80	500–2000
Tanning, hide	lb (kg)	8–12	30–45	900	6000
Textiles					
Synthetic	lb (kg)	12–25	45–95	1500–6000	500
Wool	lb (kg)	70	265	900	100
Vegetable oil	gal (L)	22	83	3050	900

to minimize the waste problem, (5) establish the basis for wastewater treatment, and (6) evaluate effect of wastes on the receiving stream.

Composition of wastewater varies with the amount of impurities initially present in water and the chemical analysis of any pollutants that are added. While domestic sewage has a fairly uniform composition, industrial wastes have an almost infinite variety of characteristics, as shown in Table 2-2. Wastewaters should be analyzed for at least BOD_5, COD, color, total solids (suspended and dissolved), pH, and turbidity. Other impurities of interest will vary with the source and type of wastewater.

Concentration of pollutants must be correlated with average, minimum, and maximum flows encountered. The analysis program must also take effluent water quality standards into account (see Table 2-1) and the BOD_5 reduction required to meet them. Any toxic impurities in the wastewater that adversely affect water quality must be determined.

Sampling and Flow Measurement

The starting point in any wastewater survey is an effective program of sampling and flow measurement. To be useful, a sample of wastewater must accurately represent the source from which it is taken and be large enough to run all the laboratory tests required. This means the method of sampling must be tailored to the type and kind of wastewater flow.

A close check of each waste source will reveal whether flow is continuous or intermittent and any wide swings in flow rate.

It is also important to know if the concentration of pollutants changes drastically or is fairly constant. The presence of oil or excessive suspended solids may also cause problems.

An integral part of this program is the need to obtain flow information on various in-plant streams as well as the plant outfalls. Wastewater flows are measured for the following reasons:

1. To determine the quantity of water being discharged, as well as variations in the flow rate
2. To determine the number of pounds of constituents being discharged on the basis of the analytical data and the determined flow rate
3. To evaluate segregation possibilities
4. To determine the effect of the wastewater discharge on the receiving stream, if applicable

Measuring Rate of Flow

Rates of flow can be approximated by the methods discussed in the following paragraphs.

Water Meters on Influent Lines. Water consumption in the plant should be determined during a wastewater survey to check on wastewater-flow measurements and to compute a water balance for the plant. Meters can also be installed at particular water-using operations to obtain flow data.

Container and Stopwatch. The time required to obtain a given volume of water in a container is measured. Volume can be determined either by weight added or by a calibrated collection container. The weight of water added is divided by 8.34 lb/gal to determine the number of gallons collected. The flow is then determined by the formula

$$\frac{\text{Gal in container} \times 60}{\text{Time, s, to fill container}} = \frac{\text{gal/min}}{15.85} = \text{L s}^{-1}$$

If the container fills in less than 10 s, the accuracy of this method is questionable.

Weirs. A weir acts like a dam or obstruction, with the water flowing through the notch, which is usually rectangular or V-shaped.

To ensure accurate weir measurements:

1. The weir crest must be sharp or at least square-edged. Steel is the best construction material, but tempered wood is also used.
2. The weir must be ventilated. There must be air on the underside of the falling water.
3. Leaks around the weir plate must be sealed.
4. The weir must be exactly level.
5. Weirs should be kept clean.
6. The head on the weir should be measured at a distance of 2.5 times the head upstream from the weir.
7. The channel upstream from the weir should be straight, level, and free from disturbing influences. A stilling box may be used to quiet the water flow.
8. The weir should be sized after the flow is estimated by other methods. The head on any weir should be greater than 3 in (7.6 cm) but not more than 2 ft (61 cm).

The flow over V-notched (triangular) weirs and rectangular weirs can be taken from the nomographs shown in Fig. 2-1.

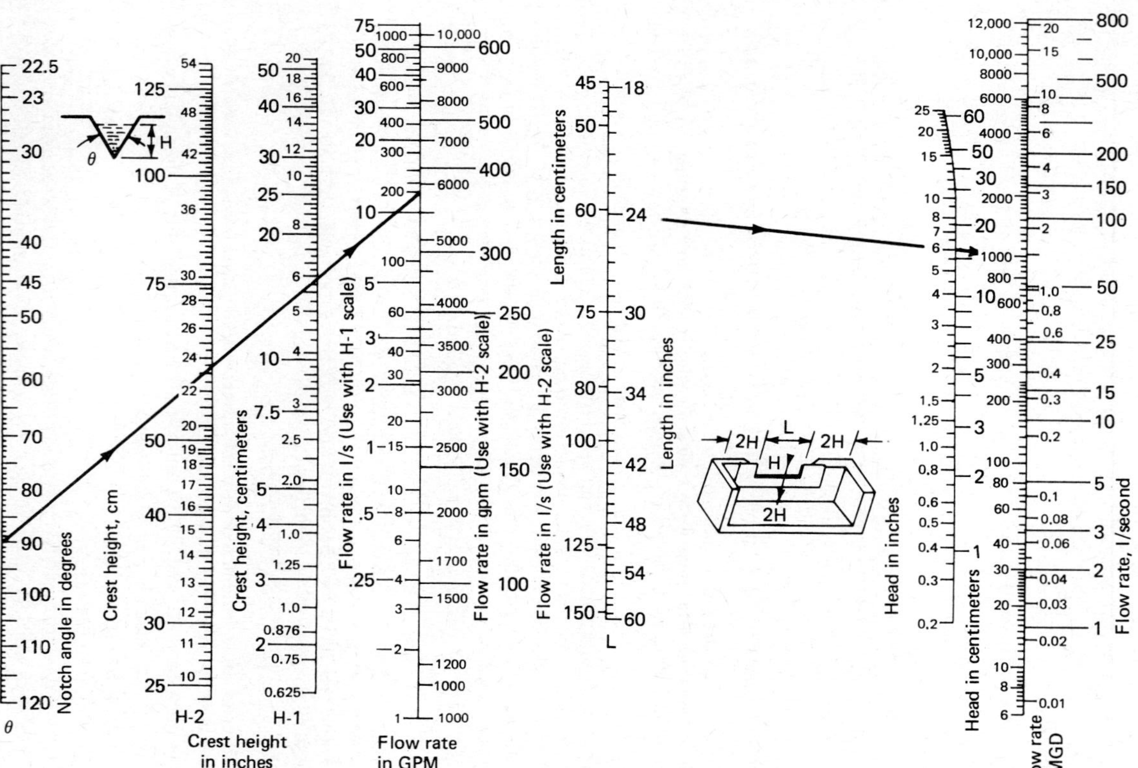

Flow for V-notch (triangular) weirs

Flow for rectangular weirs

Figure 2-1 Nomographs for measuring flow over weirs.

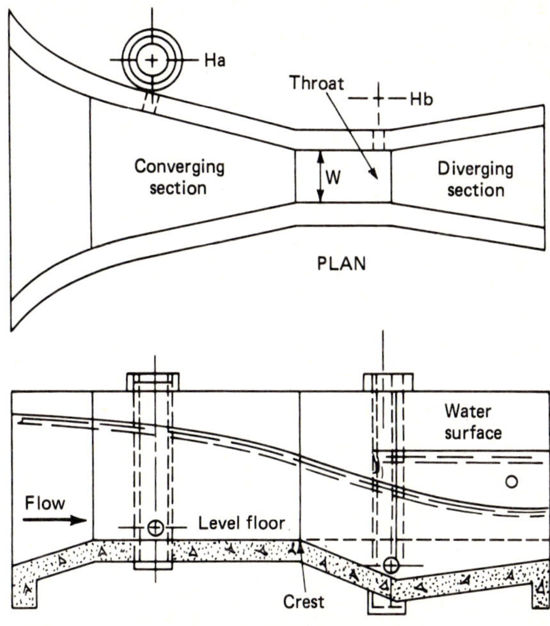

Figure 2-2 Parshall flume.

Parshall Flume. A Parshall flume (Fig. 2-2) can be used to measure flows in open channels at or near ground surface. This device is valuable when it is not possible to dam the water. It is also advisable for a permanent installation because it is self-cleaning.

Flow under submerged conditions can be calculated from readings taken at gauges. If the water surface downstream from the flume is high enough to retard the rate of discharge, submerged flow exists. When there is no backwater effect, water passing through the throat and diverging section assumes a level which corresponds to the floor of the channel. This pattern demonstrates free flow.

The flow of a free discharge from a Parshall flume is calculated by

$$Q = 4WH_a n$$

where

Q = flow, ft^3/s
W = throat width, ft
H_a = head of water above level floor, ft
$n = 1.522W^{0.026}$

or in metric units

$$Q = 8.52 \times 10^3 (11.4 \log W) H_a (1.57 + 0.09 \log W)$$

where

Q = flow, m^3/h
W = throat width, m
H_a = head of water above level floor, m

Flow under submerged conditions can be calculated from readings taken at gauges, one located at a point two-thirds the length of the converging section measured back from the crest of the flume H_a and one located near the downstream end of the throat section H_b. Degree of submergence is given by the ratio H_b/H_a.

Sample Collection and Analysis

Wastewaters are sampled and analyzed to identify those pollutants that require treatment and to select the proper treatment process.

Collecting Samples. Since a wastewater's characteristics can vary considerably, composite samples are collected to obtain a truer representation of the waste. Small samples are collected at frequent intervals during the sampling period. They are mixed together to form the composite sample.

Compositing Samples. Depending on plant operation, 8-, 16-, or 24-h composites can be collected. Daily sampling for 3 days generally constitutes a sampling program. Samples can be composited on the basis of the following criteria.

Flow. The amount of sample collected at any time during the sampling period is proportional to the flow of wastewater at that time.

Time. The same amount of sample is collected at every interval during the sampling period regardless of variations in wastewater flows.

Sample Size. The sample size collected at any one time should be at least 200 mL. Composite samples can be collected either manually or with automatic samplers. Automatic, battery-operated samplers are available for collecting composite samples on the basis of flow or time.

Amount of sample to be collected depends upon the laboratory tests to be run. The amount of sample required for each test to be performed should be determined before the sampling program is begun to ensure that sufficient sample is collected.

Analytical Determinations. Determinations that may be conducted on a wastewater sample are:

pH
Alkalinity or acidity
Total hardness
Chloride
Sulfate
Suspended solids
Volatile suspended solids
Settleable solids
Total nitrogen
Ammonia nitrogen
Total phosphate
Total solids
Volatile total solids
Copper
Nickel
Zinc
Chromium, hexavalent
Chromium, total
Iron
Manganese
Solvent soluble (oil)
Phenol
Biochemical oxygen demand (BOD_5)
Chemical oxygen demand (COD)
Total organic carbon (TOC)
Cyanide

Standard methods are available for conducting these determinations. Phases involved in a full pollution control program are:

I. Definition of the problem and development of an action plan (survey or feasibility study)
II. Detailed engineering
III. Construction and start-up

An outline of the steps involved in each of these phases is shown in Table 2-3.

TABLE 2-3 Pollution Control Program

I. Survey or feasibility study
  A. Fact finding:
    1. Develop a plant water balance for average and peak operating conditions.
    2. Inventory all industrial processes using water.
    3. Determine characteristics of the receiving waterway both upstream and downstream from plant's discharge.
    4. Determine chemical characteristics of waste streams.
    5. Study all operations using water and producing wastes.
    6. Determine local requirements with respect to pollution.
  B. Analyze data to determine:
    1. Sources of offending contaminants.
    2. Feasibility of segregating contaminated wastes requiring treatment from dilute wastes which would be acceptable without treatment.
    3. Availability of "natural" dilution waters, that is, waters employed for useful purposes but not contaminated.
    4. Quality of effluent required for compliance with discharge standards.
    5. Whether treatment is necessary.
  C. Exploit in-plant and/or process changes to minimize the problems by:
    1. Reducing wastes or waste volume at sources.
    2. Exploring the possibilities for reuse of process materials without treatment.
    3. Investigating recovery of valuable process materials.
    4. Reexamining the degrees of treatment required to meet standards.
    5. Reevaluating to decide whether treatment is necessary.
  D. Detailed report on the engineering survey:
    1. Recommend a preliminary course of action.
    2. Advise management whether a waste-treatment plant is necessary.
    3. Describe the general type of plant required.
    4. Provide preliminary estimate of construction cost.
    5. Prepare preliminary estimate of operating costs.

II. Detailed engineering
  A. Process design and evaluation:
    1. Assign liaison and engineering personnel as required.
    2. Evaluate bench scale or pilot plant data.
    3. Translate the total evaluated data into process flow diagrams and functional specifications for the treatment plant.
    4. Prepare plot plan showing layout on plant site.
    5. Assemble an engineering report for review and approval.
    6. Obtain preliminary approval of regulatory agency.
  B. Definitive engineering:
    1. Prepare detailed engineering flow diagrams which form the basis of final plant design.
    2. Obtain approval of overall plant design.
    3. Complete the definitive design.
    4. Obtain final approval and permit of regulatory agency.

III. Construction and start-up
  A. Procurement and scheduling:
    1. Prepare complete equipment specifications, bills of material, and preliminary timetable.
    2. Prepare item delivery and installation schedule.
    3. Use critical-path scheduling when warranted.
    4. Coordinate and inspect all phases of the work performed by fabricators.
  B. Facilities erection and testing:
    1. Plan, supervise, and coordinate erection of the complete wastewater-treatment plant.
    2. Conduct unit tests, after assembly, to assure proper functioning of all related facilities.
    3. Inspect, adjust, and calibrate instruments and controls to conform to high accuracy standards, with engineers performing the work.
  C. Operator training:
    1. Prepare detailed operating manuals for all unit operations in the plant.
    2. Assemble vendors' manuals for use by plant personnel in maintaining, repairing, and replacing mechanical, instrument, and electric equipment parts.
    3. Assist with training of operating crews while construction work is in the final stages.
  D. Start-up of treatment facilities:
    1. Initiate a control testing program.
      a. Operational
      b. Quality of effluent
    2. Initiate an efficiency testing program.
    3. Establish conditions for operations.
    4. Initiate a development program.
    5. Establish record-keeping procedures.
  E. Supervise operation.

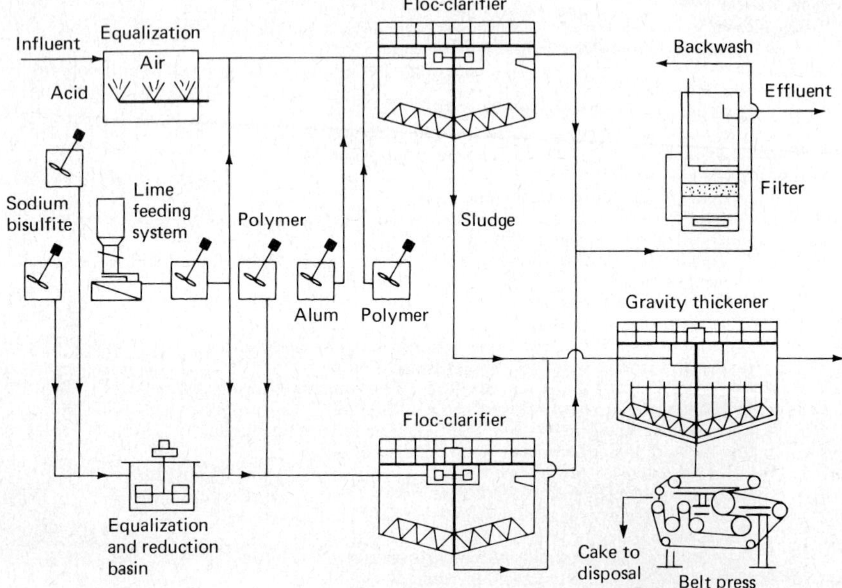

Figure 2-3 General manufacturing wastewater treatment.

TREATMENT

The necessity of effective industrial wastewater treatment must be considered an integral part of the manufacturing process, and the cost of treatment must be charged against the product.

The method of treatment depends on economic considerations and the degree of treatment required. The best alternative system for pollutant removal must be selected on the basis of a case-by-case study of efficiency and actual costs. It must be recognized that a complete system may involve several unit components and that pretreatment is required before tertiary treatment.

Figure 2-3 illustrates various treatment units combined to form a treatment system. Figure 2-4 shows the various treatment processes classified according to type, and illustrates how they can be combined to give the desired effluent quality.

Primary Treatment

The physical removal or combined chemical coagulation and physical removal of solids from wastewater is classified as primary treatment, especially if these processes are followed by biological treatment. Gravity or flotation units are used to remove suspended or coagulated colloidal material. Solids thus concentrated can be treated more economically by disposal, incineration, or biological degradation.

Oily waters are usually treated separately to remove the oil prior to mixing them with other waste streams. Chemicals can be used to enhance gravity separation of oil when emulsions are present.

Figure 2-5 presents a general guide for the design of a gravity separator based on parameters set forth by the American Petroleum Institute. It is desirable to have a minimum depth of 5 ft (1.5 m), a maximum horizontal flow-through velocity not exceeding 3 ft^3/min (0.015 L/s), and a minimum length-to-width ratio of 3:1 with a depth-to-width ratio of 0.3 to 0.5.

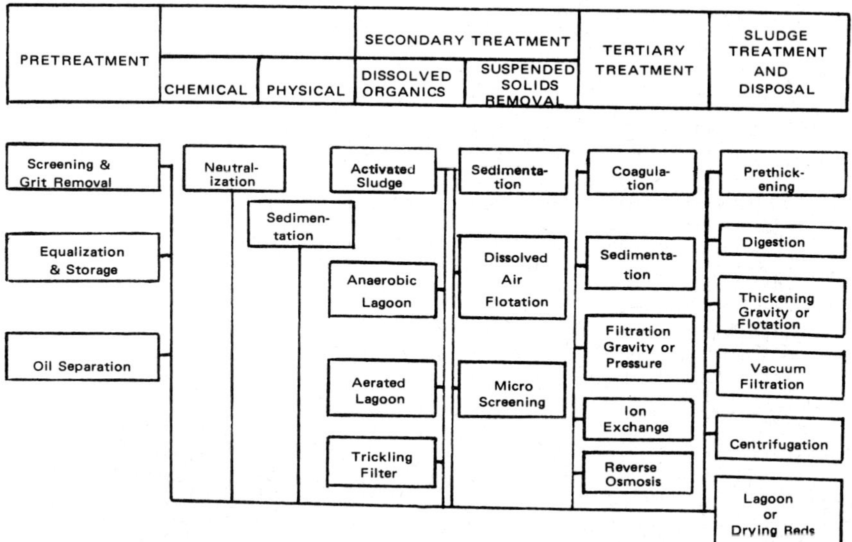

Figure 2-4 Alternatives for wastewater pollutant removal processes and how they may be combined in treatment programs.

Alkaline or acidic waste streams must be neutralized before secondary treatment or discharge.

Secondary Treatment

Secondary treatment is used to reduce soluble organic pollutants that are degradable to certain levels of organic materials remaining (the degradation products). If the effluent quality required is higher than that which can be obtained by biological treatment, tertiary processes are needed to remove the degradable with the undegradable fractions.

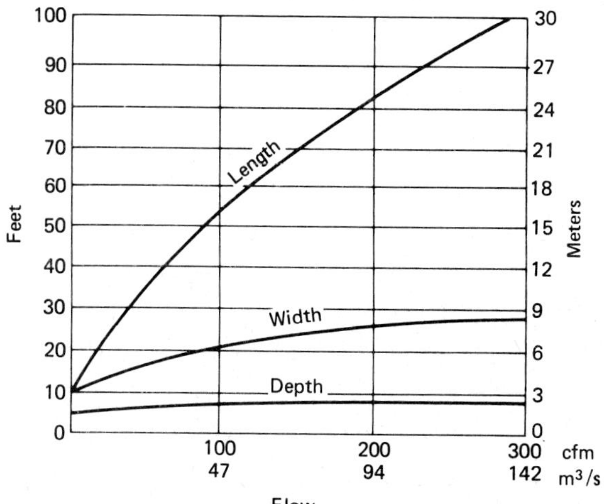

Figure 2-5 Design of gravity separator.

TABLE 2-4 Treatment-Process Removal Efficiencies

	Removal efficiency, %				
Treatment method	BOD_5	Suspended solids	Total dissolved solids	Fats, oils, & greases	Alkalinity
Screening	0–5	5–20	0	0	0
Sedimentation	5–15	15–6	0	5–15	0
Chemical precipitation	25–60	30–90	0–50	10–40	80–95
Dissolved-air flotation	10–30	70–85	10–20	80–95	10–20
Trickling filter	40–85	80–90	0–30	10–20	10–25
Nonaerated lagoon	30–70	30–70	30–80	0–40	10–20
Aerated lagoon	50–80	50–90	0–40	5–15	10–20
Activated-sludge	70–90	85–95	0–40	0–15	15–30
Filtration	80–85	30–70	10–60	0–10	10–20
Activated carbon	95–99	95–99	10	N/A	85–90
Ion-exchange	N/A	N/A	95–99	N/A	99
Reverse osmosis	N/A	N/A	99	N/A	99

Commonly used secondary-treatment processes include the completely mixed activated-sludge process, extended aeration, aerobic and aerated lagoons, trickling filters, and anaerobic and facultative waste stabilization ponds.

Table 2-4 is indicative of removal efficiencies for unit treatment processes.

EQUIPMENT

Various unit items of equipment are combined to provide the degree of treatment required.

Bar Screens

A mechanically cleaned bar screen, like the one shown in Fig. 2-6, is the simplest tool for removing debris or suspended matter which could damage equipment or disrupt the treatment process. All solids with larger than ¾- to 2-in (2- to 5-cm) bar rack openings are trapped on the upstream side and removed by moving rakes.

Clarifiers

Clarifiers used for removal of settleable solids and readily floating oils and greases are either rectangular or circular basins (Figs. 2-7 and 2-8). Clarifiers are sized on the basis of settling rate (area) and detention time (volume).

Typical overflow rates vary from 250 to 1400 gal/(day)(ft^2) [10 to 50 m^3/(day)(m^2)]. Detention time is in the range of 1 to 4 h.

Flocculation Systems

Flocculation is the agglomeration of finely divided suspended matter and floc caused by gently stirring or agitating the wastewater. The resulting increase in particle size increases the settling rate and improves suspended solids removal by providing more efficient contact between suspended solids, dissolved impurities, and chemical coagulants.

Mechanical flocculation uses paddles slowly rotating on a horizontal or vertical axis (Fig. 2-9). Peripheral speed at the paddle tip is about 1 ft/s. Various other mechanical devices are used to achieve the same result. An air flocculation system has diffusers along

Figure 2-6 Mechanically cleaned bar screen. (*Envirex.*)

one side of the basin near the bottom to produce a gentle rollover action perpendicular to flow.

The size of the required basin is determined by the detention time, which is normally in the range of 20 to 30 min at rated flow. In some cases involving industrial wastes, detention may be reduced to as little as 10 min.

Combined Equipment

Many clarifier designs, such as the solids contact unit (Fig. 2-10), combine mixing, flocculation, and coagulation in one basin. This may have economic advantages and may

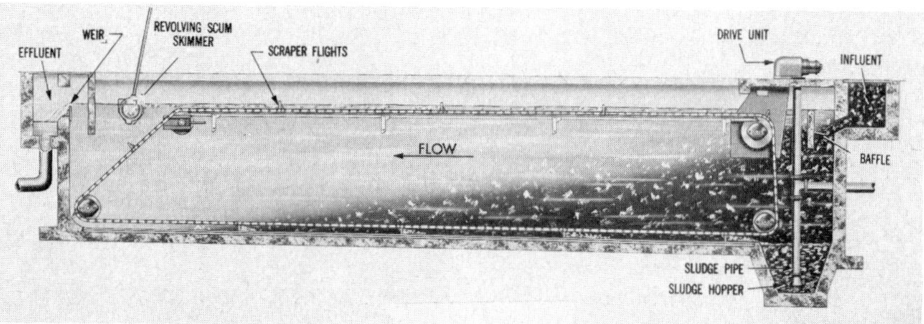

Figure 2-7 Clarifier for a rectangular basin. (*Envirex.*)

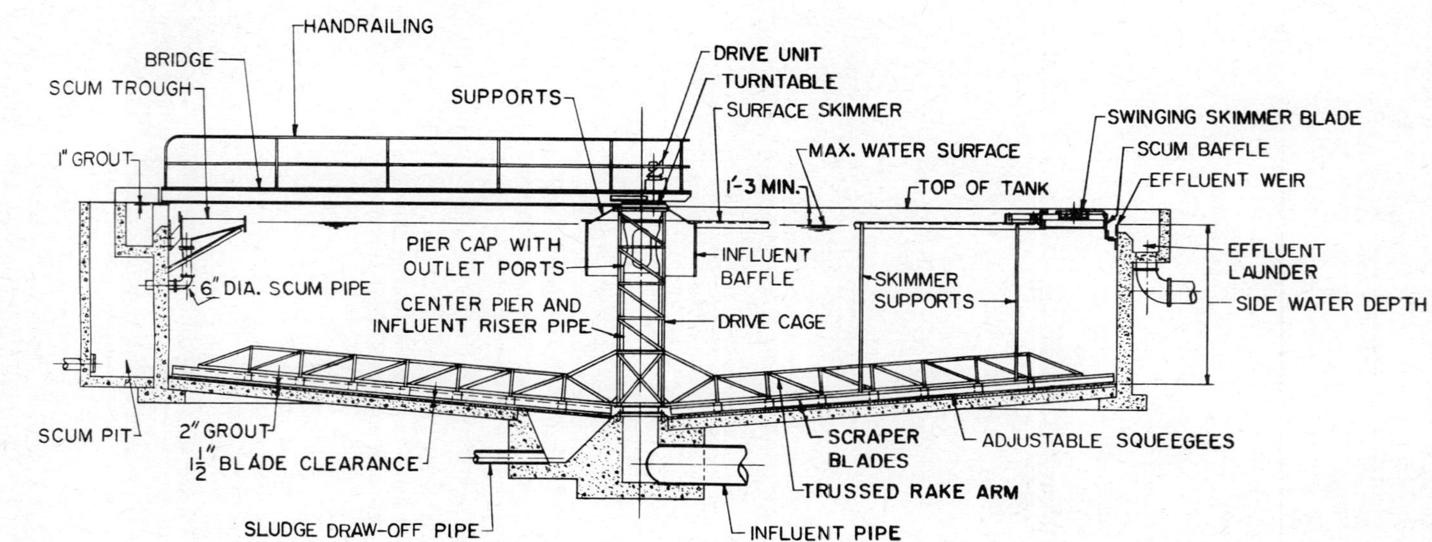

Figure 2-8 Clarifier for a circular basin.

produce better-quality effluent with shorter overall detention time than the approach using separate treatment units.

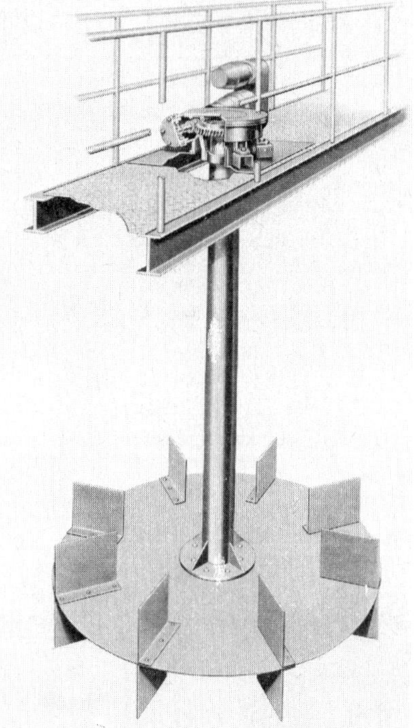

Figure 2-9 Mechanical flocculation device. (*Envirex.*)

Flotation Systems

Flotation is sedimentation in reverse to remove floatable materials and solids with a specific gravity so close to that of water that they settle very slowly or not at all. The principle of air flotation is based on the fact that when the pressure on a liquid is reduced, dissolved gases are released as extremely fine bubbles. These bubbles attach themselves to any suspended matter present and rapidly float them to the surface, where they concentrate and can be removed by skimming.

Pressure-flotation units dissolve air in the water under pressure and then release it to the atmosphere in the flotation tank.

Flotation equipment may be circular or rectangular.

Recycle Pressurization

The rectangular unit in Fig. 2-11 illustrates the use of recycle pressurization. Air is injected into the effluent recycle stream before it discharges into the inlet compartment of the flotation unit. There it is mixed with the incoming raw waste and releases the required air for flotation. The amount of effluent recycled varies from 25 to 50 percent of the forward flow. This approach has advantages when the raw waste is highly variable in composition or contains large amounts of solids.

Flotation units are normally designed for a feed-flow detention period of 15 min and an overflow rate of 1.5 to 4.0 gal/(min)(ft^2) [760 to 2025 L/(day)(m^2)]. Increased deten-

Figure 2-10 Solids contact clarifier. (*Envirex.*)

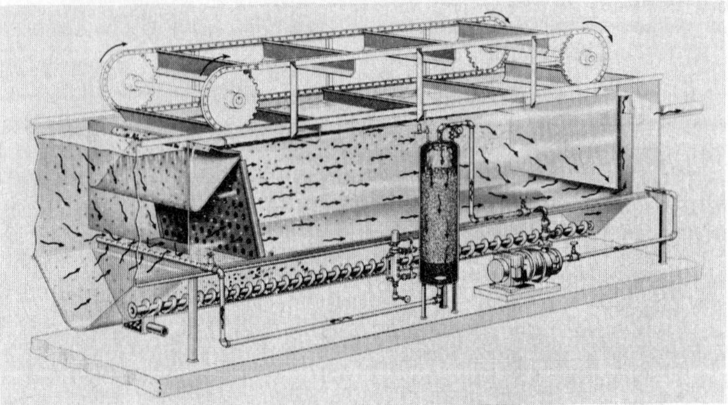

Figure 2-11 Rectangular flotation unit with recycle pressurization. (*Envirex.*)

tion time is needed if floated sludge must be thickened since surface area is based on the loading rate of solids.

High-Rate Gravity Filters

High-rate gravity filters remove suspended solids and operate in the range of 5 to 10 gal/(min)(ft^2) [2532 to 5063 L/(day)(m^2)]. They may be vertical or horizontal and filled with a variety of media such as anthracite coal, sand, and gravel.

PRODUCT RECOVERY

In many cases, industrial wastes contain valuable products such as high-value metals, acids, and other substances which can be used for manufacturing by-products, and these, when recovered, will yield high economic returns. Also obtainable are solvents, recovered with activated-carbon adsorption used for removal and recycling of solvents contained in the waste as vapors; these solvents include hydrocarbons, esters, alcohols, freons, ketones, and chlorinated or fluorinated organic compounds.

In the plating industry, rinse waters contain many of the following contaminants, usually in intolerable amounts: hexavalent chromium; sodium cyanide; complex cyanides of the heavy metals, such as cadmium, copper, zinc, and sometimes silver and gold; soluble nickel salts, strong mineral acids, and strong alkalis. Of these, the most toxic are the hexavalent chromium and cyanide ions.

Batch Treatment Method

There are several basic methods of treating such rinse waters for disposal or recovery of products. The oldest is the batch treatment method whereby the rinse waters are collected for treatment and disposal. The destruction of toxic chemicals is accomplished by oxidation of the cyanides and reduction of chomium. Chlorine (Cl_2) is used to destroy cyanides (CN^-), and the chlorinated rinse water is pumped into a reduction tank containing sulfuric acid and a reducing agent such as ferrous sulfate ($FeSO_4$), sodium bisulfite ($NaHSO_4$), or sulfur dioxide (SO_2). Chromium is reduced from its hexavalent state (Cr^{6+}) to the trivalent state (Cr^{3+}) and precipitated as a hydroxide, along with other metal hydroxides, by the addition of lime. The clear water from the clarifier is then ready for discharge into a sewer, and the sludge can be vacuum-filtered or disposed of in an acceptable manner.

Continuous-Flow Method

Another treatment of rinse water is the continuous-flow method. To be effective, this requires good instrumentation to monitor the flow and deliver the proper amount of reactant.

Ion Exchange

When the volume of chromic acid and sodium or potassium cyanide used is great enough, the recovery of the cyanides and chromic acid by ion exchange and evaporation has obvious economic advantages.

An integrated system for ion-exchange treatment of cyanide and chromium wastes has been found to be suitable for a plating operation that is split up into several parts. The system involves countercurrent rinsing, and the basic advantage of this system is that the toxic contaminants are destroyed before they can enter the rinse water, rather than having to be removed later.

Ion exchange plays an important part in chemical reclamation. The rinse waters are kept separate in the following categories:

1. Hard chrome rinse waters
2. Cyanide rinse waters containing silver, gold, or any other valuable metal
3. Miscellaneous wastes such as alkaline cleaners and acids

The hard chrome rinse waters are passed through a cation exchange and a strong-base anion exchanger. The contaminating metals such as copper, nickel, and trivalent chromium are exchanged on the cation exchanger, while the hexavalent chromium is exchanged on the anion exchanger. Upon exhaustion of the system, the cation exchanger is regenerated with an acid and the anion exchanger regenerated with caustic soda.

The regenerant effluent from the cation exchanger can be discharged to a collection chamber for neutralization. The regenerant effluent from the anion exchanger will contain sodium chromate and some caustic soda. This material can then be converted to chromic acid by passing it back through the cation exchanger so that the two exchangers may be alternated between service and conversion.

Table 2-5 shows recovery value of by-products from plating wastes.

An ion-exchange demineralization unit with a cation-anion mixed bed results in an effluent of less than 1 mg/L of total dissolved solids at a cost of roughly 5 cents per 1000 gal of waste treated. Membrane reverse-osmosis units can be used, but at substantially increased costs of five times or more than ion exchange.

Closed-Loop System

In the area of metal fabrication, parts are frequently washed or rinsed to clean away oils and other wastes. With a little care, improved wash operations cut downtime, reduce energy requirements, and cut soap costs. This is accomplished by a closed-loop wash-water treatment system.

TABLE 2-5 Economic Value of By-Product Recovery

Process	By-product	Concentration, mg/L	1980 value, $/1000 gal
Rinse water	Copper	100–500	0.85–4.26
	Nickel	150–900	1.14–9.94
	Zinc	70–350	0.57–3.40
	Cadmium	50–250	1.70–8.50
	Chromium	400–2000	4.50–21.30
	Tin	100–600	1.14–6.80
Aluminum, bright dip	Phosphoric acid	10%	630

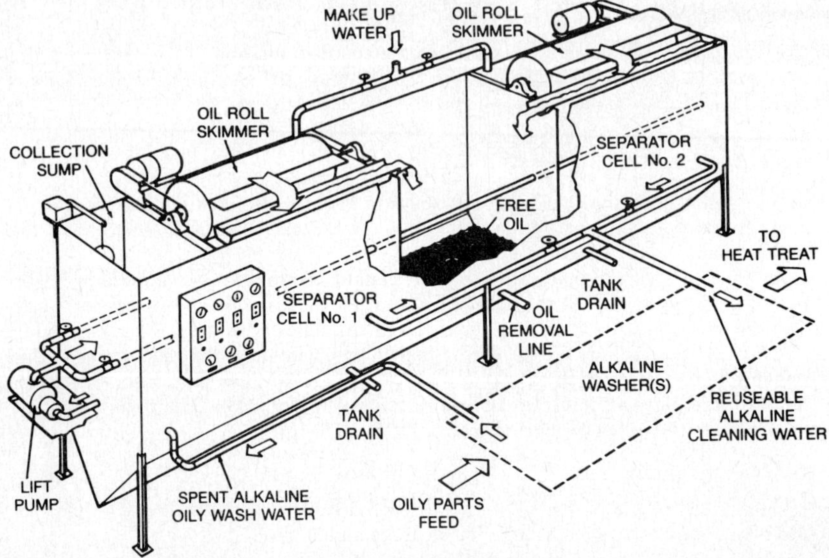

Figure 2-12 Alkaline wash system.

A typical closed-loop system (Fig. 2-12) is best described as a continuous-batch oil separator. It has dual compartments holding caustic wash solution, each equipped with an oil roll skimmer and separated by a waste tank. Piping leads from each compartment to a series of washers and back to a pump. Automated valves control flow from the pump to one of the two compartments.

One compartment continuously supplies caustic solution to a group of washers as the other stands for 24 h, allowing heavy materials to settle as oils float to the surface. Then, surface oils containing less than 0.1 percent water are skimmed off and drained into a waste tank; these may then be sold to an oil reclamation firm. While one wash solution in the first compartment undergoes treatment, the clean solution in the other compartment is circulated through the washers.

Treatment of Spent Coolants

Spent coolants can be treated for recovery of oil by acidification and centrifugation or by heat treatment at 223°F (106°C) for 22 h.

In the heat-treatment process, three distinct layers develop after there is water loss of approximately 12 percent through evaporation. The top layer consists of reusable oil, amounting to approximately 24 percent of the original scum volume; the middle layer, containing approximately 16 percent of the scum, is called the *rag layer* and consists of flocculated particulate matter; the bottom layer contains approximately 48 percent of the total scum and appears as relatively clear water. This bottom water layer is returned to the spent-coolant holding tanks for retreatment, and the rag, or intermediate, layer is combined with the swarfing dust and other heavy solids for removal by haulage.

A portion of the recovered oil can be used as the fuel for heating the floated scum, thereby making the system self-sustaining. The balance can be reprocessed for reuse as coolant. A simple product balance would show that for every 100 L of coolant treated, from 0.3 L to as much as 5 L of reusable oil can be recovered. A minimum flow of about 30,000 L of coolant per day might be selected as an economic break-even point.

At this minimum flow, taking, on an average, 3 L of recoverable oil for each 100 L of feed, 750 L per day of coolant would be recovered. Of this, approximately 50 L would be

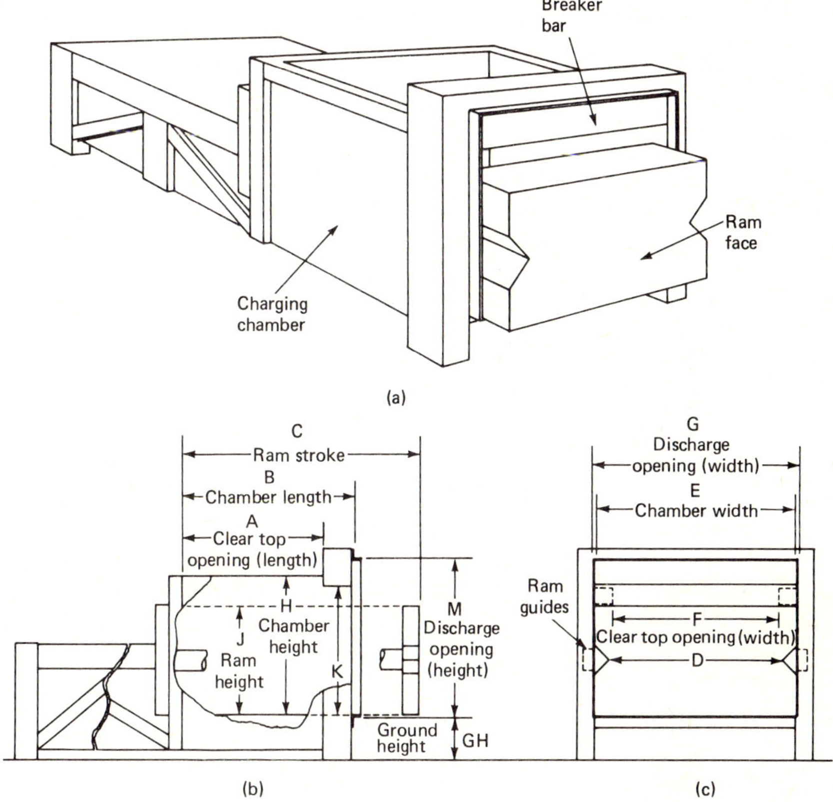

Figure 3-1 Horizontal commercial-industrial stationary compactor.

the charging chamber, (2) space for the container for the compacted wastes, and (3) space required for the pickup vehicle to maneuver, pull away the full container, replace an empty one, and then haul the full box away. Finally, sufficient room must be provided at the loading area to accommodate trash carts, containers, or even a conveyor chute leading to the hopper.

Dock space may be saved and plant security increased by installing the machine through the wall so that the charging area and the working mechanism remain inside the building, but the container is attached to the machine through an opening in the wall. Also, chutes from upper floors may run to the charging area, decreasing internal transportation costs and reducing the need for additional dock space.

Cycle Time. Cycle time is an important parameter of a container specification. It refers to the time required for a fully retracted ram face to pack the refuse from the charging box into the container and return to its original fully retracted position ready for another load of refuse.

Cycle times run from as short as 20 s to more than 1 min. Short cycle times are important if the application under consideration requires the capability of accepting refuse very quickly. A purchaser should carefully consider whether a compactor with a short cycle time is needed. Often compactors built with a rapid cycle utilize a small-diameter hydraulic cylinder which consequently exerts a relatively low force on the ram face, thereby sacrificing compaction pressure and material density.

Pressure of the Ram Face. A third important parameter of a stationary-compactor specification is the pressure of the ram face. Pressure, in pounds per square inch, is more important than total force in determining compaction density.

The ram pressure is the total force exerted on the ram, divided by the area of the ram. For a hydraulically operated compactor, the total ram force is the product of the hydraulic pressure and the hydraulic cylinder area.

For example, a compactor with a ram face measuring 30 x 60 in (0.8 x 1.6 m) has an area of 1800 in^2 (1.3 m^2). If it is acted upon by a 5-in-(12-cm) diameter hydraulic cylinder with a fluid pressure of 2000 lb/ft (14 x 10^6 N/m^2), the total ram force will be 39 200 lb (175 000 N), and the pressure exerted by the ram will be 21.7 lb/in^2 (140 000 N/m^2).

Most compactors have both a normal and maximum pressure rating. The higher pressure is used when finally packing out a container to make it easy to detach the container and ensure that the waste remains within it.

Penetration of the Ram into the Container. The penetration of the ram face into the container is an important factor in the operation of a compactor. Essentially, it represents the position of the forward ram stroke available for final compaction of the refuse load into the container. As the container begins to fill up with compacted waste, there is a tendency for the portion of refuse closest to the ram to fall back into the charging area. This is of most concern when the container is detached because waste falling from the container will litter the environment. The further a ram penetrates into the compaction container, the easier it is to load and detach a container cleanly.

Base Size. Compactor specifications are summarized as a single parameter, the base size. This is defined as the volume of waste theoretically moved through the compactor in a single stroke. Purchasers commonly use the base size as the primary specification, adding the parameters, listed above as necessary. Table 3-1 gives typical uses of compactors of various base sizes.

Rated stationary compactors manufactured by members of the WEMI carry the NSWMA Rating Seal. This seal assures performance conforming to established standards, certified by a registered professional engineer. Table 3-2 gives a hypothetical sample listing of rated compactors.

In addition to obtaining a compactor, the plant engineer will have to specify a container, which is usually provided by the refuse hauler. Before purchasing any equipment, compactor or container, the engineer should always consult the hauler to determine what type, size, and weight of containers can be handled. Some points to watch for are:

1. The container must be able to withstand the pressure of the waste without damage or distortion.
2. Roll-off and drop-off hoists, which are used to remove containers, have weight limits. The hoist will not be able to lift containers which exceed these limits.
3. State highway laws limit the amount of weight which can be carried on each chassis.
4. The length of the container affects the compaction ratio. The larger the container, generally the less the compaction.

There is no single formula for use in selecting a compactor for a plant. However, careful consideration of the tangible and intangible factors that have been outlined here will aid in avoiding costly mistakes.

TABLE 3-1 Typical Compactor Applications

Use	Base size	
	yd^3	m^3
High-rise apartment building	Up to 1	0.8
Commercial establishment	1 to 4	0.8 to 3
Industrial plants	1.5 to 7	1.2 to 5.5
Transfer stations	7 to 12	5.5 to 9

TABLE 3-2 NSWMA Commercial-Industrial Stationary Compactor Ratings

Manufacturer: Hypothetical Manufacturing, Inc. — Date: August 1977

Model number	NSWMA base size, yd^3	Clear top opening length (L) width (W), in	Chamber length, in	Ram stroke, in	Ram penetration, in	Force rating normal (N) maximum (M), (ram lb/in^2 per force, lb)	System pressures normal (N) maximum (M)	Ram face, in	Volume displacement rate, yd^3/h	Rated motor size, hp	Cycle time, s	Discharge opening width (W) height (H), in	Ground height, in	Base unit weight, lb
A	0.38	L = 24.0 W = 34.5	24.5	30.0	5.5	N = 25.3/18200 M = 27.8/20000	N = 1450 M = 1600	W = 36.0 H = 20.0	39	1	35	W = 37.0 H = 24.5	15.5	1800
B	1.06	L = 36.0 W = 51.5	39.0	50.0	11.0	N = 28.5/36200 M = 30.8/39100	N = 1850 M = 2000	W = 53.0 H = 24.0	106	10	36	W = 53.5 H = 32.5	16.0	4000
C	1.60	L = 40.0 W = 58.5	41.0	55.0	14.0	N = 21.6/38800 M = 24.3/43700	N = 2000 M = 2250	W = 60.0 H = 30.0	144	10	40	W = 61.0 H = 37.5	15.5	5100
D	1.91	L = 48.0 W = 58.5	49.0	65.0	16.0	N = 24.3/43700 M = 27.6/49600	N = 2250 M = 2550	W = 60.0 H = 30.0	146	10	47	W = 61.0 H = 37.5	15.5	5900
E	3.43	L = 84.5 W = 58.0	89.0	110.0	21.0	N = 55.8/100400 M = 61.4/110500	N = 2000 M = 2200	W = 60.0 H = 30.0	225	20	55	W = 61.5 H = 37.5	16.5	13000

Note: The National Solid Wastes Management Association presents this technical information on commercial-industrial stationary compaction equipment to assist users in the selection of equipment. The National Solid Wastes Management Association does not present these data as a warranty for equipment performance. The ratings represent data supplied by manufacturers that have been computed according to the NSWMA criteria and certified for accuracy by a registered professional engineer, selected by the manufacturer.

In-Plant Incineration

An alternative method of disposal is *incineration.* Although incinerators are more expensive to build and install than stationary compactors, their use can provide additional savings in refuse hauling costs. Energy recovery in the form of steam may also be considered as a source of additional savings.

When one is considering an incinerator, one should evaluate the following factors:

1. Any limitations imposed by air pollution emission limits in the area of the plant.
2. The physical constraints imposed by the dimensions of the plant.
3. The appropriateness in quantity and composition of the solid waste generated within the plant: a sufficiently high fired Btu value.
4. Fuel requirements. Most incinerators require supplementary fuel, either gas or oil, to control air pollution. Fuel costs can be high, and fuel availability may become uncertain. However, energy recovery can offset some of the costs.
5. By-product disposal: Adequate methods of disposal must be provided to handle the by-products of the incinerated waste.

Controlled-air incinerators (Fig. 3–2), the most frequently used incinerators today, were first commercially available in the United States in the early 1960s, but they were not really accepted until the late 1960s and early 1970s. This acceptance was mostly due to the increasing demands for high performance as measured by very low particulate emissions and a very high reduction ratio.

Most controlled-air incinerators employ two chambers. These chambers are desig-

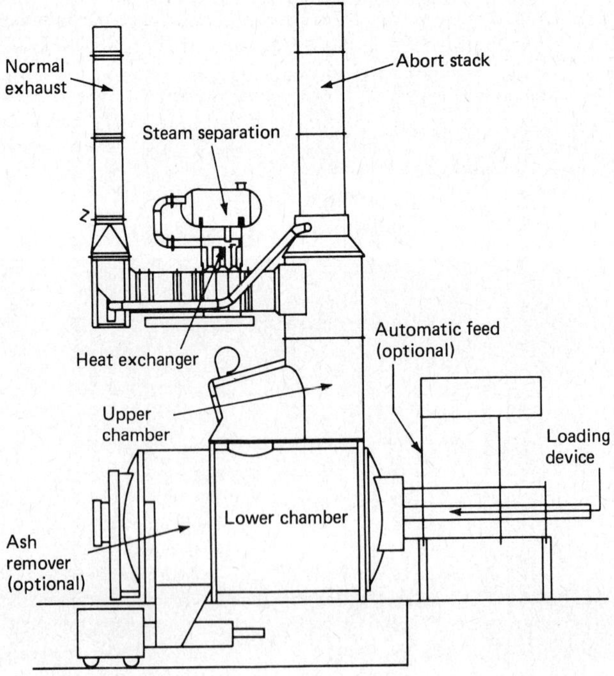

Figure 3-2 Controlled-air incinerator. (*Adapted with permission from the* American City & County Magazine.).

nated lower, or primary, and upper, or secondary. Performance of the antipollution functions of the system depends on controlling the conditions within these two chambers. The lower chamber is required to operate at low interior gas velocities and under controlled temperature conditions. This is done by limiting the air introduced into the primary chamber to less than the amount required for complete combustion (hence, the system is sometimes called "starved-air" incineration). This gives the lower chamber the operating characteristics of a partial oxidation system.

The heat released in the lower chamber is controlled by limiting the introduction of combustion air to an amount which will give partial oxidation of the waste in the chamber. The heat is sufficient to sustain the partial oxidation reactions. The gases from the lower chamber pass into the upper chamber through a turbulent zone, where additional air is added and ignition takes place to complete the oxidation reactions. The noncombustible portion of the waste and the carbonaceous residue from the reactions remain in the lower chamber. The noncombustibles are rendered sterile by the relatively high temperature while the carbonaceous material is further oxidized by the incoming air. The result is a high-quality sterile ash.

The gas velocity in the lower chamber is influenced by several factors. The gas which evolves from the chamber is a result of the interaction of the air, the auxiliary fuel, and the oxidation and volatilization products from the waste. The quantity of gas from the waste can vary substantially depending on chamber conditions of the waste and could therefore alter the gas velocity in the lower chamber significantly. The airflow controls of the upper and lower chambers are integrated in order to minimize cycling and provide a uniform flow of gases. This is important for controlling pollution performance and especially so for an efficient energy-recovery system.

When volatilization proceeds at an excessive rate as a result of the high temperatures, two distinct adverse effects ensue. First, the velocity in the lower chamber will exceed the design velocity, and particles which are too large to be oxidized properly will be carried into the upper chamber. Second, the gases will flow to the upper chamber at a rate which exceeds the capacity of the chamber and can result in excessive particulate emissions or smoking.

The function of the upper chamber is to complete the oxidation reactions of the combustible products as they are received from the lower chamber. In order to accomplish this, conditions in the chamber must be controlled within a rather narrow band, from inputs which vary rather widely. The control system is designed to maintain the required conditions by modulating both air and fuel to the system. This in effect controls the air input, auxiliary fuel input, and gas flow from the lower chamber.

The gases pass from the lower chamber into the upper chamber through a turbulent mixing region in a controlled manner. Additional air is introduced into the system and the gases are ignited, again under controlled conditions. The gas temperature at this point is somewhat higher than in the lower chamber, and the atmosphere is oxidizing (more than sufficient air for complete combustion). Temperatures in the upper chamber are limited to less than 2500°F (1400°C) in order to minimize production of nitrogen oxides and in the interest of equipment durability. On the lower-temperature side, it is recommended that at least 1800°F (1000°C) be maintained in order to stabilize an adequate reaction rate to complete the combustion process. The desired operating temperature point is adjustable but is factory preset for maximum performance. The primary means of controlling the temperature in the upper chamber is to control the quantity of combustion air. Air quantity is decreased when the temperature drops below the set point and increased when the temperature rises above the set point.

If heat is to be recovered, the gas temperature at the inlet to the heat exchanger is not allowed to exceed 1800°F (1000°C). This is done to protect the heat exchanger and is an addition to the normal safety controls. An over-temperature condition will automatically drive the hot-gas flow to the abort stack.

Compared with compaction, incineration is a more costly and more complex way of disposing of waste; however, the energy-recovery potential may in the long run prove to be a decisive factor for incineration in many plants.

Hazardous Waste Disposal

Disposal of hazardous wastes has become a serious problem for many industrial plant engineers. This problem is especially critical for chemical industries, and most such companies now have entire departments devoted to this concern. Many other industries also have hazardous wastes to dispose of, generally in lesser amounts.

Transportation, treatment, storage, and disposal of hazardous wastes are now very tightly regulated under provisions of Title C of the Solid Waste Disposal Act. Even though generators of small quantities of hazardous wastes are exempted from the full scope of the regulations, it is not a safe practice merely to intermingle hazardous wastes, even in small quantities, with other solid waste without taking special precautions.

To arrange for disposal of hazardous wastes, a plant engineer should contact a waste service company that is fully permitted to transport and dispose of those wastes. The company should provide a written statement certifying where the wastes are to be taken and how they are to be treated or disposed of. The disposal facilities should be inspected and the actual waste disposition verified.

The hazardous-waste generator will be requested to initiate a manifest for shipments of hazardous wastes going off-site for disposal and will also be required to designate where the wastes are to be taken.

Transportation of hazardous wastes, inter- or intra-state, is regulated by the Department of Transportation and the Environmental Protection Agency under both the Hazardous Materials Transportation Act and the Solid Waste Disposal Act. The plant engineer should be very sure that the transporter chosen is in full compliance with these regulations.

section 8

Plant Safety and Sanitation

chapter 8-1

Safety Considerations in Plant Operations

by

Arthur Spiegelman, P.E.
M & M Protection Consultants
A Technical Service of Marsh & McLennan Inc.
New York, New York

SAFETY

A well-organized plant safety program should encompass all phases of the plant environment and operations. From the standpoint of this handbook we are primarily concerned with safety controls and devices in the environment and only make reference to supervisory controls of the safety program.

Safeguarding industrial hazards has always been one of the principal tasks of the plant engineer. Production depends on the ability to maintain a continuous flow of materials without the interruptions caused by accidents. Engineering controls and safeguards serve to protect inexperienced workers or those workers who are distracted or who suffer from fatigue.

The basic elements of a good safety program include:

1. Management leadership
2. Assignment of responsibility
3. Maintenance of safe working environment
4. Training program
5. Record system
6. Medical follow-through

The plant engineer is basically concerned with items 1, 2, and 3, but items 4, 5, and 6 are also important.

LEGAL ASPECTS OF SAFETY

The early legal aspects of industrial safety were limited to laws which were developed in conjunction with workers' compensation acts. They were primarily aimed at providing for just compensation to the injured party, but they also established accident investigation procedures and some regulation of hazards.

The past two decades have sent a greater public demand for safety in the workplace. With it came a proliferation of federal laws and regulations which include the following.

Laws Applicable to Industrial Plants

1. Environmental Protection Agency
   - Toxic Substances Control Act of 1976
   - Marine, Protection, Research and Sanctuaries Act of 1972
   - Safe Drinking Water Act of 1974
   - Water Pollution Control Act of 1972
   - Clean Air Act of 1970
   - National Environment Policy Act of 1969
   - Atomic Energy Act of 1954
   - Clean Water Act
   - Endangered Species Act of 1973
   - Energy Supply and Environmental Coordination Act
   - Fish and Wildlife Coordination Act
2. Solid Waste Disposal Act of 1965
3. Occupational Safety and Health Act of 1971
4. Department of Transportation
   - Transportation Safety Act
   - Hazardous Materials Transportation Act of 1974
   - Ports and Waterways Safety Act of 1972
5. Resource Conservation and Recovery Act of 1976
6. Federal Insecticide, Fungicide, and Rodenticide Act of 1972
7. Consumer Product Safety Commission
   - Federal Hazardous Substances Act
   - Consumer Product Safety Act
   - Poison Prevention Packaging Act
8. Mine Safety and Health Act

The Occupational Safety and Health Act will probably have a far greater effect on business and industry than the other legislation. The law requires all employers in the private sector of business to furnish their employees with a workplace free from those recognized hazards likely to cause physical harm or death. Standards are published or referenced in the act. This 800-page document is entitled *OSHA Safety and Health Standards* (29 CFR 1910). It is available from the Superintendent of Documents, Washington, DC 20402, and is recommmended as a reference for plant engineers.

The adoption of specific federal legislation has also led to an increase in the number of civil suits. The violation of a federal standard may become prima facie evidence of negligence. Such cases have been very costly.

Recent developments at OSHA have placed greater emphasis on the health aspects of worker's compensation. What started in specialized industries, such as coal mining and asbestos, is spreading to the chemical industry and others. Much more of the social responsibility for the well being of the employee is being directed to the employer.

All employers must keep accurate records of work-related injuries, illnesses, and deaths. Any injury that involves medical treatment, loss of consciousness, restrictions of motion or work, or job transfer, must be recorded. Required information must be logged and specific information regarding each case detailed. At the end of the calendar year a summary of logged cases must be posted in each plant.

INSTRUMENTATION AND CONTROLS

Some of the most important advances in safety over the past two decades have been achieved in the area of instrumentation and controls. Standards of the Instrument Society of America (ISA) should be followed. The plant engineer is directly concerned with these devices since they are involved in controlling the operations, materials flow, and environment of the plant. Instrumentation and controls fall into the following categories:

1. Temperature systems
2. Pressure systems
3. Flow systems
4. Analytical and testing systems (i.e., gas analysis, solids and dust analysis, reaction product tests, oxygen analysis, pollution instrumentation, electronic components, and system checkout equipment)
5. Weighing, feeding, and batching systems
6. Force and power systems
7. Motion and geometric systems (i.e., speed, velocity, vibrations)
8. Humidity and moisture determinations
9. Radiation systems
10. Electrical determinations
11. Communication systems
12. Automatic process controllers (i.e., computer process control, safety in instrumentation and control systems, electronic, pneumatic and hydraulic control devices)
13. Controlling elements (i.e., valves, actuators, electric motor drives and controls, mercury switches, etc.)

The importance of instrumentation and controls cannot be overestimated. From the standpoint of safety and environmental control the following applications are dominant:

1. Detection of leaks in equipment
2. Survey of operating areas for escape of toxic materials
3. Detection of flammable or explosive mixtures in the atmosphere or process lines
4. Monitoring plant stacks and other areas for the accidental discharge of toxic gases, vapors, or smokes.
5. Analysis of waste streams for toxic or other objectionable material.
6. Control of waste-treatment or product-recovery facilities

Automatic process control, unlike manual control, provides continuous monitoring and corrective action. However, automatic controllers cannot react to new conditions, nor can they predict beyond the data programmed into them. Therefore, the ability to foresee an upset condition in a process and the capability of developing fail-safe actions may well determine whether a serious accident is averted.

Analytical and testing instrumentation provides the plant engineer with much needed information. For instance, decisions can be made in these areas:

1. Purity of raw materials entering a process. Contaminants may be hazardous.
2. Process control by automatization of sampling.
3. Process troubleshooting by continuous analysis to prevent process upsets.
4. Determination of product quality to provide an impetus to a product safety program.

PLANT ENGINEER'S FUNCTION WITHIN THE SAFETY COMMITTEE

A safety program is a management method to develop specific safety objectives, assign responsibility, and obtain desired results. The plant engineer should participate with the safety committee from the standpoint of developing adequate inspection and control procedures in order to maintain a safe work environment and control of safety devices. Safety devices may be anything from a pressure-relief valve to a machine guard.

Inspection and Maintenance

In many plants an inspection committee is appointed. A safety inspection may be done in conjunction with the regular plant inspection and maintenance program. This is an opportunity for the plant engineer to check all safety devices. The safety inspection should focus attention on those items directly concerned with accident prevention. The plant engineering staff should have a working knowledge of safety standards if their inspections are to cover safety as well as normal wear and tear of equipment. A general plant inspection should include:

1. All buildings and physical equipment
2. Inspection of new machinery before it is placed in operation
3. Inspection of walking and working surfaces and means of egress
4. Special equipment such as powered platforms, personnel lifts, and vehicle-mounted work platforms
5. Materials handling and storage facilities, elevators, cranes
6. Machinery, machinery guards, and electrical equipment.
7. Compressed gas and air equipment
8. Special equipment such as pressure vessels, drums and furnaces, and welding, cutting, and brazing equipment
9. Hand and portable powered tools and other hand-held equipment
10. Environmental control, equipment, ventilation, and pollution controls for toxic and hazardous substances

A systematic method of inspection and maintenance is preferred. Some companies require *preventive* maintenance programs which call for the replacement of *critical* equipment periodically regardless of inspection results. Others require general inspections of the entire premises, annually.

Areas of the plant that have the potential of developing into catastrophic hazards require special inspection procedures. Items that may develop into a catastrophe include: (1) structural failure, (2) fire or explosion, (3) release of hazardous gases or vapors.

Some elements of a plant may require inspection and maintenance on the basis of federal, state or local laws. Items such as elevators, boilers, and unfired pressure vessels are in this category. This equipment may require the use of specially trained and licensed inspectors either from a governmental agency or an insurance carrier.

A careful record should be kept of all inspections and recommendations. This is particularly important in the event that an accident occurs at this location which results in litigation. Some companies assign inspection tasks to maintenance personnel, electricians, and others who are charged with the job of repairing the equipment. Supervisors should continuously examine their own work areas to make sure that tools, machinery, and other types of equipment are safe to handle.

Inspection methods should be established for all new equipment and processes. Nothing should be placed into operation until all safeguards have been checked and operations evaluated by the plant engineer. Safe operating instructions should be given to all workers concerned.

Inspection Procedures

The following inspection criteria should be applied:

1. Inspectors must be familiar with the company's safety and health policies as well as the particular laws and regulations that pertain. Frequently, these regulations are

only minimum requirements and it may be necessary to exceed them to secure adequate safety.

2. Inspectors should have available an analysis of all accidents that have occurred in the plant within the past year.
3. The inspectors should utilize all aids available, including inspection checklists, report forms, and other pertinent information.

An inspection report should be divided into three areas of interest:

1. A report on imminent hazards which require immediate corrective action.
2. A routine report on unsatisfactory (nonemergency) conditions which need corrective action.
3. A general report on the overall safety conditions of the facility.

ACCIDENT PREVENTION

The four basic steps for preventing accidents are as follows:

1. Elimination of the hazard
2. Control of the hazard
3. Training of personnel to be aware of and avoid the hazard
4. Utilization of personal protective equipment

The plant engineer is primarily concerned with steps 1 and 2 by ensuring a safe design for plant, physical facilities, and machinery.

BUILDING STRUCTURE

The plant engineer is directly concerned with the inspection and maintenance of the building structures. Slippery floors, stairs, runways, ramps, and other means of access are involved in many serious plant accidents. About one-fifth of the industrial injuries result from falls. Many of these can be avoided by careful design, construction, and maintenance. Applicable standards and codes should be applied.

Basic Rules

Housekeeping is a prime consideration in minimizing the hazard of falls. Some of the basic rules in this area are as follows:

1. All places of employment should be kept clean and orderly. Floors shall be maintained in a clean and dry condition with adequate drainage.
2. All passageways including aisles, ramps, and stairways should be kept clear and in good repair.
3. Floor loading shall be kept well within prescribed limits.
4. All floor and wall openings should be guarded by standard railings or properly constructed closures.
5. Stairways should conform to acceptable standards.
6. Exits should be sufficient in number and properly located so that the building can be evacuated quickly in an emergency.
7. All ladders should conform to the applicable standards and codes and they should be properly maintained.
8. Ladders should not take the place of fixed stairways.
9. Workers should follow good practices in the use of ladders with regard to placement,

support, angle between horizontal base and vertical plane of support, and proximity to other hazards (i.e., electrical).

10. Scaffolds, which are in effect elevated working platforms, should be designed with an adequate factor of safety and protection for the workers.
11. Scaffolds must be maintained, inspected, and guarded on all exposed sides. Metal scaffolds should not be constructed near electrical equpment.

Standards

The following standards apply to building structures:

ANSI Z4.1-1979, "Requirements for Sanitation in Places of Employment"

41 CFR 50-204.2 *Code of Federal Regulations,* "Public Contracts and Property Management," General Safety and Health Standards.

ANSI A58.1-1972, Building Code Requirements for "Minimum Design Loads in Building and Other Structures"

ANSI A12.1-1973, "Safety Requirements for Floor and Wall Openings, Railings, and Toeboards"

ANSI A64.1-1968, "Requirements for Fixed Industrial Stairs"

ANSI A14.1-1975, "Safety Requirements for Portable Wood Ladders"

ANSI A14.2-1972, "Safety Requirements for "Portable Metal Ladders"

ANSI A14.3-1974, "Safety Requirements for Fixed Ladders"

ANSI A10.8-1977, "Safety Requirements for Scaffolding"

ANSI A92.1-1977, "Manually Propelled Mobile Ladder Stands and Scaffolds"

MEANS OF EGRESS FOR INDUSTRIAL OCCUPANCIES

A means of egress is a continuous and unobstructed way to exit from any point in a building or structure to a public way. It includes vertical and horizontal ways of travel and intervening room spaces and other areas. The number of exit facilities required is specified in standards and codes, and it depends on the structure, occupants, and hazard exposures.

Inspections of means of egress have shown the following items at fault in such facilities:

1. Improper number of exits and locations.
2. Poor illumination—lack of emergency lighting.
3. Lack of directional signs.
4. Poor housekeeping—obstructions.
5. Improper and hazardous floor surfaces.
6. Faulty operation of exit doors.

Standard

The following standard applies to egress for industrial occupancies:

NFPA 101 1970, "Life Safety Code"

POWERED PLATFORMS, PERSONNEL LIFTS, AND VEHICLE-MOUNTED WORK PLATFORMS

Powered platforms, personnel lifts, and vehicle-mounted work platforms are means of elevating workers on suspended operated work platforms or personnel lifts. The platforms are used for exterior building maintenance work, the personnel lifts are used to lift workers to an elevated jobsite, and the vehicle-mounted work platforms are used to position personnel to an elevated work site. All of these installations should be in conformance with the appropriate ANSI standards.

Every work platform should be tested and inspected frequently. New platforms should be tested before they are placed in service. Each installation should be inspected and tested at least every 12 months and should undergo a maintenance inspection and test every 30 days. Results of all inspections and tests should be logged, including the date, time, and inspector.

Items which require special inspections are as follows:

1. Governors
2. Initiating devices
3. Both independent braking systems
4. Interlocks and emergency electric devices
5. Electric systems
6. Emergency communication system

Maintenance is required specifically on all parts of the equipment related to safe operations. Broken or worn parts and electric devices should be replaced promptly. Gears, shafts, bearings, brakes, and hoisting drums should be maintained in proper alignment. Gears should be replaced when there is evidence of appreciable wear. All parts should be kept free from dirt. Wires or ropes should be replaced when they are damaged or in a deteriorated condition. Guardrails and toeboards are required for working platforms. Load-rating plates must be conspicuously posted and adhered to.

The personnel lifts require frequent inspections. Their use should be limited to *personnel.* They should not be used to lift construction materials. The inspections should include but are not limited to the following items:

Steps	Drive pulley
Rails and supports	Electrical systems
Belt and belt tension	Vibration—alignment
Handholds	Brake systems
Floor landing	Warning lights
Limit switches	Pulleys—clearance

Standards

The following standards apply to platforms, personnel lifts, and vehicle-mounted work platforms:

ANSI A120.1-1970, "Safety Code for Powered Platforms for Exterior Building Maintenance"

ANSI A9.2-1969, "American National Standard for Vehicle-Mounted Elevating and Rotating Work Platforms"

ANSI A90.1-1969, "Safety Code for Manlifts"

VENTILATION

Ventilation systems protect the health and environment of plant workers by removing objectionable dusts, fumes, vapors, or gases. They are essential safety devices and must be properly installed and maintained. Ventilation systems can be divided into two primary groups, local systems and general systems.

Local Systems

Local exhaust systems prevent the accumulation of toxic or flammable materials near the process unit. Due to the variety of work and the types of equipment, it is necessary to design hoods specifically developed to be as close to the operation as possible. Air-sampling devices are available to determine whether the atmosphere has been adequately cleared or constitutes a hazard. Safety standards may specify local exhaust ventilation for particular processes.

After contaminated air passes from the hood to exhaust ducts, it is sent through a cleaning system or to the outdoors. Such exhaust air must conform to the regulations of the EPA. Dusts may be collected by means of cyclone dust collectors (Fig. 1-1) or electrostatic precipitators. Many types of air-cleaning devices are used. The selection is determined by the degree of the hazard associated with the dust. Gases or vapors may be removed from the waste air by absorption or adsorption processes which dissolve or

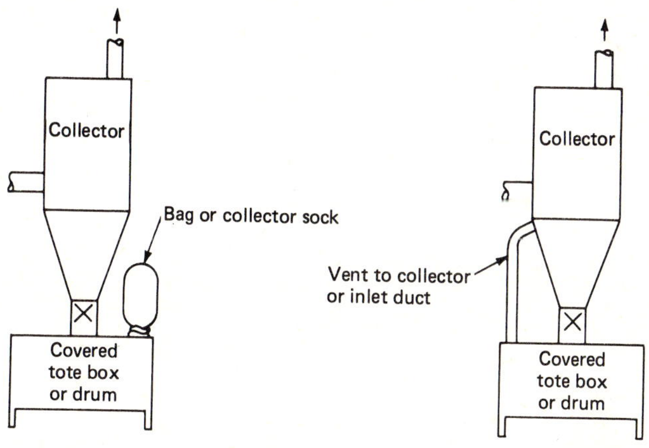

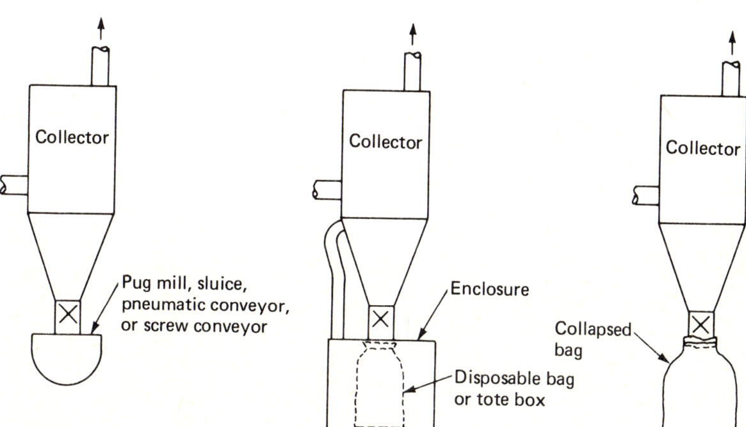

Figure 1-1 Dry-type dust collectors for dust disposal. *(American Conference of Governmental Industrial Hygienists.)*

react with the waste product chemically. Some gases and vapors are rendered harmless by passing them through a combustion chamber where they are changed to acceptable gases. In other cases condensers are utilized to liquefy toxic vapors for removal.

When ventilation is used to control potential exposure to workers it is adequate to reduce the concentration of contaminant so that the hazard is removed.

Local exhaust ventilation is used for a great many industrial operations such as anodizing, pickling, metal cleaning, and open-surface tank operations. There is a standard developed for this purpose. Other uses include spray booths, dip tanks, fume controls in electric welding grinding operations, cast-iron machining, dust control in foundries, ventilation of internal combustion engines, and kitchen range hoods.

One of the basic maintenance items in the operation of local exhaust ventilation systems is to ensure that the flow of air is unobstructed. A minimum maintained velocity must exist in order to meet the health and safety requirements. Where flammable gases or vapors are removed, the electric equipment must conform to code requirements.

At intervals of not more than 3 months the hood and duct system should be inspected for evidence of corrosion, damage, or obstruction. In any event if the airflow is found to be less than required, it should be increased to that required.

General Systems

General ventilation in the workplace contributes to the comfort and efficiency of the employees. It also serves to clear the air of hazardous contaminants and excessive heat or humidity. When local exhaust ventilation cannot be applied due to the many sources of vapor release, general ventilation is prescribed. The publication of the American Society of Heating, Refrigerating and Air Conditioning Engineers, *Handbook of Fundamentals*, defines the prescribed methods and requirements for the development of acceptable ventilation systems. The plant engineer should inspect this equipment frequently to make sure that it performs as required.

Standards

The following standards pertain to ventilation:

ANSI Z9.4-1979, "Ventilation and Safe Practices of Abrasive Blasting Operations"

ANSI Z43.1-1966, "Ventilation Control of Grinding, Polishing, and Buffing Operations"

ANSI Z9.3-1964 (R1971), "Safety Code for Design, Construction and Ventilation of Spray Finishing Operations"

ANSI Z9.1-1977, "Practices for Ventilation and Operation of Open-Surface Tanks"

41 CFR 50-204.10 *Code of Federal Regulations*, "Public Contracts and Property Management," General Safety and Health Standards: Occupational noise exposure.

41 CFR 50-204.22 *Code of Federal Regualtions*, "Public Contracts and Property Management," Radiation Standards: Exposure to airborne radioactive material.

COMPRESSED GASES

Plant engineers should be aware of the safety requirements for compressed-gas handling and storage. Pressure-relief devices for gas cylinders, portable tanks, and cargo tanks should be installed and maintained in accordance with Compressed Gas Association (CGA) pamphlets S1.1-1963 and S1.2-1963.

References

The following references pertain to compressed gases:

CGA G-1-1966, "Cylinders"

CGA G1.3-1959, "Piped Systems"
CGA G1.4-1966, "Generators and Filling Cylinders"

MATERIALS HANDLING AND STORAGE

Powered Industrial Trucks

Safety requirements for powered industrial trucks involve the following:

1. The plant engineer is directly concerned with the safety requirements relating to fire protection: design, maintenance, and use of fork trucks, tractors, platform lift trucks, motorized hand trucks, and other specialized industrial trucks.
2. All new industrial trucks should meet the requirements of ANSI B56.1-1969, "Powered Industrial Trucks."
3. In locations used for the storage of hazardous liquids or liquefied or compressed gases, only approved trucks designated as DS, ES, GS, or LPS may be used. In areas containing combustible dusts approved EX or ES trucks may be used.
4. Any power-operated industrial truck not in safe operating condition should be removed from service.

Overhead and Gantry Cranes

Safety requirements for overhead and gantry cranes include:

1. All new and existing overhead and gantry cranes should meet the design specifications of ANSI B30.2.0-1976, "Safety Standard for Overhead and Gantry Cranes (Top Running Bridge, Multiple Girder)."
2. Exposed moving parts such as gears, set screws, projecting keys, chains, chain sprockets, and reciprocating components which might constitute a hazard under normal operating conditions should be guarded.
3. Both holding brakes and control brakes should be tested to determine whether they are within required tolerances.
4. Brakes on trolleys and bridges should be examined and adjusted when necessary.
5. Inspections fall into two classes: *frequent inspections,* which fall into daily or monthly intervals, and *periodic inspections,* which fall into 1- to 12-month intervals. *Frequent inspections* should include (*a*) all functional operating mechanisms for maladjustments interfering with daily operations, (*b*) deterioration or leakage in air or hyraulic systems, (*c*) hooks and hoist chains and ropes including end connections, (*d*) all functional operating mechanisms. *Periodic inspections* should include (*a*) all items listed under frequent inspections, (*b*) deformed, cracked, or corroded members, (*c*) worn, cracked, or distorted parts, (*d*) excessive wear on brake system, (*e*) improper performance of power plants, (*f*) excessive wear of chain drive sprockets and excessive chain stretch, (*g*) defective electrical apparatus.
6. Preventive maintenance based on crane manufacturers' recommendations should be established.
7. Any unsafe conditions—including those in ropes—disclosed in the inspection requirements should be corrected before operation of the crane is allowed.

Crawler, Locomotive, and Truck Cranes

Safety requirements for crawler, locomotive, and truck cranes include:

1. All new and existing locomotive and truck cranes should meet the design specifications of ANSI B30.5-1968, "Safety Code for Crawler, Locomotive, and Truck Cranes."

2. Load ratings should not exceed the stipulated percentages for cranes with indicated types of mountings.
3. Inspections are frequent and periodic as cited under overhead gantry cranes.

Derricks

All new derricks constructed after August 31, 1971, should meet the design specifications of ANSI B30.6-1977, "Safety Code for Derricks." Those derricks constructed prior to that date should be modified to conform. The plant engineer should check the standard to assure compliance.

Inspection procedures are similar to those required for cranes under frequent or periodic inspections. Derricks not in regular use for a period of 6 months should be given a complete inspection before use.

Maintenance and repair is an important part of the plant engineer's assignment and should include the following steps:

1. Any unsafe conditions shown in the inspection should be corrected.
2. Adjustments should be made to assure correct functioning of all components.
3. Ropes and wires should be thoroughly inspected and replaced or repaired.
4. No derrick should be loaded beyond the rated load.

Helicopter Cranes

Helicopter cranes should comply with the applicable regulations of the Federal Aviation Administration. The weight of an external load should not exceed the helicopter manufacturer's rating. All equipment should be checked frequently. The cargo hooks should be tested prior to each day's operation to determine that the release functions properly, both electrically and mechanically.

Slings

All slings used in conjunction with materials handling equipment made from alloy-steel chains, wire rope, metal mesh, natural or synthetic fiber rope, and synthetic web (nylon and polypropylene) should conform to the applicable standards. They should be inspected daily for damage or defects.

Standards

The following standards pertain to materials handling and storage:

41 CFR 50-204.3 *Code of Federal Regulations,* "Public Contracts and Property Management," General Safety and Health Standards: Material Handling and Storage.

NFPA 231-1970, "General Indoor Storage"

NFPA 505-1969, "Powered Industrial Trucks"

ANSI B56.1-1975, "Safety Standard for Powered Industrial Trucks—Low Lift and High Lift Trucks"

ANSI B30.2.0-1976, "Safety Standard for Overhead and Gantry Cranes"

ANSI B30.5-1968, "Safety Code for Crawler, Locomotive, and Truck Cranes"

ANSI B30.6-1977, "Safety Code for Derricks"

MACHINERY AND MACHINE GUARDING

One or more methods of machine guarding should be provided to protect the operator and other employees in the machine area from hazards such as those created by point of operation, ingoing nip points, rotating parts, flying chips, and sparks. Examples are barrier guards, two-hand tripping devices, electronic safety devices, etc.

General

The following machines usually require point-of-operation guarding:

Guillotine cutters
Shears (Figs. 1-2 and 1-3)
Alligator shears
Power presses (Fig. 1-4)
Milling machines
Pointer saws
Jointers
Forming rolls and calenders
Revolving drums, barrels, and containers
Exposed blades

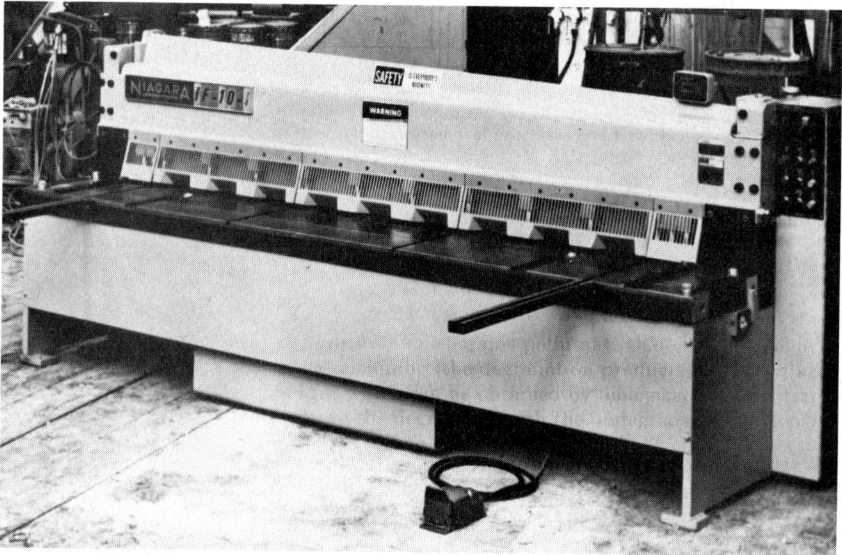

Figure 1-2 Metal shear protected by finger guards with plastic windows. *(National Safety Council.)*

Woodworking Machinery

Woodworking machinery includes automatic cutoff saws, circular saws, hinged saws, revolving double arbor saws, hand-fed ripsaws, hand-fed crosscut saws, circular resaws, swing-cut saws, radial saws, bandsaws, jointers, tenoning machines, boring and mortising machines, wood shapers, planers, lathes, sanding machines, guillotine veneer cutters, and miscellaneous woodworking machines. These machines should be properly guarded in accordance with appropriate ANSI standards. All belts, pulleys, gears, shafts, and moving parts should be guarded in accordance with requirements.

Abrasive Wheel Machinery

Abrasive wheel machinery includes cylindrical grinders, surface grinders, swing frame grinders, and automatic snagging machines. The guard design and specifications should be in accordance with appropriate ANSI standards.

Figure 1-3 Paper shear with two hand controls for clamping and shearing operations. *(National Safety Council.)*

MILLS AND CALENDERS IN THE RUBBER AND PLASTICS INDUSTRIES

All new and existing installations of mills and calenders should comply with the OSHA requirements and pertinent standards.

Mill safety controls consisting of safety trip controls, pressure-sensitive body bars, safety trip rods, safety trip wire cables, or wire center cords should be installed in all mills either singly or in combination. Fixed guards should be used where applicable.

Calender safety controls should be provided on all calenders within reach of the operator and the bite. Calenders and mills should be installed so that persons cannot normally reach over or under, or come in contact with, the roll bite. All trip and emergency switches should require manual resetting. A suitable alarm should be provided in conjunction with the safety devices.

MECHANICAL POWER PRESSES

All mechanical power presses (excluding pneumatic power presses, bulldozers, hot-banding and hot-metal presses, forging presses, hammers, riveting machines, and similar types of fastener applicators) should conform to the OSHA regulations and applicable standards. Mechanical power presses require particular emphasis as they are involved in many accidents. The machine components should be designed, secured, and covered to minimize hazards. Safeguards should include:

Figure 1-4 Swing-in safety prop for air hammer press. *(National Safety Council.)*

1. Brakes or blocks capable of stopping the motion of the slide (Fig. 1-4)
2. Foot pedals, protected to prevent unintended operation
3. Hand-operated levers to prevent premature or accidental tripping
4. Two-hand trips to have individual operator hand controls arranged to require the use of both hands to trip the press (Fig. 1-3).
5. Air-controlling equipment to be protected against foreign material
6. Brake monitoring systems installed on the press to indicate when the performance of the braking system has deteriorated
7. Point-of-operation guards installed and used in accordance with the standard's requirements
8. A regular program of periodic and regular inspections on a weekly basis of power presses
9. Reports of injuries to employees operating mechanical power presses required by OSHA, with an analysis of each accident on a 30-day basis

FORGING MACHINES

The safety requirements apply to use of lead casts and other uses of lead in the forge or die shop. All equipment should comply with OSHA requirements and the applicable standards. This should include:

1. Thermostatic control of heating elements required for melting lead

2. Industrial hygiene and personal protective equipment required due to the toxicity and heat condition of the lead
3. All presses and hammers controlled by operators who are protected by required guards
4. Devices included to lock out the power when dies are being changed

MECHANICAL POWER TRANSMISSION

This section pertains to mechanical power transmission equipment (see "Standards"), with the exception of small belts operating at a reduced speed and reduced requirements for the textile industry to prevent accumulation of combustible lint. Standard guards are usually secured with the following materials: expanded metal, perforated or solid sheet metal on a frame of angle iron, or an iron pipe securely fastened to the floor or frame of machine. The standard requirements include guards for the following:

1. Flywheels located 7 ft or less from the floor
2. Cranks and connecting rods when exposed to contact
3. Tail rods or extension piston rods
4. Shafting–horizontal, vertical, and inclined
5. Power transmission apparatus and pulleys
6. Belts, ropes, chain drums, gears, sprockets, and chains
7. Shafting, collars, couplings, belt shifters, clutches, perches, or fasteners

Periodic inspection is required of all power transmission equipment. Intervals not exceeding 60 days are recommended.

Standards

The following standards apply to mechanical power transmission:

41 CFR 50-204.5 *Code of Federal Regulations,* "Public Contracts and Property Management, General Safety and Health Standards: Machine Guarding.

ANSI O1.1-1975, "Safety Requirements for Woodworking Machinery"

ANSI B7.1-1978, "Safety Requirements for the Use, Care, and Protection of Abrasive Wheels."

ANSI B28.1-1967 (R1972), "Safety Specifications for Mills and Calenders in the Rubber and Plastic Industries"

ANSI B11.1-1971, "Safety Requirements for Construction, Care, and Use of Mechanical Power Presses"

ANSI B24.1-1971, "Safety Requirements for Forging"

ANSI B15.1-1972, "Safety Standard for Mechanical Power Transmission Apparatus"

HAND AND PORTABLE POWERED TOOLS AND OTHER HAND-HELD EQUIPMENT

Portable tools must be equipped with adequate guards and should comply with OSHA and the appropriate standards. This includes:

1. Portable circular saws, saber, scroll, and jigsaws, portable belt sanding machines, portable abrasive wheels, explosive actuated fastening tools, power lawnmowers, jacks.

2. Warning instructions should be supplied with the equipment. All equipment should be inspected periodically.

Standards

The following standards pertain to hand-held equipment:

ANSI A10.3-1977, "Safety Requirements for Power Actuated Fastening Systems"

ANSI B7.1-1978, "Safety Requirements for the Use, Care, and Protection of Abrasive Wheels"

ANSI B71.1-1972, "Safety Specifications for Power Lawn Mowers, Lawn and Garden Tractors, and Lawn Tractors"

41 CFR 50-204.4 and 50-204.8 *Code of Federal Regulations,* "Public Contracts and Property Management," General-Safety and Health Standards: Tools and equipment; Use of compressed air.

ANSI O1.1-1975, "Safety Requirements for Woodworking Machines"

ANSI B30.1-1975, "Safety Standard for Jacks"

ANSI Z9.4-1979, "Ventilation and Safe Practices of Abrasive Blasting Operations"

WELDING, CUTTING, AND BRAZING

Most welding and cutting operations are mobile and generally used for construction, demolition, maintenance, and repair. Production line welding and cutting equipment is permanently installed; the hazards can be controlled through proper design and operation. The oxygen and acetylene used for welding and cutting must be handled and stored in accordance with the pertinent standards.

Compressed-gas cylinders should be handled with care. The fusible safety plugs in acetylene cylinders melt at about 212°F (100°C). This is an obvious fire hazard if the cylinder is subjected to heat. Oxygen, on the other hand, will react strongly with oils or other hydrocarbons upon contact.

Regulators and pressure gauges should be used on the appropriate cylinders. Leaking cylinders must be removed from the building and taken away from sources of ignition. All equipment should be kept free from oily or greasy substances.

Piping systems should be tested and proved gas-tight at 1½ times the maximum operating pressure. Service piping systems should be protected by pressure-relief devices discharging upward to a safe location. Backflow protection should be provided by an approved device that will prevent oxygen from flowing into the fuel system.

Acetylene generators should be of approved construction and clearly marked with the maximum rate of acetylene production and the pressure limitations. Relief valves should be regularly operated to ensure proper functioning. Storage of all chemicals should conform to the requirements of the standards.

ARC-WELDING AND CUTTING EQUIPMENT

Arc-welding apparatus should comply with the requirements of the National Electrical Manufacturers "Standard for Electric Arc-Welding Apparatus," NEMA-EW-1 1962, and for ANSI/UL551 1976, "Safety Standards for Transformer-Type-Arc-Welding Machines." The design requirements include the following items of concern to the plant engineer:

1. Input power terminals, tip change devices, and live metal parts should be completely enclosed.
2. Terminals for welding leads should be protected from accidental electrical contact by

personnel or by metal objects. The frame or case of the welding machine should be grounded as specified.

3. Printed rules and instructions concerning operation of the equipment supplied by the manufacturer should be strictly followed.
4. All parts of the equipment should be frequently inspected. Cables with damaged insulation or exposed bare conductors should be replaced.

RESISTANCE-WELDING EQUIPMENT

All resistance-welding equipment should be installed in accordance with article 630D of the ***NEC***®.* The following items of particular interest to the plant engineer are cited:

1. Controls on all automatic or air and hydraulic clamps should be arranged or guarded to prevent accidental actuation.
2. All doors and access and control panels should be kept locked and interlocked to prevent access by unauthorized persons to live portions of the equipment.
3. All press-welding machine operations, where there is a possibility of the operator's fingers being under the point of operation, should be effectively guarded in a manner similar to that prescribed for punch-press operations.
4. Flash-welding equipment should be equipped with hoods to control flying flash and ventilation of fumes.
5. Combustible material must be removed from the welding area and the basic fire-prevention requirements in accordance with ANSI/NFPA 51B-1977. Fire watchers are required wherever a fire problem exists. A welding permit system should be instituted to control hazardous exposures.
6. Welding operators should use all of the protective equipment specified in the standard, and ventilation should be used where required.

Standards

The following standards are for resistance-welding equipment:

ANSI Z49.1-1973, "Safety in Welding and Cutting"

ANSI/NFPA-51-1978, "Oxygen–Fuel Gas Systems for Welding and Cutting"

ANSI/NFPA 51B 1977, "Cutting and Welding Processes"

41 CFR 50-204.7 *Code of Federal Regulations,* "Public Contracts and Property Management," General Safety and Health Standards: Personal protective equipment

SPECIAL INDUSTRIES

Plant engineers involved in the following special industries should consult the OSHA regulations for specific requirements pertinent to their industry.

1. Pulp, paper, and paperboard mills
2. Textiles
3. Bakery equipment
4. Laundry machinery and operations
5. Sawmills
6. Pulpwood logging
7. Agricultural operations
8. Telecommunications

****NEC***® is a Registered Trademark of the National Fire Protection Association, Quincy, MA 02269

Standards

Following are some special-industry standards:

ANSI P1.1-1969, "Safety Requirements for Pulp, Paper, and Paperboard Mills"

ANSI L1.1-1972, "Safety Requirements for the Textile Industry"

ANSI Z50.1-1977, "Safety Requirements for Bakery Equipment"

ANSI Z8.1-1972, "Safety Requirements for Commercial Laundry and Drycleaning Equipment and Operations"

ANSI O2.1-1969, "Safety Requirements for Sawmills"

ANSI O3.1-1978, "Safety Requirements for Pulpwood Logging"

ELECTRIC EQUIPMENT

All electric equipment in the plant should comply with OSHA regulations and the **National Electrical Code®** (NFPA 70-1981)*. It is particularly important for the plant engineer to be knowledgeable in the following sections:

1. Article 250, Grounding

2. Article 500, Hazardous(Classified) Locations

TOXIC AND HAZARDOUS SUBSTANCES

The exposure of employees to any of approximately 600 chemicals (see "Standards" at end of this section) should at no time exceed the ceiling value given for that material. To obtain compliance with this section of the OSHA regulations, it is necessary to develop administrative and engineering controls. When such controls are not feasible, protective equipment or other protective measures should be used.

Controls can be developed by environmental monitoring, personal monitoring, and employee observation. The plant engineer has an important part to play in the inspection and maintenance of the environmental monitoring equipment.

One method for controlling exposure is by local exhaust-ventilation and dust-collection systems. They should be constructed, installed, and maintained in accordance with ANSI Z9.2-1979, "Fundamentals Governing the Design and Operation of Local Exhaust Systems," which is incorporated by reference in this section of the handbook.

Twenty-three materials require special care in handling since they are in the category of regulated substances which present a cancer hazard. These are listed below:

Asbestos	beta-Propiolactone
Coal-tar pitch volatiles	2-Acetylaminofluorene
4-Nitrobiphenyl	4-Dimethylaminoayobenzene
alpha-Naphthylamine	*N*-Nitrosodimethylamine
Methylchloromethyl ether	Vinyl chloride
3,3-Dichlorobenzidene	Inorganic arsenic
bis(Chloromethyl) ether	Benzene
beta-Naphthylamine	Coke-oven emissions
Benzidine	Cotton dust
4-Amino diphenyl	1,2-dibromo-3-chloropropane
Ethylenimine	Acrylonitrile

*__National Electrical Code®__ is a Registered Trademark of the National Fire Protection Association, Quincy, MA 02269.

Controls on these materials are very detailed since the materials must be contained in a closed-system operation. The operating area must be restricted to authorized personnel only. Warning signs and instructions should be posted.

Standard

Refer to the following for standards on toxic and hazardous substances:

Tables 2-1, 2-2 and 2-3, "Threshold Limit Valves for Chemical Substances in Workroom Air, with Intended Changes," 1978 Conference of American Governmental Industrial Hygienists PO Box 1937, Cincinnati, OH 45201.

chapter 8-2

Fire Protection and Prevention*

by

Robert William Ryan
Assistant Director
Department of Environmental Safety
University of Maryland
College Park, Maryland

Richard Best
Senior Fire Analysis Specialist
Research and Fire Information Services
National Fire Protection Association
Quincy, Massachusetts

GLOSSARY

The following list contains the terms most often used in this chapter or in the fire protection field.

Boiling point The temperature at which the liquid boils when under normal atmospheric pressure (14.7 psia). The boiling point increases as pressure increases and is dependent on the total pressure.

Combustible A material or structure which can burn is labeled combustible. Combustible is a relative term; many materials which will burn under one set of conditions will not burn under others, e.g., structural steel is noncombustible, but fine steel wool is combustible. The term *combustible* does not usually indicate ease of ignition, burning intensity, or rate of burning, except when modified, as in *highly combustible interior finish.*

Fire prevention Measures directed toward avoiding the inception of fire.

Fireproof A misnomer for fire-resistive.

Fire load The amount of combustibles present in a given situation, usually expressed in terms of weight of combustible material per square foot. This measure is employed frequently to calculate the degree of fire resistance required to withstand a fire or to judge the rate of application and quantity of extinguishing agent needed to control or extinguish a fire.

Fire point The lowest temperature of a liquid in an open container at which vapors are evolved fast enough to support continuous combustion.

Fire resistance A relative term, used with a numerical rating or modifying adjective to indicate the extent to which a material or structure resists the effect of fire, e.g., "fire resistance of 2 h."

Fire-resistive Properties or designs to resist the effects of any fire to which a material or structure may be expected to be subjected. *Fire-resistive materials* or assemblies of materials are noncombustible, but noncombustible materials are not necessarily fire-resistive; fire-resistive implies a higher degree of fire resistance than noncombustible. *Fire-resistive construction* is defined in terms of specified fire resistance as measured by the standard time–temperature curve.

Fire-retardant Usually denotes a substantially lower degree of fire resistance than fire-resistive and is often used to refer to materials or structures which are combustible in whole or in part, but have been subjected to treatments or have surface coverings to prevent or retard ignition or the spread of fire under the conditions for which they are designed.

Flameproof, flameproofing Misleading terms and their use is discouraged in favor of *flame-retardant* or *flame-resistant.*

Flame-resistant A term that may be used more or less interchangeably with flame-retardant.

Flame-retardant Materials, usually decorative, which due to chemical treatment or inherent properties, do not ignite readily or propagate flaming under small to moderate exposure.

Flammable A combustible material that ignites very easily, burns intensely, or has a rapid rate of flamespread is called flammable. Flammable is used in a general sense without reference to specific limits of ignition temperature, rate of burning, or other property. *Flammable* and *inflammable* are identical in meaning. Flammable is used in preference to inflammable.

Flammable limits The extreme concentration limits of a combustible in an oxidant through which a flame will continue to propagate at the specified temperature and pressure. For example, hydrogen-air mixtures will propagate flame between 4.0 and 75 percent by volume of hydrogen at 21°C and atmospheric pressure. The smaller value is the lower (lean) limit, and the larger value is the upper (rich) limit of flammability. For liquid fuels in equilibrium with their vapors in air, a minimum temperature exists for each fuel above which sufficient vapor is released to form a flammable vapor-air mixture. There is also a maximum temperature above which the vapor concentration is too high to propagate flame. These minimum and maximum temperatures are referred to respectively as the lower and upper *flash points* in air. The flash-point temperatures for a combustible liquid vary directly with environmental pressure.

Flashover The phenomenon of a slowly developing fire (or radiant heat source) producing radiant energy at wall and ceiling surfaces. The radiant feedback from those surfaces gradually heats the contents of the fire area, and when all the combustibles in the space have become heated to their ignition temperature, simultaneous ignition occurs.

Flash point The lowest temperature at which the vapor pressure of a liquid will produce a flammable mixture and resultant flame. The flame will not continue to burn at this temperature if the source of ignition is removed.

Glowing combustion and flame Combustion is the process of exothermic, self-catalyzed reactions involving either a condensed-phase or a gas-phase fuel, or both. The process is usually (but not necessarily) associated with oxidation of a fuel by atmospheric oxygen. Condensed-phase combustion is usually referred to as glowing combustion, while gas-phase combustion is referred to as a flame.

Ignition temperature (autoignition temperature, autogenous ignition temperature The minimum temperature to which the substance in air must be heated in order to initiate, or cause, self-sustained combustion independently of the heating or heated element. The ignition temperature of a combustible solid is influenced by rate of airflow, rate of heating, and size and shape of the solid. Small sample tests have shown that as the rate of airflow and the rate of heating are increased, the ignition temperature of a solid drops to a minimum and then increases.

Latent heat Heat is absorbed by a substance when converted from a solid to a liquid and from a liquid to a gas. Conversely, heat is released during conversion of a gas to a liquid, or a liquid to a solid. Latent heat is the quantity of heat absorbed or given off by a substance in passing between liquid and gaseous phases (latent heat of vaporization) or between solid and liquid phases (latent heat of fusion). The high heat of vaporization of water is a reason for the effectiveness of water as an extinguishing agent.

Noncombustible Not combustible.

Nonflammable Not flammable.

Specific heat The heat, or thermal capacity, of a substance is the number of calories required to raise 1 g 1°C. The specific heats of various substances vary over a considerable range; for all common substances, except water, they are less than unity. Specific-heat figures are significant in fire protection as they indicate the relative quantity of heat needed to raise the temperature to a point of danger, or the quantity of heat that must be removed to cool a hot substance to a safe temperature.

Vapor density The weight of a volume of pure gas compared with the weight of an equal volume of dry air at the same temperature and pressure. A figure less than 1 indicates that a gas is lighter than air, and a figure greater than 1 that a gas is heavier than air. If a flammable gas with a vapor density greater than 1 escapes from its container, it may travel at a low level to a source of ignition.

Vapor pressure Because molecules of a liquid are always in motion (with the amount of motion depending on the temperature of the liquid), the molecules are continually escaping from the free surface of the liquid to the space above. Some molecules remain in space while others, due to random motion, collide with the liquid. If the liquid is in an open container, molecules (collectively called *vapor*) escape from the surface, and the liquid is said to evaporate. If, on the other hand, the liquid is in a closed container, the motion of the escaping molecules is confined to the vapor space above the surface of the liquid. As an increasing number strike and reenter the liquid, a point of equilibrium is eventually reached when the rate of escape of molecules from the liquid equals the rate of return to the liquid. The pressure exerted by the escaping vapor at the point of equilibrium is called *vapor pressure.* Vapor pressure is measured in millimeters of mercury (mm), or torr.

INTRODUCTION

Fire protection is an area in which most plant engineers can make a significant contribution to plant operations. At many facilities the chief engineer may also serve as the *fire marshal* or *fire chief.* Even at larger plants, where there is a full-time safety or fire-protection engineer responsible for plant protection and loss prevention, the plant engineer should be familiar with the fire problem, fire-prevention methods, and fire protection systems.

Fires in private industry have a great potential for significant economic and financial loss to both the industrial plant and the community. Recovery from an industrial fire includes not only replacement of equipment and facilities at higher costs, but also temporary and permanent lost business income, loss of skilled employees during the time the plant is closed, loss of profits on damaged finished goods, and extra expenses to restore operations. Many plants destroyed by fire do not reopen, contributing to local unemployment and disrupting the personal lives of employees.

This chapter is intended to give the plant engineer a basic understanding of fire protection and prevention, to provide some basic information about fire, and to identify other resources for the incorporation of fire safety in every aspect of plant operations.

In addition to the National Fire Protection Association in Quincy, Mass., and the Federal Emergency Management Agency in Washington, D.C., the following fire protection services are also available in most cities in the United States: the Society of Fire Protection Engineers, fire protection consultants, equipment manufacturers, testing laboratories, and companies specializing in special-hazard protection and control, fire investigations, fire-protection equipment testing, installation, and servicing. There are also state and regional training facilities, and the resources of municipal fire departments should not be overlooked.

This chapter does not attempt to discuss management programs which may have a significant impact on plant fire protection. Information regarding risk management, personnel training, plant fire brigades, plant emergency organization, and fire-prevention programs must be obtained from other sources. A list of sources is presented in the final section of this chapter.

THE NATURE OF FIRE

A simple method of visualizing what fire is and how burning takes place is to use a four-sided object, or tetrahedron. Each surface of the tetrahedron is used to represent one of the conditions necessary for fire to occur (see Figure 2-1).

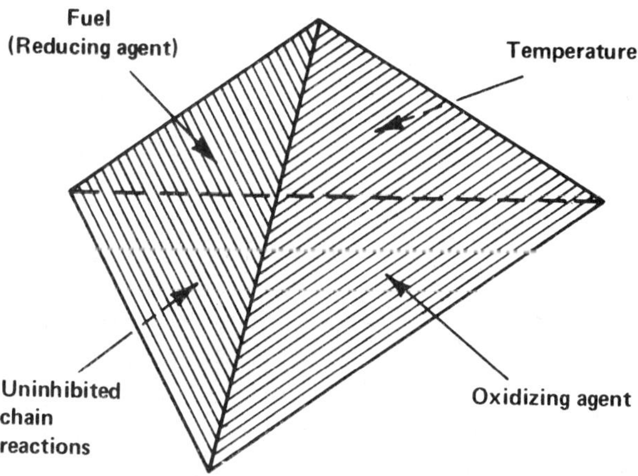

Figure 2-1 Tetrahedron fire model. *(From R. Tuve,* Principles of Fire Protection Chemistry, *NFPA, Quincy, Massachusetts, 1976.)*

The four components of this simple fire model are fuel, heat, oxidant, and the chemical chain reaction. While this model does not provide a complete scientific description of fire, it is sufficient for explaining most fire-protection concepts. In chemical terminology, the *fuel* component may also be referred to as a reducing agent. During the fire reaction the reducing agent loses electrons. The *heat* component includes both the heat which causes the fire and the heat emitted by the fire which causes it to be self-sustaining. The *oxidant* required for fire is most often provided by the oxygen in the ambient air, approximately 21 percent. Although oxygen is the most common oxidizing agent and is usually necessary for fire to occur, there are some chemicals that release oxygen and some that can burn in an oxygen-free atmosphere.

The *chemical chain reaction* of fire is a self-sustaining reaction yielding energy or products that cause further reactions of the same kind, or burning.

Vapor State of Fuel

Before a fuel can be burned it must be in a vapor state. Therefore, flammable gases are most easily ignited. Even solids and liquids, such as wood and gasoline, must be vaporized before they will burn. The decomposition of matter due to heat which generates flammable vapors is called *pyrolysis.*

The initial vaporization of the fuel may be caused by heat from a source of ignition such as a chemical reaction, electric energy, or mechanical heat energy. Ambient heat may also be sufficient to vaporize fuels which are normally liquids.

Once sufficient vaporization has occurred, the combustible vapors can be ignited by an open flame or spark or, at a sufficiently high temperature, the vapor and oxygen mixture will ignite spontaneously.

After ignition of the vaporized fuel has occurred, the heat generated by the fire will cause further vaporization of the fuel and the intensity of the fire will increase. This process is known as *radiation feedback.*

Heat Transfer

The heat which causes the fuel to ignite and the heat generated by the resulting fire can be transmitted by one or all of the following three methods: conduction, radiation, or convection.

- **Conduction.** Heat transferred by direct contact from one body to another.
- **Radiation.** Energy travel through space or materials as waves.
- **Convection.** Heat transferred by a circulating medium—either a gas or a liquid.

Products of Combustion

There are four categories of products of combustion: (1) fire gases, (2) flame, (3) heat, and (4) smoke. All of these products are produced in varying degrees by each fire. The material or materials that are involved in the fire and the resulting chemical reactions produced by the fire determine the products of combustion.

Fire Gases

The primary cause of loss of life in fires is the inhalation of heated, toxic, and oxygen-deficient gases and smoke. The amount and kind of fire gases present during and after a fire vary widely with the chemical composition of the material burning, the amount of available oxygen, and the temperature. The effect of toxic gases and smoke on people will depend on the time of exposure, the concentration of the gases in air, and the physical condition of the individual.

There are usually several gases present during a fire. Those that are most lethal are carbon monoxide, carbon dioxide, hydrogen sulfide, sulfur dioxide, ammonia, hydrogen cyanide, hydrogen chloride, nitrogen dioxide, acrolein, and phosgene.

Flame

The burning of materials in a normal oxygen-rich atmosphere is generally accompanied by flame. For this reason, flame is considered a distinct product of combustion. Burns can be caused by direct contact with flames or heat radiated from flames. Flame is rarely separated from the burning materials by any appreciable distance.

Heat

Heat is the combustion product most responsible for fire spread.

Exposure to heat from a fire will affect persons in proportion to the length of exposure and the temperature of the heat. The dangers of exposure to heat from fire range from minor injury to death. Exposure to heated air increases the heart rate and causes dehydration, exhaustion, blockage of the respiratory tract, and burns. Fire fighters should not enter atmospheres exceeding 120°F (48°C) to 130°F (54°C) without special protective clothing and masks. The maximum survivable breathing level of heat from fire in a dry atmosphere for a short period has been estimated at 300°F (148°C). Any moisture present in the air greatly increases the danger and sharply reduces the time of survival.

Smoke

Smoke is matter consisting of very fine solid particles and condensed vapor. Fire gases from common combustibles (such as wood) contain water vapor, carbon dioxide, and carbon monoxide. Under the usual conditions of insufficient oxygen for complete combustion, methane, methanol, formaldehyde, and formic and acetic acids are also present. These gases are usually evolved from the combustible with sufficient velocity to carry droplets of flammable tars which appear as smoke. Particles of carbon develop from the decomposition of these tars; they are also present in the fire gases from the burning of petroleum products, particularly from the heavier oils and distillates.

These small particles of carbon and tarlike particles that are visible and the phenomenon of fire gases rendered visible by the particles are generally defined as *smoke*.

Modes of Combustion

The combustion process may occur in two modes, *flaming* (including explosions) and *flameless* (including glow and deep-seated glowing embers). The flaming mode is characterized by relatively high burning rates. Intense and high levels of heat are usually associated with the flaming mode.

The flaming and flameless modes are not mutually exclusive; combustion may involve one or both modes. Often combustion may occur in the flaming mode and gradually make a transition to the flameless mode. At one point in this process both modes occur simultaneously.

Fire Control

Fire prevention and fire extinguishment can be described in terms of the fire model previously discussed. Fire prevention is generally a matter of keeping heat and fuel separated; or, in some processes, keeping heated fuel from combining with oxygen.

Fire extinguishment can be summarized by four methods:

1. Removal or dilution of air or oxygen to a point where combustion ceases.
2. Removal of fuel to a point where there is nothing remaining to oxidize.
3. Cooling of the fuel to a point where combustible vapors are no longer evolved or where activation energy is lowered to the extent that no activated atoms or free radicals are produced.
4. Interruption of the flame chemistry of the chain reaction of combustion by injection of compounds capable of quenching free-radical production.

Removal of Oxygen

The amount of dilution of oxygen necessary to stop the combustion varies greatly with the kind of material that is burning. Ordinary hydrocarbon gases and vapors will not burn when the oxygen level is below 15 percent. Acetylene will continue to burn unless the oxygen concentration is lowered, but will continue to glow on the surface even if the oxygen level is as low as 4 to 5 percent.

A fire in a closed space can extinguish itself by consuming the oxygen. However, incomplete combustion, which takes place when the oxygen is consumed, usually results in considerable generation of flammable gases.

A commonly used method of putting a fire out by removing or diluting the oxygen is by flooding the entire fire area with carbon dioxide or with some other inert gas.

Some burning materials react violently with water and can best be extinguished by covering them with a suitable inert material.

Fuel Removal

Fuel removal can be accomplished in a variety of ways. One of the most common examples is the practice of bulldozing a firebreak across the path of an advancing forest fire.

Fires in large coal or wood pulp piles can usually be controlled only by moving the pile out of the fire zone. Fires in large oil storage tanks have been controlled by pumping the oil out of the burning tank into an empty tank. If a gas line is ruptured and the gas ignited, shutting off the supply of gas is the only way to stop the fire.

If it is not practical to remove the fuel, extinguishment can be accomplished by shutting off the fuel vapors or by covering the burning or glowing fuel. The use of fire-fighting foams and dry powder extinguishers are effective procedures for covering or coating a fire.

Cooling

For most common combustibles such as wood, paper, and cloth, the simplest and most effective means of removing the heat of a fire is through the application of water. Water application can be varied and will depend on the fire.

Applying water to the burning fuel cools the fuel until the rate of release of combustible vapors and gases is reduced and ultimately stopped. Heat developed by a fire is carried away by radiation, conduction, and convection. This helps reduce the amount of heat and makes the use of water more effective. Only a relatively small proportion of the heat evolved needs to be cooled by the water in order to extinguish the fire.

Effective use of water cannot be accomplished if the water cannot reach the burning fuel directly. For this reason, areas where fire fighters cannot readily reach the fire with water streams, such as high-rise buildings and high-piled storage areas, must be provided with automatic sprinklers or other automatic fire-protection systems.

Interruption of Chemical Reaction

Extinguishment by cooling, by oxygen dilution, and by fuel removal is applicable to all classes of flaming and glowing fires. Extinguishment by chemical flame inhibition applies to the flaming mode only.

Note that in the fourth method of extinguishment the action occurs only during contact of the chemical agents with activated groups or with atoms being produced by the combustion process. In a sense this could be seen as a temporary extinguishment process, operating only when the agent particles are present in the flame. If activation energy continues to exist (as with ignition points for vapor or gas reignition) after withdrawal of the agent, the flame reaction will reestablish and will continue.

Summary

Combustion occurs under the following conditions:

1. An oxidizing agent, a combustible material, and an ignition source are essential for combustion.
2. The combustible material must be heated to its ignition temperature before it will burn.
3. Combustion will continue until:
   - **a.** The combustible material is consumed or removed.
   - **b.** The oxidizing agent concentration is lowered to below the concentration necessary to support combustion.
   - **c.** The combustible material is cooled to below its ignition temperature.
   - **d.** Flames are chemically inhibited.

THE PLANT FIRE PROBLEM: CAUSES AND PREVENTION

In general, the cause of most plant fires is the *exposure of a fuel to a source of heat.* Where the fuel, such as accumulations of trash or debris, is not necessary to plant operation, fires can be prevented by removal of the fuel. Where the exposed fuel, such as raw materials or finished products, is essential, the source of heat must be protected or controlled.

Sources of Heat of Ignition and Fuel

Some of the most common sources of heat and fuel that cause plant fires are heating and cooking equipment, smoking, electric equipment, burning, flammable liquids, open flames and sparks, incendiary (arson), spontaneous ignition, gas fires, and explosions. These sources of heat are summarized below:

Heating and Cooking Equipment

Defective or Overheated Equipment. This includes improperly maintained or operated furnaces, smoke pipes, vents, portable and stationary heaters, industrial and commercial furnaces, and incinerators.

Chimneys and Flues. Fire can arise from ignition of accumulated soot or inadequate separation from combustible material.

Hot Ashes and Coals. These can cause problems when improper disposal or disposal in combustible containers or with combustible debris occurs.

Improper Location. This can mean installation too close to combustibles or accumulation of combustibles near an appliance.

Smoking

Smoking in Flammable or Explosive Atmospheres

Discarding Smoking Materials in Combustible Debris

Electric Equipment

Wiring and Distribution Equipment. These include short-circuit faults, arcs, and sparks from damaged, defective, or improperly installed components.

Motors and Appliances. These include careless use, improper installation, and poor maintenance.

Burning

Trash and Rubbish. Burning trash and rubbish can furnish the fuel for accidental ignition; careless burning ignites other material.

Warming Fires. Careless burning ignites other material.

Flammable Liquids

Storage and Handling. These hazards include careless spills, leaking fuel, and overturned tanks.

Inadequate Safeguards. Fires can be started by improper storage containers or facilities, improper electrical equipment near open processes, or improper bonding and grounding of transfer processes.

Open Flames and Sparks

Sparks and Embers. Problems include ignition of roof coverings by sparks from chimneys, incinerators, rubbish fires, locomotives, etc.

Welding and Cutting. Hazards include ignition of combustibles by the arc or flame itself, heat conduction through the metals being welded or cut, molten slag and metal from the cut, or sparks.

Friction, Sparks from Machinery. Friction heat or sparks resulting from impact between two hard surfaces are a hazard.

Thawing Pipes. Open-flame devices are a hazard when used in the dangerous practice of thawing pipes.

Other Open Flames. These include ignition sources such as candles, locomotive sparks, incinerator sparks, and chimney sparks.

Lightning. This includes building fires caused by the effects of lightning.

Exposure. Exposure fires are those originating in places other than buildings, but which ignite buildings.

Incendiary, Suspicious. These are fires that are known to be or thought to have been set: fires set to defraud insurance companies, fires set by mentally disturbed persons, and fires set by malicious persons.

Spontaneous Ignition. This means fires resulting from the uncontrolled spontaneous heating of materials.

Gas Fires and Explosions. These are fires and explosions that involve gas that has escaped from piping, storage tanks, equipment, or appliances and fires caused by misuse or faulty operation of gas appliances.

PLANT FIRE HAZARDS

Texts and handbooks which describe the hazard and control technique for specific risks are listed in the last section of this chapter.

Tables 2-1 and 2-2 present a statistical profile of large-loss fire experience in the United States for 1978 and 1979. A large-loss fire is defined by the NFPA as one that caused $500,000 or more in direct fire damage. Indirect losses, such as business interruption losses, are not included.

Fire Hazards of Materials

Virtually all matter can be changed by exposure to sufficient quantities of energy. Energy in the form of heat was previously discussed. The heat energy which causes or is produced by a fire usually causes undesired changes to the material involved. The relative fire risks and products of burning of different types of materials are presented in this section.

Wood

When in contact with sufficient heat, all wood or wood-based products will ignite, the time of ignition depending on the ignition source and the length of exposure.

Wood or wood-based products can be treated with fire-retardant chemicals. When so treated, the flammability of wood and wood-based products is reduced. The flammability of these products can also be reduced when they are used in combination with other materials, such as insulation.

Moisture Content. Wood is made up primarily of carbon, hydrogen, and oxygen. Live wood cells retain considerable moisture. When the wood is dead, air replaces most of the water in the cellular structure of wood.

The fire behavior of wood and other combustible solids of the same size and shape varies greatly with the moisture content. Wet wood is harder to ignite and will not burn as fast as dry wood. Burning rate is also influenced by the moisture content in materials. See Table 2-3 for heat of combustion for various wood-based products composed with petroleum-based materials.

Even when exposed to a relatively high heat source for a prolonged period of time, ignition is generally difficult when the moisture content of wood (and similar fuels) is

TABLE 2-1 Property Uses Where Large-Loss Fire Occurred, 1980 (United States)

Property use	No. of large-loss fires	Loss	Total no. of large-loss fires for major property uses	Total loss for major property uses
Public Assembly			26	$ 60,210,000
Bowling establishments	3	$ 3,500,000		
Racetrack	1	20,000,000		
Churches	7	12,780,000		
Country clubs	3	5,800,000		
Historic buildings	2	2,750,000		
Restaurants, drinking places	7	10,650,000		
Theaters	3	4,730,000		
Educational			27	52,801,800
Public schools	22	40,113,000		
Residential schools	2	7,071,600		
Colleges	2	4,050,000		
Other educational	1	1,567,200		
Institutional			1	1,800,000
Prison	1	1,800,000		
Residential			25	93,594,500
Private dwellings	3	8,275,000		
Apartments	15	24,600,000		
Hotels	5	57,084,000		
Dormitories	2	3,635,500		
Stores and Offices			53	97,692,200
Food, beverage sales	4	4,325,000		
Clothing, fabric sales	3	3,000,000		
Furniture, appliance sales	12	18,500,000		
Drugstore	1	1,000,000		
Recreation, hobby, home repair	5	7,634,000		
Linen supply	1	1,000,000		
Vehicle sales/repair	3	4,000,000		
General-item stores	13	29,038,200		
Offices	10	27,515,000		
Other stores	1	1,680,000		
Basic Industry, Utility, Defense			17	49,570,000
Non-nuclear energy production	5	28,720,000		
Communication facility	1	2,000,000		
Energy distribution systems	2	3,000,000		
Fruit packing plants	2	4,500,000		
Forests	3	6,000,000		
Coal mine	1	1,000,000		
Mineral product manufacture	3	4,350,000		
Manufacturing			81	339,510,700
Food	6	11,635,000		
Textile	3	5,835,000		
Clothing, rubber manufacture	4	6,440,000		
Wood, furniture, paper, printing	18	47,199,900		
Chemical, plastic, petroleum	16	61,962,800		
Metal, metal products	9	33,911,500		
Other manufacturing	7	35,061,500		
Storage			72	172,950,000
Agricultural products	4	6,720,000		
Textile, textile products	4	12,083,500		
Processed food, beverage	3	8,710,000		
Petroleum, petroleum products	10	20,301,000		
Wood, paper products	23	56,316,000		
Chemical, plastic products	5	10,100,000		
Metal, metal products	3	15,023,500		
Vehicles	6	14,666,000		
General-item storage	14	29,030,000		

TABLE 2-1. Property Uses Where Large-Loss Fire Occured, 1980 (United States) (*Continued*)

Property use	No. of large-loss fires	Loss	Total no. of large-loss fires for major property uses	Total loss for major property uses
Special Properties			21	80,674,900
Under construction, unoccupied	15	35,707,500		
Outdoor properties (brush) including structures	2	40,233,900		
Rail transport	3	3,733,500		
Rail transport	1	1,000,000		
TOTAL:			323	$948,804,100

Source: Fire Journal, September 1980, National Fire Protection Quincy, Massachusetts.

TABLE 2-2 Percentage of Large-Loss Fires by Property Use Classification, 1979 (United States)

Property use	Percent of fires	Percent of loss
Public assembly	11.0	7.7
Educational	4.8	3.9
Institutional	0.5	0.4
Residential	6.6	7.0
Stores and offices	17.1	13.3
Basic industry, utility, and defense	6.0	4.6
Manufacturing	22.8	18.5
Storage	22.3	21.8
Special properties	8.9	22.8
	100.0	100.0

Source: Fire Journal, September, 1980, National Fire Protection Association, Quincy, Massachusetts.

above 15 percent. Once ignition and resultant fire have begun, heat radiation and the rate of pyrolysis reduce the importance of the moisture factor.

Plastics

There are thousands of plastic product formulations that are produced in a variety of shapes and sizes, such as solid shapes, films and sheets, foams, molded forms, synthetic fibers, pellets, and powders. They are classified into 30 major groups of plastics and polymers. In addition, most finished products contain additives such as colorants, reinforcing agents, fillers, stabilizers, and lubricants. These additives vary the chemical nature of the product still more.

Most plastics are combustible, but the degree of combustibility varies widely because of the range of chemical compositions and combinations. As a result, it is virtually impossible to assign a fire hazard or flammability limit to any general plastic group. The only method of determining the fire hazard of a particular plastic is to fire test the plastic under exact end-use conditions. See Tables 2-4 and 2-5.

Storage. The chemical composition, the physical form, and the manner and arrangement in which plastics are stored greatly affect the degree of fire hazard that is present. Large quantities of smoke are usually generated when stored plastics are involved in fire, a condition made more or less difficult by the amount of ventilation present or available in a given storage area.

Thermoplastics, such as polyethylene and plasticized polyvinylchloride, and ther-

TABLE 2-3 Heat of Combustion of Various Wood and Wood-Based Products and Comparative Substances*

Substance	Heating value, Btu/lb
Wood sawdust (oak)	8,493†
Wood sawdust (pine)	9,676†
Wood shavings	8,248†
Wood bark (fir)	9,496†
Corrugated fiber carton	5,970†
Newspaper	7,883†
Wrapping paper	7,106†
Petroleum coke	15,800
Asphalt	17,158
Oil (cottonseed)	17,100
Oil (paraffin)	17,640

*Extracted from *Kent's Mechanical Engineers' Handbook*, 12th ed., H. B. Carmichael and J. K. Salisbury, eds., Wiley-Interscience, New York, 1950.

†Dry.

mosets, such as polyesters, present severe fire hazards. Plastics in foamed material form present the most severe hazard of all. In a fire, thermoplastics will melt and break down and behave and burn like flammable liquids. Automatic sprinkler systems with high sprinkler-discharge densities are necessary for adequate fire protection.

Dusts

When some combustible solids are ground or rubbed into minute particles, the particles tend to mix with the air in much the same way that vapor or gas mixes with the air. The finer the dust particle, the more completely it will mix with the air and remain suspended in the air. Although dust particles from all combustible solids do not result in potentially explosive dust particles, a large number of combustible solids can yield explosive dust particles. See Table 2-6.

Metals

Nearly all metals will burn in air under certain conditions. See Table 2-7. Some oxidize rapidly in the presence of air or moisture, generating sufficient heat to reach their ignition temperatures. Others oxidize so slowly that heat generated during oxidation is dissipated before they become hot enough to ignite. Certain metals, notably magnesium, titanium, sodium, potassium, calcium, lithium, hafnium, zirconium, zinc, thorium, uranium, and plutonium, are referred to as combustible metals because of the ease of ignition of thin sections, fine particles, or molten metal. The same metals in massive solid form are comparatively difficult to ignite.

Hot, burning metals may react violently with the extinguishants used on fires involving ordinary combustibles or flammable liquids. A few metals, such as uranium, thorium, and plutonium, emit ionizing radiations that can complicate fire fighting and introduce a contamination problem.

Temperatures in burning metals are generally much higher than the temperature in burning flammable liquids. Some hot metals can continue burning in nitrogen, carbon dioxide, or steam atmospheres in which ordinary combustibles or flammable liquids would be incapable of burning.

Flammable and Combustible Liquids

The improper storage, handling, and use of flammable and combustible liquids has been the cause of many deaths, injuries, and disastrous fires.

TABLE 2-4 Small-Scale Tests for Combustibility of Plastics

Test method	Sample size, in	Position of sample	Ignition source, flame, in	Time and limit of exposure, s	Value reported, in/min	Usual material application
ASTM D 635	⅛ × ½ × 5	Horizontal	1	2–30	Burning rate	Rigid plastic
ASTM D 568	0.05 × 1 × 18	Horizontal	1	15	Burning rate	Films
ASTM D 229	1/32 × ½ × 5	Horizontal	1	30	Burning time for 4 samples	Elec. insulation
ASTM D 1692	½ × 2 × 6	Horizontal	1	60	Burning rate	Foam
UL 94*	⅛ × ½ × 5	Horizontal	¾	30	Burning rate	Rigid plastics
	⅛ × ½ × 5	Vertical	¾	2–10	Extinguishment time	Rigid plastics
NFPA 701	2¾ × 10	Vertical	1½	12	Length of char	Films
UL 214	2¾ × 10	Vertical	1½	12	Length of char	Films

*UL stands for Underwriters Laboratories Inc.
Source: NFPA, Quincy, Massachusetts.

TABLE 2-5 Medium- and Large-Scale Tests for Combustibility of Plastics

Test for	Number
Surface burning of building materials	NFPA 255 UL 723 ASTM E 84
Fire tests of building construction and materials	NFPA 251 UL 263 ASTM E 119
Fire tests of roof coverings	NFPA 256
Radiant panel test for flame spread	ASTM E 162
Factory mutual calorimeter test	
UL and FM corner wall tests	
Full-room burnouts—FM, UL*	
Flame-retardant films	NFPA 701 UL 214

*UL stands for Underwriters Laboratories Inc. FM for Factory Mutual Systems. See p. 14-69.

Source: NFPA, Quincy, Massachusetts.

TABLE 2-6 Common Combustible Solid Dusts Generating Severe Explosions*

Type of dust	Maximum explosion pressure		Maximum rate of pressure rise	
	psig	bar	psig/s	bar/s
Corn (processing)	95	6.55	6,000	413.7
Cornstarch	115	7.93	9,000	620.5
Potato starch	97	6.89	8,000	551.6
Sugar (processing)	91	6.27	5,000	344.7
Wheat starch	105	7.24	8,500	586.0
Ethyl cellulose plastic molding compound	102	7.03	6,000	413.7
Wood flour filler	110	7.58	5,500	379.2
Natural resin	87	6.0	10,000	689.5
Aluminum	100	6.9	10,000	689.5
Magnesium (powder)	94	6.48	10,000	689.5
Silicon (powder)	106	7.31	10,000	689.5
Titanium (powder)	80	5.52	10,000	689.5
Aluminum magnesium alloy (powder)	90	6.20	10,000	689.5

*Extracted from U.S. Bureau of Mines Investigations and Reports, Nos. 5753, RI 5971, RI 6516

It is the vapor from the evaporation of a flammable or combustible liquid when exposed to air or under the influence of heat, rather than the liquid itself, which burns or explodes when mixed with air in certain proportions in the presence of some source of ignition. There is a flammable range below which the vapor mixture is too lean to burn or explode, or above which the vapor mixture is too rich to burn or explode. (See Table 2-8.) For gasoline, the most common and widely used flammable liquid, the flammable range is 1.4 and 7.6 percent by volume. When the vapor-air mixture is near either the lower flammable limit (LFL) or upper flammable limit (UFL), the explosion is less intense than when the mixture is in the intermediate range. The violence of the explosion depends on the concentration of the vapor as well as the quantity of vapor-air mixture and the type of container. Thus, it is important in controlling the fire hazard, to store a flammable liquid in the proper type of closed container and minimize the exposure to air. When exposed to heat from a fire, a tank or other container may rupture with dangerous results if properly designed vents are not provided or if the exposed tank or con-

TABLE 2-7 Melting, Boiling, and Ignition Temperatures of Pure Metals in Solid Form*

Pure metal	Temperature, °F		
	Melting point	Boiling point	Solid metal ignition
Aluminum	1220	4445	above 1832†
Barium	1337	2084	347†
Calcium	1548	2625	1300
Hafnium	4032	9750	——
Iron	2795	5432	1706†
Lithium	367	2437	356
Magnesium	1202	2030	1153
Plutonium	1184	6000	1112
Potassium	144	1400	156†S‡
Sodium	208	1616	above 239
Strontium	1425	2102	1328†
Thorium	3353	8132	932†
Titanium	3140	5900	2900
Uranium	2070	6900	below 6900
Zinc	786	1665	1652*
Zirconium	3326	6470	2552*

*Because of the variations in available information, the values given must be considered as being approximate and for guidance only. Generally, if a metal melts and flows before burning, an ignition temperature can be determined.

†Ignition in oxygen

‡S Spontaneous ignition in moist air

Source: NFPA, Quincy, Massachusetts.

TABLE 2-8 Flash Points and Flammable Limits of Some Common Liquids and Gases

Liquid (or gas at ordinary temps.)	Flash point		Flammable limits, percent by volume
	°F	°C	
Acetylene	(Gas)		2.5–81.0+
Benzene	12	−11	1.3–7.1
Ether (ethyl ether)	−49	−45	1.9–36.0
Fuel oil			
Domestic, No. 2	100 (min.)	38	None at ordinary temps.
Heavy, No. 5	130 (min.)	54	None at ordinary temps.
Gasoline (high test)	−36	−38	1.4–7.4
Hydrogen	(Gas)		4.07–75.0
Jet fuel (A & A-1)	110 to 150	43 to 65	None at ordinary temps.
Kerosene (Fuel oil, No. 1)	100 (min.)	38	0.7–5.0
LPG (propane-butane)	(Gas)		1.9–9.5
Lacquer solvent (butyl acet.)	72	22	1.7–7.6
Methane (natural gas)	(Gas)		5.0–15.0
Methyl alcohol	52	11	6.7–36.0
Turpentine	95	35	0.8–(undetermined)
Varsol (standard solv.)	110	43	0.7–5.0
Vegetable oil (cooking, peanut)	540	282	(Ignition temp. = 833°F)

Source: NFPA, Quincy, Massachusetts.

tainer is not cooled by hose streams. The principal fire and explosion prevention measures under such circumstances are: (1) exclusion of sources of ignition, (2) exclusion of air, (3) keeping the liquid in a closed container, (4) ventilation to prevent the accumulation of vapor in the flammable range, and (5) use of an atmosphere of inert gas instead of air.

An arbitrary system of classifying liquids has generally been adopted.

Flammable Liquids. Flammable liquids are any liquid having a flash point below 100°F (38°C) and having a vapor pressure not exceeding 2068.6 mm at 100°F (38°C). Class I liquids include those having flash points below 100°F (38°C) and may be subdivided as follows:

Class IA includes those having flash points below 73°F (23°C) and a boiling point below 100°F (38°C).

Class IB includes those having flash points below 73°F (23°C) and a boiling point at or above 100°F (38°C).

Class IC includes those having flash points at or above 73°F (23°C) and below 100°F (38°C).

Combustible Liquids. Liquids with a flash point at or above 100°F (38°C) are referred to as combustible liquids. They are:

Class II liquids include those having flash points at or above 100°F (38°C) and below 140°F (60°C).

Class IIIA liquids include those having flash points at or above 140°F (60°C) and below 200°F (93°C).

Class IIIB liquids include those having flash points at or above 200°F (93°C).

Some typical liquids would be classed as follows:

Denatured Alcohol	Class IB
Fuel oil	Class II
Gasoline	Class IB
Kerosene	Class II
Peanut oil	Class IIIB
Turpentine	Class IC
Paraffin wax	Class IIIB

Storage and Handling. Proper storage and handling of flammable and combustible liquids are necessary to prevent fire or explosion. Ventilation to prevent accumulations of flammable vapors is of primary importance because there is the possibility of breaks or leaks in the storage and handling in a closed system. It is important to eliminate possible sources of ignition in an area where flammable liquids are stored, handled, or used.

Ventilation of an area where flammable liquids are manufactured or used can be accomplished by natural or mechanical means. Wherever possible, equipment such as compressors, stills, and pumps should be located in a spacious, open area. Most flammable liquids produce heavier-than-air vapors that flow along the ground or floor and settle in depressions. These can travel long distances and be ignited and flash back from a point remote from the origin of the vapors. NFPA 30, "Flammable and Combustible Liquids Code," is the accepted national standard for fire protection of such liquids.

NFPA standards specify the construction, installation, spacing, venting, and diking of above-ground and underground storage tanks, container storage in buildings, loading and unloading practices, safeguards for dispensing the liquids, and standards for transporting the liquids in trucks, ships, or pipelines (Tables 2-9 and 2-10).

Gases

There are many kinds of materials that exist in the form of gas. In general, gases are thought of and described when the substance exists in a gaseous state at normal temperature and pressure 70°F (21°C) and 14.7 psia (101430 N/m^2).

Classification by Chemical Properties. Gases can be broadly classified according to chemical properties, physical properties, or usage. Classification by chemical properties helps to define the hazards of gases to people and in fires.

TABLE 2-9 Storage Limitations for Inside Storage Rooms

Fire protection* provided	Fire resistance, h	Maximum size, ft^2	Allowable loading, gal/ft^2 floor area
Yes	2	500	10
No	2	500	4
Yes	1	150	5
No	1	150	2

*Fire-protection system of sprinkler, water spray, carbon dioxide, dry chemical, halon, or other acceptable type of system.

Source: NFPA, Quincy, Massachusetts.

TABLE 2-10 Storage Limitations for Warehouses or Storage Buildings

		Protected storage* maximum per pile height		Unprotected storage maximum per pile height	
Class liquid	Storage level	gal	ft	gal	ft
IA	Ground & upper floors	2,750	3	660	3
		(50)	(1)	(12)	(1)
	Basement	Not permitted		Not permitted	
IB	Ground & upper floors	5,500	6	1,375	3
		(100)	(2)	(25)	(1)
	Basement	Not permitted		Not permitted	
IC	Ground & upper floors	16,500	6	4,125	3
		(300)	(2)	(75)	(1)
	Basement	Not permitted		Not permitted	
II	Ground & upper floors	16,500	9	4,125	9
		(300)	(3)	(75)	(3)
	Basement	5,500	9	Not permitted	
		(100)	(3)		
Combustible	Ground & upper floors	55,000	15	13,750	12
		(1,000)	(5)	(250)	(4)
	Basement	8,250	9	Not permitted	
		(150)	(3)		

*A sprinkler or equivalent fire-protection system installed in accordance with the applicable NFPA Standard. (Numbers in parentheses indicate corresponding number of 55-gal drums.) *Note 1:* When two or more classes of materials are stored in a single pile, the maximum gallonage permitted in that pile is the smallest of the two or more separate maximum gallonages. *Note 2:* Aisles are provided so that no container is more than 12 ft from an aisle. Main aisles shall be at least 8 ft wide and side aisles at least 4 ft wide. *Note 3:* Each pile is separated from each other pile by at least 4 ft. When stored on suitably protected racks or when the storage is suitably protected, containers may be piled up but no closer than 3 ft to the nearest beam, chord, girder, or other obstructions. Good practice is to maintain 3 ft clearance below sprinkler deflectors or discharge orifices or other overhead fire protection systems.

Source: NFPA, Quincy, Massachusetts.

Flammable Gases. Any gas that will burn in the normal concentrations of oxygen in the air is a flammable gas. Like flammable liquid vapors, the burning of this gas in air is in a range of gas-air mixture (the flammable range).

Nonflammable Gases. Nonflammable gases will not burn in air or in any concentration of oxygen. A number of nonflammable gases, however, will support combustion. Such gases are often referred to as "oxidizers" or oxidizing gases. Common oxidizers are oxygen or oxygen in a mixture with other gases.

Nonflammable gases that will not support combustion are usually called *inert gases.* Common inert gases are nitrogen, carbon dioxide, and sulfur dioxide.

Toxic Gases. Toxic gases endanger life when inhaled. Gases such as chlorine, hydrogen sulfide, sulfur dioxide, ammonia, and carbon monoxide are poisonous or irritating when inhaled.

Reactive Gases. Reactive gases react with other materials or within themselves by a reaction other than burning. When exposed to heat and shock, some reactive gases rearrange themselves chemically. Such gases can produce hazardous quantities of heat or reaction products. Fluorine is a highly reactive gas. At normal temperatures and pressures it will react with most organic and inorganic substances, often fast enough to result in flaming. Other examples of reactive gases are acetylene and vinyl chloride.

Classification by Physical Properties. Gases can also be classified by their physical properties. They can be compressed, liquefied, or cryogenic.

Compressed Gases. A compressed gas is at normal temperature inside a gas container and exists solely in the gaseous state under pressure; common compressed gases are hydrogen, oxygen, acetylene, and ethylene.

Liquefied Gases. Liquefied gases can be liquefied relatively easily and stored at ordinary temperatures at relatively high pressure. Liquefied gas exists in both liquid and gaseous states; at storage pressure, both the liquid and gas in the liquefied-gas container are in equilibrium and will remain so as long as any liquid remains in the container. Liquefied gas is more concentrated than compressed gas.

Cryogenic Gases. Cryogenic gases are stored in a completely liquid state. They must be maintained in their containers as low-temperature liquids at relatively low pressure. Cryogenic gases must be stored in special containers that allow the gas from the liquid to escape in order to prevent a pressure buildup caused by the production of the gaseous state within the container, which would result in container failure.

Classification by Usage. An understanding of gases as they are classified by usage is important to those involved in fire protection because the terms of these classifications are used in codes, standards, and general industrial and medical terminology (Table 2-11).

Fuel Gases. These gases are customarily used for burning with air to produce heat which in turn is used as a source of heat (comfort and process), power, or light; the principal and most widely used fuel gases are natural gas and the liquefied petroleum gases, butane and propane.

Industrial Gases. These are classified by chemical properties customarily used in industrial processes, for welding and cutting, heat treating, chemical processing, refrigeration, water treatment, etc.

Medicinal Gases. These are for medical purposes such as anesthesia and respiratory therapy; cyclopropane, oxygen, and nitrous oxide gases are common medical gases.

Hazardous Materials

Corrosive Chemicals. Corrosive chemicals are usually strong oxidizing agents that can increase fire hazards. Caustics, which are classified as water- and air-reactive chemicals, are also corrosive.

Inorganic Acids. Concentrated aqueous solutions of inorganic acids are not in themselves combustible; however, in addition to their corrosive and destructive effect on living tissue, their chief fire hazard results from the possibility of their mixing with combustible materials or other chemicals, which could result in fire or explosion; almost all corrosive chemicals are strong oxidizing agents.

Halogens. These salt-producing chemicals are very active. They are noncombustible but will support combustion; presence of halogens cause turpentine, phosphorus, and finely divided metals to ignite spontaneously. The fumes are poisonous and corrosive.

Storage and Fire Protection for Corrosive Chemicals. Storage of corrosive chemicals should be provided with two considerations in mind: (1) protection against the dam-

TABLE 2-11 Combustion Properties of Common Flammable Gases

Gas	Btu/ft³ (Gross)	Limits of flammability percent by volume in air: Lower	Upper	Specific gravity (air = 1.0)	Air needed to burn 1 ft³ of gas, ft³	Ignition temp., °F
Natural gas						
High inert type*	958–1051	4.5	14.0	0.660–0.708	9.2	——
High methane type†	1008–1071	4.7	15.0	0.590–0.614	10.2	900–1170
High Btu type‡	1071–1124	4.7	14.5	0.620–0.719	9.4	——
Blast furnace gas	81– 111	33.2	71.3	1.04–1.00	0.8	——
Coke oven gas	575	4.4	34.0	.38	4.7	——
Propane (commercial)	2516	2.15	9.6	1.52	24.0	920–1120
Butane (commercial)	3300	1.9	8.5	2.0	31.0	900–1000
Sewage gas	670	6.0	17.0	.79	6.5	——
Acetylene	1499	2.5	81.0	.91	11.9	581
Hydrogen	325	4.0	75.0	.07	2.4	752
Anhydrous ammonia	386	16.0	25.0	.60	8.3	1204
Carbon monoxide	314	12.5	74.0	.97	2.4	1128
Ethylene	1600	2.7	36.0	.98	14.3	914
Methyl acetylene, propadiene, stabilized§	2450	3.4	10.8	1.48	——	850

*Typical composition CH_4, 71.9–83.2%; N_2, 6.3–16.20%.
†Typical composition CH_4, 87.6–95.7; N_2, 0.1–2.39.
‡Typical composition CH_4, 85.0–90.1; N_2, 1.2–7.5.
§MAPP® gas.
Source: NFPA, Quincy, Massachusetts.

aging effect of corrosive chemicals on living tissue and (2) guarding against any fire and explosion hazard that might be associated with the corrosive chemical.

Inorganic Acids. These should be stored in cool, well-ventilated areas that are not exposed to the sun or other chemical and waste materials; they should be protected from freezing temperatures. Water in spray form is the recommended procedure for fighting fires in inorganic acid storage areas; in fires involving perchloric acid, extra care should be taken since this may mix with other organic materials and result in an explosion.

Halogens. Fluorine and chlorine should be stored in special containers; fluorine may be safely stored in nickel or Monel cylinders. Impurities or moisture in the cylinder may cause an explosive reaction; chlorine, a serious inhalation hazard, should be stored in areas where ventilation is a prime consideration. Where chlorine leakage is suspected, use a self-contained breathing apparatus; fluorine requires protection of the self-contained breathing apparatus and special protective clothing.

Corrosive Vapors. Very often, corrosive vapors are a by-product of an industrial or chemical process. Ducts must be used to carry these vapors safely from the area. The type of duct used is determined by the vapor. Heavier gauge metal may be sufficient, although a protective coating or special lining may be required in the ducts. Stainless steel, asbestos cement, and plastic linings have been used with success depending on the corrosive vapor.

Radioactive Materials

Fire Protection for Radioactive Materials. The main concern in fire protection of radioactive materials is to prevent the release (or control the release) of these materials during fire extinguishment. Although fire protection operations are similar to those used when nonradioactive materials are involved, fire involving buildings or areas containing radioactive materials presents two additional considerations: (1) the presence of harmful radioactive materials might necessitate that normal fire-fighting procedures be changed and (2) because of the presence of radioactive matter, delay in salvage and resumption of normal operations may occur.

DESIGN AND CONSTRUCTION FOR FIRE SAFETY

The architectural design and building methods and materials used for a structure will often determine how a fire will be confined or will spread. Firesafety objectives must be determined before a facility is designed. Often *code compliance* is not a sufficient standard to meet the level of risk acceptable to a particular organization.

Fire safety design decisions are necessary in at least three objective areas: life safety, property protection, and continuity of operations. Systems-analysis techniques are used by fire protection engineers to determine the level of protection to meet design objectives.

Building and Site Planning for Fire Safety

Two categories of decisions should be made in the design process to provide effective firesafe design: interior building functions and exterior site planning. Building fire defenses, active and passive, should be designed so that the building assists in the suppression of fire.

Firesafety Planning for Buildings

Interior layout, circulation patterns, finish material, and building services are all important fire-safety considerations in building design. Building design also has a significant influence on the efficiency of fire-department operations. Manual fire-suppression activities should be considered during architectural design.

Fire Fighting Accessibility to Building's Interior. This includes access to the building itself as well as access to the interior of the building. Spaces in which fire fighting access and operations are restricted because of architectural, engineering, or functional requirements should be provided with effective protection. A complete automatic sprinkler system is often the best solution. Other methods which may be used include access panels in interior walls and floors, fixed nozzles in floors with fire department connections, and roof vents and access openings.

Ventilation. This is of vital importance in removing smoke, gases, and heat. Appropriate skylights, roof hatches, emergency escape exits, and similar devices should be provided when the building is constructed.

Ventilation of building spaces performs the following important functions:

1. Protection of life by removing or diverting toxic gases and smoke
2. Improvement of the environment in the vicinity of the fire by removal of smoke and heat; enables fire fighters to advance close to the fire
3. Control of the spread or direction of fire by setting up air currents that cause the fire to move in a desired direction
4. Provision of a release for unburned, combustible gases before they acquire a flammable mixture, thus avoiding a backdraft explosion

NFPA 204, "Smoke and Heat Venting Guide" recommends automatic smoke venting of large industrial buildings.

Curtain Boards as Venting Aids. In large-area buildings, unless vented areas are subdivided by means of walls or partitions, curtain boards are essential. The function of curtain boards is to delay and limit the horizontal spread of heat by providing the horizontal confinement needed to obtain the desired *stack* action. The depth of such curtain boards largely determines the height of the stack which affects the capacity of the vent. If an area is protected by automatic sprinklers, curtain boards have added values: confinement of heat to speed up operation of sprinklers over the fire and obstructed lateral spread of heat to minimize the operation of an excessive number of sprinklers. See Fig. 2-2.

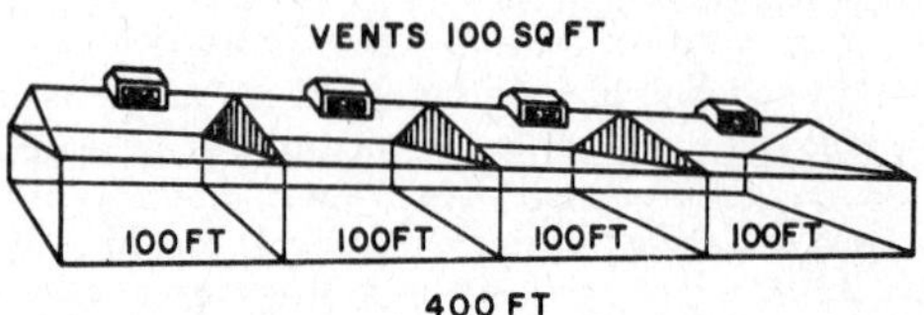

Figure 2-2 Curtain boards and roof vents. *(From* Fire Protection Handbook, *14th ed., G. P. McKinnon (ed.), NFPA, Quincy, Massachusetts, 1976.)*

Firesafety Planning for Sites

Proper building design for fire protection should include a number of factors outside the building itself. The site on which the building is located will influence the design. Among the more significant features are traffic and transportation conditions, fire department accessibility, and water supply. Inadequate water mains and poor spacing of hydrants have contributed to the loss of many buildings.

Traffic and Transportation. Fire department response time is a vital factor in building design considerations; traffic access routes, traffic congestion at certain times of the day, traffic congestion from highway entrances and exits, and limited-access highways have significant effects on fire department response distances and response time.

Fire Department Access to the Site. Is the building easily accessible to fire apparatus? Ideal accessibility occurs where a building can be approached from all sides by fire-department apparatus; congested areas, topography, or buildings and structures located appreciable distances away from the street can cause difficulty and prevent effective use of fire apparatus. Inadequate attention to site details can place the building in an unnecessarily vulnerable position; if fire defenses are compromised by preventing adequate fire department access, the building itself must make up the difference in more complete internal protection.

Water Supply to the Site. Are the water mains adequate, and are the hydrants properly located? The number, location, and spacing of hydrants and the size of the water mains are vital considerations; consult local standards and insurance requirements.

Interior Finish

Interior finish is defined as those materials that make up the exposed interior surface of wall, ceiling, and floor constructions. The common interior-finish materials are wood, plywood, plaster, wallboards, acoustical tile, insulating and decorative finishes, plastics, and various wall coverings.

Some building codes include floor coverings under their definition of interior finishes.

Fire Tests

It is possible to estimate the damage that fire can cause to a building by studying: (1) the amount and kind of combustible materials in the buildings and (2) the way they are distributed throughout the building. These two factors indicate the rate of combustion, the duration of the fire, and the degree of difficulty to extinguish the fire.

The effects of fire on the components of a building (such as the columns, floors, walls, partitions, and ceiling or roof assemblies) are tested against both time and temperature.

Results of the tests are recorded in hours or minutes and indicate the duration of fire resistance.

Ratings for flame spread of interior finish materials have been established with the use of the 25 ft (7.6 m) tunnel developed by A. J. Steiner at Underwriters Laboratories Inc. These ratings are used in the NFPA **Life Safety Code®** and in other codes to indicate the areas in which finishes of varying flame spread characteristics may be used (Table 2-12). The five classifications used in the **Life Safety Code®** are:

Class	*Flame-Spread Range*
A	0–25
B	26–75
C	76–200
D	201–500
E	over 500

The higher the flame spread, the greater the hazard. For example, in a new hospital, Class A materials would be required for most areas, and Class D or E materials would not be permitted at all.

Materials are measured on a relative scale with cement asbestos board rated 0 and red oak flooring rated 100. Some highly combustible wallboards have received ratings as high as 1500.

One of the best sources of information showing the wide variety of building assemblies and giving the fire-resistance ratings of beams, columns, floors, walls, and partitions is the Underwriters Laboratories Inc. *Fire Resistance Index.*

Confinement of Smoke and Fire

Design criteria for plant facilities are generally based on estimated fire severity (Table 2-13). Specific industrial-hazard fire-severity data may be obtained from insurance organizations (Table 2-14 and Fig. 2-3).

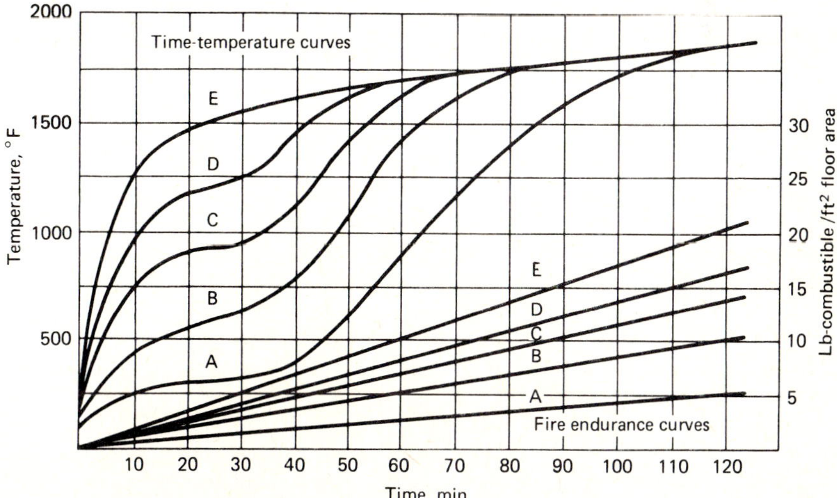

Figure 2-3 Possible classification of building contents for fire severity and duration. The straight lines indicate the length of fire endurance based upon amounts of combustibles involved. The curved lines indicate the severity expected for the various occupancies (see Table 2-14). There is no direct relationship between the straight and curved lines, but, for example, 10 lb of combustibles per square foot will produce a 90-min fire in a C occupancy, and a fire severity following the time-temperature curve C might be expected. *(From R. Tuve,* Principles of Fire Protection Chemistry, *NFPA, Quincy, Massachusetts, 1976.)*

TABLE 2-12 Summary of Life Safety Code Requirements for Interior Finish

Occupancy	Class of interior finish[a]		
	Exits	Access to exits	Other spaces
Places of assembly—class A[b]	A	A	A or B
Places of assembly—class B[c]	A	A	A or B
Places of assembly—class C[d]	A	A	A, B, or C
Educational	A	A	A, B, or C
Educational—unsprinklered open-plan buildings[e]	A	A or B }	A or B
Flexible-plan buildings[f]	A	A }	C or low height partitions
Institutional, existing—hospitals, nursing homes, residential-custodial care	A or B	A or B	A or B
Institutional, new—hospitals, nursing homes, residential-custodial care	A	A	A B in individual room with capacity-not more than 4 persons
Residential, new—apartment houses	A or B	A or B	A, B, or C
Residential, existing—apartment houses	A or B	A, B, or C	A, B, or C
Residential—dormitories	A or B	A, B, or C	A, B, or C
Residential, new—1- and 2-family, lodging or rooming houses			A, B, or C
Residential, existing—1- and 2-family, lodging or rooming houses			A, B, C, or D
Residential, new—hotels	A or B	A or B	A, B, or C
Residential, existing—hotels	A or B	(1) A or B if required path of exit travel; (2) A, B, or C if not used as required path of exit travel	A, B, or C
Mercantile—class A[g]	A or B		Ceilings—A or B Walls—A, B, or C
Mercantile—class B[h]	A or B		Ceilings—A or B Walls—A, B, or C
Mercantile—class C[i]	A or B		A, B, or C
Office	A or B	A or B	A, B, or C
Industrial	A, B, or C	A, B, or C	A, B, or C
Towers	A or B		A or B

[a]There are five classes of interior finish: Class A, flame spread 0–25; Class B, flame spread 25–75; Class C, flame spread 75–200; Class D, flame spread 200–500; and Class E, over 500. Where a standard system of automatic sprinklers is installed, an interior finish with a flame spread rating not over Class C may be used in any location where Class B is normally specified, and with a rating of Class B in any location where Class A is normally specified, unless specifically prohibited elsewhere in the *Life Safety Code*.

[b]Class A Places of Assembly—1000 persons or more.

[c]Class B Places of Assembly—300 to 1000 persons

[d]Class C Places of Assembly—100 to 300 persons

[e]Open plan buildings—includes all buildings where no permanent partitions are provided between rooms or between rooms and corridors.

[f]Flexible plan buildings have movable corridor walls and movable partitions of full height construction with doors leading from rooms to corridors.

[g]Class A Mercantile Occupancies—stores having aggregate gross area of 30,000 ft^2 or more, or utilizing more than 3 floor levels for sales purposes.

[h]Class B Mercantile Occupancies—stores of less than 30,000 ft^2 aggregate gross area, but over 3,000 sq ft,, or utilizing any floors above or below street floor level for sales purposes, except that if more than 3 floors are utilized, store shall be Class A.

[i]Class C Mercantile Occupancies—stores of 3000 ft^2 or less gross area, used for sales purposes on street level only. (A single balcony or mezzanine floor with less than half the area of the street level floor and which is used for sales purposes is not counted as another floor.)

Source: NFPA, Quincy, Massachusetts.

TABLE 2-13 Estimated Fire Severity for Offices and Light Commercial Occupancies*

Combustible content, total, including finish, floor and trim, lb/ft²	Heat potential assumed Btu/ft²†	Equivalent fire severity, approximately equivalent to that of test under standard curve for the following periods
5	40,000	30 min
10	80,000	1 h
15	120,000	1½ h
20	160,000	2 h
30	240,000	3 h
40	320,000	4½ h
50	380,000	7 h
60	432,000	8 h
70	500,000	9 h

*From Gordon P. McKinnon (ED.) *Fire Protection Handbook,* NFPA, Quincy, 1976. Data applying to fire-resistive buildings with combustible furniture and shelving.

†Heat of combustion of contents taken at 8000 Btu/lb up to 40 lb/ft²; 7600 Btu/lb for 50 lb, and 7200 Btu for 60 lb and more to allow for relatively greater proportion of paper. The weights contemplated by the tables are those of ordinary combustible materials, such as wood, paper, or textiles.

TABLE 2-14 Fire Severity Expected by Occupancy (See Fig. 2-3)

Temperature curve A (slight)
- Well-arranged office, metal furniture, noncombustible building
- Welding areas containing slight combustibles
- Noncombustible power house
- Noncombustible buildings, slight amount of combustible occupancy

Temperature curve B (moderate)
- Cotton and waste-paper storage (baled) and well-arranged, noncombustible building
- Paper-making processes, noncombustible building
- Noncombustible institutional buildings with combustible occupancy

Temperature curve C (moderately severe)
- Well-arranged combustible storage, e.g., wooden patterns, noncombustible buildings
- Machine shop having noncombustible floors

Temperature curve D (severe)
- Manufacturing areas, combustible products, noncombustible building
- Congested combustible storage areas, noncombustible building

Temperature curve E (standard fire exposure—severe)
- Flammable liquids
- Woodworking areas
- Office, combustible furniture and buildings
- Paper working, printing, etc.
- Furniture manufacturing and finishing
- Machine shop having combustible floors

Source: NFPA, Quincy, Massachusetts.

Fire Doors

Fire doors are the most widely used and accepted means of protecting vertical and horizontal openings. Suitability of fire doors is determined by nationally recognized testing laboratories; doors that have not been tested cannot be relied upon for effective protection. The doors are tested as they are installed in the field, that is, with the frame, hardware, wired-glass panels, and other accessories necessary to complete the installation.

Nearly all building codes use NFPA 80, "Standard for Fire Doors and Windows." This standard establishes the minimum ratings for the five most commonly encountered types of openings in walls. They are as follows:

1. Class A openings are in walls separating buildings or dividing a single building into fire areas. Doors for the protection of these openings have a fire protection rating of 3 h.

2. Class B openings are in enclosures of vertical communication through buildings (stairs, elevators, etc.). Doors for the protection of these openings have a fire protection rating of 1 or 1½ h.
3. Class C openings are in corridor and room partitions. Doors for the protection of these openings have a fire protection rating of ¾ h.
4. Class D openings are in exterior walls which are subject to severe exposure from the outside of the building. Doors and shutters for the protection of these openings have a fire protection rating of 1½ h.
5. Class E openings are in exterior walls which are subject to moderate or light fire exposure from outside of the building. Doors, shutters, or windows for the protection of these openings have a fire-protection rating of ¾ h.

It is important to note that this classification applies to the various types of openings and not to the fire door itself. A fire door is not a Class A fire door. It ia a door that is suitable for a Class A opening.

The following excerpt from the NFPA *Fire Protection Handbook** is a description of the various types of construction for fire doors:

> *Composite Doors.* These are of the flush design and consist of a manufactured core material with chemically impregnated wood-edge banding and untreated wood-face veneers, or laminated plastic faces, or surrounded by and encased in steel.
>
> *Hollow-Metal Doors.* These are of formed steel of the flush and paneled designs of No. 20 gage or heavier steel.
>
> *Metal-Clad (Kalamein) Doors.* These are of flush and paneled design consisting of metal covered with steel of 24 gage or lighter.
>
> *Sheet-Metal Doors.* These are of formed No. 22 gage or lighter steel and of the corrugated, flush, and paneled designs.
>
> *Rolling Steel Doors.* These are of the interlocking steel slat design or plate steel construction.
>
> *Tin-Clad Doors.* These are of two- or three-ply wood-core construction, covered with No. 30 gage galvanized steel or terneplate (maximum size 14 by 20 in.) or No. 24 gage galvanized steel sheets not more than 48 in. wide.
>
> *Curtain-Type Doors.* These consist of interlocking steel blades or a continuous formed-spring-steel curtain in a steel frame.
>
> The suitability of a fire door should be judged on the class of opening in which it is to be installed, not on the fire-resistance rating of the wall in which it is to be installed....
>
> If the opening is in a wall dividing a building into separate fire areas, the door should be suitable for installation in a Class A opening (3-h fire protection rating). The same door can be used whatever the fire resistance rating of the wall. If a wall encloses a vertical communication, the door should be suitable for a Class B opening.

The NFPA fire doors and windows standard† gives recommendations on the installation of suitable approved doors, windows, and shutters, and it also specifies how the opening should be constructed and how the door or window should be mounted, equipped, and operated.

FIRE-DETECTION AND-ALARM SYSTEMS

There are several general systems and many devices which can be effectively used to detect fire and transmit a warning. This section briefly describes this equipment.

Heat Detectors

Heat-detection devices are categorized in two ways: (1) those that respond when the detection element reaches a predetermined temperature (fixed-temperature-types), and

**Fire Protection Handbook,* 14th ed., NFPA, Quincy, Massachusetts, 1976, p. 6-84.
†NFPA 80, "Standard for Fire Doors and Windows," NFPA, Quincy, Massachusetts, 1979.

(2) those that respond to an increase in heat at a rate greater than some predetermined value (rate-of-rise-types). Some devices combine both principles. The same principles apply whether the devices are of the spot-pattern type, in which the thermally sensitive element is a unit, or the line-pattern type, in which the element is continuous along a line or a circuit.

Fixed-Temperature Detectors

Thermostats are the most widely used fixed-temperature heat detectors in signaling systems.

Bimetallic Thermostats. The common form of thermostat is the bimetallic type that utilizes the different coefficients of expansion of two metals under heat to cause a movement resulting in closing of electrical contacts (Fig. 2-4).

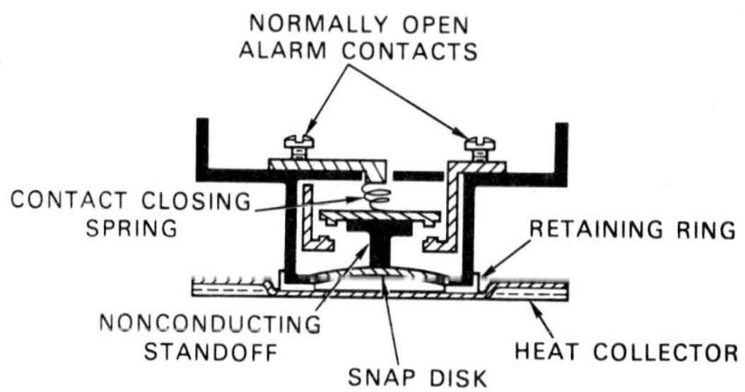

Figure 2-4 Spot-type, fixed-temperature snap-disk detector.

Snap-Action Disk Thermostats. A metal disk goes from concave to convex when the temperature rating of the thermostat is reached. One special advantage of these thermostats is that when the temperature goes down, they are restored to their original condition.

Line Thermostats. The *thermostat cable* is a line type of thermostat. The cable is made up of two metals separated from each other by a heat-sensitive covering applied directly to the wires. When the rated temperature is reached, the covering melts and the two wires come in contact to initiate an alarm. The section of wire affected must be replaced after operation.

Other Types. Other forms of fixed-temperature heat detectors are the *fusible link,* occasionally employed to restrain operation of an electrical switch until the point of fusion is reached, and the *quartzoid bulb thermostat,* which depends on removal of the restriction by breaking the bulb. Both of these units require replacement after operation.

Rate-of-Rise Detectors

Fire detectors that operate on the rate-of-rise principle function when the rate of temperature increase exceeds a predetermined rate. Detectors of this type combine two functioning elements, one of which initiates an alarm on a rapid rise of temperature, while the other acts to delay or prevent an alarm on a slow temperature rise. Advantages to rate-of-rise devices are: (1) they can be set to operate more rapidly under most conditions than can fixed-point devices; (2) they are effective across a wide range of ambient temperatures; (3) they recycle rapidly and are usually readily available for continued service; (4) they tolerate slow increases in ambient temperature without giving an alarm. The disadvantages of rate-of-rise detectors for some applications are their susceptibility to false alarms where there is a rapidly increasing temperature and their possible failure to respond to a fire that propagates very slowly.

Pneumatic-Tube Detectors. These operate on the rate-of-rise principle. When the temperature increases at a certain rate, the air in the tube expands and causes a diaphragm to move and close a cirucit, thus causing an alarm. The device will not cause an alarm if the temperature rise is too slow.

Combined Rate-of-Rise and Fixed-Temperature Detectors

Thermostats have been developed to take advantage of the rate-of-rise feature to sense a fast-developing fire; the fixed-temperature part takes care of a fire whose growth is slow. The typical form of the rate-of-rise thermostat is a vented air chamber that heats up in a flexible diaphragm carrying electric contacts (Fig. 2-5). Heat outside the chamber causes air within the chamber to expand. When such expansion exceeds the capacity of the vent to relieve pressure, the diaphragm is flexed, thus closing the electric contacts. Slow changes in ambient temperature near the chamber allow it to "breathe" through its vent, and the diaphragm is not moved sufficiently to cause an alarm.

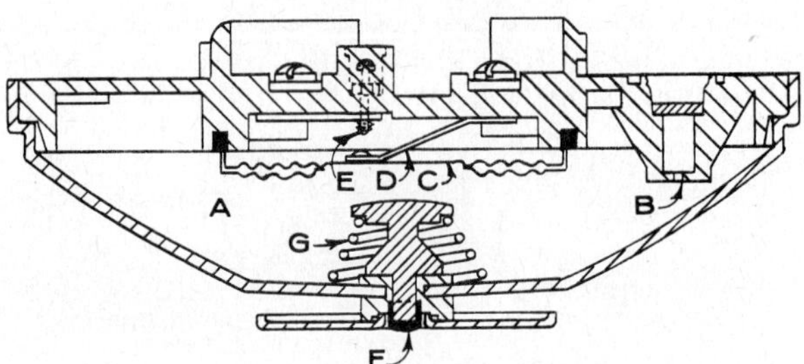

Figure 2-5 A spot-type combination rate-of-rise, fixed-temperature device. The air in chamber A expands more rapidly than it can escape from vent B. This causes pressure to close electrical contact D between diagragm C and insulated screw E. Fixed-temperature operation occurs when fusible alloy F melts releasing spring G which depresses the diaphrahm closing the contact points.

Rate-Compensation (Anticipation and Differentiation) Devices

These provide an assured actuation at some predetermined maximum temperature and compensate for changes in rates of temperature rise (Fig. 2-6).

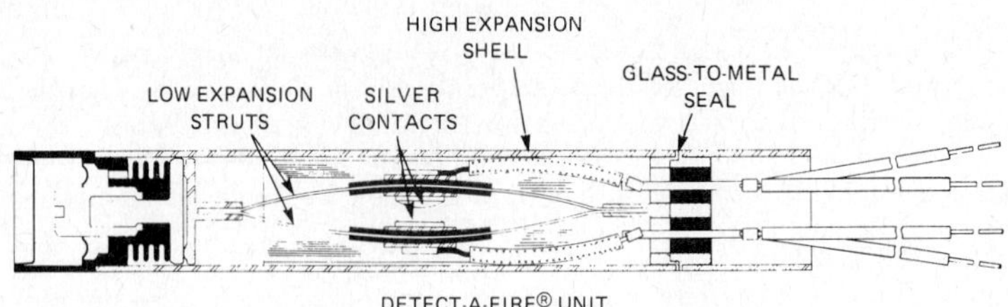

Figure 2-6 Rate-compensation heat detector. *(From* Fire Protection Handbook, *14th ed., G. P. McKinnon (ed.), NFPA, Quincy, Massachusetts, 1976.)*

Smoke Detectors

There are four types of smoke detectors: (1) photoelectric, (2) beam-type, (3) ionization, and (4) sampling detectors.

Photoelectric Detectors

These detectors operate on a beam of light. The smoke either obscures a beam of light directly, or enters a refraction chamber where the smoke reflects the light into the photocell. The change in electric current resulting from either partial obscuring of a photoelectric beam by smoke particles, or the scattering of light onto a photosensitive device, causes an alarm sound when the smoke reaches a sufficient density. (Fig. 2-7 *a* and *b*, respectively)

Beam-Type Detectors

These employ a light beam that is carried between elements at extreme ends or sides of the protected area and crosses the area to be protected. The beam is projected into a photosensing cell. Smoke between the light source and the receiving photocell reduces the light that reaches the cell, activating the alarm (Fig. 2-8).

Ionization Detectors

These detectors consist of one or more ionization chambers and the necessary related amplification circuits. The ionization detector has as a sensing element, the ionization chamber, in which air is made electrically conductive (ionized) by a minute source of radioactive material. A voltage applied across the ionization chamber causes a very small electric current to flow as the ions travel to the electrode of opposite polarity. When smoke particles enter the chamber, they attach themselves to the ions and cause a reduction in mobility and thus a reduction in current flow. The reduced current flow increases the voltage on the electrodes which, when reaching a predetermined level, results in an alarm (Fig. 2-9).

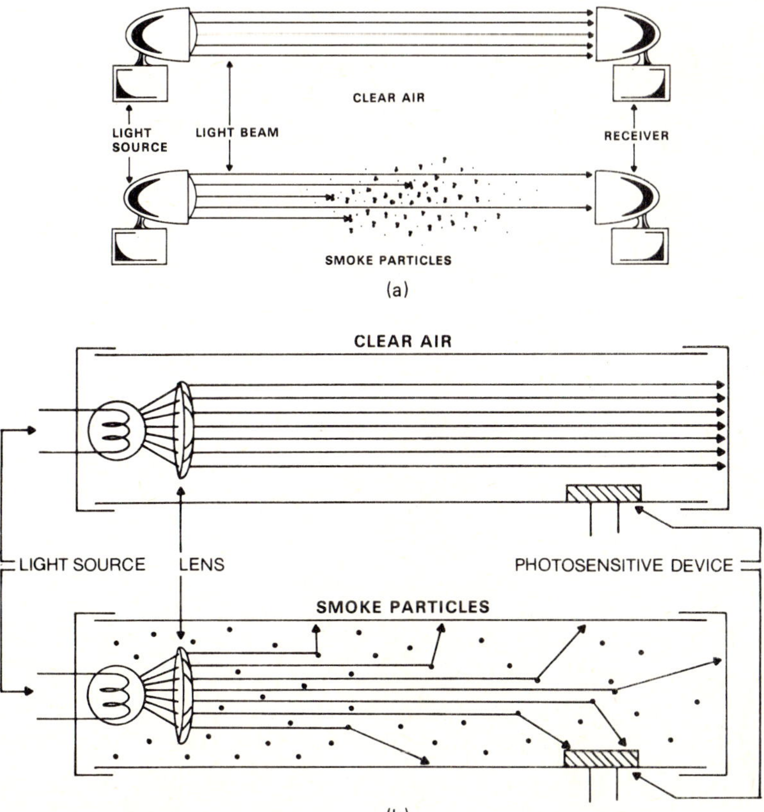

Figure 2-7 Principle of operation for (*a*) a photoelectric obscuration smoke detector, and (*b*) a photoelectric scattering smoke detector.

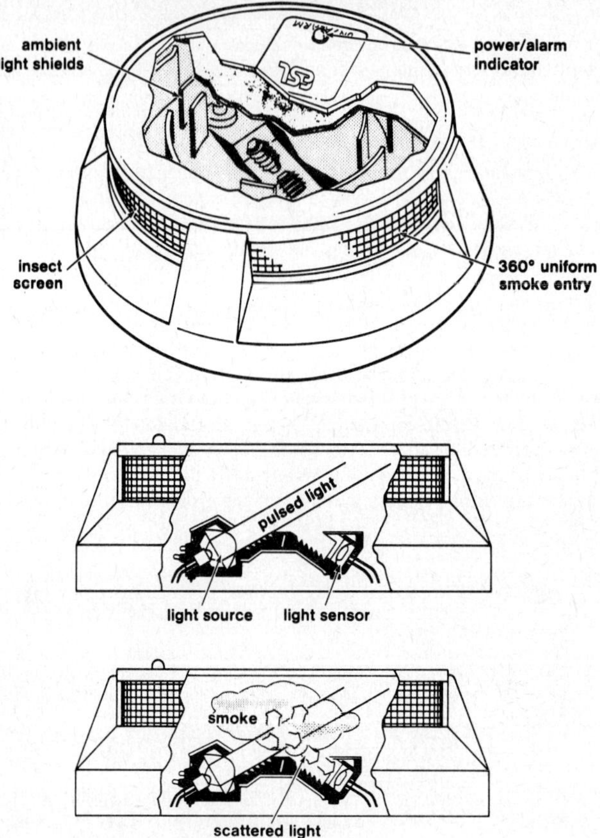

Figure 2-8 Cross-sectional view of a photoelectric light-scattering smoke detector. *(Electro Signal Lab., Inc.)*

Sampling Detectors

These consist of tubing distributed from the detector unit into the area(s) to be protected. An air pump draws air from the protected area back to the detector. A cloud-chamber smoke detector is a form of sampling detector. The air pump draws a sample of air into a high-humidity chamber; the pressure is lowered slightly. If smoke particles are present, the moisture in the air condenses on them forming a cloud in the chamber. The density of this cloud is measured by the photoelectric principle. When the density is greater than a predetermined level, the detector responds to the smoke and the alarm is activated.

Flame Detectors

There are four basic types of flame detectors: (1) infrared, (2) ultraviolet, (3) photoelectric, and (4) flame flicker.

Infrared and Ultraviolet

These detectors have sensing elements responsive to radiant energy outside the range of human vision.

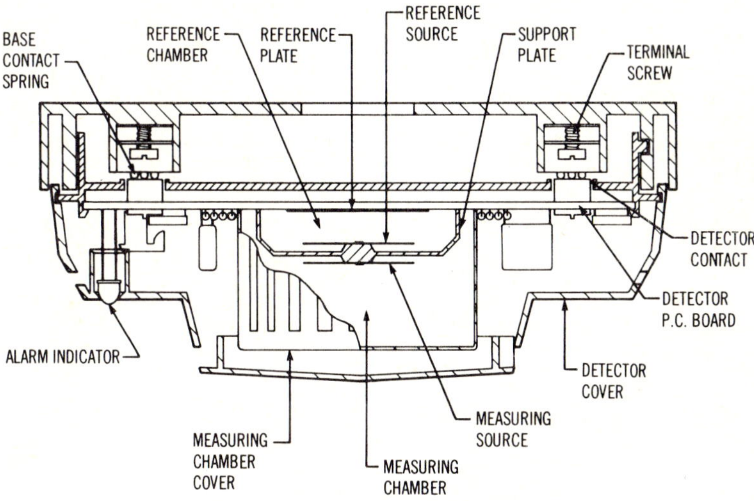

Figure 2-9 Cross-sectional view of an ionization smoke detector. *(Pyrotronics, Inc.)*

Photoelectric

This type employs a photocell that either changes its electric conductivity or produces an electric potential when it is exposed to radiant energy.

Flame Flicker

This is a photoelectric type of detector that includes means to prevent response to visible light unless the observed light is modulated at a frequency characteristic of the flicker of a flame.

Gas Detectors

Various detection devices will monitor the amount of flammable gases or vapors in an area. Portable gas detectors are used to detect the presence of combustible gas or vapor in basements, sewers, manholes, etc. Other devices will analyze the air samples brought into the device from various points. Gas and vapor testing equipment is valuable for preventing fires and explosions in petroleum and chemical plants and in industries where combustible vapors may be generated.

Fire-detection devices are usually installed in systems which combine manually activated fire-alarm stations and audible and visual warning devices. They may also be connected to fire-suppression systems in some hazardous areas.

Alarm Systems

The detection and alarm systems in a plant building should be connected to a constantly supervised monitoring system. The most common systems are described below.

Local Systems

These systems produce a signal manually or automatically at the protective premises for an alarm of fire and for required supervisory services, including supervision of a security guard's rounds, supervision of sprinkler water-flow alarm service and of sprinkler systems, etc. Local systems are used for the protection of property and for the protection of life by indicating the necessity for evacuation of the building.

Proprietary Systems

Proprietary systems are used for individual properties where the system is under constant supervision by competent and experienced personnel in a central supervisory station at the property protected. Such systems are usually found in large industrial plants; signals are received at a central supervisory station where experienced operators are on duty at all times. The central supervisory station is under the control of the owner or occupant of the protected property and is usually on or near that property.

Remote-Station Systems

These are usually used to protect premises on which there is frequently no one present. The signal is received at fire-alarm headquarters or at the office of a communications agency, usually located at a distance from the protected property. Signals are transmitted and received on privately owned equipment; the agency receiving the signals may be a municipal fire department or a communications agency capable of receiving the signals and acting upon them.

Auxiliary Systems

This type connects devices in the protected plant with the municipal fire alarm system. Alarms are received at fire-alarm headquarters on the same equipment and by the same alerting methods as alarms transmitted from municipal street boxes. Signals are recorded at a municipal fire department; connecting facilities between the protected property and the fire department are part of the municipal fire alarm system. Devices in the protected plant are customarily owned and maintained by the property owner. Equipment that connects the devices to the city's circuits is owned and maintained by the municipality, or leased by it, as part of the municipal alarm system and limited to alarm service only.

Central Station Systems

Central station systems are operated by firms whose principal business is the furnishing and maintaining of supervised signaling service. The central station services properties subscribing to the service; alarm and signaling devices on the subscribers' property are connected to the central station where operators are on hand to receive the signal and take the appropriate action. Central station operators retransmit alarms to the fire department.

Standards for the installation and maintenance of fire detection and alarm equipment are listed in the last section of this chapter.

FIRE-SUPPRESSION SYSTEMS

This section describes systems which extinguish or control fires.

Water Supply

The most common type of fire-suppression systems rely on water. Therefore it is essential that adequate supplies of water be provided and maintained. A thorough discussion of water-supply systems is presented in Sec. 6 of this handbook.*

The plant water-supply system or nearby public water supply will be the primary fire-suppression system used by the plant fire brigade or public fire department. Water must be provided in quantity and pressure sufficient for supplying automatic sprinkler systems and fire hoses, in addition to normal plant requirements. When the public water supply is inadequate for plant protection, supplemental private supplies are necessary.

*See Rosaler and Rice: *Standard Handbook of Plant Engineering*, 1983.

Pipe Networks

The minimum recommended pipe size for fire protection is 6 in; the pipe network is looped in a grid pattern, and no leg is more than 600 ft long.

It is usually cost-effective to use larger pipe sizes since the installation costs are relatively the same (Table 2-15).

TABLE 2-15 Comparison of Pipe Capacity

Size of pipe, in	Relative capacity
6	1.0
8	2.1
10	3.8
12	6.2
14	9.3
16	13.2

Source: NFPA, Quincy, Massachusetts.

Wherever possible pipe networks should be arranged in loops to provide water flow from two directions when necessary (Fig. 2-10).

A pipe 8 in or larger is necessary to supply:

- More than one hydrant on a dead-end main
- A dead-end main exceeding 500 ft
- Two hydrants on a loop exceeding 1500 ft
- Three hydrants on a loop exceeding 1000 ft
- Four or more hydrants
- Hydrants with three or four outlets
- Water at low pressure

Fire Hydrants

Fire hydrants are provided on public mains to allow the fire department to draw water with mobile pumpers to supply sprinkler and standpipe systems, or hose streams. Fire hydrants are provided on private mains to allow the fire brigade or fire department to supply hose streams and where the supply is adequate to support sprinkler and standpipe systems with mobile pumpers.

Hydrants are available in wet-barrel (California, Fig. 2-11) and dry-barrel (base valve, Fig. 2-12) types. Base-valve hydrants are necessary where there is any chance of freezing.

Hydrants on plant pipe networks should be located every 250 ft and about 50 ft from the buildings protected. They must be protected from damage by vehicles or machinery.

The available water flow for fire suppression is determined by flow testing the hydrant system. The water flow at 20 psig is calculated since this is the minimum pressure required by fire department pumpers.

Valves in pipe lines supplying fire-protection water are generally required to be indicating valves. These include underground gate valves with indicator post, underground butterfly valves with indicator post, and outside screw and yoke (OS&Y) gate valves.

Incorrectly shut valves have been the primary cause of sprinkler systems failing to control fires.

Fire Pumps

Fire pumps are essentially the same as water-supply pumps discussed in Sec. 6.* Additional considerations required for fire pumps include:

- Use of equipment approved for fire pumps
- Use of approved accessories

*See Rosaler and Rice: *Standard Handbook of Plant Engineering,* 1983.

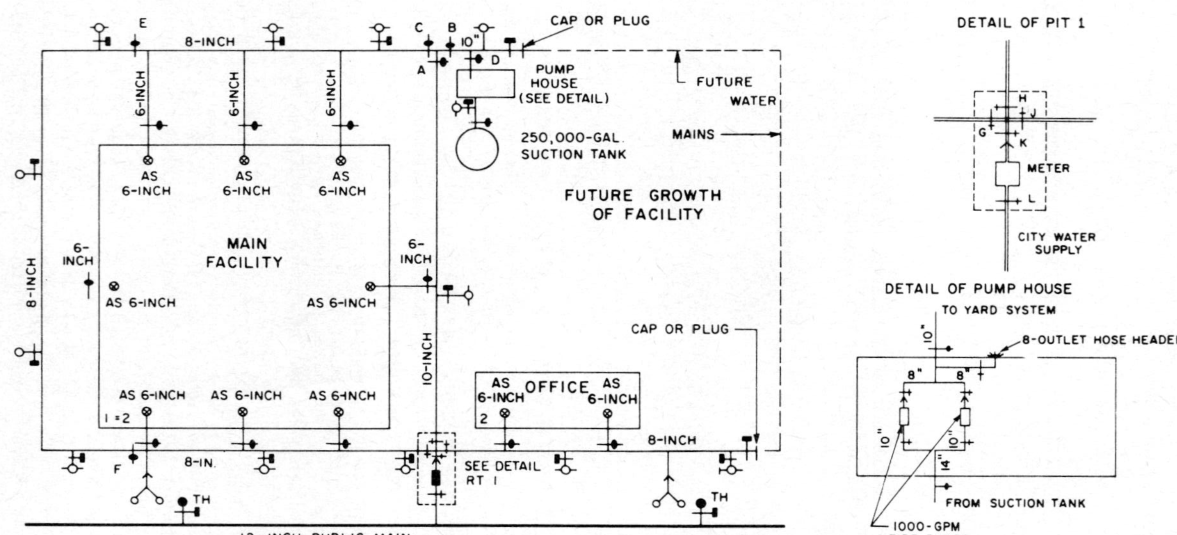

Figure 2-10 Water-pipe network. *(From* Fire Protection Handbook, *14th ed., G. P. McKinnon (ed.), NFPA, Quincy, Massachusetts, 1976.)*

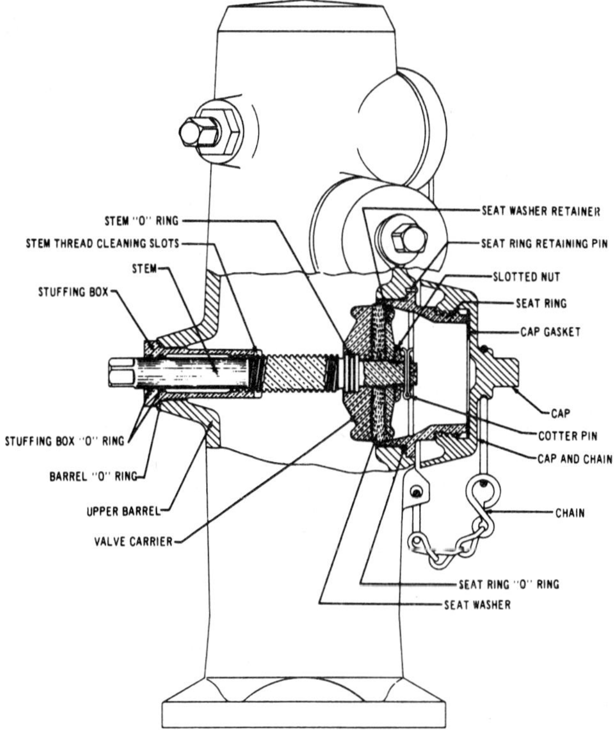

Figure 2-11 Wet-barrel hydrant. *(Mueller Co.)*

- Adequate capacity to meet fire-flow demands
- Selection of fire pump driver based on reliability, adequacy, economy, and safety of power source
- Automatic operation
- Safe location for uninterrupted service
- Annual testing
- Maintenance

Sprinkler Systems

A sprinkler system, for fire-protection purposes, is an integrated system of underground and overhead piping designed in accordance with fire-protection engineering standards. The installation includes a water supply, such as a gravity tank, fire pump, reservoir or pressure tank and/or connection by underground piping to a city main. The portion of the sprinkler system above ground is a network of specially sized or hydraulically designed piping installed in a building, structure, or area, generally overhead, and to which sprinklers are connected in a systematic pattern. The system includes a controlling valve and a device for actuating an alarm when the system is in operation. The system is usually activated by heat from a fire and discharges water over the fire area.

Wet-Pipe Systems

These are under water pressure at all times; water will be discharged immediately when an automatic sprinkler operates (Fig. 2-13).

Water flowing from the wet-pipe sprinkler system actuates an alarm valve that gives off a signal. Figure 2-13 illustrates the total concept of the wet-pipe automatic sprinkler system.

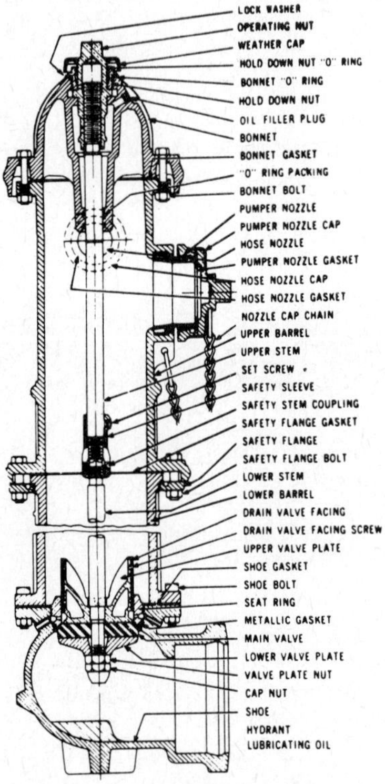

Figure 2-12 Dry-barrel hydrant. *(Mueller Co.)*

The essential features of wet-pipe sprinkler systems, which represent about 75 percent of sprinkler installations, include provisions for water supplies, piping, and location and spacing of sprinklers. This system is generally used wherever there is no danger that the water in the pipes will freeze and wherever there are no special conditions requiring one of the other systems. Inspection of the wet-pipe sprinkler system at regular intervals is essential, and weekly inspection of all water-control valves and alarm-control valves is recommended.

Where subject to temperatures below freezing, the ordinary wet-pipe system cannot be used. There are two recognized methods of maintaining automatic sprinkler protection in such locations: (1) through the use of systems where water enters the sprinkler piping only after operation of a control valve (dry-pipe, pre-action, etc.) and (2) by the use of anti-freeze solution in a portion of the wet-pipe system.

Regular Dry-Pipe Systems

In locations where there is danger of freezing, it is the usual practice to install a dry-pipe system. In regular dry-pipe systems, the sprinkler piping contains air or nitrogen under pressure instead of water, and admission of the water is controlled by a dry-pipe valve. When a sprinkler is opened by heat from a fire, the pressure is reduced, a dry-pipe valve is opened by water pressure, and water flows out of any opened sprinklers.

Pre-Action Systems

Systems in which the air in the piping may or may not be under pressure are called pre-action systems. These systems are designed primarily to protect properties on which the danger of water damage from broken sprinklers or piping could be serious. The water-supply valve is actuated independently of the opening of the sprinkler heads by an automatic fire-detection system; the valve is opened sooner than with the dry-pipe system, and the alarm is given when the valve is opened.

The pre-action system has several advantages over a dry-pipe system. The valve is opened sooner because the fire detectors have less thermal lag than sprinklers. The detection system also automatically rings an alarm. Fire and water damage is decreased because water is on the fire sooner, and the alarm is given when the valve is opened. Sprinkler piping is normally dry; thus, pre-action systems are nonfreezing and applicable to dry-pipe service.

Deluge Systems

Deluge systems are used for areas of extra occupancies. All sprinkler heads are open at all times so that when the water comes on, the entire area is flooded; when heat from a

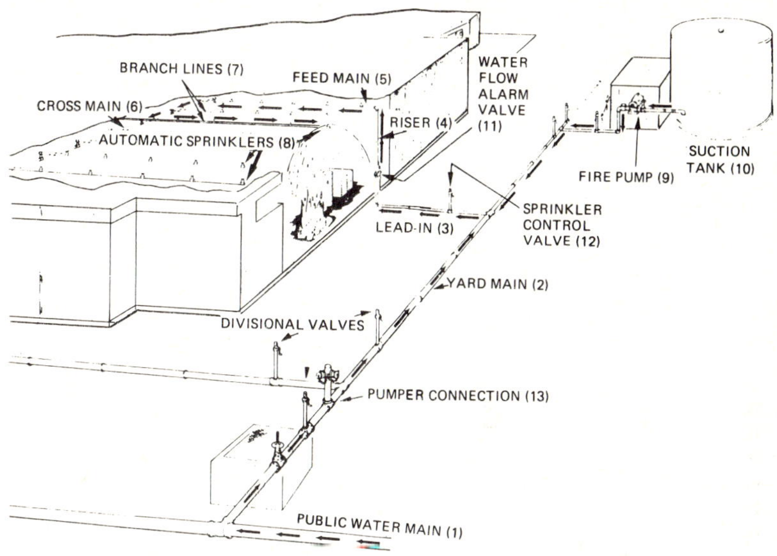

Figure 2-13 Sprinkler system. *(Factory Mutual System.)*

fire actuates the fire-detecting device, water flows to and is discharged from all sprinklers on the piping system, thus "deluging" the protected areas. These systems are often used in airplane hangars and in areas where flammable liquids are handled or stored.

By using sensitive thermostatic controls operating on the rate-of-rise or fixed-temperature principle, or controls designed for individual hazards, it is possible to apply water to a fire more quickly than with systems in which operation depends on opening of sprinklers only as the fire spreads.

Sprinkler Heads

Sprinkler heads are designed with temperature ratings ranging from 135°F (57°C) to as high as 500°F (260°C). Ratings of 165°F (74°C) are usual for use in buildings that are maintained at normal, constant temperatures.

The location and spacing of sprinkler heads depends on the degree of hazard and type of construction. The "Standard for the Installation of Sprinkler Systems," NFPA 13, provides detailed design and maintenance specifications (Table 2-16).

Standpipes

The four generally recognized standpipe system concepts are described in (Table 2-17):

1. A wet-standpipe system, having supply valve open and water pressure maintained at all times. This is the most desirable type of system.
2. A dry-standpipe system arranged to admit water to the system through manual operation of approved remote-control devices located at each hose station. The water-supply control mechanism introduces an inherent reliability factor that must be considered.
3. A dry-standpipe system in an unheated building. The system should be arranged to admit water automatically by means of a dry-pipe valve or other approved device. The depletion of system air at the time of use introduces a delay in the application of water to the fire and increases the level of competency required to control the pressurized hose and nozzle assembly during the charging period.

TABLE 2-16 Summary of Spacing Rules

Type of construction	Maximum distance of deflectors below ceiling, in			
	In bays*		Under beams	
	Comb.	Noncomb.	Comb.	Noncomb.
Smooth ceiling	10	12	14	16
Beam and girder	16	16	20	20
Panel up to 300 ft^2	18	18	22	22
Bar joists	10	12	—	—
Open wood joists—center 3 ft or less	6	—	—	—

Minimum below ceiling is 1 in.
Minimum below beams 1 in, maximum 4 in. Do not exceed maximum below ceiling.

Maximum coverage per sprinkler

Light hazard	200 ft^2 smooth ceiling and beam and girder construction
	130 ft^2 open wood joist
	168 ft^2 all other types of construction
Ordinary hazard	130 ft^2 all types of construction except
	100 ft^2 high piled storage
Extra hazard	90 ft^2 all types of construction

Direction of lines: Either direction to facilitate hanging except: across beams for beams on girders 3 ft to 7½ ft on centers and across joists for wood joists (open or sheathed) and bar joists (through or under)

Maximum spacing between lines and sprinklers

Light and ordinary hazard	15 ft except 12 ft for high piled storage
Extra hazard	12 ft

*Not more than 4 in below beams where lines run across beams.
Source: NFPA 13, "Installation of Sprinkler Systems," Quincy, Massachusetts.

4. A dry-standpipe system having no permanent water supply. This type would be used for reducing the time required for fire departments to put hose lines into action on upper floors of tall buildings. This type of system might also be used in buildings during construction, where allowed in lieu of the wet standpipe in unheated areas.

Water-Spray Fixed Systems

Water-spray fixed systems are generally used to protect flammable liquid and gas tankage; piping and equipment; electrical equipment such as transformers, oil switches, and rotating electrical machinery; and openings in firewalls and floors through which conveyors pass. The type of water spray required for any particular hazard depends, of course, upon the nature of the hazard and the purpose for which the protection is provided.

NFPA 15, "Standard for Water Spray Fixed Systems for Fire Protection," calls for piping, valves, pressure gauges, and detection systems of an approved type. The spray nozzles generally used in these systems are open, and the pipes, especially outdoor ones that are subject to freezing temperatures, are usually dry.

Foam Extinguishing Systems

Foam extinguishing systems have been used extensively for many years, especially in the petrochemical industry, for the extinguishment of flammable liquid fires. The principal

TABLE 2-17 Summary of National Fire Protection Association Standpipe Standards*

Type	Intended use	Size hose and distribution	Minimum size pipe	Minimum water supply
Class I	Heavy streams Fire department Trained personnel Advanced stages of fire	2½-in connections All portions of each story or section within 30 ft of nozzle with 100 ft of hose	4 in up to 100 ft 6 in above 100 ft (275 ft maximum unless pressure regulated)	500 gal/min 1st standpipe 250 gal/min each additional 2500 gal/min maximum) 30-min duration 65 lb/in² at top outlet with 500 gal/min flow
Class II	Small streams Building occupants Incipient fire	1½-in connections (distribution same as class I)	2 in up to 50 ft 2½ in above 50 ft	100 gal/min per building 30-min duration 65 lb/in² at top outlet with 100 gal/min flowing
Class III	Both of above	Same as class I with added 1½-in outlets or 1½-in adapters and 1½-in hose.	Same as class I	Same as class I

*From NFPA 14, "Standard for the Installation of Standpipe and Hose Systems," Quincy, Massachusetts.

kinds of foam are chemical and mechanical (determined by how they are generated), and these classes are further subdivided.

Special compatible foam concentrates result in the generation of a foam that does not break down as readily as ordinary foam when mixed with dry chemical. Other special foams are available for application on fires in alcohols, esters, ketones, and ethers (called water-soluble or polar liquids). This concentrate produces a foam that does not deteriorate like ordinary foam when in contact with water-miscible solvents.

Carbon Dioxide Systems

Carbon dioxide is a noncombustible gas that has been effectively used to extinguish certain types of fires. It acts to reduce the oxygen in the fire area to a point where it will no longer support combustion (Table 2-18). Because carbon dioxide is stored under pressure, it can readily be discharged from its cylinder or extinguisher. Carbon dioxide is inert and will not conduct electricity. It can be used safely on energized electric equipment fires without causing damage to the equipment.

Because carbon dioxide does little or no damage to equipment or materials with which it comes in contact, it is very useful for protection of rooms with contents of high value and contents subject to water damage. Typical of such occupancies are fur vaults, record storage rooms, computer rooms, and rooms housing live electric equipment. Carbon dioxide is also widely used for extinguishing flammable liquid fires because the carbon dioxide gas rapidly spreads above the surface of the burning liquid and shuts off the oxygen.

Halogenated Agents and Systems

One of the earliest halons used for fire extinguishing was carbon tetrachloride. Concern about the toxic properties of carbon tetrachloride began to manifest itself about 1960, and the *NFPA and the extinguisher-testing laboratories discontinued any recommendation of carbon tetrachloride as a fire extinguishing agent.*

TABLE 2-18 Minimum Carbon Dioxide Concentrations for Extinguishment*

Material	Theoretical min. CO_2 concentration, (%)
Acetylene	55
Acetone	26†
Benzol, benzene	31
Butadiene	34
Butane	28
Carbon disulfide	55
Carbon monoxide	53
Coal gas or natural gas	31†
Cyclopropane	31
Dowtherm	38†
Ethane	33
Ethyl ether	38†
Ethyl alcohol	36
Ethylene	41
Ethylene dichloride	21
Ethylene oxide	44
Gasoline	28
Hexane	29
Hydrogen	62
Isobutane	30†
Kerosene	28
Methane	25
Methyl alcohol	26
Pentane	29
Propane	30
Propylene	30
Quench, lubricating oils	28

*From Gordon P. McKinnon (ed.) *Fire Protection Handbook,* NFPA, Quincy, Massachusetts, 1976.

†The theoretical minimum extinguishing concentrations in air for the above materials were obtained from Bulletin 503, U.S. Bureau of Mines, Washington, 1952. Those marked † were calculated from accepted residual oxygen values.

A halon is a hydrocarbon (hydrogen and carbon) in which some of the hydrogen atoms have been replaced by such elements as bromine, chlorine, or fluorine, or by combinations of these (Table 2-19). A number of halons are toxic, thus making them undesirable for general use; two of them, Halon 1301 and Halon 1211, have acceptable levels of toxicity and excellent flame extinguishment properties.

Halon 1211 and Halon 1301 are the only two agents recognized by the NFPA Technical Committee on Halogenated Fire Extinguishing Agent Systems. Both Halon 1211 and 1301 are widely used for protection of electric equipment (both are nonconductors of electricity), airplane engines, and computer rooms. As both of these halons rapidly vaporize, they leave little corrosive or abrasive residue to clean up and do not interfere as much with visibility during fire fighting as foam or carbon dioxide. These halons are used for hand extinguishers and fixed systems.

Dry-Chemical Extinguishing Systems

Dry-chemical extinguishing agents consist of finely divided powders that effectively extinguish a fire when applied to the fire by portable extinguishers, hose lines, or fixed systems. The original dry powder was sodium bicarbonate (ordinary baking soda). Potassium bicarbonate and other chemical powders, with additives to make the powders free flowing and more moisture resistant, are now in use. Dry chemical has been found to be

TABLE 2-19 Some Physical Properties of the Common Halogenated Fire Extinguishing Agents

Agent	Chemical formula	Halon no.	Type of agent	Approx. boiling point, °F	Approx. freezing point, °F	Specific gravity of liquid at 68°F (water = 1)	Approx. critical temp., °F	Estimated pressure, psig		Latent heat of vaporization, cal/g water = 540 cal/g CO_2 = 138 cal/g
								At 130°F	At critical temp.	
Carbon tetrachloride	CCl_4	104	Liquid	170	−8	1.595	—	—	—	46
Methyl bromide	CH_3Br	1001	Liquid	40	−135	1.73	—	—	—	62
Bromochloromethane	$BrCH_2Cl$	1011	Liquid	151	−124	1.93	—	—	—	—
Dibromodifluoromethane	Br_2CF_2	1202	Liquid	76	−223	2.28	389	23	585	29
Bromochlorodifluoromethane	$BrCClF_2$	1211	Liquefied gas*	25	−257	1.83	309	75	580	32
Bromotrifluoromethane	$BrCF_3$	1301	Liquefied gas	−72	−270	1.57	153	435	560	28
Dibromotetrafluoroethane	BrF_2CCBrF_2	2402	Liquid	117	−167	2.17	—	3.8	—	25

*May be kept as a liquid at reduced temperatures.

Source: NFPA, Quincy, Massachusetts.

an effective extinguishing agent for fires in flammable liquids and in certain types of ordinary combustibles and electric equipment, depending upon the type of dry chemical used.

Dry-chemical extinguishing systems are used to protect flammable-liquid storage rooms, dip tanks, kitchen range hoods, deep-fat fryers, and similar hazardous areas and appliances. Because dry chemical is nonconductive, these systems are useful in the protection of oil-filled transformers and circuit breakers. Dry-chemical systems are not recommended for telephone-switchboard or computer protection. Dry-chemical hose-line systems are used in crash trucks and for the protection of aircraft hangars. Dry chemicals are used in portable fire extinguishers.

Combustible-Metal Extinguishing Systems

A number of metals and metal powders found in industrial situations and in transport will burn. Some metals burn when heated to high temperatures by friction or exposure to external heat. Others burn from contact with moisture or in reaction with other materials. These metals and metal powders require special extinguishing agents and special fire-fighting techniques. Some result in explosions and very high temperatures, and some react violently with water. Still others give off toxic fumes when burning.

Some combustible metal extinguishing agents' success in handling metal fires has led to the terms *approved extinguishing powder* and *dry powder*. Such terms have been accepted in describing extinguishing agents for metal fires, and should not be confused with the name *dry chemical,* which normally applies to an agent suitable for use on flammable-liquid and live electric equipment fires. Graphite powder, talc, and sand have all been used to smother metal fires.

Portable Fire Extinguishers

Portable fire extinguishers are required in most plants by local, state, and federal regulations and insurance requirements. Where there are trained personnel available to use the proper extinguisher on a small incipient fire, extinguishers may prove useful in preventing a larger, more devastating fire.

The limitations of extinguishers, personal exposure to fire and smoke, capacity, range, selectivity, and availability necessitate that training be provided if they are expected to be effective. Use of extinguishers should be simultaneous with notification of the fire brigade or department.

Types of Portable Fire Extinguishers

The kind and number of extinguishers needed for particular types of fires are specified in NFPA 10, "Standard for Portable Fire Extinguishers." The most common types of extinguishers in use are the pressurized water, carbon dioxide, and multipurpose dry chemical. Other extinguishers commonly used are water pump tanks, liquefied gases such as Halon 1211, and combustible-metal-type dry powder.

Application of Portable Fire Extinguishers

NFPA 10, "Standard for Portable Fire Extinguishers," classifies fires in four ways:

Class A. Fires involving ordinary combustible materials (wood, cloth, paper, rubber, and many plastics) requiring the heat-absorbing (cooling) effects of water, water solutions, or the coating effects of certain dry chemicals that retard combustion (Table 2-20).

Class B. Fires involving flammable or combustible liquids, flammable gases, greases, and similar materials where extinguishment is most readily secured by excluding air (oxygen), inhibiting the release of combustible vapors, or interrupting the combustion chain reaction (Table 2-21).

Class C. Fires involving live electric equipment where safety to the operator requires the use of electrically nonconductive extinguishing agents. (*Note:* When electric equipment is de-energized, the use of Class A or B extinguishers may be indicated.)

TABLE 2-20 Fire Extinguisher Size and Placement for Class A Hazards

Basic minimum extinguisher rating for area specified	Maximum travel distances to extinguishers, ft	Areas to be protected per extinguisher		
		Light hazard occupancy, ft^2	Ordinary hazard occupancy, ft^2	Extra hazard occupancy, ft^2
1-A	75	3,000	†	†
2-A	75	6,000	3,000	†
3-A	75	9,000	4,500	3,000
4-A	75	11,250	6,000	4,000
6-A	75	11,250	9,000	6,000
10-A	75	11,250*	11,250*	9,000
20-A	75	11,250*	11,250*	11,250*
40-A	75	11,250*	11,250*	11,250*

*11,250 ft^2 is considered a practical limit.

†Protection requirements may be fulfilled by several extinguishers of the minimum specified rating with the approval of the authority having jurisdiction.

Source: NFPA, Quincy, Massachusetts.

TABLE 2-21 Fire Extinguisher Size and Placement for Class B Hazard Excluding Protection of Deep Layer Flammable Liquid Tanks (For Extinguishers Labeled After June 1, 1969)

Type of hazard	Basic minimum extinguisher rating	Maximum travel distance to extinguishers, ft
Light	5-B	30
	10-B	50
Ordinary	10-B	30
	20-B	50
Extra	20-B	30
	40-B	50

Source: NFPA, Quincy, Massachusetts.

Class D. Fires involving certain combustible metals (such as magnesium, titanium, zirconium, sodium, potassium, etc.) requiring a heat-absorbing extinguishing medium not reactive with the burning metals.

Figures 2-14 through 2-16 illustrate fire extinguishing agents, classifications, and symbols.

CODES AND STANDARDS

Model Fire Prevention Codes

Currently, there are five "model" fire prevention codes available in the United States. Each of these five model codes utilizes NFPA standards as the basis for the technical details of its fire-prevention and fire-control measures. These five codes are: (1) the NFPA's *Fire Prevention Code,* which was developed by a committee representing state and city fire marshals and other concerned interests, and which was adopted after going through the usual NFPA standards-making process and submission procedure at an annual meeting of the association; (2) the American Insurance Association's *Fire Prevention Code,* which was the first model fire prevention code issued, and which has been adopted by many communities; (3) the *Uniform Fire Code,* published by the International Conference of Building Officials in cooperation with the Western Fire Chiefs

1. Extinguishers suitable for "Class A" fires should be identified by a triangle containing the letter "A." If colored, the triangle shall be colored green.*

2. Extinguishers suitable for "Class B" fires should be identified by a square containing the letter "B." If colored, the square shall be colored red.*

3. Extinguishers suitable for "Class C" fires should be identified by a circle containing the letter "C." If colored, the circle shall be colored blue.*

4. Extinguishers suitable for fires involving metals should be identified by a five-pointed star containing the letter "D." If colored, the star shall be colored yellow.*

Figure 2-14 Fire extinguisher identification.* Recommended colors per PMS (Pantone Matching System): GREEN—*Basic Green*, RED—*192*, BLUE—*Process Blue*, YELLOW—*Basic Yellow. (From NFPA 10, "Standard for Portable Fire Extinguishers," 1978.)*

Association; (4) the *BOCA Basic Fire Prevention Code,* published by the Building Officials and Code Administrators International, Inc.; and (5) the *Southern Standard Fire Prevention Code* of the Southern Building Code Congress.

Model Building Codes

When a code commission or building official is revising an outdated building code, there are four model codes that are usually used as guides. These four model codes are: (1) the *National Building Code,* (2) the *Uniform Building Code,* (3) the *Standard Building Code,* (4) the *BOCA Basic Building Code.*

Fire Safety Standards-Making Organizations

American National Standards Institute (ANSI)

ANSI sets public requirements for national standards and develops and publishes them on a wide range of subjects. In order to achieve uniformity in voluntary and mandatory state and federal standards, it coordinates voluntary standardization activities of concerned organizations.

ANSI standards cover a variety of products, materials, and equipment that is used both in highly specialized fields and in nearly all other areas of modern life. The ANSI publishes standards on ceramic tiles, chemical process equipment, home appliances, electronics equipment, motion picture film and equipment, acids, refractory materials, oil burners, office machines and supplies, hospital supplies, and combustion engines.

Fire Extinguisher/Agent Characteristics

SUITABLE FOR USE ON TYPE OF FIRE	AGENT CHARACTERISTICS	Available Sizes	Horizontal Range	Discharge Time
DRY CHEMICAL B C	Sodium Bicarbonate or Potassium Bicarbonate or Potassium Chloride. Discharges a white or bluish cloud. Leaves residue. Non-freezing.	1 to 30 lbs.	5 to 20 ft.	8 to 25 Sec.
MULTIPURPOSE DRY CHEMICAL A B C OR B C A CAPABILITY	Basically Ammonium Phosphate. Discharges a yellow cloud. Leaves residue. Non-freezing. Some extinguishers utilizing this agent do not have an "A" rating – however, they are designated as having "A" capability.	2 to 30 lbs.	5 to 20 ft.	8 to 25 Sec.
FOAM B	Basically water and detergent. Discharges a foamy solution After evaporation, leaves a powder residue. Protect from freezing.	21 oz.	4 to 6 ft.	24 Sec.
CARBON DIOXIDE B C	Basically an inert gas that discharges a cold white cloud. Leaves no residue. Non-freezing.	2½ to 20 lbs.	3 to 8 ft.	8 to 30 Sec.
HALON 1211 A B C	Basically halogenated hydro-carbons. Discharges a white vapor Leaves no residue Non-freezing.	2 to 9 lbs.	8 to 15 ft.	8 to 15 Sec.
WATER ▲ A	Basically tap water. Discharges in a solid or spray stream. Protect from freezing!	2½ Gal.	30 to 40 ft.	1 Minute

▲ NOTE: Pump tanks available.

NOTE: A garden hose connected to a suitable weather protected hose connection is advisable for use in fighting Class A fires. This should not be considered as a replacement for extinguishers.

NOTE: 1 ft. = 0.305 m: 1 lb. = 0.454 kg: 1 gal. = 3.785*l*.

Figure 2-15 Fire extinguisher classification. *(From NFPA 10, "Standard for Portable Fire Extinguishers," 1978.)*

American Society for Testing and Materials (ASTM)

ASTM develops and publishes standards on finished products and on materials used in manufacturing and construction. Because some products and materials are used only within certain companies, industries, and government agencies, not all ASTM standards are developed by the full-consensus system. However, standards that deal with commod-

Typical Pictorial Extinguisher Marking Labels

*NOTE: Recommended colors, per PMS (Pantone Matching System):
(BLUE–*299*)
(RED–*Warm Red*)

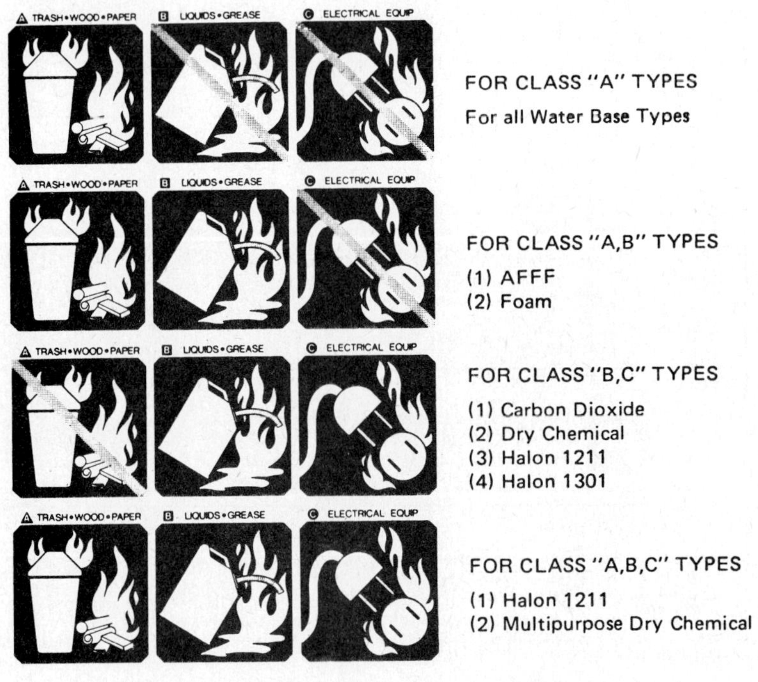

Color Separation Identification (picture symbol objects are white; background borders are white)

BLUE *	–	background for "YES" symbols
BLACK	–	background for symbols with slash mark ("NO")
RED *	–	slash mark for black background symbols

Figure 2-16 Fire extinguisher symbols. *(From NFPA 10, "Standard for Portable Fire Extinguishers," 1978.)*

ities used by the general public are developed by a full-consensus procedure, wherein all interested parties are fairly represented in the committee writing the standard. The standard committee is made up of anyone technically qualified or knowledgeable in the area of the committee's scope.

National Fire Protection Association (NFPA)

NFPA codes and standards encompass the entire scope of fire prevention, fire protection, fire fighting, and fire hazards, ranging from the **National Electrical Code,*** believed to be the most widely adopted set of safety requirements in the world, to codes or standards of specific limited areas that are nevertheless important in controlling a life or fire hazard.

**National Electrical Code®* is a Registered Trademark of the National Fire Protection Association, Quincy, MA 02269.

Once a code or standard has been adopted by the NFPA, it becomes available for adoption by any organization or jurisdiction having enforcement authority. A number of NFPA standards are widely used and commonly referenced in fire legislation.

Fire Testing and Research Laboratories

There are many laboratories in the United States capable of performing, in varying degrees, fire tests of materials and/or equipment; many of these same laboratories, as well as other laboratories, have facilities for conducting fire-related research work. Generally, these laboratories can be classified into three categories: (1) private and industrial laboratories, (2) university laboratories, and (3) government laboratories.

Private and Industrial Laboratories. In the United States there are approximately 65 private and industrial laboratories that perform a wide range of fire tests. Space does not permit that each be described in detail. However, there are two, Underwriters Laboratories Inc. and Factory Mutual Laboratory Facilities, whose work warrants particular emphasis.

Underwriters Laboratories Inc. (UL). Annually UL publishes lists of manufacturers whose products, when tested, have proved acceptable under appropriate standards and which are subjected to one of the follow-up services provided by the laboratories as a counter-check. The word "listed" appears on UL labels attached to these products as authorized evidence that these products have been found to be in compliance with the laboratories' requirements.

Factory Mutual Systems (FM). Factory Mutual maintains testing facilities in Norwood, Mass., and also conducts large-scale applied research in its 1-acre, 60-ft-high FM test center in West Gloucester, R.I. Factory Mutual Laboratory facilities are available on a contract basis through Factory Mutual Research.

University and Federal Government Laboratories. More than forty American colleges and universities are equipped with laboratories for fire testing and research. In addition to the colleges and universities that serve primarily as institutions for fire science training and educations, there are others, both private and state-supported, whose engineering, physics, or science departments engage in such activities.

Several departments of the federal government—Agriculture, Air Force, Army, Commerce, Navy, and Transportation—as well as independent agencies also have research laboratories located throughout the country. These facilities are a direct result of an increasing national interest in fire safety as well as other safety- and health-related issues.

Insurance Organizations

Many important groups perform varied fire protection and inspection services on behalf of the insurance industry and its insureds. For example, the Association of Mill and Elevator Mutual Insurance Companies serves the mill and elevator industry's needs; the American Institute of Marine Underwriters is organized to serve the marine underwriters and to promote, advance, and protect their interests.

There are five large insurance organizations, however, that serve a wide range of casualty and property insurers and contribute to fire protection in many ways. They are: (1) American Insurance Association, (2) American Mutual Insurance Alliance, (3) Factory Mutual System, (4) Industrial Risk Insurers, and (5) Insurance Services Office.

BIBLIOGRAPHY

Bryan, John L.: *Automatic Sprinkler & Standpipe Systems,* NFPA, Quincy, Massachusetts, 1976.
Bryan, John L.: *Fire Suppression and Detection Systems,* Glencoe Press, Beverly Hills, 1974.
Bugbee, Percy: *Principles of Fire Protection,* NFPA, Quincy, Massachusetts, 1978.
Factory Mutual: *The Handbook of Property Conservation,* Factory Mutual System, Norwood, Massachusetts.
Factory Mutual: *Loss Prevention Data Books,* Factory Mutual System, Norwood, Massachusetts.

Factory Mutual: *Property Conservation Workbook,* Factory Mutual System, Norwood, Massachusetts 1979.

Kimball, Warren Y.: *Fire Department Terminology,* NFPA, Quincy, Massachusetts, 1970.

Magison, Ernest C.: *Electrical Instruments in Hazardous Locations.* 3d ed., Instrument Society of America, Triangle Park, North Carolina, 1978.

McKinnon, Gordon P. (ed.): *Fire Protection Handbook,* 15th ed., NFPA, Quincy, Massachusetts, 1981.

McKinnon, Gordon P. (ed.): *Industrial Fire Hazards Handbook,* NFPA, Quincy, Massachusetts, 1979.

NFPA: *Guide to OSHA Fire Protection Regulations,* NFPA, Quincy, Massachusetts.

NFPA: *Industrial Fire Brigades,* NFPA, Quincy, Massachusetts, 1978.

NFPA: *Introduction to Fire Protection,* NFPA, Quincy, Massachusetts, 1982.

NFPA: *National Fire Codes,* NFPA, Quincy, Massachusetts, annual.

Planer, Robert G.: *Fire Loss Control, A Management Guide,* Marcel Dekker, New York, 1979.

Roytman, M. Ya.: *Principles of Fire Safety Standards for Building Construction,* Amerind, New Delhi, India, 1975.

Tuck, Charles A. (ed.): *NFPA Inspection Manual,* NFPA, Quincy, Massachusetts, 1976.

Tuve, Richard C.: *Principles of Fire Protection Chemistry,* NFPA, Quincy, Massachusetts 1976.

U.S. Department of Labor: *General Industry Standards,* part 1910, title 29, Code of Federal Regulations, Occupational Safety and Health Administration.

Williams, C. A., Jr., and Heins, R. M.: *Risk Management and Insurance,* McGraw-Hill, New York, 1976.

Zajic, J. E., and Himmelmann, W. A., *Highly Hazardous Materials Spills and Emergency Planning,* Marcel Dekker, New York, 1978.

chapter 8-3

Electrical Hazard Protection and Prevention

by

Robert L. Smith, Jr.
George W. Walsh
General Electric Company
Schenectady, New York

GLOSSARY

Basic impulse insulation level (BIL) A reference level expressed in impulse crest of a standard 1.2×50-μs wave. The withstand of apparatus insulation, as demonstrated by suitable tests, shall be equal to or greater than the basic impulse insulation level.

Chopped wave An impulse voltage wave that is suddenly reduced substantially to zero value by the sparkover of an air gap.

Crest value of a wave The maximum value (voltage or current) attained by a wave.

Discharge current The surge current that flows through an arrester.

Discharge voltage The voltage that appears across the terminals of an arrester during the passage of discharge current.

Follow current The power-frequency current that flows through an arrester during and following the passage of discharge current.

Front of wave sparkover Voltage at which arrester-gap sparkover occurs on the front of a wave rising at the rate of 100 kV/μs for each 12 kV of arrester rating for arresters rated from 3 to 240 kV and 2000 kV/μs for arresters rated above 240 kV. For arresters rated less than 3 kV, 10 kV/μs; and all rotating machine arresters, 10 ± 3 kV/μs.

Full wave An impulse voltage wave that rises to crest value and then decays in a normal manner without modification by the sparkover of an air gap or similar occurrence (contrast with "chopped wave").

Impulse A surge of unidirectional polarity.

Impulse sparkover voltage The highest value of voltage attained prior to the flow of discharge current when an impulse of a given wave shape and polarity is applied across the line and ground terminals of an arrester.

Microsecond One-millionth of a second (abbreviated μs).

Sparkover A disruptive discharge between electrodes of a measuring gap, voltage-control gap, or protective device.

Surge A transient variation in the current or potential at a point in a circuit.

Surge arrester A protective device for limiting surge voltages on equipment by discharging a surge current. It prevents continued flow of follow current and is capable of repeating these functions.

Switching surge (slow-front) Impulse of front time of 30 to 2000 μs, generally, with time to half-crest on the tail appreciably longer than twice the time to crest.

Traveling wave The resulting wave when the electric variation in a circuit takes the form of energy translation along a conductor, such energy being always equally divided between current and potential forms.

Valve-type lightning arrester An arrester having an element consisting of a resistor with a nonlinear voltampere characteristic which limits the follow current to a value the series gap can interrupt. If the arrester has no series gap, the characteristic element limits the follow current to a magnitude which does not interfere with normal operation of the system.

Voltage rating of an arrester The highest permissible rms alternating voltage that may be present between the line and ground terminals of the arrester while it is performing its operating duty cycle.

Wave The variation of current and/or voltage in an electric circuit with respect to time or distance.

Wave front That part of a surge or impulse which occurs prior to reaching the crest value.

Wave shape The variation of voltage or current of an impulse with time. For test purposes it is expressed as a combination of two numbers. The first, an index of wave-front steepness, is the time in microseconds from a virtual zero to the instant at which the crest value is reached. The second, an index of duration, is the time in microseconds from the virtual zero to the instant at which one-half of the crest value is reached on the wave tail. Examples are 1.2 × 50 and 8 × 20 waves.

Wave tail That part between the crest value and the end of an impulse.

OVERLOAD AND FAULT PROTECTION

Medium- and low-voltage systems require both *overload* and *fault* protection. An overload is circuit operation in excess of its capability for a time long enough to cause damage or dangerous overheating. A fault is either a short circuit or an open circuit. A bolted short circuit is a solid connection between two phase conductors, a phase conductor and ground, or all three phase conductors. Protective devices activated by overloads or faults include fuses which melt as well as relays and direct-acting trip devices which open circuit breakers, open contactors, or sound alarms.

OVERLOAD PROTECTION

Overloads may cause damaging elevated temperatures to persist for extended periods of time. Resistance temperature detectors are overload-protective devices which sense temperature directly. They are sometimes used to protect the most vulnerable parts of major equipment. Other overload-protective devices measure current in a circuit element and operate after a specific overcurrent persists for a time inversely related to current magnitude. Some current-detecting devices are ambient-compensated to make the protection sensitive to total hot-spot temperature. Since current unbalances resulting from unbalanced voltages cause rotating machine overheating, some protective devices sense current unbalance; others sense voltage unbalance. See Table 3-1 for a list of common overload-protective devices.

FAULT PROTECTION

Short-Circuit Fault Protection

Short-circuit currents of large magnitude flow when circuit insulation suddenly fails. These currents usually are much greater than normal load or overload currents. Sudden

TABLE 3-1 Overload Protective Devices

Device	Device function number	Measured quantity	Sensor	Usual application
Fuse	—	Current	Fusible element	Low- or medium-voltage circuits
Direct-acting trip device	—	Current	Trip device sensor	Low-voltage circuits
Thermal overcurrent relay	49	Current	Relay element	Low-voltage motor controller size 4 and smaller
			Current transformer	Low-voltage motor controller size 5 and larger
			Current transformer	Medium-voltage motor controller or circuit breaker
Time-delay magnetic or solid-state overcurrent relay	51	Current	Current transformer	Medium-voltage circuits
Temperature relay	49	Temperature	Resistance temperature detectors	Motors 1500 hp and over Transformers over 10 MVA
Combination solid-state relay	49	Current and temperature	Current transformers and resistance temperature detector	Motors 1500 hp and over
Solid-state or magnetic voltage balance relay	60	Voltages	Potential transformers	Motor buses
Solid-state or magnetic current balance relay	46	Currents	Current transformers	Motors 1500 hp and larger

failure of insulation results from either gradual deterioration or sudden overvoltage. Short-circuit current flow usually damages adjacent circuit components or equipment by arcing, explosive heating, or fire. Under short-circuit conditions, currents much greater than load currents flow in all circuit components between the source and the fault. A system short-circuit study can be conducted to reveal the magnitudes of short-circuit current throughout a system for important fault locations and can guide selection of system components rated for the calculated duties.

Short-circuit protective devices such as fuses, overcurrent trip devices, or relays operate in durations that can be preset or preselected depending on the magnitude of fault current. Series overcurrent devices between a source and a fault all experience the same short-circuit current, changed in magnitude only by the presence of transformers. This fact can be used to provide a completely selective protective device system in which the system device closest to the point of fault operates first and other series devices operate later. When several devices exist between the point of fault and the source, device characteristics other than progressively longer times are appropriate, because delays to attain complete time interval selectivity tend to get too long for source end devices.

If more than one source supplies a system, each source circuit breaker usually requires directional as well as nondirectional relays for best continuity of service. See References 8 and 9 for description of this type of relaying.

Important system components require rapid removal of a short circuit in order to minimize fault damage. Differential relaying accomplishes this without compromising system selectivity. Differential relaying compares the currents entering and leaving a circuit element and operates if the difference exceeds the setting or sensitivity of the relay. This relaying requires separate current transformers for both the incoming and outgoing conductors to a circuit element in order to obtain the proper sensitivity and accuracy. A special exception for motor circuits uses three current transformers, one for *both ends* of *each* motor phase winding.

Fault pressure relays provide excellent protection for internal transformer tank faults by detecting the sudden change in internal tank pressure which accompanies the initial insulation breakdown when a fault occurs. These completely selective relays usually are applied to main stepdown liquid-filled transformers. They can also be applied to smaller liquid-filled transformers.

Occasionally, distance relays instead of overcurrent relays are used to provide selective operation for large industrial systems. These relays measure the impedance from the relay to some system location. They operate only for faults within the protected zone.

Providing a reasonably selective system with appropriate system component protection involves compromises between overcurrent selectivity and protection. A system coordination study performed by a qualified power-system protection engineer incorporates the appropriate compromises to assure optimum system selectivity and protection. See Table 3-2 for a list of common short-circuit fault-protective devices.

Open-Circuit Fault Protection

An open-circuit fault condition results from the opening of any or all phase conductors at any point between the source and the load. Single-phase or three-phase undervoltage devices or relays easily detect this condition instantaneously on systems with no motors or generators connected. Protective devices consist of instantaneous or time-delay circuit-breaker undervoltage trip devices, instantaneous undervoltage relays with or without a timer, or time-delay undervoltage relays. These devices may be connected directly to a circuit or connected through potential transformers. They may operate to open a source or feeder circuit breaker, initiate an automatic throwover to an alternative source, or start an automatic sequence to energize the system from a standby generator.

Local generation can maintain both voltage and frequency at normal system values after a system disconnection, provided the generation is not overloaded. Overloading causes both of these quantities to decay.

TABLE 3-2 Short-Circuit Fault Protective Devices

Device	Device function number	Measured quantity	Detection devices	Usual application
Fuse	—	Current	Fuse element	Low- or medium-voltage circuits
Direct-acting trip device	—	Current	Trip device sensor	Low-voltage circuits
Overcurrent relay		Current	Current transformer	Medium-voltage circuits:
	50/51			Feeder-phase-time and instantaneous
	50/51N			Feeder residual ground time and instantaneous
	50GS			Feeder-ground sensor, instantaneous
	51GS			Feeder-ground sensor, time
	51N			Incoming or tie-residual ground, time
	51G			Neutral ground, time
	51			Incoming or tie phase, time
Differential relay		Current	Current transformer	Medium-voltage circuits:
	87T			Transformer over 10 MVA
	87T-G			Transformers, resistance grounded, overcurrent 10 MVA
	87M			Motors 1500 hp and over
	87B			Buses with 750 MVA or over, short circuit
	87L			Lines, ties to subbuses
	87G			Generators, any rating
Fault pressure relay	63FP	Pressure	Tank-mounted relay	Medium-voltage liquid-filled transformers over 10 MVA
Directional overcurrent relays	67 67N	Current and voltage	Current and potential transformers	Medium-voltage sources connected to a common bus
Distance relays	21	Current and voltage	Current and potential transformers	Large industrial medium- or high-voltage circuits

Motors connected to an electric system maintain the system voltage for some short time after the source system is disconnected. The length of time this voltage is maintained and its frequency depend on motor size, the amount of connected motor load, and motor plus load inertia.

Out-of-phase reclosing may ensue if power is lost for systems without local generation or if the loss overloads local generation and system voltage decay is not rapid enough to cause undervoltage relays to operate and disconnect the local system from the source before the source is automatically reclosed. This can result in extensive motor and associated equipment damage. To mitigate this possible event, various arrangements of sensitive reverse power, under- or overfrequency, or synchronizing check relays are available.

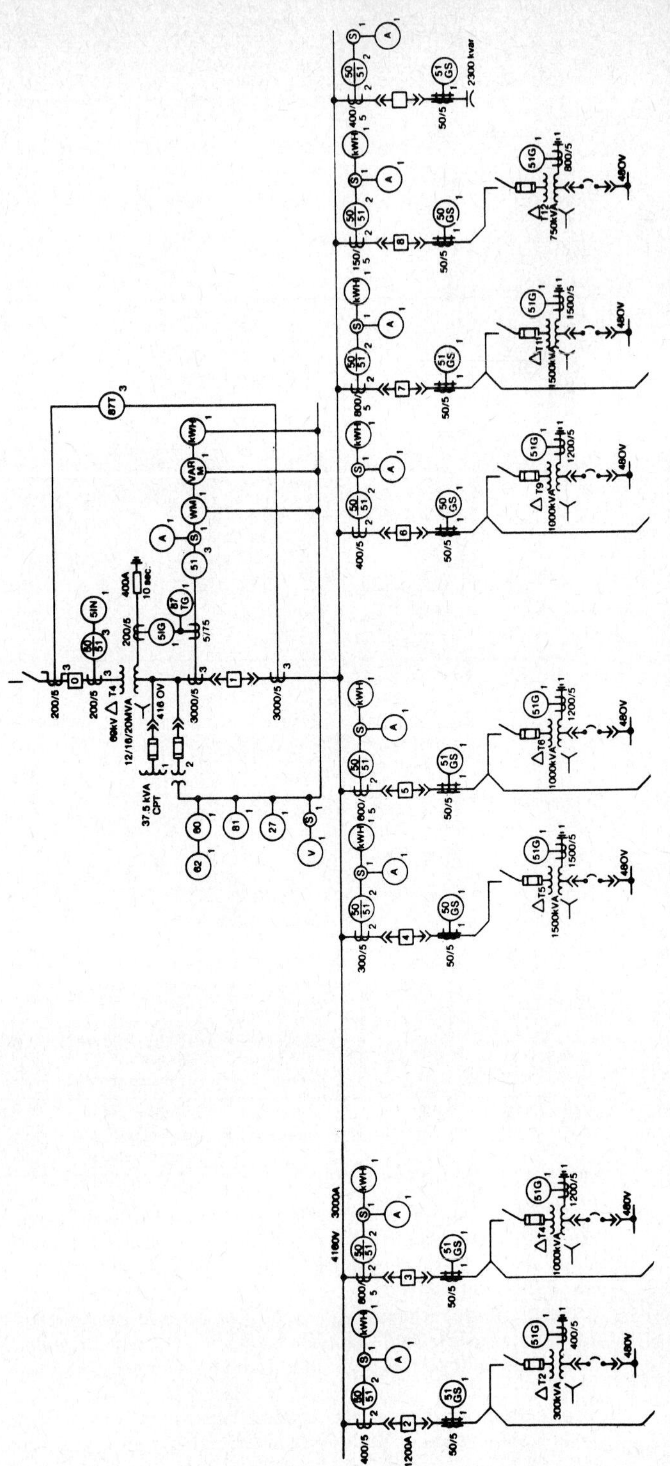

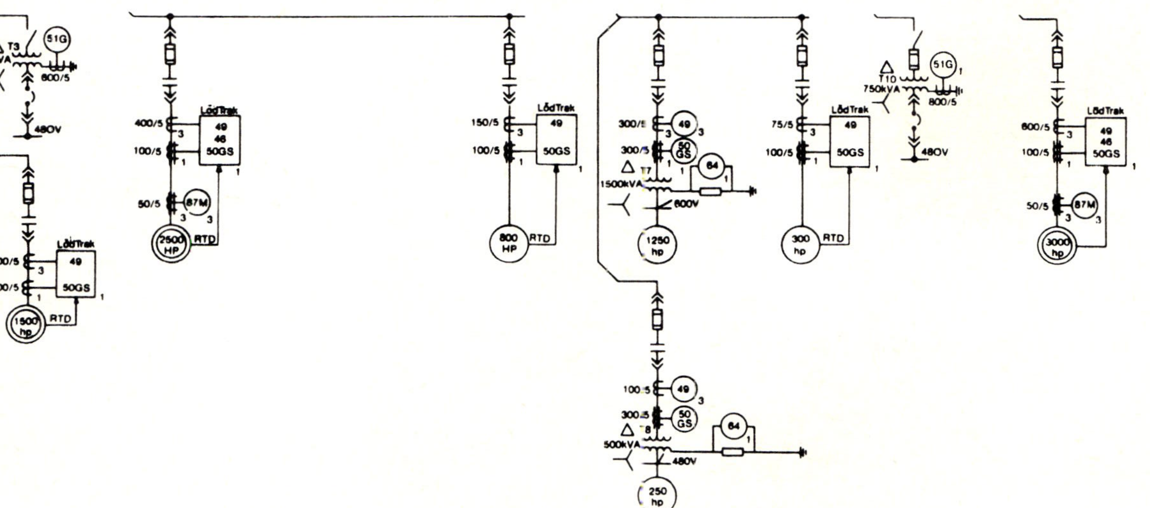

Figure 3-1 Typical industrial-plant one-line diagram. *(From* Industrial Power Systems Magazine, *General Electric Co., Schenectady, N.Y., March 1980, pp. 8–9.)*

Over-undervoltage relays connected to a single line-to-ground potential transformer detect line-to-ground faults on an incoming line that is disconnected from a grounded system at the source end. Such relays also can detect the return of normal voltage to a de-energized source and initiate automatic reclosing of the incoming-line circuit breaker.

Selection of appropriate relay settings for other than overcurrent relays is an important part of any system coordination study. Such selection may depend on the results of load flow, load shedding, or stability studies. The selection of the type of study necessary is at the discretion of the power-system protection engineer. See Table 3-3 for a list of common open-circuit fault-protective devices.

Figure 3-1 shows a typical industrial plant one-line diagram. Device function numbers identify the protective devices included. Figure 3-2 compares the time-current characteristics of some protective devices.

Figure 3-3 is a time-current curve for a portion of a system similar to Fig. 3-1. It is of utmost importance to identify each device represented by manufacturer and model number or other manufacturer's nomenclature when making a corodination study for selecting device settings.

Additional discussion of fault protection will be found in Sec. 3, Chap. 1, "Power Distribution Systems."*

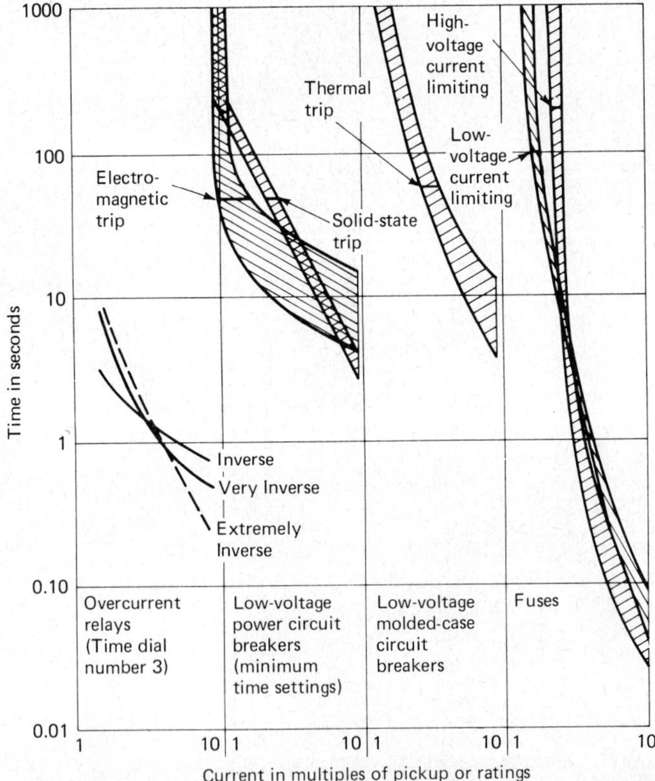

Figure 3-2 Operating time-current characteristics of four most commonly used power-system protective devices; note that plots have been developed by plotting on log-log paper on a common time basis. The more common relays normally applied on high-voltage systems operate considerably faster than the low-voltage devices with which selectivity is desired. *(Reproduced with permission from* Plant Engineering *Magazine, February 7, 1974, p. 87.)*

*See Rosaler and Rice: *Standard Handbook of Plant Engineering,* 1983.

TABLE 3-3 Open-Circuit Fault Protective Devices

Device	Device function number	Measured quantity	Sensor	Usual application
Undervoltage device	—	Voltage	Potential transformer	Low-voltage motor circuits
Single-phase undervoltage relay	27	Single-phase voltage	Potential transformer	Low or medium voltage or incoming line
Three-phase undervoltage and phase-sequence relay	47	Three-phase voltage	Potential transformers	Low or medium voltage or incoming line
Directional power relay	32	Current and voltage	Current and potential transformers	Low- or medium-voltage incoming line or generator circuits
Frequency relay	81	Frequency	Potential transformer	Generator bus
Under-overvoltage relay	27/59	Voltage	Potential transformer	Incoming line circuits
Synchronizing check relay	25	Voltage and phase angle	Potential transformer	Nonsynchronous source tie circuits

SURGE-VOLTAGE PROTECTION OF ELECTRIC-POWER-SYSTEM EQUIPMENT

APPARATUS-INSULATION CAPABILITY

Surge protection is intended to avoid surge stresses beyond the fairly well standardized surge capabilities of power-system-apparatus insulation. Apparatus-insulation capability and surge-arrester protective ability are both established by tests with various elevated voltage and/or current waveshapes.

Tables 3-4 to 3-7 list surge capabilities of basic power-system equipment and ac rotat-

TABLE 3-4 Impulse Test Levels for Liquid-Filled Transformers*

Insulation class and nominal bushing rating, kV	Windings				Bushing withstand voltages		
	High-pot. tests, kV	Chopped wave: Minimum time to flashover, kV	Chopped wave: Minimum time to flashover, μs	BIL full wave (½ × 50), kV	60-cycle 1-min dry, kV (rms)	60-cycle 10-s, kV (rms)	BIL impulse full wave (1.5 × 40) kV (crest)
1.2	10	54(36)	1.5(1)	45(30)	15(10)	13(6)	45(30)
2.5	15	69(54)	1.5(1.25)	60(45)	21(15)	20(13)	60(45)
5.0	19	88(69)	1.6(1.5)	75(60)	27(21)	24(20)	75(60)
8.7	26	110(88)	1.8(1.6)	95(75)	35(27)	30(24)	95(75)
15.0	34	130(110)	2.0(1.8)	110(95)	50(35)	45(30)	110(95)
25.0	50	175	3.0	150	70	70(60)	150
34.5	70	230	3.0	200	95	95	200
46.0	95	290	3.0	250	120	120	250
69.0	140	400	3.0	350	175	175	350
92.0	185	520	3.0	450	225	190	450
115.0	230	630	3.0	550	280	230	550
138.0	275	750	3.0	650	335	275	650
161.0	325	865	3.0	750	385	315	750

*Values in parentheses are for distribution transformers, instrument transformers, constant-current transformers, step- and induction-voltage regulators, and cable potheads for distribution cables. Data taken from industry standards.

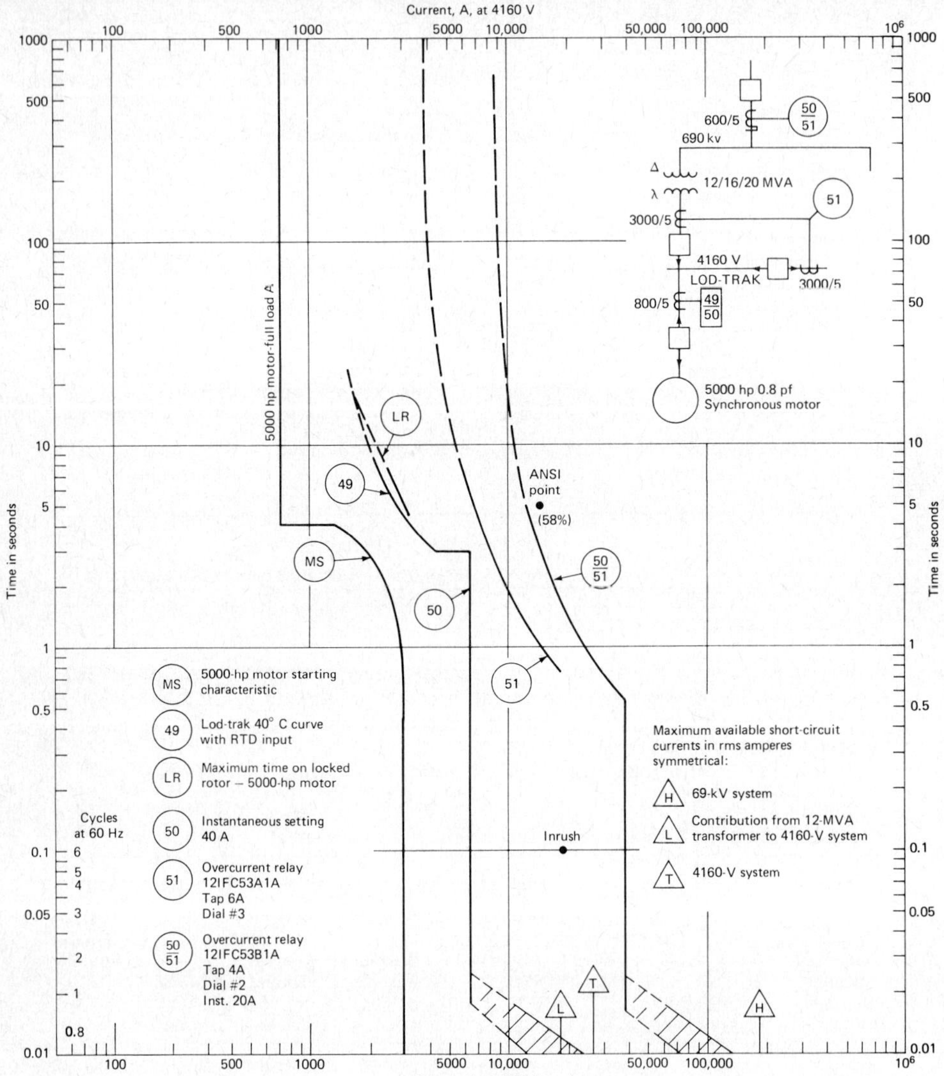

Figure 3-3 Time-current, phase-overcurrent protection, for a portion of a system similar to Fig. 3-1. *(From "Cement Plant Power Distribution," GET-2672C, General Electric Co., Schenectady, N.Y., December 1979.)*

ing machines. Impulse withstand of open-wire lines varies somewhat in specific value, depending upon construction, maintenance, weather, etc., but is generally considered well above that of associated transformers. An open-wire 13.8-kV distribution circuit, for example, is typically considered to have a 400-kV BIL. While cables do not have assigned BILs (basic impulse insulation levels), they too have impulse capability significantly higher than associated liquid-filled transformers.

TABLE 3-5 Impulse Test Levels for Power Circuit Breakers, Switchgear Assemblies, and Metal-Enclosed Buses

Voltage rating, kV	BIL full wave (1.2 × 50), kV, crest	Voltage rating, kV	BIL full wave (1.2 × 50), kV, crest	Voltage rating, kV	BIL full wave (1.2 × 50), kV, crest
2.4	45	23	150	115	550
4.6	60	34.5	200	138	650
7.2	75*	46	250	161	750
13.8	95	69	350	230	900
14.4	110	92	450	345	1300

*95 kV for metal-clad switchgear with power circuit breakers.

TABLE 3-6 Impulse Test Levels for Dry-Type Transformers*

Nominal equipment voltage		High pot. test, kV (rms)	Standard BIL (1.2 × 50), kV, crest
Delta or ungrounded wye	Grounded wye		
120–1,200		4	10
	1,200/693	4	10
2,520		10	20
	4,360/2,520	10	20
4,160–7,200		12	30
	8,720/5,040	10	30
8,320		19	45
12,000–13,800		31	60
	13,800/7,970	10	60
18,000		34	95
	22,860/13,200	10	95
2,300		37	110
	24,940/14,400	10	110
27,600		40	125
	34,500/19,920	10	125
34,500		50	150

*Data from Ref. 11, ANSI C57.12.01, "General Requirements for Dry-Type Distribution and Power Transformers."

TABLE 3-7 AC Rotating Impulse Levels

Motors			Generators		
Voltage rating, kV	High pot. test, kV (rms)	Impulse capability,* kV, crest	Voltage rating, kV	High pot. test, kV (rms)	Impulse capability,* kV, crest
2.3	5.6	9.9	2.4	5.8	10.2
4.0	9.0	15.9	4.16	9.3	16.5
6.6	14.2	25.1	6.9	14.8	26.2
13.2	27.4	48.4	13.8	28.6	50.6

*Equivalent BIL, commonly accepted values. Not standardized.

SURGE-PROTECTION TECHNIQUES

Electric-power-system components are protected against lightning and switching-produced surges by *interception* and *diversion* of the surges to ground. Shielding via overhead static wire and by conducting structures is used to greatly reduce the probability of significant lightning penetration of the system. Surge arresters are applied between

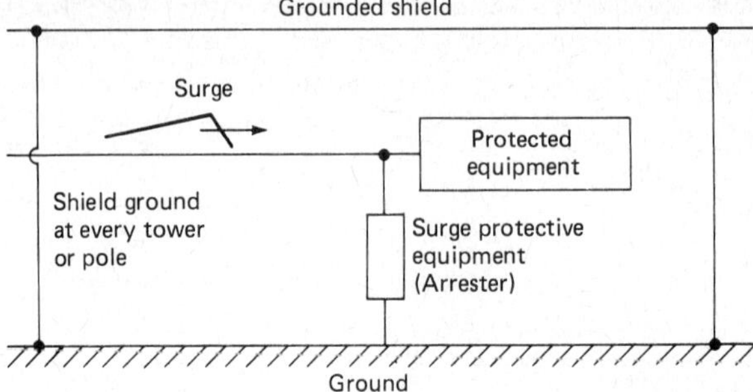

Figure 3-4 Schematic diagram of basic lightning-protection philosophy. Grounded shield intercepts a very large percentage of direct strokes. Surges that appear on incoming circuits are diverted to ground by surge arresters located at or near terminals of equipment to be protected.

line and ground at selected locations to divert surges from important and sensitive power-system and utilization equipment. Additionally, surge capacitors are sometimes applied at the terminals of motors and generators to reduce the surge voltage gradient. See Fig. 3-4.

SHIELDING

Lightning exposure to plant power systems is principally through outdoor supply substations and associated overhead lines. Effectively shielded installations are desirable. These are described in standards[13] as having shielding against direct strokes for the station and for all connected lines, at least for ½ mi from the station (line end protection).

Shielding of outdoor substation areas is usually provided by masts, or equivalent, which are designed to form a protective zone (not likely to be entered by lightning) within which all vulnerable parts will lie. With a single mast the protective zone is usually considered to be a cone having its apex at the top of the mast and whose sides make an angle with the vertical of 30 to 45°. With two or more masts the protective zone of each is increased somewhat in the area between them. With the usual spacings between masts, this shielding angle may increase to 60°. See Fig. 3-5.

Incoming line shielding is provided by overhead ground wire(s) which should be grounded at each pole or tower, through as low a ground resistance as it is practicable to

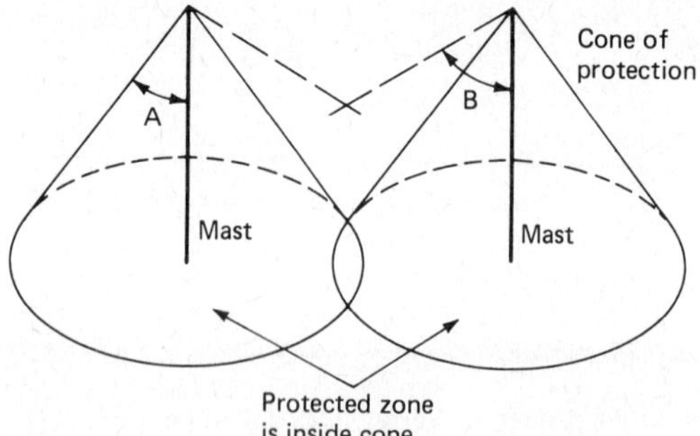

Figure 3-5 Shielding by masts. Protected zone with single mast is cone as illustrated with angle *A* of 30° to 45°. Cone of protection between masts extends up to 60°, angle *B*.

obtain, and should be connected to the ground bus at the substation. Low ground resistance is particularly important for the ground connection at the first few poles or towers adjacent to the substation. The associated protective zone is quite commonly considered to be of triangular cross section perpendicular to the shield wire with apex at the shield wire and with base at ground level extending a distance of twice the shield wire height on *each* side of the pole or tower. See Fig. 3-6.

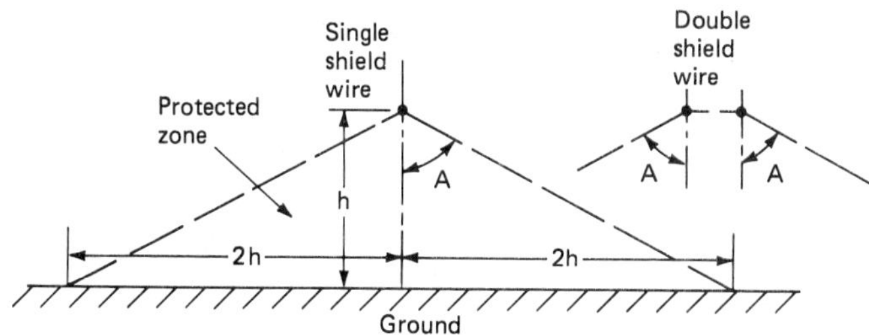

Figure 3-6 Commonly considered zone of protection provided by overhead shield wires for transmission (230 kV and below), subtransmission, and distribution lines. Specific shielding practices vary widely.

Outdoor substations not meeting the above criteria are generally classified as noneffectively shielded. Since noneffectively shielded applications entail a much higher surge exposure, it is important that the lowest practical ground resistance be obtained to minimize voltage gradients at the substation caused by surge current discharge to ground and to enhance arrester-protective performance.

Proper connection to earth is vital to the satisfactory performance of shielding and surge arresters. Resistance to ground should not exceed 1 Ω for large substations and generating stations, and 5 Ω for small substations and industrial plants. The ***National Electrical Code***[5] sets maximum resistance to earth at 25 Ω and requires arrester ground lead size be not less than No. 6 AWG.

SURGE ARRESTERS

Valve-type arresters are used for surge protection of plant power systems. Two basic valve material designs are employed to provide the necessary nonlinearity: (1) the zinc oxide arrester and (2) the gapped silicon carbide arrester. Zinc oxide–based valve elements have sufficient nonlinearity to accommodate both surge and nonsurge (normal operating) conditions. However, the silicon carbide–based valve elements require a series gap which effectively isolates them from continuous power frequency voltage. The gap sparks over at a predetermined elevated voltage which engages the valve element.

Arrester-Protective Characteristics

Three classes of arresters are recognized by standards[12] for medium-voltage and high-voltage systems—station class, intermediate class, and distribution class. Low-voltage-system (below 1000 V) arresters are classified as secondary arresters. Arrester manufacturers list arrester-surge-protective performance characteristics by arrester class. These performance characteristics are based on surge-voltage and surge-current test waveshapes also specified by standards. See Fig. 3-7. The two surge-voltage tests most frequently used for such listing are the front-of-wave sparkover and the 1.2 × 50-μs wave test. Equivalent front-of-wave test data is also listed for the nongapped zinc oxide arrest-

****National Electrical Code***® is a Registered Trademark of the National Fire Protection Association, Quincy, MA 02269.

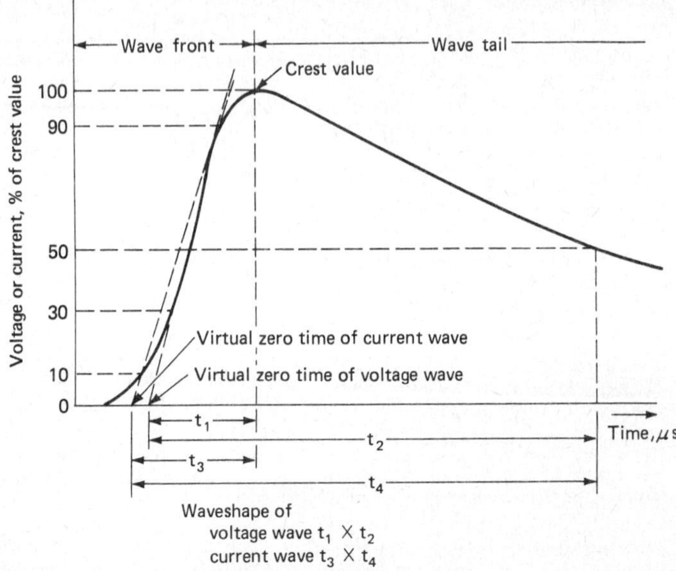

Figure 3-7 Wave geometry used to designate waveshape for purposes of standardizing surge-arrester protective characteristics and specifying insulation surge capability.

ers. The standardized 8 × 20-μs current wave is the basis for listing arrester-discharge-voltage levels. See Tables 3-8 to 3-11. As such, the listed characteristics usually provide the *maximum* surge voltage permitted (between arrester terminals) at the arrester location for the specified voltage and current surge duties.

Zinc oxide valve element technology is relatively new to power-system surge-arrester designs. It promises to affect future improvements in intermediate-class and distribution-class arresters in similar fashion to that already initiated in station-class designs. Accordingly, it is recommended that the most up-to-date vendor data and standards be used in selection of arresters.

Application of Arresters

The application of arresters depends upon three basic selections: (1) selection of arrester rating, (2) selection of arrester class, and (3) selection of arrester location.

Selection of Arrester Rating

A surge arrester should be selected with the *minimum* rating that will have a satisfactory service life on the power system. This will provide the greatest practical margin of safety for the protected insulation and at lowest cost. Expressed as a percentage this protective margin is defined as

$$100 \times \left(\frac{\text{apparatus insulation withstand}}{\text{surge-arrester protective level}} - 1 \right)$$

The generally recommended protective margin is 20 percent for impulse coordination (front-of-wave, full wave) and 15 percent for switching surge coordination.

Although protective margins are increased by reducing arrester ratings, there is a practical limit to such reduction because arresters are rated in relation to their ability to withstand power-frequency voltage during and following surge current discharge. The

TABLE 3-8 Surge-Protective Characteristics of Station-Class Arresters* for Zinc Oxide Valve Elements

Arrester rating, kV, rms	Equivalent front-of-wave protective level, kV, crest	Maximum switching surge protective level, kV, crest	Maximum discharge voltage, kV, crest, at indicated impulse current using an 8 × 20-μs current wave						
			1.5 kA	3.0 kA	5 kA	10 kA	15 kA	20 kA	40 kA
2.7	8.0	5.5	5.9	6.2	6.5	6.9	7.3	7.6	8.8
3.0	9.4	6.4	6.9	7.3	7.6	8.1	8.5	8.9	10.4
4.5	13.6	9.3	10.0	10.5	11.0	11.7	12.3	12.9	15.0
5.1	15.2	10.4	11.2	11.8	12.3	13.1	13.8	14.5	16.8
6.0	18.6	12.7	13.7	14.4	15.0	16.0	16.9	17.7	20.5
7.5	22.0	15.0	16.3	17.1	17.8	19.0	20.0	21.0	24.3
8.5	25.0	17.1	18.5	19.4	20.3	21.6	22.8	23.9	27.7
9.0	27.6	18.8	20.4	21.4	22.4	23.8	25.1	26.3	30.5
10	30.6	19.4	22.6	23.8	24.8	26.4	27.9	29.2	33.8
12	36.8	25.1	27.1	28.5	29.8	31.7	33.4	35.1	40.6
15	45.8	31.2	33.8	35.6	37.1	39.5	41.6	43.7	50.6
18	55.0	37.5	40.6	42.7	44.5	47.4	50.0	52.4	60.7
21	61.7	42.1	45.6	47.9	50.0	53.2	56.1	58.8	68.1
24	70.5	48.1	52.1	54.7	57.1	60.8	64.1	67.2	77.8
27	79.1	53.9	58.4	61.4	64.0	68.2	72.0	75.3	87.3
30	88.2	60.1	65.1	68.4	71.4	76.0	80.2	84.0	97.3
36	106	72.1	78.1	82.1	85.6	91.2	96.2	101	117
39	115	78.0	84.5	88.8	92.7	98.7	104	109	126
45	132	90.2	97.6	103	107	114	120	126	145
48	140	95.7	103.6	109	114	121	128	134	155
54	142	106	103	111	115	122	128	135	157
60	158	118	114	123	128	136	143	150	174
66	174	130	125	135	141	150	157	165	191
72	189	142	137	147	153	163	172	180	209
90	237	177	171	184	191	204	215	225	261
96	253	189	182	196	204	218	229	240	278
108	285	213	205	220	230	245	258	270	313
120	316	236	228	245	255	272	286	300	348
132	347	260	251	269	280	299	315	330	383
144	379	283	273	294	306	326	344	360	417
168	442	331	319	343	357	381	401	420	487
172	453	339	326	350	365	390	410	430	499
180	474	354	341	367	382	408	430	450	522
192	505	378	364	391	408	435	458	480	556
228	599	449	432	465	484	516	544	570	661
240	630	472	455	489	510	543	572	600	696

*Data in this table courtesy General Electric Co.
†Based on 500 A for 2.7–54-kV ratings and 3000 A for ratings 60 kV and above.

circumstance of a *line to ground* on the system and the associated *line to ground* voltage on the unfaulted phases is the prime criterion for selection of minimum arrester rating. This depends upon the degree of system grounding. Standards[12] define *coefficient of grounding* as " . . . The ratio E_{LG}/E_{LL}, expressed as a percentage, of the highest rms line-to-ground power frequency voltage E_{LG} on a sound phase, at a selected location, during a fault to ground affecting one or more phases to the line-to-line power-frequency voltage E_{LL} which would be obtained, at the selected location, with fault removed." Thus, the *minimum required arrester rating is the maximum line-to-line operating voltage times the coefficient of grounding.*

Standards[13] contain various aids to determine coefficient of grounding. The majority of industrial-plant systems in the 2.4- to 13.8-kV range are resistance-grounded with a 100 percent coefficient of grounding, as is also the case for ungrounded systems. Accord-

TABLE 3-9 Surge-Protective Characteristics of Station-Class Arresters* for Silicon Carbide Valve Elements

Arrester rating, kV, rms	Maximum ANSI std. front-of-wave sparkover, kV, crest †	Maximum 1.2 × 50 μs, sparkover, kV, crest	Maximum switching surge protective characteristic, kV, crest	Minimum 60 Hz sparkover, kV, rms	Maximum discharge voltage, crest, kV, at indicated impulse current, 8 × 20 μs					
					1.5 kA	5.0 kA	10.0 kA	15.0 kA	20.0 kA	40.0 kA
3	11	10	8.25	4.5	5.0	6.4	7.3	7.8	8.3	10.2
4.5	16.5	15	12.4	6.8	7.4	9.5	10.8	11.6	12.3	15.1
6	19	16	15.5	9.0	9.8	12.6	14.3	15.3	16.3	19.9
7.5	24	20	19.5	11.3	12.2	15.7	17.7	19.0	20.3	24.8
9	28.5	24	23.5	13.5	14.6	18.8	21.2	22.7	24.3	29.6
12	37	32	31	18	19.4	24.9	28.1	30.2	32.1	39.2
15	46.5	40	39	22.5	24.2	31.0	35.0	37.5	40.0	48.8
18	55.5	48	46.5	27	28.9	37.1	41.8	44.8	47.8	58.5
21	65	56	55.5	31.5	33.7	43.2	48.7	52.3	55.5	68.0
24	74	64	62	36	38.4	49.2	55.5	59.5	63.5	77.5
27	83	72	70	40.5	43.1	55.3	62.5	67.0	71.2	87.0

30	92	80	78	45	47.8	61.5	69.5	74.5	79.0	96.6
36	111	96	93	54	57.5	73.5	83.0	89.2	94.5	115.0
39	120	104	101	58.5	62.5	79.5	89.5	96.0	102.0	125.0
48	148	128	124	72	76.0	97.5	110	117	125	153
60	170	141	136	84	95.0	122.0	137	147	156	190
72	204	169	163	100	114	146	164	170	187	227
78	220	184	177	109	123	158	178	191	202	246
84	237	198	191	117.5	133	170	191	205	217	265
90	254	212	204	126	142	182	204	219	232	283
96	270	226	218	134.5	151	194	218	234	248	302
108	304	254	245	151	170	218	245	262	278	339
120	338	282	272	168	183	241	272	292	309	376
132	372	310	299	185	207	262	294	315	333	402
144	405	338	326	201.5	226	287	321	344	363	439
168	483	395	381	235	263	334	374	401	422	510
180	521	400	400	243	281	358	400	424	452	550
192	559	427	426	260	300	382	427	457	482	585
228	677	510	506	308	355	452	510	546	575	695
240	718	535	533	324	374	476	535	574	605	730
258	775	575	573	349	402	515	575	616	650	785

*Data in this table courtesy General Electric Co.

†ANSI/IEEE 28-1979, "Surge Arresters for Alternating-Current Power Circuits."

TABLE 3-10 Surge-Protective Characteristics of Intermediate-Class Arresters* for Silicon Carbide Valve Elements

Arrester rating, kV, rms	Maximum ANSI front-of-wave sparkover, kV crest	Maximum 1.2 × 50 μs sparkover, kV, crest	Maximum discharge voltage, kV, crest, at indicated impulse current 8 × 20 μs wave				
			1.5 kA	3 kA	5 kA	10 kA	20 kA
3	11	11	6.5	7	7.4	8.3	9.5
4.5	16	15	9.5	10.1	10.8	12.0	14.0
6	21	19	12.5	13.3	14.5	16.0	18.5
7.5	26	23.5	15.8	16.7	17.8	20.0	22.8
9	31	27.5	19.0	20.0	21.0	23.5	27.0
10	35	31	21.8	23.0	24.5	27.2	31.5
12	40	35.5	24.7	26.1	28.0	31.0	36.0
15	50	43.5	31.0	32.5	35.0	39.0	45.0
18	59	51.5	37	39	42	47	54
21	68	59	43.0	45.5	49.0	55.0	63.0
24	78	67	49.0	52.0	56.0	62.0	72.0
30	97	81	61.5	65.0	70.0	78.0	90.0
36	116	95	74.0	78.0	83.0	94.0	108
39	126	102	80.0	84.5	90.0	102	117
45	144	116	92.0	98.0	104	117	135
48	154	123	98.0	104	111	125	145
60	190	153	123	130	139	156	180
72	228	180	148	156	166	187	216
90	282	223	184	195	208	233	270
96	300	236	197	208	222	249	288
108	335	263	221	234	249	281	324
120	370	290	246	260	277	311	360

*Data in this table courtesy General Electric Co.

ingly, a large majority of plant medium-voltage systems require minimum arrester ratings of at least 100 percent of the maximum operating voltage of the system (100 percent arresters). Grounded systems typically require 75 to 80 percent arresters. See Table 3-12. Multigrounded (four-wire) distribution systems and some high-voltage transmission systems (69 kV and above) may safely use lower rated arresters. These systems require individual determination of their coefficients of grounding to ensure the most economical, secure arrester rating selection.

Selection of Arrester Class

In order of cost, protective efficiency, and durability, the three classes of arresters are: (1) station class, (2) intermediate class, and (3) distribution class. As a general guide to arrester-class usage vs. equipment size, the following may be considered typical practice:

Station Class. Component protection of 7.5 MVA and above substations and large or essential rotating machines

Intermediate Class. Component protection of 1 to 20 MVA substations, overhead lines, and rotating machines

Distribution Class. Distribution-class apparatus, dry-type transformers, and small rotating machines

These classes overlap considerably. There is a tendency to use higher-class arresters at higher voltages. In some cases of distribution-class arrester application, low sparkover models are recommended, and, in fact, are necessary. Limitations on the available range of ratings of distribution-class and intermediate-class arresters eliminate them as a choice in higher-voltage applications.

TABLE 3-11 Surge-Protective Characteristics of Distribution-Class Arresters* for Silicon Carbide Valve Elements

Arrester rating, kV, rms	Maximum ANSI front-of-wave sparkover, kV, crest	Maximum discharge voltage, kV, crest at indicated 8 × 20 μs impulse current wave				
		1500 A	2500 A	5000 A	10,000 A	20,000 A
3	15	9.5	10	11	12	13.5
	11†	9.5	10	11	12	13.5
4.5	17†	14	15	17	19	20.5
6	29	19	20	22	24	27
	21†	19	20	22	24	27
7.5	26.5†	23	25	28	31	34
9	40.5	28	30	33	36	40
	32†	28	30	33	36	40
10	44.5	28	30	33	36	40
	32†	33	35	39	43	47.5
12	56	37	40	44	48	54
	39.5†	37	40	44	48	54
15	63	42	45	50	54	61
	46†	46	49	55	60	67
18	75	50	55	61	66	74
21	89	60	64	72	78	88
27	98	73	79	87	96	107

*Data in this table courtesy General Electric Co.
†Low-sparkover model data.

TABLE 3-12 Voltage Ratings of Arresters Usually Selected for Three-Phase Systems

Nominal system voltage, kV	Voltage rating of arrester, kV: System neutral ungrounded or resistance-grounded	System neutral solidly grounded
0.120/0.208Y	0.65	0.175
0.240	0.65	0.65
0.480	0.65	0.65
0.600	0.65	0.65
2.4	3 (2.7)	3 (2.7)
4.16	4.5	3 (4.5)
4.8	6.0 (5.1)	4.5 (5.1)
6.9	7.5	6.0
12.47	15, 12	10, 9
13.8	15	12, 10
23	24	24, 21, 18
34.5	39, 36	30, 27
46	48	39
69	72	60, 54
115	120	90, 96, 108
138	144	108, 120
161	—	120, 132, 144
230	—	172, 180, 192

While actual lightning-protective practices may necessarily vary from one type of installation to the next, all installations are either effectively shielded or noneffectively shielded. There are different degrees of jeopardy for each of these two basic categories, and the degree may vary with a change in system arrangement or operating mode.

Excessive arrester discharge currents in noneffectively shielded installations are a prime cause of arrester failure, and inadvertent loss of protection may occur when arrester discharge voltages exceed anticipated levels. In effectively shielded systems, arrester discharge currents are unlikely to exceed 5000 A as a conservative maximum, but 20,000 A is a conservative maximum for noneffectively shielded systems. This encourages use of station-class and intermediate-class arresters in noneffectively shielded installations.

Location of Arresters

The ideal location for surge arresters, from the standpoint of protection, is directly at the terminals of equipment to be protected. Usually, low-BIL apparatus (certain dry-type transformers and rotating machines) requires surge-protective devices in direct shunt with associated insulation. Otherwise practical circumstances dictate often that arresters be remote, requiring that one set of arresters protect more than one piece of apparatus.

For remote arresters it is necessary to estimate the depreciation in protection caused by separation distances between the arresters and the protected equipment. These determinations require the use of traveling wave mechanics by a practiced surge protection engineer. The following are *general guides* for locating arresters relating to typical components and their arrangements.

Effectively Shielded Substations. Arresters are required on each overhead line as it enters the substation, for protection of disconnect switches, buses, etc. Separation distances of 75 to 200 ft, and sometimes more, between arresters and full-BIL equipment can be tolerated at 23 kV and above. At 15 kV and below, practice usually avoids appreciable separation distance. Usually arresters are applied at transformer incoming line terminals.

Noneffectively Shielded Substations. Arresters should be applied at or very close to terminals of transformers and breakers. A minimum separation distance between arresters and protected equipment for overcurrent protection equipment is permissible.

Metal-Clad Switchgear. Metal-clad switchgear installed in substations (as above) should be protected in the same fashion as transformers. In effectively shielded substations where continuous metallic-sheathed cable (or equivalent) intervenes between the switchgear and exposed line, an arrester at the junction of line and cable suffices if it is station class, intermediate class, or special distribution class. When exposure is through a power transformer protected on the exposed side, generally arresters are not necessary at the switchgear.

Dry-Type Transformers. Full-BIL dry-type transformers should be protected as described above for full-BIL equipment. Reduced-BIL dry-type transformers (with BILs below comparably rated liquid-filled distribution transformers) require arresters at their terminals when connected directly to overhead lines. When exposure is through continuous metallic-sheathed cable (or equivalent), a line-cable junction arrester may or may not protect the low-BIL dry-type transformer. In such cases, a special (low sparkover) distribution-class arrester at the transformer terminals will suffice. If exposure is through another transformer, usually arresters are not required at the dry-type transformer.

Cable. Cable should be surge-protected the same as full-BIL transformers.

Aerial Cable. Arresters are required at junctions of open-wire line and aerial cable. Messenger and sheath should be grounded through a low value of ground resistance at every pole. Aerial cable should be considered the same as open-wire line for protection of terminal equipment.

ROTATING-MACHINE PROTECTION

Medium-voltage (2.3- to 13.8-kV) rotating machines require: (1) a strictly effectively shielded environment, (2) arresters at the terminals of the machine, (3) surge capacitors at the terminals of the machine, and (4) strict adherence to good grounding practices. The surge capacitors are used to reduce the voltage gradient of surges that may be damaging to machine turn insulation. Such capacitors should be connected in the closest possible shunt relation to the machine line-to-ground insulation. Table 3-13 lists commonly available surge-protective capacitors. Note surge-protective capacitors are rated on a line-to-line (rms) voltage basis with associated designated capacitance per pole (terminal to case).

TABLE 3-13 Ratings and Sizes of Surge-Protective Capacitors

Voltage rating, rms, V, L-L	Maximum voltage, rms, V, L-L	Poles per unit	μF per pole
0–650	715	3	1.0
2,400	2,640	3	0.5
4,160	4,576	3	0.5
6,900	7,590	1	0.5
13,800	15,180	1	0.25
24,000	26,400	1	0.125

Switching events within the distribution system (e.g., motor switching, fault inception, fault removal, insulation flashover, capacitor switching, etc.) present surge exposure possibilities to motors and generators. In practice, a high percentage of machines above 4 kV have arresters and surge capacitors. At least half of the 4-kV motor installations are so protected, while at 2.3 kV only a minority are so protected. Low-voltage motors and generators (below 1000 V) seldom have such surge protection, except in exposed pumping applications.

REFERENCES AND BIBLIOGRAPHY

Standards (Use Latest Issue)

1. ANSI C42.100-(year), "IEEE Standard Dictionary of Electrical and Electronics Terms."

IEEE Recommended Practices

2. IEEE 141-(year), "IEEE Recommended Practice for Electric Power Distribution for Industrial Plants."
3. IEEE 142-(year), "IEEE Recommended Practice for Grounding of Industrial and Commercial Power Systems."
4. IEEE 242-(year), "IEEE Recommended Practice for Protection and Coordination of Industrial and Commercial Power Systems."
5. NFPA 70-(year), ***National Electrical Code.***®
6. NFPA 70E-(year), "Electrical Safety Requirements for Employee Workplaces."
7. NFPA 70B-(year), "Recommended Practice for Electric Equipment Maintenance."

ANSI Standards for Equipment

8. ANSI C19 Series, "Package Control Equipment."
9. ANSI C37.2-1970, "Switchgear."

10. ANSI C51:1-1978, "Safety Standards for Construction and Guide for Selectors, Installation and Use of Electric Motors and Generators."
11. ANSI C57 Series, "Transformers, Regulators, Reactors."
12. IEEE Std 28-(1974); also ANSI C62.1-(1974), "Surge Arresters for Alternating-Current Power Circuits."
13. ANSI C62.2-(1969), "Guide for Application of Valve Type Lightning Arresters for Alternating-Current Systems."

Books

14. Beeman, D. L. (ed.): *Industrial Power Systems Handbook*, McGraw-Hill, New York, 1955.
15. *Electrical Transmission and Distribution Reference Book*, Westinghouse Electric Corp., Pittsburgh, 1950.
16. *Applied Protective Relaying*, Westinghouse Electric Corp., Pittsburgh, 1976.
17. Smeaton, R. W. (ed.): *Switchgear and Control Handbook*, McGraw-Hill, New York, 1977.

chapter 8-4

Security Equipment

by
A. J. Grosso
Vice President, Engineering
American District Telegraph Co.
New York, New York

INTRODUCTION

The term *security* may be defined as the protection of life and property. Two of the basic hazards are fire (see Chap. 8-2) and crime. Crime hazards, which will be covered here, consist of intruders who enter a premise to steal physical property or secret information or to commit sabotage. The purpose of crime security is to deter the entrance of potential criminals or vandals, to detect intruders, and to alert a guard force as promptly as possible. A basic premise of good security is protection in depth—the provision of more than one line of defense to discourage incursions, to slow the advance of knowledgeable intruders, and to provide secondary defenses in case of penetration beyond the first line.

Security is much more than an alarm system and must include:

1. A complete plan, attitude, and purpose by the persons involved in "securing" life and property
2. Physical security

3. Detection and reporting methods
4. Signal transmission
5. Signal handling
6. Continued review of quality, value, and changing conditions and the intent to amend and upgrade as necessary

TYPES OF SYSTEMS

The alarm industry recognizes four types of alarm systems:

1. **Local Alarm** A system which produces an audible and/or visible signal at the protected premises only as a result of an alarm condition or fault (capability degradation).
2. **Direct-Connect (or Headquarters)** A system in which signals are transmitted to police or fire headquarters is sometimes combined with a local alarm system.
3. **Central Station** A system in which signals of various types are transmitted to an independent control center where trained personnel are maintained continuously to supervise the status of protected premises and take appropriate action upon the receipt of signals. The central station is generally considered to be the preferred form of protection.
4. **Proprietary** This system is similar to the central station system except that the control center is staffed and maintained by the owner of the protected properties.

In practice, fire-protection and crime-protection systems are closely related, and the equipment and services needed for both types of systems are usually provided by the same equipment manufacturers, installers, and service companies.

Modern practice, especially for proprietary systems, in large plants, allows for the integration of the fire- and burglar-alarm systems with the building automation services. Air conditioning, energy management, closed-circuit television, access control, and communications, together with the protection services, are controlled from a central location under the direction of a computer that has been programmed to initiate the proper actions and responses required under a wide variety of operating and emergency conditions.

SECURITY PLANNING

The planning of a plant-wide security system is a technical matter that may require the services of a specialist, either an independent consultant or the representative of a security company. Consultation should begin before the plans for a new building are completed as there are aspects of architecture and design which affect security.

With the aid of the expert, a security system can be designed to meet the sometimes conflicting requirements for:

1. **Likelihood of Attack** A precious-metals refinery is more likely to be attacked than a junk storage area. Criminals prefer readily portable articles of high value. The local crime rate is also an important factor.
2. **Police Response** A large, well-equipped police force located within a few minutes travel time establishes a smaller security risk than a remotely located plant in a rural setting.
3. **Building Construction** A modern masonry structure is more resistant to attack than an old wooden building or, in an extreme case, an air house.
4. **Insurance** The security expert can be a guide in the selection of equipment and techniques for both security and fire-alarm systems that will meet the requirements

of such approval agencies as the Underwriters Laboratories, Inc., Factory Mutual, and National Fire Protection Association which lead to insurance premium discounts.

5. **Technical Considerations** These include a variety of technical considerations leading to reliable security systems which produce a minimum of nuisance alarms and other trouble conditions and assure compliance with any local codes and ordinances.

PHYSICAL SECURITY

Physical security is passive in nature as opposed to the active methods for the detection and reporting of intruders. It is often the first line of defense and can be seen in the form of fences, walls, and moats. It includes visual definition of boundary lines as psychological barriers; for example, low fences, a line of gardens, well-kept lawns, and a change of walkway materials, e.g., from concrete to brick.

National records indicate that crime against property averages 90 percent of all reported crime and that, in the majority of all burglary cases, entry was gained through doors or windows. In spite of this knowledge, there is a consistent use of inferior, low-security hardware and other materials including locks.

In some localities in the United States there are codes and ordinances which set forth standards for door frames, doors, door locks, windows, window locks, elevators, and other openings such as skylights, hatchways, air ducts, vents and transoms. In the absence of local standards, or the presence of outdated codes, reference should be made to the Law Enforcement Assistance Administration, National Institute of Law Enforcement and Criminal Justice Standards (NILECJ; See Bibliography).

Physical security is dependent on good housekeeping. The best security equipment and alarm systems lose effectiveness when doors are not locked at night.

DETECTION AND REPORTING METHODS

Outdoor Perimeter Protection

The protection of an outdoor perimeter is the first line of defense in a protection-in-depth security plan. It can start with a masonry wall or a fence constructed of wood. A widely used construction is the familiar chain-link (wire mesh) fence six or more feet in height and topped with an outward slanting barbed-wire or barbed-tape-coils section to discourage fence climbers. While difficult to climb for the inexperienced, such a barrier is susceptible to attack with wire cutters, or by jacking. Thus, it may be necessary to provide a patrolling guard, with or without dogs.

An alternative to the expense of a guard force is the use of an electronic barrier consisting of fence-mounted transducers to detect and annunciate the mechanical vibrations caused by attempts to climb or cut the fence.

Another form of electronic barrier comprises a series of posts strung with a number of wires in telephone-line style. This type of fence presents no physical barrier to an intruder but the electromagnetic field set up by the current flowing through the wires is disturbed by the presence of an intruder. The field disturbance is detected electronically and annunciated at a guard station.

In an effort to reduce the costs of installing and maintaining a long line of posts and wires, projected energy beams are sometimes used. A beam of energy (which may be light in the invisible infrared frequency spectrum or microwave energy) is projected from a source to a receiver and, when interrupted by the passage of an intruder, an alarm is annunciated at the guard station. (See Fig. 4-1.) *Note:* This type of system is best adapted to flat terrain where the beam length can be relatively long. When the terrain is hilly, or very irregular in outline, the fence-post type of system is preferred to reduce equipment cost.

All types of electronic intrusion barriers require both means to keep animals at a safe distance and constant maintenance to control the growth of vegetation to prevent an

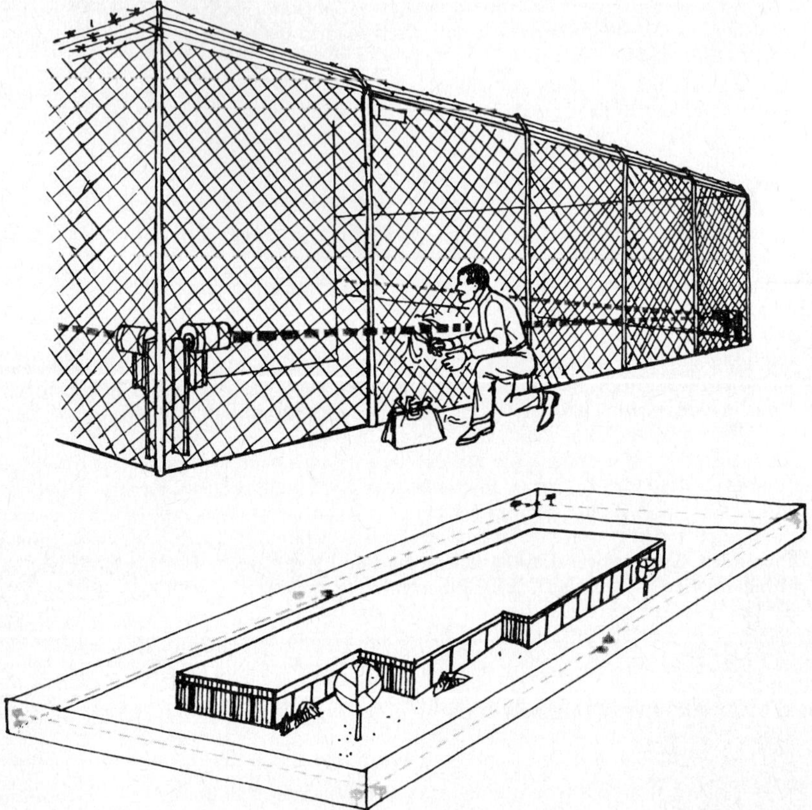

Figure 4-1 Use of light beams to protect restricted area.

excessive rate of nuisance alarms. Erosion of the earth under the barrier must be controlled to prevent access paths.

Portal Protection

Outdoor perimeter fences as well as building entrances must have portals that are guarded in such a manner that authorized persons may be admitted and others excluded. This may be done by means of a guard who provides other services such as directing visitors.

The cost of guard service may be reduced by an electrical lock on the door and a voice intercom connection as is commonly provided in apartment house lobbies. Where voice identification is not sufficient, a closed-circuit television system will permit visual inspection before admission. Obviously, a single guard at a central location can monitor many portals.

The simplest means of admission control is to provide authorized persons with a key to the door. Pushbutton locks avoid the problems of lost and stolen keys since it is easier and cheaper to change the combination when necessary than to provide a new lock and set of keys.

Card access systems avoid the problems associated with metallic keys and provide other advantages. In its simplest form, the card access system is the equivalent of a metallic door key. When a photograph plus other information is applied to the card, it becomes an identification card presented to building guards when needed. With additional coding, a zoned system may be established in which many persons may be allowed to enter a first door, a lesser number to enter another door, and only a select few to pass through a third door.

Computer-based systems provide such sophisticated functions as sounding an alarm when the wrong card is presented, keeping a record of who entered what door at what time, and establishing control according to the time of day.

Building Perimeter Protection

Experience has shown that the large majority of burglarious attacks are made on doors and windows. The basic device employed for their protection is known as the *burglar-alarm contact* and has evolved from clumsy mechanical contrivances to the modern magnetically operated switch. Areas of glass have been protected for almost a century by the familiar current-carrying foil strip whose rupture initiates an alarm signal. Foil is now being replaced by transducer devices cemented to the glass and tuned to respond to the characteristic sound frequencies produced when the glass is broken.

An older form of protection for windows and other openings is the *burglar-alarm screen* consisting of a small-diameter, current-carrying wire embedded in a frame of wooden dowels. To pass such a barrier, the intruder must break at least some of the dowels (and the embedded wire) and so actuate the alarm.

Since it is also possible to effect entry by penetration of the walls, floors, and ceilings, such surfaces are protected by *pads* of wire or foil mounted on building-material panels attached to the surface to be protected. Figure 4-2 illustrates these as well as other devices described below.

Interior Protection

In accordance with the principle of protection in depth, *traps* are employed to reveal the intruder who has penetrated the perimeter. A simple and effective trap is the pressure mat hidden under the carpet which produces a signal when stepped on by the intruder. The *floor trap* is a trip-wire device in which the wire need not necessarily be broken; its displacement alone will initiate the alarm. A photoelectric beam made invisible by the use of infrared light is a more advanced form of trip wire.

Discrete objects such as safes and file cabinets may be protected by contacts to give warning of their being opened but it is better to have an alarm prior to that event. For this purpose, capacitance alarm systems which produce an electric field around the protected object will give the alarm signal upon the approach of the intruder before he or she touches the object.

An earlier form of protection still in use is the "cabinet" made of wood lined with foil or fine wire surrounding the object to be protected. Penetration of the cabinet, as in the case of the burglar alarm screen, initiates an alarm signal.

Space Protection

About 1950, the concept of space protection was introduced in the form of the ultrasonic motion detector. The space to be protected is flooded with sound energy in the ultrasonic range, and the motions of the intruder cause a Doppler shift in the frequencies detected at the receiver which is employed to actuate the alarm signal. See Fig. 4-3. Problems in some situations caused by air turbulence led to the introduction of generally similar systems operating on energy in the microwave range (Fig. 4-4) and, most recently, to passive infrared systems which detect the warmth of the intruder's body.

Designated areas may be kept under constant surveillance by means of closed-circuit television cameras or a record of activities established by constantly or intermittently operated photographic cameras. Alternatively, surveillance may be provided by microphones which relay the sounds made by an intruder to a listening post.

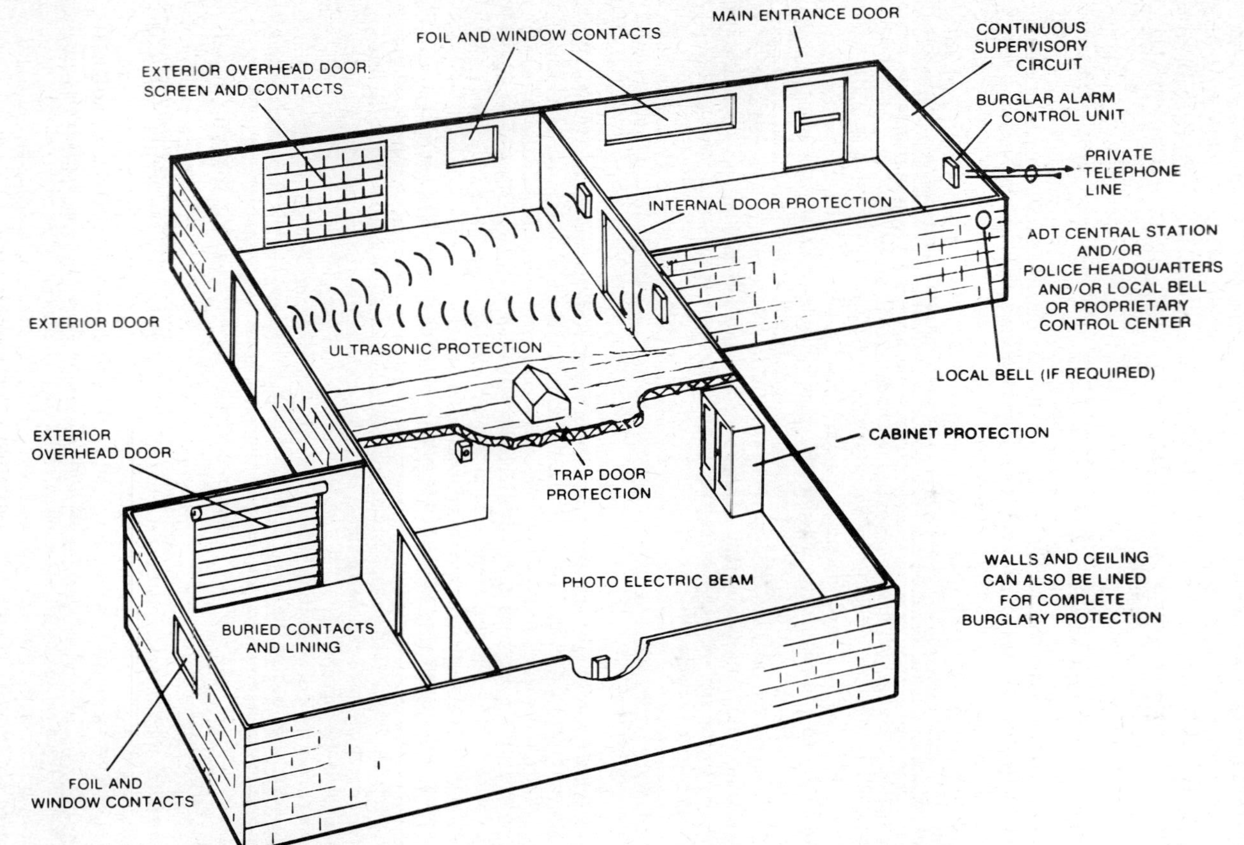

Figure 4-2 Typical perimeter and interior protection plan. *(American District Telegraph Co.)*

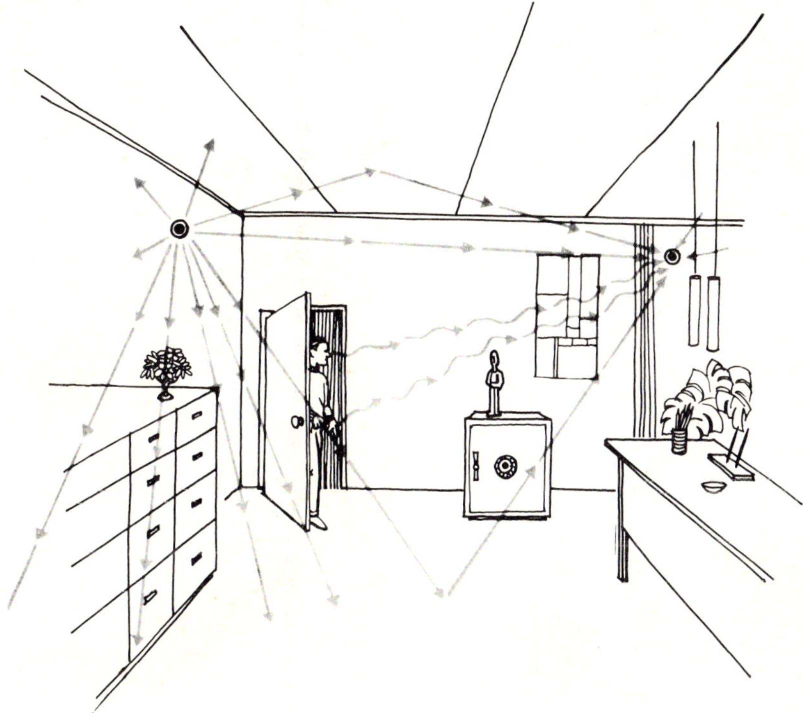

Figure 4-3 Diagram shows how system utilizes Doppler effect, a scientific phenomenon which causes sound waves to change tone when hitting a moving surface. The change triggers an alarm.

A relatively new device is the rail television system in which a mobile camera travels on a track-type rail to provide coverage of an area which would otherwise require multiple cameras.

Vaults like those used for safe deposit, fur, and narcotics are often protected by alarm systems responsive to the sounds created during an attack which range from the single, high-intensity burst of a dynamite blast to the low-level noise produced by the scraping away of mortar with a pointed tool such as a screwdriver. Two basic types of system are employed: one is responsive to airborne sounds and the other to sounds transmitted through the structure of the vault.

Attacks conducted with oxyacetylene torches or burning bars are detected by heat and/or smoke detectors adapted from fire-protection technology.

Night depositories can be protected by lead-sheathed, current-carrying cables embedded in the concrete at the time of construction. Rupture of the cable during the course of an attack results in an alarm.

Holdup Alarms

It is often advisable to provide a holdup alarm system for payroll departments and other high-risk areas so that assistance may be summoned rapidly either during or immediately after a raid. This system could also actuate automatic cameras, photographic or television. The initiating devices are concealed hand- or foot-operated switches and short-range radio transmitters which are carried on the belt or in a pocket.

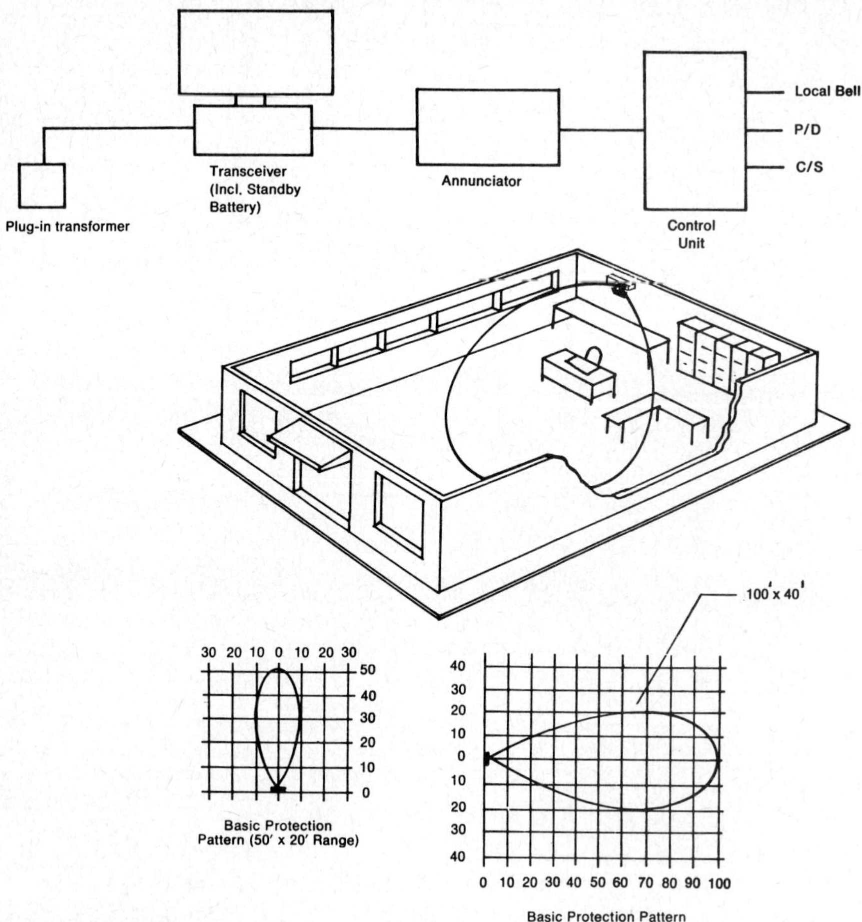

Figure 4-4 Typical protection patterns with microwave beams.

There are also a variety of *cash-drawer* devices intended to permit the surreptitious initiation of a signal: pressure pads which can be pressed by the cashier when removing money from the drawer; a money clip which is a switch whose contacts are held apart by one or more pieces of paper currency whose withdrawal allows the contacts to close and actuate a silent alarm.

A recent addition is the exploding-money package. A bundle of what appears to be currency contains tear gas and marker dye. It is actuated by an electric field at the exit door as the perpetrator departs. After a short delay, the package ruptures to scatter the tear gas and dye. This assists in the capture and identification of the bandit.

SIGNAL TRANSMISSION

The basic burglar-alarm circuit is that of the doorbell; i.e., operate a switch and the bell rings. In practice, such a simple, normally open circuit is not acceptable because loss of the power supply or a break in the line (accidental or intentional) would not be discovered until the system failed to operate when needed.

It is, therefore, the general practice to monitor a small current continually passing through the circuit and initiate alarm signals when its value rises above or falls below specified levels. Means are sometimes provided for the operator to distinguish between true alarms and circuit faults.

In central-station service, two basic types of circuit are employed. Direct-wire service employs a dedicated telephone connection from each subscriber to an individual readout device at the central station and is the equivalent of a private-wire telephone connection. Circuit burglar-alarm service connects a number of subscribers to a common circuit path, as in party-line telephone service, and the individuals are identified at the central station by means of coded signals. The McCulloh circuit arrangement is usually employed in this type of service to permit continued operation when faults appear on the line.

Where high-security risks are involved, some form of line security should be provided to detect attempts to compromise the transmission lines. The technical sophistication of the modern criminal element poses a substantial threat to security.

Traditionally, the central-station services have been provided by dc circuits which have distance limitations. As more and more subscribers leave the city for suburban locations, the central stations have turned to ac signal transmission to relieve the problem. Interrogate-response polling techniques enhance the reliability and security of the service.

An intermediate step is the use of automatic telephone dialers connected to the switched network which eliminates the need to wait for (and pay for) the availability of a dedicated telephone line. The best form of this service is provided by digital dialers which check for the completion of a connection before transmission of the signal. Tape-recorded voice message dialers have presented service problems which have led to their being prohibited by the police in certain locations. Dialers are sometimes used as backup for other forms of transmissions.

The radio transmission of alarm signals has been impeded by cost and federal regulations. Nonetheless, the transmission of signals by radio can be a valuable method when other facilities are unreliable or nonexistent and when used as a backup for the conventional transmission facilities.

SIGNAL HANDLING

A traditional problem in the monitoring of security systems has been the processing of the large number of signals that do not represent an emergency situation. These signals are caused by the opening and closing of protected premises at the proper time of day and the signals received from patrolling watchmen according to schedule.

Carefully programmed computers are now used to process routine signals automatically with the added advantage of automatically providing printouts of desired records. Computer-based operations relieve the pressure on the operators during busy periods and generally improve the efficiency and reliability of security services.

Modern microprocessor technology has been especially effective in proprietary systems where it provides such assistance to the operators as illuminated maps showing the exact location of the alarm source and simultaneously presenting detailed instructions as to the actions to be taken by the operator.

REVIEW AND UPGRADING

Security is a never-ending procedure because it must be constantly reviewed and upgraded to allow for changes in the physical structure of the protected premises as well as their content and the environment, as discussed under "Security Planning." In addition, consideration must be given to the replacement of aging equipment with new-technology devices offering superior performance.

BIBLIOGRAPHY

National Institute of Law Enforcement and Criminal Justice (NILECJ)-STD-0306.00, May 1976, "Physical Security of Door Assemblies and Components," National Institute of Law Enforcement and Criminal Justice, Law Enforcement Assistance Administration, U.S. Department of Justice.

NILECJ-STD-0316.00, March 1979, "Physical Security of Window Units," National Institute of Law Enforcement and Criminal Justice, Law Enforcement Assistance Administration, U.S. Department of Justice.

Security World Publishing Company, P.O. Box 272, Culver City, CA 90230, specializes in books and periodicals relating to security.

chapter 8-5

Toxic Substances and Radiation Hazards

by
Jaswant Singh, Ph.D., C.I.H.
Vice President/Technical Director
Clayton Environmental Consultants, Inc.
Southfield, Michigan

GLOSSARY

Aerosols Liquid droplets or solid particles dispersed in air that are of fine enough particle size (0.01 to 100 μm) to remain so dispersed for a period of time.

Alveoli Tiny air sacs of the lungs, formed by a dilation at the end of a bronchiole; through the thin walls of the alveoli, the blood takes in oxygen and gives up its carbon dioxide through respiration.

Anthrax A highly virulent bacterial infection picked up from infected animals and animal products.

Aplastic anemia A condition in which the bone marrow fails to produce an adequate number of red blood corpuscles.

Asbestos A hydrated magnesium silicate in fibrous form.

Asbestosis A disease of the lungs caused by the inhalation of fine airborne asbestos fibers.

Asphyxia Suffocation from lack of oxygen. *Chemical asphyxia* is produced by a substance, such as carbon monoxide, that combines with hemoglobin to reduce the blood's capacity to transport oxygen. *Simple asphyxia* is the result of exposure to a substance, such as carbon dioxide, that displaces oxygen.

Bronchi The two main branches of the trachea that go into the right and left lung.

Bronchiole The smallest of the many tubes that carry air into and out of the lungs.

Byssinosis Disease occurring to those who experience prolonged exposure to heavy air concentrations of cotton dust.

Carcinoma Malignant tumors derived from epithelial tissues, that is, the outer skin, the membranes lining the body cavities, and certain glands.

Cesium 137 An isotope of the element cesium having an atomic mass number of 137. One of the important fission products.

Chloracne A disease caused by chlorinated polyphenyls and naphthalenes acting on sebaceous glands and the liver.

Chronic bronchitis An inflammation of the bronchial tubes occurring over a long period of time and/or frequently.

Cilia Tiny hairlike "whips" in the bronchi and other respiratory passages that aid in the removal of dust trapped on these moist surfaces.

Conjunctivitis Inflammation of the delicate mucous membrane (conjunctiva) that lines the eyelids and covers the front of the eyeball.

Contact dermatitis Dermatitis caused by a primary irritant.

Cristobalite A crystalline form of free silica. Quartz in refractory bricks and amorphous silica in diatomaceous earth are altered to cristobalite when exposed to high temperatures (calcined).

Curie A measure of the activity, or the rate at which a radioactive material throws off particles. The radioactivity of one gram of radium is a curie. One curie corresponds to 37 billion disintegrations per second or 37 becquerels. The becquerel (abbreviated Bq) is the SI unit that supersedes the curie (abbreviated Ci).

Cutie-pie A portable instrument equipped with a direct-reading meter used to determine the level of radiation in an area.

Cyclone As used in industrial-hygiene monitoring, a particle-size selector whose operation is based on imparting sufficient tangential velocities to relatively larger (heavier) particles sufficient to cause impaction on the walls of a conical chamber, while permitting smaller (respirable) particles to remain entrained in the air system.

Diatomaceous earth A soft, gritty, amorphous silica composed of small aquatic plants. Used in filtration and decolorization of liquids. Calcined and flux-calcined diatomaceous earth contains appreciable amounts of cristobalite.

Dose The total amount of a substance taken into the body by all routes of exposure during some recognized time period.

Dyspnea Shortness of breath, difficult or labored breathing.

Edema A swelling of body tissues as a result of being waterlogged with fluid.

Electromagnetic radiation The propagation of varying electric and magnetic fields through space at the speed of light, exhibiting the characteristics of wave motion.

Emphysema A lung disease, in which the walls of the alveoli have been stretched too thin and broken down; frequently accompanied by impairment of the heart action.

Epidemiology The science of correlating incidence and distribution of disease with causative factors or agents.

Etiology The study or knowledge of the causes of disease.

Fibrosis A growth of fibrous tissue in an organ in excess of that naturally present. A condition marked by increase of interstitial fibrous tissue.

Forced vital capacity (FVC) The maximum volume of air that can be expelled with maximum effort after a full inspiration.

Free crystalline silica Silicon dioxide with the SiO_2 molecule oriented in a fixed tetrahedral (crystalline) pattern, the most prevalent forms being quartz, cristobalite, and tridymite.

Hemoglobin The red coloring matter of the blood which carries the oxygen.

Ionizing radiation Refers to *(a)* electrically charged or neutral particles, or *(b)* electromagnetic radiation which will interact with gases, liquids, or solids to produce ions.

Industrial hygiene The science of recognizing, evaluating, and controlling environmental stresses arising in or from the workplace.

Inertial impaction The forceful impingement on, or striking of, a particle on a surface with resulting adherence.

LD_{50} Abbreviation of lethal dose 50, the dose which is required to produce death in 50 percent of the exposed species. Death is usually reckoned as occurring within the first 30 days.

Leukemia A blood disease distinguished by overproduction of white blood cells. It may result from overexposure to radiation or it may generate spontaneously.

Lymph A clear, colorless fluid which circulates through the vessels of the lymphatic system.

Maser Microwave amplification by stimulated emission of radiation.

Metastasis Spread of malignancy from the site of primary cancer to secondary sites due to transfer through the lymphatic or blood system.

mg/m^3 Milligrams of substance per cubic meter of air; a common unit of exposure concentration.

Milliroentgen One one-thousandth of a roentgen. A roentgen is a unit of radioactive dose.

mppcf Millions of particles per cubic foot of air; a common unit of exposure concentration for mineral dusts.

Narcosis Stupor or unconsciousness produced by chemical substances.

Necrosis Destruction of body tissue.

Papilloma A small growth or tumor of the skin or mucous membrane.

Phagocyte A cell in the body that characteristically engulfs foreign material and consumes debris and foreign bodies, bacteria, and other cells.

Pharynx A part of the alimentary canal located between the mouth and the esophagus.

Pneumoconiosis A disease of the lungs caused by irritation of dusts and other particles.

Pulmonary function tests Measurement of ventilatory capacity of the lung by a series of tests, such as forced expiratory volume (FEV) and forced vital capacity (FVC).

Quartz The most prevalent, naturally occurring form of free silica, the basic raw material for the industrial sand industry.

Rad Standard unit of radioactive dose. It supersedes the roentgen.

Radioactivity Emission of energy in the form of alpha, beta, or gamma radiation from the nucleus of an atom.

Radiologist A specialist in the diagnostic and therapeutic use of x-rays and other forms of ionizing radiant energy.

Respirable Capable of penetrating into the lower respiratory tract, generally regarded as requiring a particle size of 10 μm or less.

Respirable mass That portion of total suspended particulate matter capable of penetrating into the lower respiratory tract.

Roentgen A unit of radioactive dose or exposure is called a roentgen (abbreviated R). A roentgen is that amount of x- or gamma radiation that will produce one electrostatic unit of charge, of either sign, in one cubic centimeter of dry air at standard temperature and pressure. It is equivalent to 2.58×10^{-4} C/kg in air.

Roentgenogram The shadow picture formed on a sensitized film or plate by x-rays passing through a body.

Sensitizer A chemical that, at first exposure, may or may not cause irritation. After extended or repeated exposure, some individuals develop an allergic type of skin irritation called sensitization dermatitis.

Siderosis Lung disease resulting from inhalation of iron oxide.

Silicosis A lung disease resulting from fibrosis of the lungs due to inhalation of silica dust.

Spirometry The measurement of air movement in or out of the lungs with the use of a spirometer.

Synergism Combined action of substances whose total effect is greater than the sum of their separate effects.

Talc A hydrous magnesium silicate used in ceramics, cosmetics, paint, pharmaceuticals, and soap.

TLV Threshold-limit value. An exposure level under which most people can work consistently for 8 h/day, day after day, with no harmful effects. A table of these values and accompanying precautions is published annually by the American Conference of Governmental Industrial Hygienists (ACGIH); TLV is a registered trademark of ACGIH.

Time-weighted average concentration A calculated average obtained by dividing the sum of products of concentration times time for all activities by the total time of exposure.

Trachea The cartilaginous and membranous tube (windpipe) by which air passes to and from the lung.

Tridymite A vitreous, colorless form of free silica formed when quartz is heated to 870°C (1598°F). A form of crystalline silica rarely found in naturally occurring deposits.

X-ray diffraction Since all crystals act as three-dimensional gratings for x-rays, the pattern of diffracted rays is characteristic for each crystalline material. This method is of particular value in determining the presence or absence of crystalline silica in industrial dusts.

INTRODUCTION

The practical considerations of occupational health programs in general, and control of workplace contaminants in particular, took on a new cast with the passage by Congress in 1970 of the Occupational Safety and Health Act, which requires all employers to comply with strict health and safety standards. Responsibility for enforcing the act was assigned by Congress to the Occupational Safety and Health Administration (OSHA), an agency established under the Department of Labor. In addition, a National Institute for Occupational Safety and Health (NIOSH) was created within the Department of Health, Education, and Welfare (now Health and Human Services) to conduct research, establish educational programs, and make recommendations to OSHA for health and safety standards.

Enactment of occupational health standards by OSHA has forced many plant engineers to become involved in the evaluation and control of industrial hygiene problems not only in existing facilities, but in the design of new facilities as well. For example, industrial hygiene considerations play a dominant role when considering ventilation systems (local and general), makeup air, enclosure and/or isolation of processes using toxic chemicals, handling and storage of toxic materials, and cleanup of exhaust emissions. Adequate disposal of toxic or hazardous wastes is also important to prevent environmental contamination and exposure of workers handling them. Plant engineers, therefore, need to be aware of toxic chemical and radiation hazards and how to effectively deal with them.

Industrial hygiene is defined by the American Industrial Hygiene Association (AIHA) as "that science and art devoted to the recognition, evaluation, and control of those environmental factors or stresses, arising in or from the workplace, which may cause sickness, impaired health, and well-being, or significant discomfort and inefficiency." An industrial hygienist is someone trained in "engineering, chemistry, physics, or medicine or related biological sciences (augmented by) special studies and training in all of the above cognate sciences" to practice industrial hygiene as defined above. A fully staffed industrial-health team also includes, at minimum, an *industrial physician,* an *industrial toxicologist,* a *lab-analysis technician,* and an *environmental* or *process engineer.*

TOXIC SUBSTANCES

Identifying Hazards: The Workplace Survey

Identifying potential health hazards is often relatively straightforward. For example, it is safe to assume that petrochemical workers will probably be exposed to one or more of a variety of organic vapors, that sandblasters risk overexposure to respirable dusts (crystalline silica, etc.), or that radiation is a potential hazard affecting workers in nuclear power plants.

In many other situations, the presence of a hazard may only be suspected, as when employee complaints of disease symptoms fall into an identifiable pattern or when strange odors are detected in the workplace. These cases call for a more rigorous type of exploratory survey, to both identify the hazard and measure its extent.

Whether the identity of the contaminant is known or not, a workplace survey should be conducted that includes an inventory of the processes taking place in the work area or entire plant, the chemicals used in these processes, the raw materials, by-products, etc. Generally, a comprehensive investigation requires the services of a highly experienced professional, such as a certified industrial hygienist. When the nature of the contaminant is indeterminate, the industrial hygienist should work with an industrial physician and a process engineer.

Many times in an investigation, visibility is a positive indicator of a hazard. The absence of visibility of dusts and fumes, however, does not mean that there are not dangerous levels of contaminants present, since many contaminants constitute a hazard at levels much too low to be visible.

In the same way, the investigator uses the sense of smell to pinpoint many contaminants. For example, it is possible to distinguish between the haylike odor of phosgene and the fishlike odor of trimethylamine. Here again, however, certain precautions must be observed. For example, although hydrogen sulfide has a very distinct rotten-egg-like odor, it can also dull the sense of smell after prolonged exposure, and thus effectively mask heavy concentrations of many other substances.

Tables 5-1 and 5-2 provide examples of typical occupational exposure to particulate and gaseous toxic agents.

TABLE 5-1 Examples of Occupational Exposure to Particulate Toxic Agents

Contaminant	Physical state	Occupation
Asbestos	Dust, fiber	Fireproofers, insulation strippers, asbestos-cement workers, auto-garage mechanics, construction workers, shipbuilding and repair workers, gasket manufacturing workers, rubber compounders, vinyl-tile workers
Silica	Dust	Abrasive blast cleaners, pottery makers, glass makers, cement workers, coal miners, construction workers, enamellers, foundry workers, smelters
Lead	Dust, fume	Babitters, glass makers, foundry workers, pottery makers, printers, paint manufacturers and sprayers, can and dye makers
Chromic acid	Mist	Electroplaters, picklers, colored glass, ink, and refractory makers
Arsenic	Dust, fume, or gas (arsine)	Copper smelters, brass manufacturing workers, ceramic and glass makers, insecticide manufacturing workers, electroplaters
Beryllium	Dust, fume	Beryllium metal and alloy manufacturing workers, alloy machinists, glass and neon tube makers, rocket fuel manufacturing workers
Coal-tar pitch volatiles	Dust, fume	Metallurgical operations workers, metal casters, petroleum refinery workers, coking operations workers

TABLE 5-2 Examples of Occupational Exposure to Gaseous Toxic Agents

Contaminant	Physical state	Occupation
Carbon monoxide	Gas	Cokers, smelters, metal casters, forklift drivers, garage operators, heat treaters, coal conversion workers, pottery makers
Nitrogen oxides	Gas	Welders, electroplaters, forklift drivers, fertilizer manufacturing workers, explosive manufacturers, dye workers
Fluorocarbons	Gas or vapor	Food processers, storage workers
Sulfur dioxide	Gas	Brewers, copper smelters, ore roasters, petroleum refiners, glass makers, powerplant operators, paper makers
Benzene	Vapor	Petroleum refiners, coke oven workers, organic chemical manufacturing workers, gasoline station (coke ovens) attendants, ink makers, insecticide makers, lithographers, paint makers, rubber makers
Toluene	Vapor	Core makers (foundry operation), polyurethane foam makers and users, spray painters, ship welders

Work and Process Inventory

The first step is to obtain descriptions of job functions within the company. This information may be provided by the personnel department and/or manufacturing staff, and should include the following:

- Nature of the job
- Description of the process
- Work duration
- Nature of potential exposure (if known)

This information allows the investigator to group various workers according to similarity of exposure to chemical and physical stresses. This grouping is essential later in selecting representative workers who may need to be sampled for exposure levels.

Chemical Inventory

It is important that the investigator conducting the industrial hygiene survey be knowledgeable about the chemicals that are used in the plant. In preparing a chemical inventory (preferably with the help of the purchasing department), all processes using chemicals are taken into consideration, such as waste treatment, boiler operations, air conditioning, and any other pertinent sources including those involving raw materials, intermediate products, finished products, and cleaning compounds.

A major problem in recognizing chemical hazards is the many chemical formulations of different suppliers and manufacturers. Industry uses a large variety of materials sold under various trade names, and obtaining sufficient or accurate information on the composition of each one is often difficult. Obtaining *material-safety data sheets* (MSDSs) from the supplier can be helpful in this regard. Updating MSDSs is recommended using various relevant publications and fact sheets, including the hygiene guides prepared by the American Industrial Hygiene Association and the chemical-safety data sheets supplied by various organizations, including the Chemical Manufacturers Association.

Consideration must also be given to toxic materials that may be present as impurities in relatively safe materials—diatomaceous earth, for example, which is supposedly 100 percent amorphous silica and thus generally considered a safe material. Diatomaceous earth has extensive applications, including use as a filtering aid and as a filler in cosmetics, detergents, and other household products. However, certain varieties of diatomaceous earth used mostly as filtration aids have been found to contain substantial quantities of crystalline silica, a widely recognized health hazard.

Moreover, some materials that apparently exist at subtoxic levels in the workplace can substantially appreciate in toxicity by undergoing chemical reaction within the body. For example, a metabolic reaction occurs in the case of benzidine dyes. The carcinogenic nature of benzidine has been recognized for many years. Benzidine-based dyes have been considered relatively safe on the assumption that they contain very little free or unreacted benzidine. Studies of analyses of urine samples collected from workers exposed to these dyes, however, indicate that in work areas where these dyes are in extremely small or even undetectable concentrations, metabolization of benzidine dye results in significantly elevated benzidine levels within the body.

Measurement of Worker Exposure

A variety of techniques are used in measuring exposure of workers to toxic chemicals. Techniques used for measuring particulates differ somewhat from those used to measure gases and vapors.

Sampling to determine worker exposure to particulates is most often done by filtration, which uses a battery-operated pump that draws air through a filter, usually 37 mm in diameter and made of paper, fiberglass, or synthetic materials such as mixed cellulose-ester, polyvinyl chloride, or polycarbonate. An impinger is recommended in sampling for some particulates. The solution within the impinger can also collect gases and vapors and, in fact, impingers are used more for this purpose than for sampling for particulates.

Route of Entry

Contaminants enter the body principally in three ways:

- Inhalation (through the respiratory tract)
- Skin absorption (through the skin)
- Ingestion (through the digestive tract)

Inhalation is by far the most common access for airborne contaminants to the body because of the continuous need to oxygenate the tissue cells and because of intimate contact with the body's circulatory system.

The effect of exposure to toxic agents is usually classified as acute or chronic.

Acute Exposure

Acute exposure is characterized by exposure to high concentrations of the toxic material over a short period. The exposure occurs quickly and can result in immediate damage to the body. For example, inhaling high concentrations of carbon monoxide gas or carbon tetrachloride vapors will produce acute poisoning.

Chronic Exposure

Chronic exposure occurs when there is continuous absorption of small amounts of contaminants over a long period. Each dose, taken independently, would have little toxic effect, but the quantity accumulated over a number of years can result in serious damage. Chronic poisoning can also be produced by exposure to small amounts of harmful material that produces irreversible damage to tissues and organs so that the injury rather than the poison accumulates. An example of such a chronic effect is silicosis, a disease produced by inhaling crystalline silica dust over a period of years.

Nature of Contaminants

Airborne contaminants can be present as liquids or solids, as gaseous material in the form of a true gas or vapor, or in combination of both gaseous and particulate matter. Most often, airborne contaminants are classified according to physical state and physiological effect on the human body. Knowledge of these classifications is necessary for proper evaluation of the work environment. One must also consider the route of entry and action of the contaminant.

Physiological Classification of Toxic Effects

Irritants

Irritants cause inflammation of the moist mucous surfaces of the body. Irritants are corrosive, but inflammation of tissues may result from concentrations well below those needed to produce corrosion. Examples of irritant materials include aldehydes, alkaline and acid mists, and ammonia. Materials that affect both the upper respiratory tract and lung tissues are chlorine and ozone. Irritants that affect primarily the terminal respiratory passages are nitrogen dioxide and phosgene.

Asphyxiants

Asphyxiants deprive the tissues of oxygen. They are generally divided into two classes—simple and chemical.

Simple Asphyxiants. These are physiologically inert gases that deprive the tissue of oxygen by diluting the available atmospheric oxygen. Examples include nitrogen, carbon dioxide, hydrogen, helium, and aliphatic hydrocarbons such as methane.

Chemical Asphyxiants. These prevent either oxygen transport in blood or normal oxygenation of the tissues. Chemical asphyxiants are active far below the level required for damage from simple asphyxiants. Examples include carbon monoxide, hydrogen cyanide, and nitrobenzene.

Primary Anesthetics

Anesthetics depress the central nervous system, particularly the brain. Examples include ethylene and ethyl ether.

Systemic Poisons

Systemic poisons cause injury to particular organs or body systems. The halogenated hydrocarbons (such as carbon tetrachloride) can damage the liver and kidneys, whereas benzene, aniline, and phenol may cause damage to the blood-forming system. Examples of materials classified as neurotoxic agents include carbon disulfide, methyl alcohol, tetraethyl lead, and organic phosphorus insecticides. Examples of metallic systemic poisons include cadmium, lead, manganese, and mercury.

Chemical Carcinogens and Teratogens

Chemical carcinogens can cause tumors in mammalian species. Carcinogens may induce a tumor type not usually observed, or induce an increased incidence of a tumor type normally seen, or induce such tumors at an earlier time than would otherwise be expected. In some instances, the worker's initial stages of exposure to the carcinogen and the tumor appearance are separated by a latent period of 20 to 30 years. Examples of chemical carcinogens include benzo(a)pyrene, beta-naphthylamine, vinyl chloride, and chromates.

Chemical teratogens are chemicals that produce malformation of developing cells, tissues, or fetal organs. These effects may result in growth retardation or in degenerative toxic effects. Examples of chemical teratogens include acetamide and methyl mercury.

Particulate Size

Airborne particulate matter varies in size from less than 0.01 to more than 25 μm (one micrometer equals approximately 1/25,000 inch). These particles are invisible to the naked eye.

Nonrespirable particulates consist of particles that are either too large to escape the respiratory tract's defenses before they can reach the lungs or that are too small to be retained in the lungs even if they get that far. Nonrespirable particulates present other types of health problems. Toxic fumes and nonrespirable dusts that are inhaled or ingested (or, less commonly, that come into contact with the skin) may be eventually absorbed into the bloodstream and cause systemic poisoning. Other nonrespirable dusts, inert "nuisance" substances like limestone and gypsum, can severely irritate the upper respiratory tract, endangering health as well as causing great discomfort.

Particulates that are small enough to find their way into the lungs and remain there, including dusts fine enough to be classified as *respirable dusts,* can produce serious chronic conditions such as the group of diseases known as *pneumoconiosis*—lung diseases like coal miners' "black lung disease" and silicosis.

In view of the above, it is clear that the measurement of dust exposure in many cases needs to be limited to that fraction of inhaled particles small enough to be deposited in alveolar spaces. Thus, a size-selective air-sampling method is required that will separate this "respirable fraction" from the coarser material that is deposited in the upper respiratory tract. The size selector most commonly used in respirable dust sampling has the following characteristics:

Aerodynamic diameter, μm (unit density sphere)	% passing selector
2.0	90
2.5	75
3.5	50
5.0	25
10.	0

Time-Weighted Average Exposures

Threshold-limit values usually refer to time-weighted concentrations for a 7- or 8-h workday and 40-h workweek, although recent efforts of NIOSH and OSHA have been aimed at defining permissible exposure limits for up to 10-h workdays.

Short-term exposures to concentrations above the threshold limit are permitted provided they are compensated for by equivalent excursions below the limit during the workday. For example, if the permissible exposure limit is 10, a worker could be permitted to work in concentrations as high as 15 for 4 h, provided that the remainder of this 8-h shift did not result in exposure to concentrations above 5, thus yielding an average 8-h weighted concentration of 10.

Stated mathematically, the time-weighted average concentration, based on an 8-h shift, is defined as

$$\text{TWA} = \frac{1}{8}\sum_{i=1}^{N} \overline{C}_i t_i$$

where $\overline{C}$ is the average concentration of the contaminant at location i, N is the total number of work locations, Σ refers to the summation of the products of concentration $\overline{C}$ in percentage and t is the time in house spent at location i. Expressed otherwise,

$$\text{TWA} = \frac{(\text{exposure time})(\text{con. } A) + (\text{exposure time})(\text{con. } B) + \cdots}{\text{total work time per shift}}$$

Obviously, this approach requires reliance upon personal-breathing-zone sampling or detailed job analyses for all classifications studied, and a comprehensive program of sampling sufficient to establish average airborne concentrations at all work sites.

Whom to Sample

Sampling programs are based upon statistical considerations, as it may be impractical and is often unnecessary to sample every worker's exposure. Thus, various statistical approaches have been devised to determine the minimum number of workers to be sampled to achieve adequate representation. A simple approach recommended by NIOSH is summarized as follows:

Number of employees exposed	Minimum number of employees whose individual exposures shall be determined
1–20	50% of the total number of exposed employees
21–100	10 plus 25% of the excess over 20 exposed employees
Over 100	30 plus 5% of the excess over 100 exposed employees

Occupational Safety and Health Standards

Measured results are compared to OSHA health standards or other widely accepted standards such as the American Conference of Governmental Industrial Hygienists (ACGIH) threshold-limit values.

Federal Standards (29 CFR 1910 Subpart Z)

The first compilation of health and safety standards promulgated by OSHA was based on the existing federal and national consensus standards.

Table 5-3 lists the standards in Sections 1910.1001 through 1910.1046 of Title 29, Chapter XVII, Section 1910.1000 (available through OSHA as Publication 2206). These standards are detailed sets of regulations for individual substances. In addition to setting exposure limits, they require exposure monitoring and medical surveillance of exposed employees. OSHA plans to promulgate many more complete standards.

Private Organizations

The first influential American organization in the field of occupational safety was the American Conference of Governmental Industrial Hygienists (ACGIH). The ACGIH publishes each year its revised and updated list of occupational exposure limits, commonly referred to as TLVs (threshold-limit values). This listing is more extensive than

TABLE 5-3 Individual Chemical Substance Standards

29 CFR 1910	Chemical name	Exposure limits		
		TWA	Ceiling	Action level
1000	Air contaminants	——		
1001	Asbestos	2 fibers/cm³	10 fibers/cm³	
1003	4-Nitrobiphenyl	*		
1004	α-Naphthylamine	*		
1005	Moca (deleted)			
1006	Methyl chloromethyl ether	*		
1007	3, 3′-Dichlorobenzidine	*		
1008	bis (Chloromethyl) ether	*		
1009	β-Naphthylamine	*		
1010	Benzidine	*		
1011	4-Aminodiphenyl	*		
1012	Ethylenimine	*		
1013	β-Propiolactone	*		
1014	2-Acetylaminofluorene	*		
1015	4-Dimethylaminoazobenzene	*		
1016	*N*-Nitrosodimethylamine	*		
1017	Vinyl chloride	1 ppm	5 ppm/15min	0.5 ppm
1018	Inorganic arsenic	10 $\mu g/m^3$		5 $\mu g/m^3$
1025	Lead	50 $\mu g/m^3$		30 $\mu g/m^3$
1028	Benzene	1 ppm	5 ppm/15 min	0.5 ppm
1029	Coke-oven emissions	150 $\mu g/m^3$		
1043	Cotton dust	200 $\mu g/m^3$ (yarn mfr.) 750 $\mu g/m^3$ (weaving) 500 $\mu g/m^3$ (nontextile)		
1044	1,2-Dibromo-3-chloropropane	1 ppb		
1045	Acrylonitrile	2 ppm	10 ppm/15 min	1 ppm
1046	Cotton dust (gins)	——		

*These carcinogen standards require that these chemicals be handled only in completely enclosed systems with exposures reduced to the lowest feasible level.

the OSHA listing of air contaminants in CFR 29 1910.1000. It should be noted that the term TLV is a registered trademark of the ACGIH and should not be used to refer to OSHA permissible exposure limits. Another organization involved in the promulgation of such standards, although its concerns extend into several other areas, is the American National Standards Institute (ANSI). Two younger organizations are the American Board of Industrial Hygienists (ABIH) and the American Industrial Hygiene Association (AIHA), both established in the 1950s; among the services they provide are the certification of industrial hygienists and of laboratories who analyze workplace samples.

Control Strategy—Reducing or Eliminating Hazards

A strategy for reducing or eliminating workplace hazards can include one of two approaches (or both): engineering or administrative controls.

Engineering Controls. Some examples are: substituting relatively safer chemicals for more toxic ones; altering a process in such a way as to reduce worker exposure to contaminants; changing or upgrading ventilation systems; isolating or enclosing contaminated areas.

Administrative Controls. An example is adjusting work schedules so that workers receive only a fraction of their present exposure to contaminants.

As an example of an engineering control, if asbestos is used as insulation, it may be possible to replace it with fiberglass or calcium silicate, materials which do not present the same degree of hazard. Here again, however, it should never be assumed that such a substitution completely eliminates the hazard. The hazardous potential for fiberglass is not completely known, for example, while certain varieties of calcium silicate, generally regarded as a relatively safe substance, have been found to contain measurable amounts of asbestos fibers and/or free crystalline silica, another highly hazardous material.

A control strategy must take into account work *practices,* which can be a major factor in exposure. Many poor work practices result from unawareness of the potential hazards of a toxic material. For example, in one paint factory, workers were observed dusting off their clothes with a compressed-air line, starting from the shoes and proceeding upward. The raw material in the plant's primary process included, among other substances, lead chromate. Sampling measurements in the plant indicated that worker exposures to lead and hexavalent chromium (both highly toxic materials) were not excessive under normal working conditions. It was determined that *this dusting procedure, however, exposed workers to higher concentrations of lead and chromium during a 2- to 4-min period than they were exposed to during the remainder of the day* when handling these materials frequently. Vacuuming of dusty clothing (rather than blowing the dust off) could easily prevent this exposure. In such cases, employee-awareness programs can be highly useful in preventing work practices that may result in unnecessary exposure.

The evaluation of existing and anticipated engineering controls is a major goal of an effective industrial-hygiene program. OSHA considers engineering controls to be the primary and most effective means of reducing or eliminating exposure to toxic substances in the workplace. The extent of engineering controls needed can be better established after sampling measurements have been taken, other aspects such as work practices, housekeeping, etc., have been evaluated, and corrective actions have been instituted. Examples of poor engineering controls are all too plentiful.

Exposure Surveillance Programs

A comprehensive ongoing occupational health or industrial hygiene program includes a monitoring program that ensures surveillance of worker health in addition to determining compliance with the applicable regulatory standards. The *surveillance program* consists not only of ongoing sampling of worker exposure to airborne concentrations of chemicals but also monitoring for exposure to such physical agents as noise, radiation, and heat stress. The frequency of sampling will depend upon the levels of various contaminants, the toxicity of the material, and the specific requirements under the applicable codes, and may vary from weekly to biannually.

Biological monitoring of worker exposure to chemical hazards through analysis of samples of blood, urine, expired air, etc., is required in those cases where measurement of airborne concentration alone is not a reliable indicator of hazard potential. In many instances, the metabolite (chemical produced within the body) of a toxic substance rather than the substance itself may be measured in the biological fluids.

Medical surveillance of exposed workers is also a highly useful and necessary tool for protection against toxic exposures. Pre-employment and periodic medical examinations can reveal the presence of toxic effects at an early stage, when a cure is often possible. Medical examinations should include specific organ functions to detect changes relative to the specific contaminants to which the worker is exposed.

RADIATION HAZARDS

Radiation is energy which is emitted, transmitted, or absorbed in wave or energetic particle form. The electromagnetic (EM) waves consist of electric and magnetic forces. When these forces are disturbed, EM radiation results. Figure 5-1 is an arrangement of known EM radiations according to their frequency and/or wavelength. It includes microwave, infrared, visible, and ultraviolet radiation. The range of biological effects of exposure within the electromagnetic spectrum is, therefore, extremely broad and diverse.

Radiation is generally divided into two categories: *ionizing* and *nonionizing*. Ionizing radiation includes x-rays, alpha particles, and beta, gamma, and neutron rays. Nonionizing radiation includes ultraviolet, visible, infrared, microwave, and radio-frequency. Laser radiations fall into most of these bands. Examples of occupational exposure to ionizing and nonionizing radiation are shown in Table 5-4.

Ionizing Radiation

Of the type of ionizing radioactivity, alpha particles are the least penetrating—paper and skin will ordinarily stop alpha particles.

Beta radiation has considerably more penetrating power than alpha particles. X-rays and gamma rays both have very good penetrating power and require the use of heavy

TABLE 5-4 Example of Occupational Exposure to Ionizing and Nonionizing Radiation

Type of radiation	Occupation
Ionizing radiation (radioactive isotopes, x-rays)	Nuclear powerplant workers, food preservers, electron microscopists, biologists, food sterilizers, high-voltage repairmen, ceramic workers, drug makers
Ultraviolet radiation	Meat curers, movie projectionists, pipeline workers, paint curers, nurses, bacteriologists, dentists, food preservers, lithographers, laboratory workers, welders, textile inspectors, plastic curers, printers
Infrared radiation	Bakers, electricians, furnace workers, glass blowers, heat treaters, solderers, welders, firemen, steel workers
Microwave, radio-frequency radiation	Automotive workers, paper product workers, plastic heat-sealing workers, rubber-product workers, textile workers, tobacco workers, electronics workers, advertising sign workers
Laser	Medical technicians, surgeons, aerospace workers, semiconductor workers

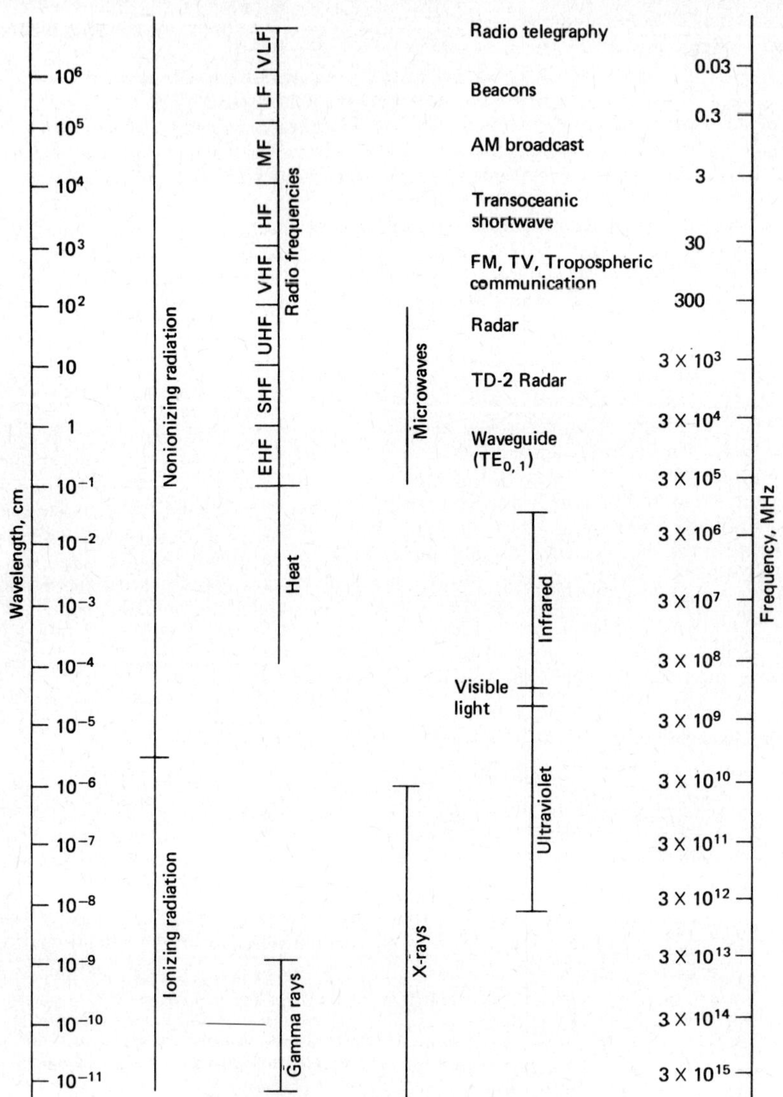

Figure 5-1 Electromagnetic spectrum.

shielding material (lead). Neutrons are very penetrating and require shielding with materials of higher hydrogen atom content rather than the use of mass alone.

Alpha emitters are an internal hazard because they do not have the ionizing ability to travel very long distances. One must take precaution against breathing or ingesting the alpha emitters. Beta emitters are also generally considered an internal hazard.

Regulations Regarding Ionizing Radiation

Jobs involving exposure to ionizing radiation fall under the provisions of Nuclear Regulatory Commission (NRC) standards (10 CFR Part 20). OSHA standards (Section

1910.96) on ionization radiation apply in cases where employees may not be protected under the NRC standards.

Table 5-5 summarizes the most frequent types of radiation encountered in health physics surveys and the type of detection most commonly used.

A photographic film badge can be used for determining beta and gamma doses. Such film badges give a sufficiently quantitative indication of the integrated weekly doses of individuals. These should be recorded along with other records of radiation levels.

TABLE 5-5 Types of Detectors Used for Various Types of Radiation

Type of detector	Type of radiation
Proportional or scintillation counter	Alpha
Geiger-Mueller tube or proportional counter	Beta
Ionization chamber	X and gamma
Proportional counter	Neutron

Nonionizing Radiations

Ultraviolet Radiation

For purposes of assessing the biological effects of ultraviolet radiation, the wavelengths of interest can be restricted to 0.1 to 0.4 μm. The major source of ultraviolet radiation is the sun; common constructed sources are mercury discharge lamps, welding and plasma torches, xenon discharge lamps, and lasers. The symptoms of overexposure to ultraviolet radiation are those characteristic of a severe sunburn.

Currently, there are no OSHA standards for exposure to ultraviolet radiation. OSHA Section 1910.97 concerning nonionizing radiation includes exposure levels and warning signs. However, this is an advisory and not a mandatory standard.

NIOSH has a recommended standard for occupational exposure to ultraviolet radiation. The American Conference of Governmental Industrial Hygienists (ACGIH) has developed TLVs (threshold-limit values) for ultraviolet radiation in the spectral region between 200 and 400 nm (see Table 5-6).

Infrared Radiation

Infrared radiation (ir) can be considered associated with the heat given off by all bodies that radiate heat and covers wavelengths from 0.75 μm to about 1 mm. Infrared rays are

TABLE 5-6 Permissible Ultraviolet Exposures

Duration of exposure per day	Effective irradiance, E_{eff}, $\mu W/cm^2$
8 h	0.1
4 h	0.2
2 h	0.4
1 h	0.8
30 min	1.7
15 min	3.3
10 min	5.
5 min	10.
1 min	50.
30 s	100.
10 s	300.
1 s	3,000.
0.5 s	6,000.
0.1 s	30,000.

mostly absorbed by the skin and burn the skin similar to the way the sun does. Like ultraviolet radiation, infrared radiation is invisible and can seriously damage the eyes. Although infrared radiation may not be felt, over the years it can cause permanent eye damage. Water can be used as a barrier, as it can absorb most ir waves. Regular clothing protects the skin against ir, but goggles should be worn when there is a potential for overexposure.

Except for thermal burns (below wavelengths of 1.5 μm), infrared radiation is insignificant as a health hazard. However, when highly intense and compacted sources of radiant energy are being used, as with lasers, injury can occur in fractions of a second, before pain is evident.

Microwave Radiation

There has been a great deal of exposure to microwaves among people in the armed forces because microwave radiation is used for radar. With the advent of microwave ovens, however, this type of radiation is now encountered in the home. Microwaves have wavelengths of 3 m to 3 mm and their effect is related to power intensity and time of exposure as well as to the wavelength. Their ability to heat the body allows them to be used for medical treatment. This ability, however, can also be a hazard to the overexposed, unprotected worker. Microwaves penetrate deeply into the body and cause body temperature to rise; if the body temperature rises high enough, the person can go into a coma and die.

Before personnel are assigned to work in or about radar equipment, they should be given a complete physical including eye and blood examinations. Workers should be instructed never to look directly at a radar beam and should be given physicals periodically and whenever exposure is indicated. Most microwave measuring devices are based on (1) bolometry, (2) colorimetry, (3) voltage and resistance changes in detectors, and (4) radiation pressure on a reflecting surface. The bolometry method is one of the most widely used in commercially available power meters.

One troublesome fact in the measurement of microwave radiation is that the near field (reactive field) of many sources may produce unpredictable radioactive patterns. Energy density rather than power density may be a more appropriate means of expressing hazard potential in the near field.

Radio-Frequency Radiation

Radio-frequency (rf) electromagnetic radiation obeys the general laws of electromagnetic radiation and is characterized by the following basic parameters: frequency f in hertz (1 Hz = 1 cycle per second), propagation time T, wavelength λ in meters, and velocity c, which in free space under normal conditions is equal to the velocity of light, i.e., 300,000 km/s. These quantities are interrelated according to the formula

$$\lambda = cT$$

Radio-frequency radiation covers wavelengths from 300 m to 1 mm. For measurement, either electric or magnetic field strength meters or power meters are used.

Using power meters, the rf energy causes a change of a temperature-sensitive element which is monitored by a bridge circuit and meter calibrated in $\mu W/m^2$ or mW/m^2.

All effects of rf radiation may be classified as either thermal or nonthermal, leading to heat stress or disturbance of the nervous system, respectively.

Nonionizing Radiation Instruments

Ultraviolet and Infrared Radiation Monitoring Devices

A variety of instruments are commonly used to measure ultraviolet and infrared radiation. They are classified according to the type of detector used, which is generally one of two types: thermal detectors or photoelectric detectors. Thermal detectors are those in which the absorbed radiation is degraded to heat and subsequently coverted to an electric signal by changing the electric resistance of a filament. Photoelectric detectors are based on the principle that the absorbed photons eject electrons from a photo-emissive material. Most of these instruments are precalibrated by the manufacturer, but should be routinely checked prior to field use.

Microwave Monitoring Devices

Most microwave radiation detectors consist of the following components: the test antenna, the attenuator, a bolometer or thermistor, and a power meter. These detectors are usually in the form of a single integrated unit. The test antenna is specific to certain wavelengths and, therefore, interchangeable. These units must be calibrated at frequencies throughout the band to ensure accuracy.

Laser Monitoring Devices

A wide variety of laser radiation detectors are commercially available to fulfill the diversified needs due to different wavelengths, pulse durations, and power and energy densities of various laser instruments. Laser radiation detectors are generally based upon two basic principles—photon and thermal detection.

Lasers

Laser technology has increased rapidly since its inception in 1960. Presently it involves virtually every scientific field in one way or another.

Laser is an acronym for light amplification by stimulated emission of radiation. A laser is a source of coherent energy which may appear in the near and far infrared, long and short ultraviolet, and visible regions. The properties of lasers are similar to those for various regions of the electromagnetic spectrum, except that the laser normally achieves great power densities. A laser beam does not rapidly diverge since a laser operates at specific frequencies in the electromagnetic spectrum. It travels in one direction in straight lines.

There are four types of laser beams, classified by generating medium:

- Solid-state (the ruby crystal is the most common)
- Gaseous-state (the helium-neon is the most common)
- Semiconductor or junction
- Liquid-state, using organic dyes as a medium

Some lasers operate continuously while others operate in pulses, sometimes as short as 10^{-11} s. Output levels range from milliwatts to kilowatts for continuous operation and up to gigawatts in pulse operations.

The two main areas of potential health effects from laser radiation are the eye and the skin. Associated electrical, chemical, and explosive hazards are also possible. Knowledge of and continued interest in laser safety are necessary in developing applications of lasers.

Ocular Hazards

Protective eyewear should be worn whenever hazardous conditions may result from operation of a laser product. Laser radiation in the infrared region is highly absorbed at the surface of the cornea and could induce opacities in the cornea and destruction of the protective epithelial layer. In the ultraviolet region, exposure may result in extreme discomfort, but a moderate exposure is not thought to produce permanent damage. In the region of the spectrum from 320 to 1500 nm (the visible range is normally 380 to 760 nm), eye hazards are confined primarily to the retina and choroid. The most critical area for vision is the fovea. Thus, safety guidelines for laser radiation are designed to protect against foveal area damage. There is an often undetected hazard from diffused laser radiation in the visible region due to the ability of the eye to focus such radiation to a very small spot on the retina. If the laser beam is incident upon a diffuse surface, the illuminated diffuse surface can serve as a secondary extended source. In this case, the retinal power density is unaltered regardless of viewing angle and how far away the surface is from the viewer.

Skin Hazards

The effects of laser radiation on the skin may vary from mild reddening (erythema), to blistering and charring, depending on the amount of energy absorbed, the wavelengths of the radiation, skin pigmentation, individual sensitivity, and duration of exposure.

Electrical and Explosion Hazards

Live parts of circuits and components with peak open-circuit potentials over 42.5 V are considered hazardous, unless limited to less than 0.5 mA circuit components of combustible materials. For example, transformers are potential fire hazards unless individual noncombustible enclosures are provided.

In the event of a tube or lamp failure, components such as electrolytic capacitors may explode if subjected to voltages higher than their ratings. A misdirected laser beam, in this case, could steam the coolant of a high output laser system to trigger an explosion.

Chemical Hazards

A misdirected laser beam could decompose nearby chemical compounds to give off possibly toxic components.

Recently liquid lasing mediums (mostly complex organic dyes) have been gaining popularity due to the advantage of tunable frequency. However, many of these organic dyes are toxic themselves and, therefore, pose a potential exposure problem for laser handlers.

Control Measures

OSHA currently has no occupational health standards for laser products. Threshold limits established by the American Conference of Governmental Industrial Hygienists (ACGIH) are the recommended values to use. All laser products should meet specifications of the ***National Electrical Code***® (NFPA 70-1981),* Articles 300 and 400. Adherence to control procedures required under this standard should eliminate any significant probability of detrimental health effects from laser product operations. Therefore, an extensive, medical surveillance protocol for laser operators is not required under this standard.

BIBLIOGRAPHY

American Industrial Hygiene Association Journal, 66 South Miller Road, Akron, OH 44313, 1940–

American Journal of Public Health, American Public Health Association, 1015 Eighteenth Street, N.W., Washington, DC 20036, 1911–

Archives of Environmental Health, American Medical Association, 535 North Dearborn Street, Chicago, IL 60610, 1950.

Cember, H.: *Introduction to Health Physics.* Pergamon, New York, 1969.

Clarke, A. M.: *Ocular Hazards from Lasers and Other Optical Sources,* CRC Press, Boca Raton, Florida, 1970.

Clayton, F. E., and G. D. Clayton (eds.): *Patty's Industrial Hygiene and Toxicology: General Principles,* vol. I, 3d ed., 1978. *Toxicology,* vol. IIA, 3d ed., Wiley-Interscience, New York, 1981.

Cleary, S. F.: *The Biological Effects of Microwave and Radiofrequency Radiations,* CRC Press, Boca Raton, Florida, 1970.

Cralley, L. J., and L. V. Cralley (eds.): *Patty's Industrial Hygiene and Toxicology: Theory and Rationale of Industrial Hygiene Practice,* vol. III, Wiley-Interscience, New York, 1979.

****National Electrical Code*** is a trademark of the National Fire Protection Association, Quincy, MA 02269.

Cralley, L. V., G. D. Clayton, and J. A. Jurgill (eds.): *Industrial Environmental Health: The Worker and the Community.* Academic, New York, 1972.

Documentation of the Threshold Limit Values for Substances in Workroom Air, 4th ed, American Conference of Governmental Industrial Hygienists, Cincinnati 1980.

Dreisbach, R. H.: *Handbook of Poisoning: Diagnosis and Treatment,* 8th ed., Lange Medical Publications, Palo Alto, California.

Drinker, P., and T. Hatch: *Industrial Dust: Hygienic Significance, Measurement and Control,* 2d ed. McGraw-Hill, New York, 1954.

Engineering Manual for Control of In-Plant Environment in Foundries, American Foundrymen's Society, Des Plaines, Illinois, 1956.

Environmental Health Monitoring Manual, United States Steel Corporation, 1973.

Hamilton, A., and H. L. Hardy: *Industrial Toxicology,* 3d ed. Publishing Sciences Group, Acton, Massachusetts, 1974.

Hunter, D.: *The Diseases of Occupations,* 5th ed., Little, Brown, Boston, 1974.

The Industrial Environment, Its Evaluation and Control, 3d ed., National Institute for Occupational Safety and Health, Rockville, Maryland, 1973.

Industrial Ventilation, a Manual of Recommended Practice, 13th ed., Committee on Industrial Ventilation, American Conference of Governmental Industrial Hygienists, Lansing, Michigan, 1974.

Intersociety Committee: *Methods of Air Sampling and Analysis,* American Public Health Association, Washington, D.C., 1972.

Kinsman, S. (ed.): *Radiological Health Handbook.* U.S. Bureau of Radiological Health, Washington, D.C., 1970.

NIOSH Manual of Analytical Methods, National Institute for Occupational Safety and Health, Cincinnati, vol. I to V, 1979.

NIOSH Manual of Sampling Data Sheets, National Institute for Occupational Safety and Health, Cincinnati, 1974.

Hemeon, W. (ed.): *Plant and Process Ventilation,* 2d ed., Industrial Press, New York, 1963.

Plunkett, E. R.: *Handbook of Industrial Toxicology,* 2d ed., Chemical Publishing, New York, 1976.

Sax, N. I.: *Dangerous Properties of Industrial Materials,* 5th ed., Van Nostrand Reinhold, New York, 1979.

Norwood, W. D.: *Health Protection of Radiation Workers.* Charles C. Thomas, Springfield, Illinois, 1975.

Schwartz, L., L. Tulipen, and D. J. Birmingham (eds.) *Occupational Diseases of the Skin,* 3d ed., Lea and Febiger, Philadelphia, 1957.

Olishifski, J. B., and F. E. McElroy: *Fundamentals of Industrial Hygiene,* National Safety Council, Chicago, 1971.

Shreve, R. N. and J. Brink: *Chemical Process Industries,* 4th ed. McGraw-Hill, New York, 1977.

Wendholz, M. (ed.): *The Merck Index of Chemicals and Drugs,* 9th ed. Merck & Co., Rahway, N. J., 1976.

chapter 8-6

Sanitation Control and Housekeeping

by

Don Williams
Technical Service Director
Huntington Laboratories, Inc.
Huntington, Indiana

GLOSSARY

Acid A compound that gives a pH below 7, produces hydrogen ions in water, and reacts with or neutralizes alkalis. Most soils including oils, greases, and waxes are acids.

Alkali A compound that gives a pH between 7 and 14, produces hydroxide ions in water, and reacts with or neutralizes acids. Hard-water films, carbonates, potash, and phosphates are alkaline.

Deodorize Destroying, masking, or modifying foul and unpleasant odors. May be accomplished by killing bacteria that make foul odors.

Detergent Any product that cleans. Usually a synthetic detergent but may be a soap or even an abrasive material.

Disinfectant Product used on inanimate surfaces, which destroys microorganisms but not necessarily spores. Also called germicide.

Finish A protective coating used as a top coat.

Germicide See Disinfectant.

Hard floors Concrete, terrazzo, ceramic, quarry, slate, etc.

Polymer Usually a synthetic plastic used as a floor seal or finish. Examples are acrylic, styrene, and urethane.

Resilient floors Vinyl, vinyl asbestos, asphalt, rubber, linoleum, etc.

Sanitary Relating to health or to the preservation of or restoration of health and hygiene.

Sanitation Use of sanitary measures to clean and maintain a building.

Sanitize Reduce bacterial counts to safe levels as determined by health requirements.

Sealer A product used to prevent excessive absorption of finish coats into porous surfaces. An undercoat.

INTRODUCTION

Proper housekeeping procedures play an important role in the total plant operation. Custodial employees work with a great variety of chemicals and equipment and perform unique procedures. They work in a building that has taken considerable money to construct, and improper procedures can shorten the normal life of various parts of it. In a relatively few years, the owners will have *spent more money on housekeeping than the initial cost of the building and its furnishings.* Custodians should be well-trained to perform their tasks. Table 6-1 gives average cleaning times for various duties so that an estimate of efficiency can be determined. In addition to maintaining the appearance and prolonging the life of the building, good housekeeping provides immeasurable safety, hygienic, and sanitary benefits.

Safety is a key reason for good custodial procedures. The Occupational Safety and Health Administration of the U.S. Department of Labor states in its Standards and Interpretations, 1910.22(a) Housekeeping:

> (1) All places of employment, passageways, storerooms, and service rooms shall be kept clean and orderly and in a sanitary condition.
>
> (2) The floor of every workroom shall be maintained in a clean, and so far as possible, a dry condition. Where wet processes are used, drainage shall be maintained, and false floors, platforms, mats, or other dry standing places should be provided where practicable.
>
> (3) To facilitate cleaning, every floor, working place, and passageway shall be kept free from protruding nails, splinters, holes, or loose boards.

Other benefits from the results of a well-trained custodial work force are:

1. A clean, sanitary environment contributes to the health of all the personnel.
2. Training improves the morale and mental attitude of each worker.
3. Each worker can have a feeling of pride in a job well done.
4. A well-maintained building is easier to keep in shape, thus saving time and money.
5. A well-maintained building and a crew of employees that enjoy their work contribute to goodwill and public relations.
6. The life of the various parts of the building and the building itself can be greatly extended by proper housekeeping procedures.

TRAINING TIPS

Some key tips to use in training are:

1. The longer a soil, stain, or coating is allowed to remain on the surface, the harder it will be to remove.
2. Quality products made by reputable manufacturers are usually best in the long run.

TABLE 6-1 Cleaning Operation Time Estimates

Floor operations	1000 ft², min
Sweeping	
Halls & corridors	10
General rooms	20
Mopping	
Dust mop (unobstructed)	5
Dust mop (obstructed)	10
Damp mop (unobstructed)	16
Damp mop (obstructed)	30
Wet mop and rinse (unobstructed)	30
Wet mop and rinse (obstructed)	50
Scrubbing	
Hand scrub 12″ brush	300
Deck scrub	100
Machine scrub	
Machine scrub 12″ diam	48
Machine scrub 14″ diam	40
Machine scrub 16″ diam	36
Machine scrub 18″ diam	31.5
Machine scrub 19″ diam	30
Machine scrub 21″ diam	27
Machine scrub 23″ diam	25
Machine scrub 24″ diam	24
Machine scrub 32″ diam	18
Machine scrub 36″ diam	16
Automatic scrub machine (24″)	6
Vacuum pickup	
Vacuum pickup (unobstructed)	20
Vacuum pickup (obstructed)	30
Wax	
Wax	30
Machine polish (19″ machine)	15
Rectangular machine (48″ plate)	3.5
Buff with steel wool	20
Strip and rewax (1 operator)	150
Dry strip and rewax (1 operator)	120
Spray buffing	
Spray buffing (unobstructed)	30
Spray buffing (obstructed)	45
Carpeting	
Vacuuming (unobstructed)	20
Vacuuming (obstructed)	30
Spot vacuuming	16
Shampoo (dry-foam)	60
Pile lift	30

Furniture and fixtures operations	per unit, min
Dusting	
Air conditioners	0.30
Ashtrays (desk)	0.25
Bookcases (3-tier sect.)	0.30
Chairs	0.30
Cigarette stands	0.40
Couch	0.25
Desks	0.80
Desk trays	0.15
File cabinets (4 drawer)	0.40
Lockers	0.20
Radiators	0.30
Tables (medium)	0.50
Telephones	0.15
Towel dispensers	0.12
Towel disposal cans	0.40
Typewriter & stand	0.50
Wash basin (office)	0.60
Wastebasket	0.50
Window sill	0.20
Venetian blinds std. size	3.50

Washrooms	1000 ft², min
Cleaning	
Cleaning commode	3.83
Door (spot wash both sides)	0.83
Mirrors	0.66
Sanitary napkin dispenser	0.16
Urinals	3.00
Washbasin—soap dispenser	3.00
General washroom cleaning	
General cleaning per 1000 ft²	120.00

Miscellaneous operations	1000 ft², min
Wall washing	
Painted walls (manual)	240
Painted walls (machine)	150
Marble walls (manual)	92
Ceiling washing	
Ceiling washing (manual)	300
Ceiling washing (machine)	180
Window washing	
Single pane	125
Multipane	170
Frosted single pane	190
Opaque glass	50
Plate glass	35
Office partitions (glass)	110

TABLE 6-1 Cleaning Operation Time Estimates (*Continued*)

Miscellaneous operations	per unit, min	Miscellaneous operations	per unit, min
Dusting lamps & light fixtures		Whisk or vacuum armchair	1
Wall fluorescent fixtures	0.13	Whisk or vacuum couch	2
Desk fluorescent lamp	0.30	Shampooing armless chair	4
Table lamp & shade	0.58	Shampooing armchair	7
Floor lamp & shade	0.58	Shampooing couch	20
Washing fluorescent light fixtures		Stairway cleaning	
Ceiling fixture (eggcrate) 4′ ea.	9	Sweep and dust 1 flight 15 steps	6
Ceiling fixture (eggcrate) 8′ ea.	12	Damp mop 1 flight 15 steps	5
Fabric upholstery cleaning		Scrubbing (hand)	20
Whisk or vacuum armless chair	0.50		

*Reprinted with permission from International Sanitary Supply Association. These cleaning time estimates represent average cleaning times. Layout, obstacles, maintenance level desired, environmental conditions, etc., will affect cleaning time.

3. Chemicals should be measured and applied properly when required.
4. Two thin coats of a finish are better than one thick coat.
5. Always follow directions on labels or given in training, for safety as well as effectiveness.
6. Do not depend upon perfumed products to cover up odors. Clean thoroughly and kill germs with disinfectants or sanitizers and there will not be putrefaction odors.
7. Surfaces vary considerably and aging often changes them.
8. Use products and procedures that will do the best job and that are made for that particular surface.
9. Clean and maintain equipment in proper working order.
10. Learn how to use equipment to the best advantage.
11. Use products with the best blend of properties to do the job intended.
12. Read supplier directions, literature, trade books, and magazines for generating ideas and becoming aware of new products.
13. Talk with fellow custodians about common problems and situations.
14. Attend local, regional, or state training programs where possible.
15. Practice cooperating with other departments, and they will cooperate with you.

PLANT DESIGN

The actual design of the plant can play a key role in contributing to simpler maintenance. Too often, those responsible for maintaining the building are never consulted during the early planning of the building design. If given a chance, many experienced custodians can provide some good ideas in planning or rearranging plant areas. Following are some basic suggestions:

1. Custodial service closets should be provided on every floor and be centrally located to save walking time.
2. The closet should have sufficient shelving, floor space, a floor-level sink, hard sur-

face floors and wall racks for mops and brooms, and other facilities that fit the needs.

3. Keys for doors, cabinets, machines, etc., should be planned and organized to make locking and unlocking by the custodians quick and easy as well as secure.
4. Access through one room, closet, or office to another should not be permitted.
5. Restrooms should be planned to make use and cleaning easy. (Wall-hung fixtures, placement of fixtures for sequential use, glazed walls, ceramic floors, and floor drains are good choices.)
6. Kick plates and push plates can help the appearance and cleanability of the custodial closet, restroom, and other doors.
7. Wall and furniture surfaces should be easily washable.
8. Entrance mats should be either recessed or integrated so that tracked-in soils are reduced.
9. Light-colored carpeting should not be laid where soiling will show quickly.
10. OSHA states in Standards and Interpretations 1910.22(b) aisles and passageways:

> (1) Where mechanical handling equipment is used, sufficient safe clearances shall be allowed for aisles, at loading docks, through doorways and wherever turns or passage must be made. Aisles and passageways shall be kept clear and in good repair, with no obstruction across or in aisles that could create a hazard.
>
> (2) Permanent aisles and passageways shall be appropriately marked.

EQUIPMENT

Cleaning equipment may vary somewhat according to type of surfaces, size of area, and other factors. Though it is tempting to order large versions of all equipment to apparently save time in covering large surface areas, there are many occasions when the larger equipment version will be too clumsy, heavy, and hard to adapt to the inevitable smaller areas. When it is impractical to own a variety of sizes to fit every need, consider a compromise size. Many units have attachments to broaden their usage in various areas; they may combine two or more functions and may be of the rider type. Details and demonstrations should always be obtained from equipment suppliers prior to commitment. All equipment should be cleaned after each use and kept in working order. Standardization of purchases can make training easier, requires stocking of fewer replacement parts, and makes possible larger quantity discounts. Following are some of the basic items:

1. Rotary *floor machines* for use in scrubbing, stripping, scouring, and polishing resilient and hard-surface floors as well as shampooing carpets. A 20-in model is most efficient.
2. Wet or dry *vacuums* are available in a number of different varieties. They may use a wand or squeegee, be lightweight backpack models, be extra quiet, contain filters, and be of various sizes or suction powers. Sweeper versions may be of the rider type.
3. *Automatic scrubber vacuums* to scrub soils and pick up the scrub solution.
4. Pressure sprayers for fast and effective cleaning.
5. Wall-washing equipment with many variations.
6. *Carpet extractors* for cleaning and rinsing on location.
7. *Compactors* for trash disposal.
8. Mops, buckets, wringers, carts, pads, sprayers, brushes, dispensers, and many other miscellaneous items.

CHEMICALS

Understanding some of the basic properties of the products used for sanitation should become one of the most important goals of maintenance personnel. Each product will

TABLE 6-2 Solving Floor-Finish Problems*

SOLUTIONS	Scratching	Black marking	Dirt pickup	Furniture sticking	Tackiness	Discoloring of light floor	Poor initial gloss	Slipperiness	Streaking, uneven, or dull film	Unpleasant odor	Powdering or dusting	Marring, smearing, or scuffing	Poor water resistance	Poor detergent resistance
Do not pour used coating back into container					*	*	*		*	*		*	*	*
Apply more coats or use sealer		*					*							
Apply thin, continuous films			*	*	*				*		*	*		
Remove all previous coating or factory finish and recoat	*	*	*	*	*	*	*	*	*		*	*	*	*
Check freezing damage	*	*	*				*		*		*	*	*	*
Use fine or medium scrubbing pads rather than coarse pads	*						*		*					
Allow adequate drying time between coats				*	*		*		*		*			
Wipe, do not rub while applying							*		*					
False impression—check natural color of flooring						*								
Use clean equipment and applicator						*			*				*	*
Remove excessive buildup on surface			*			*					*			
Remove dust and loose soil regularly and use carpet at entranceways	*							*			*			
Spray clean and buff or recoat	*	*					*							
Allow adequate drying time before resetting furniture—ventilate				*	*									
Avoid applying in hot and humid weather—ventilate well				*	*								*	
Avoid or remove soap or oil films on floor					*		*	*	*			*		
Floor too cold—apply closer to room temperature, 70° (21°C)	*								*		*			
Change to metal interlocked polymer														*
Buff dry stripped floor before applying coating							*		*					
Do not apply polymer finish over a wax	*							*			*			
Change to hard polymer finish		*	*									*		
Unstable plasticizer in vinyl tile—allow curing time		*	*		*									

Source Huntington Laboratories, Inc., Huntington, Indiana.

*Most waxes and polymer finishes are quality formulations, made of the finest raw materials, under most careful laboratory control. Yet, many outside factors may cause the finest product to do a poor job. This chart explains some of the problems encountered and their solutions.

perform certain tasks if used properly but have the potential to harm surfaces or personnel if improperly applied. Use only where and in the manner recommended. Observe the cautions on labels to preclude dangerous situations. Use the mildest abrasive, coupled with the mildest chemical possible, to complete the task and protect the surface.

A cleaning product may be either all-purpose for a variety of tasks or highly specialized. Mild *alkaline cleaners* will remove oily, greasy soils from any surface not harmed by water alone. Special alkaline cleaners are recommended for specific tasks such as strippping of floor coatings, window washing, carpet cleaning, wall washing, etc. Strong alkaline cleaners and degreasers are for heavy industrial soil buildups. *Acid cleaners* remove lime and hard-water deposits or rust stains from hard surfaces such as toilet bowls, metals, ceramics, and concrete. *Solvents* are generally used solely for spot removers in housekeeping.

Abrasive or mechanical action of some sort is usually necessary for the best cleaning procedures. Scouring powders, abrasive hand pads, abrasive floor pads, brushes, pressure sprayers, and simple "elbow grease" rubbing are supplements to the chemicals.

Germicides combined in certain cleaners can help control harmful microorganisms and unpleasant odors. This results in a healthful as well as a pleasant environment.

Floor coatings are of two types, sealer or finish. Sealers are generally designed for porous surfaces to smooth them out and make them easier to maintain. Concrete and terrazzo sealers stop dusting and may be used to provide a good base for a finish. Wood sealers bring out the natural beauty of wood, prevent penetration of stains, and provide gloss, nonslip, and other desirable properties. Resilient-tile sealers are mainly for the porous tile and provide a good base for a finish. Most sealers are solvent-based and very hard to remove, often requiring special removers. Note that some products may perform both as a sealer and finish.

Finishes generally are water-based and easily removed with a detergent stripping solution. The first finishes were made of natural waxes that required much dry buffing to retain a smooth, glossy appearance. The relatively recent chemical revolution has brought about many new coatings. At first synthetic *waxes* replaced the natural waxes, but more recently hard *polymers* have prevailed. These polymers dry hard and glossy without buffing, require less maintenance, and perform best on the light floors in use today. Table 6-2 outlines specific solutions to floor finish problems.

REFERENCES

Edwards, J. K. P.: *Floors and Their Maintenance*, Butterworths, London, 1972.

Feldman, Edwin B.: *Building Design for Maintainability*, McGraw-Hill, New York, 1975.

Feldman, Edwin B.: *Housekeeping Handbook for Institutions, Business and Industry*, Frederick Fell Publishers, New York, 1978.

General Industry Standards and Interpretations, U.S. Department of Labor, Occupational Safety and Health Administration, 1977.

Meyers, Earl M.: "Standardized Housekeeping Program Improves Utilization of Manpower and Materials," *Maintenance Engineering*, October 1972.

Ruhlin, Robert R.: "Work Control: A Sure Way to Improve Building Maintenance," *Buildings*, February 1974.

Sack, Thomas F.: *A Complete Guide to Building and Plant Maintenance*, Prentice-Hall, Englewood Cliffs, N.J., 1971.

Sipes, Sherrill F., Jr.: "Plan to Manage Your Maintenance Program," *Buildings*, May 1971.

Index

ABOUT THE EDITORS

Editor in Chief ROBERT C. ROSALER, P.E. is Vice President of James O. Rice Associates, Inc., organizers of practical engineering conferences and seminars. He is a member of the American Institute of Plant Engineers and the Plant Engineering Division of the American Society of Mechanical Engineers. He has over 40 years of experience as an engineer executive with both large and small corporations.

Associate Editor JAMES O. RICE has been identified with the professional management movement since 1935. A former Vice President and General Manager of the American Management Association, where he introduced the seminar system of education, Mr. Rice is the founder and President of James O. Rice Associates, Inc.